D1573717

The Integrative Action of the Autonomic Nervous System

Almost all bodily functions are dependent on the functioning of the autonomic nervous system – from the cardiovascular system, the gastrointestinal tract, the evacuative and sexual organs, to the regulation of temperature, metabolism and tissue defense. Balanced functioning of this system is an important basis of our life and well-being. The Integrative Action of the Autonomic Nervous System: Neurobiology of Homeostasis gives a detailed description of the cellular and integrative organization of the autonomic nervous system, covering both peripheral and central aspects. It brings to light modern neurobiological concepts that allow understanding of why the healthy system runs so smoothly and why its deterioration has such disastrous consequences. This broad overview will appeal to advanced undergraduate and graduate students studying the neurobiology of the autonomic nervous system within the various biological and medical sciences and will give access to ideas propagated in psychosomatic disease and alternative medicines.

Wilfrid Jänig is Professor of Physiology at the Christian-Albrechts University in Kiel, Germany.

The Integrative Action of the Autonomic Nervous System

Neurobiology of Homeostasis

Wilfrid Jänig
Physiologisches Institut, Christian-Albrechts-Universität zu Kiel, Kiel, Germany
Supported by the German Research Foundation and the Max-Planck Society

CAMBRIDGE UNIVERSITY PRESS
Cambridge, New York, Melbourne, Madrid, Cape Town, Singapore, São Paulo

Cambridge University Press
The Edinburgh Building, Cambridge CB2 2RU, UK

Published in the United States of America by Cambridge University Press, New York

www.cambridge.org
Information on this title: www.cambridge.org/9780521845182

© W. Jänig 2006

This publication is in copyright. Subject to statutory exception
and to the provisions of relevant collective licensing agreements,
no reproduction of any part may take place without
the written permission of Cambridge University Press.

First published 2006

Printed in the United Kingdom at the University Press, Cambridge

A catalog record for this publication is available from the British Library

ISBN-13 978-0-521-84518-2 hardback
ISBN-10 0-521-84518-1 hardback

Cambridge University Press has no responsibility for
the persistence or accuracy of URLs for external or
third-party internet websites referred to in this publication,
and does not guarantee that any content on such
websites is, or will remain, accurate or appropriate.

Contents

Foreword
Elspeth McLachlan   *page* ix
Preface   xv
List of abbreviations   xviii

Introduction   1

    The autonomic nervous system and the regulation of
        body functions   1
    Organization and aims of the book   7

Part I The autonomic nervous system: functional anatomy
        and visceral afferents   11

Chapter 1 Functional anatomy of the peripheral sympathetic
        and parasympathetic nervous system   13

1.1 Definitions and limitations   13
1.2 Gross anatomy of the peripheral sympathetic
    and parasympathetic nervous system   14
1.3 Reactions of autonomic target organs to activation
    of sympathetic and parasympathetic axons   24
1.4 Neuropeptides in autonomic neurons and the idea of
    "neurochemical coding"   29

Chapter 2 Visceral afferent neurons and autonomic
        regulations   35

2.1 Visceral afferent neurons: general characteristics   37
2.2 Visceral primary afferent neurons as interface between
    visceral organs and brain   42
2.3 Receptive functions of visceral afferent neurons   45
2.4 Role of visceral afferent neurons in visceral nociception
    and pain   54
2.5 Relation between functional types of visceral afferent
    neurons, organ regulation and sensations   60
2.6 Central ascending pathways associated with autonomic
    regulations and visceral sensations   65

Part II Functional organization of the peripheral autonomic
        nervous system   85

Chapter 3 The final autonomic pathway and its analysis   87
3.1 The final autonomic pathway   88

3.2 Functions of the autonomic nervous system and levels of integration — 90
3.3 Activity in peripheral autonomic neurons reflects the central organization — 92
3.4 Reflexes in autonomic neurons as functional markers — 95
3.5 Some methodological details about recording from peripheral autonomic neurons in vivo — 96
3.6 Confounding effects of anesthesia in animal experiments — 104

Chapter 4 The peripheral sympathetic and parasympathetic pathways — 106

4.1 Sympathetic vasoconstrictor pathways — 107
4.2 Sympathetic non-vasoconstrictor pathways innervating somatic tissues — 123
4.3 Sympathetic non-vasoconstrictor neurons innervating pelvic viscera and colon — 136
4.4 Other types of sympathetic neuron — 141
4.5 Adrenal medulla — 143
4.6 Sympathetic neurons innervating the immune tissue — 148
4.7 Proportions of preganglionic neurons in major sympathetic nerves — 151
4.8 Parasympathetic systems — 153

Chapter 5 The enteric nervous system — 168

5.1 Anatomy, components and global functions of the enteric nervous system — 169
5.2 The enteric nervous system is an autonomic nervous system in its own right — 178
5.3 Regulation of motility and intraluminal transport in the small and large intestine: the neural basis of peristalsis — 181
5.4 Integration of enteric neural, pacemaker and myogenic mechanisms in generation of motility patterns — 188
5.5 Regulation of secretion and transmural transport — 194
5.6 Defense of the gastrointestinal tract and enteric nervous system — 196
5.7 Control of the enteric nervous system by sympathetic and parasympathetic pathways — 200

Part III Transmission of signals in the peripheral autonomic nervous system — 209

Chapter 6 Impulse transmission through autonomic ganglia — 211

6.1 Morphology, divergence and convergence in autonomic ganglia — 213

6.2 Strong and weak synaptic inputs from preganglionic neurons — 217
6.3 The autonomic neural unit: structural and functional aspects — 223
6.4 Electrophysiological classification, ionic channels, functions and morphology of sympathetic postganglionic neurons — 225
6.5 Different types of autonomic ganglia and their functions in vivo — 230
6.6 Non-nicotinic transmission and potentiation resulting from preganglionic stimulation in sympathetic ganglia — 240

Chapter 7 | Mechanisms of neuroeffector transmission — 251

7.1 Transmitter substances in postganglionic neurons — 252
7.2 Principles of neuroeffector transmission in the autonomic nervous system — 255
7.3 Specific neuroeffector transmissions — 264
7.4 Integration of neural and non-neural signals influencing blood vessels — 273
7.5 Unconventional functions of sympathetic noradrenergic neurons — 276

Part IV Central representation of the autonomic nervous system in spinal cord, brain stem and hypothalamus — 289

Chapter 8 | Anatomy of central autonomic systems — 293

8.1 Tools to investigate the anatomy of the central autonomic systems — 293
8.2 Morphology and location of preganglionic neurons — 297
8.3 Nucleus tractus solitarii — 311
8.4 Sympathetic and parasympathetic premotor neurons in brain stem and hypothalamus — 317

Chapter 9 | Spinal autonomic systems — 331

9.1 The spinal autonomic reflex pathway as a building block of central integration — 332
9.2 Spinal reflexes organized in sympathetic systems — 336
9.3 Sacral parasympathetic systems — 349
9.4 The spinal cord as integrative autonomic organ — 362

Chapter 10 | Regulation of organ systems by the lower brain stem — 375

10.1 General functions of the lower brain stem — 377
10.2 Sympathetic premotor neurons in the ventrolateral medulla oblongata — 378

10.3 Baroreceptor reflexes and blood pressure control	398
10.4 Arterial chemoreceptor reflexes in sympathetic cardiovascular neurons	410
10.5 Sympathetic premotor neurons in the caudal raphe nuclei	414
10.6 Coupling between regulation of autonomic pathways and regulation of respiration	420
10.7 Vagal efferent pathways and regulation of gastrointestinal functions	440

Chapter 11 | Integration of autonomic regulation in upper brain stem and limbic-hypothalamic centers: a summary — 459

11.1 Functions of the autonomic nervous system: Cannon and Hess	460
11.2 General aspects of integrated autonomic responses	469
11.3 Autonomic responses activated quickly during distinct behavioral patterns	474
11.4 Emotions and autonomic reactions	491
11.5 Integrative responses and the hypothalamus	498
11.6 Synopsis: the wisdom of the body revisited	507
11.7 Future research questions	510
References	519
Index	600

Foreword

Elspeth M. McLachlan, Prince of Wales Medical Research Institute and the University of New South Wales, Sydney, NSW, Australia.

The autonomic nervous system carries the signals from the central nervous system to all organs and tissues of the body except skeletal muscle fibers. It is made up of preganglionic and postganglionic neurons linked together in functionally distinct pathways. The postganglionic terminals have specific relationships with their target tissue. As well as distributing centrally derived command signals, this system can also integrate reflex interactions between different parts of the peripheral nervous system, even without involving the spinal cord. All of these activities are specific for each organ system and attempts to generalize have often proved incorrect. The breadth and scope of involvement of this system in body function are obvious. The autonomic nervous system controls not only the quantity and quality of tissue perfusion in response to varying needs, and the maintenance of secretions for protection of the body's orifices and the lining of the gastrointestinal tract, but it also regulates the usually intermittent but complex functions of the abdominal viscera and pelvic organs, the mechanical aspects of the eye and the communication between the nervous system and the immune system. Many autonomic pathways are continuously active but they can also be recruited when the environmental and/or emotional situation demands it. This system is essential for homeostasis – hence the subtitle of this book.

Despite its enormous importance for the maintenance of normal physiology in all vertebrate species, and for the understanding of many clinical symptoms of disease, the autonomic nervous system has not, even transiently, been the center of attention in neuroscience research internationally over the past 40 years. Many seem to think that this system has been worked out and there is nothing new to investigate. The discovery of neuropeptides as putative transmitters was probably the only interlude that triggered widespread excitement. Others simply forget that the system exists except for emergencies.

Two views about the autonomic nervous system are often encountered:

1. that this system is similar to the endocrine system and its functions can all be explained by the pharmacological actions of the major neurotransmitters, noradrenaline and acetylcholine, possibly involving modulation by cotransmitters and neuropeptides, or
2. that the functions of this system are not important as life continues without them.

For anyone who thinks about it, at least the latter of these concepts is obviously not true. Life can be maintained in a cocoon in individuals with autonomic failure but the ability to cope with external stressors severely compromises their quality of life. The extent to which the practical difficulties of daily life for people with spinal cord injury, which disrupts the links between the brain and the autonomic control of the body's organs, absorb personal energy and resources should not be underestimated by those who take their bodies for granted. Elderly people face similar problems as some of their autonomic pathways degenerate.

On the other hand, the former of the above two concepts dominates almost all current textbooks of physiology and neuroscience. It is true that some of the effects of autonomic nerve activity can be mimicked by the application of neurotransmitter substances locally or systemically. However, the mechanisms by which the same substances released from nerve terminals produce responses in the target tissue have proved to be quite different in most cases so far analyzed. This helps to explain the failure of many pharmaceutical interventions based on this simplistic idea as outlined above. What is important here is that the present volume collates the evidence against both these ideas and develops the factual and conceptual framework that describes how an organized system of functional nerve connections that operate with distinct behaviors is coordinated to regulate the workings of the organ systems of each individual.

Nevertheless, over the past 40 years, there have been remarkable strides in our understanding. Technical problems limit how the complexities of this system can be unraveled. There are enormous challenges involved in investigating a complex interconnected system made up of small neurons that are not always packaged together in precisely the same way between individuals. Even in the spinal cord, the neuroanatomical distribution and apparent imprecision have been daunting. To study this system requires patience and persistence in the development of manipulative and analytical skills. These attributes are relatively rare.

Fortunately, over this period, a small but steady stream of researchers has persisted in their endeavors to clarify how this functionally diverse system works. One of the most significant players has been Wilfrid Jänig. Wilfrid and his many students and collaborators at the Christian-Albrechts-Universität in Kiel have pursued a major and uniquely productive approach to understanding how sympathetic pathways work. This has been to apply the technique of extracellular recording from single identified axons dissected from peripheral nerves projecting to particular target tissues and therefore acting in known functional pathways. Over the 40 years, this work, originally in cats and latterly in rats, has revealed the principles underlying reflex behavior of sympathetic axons in the anaesthetized animal. The characteristic behavior of pre- and postganglionic neurons in over a dozen functional pathways has been defined. As the reader progresses through the book, it will become clear that many of these reflexes are also present in humans. The parallel technique of microneurography, pioneered by Hagbarth, has been implemented over a similar period in the sympathetic pathways of conscious humans by Gunnar Wallin and his colleagues in Göteborg. While pathways to the viscera are currently too hard to study in humans because they are less accessible, the principles of their organization can be deduced from Wilfrid's data on pre- and postganglionic discharge patterns and from the analyses of ganglionic and neuroeffector transmission conducted by him and others.

Over the 40 years, Wilfrid's various interests have been broad but always focussed. They have taken him to many places to answer questions about the structure and function of sympathetic pathways. His earliest training in single unit recording was in sensory neurophysiology and this background has been the basis of his parallel studies of visceral afferent behavior and nociception. Early in his career, he was interested in integrative autonomic control at the higher levels of the nervous system and developed a passion to follow on the work of Philip Bard. After returning to Germany from New York in the 1970s, he conducted experiments on decorticate and decerebrate cats in which he created behaviors such as sham rage during which he planned to record and analyze the sympathetic outflow. These experiments did not progress because of limited resources, but instead he undertook a most detailed analysis of the distinctive behavior of skin and muscle sympathetic vasoconstrictor axons. These results provided evidence that strongly rejected popular ideas that a general level of "sympathetic tone" was the determinant of peripheral vascular resistance. It was clear that the reflex connectivity of the pathways involved in cutaneous and skeletal muscle blood flow are largely independent. This concept was more dramatically confirmed in recordings from humans where it is possible to demonstrate the strong emotional drive that modulates cutaneous vasoconstrictor activity (see Subchapter 4.1.2 in this book). Subsequently Wilfrid's laboratory has extended this type of analysis to over a dozen different pathways that they have studied in anesthetized animals.

I first met Wilfrid in 1979 when he came to give a seminar in Edinburgh where I was on sabbatical leave at the time. As my original background was in cardiovascular physiology, I had naturally read his work on vasoconstrictor discharge patterns and had lots of questions to ask him. Wilfrid invited me to visit Kiel on my way home (it was very cold and wet in November) and then he came to Melbourne to work with me to trace the peripheral sympathetic pathways quantitatively. In my laboratory at Monash University, I had established the retrograde tracing technique using horseradish peroxidase to identify the location of preganglionic neurons in the spinal cord as a prelude to recording intracellularly from them. He worked hard with me cutting and mounting thousands of sections and soon after I spent a similar period in Kiel helping his group establish the technique there. This quantitative work dovetailed well to explain how the axons that his group sampled in their recordings related to the entire population.

It has been my great privilege to continue to work with Wilfrid and his colleagues, particularly up to the early 1990s, undertaking studies for which he and I received the Max-Planck Forschungspreis for international collaboration in 1993. Since that time, and in various parts of Australia as I have moved between Universities, we have worked together and in parallel on aspects of the interactions between the sympathetic and sensory systems that may be involved in neuropathic pain after nerve injury. We have continued to communicate frequently and his younger colleagues, notably Ralf Baron, Ursula Wesselmann and Joachim Häbler, have spent time in Australia working in my laboratory. I hope and expect that these interactions will continue.

Wilfrid's early studies of sympathetic activity were made when Robert Schmidt was in Kiel and were conducted in parallel with studies of somatosensory, particularly nociceptive, afferents. This anteceded his interest in visceral afferent function to which he applied the same technical expertise to unravel the behavior of these neurons, particularly in pelvic organ reflexes. His interests in nerve injury were pursued in part with Marshall Devor in Jerusalem. This involved extended studies of the ectopic activity of sensory neurons after peripheral nerve lesions and the role of sympathetic activity in triggering this. His laboratory has also conducted a wide range of studies on the effects of various nerve lesions on the properties of sympathetic and afferent axons. As Wilfrid appreciated that the problem of neuropathic pain was probably related to inflammation, he sought out Jon Levine in San Francisco where he was exposed to a very strong research community involved in pain and inflammation research. He has a prodigious output from Jon's laboratory deciphering the components of the neuroimmune interactions using rigorous and systematic approaches to identify the pathways and sites at which the hypothalamo–pituitary–adrenal axis (HPA) intervenes in inflammation and in nociception, in some cases with sympathetic involvement. More recently, Wilfrid and Ralf Baron have worked with the clinical community worldwide on clarifying the misnamed concept of "reflex sympathetic dystrophy" and developing the newer definitions of various "complex regional pain syndromes" to help to clarify the diagnosis of the mechanisms underlying chronic neuropathic pain.

When visiting my laboratories at Monash and the Baker Institute in Melbourne, and subsequently at the Universities of New South Wales and Queensland, and more recently at the Prince of Wales Medical Research Institute in Sydney, Wilfrid has been able to visit many neuroscientists around Australia where research on the neurobiology of the autonomic nervous system and on central cardiovascular control is prolific by world standards. He has seized upon these opportunities to learn what the community of Australian autonomic researchers is doing and has established strong relationships with the leaders of many active laboratories including those of David Hirst, Ian Gibbins and Judy Morris, John Furness, Marcello Costa,

Christopher Bell, Janet Keast, Sue Luff, James Brock, Roger Dampney, Robin McAllen, Bill Blessing, Paul Korner, Dick Bandler, Paul Pilowsky, and Dirk van Helden. This extensive Australian involvement in autonomic neuroscience arose in part from the students who trained with Geoff Burnstock and Mollie Holman in Melbourne in the 1960s and 1970s and who have taken their skills across the country and have been training the next generations since that time. Despite the divergence of their specific interests, this community continues to be one of the largest internationally working in the autonomic nervous system. Wilfrid's exposure to the cellular, pharmacological and neuroanatomical aspects of ganglionic and junctional transmission in the peripheral pathways gave him a very wide view of autonomic effector systems, which he has so cleverly incorporated into this book.

Throughout these years, Wilfrid has been a prodigious author of textbook chapters and review articles. Although many of the former have been written in German, he has also developed and expounded his ideas about neural control of vasoconstriction, pain and the sympathetic nervous system, the consequences of nerve injury, the involvement of the HPA axis in inflammation and nociception, and on clinical aspects of these topics. This book arises from this lifetime of synthetic writing and from his reflection on the wider issues of this area of science. It also is the product of his frustration, which I share, with the limited availability of publications that summarize the scientific background and present the current status of our understanding of how the autonomic nervous system works. As in his experiments, he has dissected the system into the major functional pathways in which reflex behavior and cellular mechanisms have been well investigated. He reviews and synthesizes the available information on the spinal cord and brain stem components of autonomic reflexes and then resynthesizes these output systems into a complex package that includes the control of autonomic discharge patterns from the midbrain and higher centers. He has extracted the key information yielded by both classical and modern technical approaches used to study these components of the nervous system. He has incorporated the conceptual background behind each area of research. Finally, he discusses how the old "unifying" concepts of Cannon and Hess misrepresented the diversity of autonomic outflow patterns that the brain recruits during the various behaviors that function to conserve the body in a range of environmental circumstances. This philosophical base needs to replace the widely held views mentioned earlier if we are to progress our understanding of this important set of control systems. The present *tour de force* has involved discussion and input from many of Wilfrid's collaborators and colleagues around the world whose contributions have ensured that the final product really contains the most up-to-date summary of our current knowledge of autonomic function.

Despite, or because of, this diversity of inputs, this book provides Wilfrid Jänig's unique overview of the autonomic nervous system.

Without his driving fascination with how the whole autonomic system works in the body, this book would never have been written. No-one else currently has the conceptual breadth and capacity to integrate so many aspects to compose this amalgam. He has collected all the available data from the past and the present and fitted them together with what is known of the central control and spinal integration that determine the activity patterns in each outflow pathway. He has taken the knowledge from Langley's time, through Cannon, Hess and Bard, Burnstock and Holman, to the recent application of cellular biology and molecular genetics to collate a truly comprehensive compendium. I am delighted that he has committed himself to drawing together so many diverse aspects of autonomic function in one place and to give us a truly integrated overview of what is known at the beginning of the twenty-first century. He has made very clear what he feels are the major questions that remain to be answered. I know that Wilfrid will contribute to many of those answers.

Preface

In the late 1960s, while I was working in Robert F. Schmidt's laboratory in the Department of Physiology of the University of Heidelberg, conducting experiments on cutaneous primary afferent neurons and presynaptic inhibition in the spinal cord, Robert introduced me to the sympathetic nervous system. We worked on somato-sympathetic reflexes and other spinal reflexes, some of the work being conducted with Akio Sato. At this time, I tried to understand *The Wisdom of the Body* by Walter Bradford Cannon (Cannon 1939) and *Vegetatives Nervensystem* by Walter Rudolf Hess (Hess 1948). However, from 1971 to 1974, I continued with my experimental work on the somatosensory system and concentrated with Alden Spencer on the cuneate nucleus and thalamus in the Department of Neurobiology and Behavior of the Public Health Institute of the City of New York (directed by Eric Kandel).

While working in New York I came into contact with Chandler McCuskey Brooks (Downstate Medical Center, State University of New York). He invited me to attend the Centennial Symposium "The Life and Influence of Walter Bradford Cannon, 1871–1945: The Development of Physiology in this Century" (Brooks *et al.* 1975). Chandler encouraged me to concentrate scientifically on the autonomic nervous system; he remained very supportive until his death seventeen years later. This influence and particularly the books of Cannon and Hess led to my decision to leave the somatosensory field and redirect my research, after my return to Germany, to investigations of the sympathetic nervous system. The books by Cannon and Hess, and the published papers on which they are based, aroused from the beginning my opposition on the one hand and my secret admiration for the authors on the other. This ambiguity in my scientific attitude towards Cannon and Hess has always been in the background of the scientific activities in my laboratory, of my teaching and of my writing on the autonomic nervous system.

I am particularly grateful to two persons who have kept me going on the scientific path amidst trials and tribulations to unravel some of the mysteries of the autonomic nervous system. Robert Schmidt has made me invest time in writing textbook chapters since 1971. Elspeth McLachlan has always been extremely supportive and virtually carried me through some periods of doubt and despondency throughout the 25 years we have worked experimentally together. She introduced me to the Australian Autonomic Neuroscience and is responsible for this book being in some ways an Australian book (see below). Finally, the many young students in my laboratory, some now professors, influenced me by their enthusiasm despite my being entirely uncompromising, which was sometimes hard for them to digest.

The German Research Foundation has fully supported my research over more than 30 years. Without this continuous funding, for research that was often methodologically, and as regards content, not in the mainstream, I never would have been able to continue my research on the autonomic nervous system for so many years. So I am deeply grateful to the many anonymous referees of the German Research Foundation for their fair judgment.

Finally, and most important, on a very personal level, the research over these decades, and still continuing, could never have happened without the never-ending tolerance and support of my wife Ute and our sons Nils and Volker. My family life has sustained my research more than anything else.

I want this book to be a forum for ongoing discussion. I strongly encourage young scientists to invest their time in research on the autonomic nervous system. While writing the book I was in continuous discourse with many scientists in Australia, Europe and the United States addressing various scientific aspects of the book. These scientists have made a major contribution.

- I would particularly like to thank those who undertook to read and comment on chapters or subchapters of the book. *AUSTRALIA*: Richard Bandler (Sydney), Bill Blessing (Adelaide), James Brock (Sydney), Roger Dampney (Sydney), John Furness (Melbourne), Ian Gibbins (Adelaide), Elspeth McLachlan (Sydney), Judy Morris (Adelaide), Paul Pilowsky (Sydney), Terry Smith (Reno, Nevada, USA), Dirk van Helden (Newcastle). *EUROPE*: Niels Birbaumer (Tübingen, Germany), John Coote (Birmingham, England), Sue Deuchars (Leeds, England), Peter Holzer (Graz, Austria), Winfried Neuhuber (Erlangen, Germany), Julian Paton (Bristol, England), Hans-Georg Schaible (Jena, Germany), Gunnar Wallin (Göteborg, Sweden). *NORTH AMERICA*: Gerlinda Hermann (Baton Rouge, Louisiana), Arthur Loewy (St. Louis, Missouri), Orville Smith (Seattle, Washington), Larry Schramm (Baltimore), Alberto Travagli (Baton Rouge, Louisiana).
- I am thankful to those who never hesitated to think about and answer my questions. *AUSTRALIA*: Andrew Allen (Melbourne), David Hirst (Melbourne), Ida Llewellyn-Smith (Adelaide), Susan Luff (Melbourne), Vaughn Macefield (Sydney), Robin McAllen (Melbourne). *EUROPE*: Jean-François Bernard (Paris, France), Uwe Ernsberger (Heidelberg, Germany), Björn Folkow (Göteborg, Sweden), Rupert Hölzl (Mannheim, Germany), Andrew Todd (Glasgow, Scotland). *NORTH AMERICA*: Susan Barman (East Lansing, Michigan), Bud Craig (Phoenix, Arizona), Lori Birder (Pittsburgh), Chet de Groat (Pittsburgh), Patrice Guyenet (Charlottesville, Virginia), John Horn (Pittsburgh), Kevan McKenna (Chicago), Shaun Morrison (Beaverton, Oregon), Michael Panneton (St. Louis, Missouri), Terry Powley (West Lafayette, Indiana), Clifford Saper

(Boston), Ann Schreihofer (Augusta, Maine), Ruth Stornetta (Charlottesville, Virginia), Holly Strausbaugh (San Francisco).

Eike Tallone (Kiel) has my very special thanks. She is responsible for the graphical work in almost all figures of this book. Without her, the figures would not be of such high quality.

Abbreviations

The main abbreviations used are listed below. Special abbreviations related to anatomical structures in the lower brain stem or hypothalamus are listed in the legends of the figures and tables, particularly Figures 10.2, 11.13, 11.14 and Tables 8.2 and 8.3.

ACC	anterior cingulate cortex
ACh	acetylcholine
AgAgCl	silver-silver chloride
AP	area postrema
ATP	adenosine triphosphate
BAT	brown adipose tissue
BDNF	brain-derived neurotrophic factor
BNST	bed nucleus of the stria terminalis
BP	blood pressure
c, C	cervical (segment)
CA	central autonomic nucleus
Ca^{2+}	calcium
cAMP	cyclic adenosine monophosphate
CCK	cholecystokinin
CGRP	calcitonin gene-related peptide
ChAT	choline acetyltransferase
CL	centrolateral nucleus (thalamus)
CM	circular musculature (gastrointestinal tract)
CN	cuneiform nucleus
CNS	central nervous system
CPA	caudal pressure area
CRG	central respiratory generator
CRH	corticotropin-releasing hormone
CRPS	complex regional pain syndrome
CSN	carotid sinus nerve
CST	cervical sympathetic trunk
CTb	cholera toxin subunit B
CVC	cutaneous vasoconstrictor (neuron)
CVD	cutaneous vasodilator (neuron)
CVLM	caudal ventrolateral medulla
cVRG	caudal ventral respiratory group
DC	dorsal column (spinal cord)
DCN	dorsal commissural nucleus (spinal cord)
DH	dorsal horn (spinal cord, trigeminal)
DMNX	dorsal motor nucleus of the vagus
DR	dorsal root
DRG	dorsal root ganglion
DVC	dorsal vagal complex
DYN	dynorphin
DβH	dopamine-β-hydroxylase

ECG	electrocardiogram
EJC	excitatory junction current
EJP	excitatory junction potential
ENK	enkephalin
EPSP	excitatory postsynaptic potential
EUS	external urethral sphincter
FB	Fast Blue
FG	Fluoro-gold
FRA	flexor reflex afferent
GABA	γ-aminobutyric acid
GAL	galanin
GALT	gut-associated lymphoid tissue
GIT	gastrointestinal tract
GLP-1	glucagon-like peptide 1
GnRH	gonadotropin-releasing hormone
GR	grey ramus
GSR	galvanic skin response
HGN	hypogastric nerve
HR	heart rate
5-HT	5-hydroxytryptamine (serotonin)
HRP	horse radish peroxidase
HVPG	hypothalamic visceral pattern generator
IC	intercalate spinal nucleus
ICC	interstitial cell of Cajal
IGLE	intraganglionic laminar ending (enteric nervous system)
IJP	inhibitory junction potential
IL	interleukin
ILF	funicular part of the intermediolateral nucleus
ILP	principal part of the intermediolateral nucleus
IMA	intramuscular array (enteric nervous system)
IMG	inferior mesenteric ganglion
IML	intermediolateral nucleus (spinal cord)
INA	integrated nerve activity
INSP	inspiratory
IPANs	intrinsic primary afferent neurons (enteric nervous system)
IP_3	inositol 1,4,5-triphosphate
IPSP	inhibitory postsynaptic potential
IRH	inhibitory releasing hormone
IVLM	intermediate ventrolateral medulla
l,L	lumbar (segment)
LAH	long afterhyperpolarizing
LH	lateral hypothalamus
LHRH	luteinizing hormone-releasing hormone
LM	longitudinal musculature (enteric nervous system)
lPMC	lateral pontine micturition center
LRN	lateral reticular nucleus
LSN	lumbar splanchnic nerve
LST	lumbar sympathetic trunk

LTF	lateral tegmental field
MC	mast cell
MCC	middle cingulate cortex
MDvc	ventral portion of the medial dorsal nucleus (thalamus)
MMC	migrating myoelectric complex (enteric nervous system)
mPMC	medial pontine micturition center
MP	myenteric plexus (enteric nervous system)
MR	motility-regulating (neuron)
mRNA	messenger ribonucleic acid
MVC	muscle vasoconstrictor (neuron)
MVD	muscle vasodilator (neuron)
NA	nucleus ambiguus
NAd	noradrenaline
NANC	non-adrenergic non-cholinergic
NGF	nerve growth factor
NK1	neurokinin 1
NMDA	N-methyl-D-aspartate (acid)
NO	nitric oxide
NOS	nitric oxide synthase
NTS	nucleus tractus solitarii
NPY	neuropeptide Y
OVLT	organum vasculosum laminae terminalis
PAF	platelet-activating factor
PAG	periaqueductal grey
PBC	parabrachial complex
Pf	parafascicular nucleus (thalamus)
PGE	prostaglandin E
PHA-L	phaseolus vulgaris leuco-agglutinin
PHR	phrenic nerve
PKA	protein kinase A
PMC	pontine micturition center
PMNL	polymorphonuclear leukocyte
PO	posterior nucleus (thalamus)
PRV	pseudorabies virus
PSDC	postsynaptic dorsal column
PVH	paraventricular nucleus of the hypothalamus
REM	rapid eye movement
RH	releasing hormone
RVC	renal vasoconstrictor (neuron)
RVLM	rostral ventrolateral medulla
rVRG	rostral ventral respirator group
s,S	sacral (segment)
SCG	superior cervical ganglion
sEJP	spontaneous excitatory junction potential
SG	stellate ganglion
SM	sudomotor (neuron)
SMC	smooth muscle cell
SMP	submucosal plexus (enteric nervous system)

SOM	somatostatin
SP	substance P
SPN	sacral parasympathetic nucleus
STT	spinothalamic tract (neuron)
t,T	thoracic (segment)
TH	tyrosine hydroxylase
TrkA	tyrosine kinase A (receptor)
TNF	tumor necrosis factor
TRH	thyrotropin-releasing hormone
VH	ventral horn (spinal cord)
VIP	vasoactive intestinal peptide
VLM	ventrolateral medulla
VMM	ventromedial medulla
VMb	basal part of the ventromedial nucleus (thalamus)
VMpo	posterior part of the ventromedial nucleus (thalamus, primate)
VPI	ventral posterior inferior nucleus
VPL	ventral posterior lateral nucleus (thalamus)
VPM	ventral posterior medial nucleus (thalamus)
VPpc	ventral posterior parvicellular nucleus of the thalamus (rat)
VR	ventral root
VRG	ventral respiratory group
VVC	visceral vasoconstrictor (neuron)
WHBS	working-heart-brain-stem preparation
WR	white ramus
X	vagus nerve/nucleus

Introduction

The autonomic nervous system and the regulation of body functions

Somatomotor activity and adjustments of the body

All living organisms interact continuously with their environment. They receive multiple signals from the environment via their sensory systems and respond by way of their somatomotor system. Both sensory processing and motor actions are entirely under control of the central nervous system. Within the brain are representations of both extracorporeal space and somatic body domains, the executive motor programs and programs for the diverse patterns of behavior which are initiated from higher levels. The brain generates complex motor commands on the basis of these central representations; these lead to movements of the body in its environment against different internal and external forces. The tools for performing these actions are the effector machines, the skeletal muscles and their controlling somatomotoneurons.

The body's motor activity and behavior are only possible when its internal milieu is controlled to keep the component cells, tissues and organs (including the brain and skeletal muscles) maintained in an optimal environment for their function. This enables the organism to adjust its performance to the varying internal and external demands placed on the organism. In the *short term* the mechanisms involved include the control of:

- fluid matrix of the body (fluid volume regulation, osmoregulation),
- gas exchange with the environment (regulation of airway resistance and the pulmonary circulation),
- ingestion and digestion of nutrients (regulation of the gastrointestinal tract, control of energy balance),
- transport of gases, nutrients and other substances throughout the body to supply organs, including the brain to maintain consciousness (regulation of blood flow and blood pressure by the cardiovascular regulation),
- excretion of substances (disposal of waste),
- body temperature (thermoregulation),
- reproductive behavior (mechanics of sexual organs),
- defensive behaviors,

and in the *long term* the control of:

- body recovery (control of circadian rhythms, of sleep and wakefulness),
- development and maintenance of body organs and tissues,

- body protection at the cellular and systems level (regulation of inflammatory processes, control of the immune system).

These body functions responsible for maintaining the internal milieu are controlled by the brain. The control is exerted by the autonomic nervous system and the endocrine system. Specifically, the brain acts on many peripheral target tissues (smooth muscle cells of various organs, cardiac muscle cells, exocrine glands, endocrine cells, metabolic tissues, immune cells, etc). The *efferent signals* from the brain to the periphery of the body by which this control is achieved are *neural* (by the autonomic nervous systems) and *hormonal* (by the neuroendocrine systems). The time scales of these controls differ by orders of magnitude: autonomic regulation is normally fast and occurs within seconds, and neuroendocrine regulation is relatively slow (over tens of minutes, hours or even days). The *afferent signals* from the periphery of the body to the brain are neural, hormonal (e.g., hormones from both endocrine organs and the gastrointestinal tract, cytokines from the immune system, leptin from adipocytes) and physicochemical (e.g., blood glucose level, temperature, etc.).

The maintenance of physiological parameters such as concentration of ions, blood glucose, arterial blood gases, body core temperature in a narrow range (but around predetermined "set points") is called *homeostasis*. Homeostatic regulation involves autonomic systems, endocrine systems and the respiratory system. The concept of homeostasis has been formulated by Walter B. Cannon (1929a) based on the idea of the fixity of the internal milieu of the body (Bernard 1957, 1974).[1] However, this concept is too narrow and static to understand how the organism is able to maintain the parameters in the body stable and to temporarily adapt them during environmental changes. To understand how the internal milieu is maintained stable during changes in the body and in the environment the concept of allostasis has been developed. This concept extends the concept of homeostasis. It distinguishes between systems that are essential for life (e.g., the concentration of ions and pH in the extracellular and intracellular fluid compartments; homeostasis in the narrow sense) and systems that maintain these systems in balance ("allostasis") as the environment changes (McEwen and Wingfield 2003). This is achieved by the autonomic nervous systems and the endocrine systems and is fully dependent on a functioning central nervous system, notably the hypothalamus and the cerebral hemispheres (Sterling and Eyer 1988; McEwen 2001b; McEwen and Wingfield 2003; Schulkin 2003a, b).

Autonomic nervous system, behavior and brain

By analogy with the organization of the somatomotor system and the sensory systems in the brain, it is natural to accept the concept that the regulation of autonomic and endocrine homeostatic mechanisms is also represented in the brain (i.e., in the hypothalamus, brain stem and spinal cord) and that these systems are under the control of the forebrain and integrated there with the somatomotor and sensory

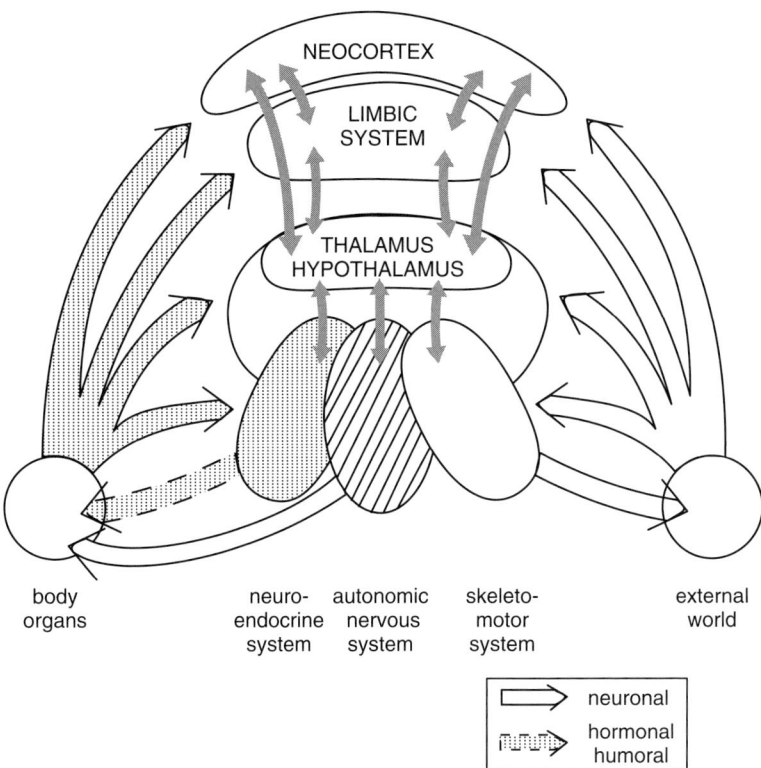

Figure 1 Autonomic nervous system, brain and body. *Right*, somatic nervous system (motor system and sensory systems) and environment. *Left*, autonomic nervous system, neuroendocrine system and body organs. In the *middle*, spinal cord, brain stem, hypothalamus' limbic system and neocortex. The afferent feedback from the body is neuronal, hormonal and humoral (physicochemical; e.g., glucose concentration, osmolality) and of other types (e.g., body temperature). *Solid line arrows*, neuronal; *dashed line arrow*, hormonal; *dotted*, neuroendocrine system, hormonal and humoral feedback. Limbic system is anatomically descriptive and a collective term denoting brain structures common to all mammals that include hippocampus, dentate gyrus with archicortex, cingulate gyrus, septal nuclei and amygdala. These forebrain structures are functionally heterogenous and not a unitary system (as the term "limbic system" may imply). They are involved in the generation of emotional and motivational aspects of behavior (see LeDoux [1996]). Note the reciprocal communication between hypothalamus, limbic system and neocortex (symbolized by the shaded arrows) indicating that the centers of the cerebral hemispheres have powerful influence on all autonomic regulations. Modified from Jänig and Häbler (1999) with permission.

representations. Thus the brain contains autonomic "sensorimotor programs" for coordinated regulation of the internal environment of the body's tissues and organs and sends efferent commands to the peripheral target tissues through the autonomic and endocrine routes (see Figure 1). Integration, within the brain, between the areas that are involved with the outputs of the autonomic, endocrine, somatomotor, and sensory systems is essential for the coordination of behavior of the organism within its environment, for the expression and development of normal feelings and emotions, and for the perception of the body and self as a unified being.

The essential role of the autonomic nervous system in these integrative homeostatic and allostatic programs is primarily to

Figure 2 Functional organization of the nervous system to generate behavior. The motor system, consisting of the somatomotor, the autonomic (visceromotor) and the neuroendocrine systems, controls behavior. It is hierarchically organized in spinal cord, brain stem and hypothalamus. The motor system receives three general types of synaptic input: (a) from the sensory systems monitoring processes in the body or the environment to all levels of the motor system generating reflex behavior (r); (b) from the cerebral hemispheres responsible for voluntary control of the behavior (c), cognitive based on neural processes related to cognitive and affective-emotional processes; (c) from the behavioral system controlling attention, arousal, sleep/wakefulness, circadian timing (s). The three general input systems communicate bidirectionally with each other (upper part of the figure). Modified from Swanson (2000, 2003).

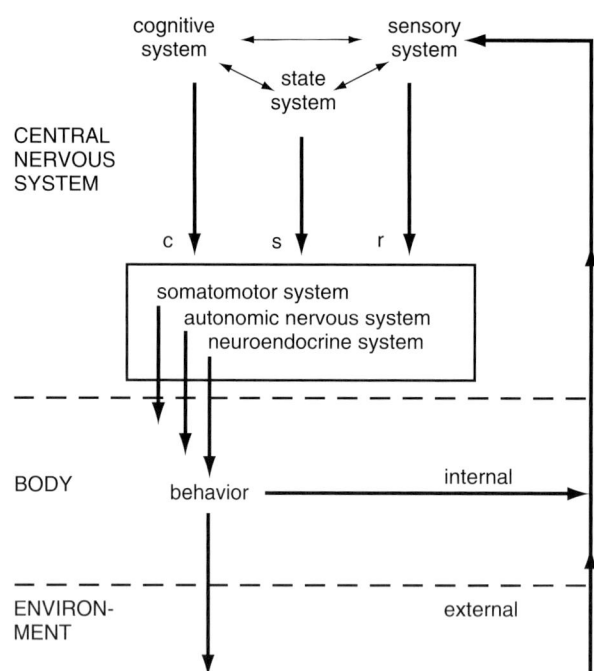

distribute specific signals generated in the central nervous system to the various target organs. In order to achieve overall coordination, the signals need to be precisely patterned to implement appropriate reactions in each target tissue or organ. Some of these signals pass continuously to the periphery in the resting state; others are recruited (or switched off) during particular body behaviors and states of the internal milieu.

The model in Figure 2 outlines the role of the autonomic nervous system in the generation of behavior. Behavior is defined as the purposeful motor action of the body in the environment. It is generated by coordinated activation (1) of somatomotor neurons to move the body in the environment and (2) of autonomic and neuroendocrine motor neurons to prepare and adjust the internal milieu and body organs enabling the body to move. Thus, under motor system we subsume the three divisions, the somatic, the autonomic and the neuroendocrine motor systems (Swanson 2000, 2003; Watts and Swanson 2002):

- The divisions of the motor system are closely integrated in spinal cord, brain stem and hypothalamus. Both somatomotor and autonomic motor systems are hierarchically organized, their integration occurs at each level of the hierarchy. The neuroendocrine motor system is represented at the top of this hierarchy (in the hypothalamus).
- The motoneuron pools (final motor pathways) extend from the midbrain to the caudal end of the spinal cord for the somatomotor system and the autonomic system (the latter with gaps in the

cervical and lower lumbar spinal cord). The neuroendocrine motor neurons are located in the periventricular zone of the hypothalamus.
- The activity of the motor system generating behavior is dependent on three major classes of input: (1) the sensory systems, (2) the cortical system that generates voluntary behavior and (3) the behavioral state system.
- The sensory systems monitor events in the body (*internal* in Figure 2) or in the environment. They are closely welded to the motor hierarchy and generate on all levels of this hierarchy reflex behavior (*r* in Figure 2).
- The cerebral hemispheres initiate and maintain behavior based on cognition and affective-emotional processes (*c* in Figure 2).
- The behavioral state system consists of intrinsic neural systems that determine the state of the brain in which it generates motor behavior. The behavioral state system controls sleep and wakefulness, arousal, attention, vigilance and circadian timing (*s* in Figure 2).
- The three global input systems to the motor system interact with each other too.

This way of looking at the autonomic nervous system shows that the activity in the autonomic neurons is dependent on the intrinsic structure of the sensorimotor programs of the motor hierarchy and on its three global input systems. Thus any change in these input systems should be reflected in the activity of the final autonomic pathways and therefore in the autonomic regulations.

Precision of autonomic regulations and its failure in disease

The precision and biological importance of the control of peripheral target organs by the autonomic nervous system are silently accepted, but the mechanisms by which they come about are not generally understood. Both of these aspects become quite obvious when the regulatory functions fail so that even the simplest actions of the body, such as standing up, may become a burden. Thus, as the blood collects in the venous system due to the effect of gravity, the control of the level of pressure perfusing the brain and other organs fails, leading to decreased cerebral perfusion and loss of consciousness. Such failure of autonomic control may occur at a time when the somatomotor and the sensory systems are functioning normally (Low 1993; Appenzeller 2000; Mathias and Bannister 2002). Failure may develop:

- when the peripheral (efferent) autonomic neurons are damaged (e.g., as a consequence of metabolic disease, such as long-term diabetes in which peripheral autonomic neurons are destroyed), resulting in failure of regulation of the cardiovascular system, the gastrointestinal tract, pelvic organs (sexual organs, urinary bladder, hindgut) or other organs;

- when certain types of peripheral autonomic neurons are inherently absent such as the rare cases of *pure autonomic failure* in which most neurons in the autonomic ganglia are absent or in which one enzyme for synthesis of the transmitter noradrenaline, dopamine-β-hydroxylase, is deficient or absent (Mathias and Bannister 2002); or in *Hirschsprung's disease* in which some of the inhibitory neurons of the enteric nervous system of the gut are missing (Christensen 1994);
- when the *spinal cord* is traumatically *lesioned* (leading to interruption of the control of these spinal autonomic circuits by supraspinal autonomic centers to the spinal autonomic circuits);
- when the efferent sympathetic and afferent pathways in the peripheral somatic tissues are disrupted after trauma (with or without nerve lesions), leading to abnormal relationships between sympathetic and afferent systems and consequently to pain syndromes such as reflex sympathetic dystrophy (complex regional pain syndrome type I) or causalgia (complex regional pain syndrome type II) (Stanton-Hicks *et al.* 1995; Jänig and Stanton-Hicks 1996; Harden *et al.* 2001; Jänig and Baron 2002, 2003);
- when hypothalamic functions are impaired (e.g., in anorexia nervosa or as a consequence of a tumor or trauma);
- during severe infectious diseases, probably since central regulation of the cardiovascular system or gastrointestinal tract fail;
- quite commonly in *old age* when many peripheral autonomic neurons may die and autonomic regulation may be reduced in effectiveness.

When the autonomic system fails for any of the reasons described above the potential range of an individual's behavior may be considerably reduced. In the extreme case, a patient with failure of blood pressure regulation may have to be confined permanently to bed in order to maintain adequate perfusion of the brain. Patients with distinct diabetic autonomic neuropathy may have problems with regulation of their pelvic organs, cardiovascular system or gastrointestinal tract. Finally, and quite commonly, life may become miserable in old age because autonomic systems decrease in their range of functioning without obvious disease triggered from within or without the body.

Functional diseases, involving the autonomic nervous system and neuroendocrine systems, may also develop when allostatic responses, which are physiologically rapidly mobilized during external and internal perturbations ("stress") and then turned off when no longer needed, remain active over a longer time. These maintained allostatic responses are called *allostatic load* or *overload* and are believed to contribute to various types of diseases, such as hypertension, myocardial infarction, obesity, diabetes type II, atherosclerosis, and metabolic syndrome (Folkow *et al.* 1997; McEwen 2001a; McEwen and Wingfield 2003).

Although we can principally live without the function of large parts of the autonomic nervous system, the life style of an individual in such a state becomes severely constrained so that many of the large range of activities for which our biology equips us, such as being sexually active, playing tennis, running a marathon, climbing mountains, diving in the sea, living in the tropics or in arctic climates or being involved in intellectual activities, are not possible without a normally functioning autonomic nervous system. Finally, all vertebrates are endowed with autonomic systems in order to meet extreme environmental challenges. Thus we can adapt, e.g., to very cold climates or very hot climates, to high altitudes, to extreme persistent exercise, to extreme states of starvation or to extreme dry climates. These examples illustrate the dynamic plasticity of autonomic regulations. This plasticity may have been essential for the evolution of mammals, humans probably being the most adaptable mammal. Development, anatomical differentiation and functional differentiation of the autonomic nervous system evolved in association with the behavioral repertoire of the different vertebrate species. Thus the complexity of autonomic regulation increased with the complexity of the behavioral repertoire (Nilsson 1983; Nilsson and Holmgren 1994). This is fully in line with the concept of functioning of the hierarchically organized motor system that includes the somatomotor, the autonomic and neuroendocrine systems as outlined above (Figure 2).

Autonomic regulation of body functions requires the existence of specific neuronal pathways in the periphery and specific organization in the central nervous system; otherwise it would not be possible to have the precision and flexibility of control that higher vertebrates possess for rapid adjustments during diverse behaviors. This implies that the various autonomic systems must be centrally integrated and have multiple, but distinct, peripheral motor pathways. These pathways are defined according to the function they mediate in the target cells they innervate. The effector cells of the autonomic nervous system are very diverse while those of the somatic efferent system are not. From this point of view, it is clear that the autonomic nervous system is the major efferent component of the peripheral nervous system and outweighs the somatic efferent pathways in the diversity of its functions and its size.

Organization and aims of the book

In this book, I will describe the properties of autonomic circuits and of single autonomic neurons or parts of autonomic neurons in the context of their biological functions in vivo so that the reader understands the principles of organization of the autonomic neurons and their interrelationships. This description should also help explain why the brain is able to adapt and coordinate the different functions of the body so precisely during our daily activities, during extreme exertion and physiological stress, as well as during various mental activities and while we experience emotions. The book is organized in five sections:

1. *The autonomic nervous system: functional anatomy and visceral afferents.* I describe the neuroanatomical basis for the precise regulation of the peripheral autonomic target organs in higher vertebrates. It provides definitions and lays the groundwork for this book. It includes a general description of the anatomy and physiology of *visceral afferent neurons*, which are closely associated with the functioning of the autonomic nervous system, with the sensations and general feelings triggered from the internal organs and, finally, also with emotions.
2. *Functional organization of the peripheral autonomic nervous system.* This section describes the functional organization of sympathetic and parasympathetic pathways in the periphery and the principles of the organization of the enteric nervous system.
3. *Transmission of signals in the peripheral autonomic nervous system.* I describe the "tools" by which the signals generated in the brain are transmitted by the autonomic pathways to the effector cells. This description includes the neurotransmitters and their receptors, ganglionic transmission and integration and neuroeffector transmission.
4. *Central representation of the autonomic nervous system in spinal cord and brain stem.* I describe some principles of organization of the autonomic control systems in the spinal cord and lower brain stem. This description will have its reference in the functional organization of the peripheral autonomic pathways that has been extensively described in the two sections before.
5. The last chapter is the *summary* and *synopsis* of the book. I will discuss how functions of autonomic systems that are represented in spinal cord and lower brain stem are integrated in complex regulatory circuits involving the somatomotor, the autonomic and the neuroendocrine systems (Figure 2) that are represented in the mesencephalon, the hypothalamus and the forebrain. I will reflect back on the concepts of how the autonomic nervous system works as those developed in the first half of the last century still direct our thinking. I will paraphrase Walter Bradford Cannon as expressed in his most influential book *The Wisdom of the Body* (Cannon 1939).

For the detailed physiology of the various autonomic control systems, as well as their specific organization in the spinal cord, brain stem and hypothalamus the reader is referred, *first*, to the series of volumes on the Autonomic Nervous System edited by Burnstock and by several volume editors (see under volume editors),[2] *second*, to Randall (1984), Loewy and Spyer (1990b), Ritter *et al.* (1992), Low (1993), Rowell (1993), Greger and Windhorst (1996), Shepherd and Vatner (1996), Appenzeller and Oribe (1997), Blessing (1997), Mathias and Bannister (2005), and Appenzeller (1999) and, *third*, to special review articles published recently (Jänig 1985a, 1986, 1988a, 1995a,

1996a; Dampney 1994; Häbler *et al.* 1994a; Spyer 1994; Kirchheim *et al.* 1998; Jänig and Häbler 1999, 2003; Folkow 2000). The approach used here should also lead to a better understanding of primary disorders of the autonomic nervous system and of autonomic disorders that are secondary to, rather than causative of, various diseases (Low 1993; Robertson and Biaggioni 1995; Appenzeller and Oribe 1997; Appenzeller 2000; Mathias and Bannister 2002). Thus, this book is not intended to discuss the pathophysiology of the autonomic nervous system; however, it is the basis to understand pathophysiological changes of autonomic functions.

Notes

1. Claude Bernard was the first to formulate the idea of the fixity of the internal milieu of the body. This formulation was not so much based on experimental observations, however, it related to the experiments he had conducted over tens of years (Bernard 1878). Physiologists became aware of the universal importance of this concept in the frame of experimental physiology and medicine when it was applied to regulation of acid–base balance in the first half of the last century.
2. Burnstock and Hoyle (1992); Hendry and Hill (1992); Maggi (1993); Nilsson and Holmgren (1994); McLachlan (1995); Bennett and Gardiner (1996); Shepherd and Vatner (1996); Unzicker (1996); Barnes (1997); Jordan (1997b); Morris and Gibbins (1997); Burnstock and Sillito (2000); Brookes and Costa (2002).

Part I

The autonomic nervous system: functional anatomy and visceral afferents

Chapter 1

Functional anatomy of the peripheral sympathetic and parasympathetic system

1.1 Definitions and limitations *page* 13
1.2 Gross anatomy of the peripheral sympathetic
 and parasympathetic nervous system 14
1.3 Reactions of autonomic target organs to activation
 of sympathetic and parasympathetic axons 24
1.4 Neuropeptides in autonomic neurons and the idea of
 "neurochemical coding" 29

Before developing the general concepts of the function of the autonomic nervous system and describing its neurobiology I will describe some aspects of the anatomy and function of this system on the macroscopic level and explain the limitations of our present understanding. Furthermore, I will provide some definitions. This conventional approach is necessary to help to convey what is meant when we speak of how the peripheral autonomic nervous system behaves and to lay the groundwork for the description of the functions in which the autonomic nervous system is involved.

1.1 Definitions and limitations

Langley (1900, 1903a, b, 1921) originally proposed the generic term *autonomic nervous system* to describe the system of nerves that regulates the function of all innervated tissues and organs throughout the vertebrate body except striated muscle fibers; that is, the innervation of the viscera, vasculature, glands, and some other tissues. This term is synonymous with the term *vegetative nervous system*, which has become obsolete in Anglo-American countries. However, the latter is still the preferred term in German ("vegetatives Nervensystem"; Jänig 2005a), French ("système végétative"), Italian ("systema neurovegetativo"), Spanish ("sistema nervioso vegetativo") and Russian ("vegetativnaia nervnaia systema"). Langley (1921) divided the autonomic nervous system into three parts: the parasympathetic nervous system, the sympathetic nervous system and the enteric nervous system. This definition has stood the test of time and is now universally used in descriptions of the autonomic nervous system in vertebrates (Nilsson, 1983; Gibbins, 1994).

The definition of the *sympathetic* and the *parasympathetic nervous systems* is based on the specialized neuroanatomical arrangement of the autonomic outflow from the central nervous system to the peripheral target tissues (see Figure 1.1). This outflow is separated into a tectal (mesencephalon), bulbar (medulla oblongata) and sacral system (collectively called craniosacral system = parasympathetic system; see Figures 1.1 and 1.4) and a thoracolumbar system (sympathetic system). The separation made by Langley was based on several criteria: the distribution of innervated target organs, the opposing effects of nerve stimulation, embryological development and the effects of exogenously applied substances (e.g., adrenaline, pilocarpine, atropine) on effector organs (Langley 1903a). The main feature distinguishing sympathetic and parasympathetic spinal outflow is their separation by the cervical and lumbar enlargements of the spinal cord (which contain the motor innervation and neural circuits related to the upper and lower limbs, respectively), and so the definition is primarily an *anatomical* one. In rats, cats and some other mammalian species the spinal parasympathetic outflow to the pelvic organs originates also from the last lumbar spinal segment. In some lower vertebrates, the distinction between sympathetic outflow and sacral parasympathetic outflow to pelvic organs is not clear. Therefore it has been proposed to use the terms *cranial autonomic outflow* and *spinal autonomic outflow* rather than the terms *sympathetic* and *parasympathetic* (Nilsson, 1983). Although there is some merit in this idea, I will not use it in this book.

Throughout the book I will use the terms *sympathetic* and *parasympathetic* as defined *anatomically* by Langley (1903a, b, 1921). When peripheral autonomic neurons or neuron populations cannot unambiguously be assigned to one of these systems (as is the case for some pelvic ganglia), I will speak of autonomic neurons. I will not use the terms *sympathetic* and *parasympathetic* in a global sense, as may be inferred from the work of Cannon (1939), Hess (1948) and others and is still commonly done in the literature, because it leads to generalizations that are not justified in view of the neurobiology and differentiation of the autonomic nervous system. Thus, to speak of sympathetic or parasympathetic "functions" generates misunderstandings and gives a wrong impression of how these systems work. This point will be discussed in more detail in Subchapter 11.1 and will be clarified in Part II. Finally, I will not apply these terms to visceral primary afferent neurons at equivalent spinal levels nor to vagal afferents (see Subchapter 2.1).

1.2 Gross anatomy of the peripheral sympathetic and parasympathetic nervous system

The basic feature of the peripheral sympathetic and parasympathetic systems is that each consists of two populations of neurons, which are arranged in series and which are connected at synapses in the periphery. The sympathetic and parasympathetic neurons that innervate the target tissue lie entirely outside the central nervous system. The cell bodies of these autonomic neurons are grouped in structures called *autonomic ganglia*. Their axons project from these ganglia to the target organs. Therefore, these neurons are called *ganglion cells* or *postganglionic neurons*. The efferent neurons that send axons from the spinal cord or brain stem into the ganglia and form synapses on the dendrites and cell bodies of the postganglionic neurons are called *preganglionic neurons*. The cell bodies of the preganglionic neurons lie in the spinal cord or in the brain stem. In subchapters 1.2 and 1.3 some aspects of the macroanatomical organization and function of the two autonomic systems are described. Details of the anatomical organization are described in the literature (Pick 1970; Gray's Anatomy 1995; Brodal 1998).[1]

Langley defined the spinal levels of the functional outflow to different organs by examining organ functions in response to stimulation of the ventral root at each segmental level. He then localized ganglionic synapses in each pathway by direct application of nicotine, a substance that excites postganglionic cell bodies. In this way, all the major efferent functional pathways to the peripheral tissues of several species of laboratory animal (cat, dog and rabbit) were defined. It was evident that every organ receives a supply from one or both of the sympathetic and parasympathetic outflows, and that the effects on each organ system are discrete and appear in some cases to be opposing each other. These experiments led to the concept of control of function of the autonomic target organs by independent peripheral autonomic pathways.

By the end of his life, Langley had also described the *enteric nervous system*, which is intrinsic to the wall of the gastrointestinal tract and is distinct from the sympathetic and parasympathetic nervous systems (see Chapter 5).

The transmission of signals from preganglionic neurons to postganglionic neurons and from postganglionic neurons to the effector cells is chemical. All preganglionic neurons are cholinergic and use *acetylcholine* as transmitter. All postganglionic parasympathetic neurons are potentially cholinergic in that they contain the enzyme choline acetyltransferase (ChAT).[2] However, not all postganglionic parasympathetic neurons seem to use acetylcholine as transmitter. The major effects of some postganglionic parasympathetic pathways appear to be mediated by cotransmitters such as nitric oxide (NO) and vasoactive intestinal polypeptide (VIP), which are synthesized in the same neurons. Whether acetylcholine present in the same neurons has another effect as yet unidentified remains to be clarified (see Subchapters 1.4, 4.8 and 7.1 and Table 1.2). Most postganglionic sympathetic neurons are adrenergic and use *noradrenaline* as transmitter; some sympathetic postganglionic neurons are cholinergic. Synaptic transmission in the peripheral sympathetic and parasympathetic systems will be described in Chapters 6 and 7.

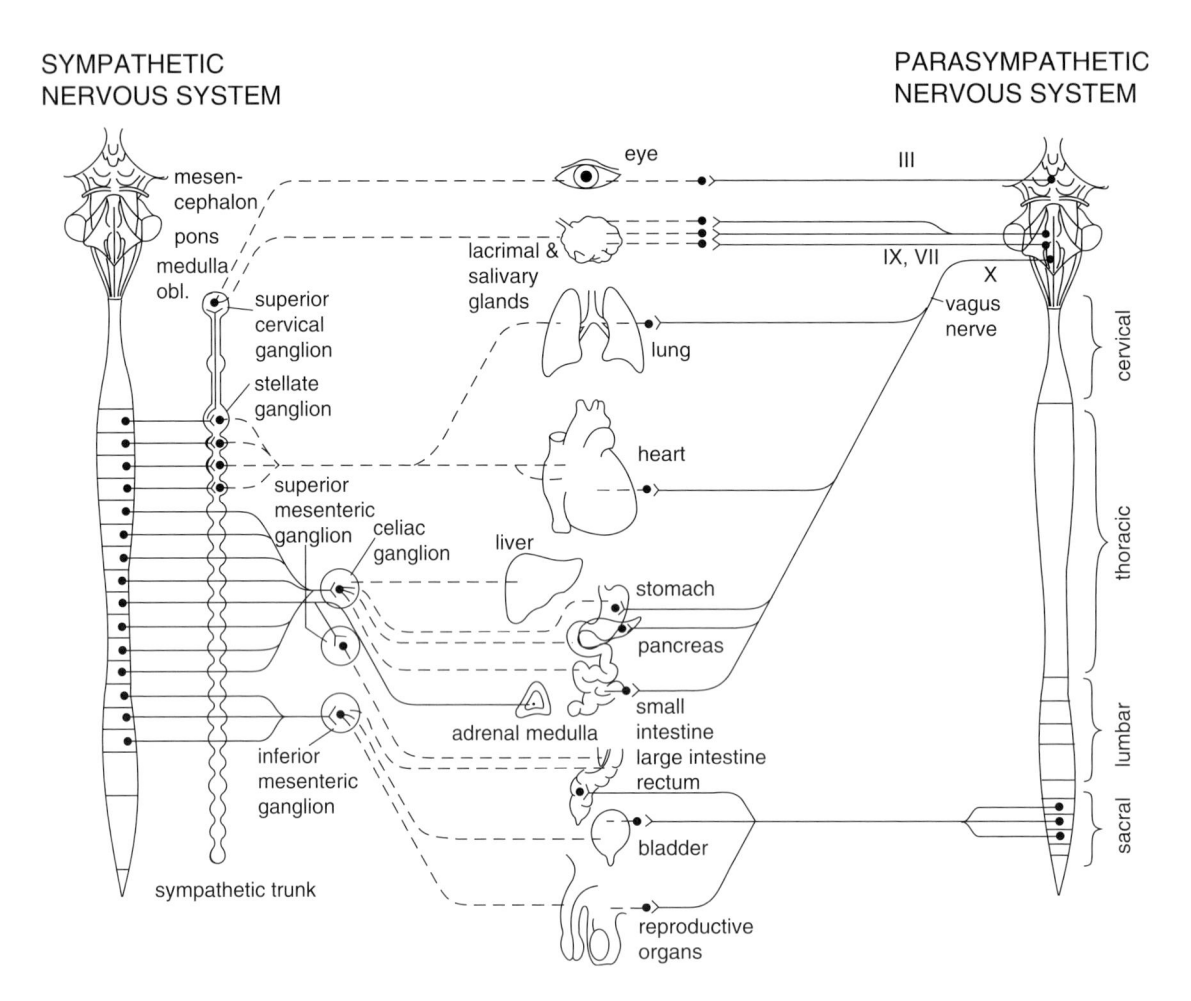

Figure 1.1 The sympathetic and parasympathetic nervous system in humans. Continuous lines, preganglionic axons; dotted lines, postganglionic axons. The sympathetic outflow to skin and deep somatic structures of the extremities (upper extremity, T2 to T5; lower extremity T12 to L3) and of the trunk are not shown. III, oculomotorius nerve; VII, facial nerve; IX, glossopharyngeal nerve; X, vagus nerve (with permission from Jänig 2005a).

1.2.1 Sympathetic nervous system

Sympathetic postganglionic neurons

Most sympathetic postganglionic neurons are aggregated in the paravertebral ganglia or in the prevertebral ganglia. The sympathetic ganglia are usually distinct in laboratory and smaller domestic animals but some may become more like plexuses in larger species like humans (particularly the prevertebral ganglia and the lumbosacral sympathetic chains). This plexus-like organization probably contributed to the erroneous belief that the sympathetic nervous system is diffusely organized.[3]

The *paravertebral ganglia* are interconnected by nerve trunks to form a chain on either side of the vertebral column, extending from the base of the skull to the sacrum. These chains are called *sympathetic trunks* (or *sympathetic chains*). The *prevertebral ganglia* (celiac, aorticorenal, superior mesenteric and inferior mesenteric)[4] lie in front of the vertebral column. They are unpaired (although they consist of right and left lobes) and mostly arranged around the origin of the major branches of the abdominal aorta. All sympathetic ganglia lie remote from the organs they supply, so that their postganglionic axons are long. The *axons of postganglionic neurons* are unmyelinated and conduct action potentials at less than 1 m/s; a few to the eye are myelinated. Sympathetic postganglionic axons do not branch in the projection course to their target tissue. They exhibit multiple branches close to their target cells. These terminal branches have up to thousands of varicosities (small swellings), which contain the biochemical machineries for synthesis, inactivation, storage, release and reuptake of the transmitter(s).

- Overall, there is approximately one pair of paravertebral ganglia per thoracic, lumbar and sacral segment. Macroscopically both sympathetic chains are sometimes fused at the lumbar level.
- The paravertebral ganglia are connected to the spinal nerves by white and grey rami.[5] Preganglionic neurons project exclusively through the white rami, which connect the ventral roots with the sympathetic trunk. Most postganglionic neurons in the paravertebral ganglia project through the grey rami to the spinal nerves, some project through the white rami to the spinal nerves and some through splanchnic nerves to viscera.
- The paravertebral ganglia are named according to the spinal nerve from which the white ramus comes. For the thoracic chain ganglia T1 to T10, the grey rami project to the same spinal nerve (e.g., grey ramus T5 to spinal nerve T5). Due to some rearrangement of white and grey rami in the thoracic segments T10 to T12, the grey rami of the lumbar and sacral chain ganglia project to the spinal nerve one segment further caudal (e.g., grey ramus from paravertebral ganglion L3 projects to spinal nerve L4). Very often there are two or more white and/or grey rami per segment. Other variations in the arrangement also exist (Pick 1970; Baron *et al.* 1985a, 1988, 1995).
- At the rostral end of the cervical sympathetic trunk (CST) lies the *superior cervical ganglion* (SCG), which contains the postganglionic neurons projecting to the head and the upper two or three cervical segments. This ganglion results from the fusion of several cervical ganglia. A small middle or intermediate cervical ganglion may exist between the SCG and the stellate ganglion, but this is inconsistent across species. In large animals, including human beings, a middle cervical ganglion is present along the CST.
- At the rostral end of the thoracic sympathetic trunk lies the *stellate ganglion* (SG), which is a combination of the lower cervical and the two or three most rostral thoracic paravertebral ganglia. This ganglion contains the postganglionic neurons projecting to the upper extremity in the brachial plexus and to the thoracic organs. These postganglionic neurons project through cardiac branches to the heart, through other branches to the lung, and through grey rami to the cervical spinal nerves C4 to C8. These grey rami are sometimes called vertebral nerve. There is no evidence that postganglionic neurons in the stellate ganglion project to intra- and extracranial target tissues (Pick 1970).

- At the caudal end, both sympathetic trunks join and sometimes form the ganglion "impar". In animals with tails, cells in the ganglion impar supply the vasculature of the distal tail.
- Most postganglionic neurons in the paravertebral ganglia project through the grey rami, the respective spinal nerves and the peripheral nerves to the effector cells in the somatic tissues. There is no experimental or clinical evidence that postganglionic sympathetic neurons in paravertebral ganglia project to the extremities along the major blood vessels, such as the subclavian and iliac arteries, as is erroneously shown in some anatomy texts (Brodal 1998). The projection of postganglionic neurons in paravertebral ganglia is shown for the innervation of the head, upper extremity and lower extremity in Figure 1.2. However, the postganglionic axons may project for a few centimeters (depending on the size of the animal species) along a blood vessel before branching into terminals on that vessel. Some postganglionic neurons in the paravertebral ganglia project through the splanchnic nerves to the viscera (Baron et al. 1985c).
- Postganglionic neurons in the SCG that project to the target organs in the head travel in perivascular bundles (with the internal and external carotid and pharyngeal arteries) and join the nerves that project to the target organs in the head. Some postganglionic neurons in the SCG project through special grey rami to the upper three or four cervical spinal nerves (Lichtman et al. 1979) or through several small branches to some other structures in the neck.
- Postganglionic neurons in the prevertebral ganglia project in nerve bundles that accompany the relevant blood vessels (e.g., the branches of the celiac, superior and inferior mesenteric arteries) or sometimes in special nerves (e.g., the hypogastric nerves) to the organs in the abdominal and pelvic cavity.

Figure 1.2 Innervation of head, neck, upper extremity and lower extremity by sympathetic postganglionic neurons and preganglionic neurons in humans. Preganglionic neurons in the thoracolumbar spinal cord project through the ventral roots and white rami (see Figure 1.3) to postganglionic neurons in the sympathetic trunk. Postganglionic neurons project through the corresponding grey rami (see Figure 1.3) to the spinal nerves or along the internal carotid artery to the head. C1, C5, T1 etc., segmental ventral root or spinal nerve. For the lower extremity only the projections in the sciatic nerve are shown. MCG, middle cervical ganglion; pre, preganglionic; post, postganglionic; SCG, superior cervical ganglion; SG, stellate ganglion. Highly schematized. Modified from Brodal (1998) with permission.

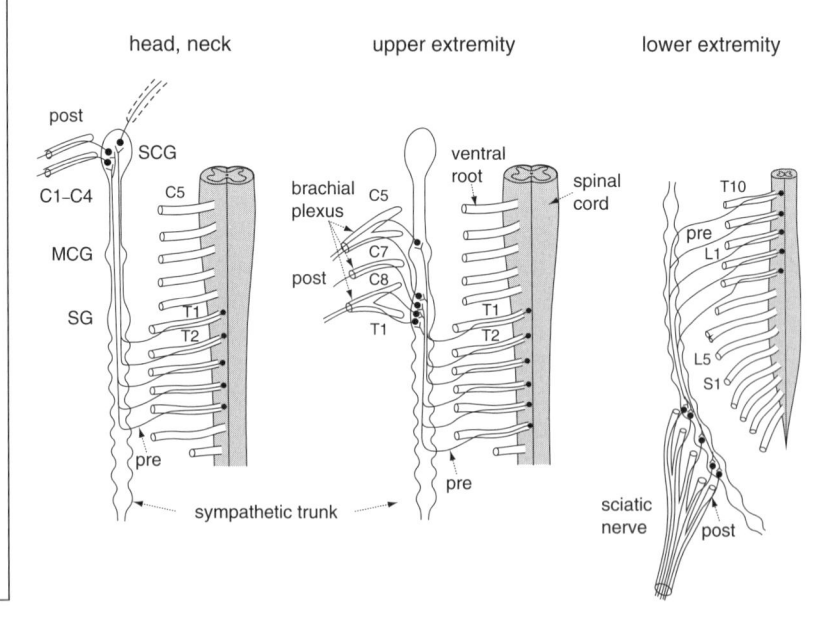

- Some aggregations of sympathetic postganglionic cell bodies are found more peripherally in ganglia in the vicinity of the pelvic organs (e.g., bladder, rectum, vas deferens, seminal vesicle, prostate). These ganglia belong to the pelvic plexus, which includes neurons that are innervated by *either* sympathetic *or* parasympathetic preganglionic axons or probably rarely by both (i.e., these latter neurons are simply autonomic since they receive convergent synaptic input from preganglionic sympathetic and parasympathetic neurons) (Keast 1995a). The preganglionic sympathetic axons to these postganglionic neurons project through the hypogastric nerves (or plexuses).

The *effector cells and organs* of the sympathetic nervous system are the smooth musculature of all organs (blood vessels, erector pili muscles, pupil, lung, evacuative organs, sphincters of the gastrointestinal tract), the heart and some glands (sweat, salivary and digestive glands). In addition, sympathetic postganglionic fibers innervate adipose tissue (white and brown), liver cells, the pineal gland and lymphatic tissues (e.g., thymus, spleen, lymph nodes and Peyer's patches in the gastrointestinal tract). Finally, activity of the enteric nervous system is modulated by postganglionic sympathetic neurons. Thus, the effects of sympathetic postganglionic neurons on gastrointestinal motility and secretion are usually mediated by at least three neurons in series.

Cells in the *adrenal medulla* are ontogenetically homologous to sympathetic postganglionic neurons. These cells are synaptically innervated by thoracic preganglionic neurons that project through the splanchnic nerves bypassing the celiac ganglion. The adrenal medulla is an endocrine gland made up of cells that release either adrenaline or noradrenaline directly into the blood, which circulates to reach tissues throughout the body. The responses of these tissues depend on the characteristic adrenoceptors and post-receptor events that are present in each, and their interaction with the innervation. However, whether this is important under physiological conditions for tissues innervated by sympathetic postganglionic fibers is questionable (see Subchapter 4.5).

Sympathetic preganglionic neurons
The cell bodies of the preganglionic sympathetic neurons lie in the *intermediate zone* of the *thoracic* and *upper lumbar spinal cord* (the upper two to four lumbar segments, depending on the species). The axons of these neurons are either myelinated or unmyelinated and conduct action potentials at 0.5 to 15 m/s. They leave the spinal cord in the ventral roots and white rami communicantes, and project (Figure 1.3):

1. either through the sympathetic chains, terminating in either of the *bilateral paravertebral ganglia*;
2. through the sympathetic chains and various splanchnic nerves (major, lesser, minor, lumbar) and terminate in the (largely

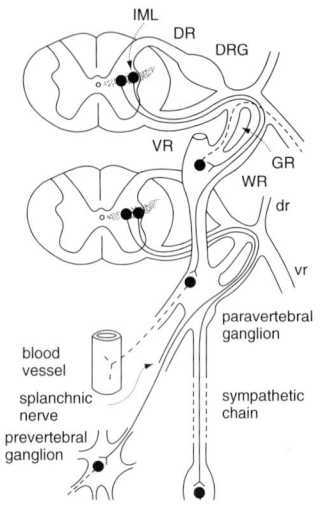

unpaired) *prevertebral ganglia* in the abdomen (celiac, aorticorenal, superior mesenteric and inferior mesenteric ganglion or plexus); or
3. further through the hypogastric nerves, terminating in the *pelvic splanchnic ganglia*.

A sympathetic preganglionic axon may travel through several paravertebral ganglia before making its main synaptic contacts (e.g., with postganglionic neurons in the SCG to the head or in lumbar paravertebral ganglia to the hindlimb). However, each preganglionic axon usually may form some synapses with postganglionic neurons in several paravertebral ganglia. There is no indication that individual preganglionic axons that innervate postganglionic neurons to the head or extremities project both up and down in the sympathetic chain, although they may do so in the midthoracic region where they innervate the upper extremities or the lower extremities. Furthermore, there is no evidence that individual preganglionic axons branch to form synapses on postganglionic neurons in paravertebral as well as prevertebral ganglia. Finally, almost all preganglionic neurons projecting to paravertebral ganglia terminate only in ipsilateral ganglia. Preganglionic neurons projecting to prevertebral ganglia terminate to about 80% on the ipsilateral side and to about 20% on the contralateral side (Baron *et al.* 1985b, c, d; Jänig and McLachlan 1986a, b).

Figure 1.3 Projection of sympathetic preganglionic neurons to postganglionic neurons in para- and prevertebral ganglia. Dotted area, intermediate zone; IML, intermediolateral nucleus; DR, dorsal root; DRG, dorsal root ganglion; VR, ventral root; GR, grey ramus; WR, white ramus; dr, dorsal ramus of spinal nerve; vr, ventral ramus of spinal nerve.

Location of sympathetic pre- and postganglionic neurons with respect to organs

Pre- and postganglionic neurons exhibit distinct topographical locations with respect to different organs. This has been detailed for various organs, skin and skeletal muscle of head and neck, the upper extremity and the lower extremity in humans in Table 1.1. The locations of these neurons in humans have been worked out with classical anatomical techniques. These locations have been fully confirmed in anatomical studies on animals (mostly cat and rat) using modern tracer techniques (see Chapter 8 and Tables 8.1, 8.2 and 8.3).

Approximately 90% of the cell bodies of sympathetic postganglionic neurons projecting through the dorsal cutaneous nerves to the dorsal skin of the trunk or through the intercostal nerves are located in the paravertebral ganglion corresponding to the dorsal root ganglion and in the next caudal paravertebral ganglion (Baron *et al.* 1995).

1.2.2 Parasympathetic nervous system

The cell bodies of preganglionic parasympathetic neurons are situated in the *brain stem* (dorsal motor nucleus of the vagus, nucleus ambiguus, superior and inferior salivary nuclei, visceral efferent oculomotor [Edinger–Westphal] nucleus[6]) and in the *intermediate zone* of the *sacral spinal cord* (Table 1.1). They project through the third (oculomotor nerve), seventh (facial and intermediate nerve) or ninth (glossopharyngeal nerve) cranial nerves to the parasympathetic ganglia associated with the intraocular smooth muscles and glands of

Table 1.1 Location of sympathetic and parasympathetic preganglionic and postganglionic neurons in humans

Organ	The sympathetic system		The parasympathetic system	
	Preganglionic neurons	Postganglionic neurons	Preganglionic neurons	Postganglionic neurons
Eye	T1–2	Superior cervical ganglion	Edinger–Westphal nucleus	Ciliary ganglion
Lacrimal gland	T1–2	Superior cervical ganglion	Superior salivary nucleus	Pterygopalatine ganglion
Submandibular & sublingual glands	T1–2	Superior cervical ganglion	Superior salivary nucleus	Submandibular ganglion
Parotid gland	T1–2	Superior cervical ganglion	Inferior salivary nucleus	Otic ganglion
Heart	T1–4 (T1–T5)	Stellate ganglion, upper thoracic ganglia (superior & middle cerv. ganglion)	Nucleus ambiguus	Cardiac plexus
Bronchi, lungs	T2–7 (T2–T4)	Stellate ganglion & upper thoracic ganglia	Nucleus ambiguus	Pulmonary plexus
Esophagus	(T5–T6)	Stellate ganglion & upper thoracic ganglia	Dorsal motor nucleus of the vagus nerve	Myenteric plexus
Stomach	T6–10	Celiac ganglion	Dorsal motor nucleus of the vagus nerve	Myenteric and submucosal plexus
Liver, gallbladder	(T7–T9)	Celiac ganglion	Dorsal motor nucleus of the vagus nerve	
Pancreas	T6–10	Celiac ganglion	Dorsal motor nucleus of the vagus nerve	Periarterial plexus?
Small, ascending large intestine	T6–10 (T9–T10)	Celiac, superior and inferior mesenteric ganglia	Dorsal motor nucleus of the vagus nerve	Myenteric and submucosal plexus
Transverse large intestine	(T11–L1)	Celiac, superior and inferior mesenteric ganglion	Dorsal motor nucleus of the vagus nerve	Myenteric and submucosal plexus
Descending large intestine, sigmoid & rectum	T11–L2 (L1–L2)	Inferior mesenteric, hypogastric & pelvic ganglia	S3–S4	Myenteric and submucosal plexus
Ureter, urinary bladder	T11–L2	Hypogastric & pelvic ganglia	S3–S4	Pelvic ganglia
Internal sexual organs, male	T11–L2	Hypogastric & pelvic ganglia	S3–S4	Pelvic ganglia
Internal sexual organs, female	(T10–L1)	Hypogastric & pelvic ganglia	S3–S4	Pelvic ganglia

Table 1.1 (cont.)

Organ	The sympathetic system		The parasympathetic system	
	Preganglionic neurons	Postganglionic neurons	Preganglionic neurons	Postganglionic neurons
Head, neck (skin and skeletal muscle)	T1–4 (T1–T5)	Superior & middle cervical ganglia[a]	No parasympathetic innervation	
Upper extremity	T3–6 (T2–T5)	Stellate & upper thoracic ganglia (projection to spinal nerves C5–C8)[b]	No parasympathetic innervation	
Lower extremity	Th10–L2	Lumbar & upper sacral ganglia (ganglia L1–S1)[c]	No parasympathetic innervation	
Trunk (innervated by dorsal rami and intercostal nerves)	Most neurons in 2 segments corresp. to paravertebral ganglia	90% neurons in two paravertebral ganglia (corresp. to dorsal root ganglion and next caudal paravertebral ganglion)[d]	No parasympathetic innervation	

Notes:
Modified from Brodal (1998) and from Gray's Anatomy (1995). Segmental locations in brackets from Gray's Anatomy (1995).
[a] The superior cervical ganglion projects to the spinal nerve C1 and C2 (possibly C3). The middle cervical ganglion projects inconstantly to the spinal nerves C3 to C6.
[b] The stellate ganglion projects through the vertebral nerve to the spinal nerves C7 and C8 and frequently also to the spinal nerves (C4) C5 to C6.
[c] In the lumbar sympathetic chain the ganglia that are named by their white ramus (rami) project with their grey ramus (rami) to the next spinal nerve further caudal, i.e. the lumbar ganglion L1 to the spinal nerve L2, the lumbar ganglion L2 to the spinal nerve L3 etc. (see Pick 1970; Baron et al. 1985a).
[d] See Baron et al. 1995.

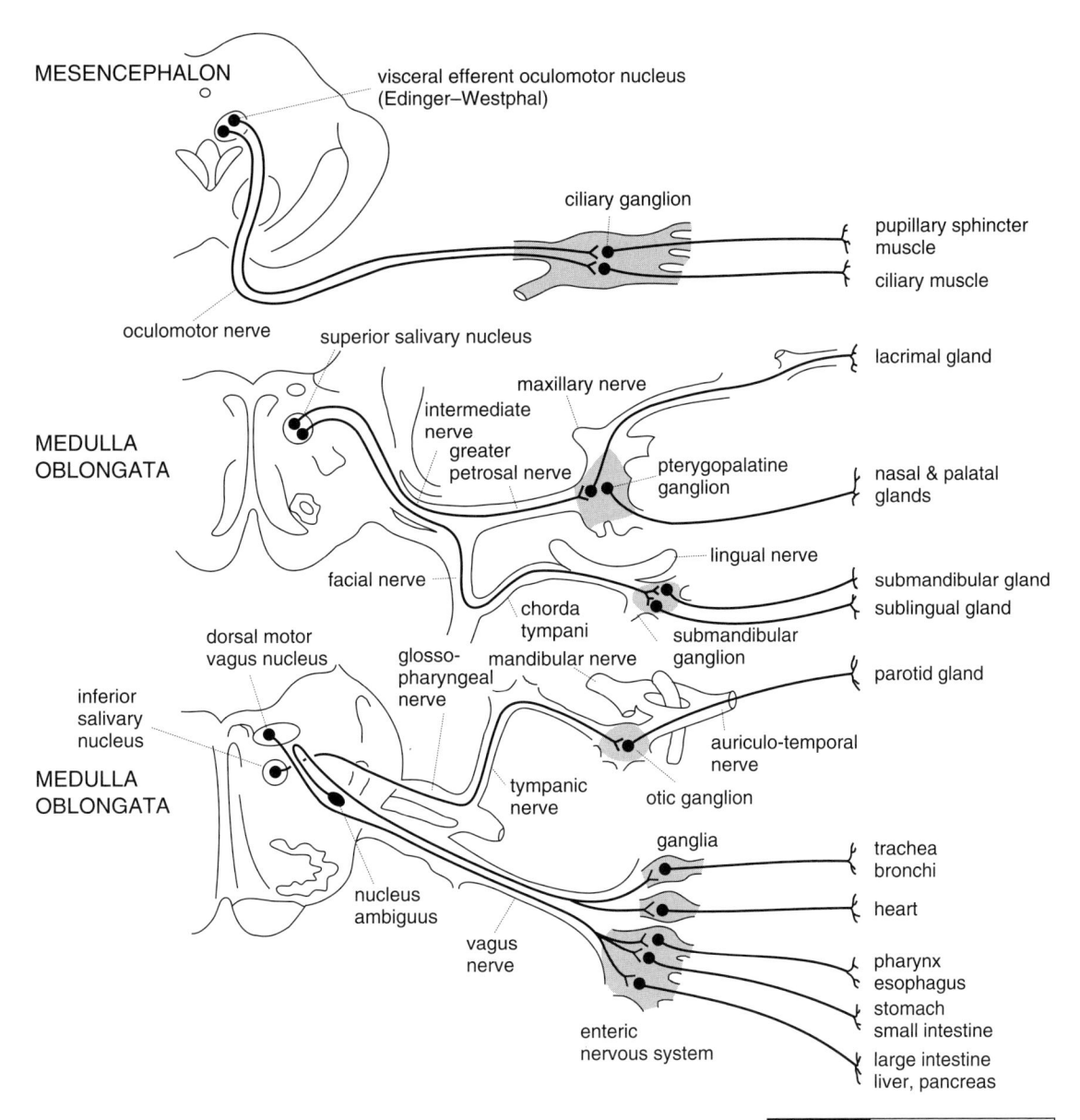

the head, through the tenth cranial (vagal) nerve to the ganglia associated with the target organs in the thoracic and abdominal cavity and through the pelvic splanchnic nerves to the pelvic ganglia associated with the pelvic organs. The preganglionic axons can be very long and are either myelinated or unmyelinated.

Discrete *parasympathetic ganglia* containing the postganglionic parasympathetic neurons are found in the head region. Figure 1.4 shows schematically these ganglia, the location of the preganglionic neurons in the brain stem, the nerves through which these preganglionic neurons project and the target organs of the postganglionic neurons (ciliary ganglion: eye; pterygopalatine ganglion: lacrimal, nasal and palatal glands; otic and submandibular ganglion: salivary

Figure 1.4 Location of cranial parasympathetic preganglionic neurons, the nerves through which the preganglionic axons project, the parasympathetic ganglia and the target organs. See also Table 9.1. Modified from Brodal (1998) with permission.

glands). Discrete ganglia are also found near or in the wall of the effector organs (heart: cardiac plexus [discrete small ganglia innervating different groups of effector cells, e.g., pacemaker cells in the sinus venosus, atrial muscle cells, atrioventricular pacemaker cells]; airways: ganglia on the membranous part; pancreas [exocrine, endocrine]; gallbladder; organs in the pelvic cavity: ganglia in the pelvic splanchnic plexus, see McLachlan [1995]). Preganglionic neurons that project to the gastrointestinal tract synapse with neurons that are part of the enteric nervous system (Figure 1.4; see Chapter 5). The sacral parasympathetic outflow to the hindgut largely consists of three populations of neurons connected synaptically: preganglionic neurons in the sacral spinal cord, postganglionic neurons in the pelvic splanchnic ganglia and neurons of the enteric nervous system (neurons in the myenteric plexus or serosal neurons) (Table 1.1).

The parasympathetic system innervates and activates the exocrine glands of the head, the intraocular smooth muscles, the smooth muscles and glands of the airways, the smooth musculature, exocrine glands and endocrine glands of the gastrointestinal tract (via the enteric nervous system), the heart (pacemaker cells and atria), the pelvic organs (lower urinary tract, hindgut, reproductive organs) and epithelia and mucosa throughout the body. Except for the helical arteries and sinusoids of the erectile tissues of the reproductive organs and some intracranial, uterine and facial blood vessels and blood vessels in the salivary glands, the parasympathetic system does not innervate blood vessels in somatic tissues and viscera. The parasympathetic supply to these vascular tissues is not involved in blood pressure regulation but is related to specific functions of each effector vessel.

1.3 | Reactions of autonomic target organs to activation of sympathetic and parasympathetic axons

Table 1.2 describes the overall reactions of the peripheral target organs and individual target tissues to activity in the sympathetic and parasympathetic neurons that innervate them. The responses have been defined by the reaction of the tissues to reflex activation or to electrical stimulation of the respective nerves, but not pharmacologically by their reactions to adrenoceptor or cholinoceptor agonists. In some cases, the apparent effects of these nerves have been identified by changes in function after removing or blocking the nerve supply. The table shows that:

- Most target tissues are innervated by only one of the autonomic systems.
- A few target tissues are innervated by both autonomic systems (e.g., iris, pacemaker cells and atria of the heart, urinary bladder, some blood vessels [erectile tissue, in salivary glands and oral mucosa, intracranial blood vessels]).

Table 1.2 | Effects of activation of sympathetic and parasympathetic neurons on autonomic target organs

Organ and organ system	Activation of parasympathetic nerves	Activation of sympathetic nerves	Adrenoceptor
Heart muscle	Decrease of heart rate, Decrease of contractility (only atria)	Increase of heart rate, Increase of contractility (atria, ventricles)	β_1 β_1
Blood vessels[a] Arteries in			
... skin of trunk and limbs	0	Vasoconstriction	α_1
... skin and mucosa of face	Vasodilation	Vasoconstriction	α_1
... visceral domain (inc. kidney)	0	Vasoconstriction	α_1
... skeletal muscle	0	Vasoconstriction	α_1
... heart (coronary arteries)	?	Vasodilation (cholinergic)[b] Vasoconstriction	α_1
... intracranial tissues	Vasodilation	Vasoconstriction	α_1
... salivary glands	Vasodilation	Vasoconstriction	α_1
Veins	0	Vasoconstriction	α_1
Gastrointestinal tract			
longitudinal and circular muscle	Increase of motility	Decrease of motility (mostly indirect)	α_2, β_1
sphincters	Relaxation	Contraction	α_1
digestive glands	Secretion	Decrease of secretion or 0 (stomach, pancreas)	
mucosa	Secretion	Decrease of secretion or 0 (small, large intestine) reabsorption[c]	
endocrine cells (e.g., gastrin by G cells)	Secretion	0	
Capsule of spleen	0	Contraction	
Kidney juxtaglomerular cells	0	Renin release	β_1

Table 1.2 (cont.)

Organ and organ system	Activation of parasympathetic nerves	Activation of sympathetic nerves	Adrenoceptor
tubuli	0	Sodium reabsorption	α_1
Urinary bladder			
detrusor vesicae	Contraction (micturition)	Relaxation (small)	β_2
trigone (internal sphincter)	0	Contraction (continence)	α_1
urethra	Relaxation	Contraction (continence)	α_1
Reproductive organs			
seminal vesicle, prostate	0	Contraction	α_1
vas deferens	0	Contraction	α_1
uterus	Vasodilation	Contraction; relaxation[d]	α_1
glands	Secretion	0	
erectile tissue	Vasodilation (helical arteries and sinusoids in penis and clitoris)	Vasoconstriction Vasodilation	α_1
vagina	Engorgement (vasodilation), transepithelial transudation of mucoid fluid	Vasoconstriction	
Eye			
dilator muscle of pupil	0	Contraction (mydriasis)	α_1
sphincter muscle of pupil	Contraction (miosis)	0	
ciliary muscle	Contraction (accommodation)	0	
blood vessels (chorioid)	Vasodilation	(vasoconstriction?)	
tarsal muscle	0	Contraction (lifting of lid)	
orbital muscle	0	Contraction (protrusion of eye)	
Tracheobronchial muscles	Contraction	Relaxation (probably mainly by circulating catecholamines)	β_2

Piloerector muscles	0	Contraction	α_1
Exocrine glands[e]			
salivary glands	Copious serous secretion	Weak mucous secretion (submandibular gland)	α_1
lacrimal glands	Secretion	0	
nasopharyngeal glands	Secretion		
bronchial glands	Secretion	?	
sweat glands	0	Secretion (cholinergic)	
Pineal gland	0	Increase in synthesis of melatonin	β_2
Brown adipose tissue	0	Heat production	β_3
Metabolism			
liver	0	Glycogenolysis, gluconeogenesis	β_2
fat cells	0	Lipolysis (free fatty acids in blood increased)	β_2
β-cells in islets of pancreas	Secretion of insulin	Decrease of secretion of insulin	α_2
α-cells in islets of pancreas		Secretion of glucagon	β
Adrenal medulla	0	Secretion of adrenaline and noradrenaline	β_2[f]
Lymphoid tissue	0	Depression of activity (e.g., of natural killer cells)	α

Notes:
The right column shows the type of adrenoceptor in the membranes of the target cells mediating the effect of activation of sympathetic neurons (see Chapter 8). 0 indicates no effect.

[a] Other blood vessels see Eye and Reproductive organs.
[b] Only in some species (e.g., cat, see Bolme et al. [1970]). Circulating adrenaline may produce vasodilation via β_2-adrenoceptors.
[c] Mostly indirect by inhibiting secretomotor neurons; also indirectly following vasoconstriction?
[d] Depends on species and hormonal state; only exogenously by noradrenaline.
[e] Other glands see Gastrointestinal tract and Reproductive organs; secretion of all exocrine glands generated by activation of secretomotor neurons is accompanied by vasodilation of the associated vasculature.
[f] Adrenaline is a metabolic hormone and acts via β_2-adrenoceptors (see Subchapter 4.5).

- Opposite reactions to the activation of sympathetic and parasympathetic neurons are more the exception than the rule (e.g., pacemaker and atria of the heart, erectile tissue of the reproductive organs, some exocrine digestive glands, insulin-producing islet cells of the pancreas).
- Most responses are excitatory even when opposing actions of the organ result; inhibition (e.g., relaxation of muscle, decreased secretion) is rare.
- Adrenaline is a metabolic hormone. It is released by cells of the adrenal medulla in the blood during synaptic activation.

Table 1.2 clearly shows that the widely propagated idea of a universal antagonism between the parasympathetic and sympathetic nervous systems is a misconception. Where there is a reciprocal effect of the two autonomic systems on some target tissues, it can usually be shown either that the systems work synergistically or that they exert their influence under different functional conditions. For example, the opposite actions of sympathetic and parasympathetic systems on the size of the pupil is a consequence of the distinct target muscles (dilator pupillae for sympathetic, sphincter pupillae for parasympathetic) supplied by each system. Moreover, in larger mammals, fast changes of heart rate, e.g., during changes of body position and emotional stress, are generated via changes in activity in the parasympathetic neurons to the pacemaker cells; the sustained increase of heart rate during exercise is mainly generated by activation of sympathetic neurons supplying the heart. In addition, it is likely that some organs which can be affected by both systems under experimental conditions are primarily under the control of only one system in vivo (i.e., under physiological conditions).

It is evident from Table 1.2 that both systems have some very specialized effects on particular organs but also generate functional effects that are similar in many tissues. For example, the parasympathetic innervation of mucosae and glands is almost always responsible for the generation of watery secretions, whereas at least part of the sympathetic innervation of most organs is associated with vasoconstriction.

In essence, Table 1.2 shows the macroscopic effects of activation of sympathetic and parasympathetic neurons on target organs; it does not show whether these responses have functional meaning nor how the autonomic systems work to regulate the behavior of these target organs. Finally, it is important to recognize that the effects of nerve activity on the autonomic target organs are not necessarily the same as the reactions of the effector organs to application of exogenous transmitter substances (see Subchapter 7.3) or to circulating adrenaline from the adrenal medulla (see Subchapter 4.5).

As already mentioned and as will be treated more extensively in Chapter 7, most sympathetic postganglionic neurons are noradrenergic (in mammals; and adrenergic in reptiles and amphibians) and all parasympathetic postganglionic neurons are cholinergic. Noradrenaline released by the postganglionic fibers reacts with adrenoceptors in the membranes of the effector cells. These adrenoceptors consist of two families each being divided into several types (Alexander et al. 2004). Some of these types of adrenoceptors mediate the nerve-induced effector responses. The type of adrenoceptor responsible for responses elicited by nerve-released noradrenaline is listed in Table 1.2. Acetylcholine released by postganglionic neurons reacts with cholinergic muscarinic receptors in all effector cells. It is important to emphasize that it is a misconception to identify the peripheral sympathetic nervous system as noradrenergic (autonomic) nervous system and the peripheral parasympathetic nervous system as cholinergic (autonomic) nervous system as it is sometimes done.

1.4 Neuropeptides in autonomic neurons and the idea of "neurochemical coding"

The presence of distinct functional subunits of the autonomic nervous system, in the sense of Langley and as outlined in this chapter, is supported by a substantial amount of histochemical evidence demonstrating that autonomic neurons may contain particular combinations of neuropeptides along with (and a few without) one of the classical transmitters, noradrenaline (NAd) or acetylcholine (ACh). The patterns of coexistence of one or more neuropeptides and a non-peptide transmitter within the same cell body in peripheral ganglia are correlated with combinations of both coexisting substances in the nerve terminals associated with target tissues (see Furness et al. [1989]; Gibbins [1990, 1995, 2004]; Morris and Gibbins [1992]). Various combinations of substances are present in the innervation of various target organs. Furthermore, there are many examples of species-specific neuropeptide expression (see Nilsson and Holmgren [1994]).

Neuropeptides in sympathetic neurons
Examples of neuropeptide/neurotransmitter combinations in sympathetic postganglionic neurons in the cat are (Table 1.3):

- Most vasoconstrictor neurons to viscera, skin and skeletal muscle contain, in addition to NAd, neuropeptide Y (NPY) and galanin (GAL).
- Muscle vasodilator neurons have, presumably in addition to ACh, vasoactive intestinal peptide (VIP).
- In addition to ACh, sudomotor neurons contain VIP, calcitonin gene-related peptide (CGRP) and substance P (SP).
- In addition to NAd, pilomotor neurons contain GAL.
- In addition to NAd, motility-regulating neurons contain somatostatin (SOM).
- In secretomotor neurons to the gastrointestinal tract or to the salivary glands no peptide has been detected so far.

Table 1.3 | *Neurochemical coding of different functional populations of postganglionic sympathetic neurons in the cat and guinea pig*

Sympathetic system	Primary transmitter	Neuropeptide(s)
Cat		
Muscle vasoconstrictor	Noradrenaline	GAL/NPY
Muscle vasodilator	Acetylcholine	VIP
Visceral vasoconstrictor	Noradrenaline	GAL/NPY
Cutaneous vasoconstrictor	Noradrenaline	GAL/NPY
Pilomotor	Noradrenaline	GAL
Sudomotor	Acetylcholine	VIP/CGRP/SP
Salivary secretomotor	Noradrenaline	Ø
Motility-regulating (to ENS)[a]	Noradrenaline	SOM/?
Guinea pig		
Muscle vasoconstrictor	Noradrenaline	NPY
Muscle vasodilator	Acetylcholine	VIP/NPY/DYN
Visceral vasoconstrictor	Noradrenaline	NPY
Cutaneous vasoconstrictor	Noradrenaline	NPY, NPY/DYN, DYN
Pilomotor	Noradrenaline	DYN
Sudomotor	Acetylcholine	VIP/CGRP
Pupillodilator	Noradrenaline	NPY/DYN
Salivary secretomotor	Noradrenaline	Ø
Motility-regulating (to ENS)[a]	Noradrenaline	Ø
Secretomotor (to ENS)[b]	Noradrenaline	SOM
Visceromotor[c]	Noradrenaline	NPY, NPY/DYN

Notes:
[a] Postganglionic neurons in the celiac ganglion innervating the myenteric plexus
[b] Postganglionic neurons in the celiac ganglion innervating the submucous plexus
[c] Sympathetic non-vasoconstrictor system which directly innervates visceral organs; this group does not include sympathetic postganglionic NA/SOM and NA/Ø neurons (see notes a and b).
Ø, no colocalized neuropeptide; CGRP, calcitonin gene-related peptide; DYN, dynorphin; ENS, enteric nervous system; GAL, galanin; NPY, neuropeptide Y; SOM, somatostatin; SP, substance P; VIP, vasoactive intestinal peptide. From Gibbins (1995).

Detailed investigation of the cutaneous vasoconstrictor neurons innervating the ear of the guinea pig has shown that different sections of the cutaneous vascular bed are innervated by neurochemically different groups of noradrenergic (NAd) neurons: (1) large distributing arteries are innervated by neurons containing NAd and NPY; (2) small arteries are innervated by neurons containing NAd, NPY and a prodynorphin-derived peptide; (3) arteriovenous anastomoses and precapillary resistance vessels are innervated by neurons containing NAd and a prodynorphin-derived peptide (but no NPY); (4) large veins are innervated by neurons containing NAd, NPY and a prodynorphin-derived peptide; (5) small veins are innervated by neurons containing NAd and a prodynorphin-derived peptide, but lack mostly NPY (Gibbins and Morris 1990; Morris 1995, 1999; Gibbins *et al.* 2003a). These findings support the hypothesis that different sections

Table 1.4 | *Neurochemically identified preganglionic and postganglionic pathways in lumbar sympathetic ganglia of female guinea pig*

Sympathetic system	Preganglionic neuron	Postganglionic neuron
Vasoconstrictor	ACh; Ø (most), CGRP, ENK, CGRP/ENK	NAd; NPY (most), NPY/DYN, Ø/DYN
Vasodilator	ACh; Ø, SP, ENK, SP/ENK	ACh; VIP/NPY/DYN
Pilomotor	ACh; Ø, ENK	NAd; DYN
Visceromotor[a]	ACh; ENK	NAd; NPY

Notes:
Primary transmitters acetylcholine (ACh) or noradrenaline (NAd). The table shows with which neuropeptide or combinations of neuropeptides the primary transmitter may be colocalized. ENK, enkephalin; for further abbreviations see Table 1.3.

[a] Sympathetic non-vasoconstrictor system that directly innervates visceral organs; this group does not include sympathetic postganglionic NAd/SOM and NAd/Ø neurons that innervate the submucous plexus or the myenteric plexus of the enteric nervous system, and are involved in regulation of secretion and motility, respectively. From Gibbins (1995).

of the cutaneous vascular bed are also innervated by functionally distinct cutaneous vasoconstrictor neurons and are consistent with the neurophysiological studies of cutaneous vasoconstrictor neurons in the cat (see Subchapter 4.1). Table 1.3 also demonstrates the difference, in neuropeptide content of functionally different types of postganglionic neurons, between cat and guinea pig to emphasize species differences.

Although only a few types of preganglionic neurons have been found to contain identified neuropeptides in the cat, this neurochemical coding is also present to a certain extent in these pathways in this species (Lundberg *et al.* 1988; Lindh *et al.* 1989, 1993; Moriarty *et al.* 1992; Shafton *et al.* 1992; Anderson *et al.* 1995; Edwards *et al.* 1996; see Gibbins [1995, 1997]).

For the guinea pig and the rat more complete sets of data on the presence and absence of peptides in pre- and postganglionic neurons of some sympathetic pathways are available (Table 1.4). These neurochemical codes are correlated with morphological data (geometry and size of soma and dendrites), electrophysiological data and anatomical data defining the projections of the sympathetic neurons. Unfortunately we do not have data about the discharge pattern of these neurons in vivo for the guinea pig and we have only limited data on the sympathetic innervation for the rat (see Chapter 4 for details).

Neuropeptides in parasympathetic neurons

Data on neuropeptides in the parasympathetic neurons of humans has been described in detail by Gibbins (2004). In the guinea pig VIP is colocalized with acetylcholine in all types of postganglionic *parasympathetic neurons to cranial targets* except for those innervating the iris and ciliary muscle. Postganglionic neurons of the parasympathetic pathways to lacrimal, nasal, parotid, sublingual and submandibular glands and to the chorioid, cerebral and facial blood vessels contain additionally nitric oxide (NO). The VIPergic parasympathetic postganglionic neurons to cranial targets are involved in secretion and vasodilation.

Postganglionic neurons of *vagal (parasympathetic) pathways* to airways (bronchomotor neurons), esophagus or gallbladder generating contraction have no known neuropeptide colocalized with acetylcholine. Postganglionic parasympathetic cardiomotor neurons (innervating the sinoatrial node, the atrioventricular node and the atrial musculature and generating bradycardia and decrease of contractility) contain SOM (in addition to acetylcholine). All other types of postganglionic neurons that elicit relaxation of gastrointestinal muscles, vasodilation or secretion contain VIP and NO. These neurons are either cholinergic or non-cholinergic, some of the latter (to the esophagus) containing NPY.

Postganglionic neurons of *sacral parasympathetic pathways* to distal colon and bladder body are cholinergic and do not contain a known neuropeptide. Other sacral postganglionic neurons contain VIP and NO (and some additionally NPY). These neurons are either cholinergic (urethra, male internal reproductive organs, erectile tissue) or non-cholinergic (urether, detrusor vesicae, uterus, vagina, internal anal sphincter). They generate either vasodilation, relaxation of visceral smooth muscle or secretion.

The concept of neurochemical coding

The term neurochemical coding has been coined for the coexistence of a classical transmitter and one or a set of neuropeptides in the autonomic neurons associated with a particular target tissue. The patterns of coding can often be used to identify different functional populations of autonomic neurons.

This concept is an interesting development, and at present it is a powerful neuroanatomical and experimental tool in the analysis of the peripheral pathways of the autonomic nervous system, in particular the sympathetic one (Gibbins 1995, 1997). The functions of most neuropeptides are unknown, although they are beginning to be revealed for a few neuropeptides (Lundberg 1981; Furness *et al.* 1992; Ulman *et al.* 1992). In much of the literature it is assumed that these neuropeptides act as transmitters as the terms transmitter, cotransmitter, neuromodulator etc. are generally applied to them. However, this assumption lacks experimental evidence and it may finally turn out that most neuropeptides do not have any obvious function as neurotransmitters in the adult organism. The principle of neurochemical coding has also been worked out and exploited to unravel the connectivity and function of the enteric nervous system in the guinea pig (Furness *et al.* 1992, 2003a; Costa *et al.* 1996; see Subchapter 5.1 and Figure 5.2).

In *conclusion*, (1) neurochemical coding of autonomic neurons can be used as a tool to unravel the organization of the autonomic pathways, (2) neurochemical coding of many functionally distinct autonomic pathways is not straightforward, (3) the correlation between target tissue and neurochemistry is by no means absolute, (4) the function of the neuropeptides is mostly unclear, and (5) neuropeptides expressed in particular functional pathways can differ markedly between species, even where a function has been demonstrated (Gibbins and Morris 1987; Potter 1991; Romano *et al.* 1991; Gibbins 1992, 1995; Ulman *et al.* 1992).

Conclusions

1. The definition of the sympathetic and parasympathetic nervous system is anatomical and based on the levels of outflow and therefore the developmental origin from the neuraxis. Visceral afferent neurons are not included in this definition, although they are indispensable to understand the functioning of both autonomic systems (see Chapter 2).
2. The sympathetic system originates from the thoracic and upper lumbar spinal segments and is therefore called the thoracolumbar system.
3. The parasympathetic system originates from the brain stem (medulla oblongata, mesencephalon) and sacral spinal cord and is therefore called the craniosacral system.
4. Both systems are efferent and consist of a chain of preganglionic and postganglionic neurons, which are synaptically connected in peripheral autonomic ganglia.
5. Sympathetic ganglia are situated remote from the target organs and organized bilaterally in the sympathetic chains and in the prevertebral ganglia. Parasympathetic ganglia are situated close to the target organs.
6. Stimulation of sympathetic neurons produces many distinct effector responses elicited from a variety of cell and tissue types including blood vessels, heart, non-vascular smooth muscles, exocrine epithelia, endocrine cells (of the pancreas), adipocytes and lymphoid tissues.
7. The adrenal medulla is an endocrine gland made up of cells that release either adrenaline or noradrenaline during synaptic activation by sympathetic preganglionic neurons. Adrenaline is a metabolic hormone.
8. Stimulation of parasympathetic neurons leads to activation of most exocrine glands, non-vascular smooth muscles, pacemaker and atria of the heart and some other target cells. It influences only a few specialized blood vessels.
9. Most target tissues react under physiological conditions to only one of the autonomic systems. Only a few react to both. Opposite reactions to sympathetic and parasympathetic inputs are more the exception than the rule.
10. The widely propagated idea of the antagonism between sympathetic and parasympathetic nervous systems is misleading.

11. Postganglionic neurons (and to some degree also preganglionic neurons) of peripheral autonomic pathways contain combinations of neuropeptides colocalized with (or some without) the classical transmitters acetylcholine or noradrenaline. The peptide content may to some extent correlate with the function of the autonomic neurons. However, there are species and organ differences and the function of most peptides is unknown.

Notes

1. A complete description of the anatomy of the peripheral sympathetic and parasympathetic nervous system in vertebrates (including humans) has been done by Pick 1970. This book documents most variations of autonomic ganglia, white and grey rami, the sympathetic chain, and the splanchnic nerves in detail.
2. Choline acetyltransferase (ChAT) catalyzes the condensation of choline and acetyl-coenzyme A, forming acetylcholine and coenzyme A. Antibodies against ChAT are used to identify cholinergic postganglionic neurons immunohistochemically (Keast *et al.* 1995).
3. The nomenclatures used to describe the macroscopy of the peripheral sympathetic nervous system are sometimes confusing, not only for the beginner but also for the experienced experimenter. This is related to differences in species, to special emphasis on the human and probably also to differences in academic teaching by anatomists and by physiologists. For example, celiac ganglion (ganglia) and celiac plexus are equivalent; the inferior mesenteric ganglion (IMG) in the human is equivalent to the proximal IMG in the cat; the superior hypogastric plexus in the human is equivalent to the distal IMG in the cat; the abdominal aortic plexus or intermesenteric plexus in the human is equivalent to the intermesenteric nerve in the cat etc. Without discussing these differences further, the reader is referred to Pick (1970) [all laboratory animals, human], Baron *et al.* (1985a) [cat], Baron *et al.* (1988, 1995) [rat], Jänig and McLachlan (1987) [lumbar sympathetic outflow to viscera, all laboratory animals and human], and Gray's Anatomy (1995) [human].
4. In books of anatomy the celiac ganglion is called celiac plexus and the inferior mesenteric ganglion superior hypogastric plexus (for details of nomenclature used in humans/primates and smaller animals see Jänig and McLachlan [1987]).
5. White rami appear macroscopically white in cadavers of humans and other higher vertebrates because many sympathetic preganglionic axons are myelinated. Grey rami appear macroscopically grey because postganglionic axons are unmyelinated.
6. Only in primates and birds. In rat, rabbit and cat most cell bodies of parasympathetic preganglionic neurons to the eye are situated ventral to the Edinger–Westphal nucleus (see Subchapter 8.2).

Chapter 2

Visceral afferent neurons and autonomic regulations

2.1 Visceral afferent neurons: general characteristics *page* 37
2.2 Visceral primary afferent neurons as interface between visceral organs and brain 42
2.3 Receptive functions of visceral afferent neurons 45
2.4 Role of visceral afferent neurons in visceral nociception and pain 54
2.5 Relation between functional types of visceral afferent neurons, organ regulation and sensations 60
2.6 Central ascending pathways associated with autonomic regulations and visceral sensations 65

The brain continuously receives messages from internal organs of the body. These afferent messages are neuronal, hormonal, chemical and physical in nature (see Figure 1). The continuous afferent feedback to the brain confers information about the state of internal organs (e.g., the degree and composition of filling of the gastrointestinal tract), the state of parameters regulated homeostatically (e.g., the level of systemic arterial blood pressure; the concentration of glucose, oxygen and bicarbonate in the blood; the size of fat stores by the concentration of leptin), the activity of endocrine glands (by the concentration of circulating hormones secreted by these glands) and the state of peripheral protective mechanisms of the body (e.g., by activity in nociceptive afferents, by signals from immune tissues [cytokines acting directly on the brain or indirectly via vagal afferents]). The multiple afferent feedback signals to the brain from various organs and tissues of the body are essential to achieve the precision of the homeostatic short- and long-term regulations in which the autonomic nervous systems are involved. This afferent (sensory) feedback connects to all levels of the autonomic motor hierarchy, to the centers of the cerebral hemispheres, which are responsible for conscious sensations and cognitive inputs to the motor hierarchy, and to the behavioral state system (Swanson 2000, 2003; see Introduction and Figure 2; see Subchapters 11.4 and 11.6).

Details about these regulations and the functional specificity of the afferent signals with respect to the autonomic regulations will be discussed in Part III of this book. In the following Chapter I will describe functional aspects of vagal and spinal visceral afferent neurons in relation to visceral sensations and autonomic regulations, with some emphasis on visceral pain and body protection. Anatomy and physiology of the afferent neurons for each organ is described in detail in the literature (see Thorén [1979]; Coleridge and Coleridge [1980, 1984]; Malliani [1982]; Mei [1983, 1985]; Cervero and Morrison [1986]; Grundy [1988]; Grundy and Scratcherd [1989]; Jänig and Koltzenburg [1990, 1993]; Ritter *et al.* [1992]; Cervero [1994]; Gebhart [1995]; Jänig [1996d]; Bielefeld and Gebhart [2005]; Undem and Weinreich [2005]).

Sherrington (1906) made the functional distinction between interoception, exteroception and proprioception. Exteroceptive afferent neurons innervating the body surface are involved in the communication between environment and body. Proprioceptive afferent neurons monitor events in the deep somatic tissues, in particular skeletal muscle and joints. Both types of afferent neuron are mechanosensitive and have large-diameter myelinated (Aα, Aβ) axons. Interoceptive afferent neurons monitor events in the viscera (originally, events at the inner surface of the body, e.g. the gastrointestinal tract, Sherrington termed the interoceptive surface of the body). These afferent neurons have Aδ- and C-axons. By the same token are processes in the brain related to these functional classes of primary afferent neurons discussed in the frame of the concept of exteroception, proprioception and interoception. Sherrington (1900) also had a clear concept about thermoreception and nociception of skin. Would he have subsumed skin senses related to thermal stimuli and to tissue damaging or potentially damaging stimuli and sensations elicited from deep somatic tissues that are related to the excitation of afferent Aδ- and C-fibers under the category interoception? From the discussion of the topic "The skin and common sensation" in his textbook chapter "Cutaneous sensations" (Sherrington 1900) I would say that the answer is in the affirmative. On p. 969 Sherrington says

> ... By common sensation is understood that sum of sensations referred, not to external agents, but to the processes of the animal body. Its "object" is the body itself – the material "me". Sensations derived from the body tissues and organs possess strong affective tone; while sensations of special sense are relatively free from affective tone.

Here I will subsume the sensory processes related to the activation of afferent receptors connected to Aδ- or C-axons under the category interoception meaning that these senses refer to the different body tissues. This view will become relevant in subchapter 2.6. It is similar or identical to that propagated by Craig and Saper. Craig (and others) claim that pain, thermal sensations, itch, muscular and visceral sensations (along with hunger, thirst, air hunger and other feelings from the body) are aspects of the representation of the physiological condition of the different body tissues and therefore belong to interoception (Craig 2003a; see below). Saper claims that nociceptive sensations are related to mechanical, thermal and metabolic stresses of deep somatic and visceral and superficial body tissues. These sensations monitor tissue integrity and are internally directed, i.e., they are concerned with that state of the body itself (Saper 2002). Saper says that it were "reasonable to describe pain per se as a visceral modality." However, I would not go as far as that.

2.1 Visceral afferent neurons: general characteristics

2.1.1 Vagal and spinal visceral afferent neurons

Visceral organs in the thoracic, abdominal and pelvic cavities are innervated by vagal and spinal visceral afferent neurons that encode physical and chemical events in the visceral organs and convey this information to spinal cord and lower brain stem. They are the interface between the visceral organs and the central nervous system. Most visceral afferent axons are unmyelinated; some are myelinated, conducting up to about 30 m/s. Visceral afferent neurons that convey information from the viscera to spinal cord and lower brain stem are distinguished from visceral afferent neurons of the enteric nervous system. These enteric neurons have their cell bodies in the wall of the gastrointestinal tract. They are called intrinsic primary afferent neurons (IPANs) of the enteric nervous system and encode mechanical and chemical events (see Chapter 5).

Visceral afferent neurons are involved in many functions (Figure 2.1):

- Organs regulations, organ reflexes, neuroendocrine regulations, related particularly to vagal afferents and sacral spinal afferents (see Chapter 10 and Subchapter 9.3).
- Multiple protective organ reflexes, particularly related to thoracolumbar spinal visceral afferents, but also to vagal afferents.
- Visceral non-painful sensations (see Table 2.1).
- Visceral discomfort and pain, related particularly to spinal visceral afferents.
- Pain referred to deep somatic tissues, other visceral organs and skin as well as the changes in the referred zones mediated by the sympathetic nervous system.
- Shaping of emotional feelings related to body states.
- Generation of sickness behavior, related particularly to the excitation of vagal afferents from the gastrointestinal tract (Watkins and Maier 1999; Jänig 2005b; Jänig and Levine 2005).
- Neural and neuroendocrine regulation of hyperalgesia and inflammation, related particularly to the excitation of vagal afferents (Jänig 2005b; Jänig and Levine 2005).

Figure 2.1 General scheme of visceral-autonomic relations. *Left*: visceral afferent inputs and efferent (autonomic) outputs of lower brain stem and spinal cord. *Right*: general functions of visceral afferent neurons. *Upper part*: medulla oblongata; cell bodies of preganglionic neurons that project in the vagus nerve are located in the dorsal motor nucleus of the vagus (DMNX) and in the nucleus ambiguus (NA, see Subchapters 8.4.2 and 10.7); X, vagus nerve; NTS, nucleus tractus solitarii. *Lower part*: spinal cord, spinal visceral afferents converge on viscero-somatic neurons with afferent input from skin and skeletal muscle in laminae I and V of the dorsal horn and deeper laminae; sensations are also referred to the segmentally corresponding body domains. Note that all efferent autonomic systems and the transmission of impulses from visceral primary afferent neurons to second-order neurons (in the NTS and in the dorsal horn of the spinal cord) are under descending control from higher brain centers (see Jänig and Morrison [1986]; Jänig and Häbler [1995]). Modified from Jänig and Häbler (1995) with permission.

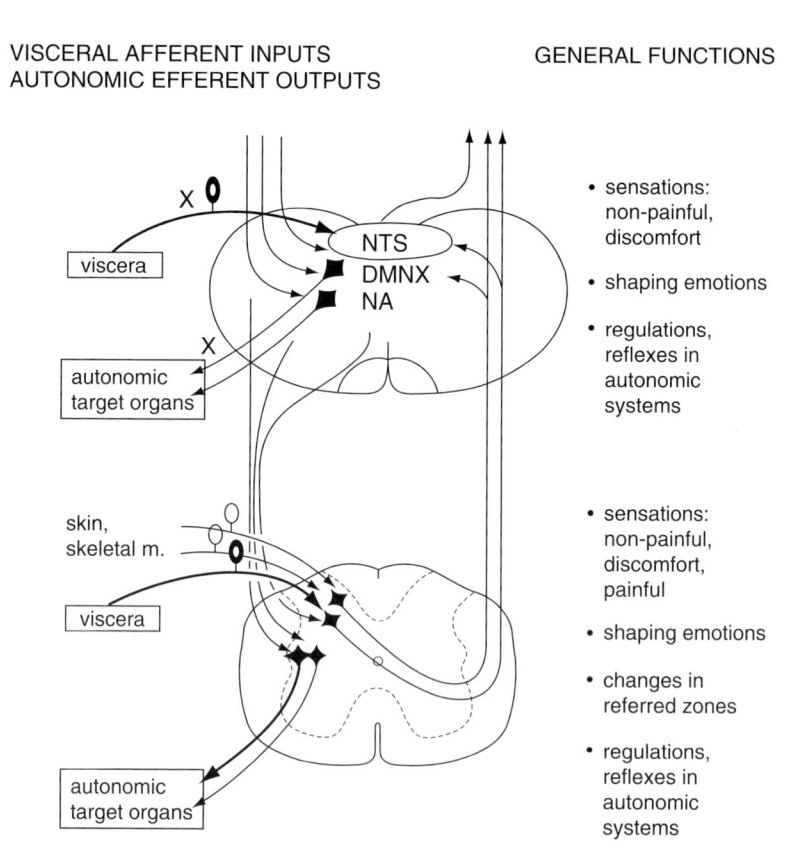

Figure 2.2 demonstrates the projections of spinal and vagal visceral afferents from the organs in the thoracic, abdominal and pelvic cavities to the spinal cord and lower brain stem:

- Thoracolumbar spinal visceral afferents project from the viscera through splanchnic nerves (cardiac nerves; major, minor, lumbar splanchnic nerves), which also contain pre- and postganglionic sympathetic axons, and the corresponding white rami to the thoracic and upper lumbar spinal cord. These afferent neurons have their cell bodies in the corresponding dorsal root ganglia (Figure 2.2 right side).
- Spinal visceral afferent neurons innervating pelvic organs (ureter, urinary bladder and urethra; internal reproductive organs; distal colon, sigmoid and rectum including anal canal) project through the pelvic splanchnic nerves and have their cell bodies in the sacral dorsal root ganglia (in the rat, also in the sixth lumbar dorsal root ganglion) (Figure 2.2 lower left).
- Some afferent neurons projecting through the thoracolumbar white rami innervate the ventral compartment of the vertebral column (Bogduk 1983; Bahns et al. 1986b; Bogduk et al. 1988). These neurons are deep somatic afferent neurons.
- There is no evidence that spinal and trigeminal afferent neurons that innervate somatic tissues (skin, skeletal muscle, joints etc.)

project with their peripheral axons along the sympathetic chain and the major distributing arteries (e.g., the subclavian, iliac, carotid artery) to the extremities, the head or – with the exception of the ventral compartment of the afferent innervations of the vertebral column – to the body trunk (see Figure 1.2 for projection of sympathetic postganglionic neurons to targets in the extremities and the head). Large blood vessels of the body trunk (such as aorta, subclavian, iliac or carotid arteries) are surrounded by terminals of primary afferent neurons that form a plexus lying outside the perivascular noradrenergic plexus. These afferents project to their target together with the visceral sympathetic innervation.

- About 1.5% to 2.5% of all spinal afferents that have their cell bodies in the dorsal root ganglia project to the viscera, the other spinal afferent neurons project to skin and deep somatic tissues (Jänig and Morrison 1986); in some dorsal root ganglia (e.g., sacral S2/S3, thoracic T8/T9 up to about 8% of all neural cell bodies may be visceral. In the cat some 22 000 to 25 000 spinal primary afferent neurons project to the viscera. This number compares to about 1 to 1.5 million spinal afferent neurons (both sides; Jänig and Morrison 1986). This illustrates that visceral organs are much less densely innervated by spinal afferent neurons than the superficial and deep somatic body tissues.

- Vagal afferents innervating thoracic organs project through the superior and recurrent laryngeal branches, the aortic nerve, cardiac branches and branches from lung and esophagus to the nucleus tractus solitarii (NTS) in the lower brain stem. Some afferents (from carotid arterial baro- and chemoreceptors) project through the carotid sinus nerve and the glossopharyngeal nerve. Abdominal vagal afferents project through the common hepatic branch (from liver, upper duodenum and pylorus), two gastric branches (from the stomach) and two celiac branches (distal duodenum, small intestines, proximal colon) to the brain stem (Berthoud et al. 1997; Berthoud and Neuhuber 2000). The cell bodies of afferent neurons projecting in the vagus nerve lie in the inferior (nodose) ganglion of the vagus nerve (and some in the superior [jugular]), those projecting in the glossopharyngeal nerve (including afferents from arterial baro- and chemoreceptors) lie in the petrosal ganglion.

- In the cat about 30 000 to 40 000 afferent neurons project through the abdominal vagal nerve. The number of vagal afferent neurons innervating thoracic visceral organs is unknown.

- Trigeminal afferent neurons innervating intracranial blood vessels and associated structures (e.g., the dura, trigemino-vascular afferents) may also be considered to be visceral. The same applies to afferents from the tongue (associated with taste), from the hard palate and the proximal part of the oropharynx that project through the facial or glossopharyngeal nerve to the NTS, taste afferents being a special class of visceral afferents.

VISCERAL AFFERENT NEURONS AND AUTONOMIC REGULATIONS

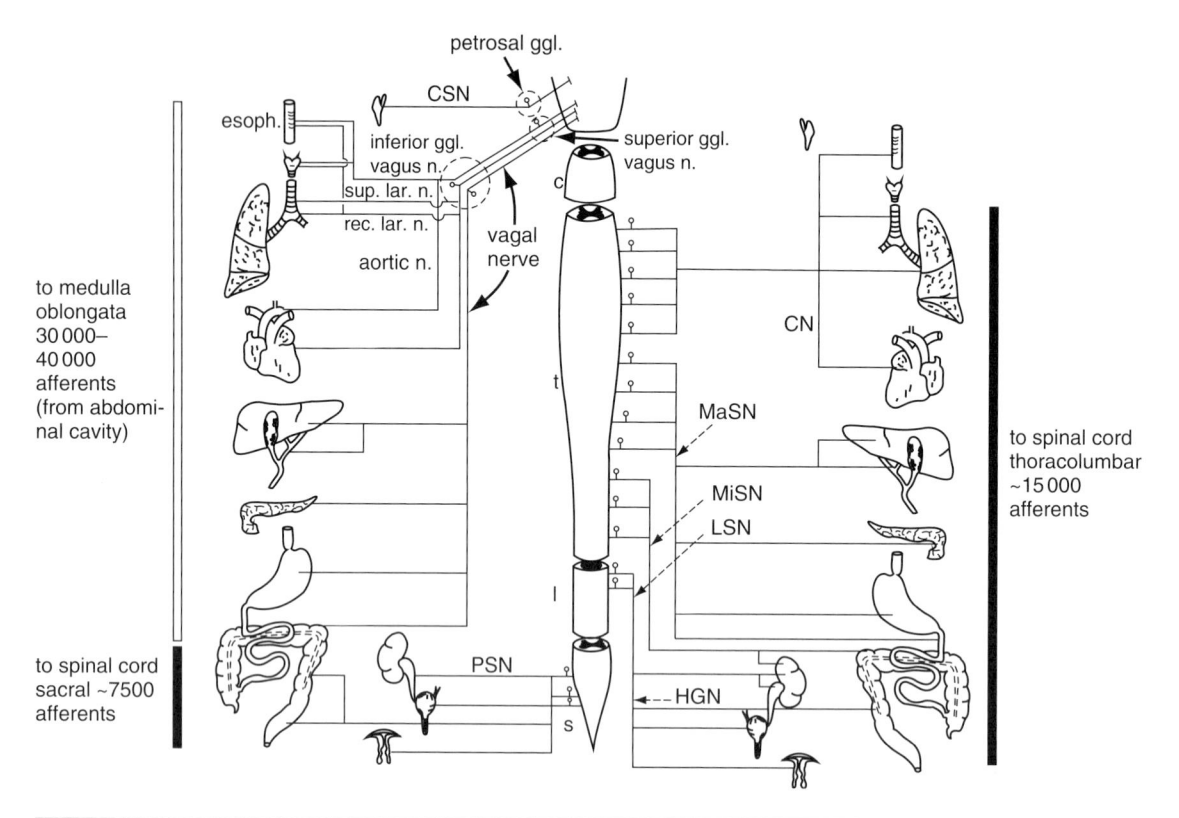

Figure 2.2 Projection of visceral afferent neurons. *Left side*: vagal visceral afferent neurons projecting to the nucleus tractus solitarii (NTS) and spinal visceral afferent neurons projecting through the pelvic splanchnic nerve (PSN) to the sacral spinal cord. Baroreceptor and chemoreceptor afferents from the carotid sinus and body project through the carotid sinus nerve (CSN) and the glossopharyngeal nerve to the NTS. *Right side*: spinal visceral afferent neurons projecting to the thoracic and upper lumbar spinal cord. Note that pelvic organs are supplied by sacral and lumbar (some lower thoracic) spinal visceral afferent neurons. The numbers of afferent neurons were estimated from counting in the cat. The number of spinal visceral afferent neurons is low compared to the number of spinal somatic afferent neurons. CN, cardiac nerves; HGN, hypogastric nerve; LSN, lumbar splanchnic nerves; MaSN, major splanchnic nerve; MiSN, minor splanchnic nerves; sup./rec. lar. n., superior/recurrent laryngeal nerve; ggl., ganglion; c, cervical; t, thoracic; l, lumbar; s, sacral. Modified from Mei (1983), Berthoud and Neuhuber (2000), Neuhuber (1989), Wank and Neuhuber (2001) with permission, Neuhuber personal communication.

2.1.2 Visceral afferent neurons and autonomic nervous systems

As described in Chapter 1, Langley defined the sympathetic and parasympathetic (cranio[bulbo-tectal]-sacral) systems on functional anatomical grounds, i.e. based on the reaction of autonomic target organs to electrical stimulation of preganglionic or postganglionic axons and on some other criteria. His definition of both autonomic systems was not based on the regulations and reflexes in which the autonomic neurons are involved. Thus, Langley never used the terms "sympathetic" and "parasympathetic" in a functional sense as it is commonly done. He did not include in this definition the visceral afferent neurons projecting through the same nerves together with the preganglionic sympathetic and parasympathetic neurons (Langley 1900, 1903a, b, 1921). In fact, for good reasons he was extremely careful in his argumentation in his paper in *Brain* (Langley 1903b) as to why afferent fibers passing through "autonomic" nerves and innervating viscera should neither be named sympathetic nor parasympathetic nor autonomic. For me the position of Langley on this issue was rather modern and is still, with some modifications, fully valid. This is best expressed in his own words:

> In what I have said so far I have dealt with what seem to me facts about the afferent fibres accompanying the efferent autonomic nerves, but I have put on one side the fundamental difficulty with regard to them, and that is: Are there any afferent nerves which deserve to be separated from afferent somatic nerves, and if so, what are the characteristics of autonomic as opposed to somatic afferent nerves? I have above tried to show that the afferent nerves of the sympathetic system are indistinguishable in form and position from those of the somatic system, and it remains to consider what other distinguishing characters may be present ...
>
> It is clear that we cannot make a like division of afferent fibres according as they run to striated muscles or to other tissue; it would lead to nothing but confusion to consider the afferent fibres of the skin as autonomic fibres and the afferent fibres of striated muscle as the only somatic afferent fibres ....
>
> Since by hypothesis the one kind of fibre gives rise to sensation, and the other does not, there must be a difference in their central connection, such that in the one the upward path to the cerebral hemispheres is absent or very slightly developed. That, and that only, so far as we can say, distinguishes autonomic from somatic afferent fibres. Further progress waits for the discovery of some distinguishing histological character. And in the meantime it is open to discussion whether the class of afferent fibres which are solely reflex in function can be properly considered as corresponding on the afferent side to the efferent fibres of the autonomic tissues.
>
> Langley 1903b

Thus the situation is still an open story, as expressed by Langley at the beginning of the twentieth century, and the putative classification of visceral (and other) afferent neurons as sympathetic or parasympathetic or autonomic awaits better, more stringent criteria.

Although visceral afferents are anatomically closely associated with either the sympathetic or parasympathetic part of the autonomic nervous systems I will call these afferent neurons *visceral* (qualified by spinal or vagal), fully in accordance with Langley. The terms *sympathetic afferent* and *parasympathetic afferent* neuron are misleading since they imply that the afferent neurons have functions that uniquely pertain to that particular part of the autonomic nervous system, the exception being that afferent neurons of the enteric nervous system are by definition *enteric afferent neurons* (IPANs, Chapter 5). No convincing functional, morphological, histochemical or other criteria exist to associate any type of visceral afferent neuron that projects to the spinal cord or brain stem with only one of the autonomic systems. By the same token should spinal and vagal visceral afferent neurons not be called autonomic. The label *sympathetic* or *parasympathetic* would lead to complications as far as the understanding of the functions of these afferents is concerned.

This point is illustrated by two examples: (1) Pelvic organs are innervated by two sets of spinal visceral afferents, one entering the spinal cord at the rostral lumbar and most caudal thoracic segmental levels and the other at sacral segmental levels, and both are involved in visceral nociception; the sacral visceral afferents are involved in organ regulations, yet not the visceral afferents projecting to the upper lumbar and lower thoracic spinal cord (Jänig and Morrison 1986; Jänig and Koltzenburg 1990, 1993; Ritter *et al.* 1992; Cervero 1994; Jänig 1996d). It makes no sense to speak of "sympathetic" and "parasympathetic" visceral nociception. (2) Arterial baro- and chemoreceptor afferents project through the aortic and glossopharyngeal nerves to the NTS. To label these afferents as parasympathetic is groundless.

Many visceral afferent neurons serve special functions related to distinct types of physiological control in which the autonomic nervous system is the efferent pathway, for example cardiovascular afferents, afferents from the respiratory tract and afferents from the gastrointestinal tract that project to the NTS or afferents from pelvic organs that project to the sacral spinal cord. These afferents monitor the inner state of the body and serve to maintain homeostasis and to adapt internal milieu and organ functions to the behavior of the organism. *In this sense do these afferents belong functionally to the autonomic nervous system.*[1] However, the second-order neurons that the visceral afferents connect to are widespread and carry signals to higher levels of integration in the brain. The brain's knowledge of the body's inner state may influence the behavior in the widest sense.

2.2 | Visceral primary afferent neurons as interface between visceral organs and brain

Many spinal afferent neurons and some vagal afferent neurons, in particular those with unmyelinated axons supplying viscera, seem to have multiple functions (Jänig 1996d). These functions are related to the impulses conducted orthodromically to spinal cord, brain stem or prevertebral sympathetic ganglia ("afferent" functions) and to the release of neuropeptides in the target tissues ("efferent" functions). The knowledge about some of these functions is well established, others are at best hypothetical. These functions do not apply to each type of visceral afferent neuron and afferent neurons may not only be specialized with respect to their receptive properties, but also with respect to the putative efferent functions (de Groat 1987; Maggi and Meli 1988; Dockray *et al.* 1989; Holzer 1992, 1998a,

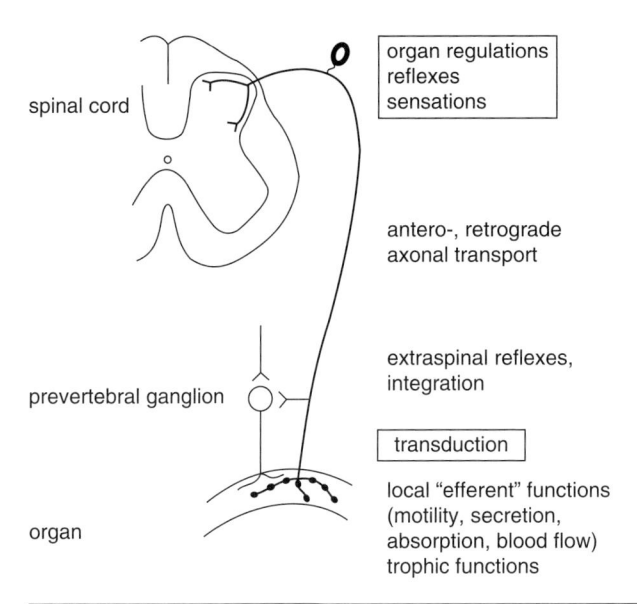

Figure 2.3 Spinal afferent neurons as interface between visceral organs and brain. Spinal primary afferents supplying viscera have multiple functions. They contain neuropeptides such as calcitonin gene-related peptide and substance P (de Groat 1989), which are ortho- and retrogradely transported between the soma and central and peripheral terminals. (1) *Afferent functions*: the prime role is to send information to the central nervous system. They can also modify the regulation of viscera by extracentral reflexes in prevertebral ganglia (see Subchapter 6.5). (2) *Local efferent (effector) functions*: they have local efferent functions by the release of neuropeptides from peripheral endings. (3) *Trophic functions*: the afferents can also signal events by slow transport of chemical substances rather than rapid electrical processes under normal conditions. These substances could have trophic effects on the peripheral tissues or might influence the synaptic connectivity between primary afferent neurons and second-order neurons in the spinal cord.

2002a, b, 2003; Holzer and Maggi 1998). Spinal visceral afferents might further be differentiated for mediating preferentially peripheral extraspinal reflexes or local regulations or trophic influences (Figure 2.3).

- The conventional function of visceral afferent neurons is to encode physical (distension, contraction) and chemical events and to signal these events by centripetal impulses to neurons in the spinal cord and brain stem leading to organ regulations, reflexes and distinct sensations. The terminals of these afferents contain the various molecules that enable the afferents to transduct and integrate the physical and chemical stimuli in their microenvironment, so as to generate impulse activity. Also involved in this transduction process are non-neural cells such as epithelia of the respiratory, gastrointestinal and urinary tract, which contain the same molecules to sense the physical and chemical stimuli in the lumen of the organs as the afferent terminals (Apodaca 2004; Birder 2005).
- Collaterals of some spinal visceral afferent fibers form peptidergic synapses with noradrenergic neurons in prevertebral ganglia (celiac, mesenteric ganglia) that are particularly involved in gastrointestinal functions (regulation of secretion and motility), establishing in this way extraspinal reflexes. The peptide transmitters

are calcitonin gene-related peptide (CGRP) and substance P (Perry and Lawson 1998). These postganglionic neurons integrate activity in preganglionic neurons, peripheral intestinofugal neurons of the enteric nervous system and spinal visceral afferents (see Subchapter 6.5; de Groat, 1987; Furness and Costa 1987; Jänig 1988b, 1995a; Dockray et al. 1989; Szurszewski and King 1989).

- Visceral afferent neurons may participate in a variety of important "efferent" functions (effector functions), by release of neuropeptides (such as CGRP and/or substance P) and other substances (e.g., ATP) within the viscera that are independent of the central nervous system and prevertebral ganglia. These functions consist of vasodilation, bronchoconstriction, secretory processes in the gastrointestinal tract, regulation of gut and urinary tract motility and modulation of the protective function of epithelia against intraluminar and otherwise toxic substances. Thus the afferents are involved to protect the gastric mucosa against acid back-diffusion (Holzer 1992, 1995, 2002a, b, 2003; Maggi et al. 1995; Santicioli and Maggi 1998) and in the protective function of the uroepithelium of the urinary tract keeping down its permeability for small molecules (water, ammonia, urea, proton ions) and toxins in the urine (Apodaca 2004; Birder 2005). Some visceral afferent neurons that have their cell bodies in the dorsal root ganglia may not project to the spinal cord at all (Häbler et al. 1990b). These afferent neurons may only have peripheral efferent functions (see also Holzer and Maggi 1998). The "efferent" functions of the afferent neurons may be particularly important under pathophysiological conditions, for example, inflammatory changes of the peripheral tissue (de Groat 1989; Dockray et al. 1989; Kumazawa 1990).[2]
- Visceral afferents may have trophic functions and could be important for the maintenance of the structure of visceral tissues (e.g., the mucosa of the urogenital tract [Apodaca 2004; Birder 2005] or of the stomach [Lundgren 1989]).
- Afferent neurons transport retrogradely neurotropic substances; this invites speculation that these substances might have long-term effects on the synaptic connections formed by primary afferent terminals in the spinal cord with second-order neurons (cf. Lewin and McMahon 1993).

This cascade of functions of spinal visceral afferents may serve the same final general aim: protection and maintenance of the integrity of visceral tissues. For example, excitation of thoracolumbar spinal visceral afferent neurons may elicit pain and discomfort, protective supraspinal, spinal and extraspinal reflexes, and peripheral changes of target organ responses (such as increase of blood flow, change of motility, and secretion). Thus many visceral afferent neurons, in particular those with unmyelinated fibers, are an active interface between visceral organs and the neuraxis. Spinal visceral afferent neurons may be differentiated into those having only peripheral (efferent) functions, those having only central functions

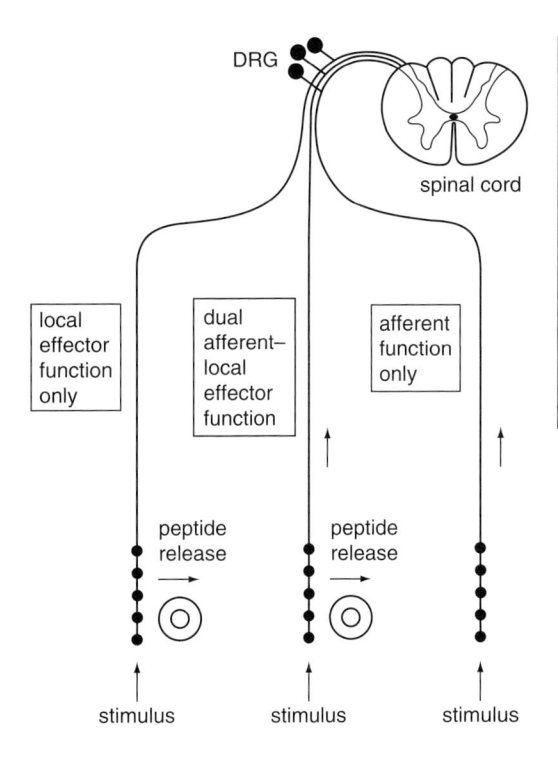

Figure 2.4 Spinal visceral afferent neurons having either afferent (central) functions (right), local effector ("efferent") functions (left) or dual functions (middle). This functional distinction of spinal visceral afferent neurons into three global types is a hypothesis. The cell bodies of each type of afferent neuron are located in the dorsal root ganglia (DRG). Modified from Holzer and Maggi (1998) with permission.

(i.e., signaling peripheral events to the spinal cord) and those having dual functions (see Figure 2.4; Holzer and Maggi 1998; Holzer 2002a, b, 2003).

2.3 | Receptive functions of visceral afferent neurons

The central nervous system receives information from the internal organs by two sets of visceral afferents (Figures 2.1, 2.2) and sends efferent impulses to the internal organs by two sets of autonomic efferents each consisting of many functionally distinct pathways (Chapter 4). Visceral afferent neurons being involved in visceral pain will be discussed separately (see Subchapter 2.4).

The degree of functional (physiological) specificity of afferent neurons is generally described by way of their quantitative responses to physical and chemical stimuli. If an afferent neuron responds preferentially to a particular physiological stimulus applied to its receptive endings at low stimulus energy but not to other physiological stimuli, then this stimulus is considered to be an adequate stimulus and the receptive ending of the afferent neuron specific for this stimulus (see Cervero 1994). Many visceral receptors are specific with respect to the various adequate stimuli occurring in the visceral domain. Table 2.1 lists, separated for different organ systems, the physiologically distinct types of visceral afferent neurons and the sensations, organ functions and regulations that are associated with the activation of these afferent neurons.

In this context I want to emphasize that the activation of a functionally particular group of afferent neuron does not *cause* a particular sensation or regulation, and the afferent signals generated in the periphery are not "carried through" to the cortical representations of these sensations and regulations in the form of a labeled-line system. We have to distinguish three classes of events involved that are not interchangeable but correlated with each other: (1) The anatomy, histochemistry and pharmacology of the neurons. (2) The physiology of the neurons and regulations and reflexes associated with them. (3) The "psychology" of the sensations and emotional feelings (Table 2.1).

2.3.1 Vagal afferent neurons

About 80% to 85% of the nerve fibers in the *vagus nerve* are *afferent*. Most of these afferents are unmyelinated, some are myelinated, but this varies between organs and between species. The fibers project viscerotopically to the NTS (Loewy and Spyer 1990a; Ritter *et al*. 1992; Barraco 1994; Paton and Kasparov 2000; see Subchapters 8.3 and 10.7). The second-order neurons in the NTS project to various sites in the lower brain stem, upper brain stem, hypothalamus and amygdala, establishing well-organized neural pathways that are the basis for distinct organ regulations and associated body perceptions (see Subchapters 2.6 and 8.3).

Receptive properties and organ regulation

Vagal afferents monitor mechanical and chemical events related to the respiratory tract, cardiovascular organs and gastrointestinal tract and therefore related to the organ regulations and reflexes associated with these organ systems. Neurophysiological recordings have shown that these afferent neurons exhibit specificity or relative (preferential) specificity with respect to the adequate mechanical (distension, contraction, shearing stimuli) and chemical stimuli (changes in blood gases, osmotic stimuli; changes of glucose, proton ions, protein products, lipids in duodenum and small intestines). This specificity is brought about by the anatomical arrangement of the receptive endings and by the transduction mechanisms. The main functional types of vagal afferents are listed in Table 2.1:

- The *respiratory tract* is innervated by various functionally specific types of vagal afferents with myelinated and unmyelinated fibers (Widdicombe 1986, 2001, 2003; Coleridge and Coleridge 1997; Lee and Pisarri 2001; Schelegle and Green 2001).
- The *cardiovascular system* (heart, ascending aorta, carotid artery) is innervated by afferents encoding in their activity arterial blood pressure and its changes (arterial baroreceptors), arterial blood gases (in particular oxygen tension, arterial chemoreceptors), right atrial pressure (volume receptors in the right atrium and

Table 2.1 | *Visceral afferent neurons, sensations, regulations and reflexes*

Organ[a]		Afferent neuron[b]	Sensation[c]	Regulation/reflex[d]
Respiratory tract				
Pharynx, larynx	V	Mechanorec. (pressure), irritant rec., cold rec., flow rec.	Rawness, irritation, desire to cough, pain, nausea	Aspiration reflex, cough reflex, swallowing, bronchodilatation, etc.
Trachea, bronchi, lung	V	Irritant rec., C-fiber (epithelial) rec.	Substernal rawness, irritation, urge to cough, tightness	Burn-cough reflex, laryngo-, bronchoconstriction, mucus secretion, hyperpnea
	V	Slowly adapting rec.	?/no	Hering–Breuer reflex
	V	J-receptor	Irritation in throat, breathlessness, discomfort, pain	Respiratory-protective reflexes
	S	yes, function?	?/no	?
Cardiovascular organs				
Large blood vessels	V	Barorec., chemorec.	No	Cardiovascular regul., reflexes
	S	Mechanorec.	Discomfort, pain	Spinal reflexes
Heart	V	Mechanorec. (atrial, ventricular)	No	Cardiovascular regul., reflexes
	S	Atrial rec.	No	Cardiovascular reflexes
	S	Ventricular, coronary	Discomfort, pain, other sensations (?)	Cardio-cardial reflexes, other cardiovascular reflexes
Gastrointestinal tract				
Esophagus	V	Mechanorec. (tension)	Fullness, thermal sensations, heartburn	Propulsive peristalsis, vomiting
	S	Mechanorec.	Discomfort, (tension) pain	?
Stomach	V	Mechanorec. (tension)[e]	Fullness/emptiness	Storage, relaxation
	V	Mucosarec. (mechano-, chemo-, thermo-)[f]	Satiety/hunger thermal sensations (?)	Secretion, peristalsis, vomiting
	S	Mechanorec. (serosal)[g]	Discomfort, pain	Intestino-intestinal reflexes
Duodenum	V	Mechanorec. (tension)[e]	?	Secretion, peristalsis
Ileum, jejunum	V	Mucosarec.[f] (mechano-, chemo-, thermo-)	?	Secretion (?), peristalsis (?)
	S	Mechanorec. (serosal)[g]	Discomfort, pain	Intestino-intestinal reflexes

Table 2.1 | (cont.)

Organ[a]		Afferent neuron[b]	Sensation[c]	Regulation/reflex[d]
Liver	V	Osmorec.	Thirst	Osmoregulation (?)
Gallbladder	S	Mechanorec.	Discomfort, pain	?
Pancreas	S	Mechanorec.	Discomfort, pain	?
Colon, rectum	S1[h]	Mechanorec. (wall)	Fullness, call to defecate, discomfort, pain	Defecation, continence reflexes
	S2[h]	Mechanorec. (serosal)[g]	Discomfort, pain	?
Anal canal	S1[h]	Mechanorec., thermorec., nociceptor (?)	Shearing sensation, thermal sensation (?). pain	Ano-rectal, ano-vesical reflexes
Urinary tract				
Kidney	S	Mechanorec., chemorec.	Pain	Reno-renal reflexes, other reflexes
Ureter	S1/S2[h]	Mechanorec.	Pain	?
Urinary bladder, urethra	S1[h]	Mechanorec.	Fullness, urge to micturate discomfort, pain	Micturition, continence reflexes
	S2[h]	Mechanorec.	Discomfort, pain	?
Spleen	S	Mechanorec.	Discomfort, pain	?
General				
All organs, inclusive blood vessels etc.	S	Mechanorec. (high threshold), chemorec. (?) (mechanoinsensitive afferents)	?; pain under pathophysiological conditions	Defensive reactions/reflexes (?), trophic functions (?)

Notes:

[a] Organ or organ system.
[b] Functional type of afferent neuron.
[c] Type of sensation(s) elicited when the respective afferent neuron(s) is (are) stimulated.
[d] Type(s) of regulation and reflexes associated with the afferent neurons.
[e] Receptors lying in the muscular wall of the gastrointestinal tract, responding to distension and contraction.
[f] Stimulus specificity unclear.
[g] Receptors lying particularly at the insertion of the mesenteries and responding to mechanical and chemical stimuli.
[h] S1, sacral afferent neurons; S2 lumbar visceral afferent neurons.

V: Visceral afferent neurons projecting through the vagal or glossopharyngeal nerves to the nucleus of the solitary tract in the medulla oblongata. The cell bodies of these afferent neurons lie in the inferior (nodose) and, superior (jugular) vagal ganglion and in the petrosal ganglion.
S: Spinal afferent neurons projecting through the splanchnic nerves to the thoracic, upper lumbar and sacral spinal cord. The cell bodies of these afferents lie in the thoracic, upper lumbar and sacral dorsal root ganglia. Data from Hertz (1911), Paintal (1973, 1986), Malliani (1982), Mei (1983, 1985), Coleridge and Coleridge (1984), Andrews (1986), Jänig and Morrison (1986), Widdicombe (1986), Grundy (1988), Jänig and Koltzenburg (1993). Modified from Jänig (1996b).

large intrathoracic veins) and ventricular mechanoreceptors (Thorén 1979; Chapleau and Abboud 2001).
- In the *gastrointestinal tract*, vagal afferents monitor mechanical and chemical events in the intestine (for review see Grundy and Scratcherd [1989]). Intramural endings consist of intraganglionic laminar endings (IGLEs) in the myenteric plexus with functional characteristics of tension receptors or intramuscular arrays (IMAs) that probably have properties of stretch receptors (Page and Blackshaw 1998; Lynn and Blackshaw 1999; Phillips and Powley 2000; Zagorodnyuk and Brookes 2000; Zagorodnyuk *et al.* 2001; Page *et al.* 2002; Lynn *et al.* 2003). Mucosal afferent endings are situated in the lamina propria but do not penetrate the basal lamina (i.e., they are not located juxtaepithelially) and do not directly come into contact with the luminal content. Exceptions are afferents innervating the squamous epithelium of the esophagus and anal canal. Responses of vagal afferents to maltose, glucose and intraluminal osmotic stimuli are mediated by enterochromaffin cells releasing 5-hydroxytryptamine (5-HT) and by 5-HT$_3$ receptors in the terminals of the vagal afferents (Zhu *et al.* 2001). Responses of vagal afferents to protein products of long-chain lipids are mediated by enteroendocrine cells releasing cholecystokinin (CCK) and the CCK$_A$ receptor in the terminals of the vagal afferent neurons. These afferents do not seem to be mechanosensitive (Richards *et al.* 1996; Lal *et al.* 2001; Beyak and Grundy 2005).

Vagal afferent neurons and body protection

Experiments on animals show that vagal afferents play an important role in general body protection (see Jänig 2005b):

- Vagal afferent neurons may be associated with the gut-associated lymphoid tissue (GALT) and excited by inflammatory and toxic processes. This excitation is probably mediated by enterochromaffin cells releasing 5-HT, by enteroendocrine cells releasing CCK and by mast cells releasing histamine and other compounds (Williams *et al.* 1997; Kirkup *et al.* 2001; Kreis *et al.* 2002). The functional specificity of these afferents, with respect to the different types of intraluminal stimuli, is unknown (see Subchapter 5.6 and Figure 5.13).
- Excitation of vagal afferents, innervating liver and upper gastrointestinal tract (proximal duodenum and distal stomach) and projecting through the hepatic branch of the abdominal vagus nerve, triggers during inflammation in the viscera (e.g., generated experimentally by intraperitoneal injection of the bacterial cell wall endotoxin, lipopolysaccharide) so-called illness responses (which include immobility, decrease of social interaction, decrease in food intake, formation of taste aversion to novel food, decrease of digestion, loss of weight [anorexia], fever, increase of sleep, change in endocrine functions, malaise,

hyperalgesic behavior) (Dantzer *et al.* 1998, 2000; Maier and Watkins 1998; Watkins and Maier 1999, 2000; Goehler *et al.* 2000). Based on lesion experiments, it is hypothesized that vagal afferents innervating the liver are activated by proinflammatory cytokines (interleukin 1 [IL-1β], IL-6, tumor necrosis factor α [TNFα]) released by activated macrophages (Kupffer cells), dendritic cells and leukocytes and signal these events to the brain resulting in the illness responses (Figure 2.5). The proinflammatory cytokines are suggested either to activate the vagal afferents directly or to bind specifically to glomus cells in the abdominal paraganglia that are innervated by vagal afferents. The physiological properties of these vagal afferents are rather unknown. Some vagal afferents innervating the hepatoportal system are excited by IL-1β (Niijima 1996). The central mechanisms underlying the sickness responses are also unknown.

- Activity in vagal afferents innervating the small intestine (i.e., projecting through the celiac branches of the abdominal vagus nerves) is important in reflex modulation of inflammatory processes (e.g., in the knee joint) and mechanical hyperalgesic behavior (by sensitization of nociceptors) in remote body tissues involving the sympatho-adrenal system and possibly the hypothalamo–pituitary–adrenal system (Green *et al.* 1995, 1997; Miao *et al.*, 1997a, b, 2000, 2001, 2003a, b; Khasar *et al.* 1998a, b, 2003; Jänig *et al.* 2000).

- Electrical stimulation of abdominal vagal afferents inhibits nociceptive impulse transmission in the spinal dorsal horn and depresses nociceptive behavior, showing that these afferents have antinociceptive function (Gebhart and Randich, 1992; Randich and Gebhart, 1992). Electrical stimulation of cervical vagal afferents in monkeys suppresses transmission of impulse activity in spinothalamic relay neurons with nociceptive function at all levels of the spinal cord whereas electrical stimulation of subdiaphragmatic vagal afferents has no effect on spinothalamic relay neurons in this species arguing that particularly cardiopulmonary vagal afferents are involved in this inhibitory control in monkeys. In the rat, indirect evidence shows that some ongoing central inhibition of nociceptive impulse transmission (occurring probably in the dorsal horn) is normally maintained by spontaneous activity in vagal afferents (Khasar *et al.* 1998a, b). The central pathways mediating the inhibitory effect are neurons in the subceruleus-parabrachial complex (noradrenergic) and neurons in the nucleus raphe magnus of the rostral ventromedial medulla (serotonergic) that project to the spinal cord (see Foreman 1989). The functional types of vagal afferents involved in this antinociception are unknown.

In *conclusion*, the animal experiments clearly indicate that cardiopulmonary and abdominal vagal afferent neurons are not only important in the context of regulation of respiratory, cardiovascular and

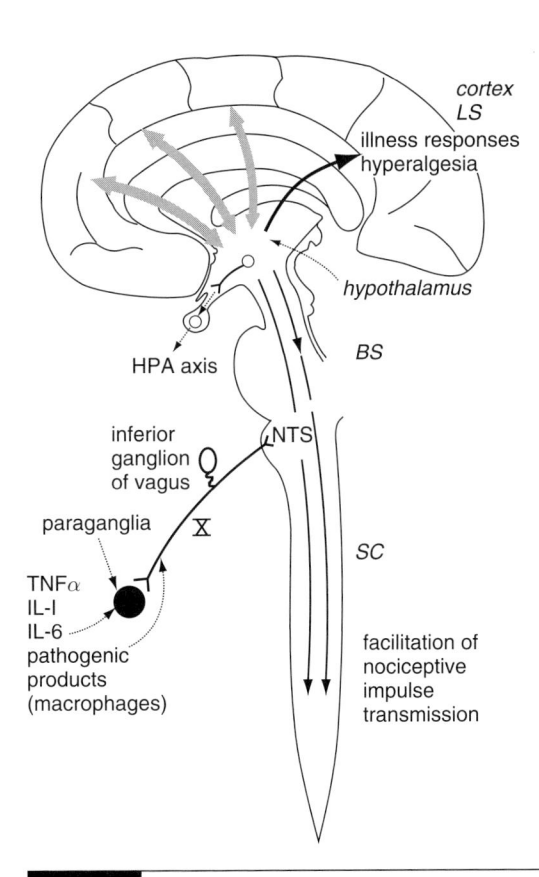

Figure 2.5 Illness responses (including hyperalgesia and pain) elicited by pathogenic stimuli in the viscera. Pathogens (bacteria, viruses, others) activate phagocytic immune cells (macrophages; Kupffer cells in the liver). These activated immune cells release proinflammatory cytokines (interleukin 1 [IL-1], IL-6, tumor necrosis factor α [TNFα]). The cytokines activate vagal afferents projecting through the hepatic branch of the abdominal vagus nerves (probably via paraganglia, but possibly also independent of the paraganglia). Stimulation of vagal afferents activates second-order neurons in the nucleus tractus solitarii (NTS) in the medulla oblongata. This leads to activation of pathways creating illness responses, which include hyperalgesia and pain. Illness responses are generated by activation of the paraventricular nucleus of the hypothalamus and structures in the limbic system (e.g., the hippocampus). Specifically, pain and hyperalgesia are generated by facilitation of nociceptive impulse transmission in the spinal cord (and probably elsewhere). This facilitation is mediated by descending pathways from the NTS (via the nucleus raphe magnus) and probably from the hypothalamus. BS, brain stem; HPA axis, hypothalamo–pituitary–adrenal axis; LS, limbic system; SC, spinal cord; X, vagus nerve. After Goehler et al. (2000).

gastrointestinal functions but also in the context of protection of the body. This latter function involves the spinal cord, brain stem, hypothalamus and limbic system structures and, as efferent pathways, neuroendocrine systems to the effector cells, such as the sympatho-adrenal system and the hypothalamo–pituitary–adrenal system (Jänig 2005). Thus, vagal afferents seem to sense events that are related to injurious events in the visceral body domain including microorganisms and toxic substances invading the body via the largest defense barrier of the body, the GALT (Shanahan 1994) in the small intestine, and via the liver. The physiological response properties of these vagal afferents are unknown; however, it is not far-fetched to predict that several different types of vagal afferents are involved (see also Subchapter 5.6 and Figure 5.13).

2.3.2 Spinal visceral afferent neurons

The projections of visceral afferent neurons to the spinal cord from different organs are segmentally organized, the projection from each organ exhibiting a wide segmental distribution (Jänig and Morrison 1986). No distinct organotopic organization of this projection is present in the dorsal horn. The afferents project to laminae I and V of the dorsal horn and to deeper laminae (laminae VI, VII and X), sparing lamina II (substantia gelatinosa Rolandi; see Figure 2.6) and laminae III and IV (nucleus proprius). Occasionally they project to the contralateral laminae V and X. Single visceral afferent neurons with unmyelinated fibers project over 4 to 5 segments and over the whole mediolateral width of the dorsal horn (Figure 2.6). This projection pattern is similar to that of small-diameter afferents from deep body structures (skeletal muscle and joints [Craig and Mense 1983; Craig et al. 1988; Mense and Craig 1988]) and contrasts with that of single cutaneous afferent neurons with unmyelinated fibers, which is spatially much more restricted (Cervero and Connell 1984; Sugiura et al. 1989; for review of spinal projection of visceral and other afferents see Willis and Coggeshall [2004a]).

The low density of the spinal visceral afferent innervation, the broad segmental projection of spinal afferents from different organs,

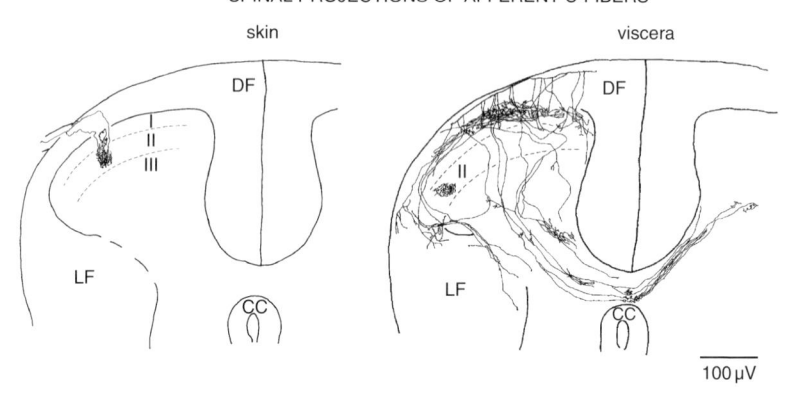

Figure 2.6 Camera lucida drawing of the central termination patterns of single unmyelinated afferent fibers innervating somatic (left) and visceral tissues (right) in the guinea pig. The tracer (*Phaseolus vulgaris* leuco-agglutinin) was injected intracellularly through a micropipette into the cell bodies of afferent neurons of the dorsal root ganglion Th13 in anesthetized animals. The afferent neurons were recorded through the same micropipette and identified as somatic or visceral by their responses to electrical stimulation of a somatic (subcostal nerve) or visceral nerve (stimulation of celiac ganglion) and the receptive fields were stimulated by noxious stimuli (pinch, heat, cold). The animals recovered from this experimental procedure. Two to four days later the animals were fixated under anesthesia and the spinal cord removed and cut into transverse 50 μm-thick parasagittal sections. The tracer was visualized in the spinal cord sections and the axon ramifications of the labeled neurons reconstructed. I, II, III indicate the laminae of the grey matter according to Rexed (1954). Transverse sections. CC, central canal; DF, dorsal funiculus; LF, lateral funiculus. For details see text. From Sugiura et al. (1989) with permission.

and the broad spinal segmental projection of individual visceral afferent neurons is one basis of the poor localization and graduation of visceral sensations mediated by spinal visceral afferent neurons.

Thoracolumbar visceral afferent neurons
The sensory receptors of *thoracolumbar visceral afferents* are situated in the serosa, at the attachment sites of the mesenteries, in the walls of some organs and in the mucosa. It is unclear whether afferents supplying hollow organs (e.g., the gastrointestinal tract or the urinary bladder) innervate the mucosa as well as serosa and wall of the organs or whether afferents innervating the mucosa are separate from those innervating the serosa and wall of the organs. Furthermore, the density of visceral afferent neurons innervating serosa, musculature or mucosa is unknown.[3] Most of these afferents seem to be mechanosensitive and react to distension and contraction of the organs. But they are also activated by chemical stimuli as occurring during inflammation and ischemia of the organs. Thus, these afferents are polymodal and probably do not signal specific events to the spinal cord except that they are associated with a particular organ; they trigger protective reflexes and regulations, pain, discomfort and local protective responses when excited (Haupt *et al.* 1983; Longhurst 1995; Pan and Longhurst 1996; Pan *et al.* 1999).

Thoracolumbar spinal visceral afferent neurons are involved in extraspinal and spinal intestino-intestinal reflexes and probably also in specific organ reflexes, for example, to the heart (cardio-cardial reflexes), the kidney (reno-renal reflexes) or bronchopulmonary system (see Malliani 1982; Jänig 1988b; DiBona and Kopp 1997; Hummel *et al.* 1997; Kopp and DiBona, 2000). However, this has not been thoroughly studied (see Subchapter 9.2). Activity in these visceral afferent neurons may be related mainly to visceral nociception, visceral pain and discomfort (see below) in addition to the peripheral ("effector") functions (see Subchapter 2.2).

Sacral visceral afferent neurons
Pelvic organs have a dual spinal visceral afferent innervation and a somewhat higher density of innervation by sacral spinal visceral afferents than the other organs, the latter probably related to the precise control of these organs by the central nervous system (see Subchapter 9.3). The sacral component of this afferent innervation is essential for the regulation of evacuation and storage functions and of the reproductive organs, as well as for the generation of non-painful and most painful sensations associated with the pelvic organs. Thoracolumbar visceral afferents are not essential for the regulation of the pelvic organs (continence function of urinary bladder and colon; emission of semen) and the associated non-painful sensations but may be important for the generation of pain (see Jänig and Morrison [1986]; Jänig and McLachlan [1987]; Jänig and Koltzenburg [1993]; Jänig and Häbler [2002]).

2.4 Role of visceral afferent neurons in visceral nociception and pain

2.4.1 Vagal afferent neurons

Vagal afferent neurons innervating abdominal organs

The general contention is that visceral pain elicited from pelvic and most abdominal organs is triggered by excitation of spinal visceral afferent neurons and not by excitation of vagal afferent neurons (Cervero 1994). This is supported by studies of behavioral responses in animals and clinical studies using stimulation and blocking techniques (Foerster 1927; Cannon 1933). However, the situation is not entirely clear for the gastroduodenal section of the gastrointestinal tract. Whether patients with complete interruption of spinal ascending impulse transmission at the thoracic level T1 or at a more rostral segmental level can experience pain from the gastroduodenal section of the gastrointestinal tract (e.g., during gastritis, a peptic ulcer or distension of the stomach) has never been systematically investigated. Patients with complete lesion of the cervical spinal cord experience abdominal hunger, dread and nausea (Crawford and Frankel 1971). Furthermore, these patients may experience vague sensations of fullness after a hot meal. But usually acid reflux or an obstructed or distended visceral organ does not generate discomfort and pain (Juler and Eltorai 1985; Strauther et al. 1999). There are occasional observations in these patients showing that gastric distension may generate pain and discomfort (Dietz, personal communication) and that an acute perforation of a duodenal ulcer is accompanied by violent pain in the right or left shoulder (page 284 in Guttmann [1976]), indicating that vagal afferents may be involved or possibly phrenic afferents projecting through the phrenico-abdominal ramus to the cervical segments C3 to C5.

Stimulation of vagal afferents innervating the stomach elicits emesis, bloating and nausea, all three being protective reactions. In the rat, acute gastric inflammation generated by acid triggers aversive responses, which are dependent on activity in vagal afferents (Lamb et al. 2003). Experiments on rats show that influx of acid into, or other chemical insults of, the gastroduodenal mucosa lead to a host of locally and centrally organized protective reactions that are mediated by spinal visceral and vagal afferent neurons. Activation of vagal afferents by these chemical stimuli leads to activation (expression of the marker protein Fos after activation of the immediate early gene *c-fos*) of neurons in the NTS, area postrema, lateral parabrachial nucleus, thalamic and hypothalamic paraventricular nuclei, supraoptic nucleus and central amygdala, but not in the insular cortex (the major central representation area of the stomach; see Subchapter 2.6) (Michl et al. 2001). Furthermore, neurons in the spinal dorsal horn do not seem to be activated. Holzer has put forward the idea that vagal afferents innervating the mucosa of the gastroduodenal section of the gastrointestinal tract are involved in chemonociception and mediate autonomic, endocrine and behavioral protective reactions. Vagal afferents are not involved in pain perception, but in the emotional aspect of pain and therefore indirectly in upper abdominal hyperalgesia (Holzer and Maggi 1998; Holzer 2002a, b, 2003). This fascinating idea needs verification by further experimentation. For example, it is unclear in which way activity in spinal visceral afferents and vagal afferents is centrally integrated in vivo so as to elicit the protective reflexes, protective behavior and pain sensations including visceral hyperalgesia. Furthermore, it is unclear in which way activity in vagal afferents is responsible for the emotional aspects of visceral pain and activity in spinal visceral afferents for the conscious perception of visceral pain (see Subchapter 2.6).

Vagal afferents innervating thoracic visceral organs
The situation is at least as complex for the thoracic visceral organs. Pain elicited from the proximal esophagus and proximal airways is probably mediated by vagal afferents innervating the mucosa of these organs. These afferents are peptidergic (i.e. they contain CGRP and/or substance P); their cell bodies are probably located in the superior [jugular] ganglion (see Berthoud and Neuhuber [2000] [cells in the inferior, nodose, ganglion of the vagus nerve are almost exclusively peptide-negative, whereas many neurons in the jugular ganglion contain peptides]). Activation of these afferents generates neurogenic inflammation in the mucosa, which has been extensively studied in the mucosa of the airways (venular plasma extravasation and vasodilation; McDonald et al. 1988; McDonald 1990, 1997). Pain elicited from the more distal sections of esophagus and airways as well as from the bronchi may be elicited by stimulation of spinal visceral afferent neurons and not of vagal afferents. However, this situation is unclear and needs further experimentation (see Hummel et al. 1997).

Cardiac pain (e.g., during ischemic heart disease) is generally considered to be mediated by spinal visceral afferents having their cell bodies in the dorsal root ganglia C8 to T9 (mainly T2 to T6). However, attempts to relieve pain associated with cardiac angina by surgical manipulations (cervico-thoracic sympathectomy, dorsal rhizotomy, injection of alcohol into the sympathetic chain) consistently showed that only 50% to 60% of patients report complete relief from angina following these interventions while the remaining patients report either partial relief or no relief at all. With the caveat that some failures to relieve pain were attributed to incomplete spinal denervation of the heart it is concluded that vagal afferents innervating particularly the inferior-posterior part of the heart may mediate cardiac pain too (for review and references see Meller and Gebhart 1992).

This conclusion is supported by neurophysiological investigations in monkeys and rats showing that some spinothalamic tract (STT) neurons in the superficial dorsal horn and deeper laminae of the cervical segments C1 to C2 (C3) can be synaptically activated by electrical stimulation of cardiopulmonary spinal and vagal afferents or by injection of algogenic chemicals in the pericardial sac via both afferent pathways. The activation of the STT neurons by vagal afferents is relayed through the NTS. These STT neurons can also be synaptically activated by mechanical stimulation of the somatic receptive fields in the corresponding segments from the head, jaw, neck and shoulder (dermatomes, myotomes). These results are fully in line with clinical observations showing that cardiac pain may be referred to neck, shoulder and jaw (Lindgren and Olivecrona 1947; White and Bland 1948; Meller and Gebhart 1992). Finally, it must be taken into account that some phrenic afferents innervate the pericard. These afferents are independent of vagal afferents and thoracic spinal afferents and project to the cervical segments C3 to C5. The same high cervical spinal segments contain neurons with similar convergent synaptic inputs from vagal, spinal visceral and somatic afferents that project to more caudal spinal thoracic, lumbar and sacral segments. These neurons are involved in the inhibitory control of nociceptive impulse transmission (for discussion and literature see Foreman [1989, 1999]; Qin et al. [2001]; Chandler et al. [2002]).

2.4.2 Spinal visceral afferent neurons

Visceral afferent neurons and peripheral mechanisms of visceral nociception

Spinal visceral afferents are involved in neural regulation of visceral organs (in particular, sacral visceral afferents), reflexes and sensations. The messages that trigger these activities must be derived entirely from the activity in the spinal visceral afferent neurons. How do central neurons decode the afferent messages in order to produce appropriate regulations, reflexes and sensations? Does the decoding process of the afferent activity depend on different types of visceral afferent neurons with respect to the different (adequate) stimuli? In other words, are there distinct functional types of spinal visceral afferent neurons that can be characterized with respect to particular functions, such as painful and non-painful sensations and organ regulations (Cervero and Jänig, 1992; Cervero 1994, 1996)? These questions have been systematically addressed for the lumbar and sacral afferent supply of the urinary bladder and colon in the cat (Blumberg et al. 1983; Haupt et al. 1983; Bahns et al. 1986a, 1987; Jänig and Koltzenburg 1990, 1991c) and rat (Sengupta and Gebhart 1994a, b, 1995; Su and Gebhart 1998), for the thoracic afferent supply of the rat stomach (Ozaki and Gebhart 2001), the heart of the cat (Lombardi et al. 1981; Malliani 1982; Malliani and Lombardi 1982), the biliary system of the ferret (Cervero, 1982) and the esophagus of the opossum (Sengupta et al. 1990), for the afferent supply of the ureter in the guinea pig (Cervero and Sann 1989), for the hypogastric and pelvic afferents innervating the female reproductive organs (Berkley et al. 1988, 1990, 1993) and for the afferent innervation of the testis (Kumazawa 1986; Kumazawa et al. 1987).

Spinal visceral afferents can be distinguished according to the visceral organs or organ sections they innervate (a visceral afferent neuron only innervates one organ or part of an organ) and according to the spatial arrangement of their terminals (e.g., in the smooth musculature, in the mucosa, in the serosa, in the myenteric plexus). These spatial arrangements also determine the adequate stimuli exciting the visceral afferents (such as distension, contraction, shearing and intraluminal chemical stimuli). Morphologically the peripheral terminals of spinal visceral afferent neurons appear rather uniform. However, this uniformity does not necessarily apply to the transduction mechanisms for physical and chemical stimuli.

Most spinal visceral afferents are polymodal and can be excited by mechanical and chemical stimuli and possibly also by thermal stimuli. In the cat, sacral visceral afferents innervating urinary bladder or colon have no spontaneous activity, but about 50% of the thoracolumbar visceral afferents do so (about 0.2 to 1 imp/s). In the rat, both populations of spinal visceral afferents may have spontaneous activity under physiological conditions. The classification of spinal visceral afferent neurons in nociceptive and non-nociceptive ones is, in view of the different types of visceral organs, difficult. Based on the experimental investigations of spinal visceral afferents innervating urinary bladder, hindgut, stomach, gallbladder and ureter, most investigators believe that the afferent neurons can be divided into low-threshold and high-threshold afferents and that the low-threshold afferents are involved in organ regulations and non-painful sensations whereas the high-threshold afferents are involved in visceral pain and protective reflexes. Figure 2.7 demonstrates a representative example for sacral afferents innervating the urinary bladder in the cat. The afferents respond in a graded way to distension of the urinary bladder (generated by slow filling or intraluminal pressure steps). Two populations of sacral visceral spinal afferents can be discriminated: a population of low-threshold ones with high maximal discharge rates and a smaller population of high-threshold ones with relatively low maximal discharge rates (marked by crosses [unmyelinated] and asterisks [myelinated]). Similar results were obtained in the rat urinary bladder, rat and cat colon and rat stomach (Cervero 1994, 1996; Gebhart 1996; Su and Gebhart 1998; Su et al. 1997a, b; Coutinho et al. 2000; Ozaki and Gebhart 2001).

The distinction of spinal visceral afferents into two groups, a high- and a low-threshold one as has been worked out particularly for

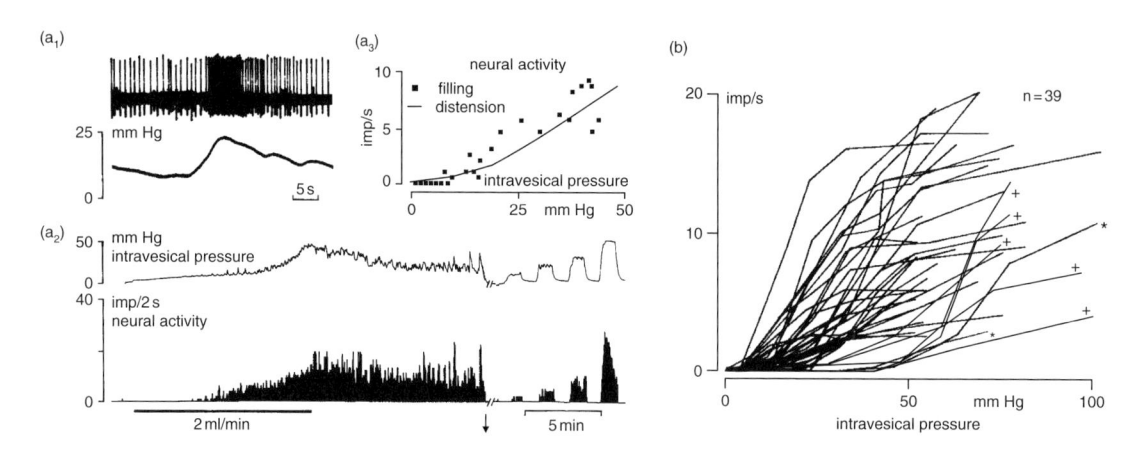

Figure 2.7 Activation of sacral visceral afferents innervating the urinary bladder by distension. (a) Activation of a single myelinated (Aδ) low-threshold afferent fiber during slow filling and to intravesical pressure steps. (a₁) Activity during filling (2 ml/min) through a urethral catheter. Lower record, intravesical pressure with isovolumetric contractions. (a₂) Activity during slow filling (bar) and during a series of intravesical pressure steps (right). Arrow, release of bladder content. The afferent fiber had no spontaneous activity with empty bladder. (a₃) Stimulus–response functions of the afferent fiber during slow filling (squares) and during intravesical pressure steps (solid-line curve). (b) Stimulus–response functions of sacral afferents innervating the urinary bladder to bladder distension by slow filling (n = 39 fibers). All low-threshold afferents were small-diameter (Aδ) fibers. Five of the seven high-threshold afferents (+) were C-fibers and two Aδ-fibers (*). After Häbler et al. (1990a, 1993a).

hollow organs, with the implication that the first group has nociceptive function and the latter one not, must be considered with caution. It is probably too simple:

- Low-threshold spinal visceral afferents encode in their activity a large range of intraluminal pressures during distension and contraction, covering the non-noxious as well as the noxious range (Bahns et al. 1987; Häbler et al. 1993a; Su et al. 1997a, b; Coutinho et al. 2000).
- High-threshold spinal visceral afferents are relatively rare and exhibit lower rates of activity to maximal stimuli (Häbler et al. 1990a; Su et al. 1997a, b; Coutinho et al. 2000).
- Many thoracolumbar spinal visceral afferents (Blumberg et al. 1983; Bahns et al. 1986a) and (in the rat) sacral visceral afferents are spontaneously active (Su et al. 1997a, b; Ozaki and Gebhart 2001).
- Low- and high-threshold afferents are polymodal and can be activated and sensitized by chemical stimuli, experimental inflammation as well as by ischemia (Haupt et al. 1983; Häbler et al. 1993a, b; Su et al. 1997a, b; Coutinho et al. 2000).
- Low- and high-threshold spinal visceral afferents cannot be distinguished by independent physiological (transduction mechanisms), anatomical (e.g., projection patterns to the spinal cord), histochemical (e.g., peptide content) or pharmacological criteria (e.g., receptors for inflammatory mediators, opioids etc.; Sengupta et al. 1996; Su et al. 1997a, b).

These results may imply that, *under biological conditions*, the same populations of afferents may well encode processes that lead to organ regulation, non-painful sensations as well as painful sensations. Visceral pain from these organs would then be associated with a high intensity of discharge in the visceral afferent neurons, with the degree of their recruitment, and with the recruitment of high-threshold afferents that contribute relatively little to the impulse activity from the urinary bladder at intraluminal pressures of >50 mmHg, which are frankly painful (Häbler *et al.* 1993b; Jänig and Häbler 1995). Also for the heart the same spinal cardiac afferent neurons may be involved in regulation of the cardiovascular system (e.g., during exercise) as well as pain (Malliani 1982; Malliani and Lombardi 1982).

As already mentioned this interpretation is not generally accepted and cannot necessarily be generalized for other organs. (1) Neurophysiological investigations of the spinal afferents supplying the biliary system in the ferret (Cervero, 1982) or the ureter in the guinea pig in vitro (Cervero and Sann, 1989) have shown that most spinal visceral afferents can only be activated at high intraluminal pressures that are normally associated with pain and an equivalent behavior in animals. These afferents probably have nociceptive function only. (2) Baker *et al.* (1980) have analyzed the spinal afferent supply of the heart and claimed to have found a small separate group of afferent neurons that respond preferentially to bradykinin (a pain producing substance) and have rather high thresholds to mechanical stimuli.

It is possible that noxious stimuli, damage and impending damage in the visceral domain are encoded in different ways by spinal afferents for different organs. Therefore care must be taken when generalizing from the investigation of one organ to the others (Cervero and Jänig, 1992). It is, however, also possible that some investigations just mentioned concentrated on populations of spinal visceral afferents that only get into action under pathophysiological conditions and are normally silent (see below).

Recruitment of normally mechanoinsensitive afferent neurons

It has been shown for the sacral afferent innervation of the urinary bladder and colon of the cat that about 90% of the visceral afferent neurons with unmyelinated fibers projecting through the pelvic splanchnic nerve are silent under normal conditions and cannot be activated at intraluminal pressures of up to 70 mm Hg in both organs (Bahns *et al.* 1987; Häbler *et al.* 1990a; Jänig and Koltzenburg 1991c), i.e. at pressures that elicit pain in humans. Some of these silent sacral visceral afferent neurons with unmyelinated fibers may innervate the internal reproductive organs, urethra or anal canal and some may innervate the mucosa of the colon or the urothelium of the urinary bladder, and distension and contraction of the hollow organs may not be the adequate stimuli for them. When the urinary bladder is inflamed some of these silent afferent neurons may develop resting activity and can be activated during normal non-noxious stimulation of the organ. The afferent neurons that are activated during evacuation under healthy conditions may develop ongoing activity after inflammation when the organ is empty and may exhibit increased responses to adequate stimulations. Thus, under pathophysiological conditions the visceral afferent supply of the urinary bladder is sensitized and novel types of visceral afferent neurons with chemo- and mechanosensitivity are recruited (Häbler et al. 1990a; Jänig and Koltzenburg 1990).

Recruitment of normally mechanoinsensitive spinal visceral afferents under pathophysiological conditions (e.g., during inflammation) may apply to all visceral organs. For example, various structures in the retroperitoneal space, such as blood vessels, nerves, lymph nodes etc., are innervated by afferents that are normally not activated, but are recruited in pathophysiological states (Bahns et al. 1986b). During angina pectoris, afferents may be sensitized and recruited, although Malliani believes that this is not the case (Lombardi et al. 1981). It is a matter of debate whether these silent and normally mechanoinsensitive visceral afferents constitute a distinct category of spinal visceral afferents that is recruited under inflamed and other pathophysiological conditions or whether these afferents are frankly extremely high-threshold nociceptive fibers (for discussion see Cervero [1996]).

The idea that tissues are innervated by silent and mechanoinsensitive afferent neurons with unmyelinated fibers that are activated in diseased states (Michaelis et al. 1996) has first been demonstrated for the afferent innervation of the joint capsule in the cat (Grigg et al. 1986; Schaible and Schmidt 1988). It probably applies to all tissues (for the skin in rat and monkey see Meyer et al. [1991], Simone et al. [1991], Kress et al. [1992], Davis et al. [1993], for the vein in cats see Michaelis et al. [1994]; for the skin in humans see Schmidt et al. [1995, 2002]). These silent and normally mechanoinsensitive (or mechanically extremely high-threshold) spinal visceral afferent neurons may overlap with the spinal visceral afferent neurons that have only peripheral (efferent) functions under physiological conditions (Figure 2.4; Holzer and Maggi 1998; Holzer 2002a, b, 2003).

2.5 | Relation between functional types of visceral afferent neurons, organ regulation and sensations

2.5.1 General considerations

Events in the visceral organs are encoded in the activity of visceral afferent neurons and by the activity of central neurons in the neuraxis, which then leads to appropriate organ regulations, reflexes and sensations. Several points need to be emphasized:

1. In healthy conditions, afferent activity elicited during regulation of visceral organs is usually *not* accompanied by distinct visceral sensations and aversive feelings. This is quite obvious for the regulation of the cardiovascular system and of most parts of the gastrointestinal tract. It is biologically meaningful.
2. Visceral sensations are dependent on the representations of the visceral organs in the limbic sensory cortex (mid/posterior insular cortex; see Subchapter 2.6). Distinct non-painful sensations can be elicited particularly from those organs that are precisely regulated in the social context of human and animal behavior, such as the entrance of organs and the evacuative organs (trachea, esophagus, stomach, hindgut, lower urinary tract). These sensations are integral components of the neural regulation of the organs, leading to appropriate behavior and suppression or enhancement of organ activity (e.g., continence and evacuation).
3. Events that endanger visceral organs and therefore also the organism may lead to discomfort and pain. These sensations can be elicited from all visceral organs or their capsules. With the exception of the afferent vagal innervation of proximal esophagus and airways, activity in spinal visceral afferent neurons is correlated with these sensations. They are integral components of protective behavior (that includes autonomic protective reflexes).
4. Emotional feelings and their expression (by the somatomotor system, e.g., in facial expression, and by the autonomic systems, e.g. in the adaptive responses of the cardiovascular system, gastrointestinal tract, evacuative organs etc.) are represented in the telencephalon. They are generated *in parallel* by the brain without peripheral afferent input from visceral and deep somatic body domains (see Figure 11.1 and Subchapter 11.3). However, activity in visceral afferent neurons (and in afferent neurons from deep somatic tissues) shapes the centrally generated emotional feelings. It is virtually impossible to disprove that emotions are evoked ("caused") by afferent activity from viscera and deep somatic structures during bodily changes as originally proposed by James and Lange (James 1884, 1994; Meyers 1986).

2.5.2 Encoding of visceral events in the activity of afferent and central neurons

Are the different qualities of sensation (including visceral discomfort and pain), organ regulations and reflexes that originate from a visceral organ elicited by the excitation of different types of afferents, or are they all subserved by a single category of afferent? Theoretically, there are several ways in which different peripheral stimuli in the visceral domain could be encoded by primary afferent and central neurons so as to elicit these sensations, regulations and reflexes:

1. Primary afferents from an organ react specifically to certain stimuli and not to others. These afferents are more or less specifically connected to second-order neurons in the spinal cord and brain stem leading, when excited, to characteristic sensations and reflexes. This concept is compatible with the specificity theory and with the idea of an "adequate stimulus" as originally described by Sherrington (1900, 1906). According to this idea, non-painful sensations, organ reflexes and regulations would be elicited by low-threshold afferents and painful sensations as well as protective reflexes by high-threshold afferents (Figure 2.8a).
2. Primary afferents from an organ are homogeneous and encode the whole stimulus range in their activity but may have different thresholds (Figure 2.8b). This is described by the "intensity theory" or "summation hypothesis" and some physiologists believe that it may apply to the sensations associated with some visceral organs (Malliani and Lombardi 1982; Jänig and Morrison 1986).
3. The way in which the afferent activity is decoded in the spinal cord is unknown. Different classes of interneurons linked to the autonomic pathways supplying the visceral organs and other target organs and to neural pathways ascending to the brain stem and thalamus (which are concerned with organ regulations and sensations) must be postulated (see Chapter 9). Each of these types of interneurons must decode the afferent messages from these visceral organs in a different way and must have different stimulus–response functions of their activity. Otherwise it would be very difficult to understand in which way the distinct regulations, sensations and reflexes are brought about by stimulation of a rather homogeneous population of visceral afferents, such as e.g. the afferents from urinary bladder

Figure 2.8 Schematic diagram portraying the two classical theories for the encoding of noxious stimuli by primary visceral afferent fibers. The *specificity theory* (a) postulates the existence of two different classes of primary afferent fibers (low threshold: LT and high threshold: HT) responsible for the separate encoding of innocuous and noxious events. The *intensity theory* (b) proposes that all peripheral stimuli are encoded by a single and homogeneous population of afferent fibers capable of encoding stimulus intensities ranging from innocuous to noxious. Although it is almost universally accepted that the specificity theory applies to peripheral mechanisms of visceral pain it cannot be excluded that low-threshold visceral afferents that encode the full range of intraluminal pressures (see Figure 2.7b) are also involved in the generation of visceral pain. From Cervero and Jänig (1992) with permission.

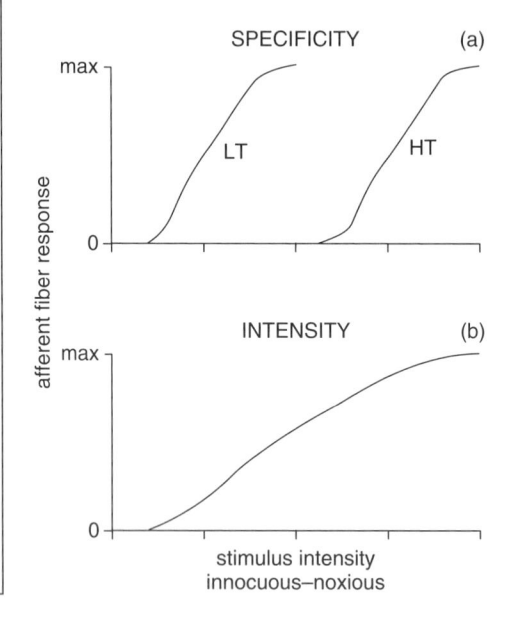

(see Subchapter 9.3) and colon. We need to know how the different populations of interneurons, or networks of interneurons, being involved in organ regulation, protective reflexes and various types of sensations, behave during physiological stimulation of distinct groups of visceral afferents (see Chapter 9).

This point of view is somewhat simplistic. Specificity is brought about by a specialization of the peripheral receptors with respect to the adequate (physical and chemical) stimuli and by the central synaptic connectivity of the central projections of the afferent neurons. The specificity applies certainly to cutaneous sensations elicited by stimulation of mechanosensitive myelinated and nociceptive unmyelinated cutaneous afferents from the distal parts of the extremities in humans. Repetitive intraneural electrical microstimulation of *single* large-diameter myelinated afferents from Pacinian corpuscles, rapidly adapting receptors or slowly adapting type I receptors in the hand elicits distinct elementary vibration, touch or pressure sensations, respectively. These elementary sensations are projected to the receptive fields of the stimulated afferent fibers. Electrical microstimulation of microbundles of cutaneous C polymodal nociceptive afferents elicits dull or burning pain sensations, which are in most cases projected to skin areas lying within 10 mm of the C nociceptor receptive fields. Thus, for cutaneous mechanical and painful sensations the "stimulus specificity" of the cutaneous afferent neurons is connected to their "modality specificity" (Ochoa and Torebjörk, 1983, 1989; Vallbo *et al.* 1984). This specificity applies to many autonomic reflexes too (see Chapters 4, 9 and 10).

2.5.3 A general model of central encoding

The perception of non-painful and painful sensations and the complex regulation of visceral organs cannot be explained by reference to the encoding of peripheral events in afferent activity. We do not know the actual neuronal mechanisms of central decoding of the activity in most visceral afferent neurons, and in particular of spinal visceral afferent neurons. Practically all second-order neurons in the spinal cord (laminae I, V and deeper) with synaptic input from visceral afferents are viscero-somatic convergent neurons that receive additionally synaptic input from skin and probably deep somatic tissues (see Subchapter 2.6). Traditionally the answer to the problem is that separate populations of afferent and central neurons are involved in the generation of painful and non-painful visceral sensations and regulatory visceral reflexes. Models explaining the mechanisms of normal painful and other sensations under biological and pathological conditions have been developed from studies of cutaneous sensations in animal models and humans (Baumann *et al.* 1991; Simone *et al.* 1991; LaMotte *et al.* 1992; Torebjörk *et al.* 1992; Willis and Coggeshall 2004a; Meyer *et al.* 2005). In analogy to these models, a model is proposed that may explain how the different sensations (including pain) and regulations are triggered from hollow viscera (such as urinary bladder or colon) in physiological and pathophysiological conditions. This model

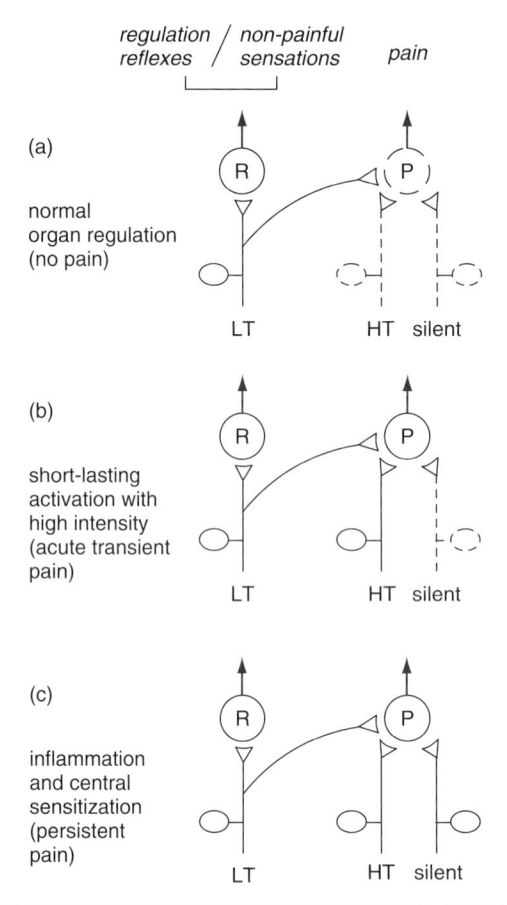

Figure 2.9 Diagrammatic representation of possible neuronal mechanisms of pain, other sensations, and regulatory reflexes elicited from visceral organs. Visceral afferents are shown to activate different central mechanisms, those mediating regulatory reflexes and non-painful sensations (left, pathway R) and that mediating pain (right, pathway P). (a) During normal organ regulation low-threshold (LT) intensity encoding afferents are active. This activity is associated with the activation of pathway R. (b) Short transient, high-intensity stimuli will evoke large responses in the low-threshold afferents. These responses may be grouped and burst-like. This afferent activity will now activate pathway P and lead to brief periods of pain. The activation of pathway P may be supported by the recruitment of some high-threshold afferents (HT), which may be present in variable percentage according to the visceral organ. (c) During prolonged forms of stimulation, including activation during inflammation, normally mechanoinsensitive nociceptors may be recruited and sensitized and high-threshold receptors sensitized. The afferent barrages in both of these may now increase the excitability of central neurons of the pathway P for input from low-threshold afferents (central sensitization). As a result, normal regulatory activity in the viscera could now be perceived as painful. Pathways being involved in regulatory reflexes may also be sensitized. Modified from Cervero and Jänig (1992) with permission.

includes low-threshold, high-threshold as well as normally silent visceral afferents (Figure 2.9):

1. Normal distension and contraction of an organ stimulate low-threshold visceral afferents. Spinal neurons (interneurons, propriospinal neurons, tract neurons projecting to the brain stem, hypothalamus and thalamus), which are involved in organ regulation (e.g., micturition, defecation and continence) or non-painful sensations, are activated (neuron R in Figure 2.9a).
2. Short intense distension or contractions of the organ with increase of intraluminal pressure above 40 to 50 mm Hg (a pressure that

elicits discomfort and pain) generate high-frequency bursts in these low-threshold visceral afferents and activate some high-threshold afferents. This afferent excitation activates additionally other spinal neurons (neuron P in Figure 2.9b), the excitation of which leads to transient pain sensation. This situation may occasionally occur in everyday life.

3. Continuous or intermittent high-frequency activation of low- and high-threshold visceral afferents and recruitment of silent normally mechanoinsensitive afferents from the visceral organs (e.g., by inflammation of the organ) sensitizes the neurons of the spinal nociceptive pathways (neuron P in Figure 2.9c; high-threshold ["nociceptive-specific"] neurons, wide dynamic range [convergent] neurons). This sensitization is either generated by all (high- and low-threshold) visceral afferents or only by high-threshold and normally silent visceral afferents. Now, activity in low-threshold afferents during normal regulation of the organs activates central sensitized neurons resulting in enhanced pain perception (Figure 2.9c).

4. For visceral organs whose spinal afferent innervation consists largely of high-threshold afferents (e.g., the ureter or the biliary system) only pathway P may be excited when these afferents are activated (Figure 2.9c).

A component, which is probably important in the generation of pain and other sensations elicited from visceral organs and which has not been explored so far, is the dual autonomic – parasympathetic and sympathetic – innervation of the organs. Neurons of these autonomic systems exhibit distinct reflexes to physiological afferent stimuli (see Chapter 4). It is possible that these autonomic neurons exhibit abnormal reflexes when the visceral afferents are continuously activated, sensitized and recruited and spinal neurons are sensitized. These abnormal reflexes may in turn enhance the afferent impulse traffic and establish a positive feedback loop consisting of visceral afferents, spinal cord and autonomic output systems to the viscera. This feedback loop may be extremely important for the understanding of mechanisms of pain (or its absence) and other sensations in coronary ischemia and myocardial infarction, in non-cardiac chest pain, in non-ulcer dyspepsia, in irritable bowel syndrome (Mayer and Raybould 1993; Mayer *et al.* 1995), in interstitial cystitis (which is believed to be the equivalent of the irritable bowel syndrome in the lower urinary tract) and in other functional diseases of the viscera.

2.6 Central ascending pathways associated with autonomic regulations and visceral sensations

Central pathways mediating the activity of spinal visceral and vagal visceral afferent neurons are related to autonomic and endocrine regulations, non-painful and painful visceral sensations and shaping of emotions (see Figure 2.1). The physiology and anatomy of these pathways are only incompletely understood. This is due to: (1) difficulties to dissect out central pathways conveying information in spinal visceral afferent neurons compared to information in primary afferent neurons with small-diameter axons (Aδ, C) from skin and deep somatic tissues; (2) difficulties to dissect out central pathways related to vagal afferent neurons projecting to the NTS (see Subchapter 8.3) and (3) species differences between primates and non-primates (in particular the rat).

Autonomic reflex pathways in spinal cord and lower brain stem linked to spinal visceral or vagal afferent neurons will be described in more detail in Chapters 8 to 10. The knowledge about central pathways mediating and representing visceral sensations and body feelings including visceral pain is based on anatomical and physiological studies conducted on animals (rat, cat, monkey), on recording and stimulation experiments conducted on patients undergoing brain surgery and on studies using brain imaging methods in humans (for review and literature see Cechetto and Saper [1990]; Cechetto [1995]; Saper [1995, 2002]; Craig [1996, 2002, 2003a, b, c]; Bernard and Bandler [1998]; Gauriau and Bernard [2002]; Willis and Coggeshall [2004a, b]). Here I will summarize what is known about ascending spinal and trigeminal tract neurons projecting to autonomic centers in the brain stem and hypothalamus and to thalamocortical systems and about ascending pathways from the NTS related to specific and general visceral sensations. This short summary does not imply that information in primary afferent neurons is relayed to the autonomic centers and thalamocortical systems in passive ascending labeled-line systems. Every synaptic transmission in these ascending systems is subject to powerful multiple descending control by the forebrain. This descending ("efferent") control is exerted by pathways that are, in number of neurons, up to five times more powerful than the ascending pathways.

2.6.1 Ascending spinal pathways conveying information from visceral afferents

Spinal visceral afferent neurons connect synaptically with second-order neurons in laminae I, V, VII, VIII and X of the spinal grey matter. They largely spare laminae II, III and IV and some projections go to the contralateral side (Figure 2.6). The second-order neurons are either (excitatory or inhibitory) segmental or propriospinal interneurons or tract neurons projecting to supraspinal centers (lower and upper brain stem, hypothalamus, thalamus). Interneurons have multiple functions mediating the effects of supraspinal centers on spinal circuits and being involved in autonomic regulations and reflexes (see Chapter 9), in regulation of movements as well as in regulation of synaptic transmission from primary afferent neurons to second-order neurons by acting pre- or postsynaptically.

In the rat by far the most common second-order neurons in the superficial and deep spinal dorsal horn are segmental interneurons or propriospinal neurons and only a few neurons project to supraspinal centers. For example, lamina I of the lumbar segment L_4 contains about 7500 to 8000 neurons. Only 5% of these neurons are tract neurons projecting to supraspinal centers and the rest interneurons (Spike *et al.* 2003; Polgar *et al.* 2004). These proportions most likely apply also to the deeper laminae of the spinal grey matter of the rat and to primates, although quantitative data are not available. In primates, the number of spinothalamic neurons is significantly higher than in the rat (Willis and Coggeshall 2004b; Dostrovsky and Craig 2005).

Functions and projections of lamina I neurons
Lamina I projection neurons of the spinal and trigeminal dorsal horn receive monosynaptic input from primary afferent neurons with small-diameter axons (Aδ, C) innervating skin, deep somatic tissues and viscera. Physiological investigations of the lamina I neurons show that they consist of several functionally distinct types. They are excited from the skin by noxious mechanical stimulation (mechanosensitive nociceptive-specific neurons) or heating, pinch and noxious cooling (polymodal nociceptive-specific neurons) or histamine ("itch" receptors) or innocuous cooling or warming or sensual touch (Figure 2.10a, b) (Bester *et al.* 2000; Craig *et al.* 2001; Andrew and Craig 2001a, b; 2002a, b; Craig 2004a; see Craig [2003a] for discussion and references). Some lamina I projection neurons are multireceptive (wide dynamic range, WDR) neurons (Willis and Coggeshall 2004a).[4]

The differentiation of lamina I projection neurons with respect to afferent inputs from deep somatic structures and from viscera has not been extensively studied. However, lamina I neurons excited from the skin may also receive additional convergent synaptic input from small-diameter primary afferent neurons innervating deep somatic tissues (such as skeletal muscle, joint) and/or viscera. A few lamina I neurons are only activated by small-diameter muscle afferents (Craig and Kniffki 1985; Wilson *et al.* 2002). However, no class of lamina I neurons receiving synaptic input from visceral afferent neurons only has so far been detected. Almost all neurons being activated by stimulation of visceral afferents are *viscero-somatic convergent (wide dynamic range) neurons*, i.e. they can be activated by stimulation of small-diameter (Aδ, C) afferents of skin, deep somatic tissues (skeletal muscle, joint, bone, fascia) and viscera (Cervero and Tattersal 1986; Cervero 1994, 1995; Katter *et al.* 1996) and by large-diameter afferents from skin. Cutaneous high-threshold (nociceptive specific) dorsal horn neurons receiving visceral afferent input seem to be rare. Thus, the neurons of the ascending pathways from lamina I are functionally not specific for the viscera.[5] This is probably not very surprising since non-painful visceral sensations, temperature sensations, sensations related to deep body tissues, and pain sensations elicited from all body domains have the commonality to be directed towards the *body* (Craig 2002, 2003b, c; Saper 2002). Furthermore, sensations projected to visceral organs have a poor spatial resolution and are referred to skin, deep somatic tissues and other visceral organs (Giamberardino 1999). Sensations elicited by noxious, temperature or chemical stimuli and projected to the body surface

Figure 2.10 Functional classes of neurons in laminae I and V of the spinal dorsal horn that can be excited by stimulation of afferent neurons with Aδ- or C-axons and project to brain stem and thalamus. (a) Nociceptive-specific (NS) neurons in lamina I. (b) Non-nociceptive neurons in lamina I. (c) Multireceptive neurons (MR) neurons (wide dynamic range [WDR] neurons) in lamina V and deeper laminae (some in lamina I). Synaptic connections from afferent neurons to most lamina I neurons are monosynaptic, those to most lamina V neurons are polysynaptic (interrupted synaptic inputs). Cold noc., cold nociceptive; HSM, high-threshold mechanical; HPC, heat pinch cold (nox); HSMH, high-threshold mechanical heat (polymodal nociceptive); Aβ, Aδ, C, neurons with large-diameter myelinated, small-diameter myelinated or unmyelinated axons. After Craig (2003a).

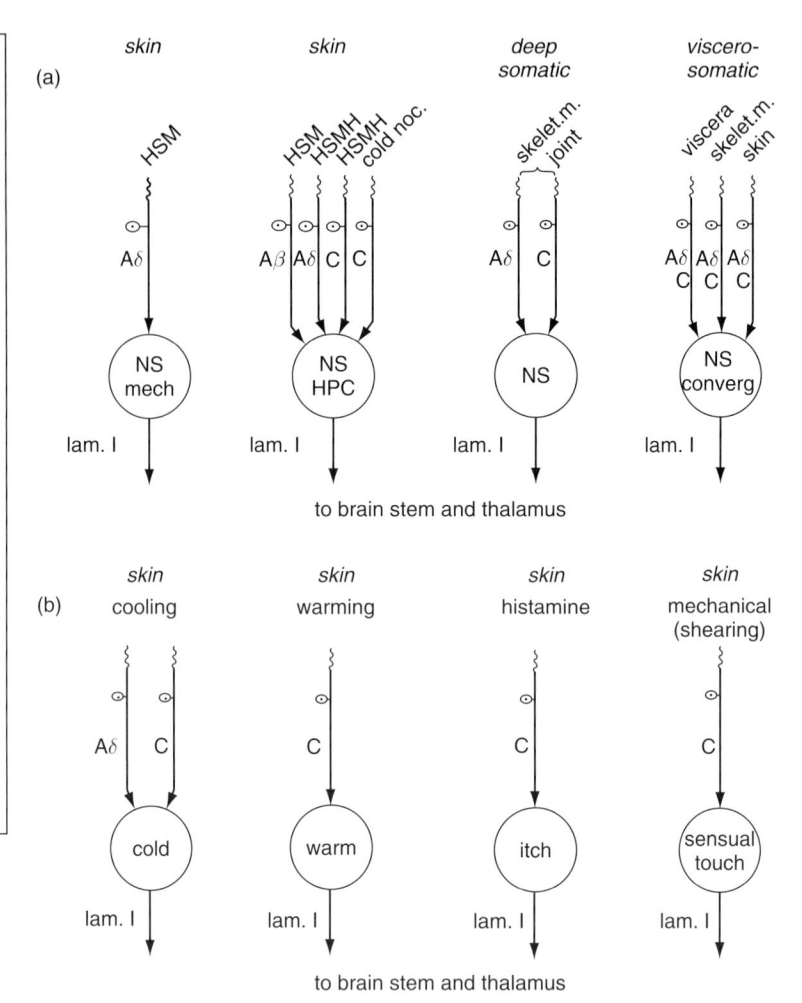

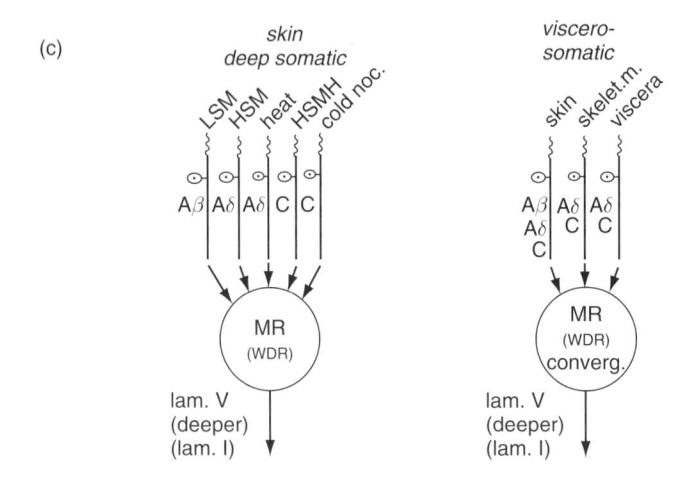

(skin) are graded and have a high spatial resolution, showing that the sensory-discriminative component is marked for these sensory submodalities (for review see Gauriau and Bernard [2002]; Craig [2003a]; Dostrovsky and Craig [2005]).

It is debated whether the uniformity of the convergent viscero-somatic lamina I neurons is an experimental artifact and related to the anesthesia and the experimental surgery, i.e. whether these neurons exhibit functional specificity under more physiological conditions. The functional specificity of viscero-somatic lamina I neurons may, e.g., be preferentially dependent on the control of synaptic transmission from primary afferent neurons to second-order neurons by descending systems. These descending control systems are represented in the rostral ventromedial medulla (nucleus raphe magnus, paragigantocellular medial nucleus, parapyramidal nucleus), dorsolateral pontomesencephalic tegmentum (area A5, area A7, nucleus ceruleus, parabrachial nucleus, cuneiform nucleus [lateral to the periaqueductal grey]) and the periaqueductal grey (Fields *et al.* 2005).

The functional characteristics of lamina I projection neurons indicate that these neurons mediate information on the state of body tissues to supraspinal centers. Most of them project through the contralateral anterolateral (ventrolateral) spinal tract to supraspinal centers in primates and some through the ipsilateral anterolateral tract (Figure 2.11a). In the rat these neurons project through the lateral and dorsolateral tract. This difference between primates and rat is related to the large pyramidal tract in primates (in particular the human) that occupies most of the lateral and dorsolateral spinal funiculus. A few lamina I neurons project ipsilaterally. Propriospinal lamina I neurons project ipsilaterally to the sympathetic preganglionic neurons in the intermediate zone. The main supraspinal projections of lamina I neurons are (Figure 2.11a; Craig 1996; Dostrovsky and Craig 2005; Villanueva and Nathan 2000; Gauriau and Bernard 2002):

- In the medulla oblongata, lamina I neurons project to the ventrolateral medulla (rostral and caudal, see Figure 10.2), to the subnucleus reticularis dorsalis and to the caudal NTS. These projections are weak. The nuclei are involved in autonomic regulation.
- In the pons and mesencephalon, lamina I neurons have a powerful projection to the lateral parabrachial area (Figure 2.12). This will be discussed below. Furthermore, they project to the lateral and ventrolateral periaqueductal grey and to the deep layers of the superior colliculus (see Subchapter 11.3.3 and Figure 11.6).
- Lamina I neurons project to various nuclei of the thalamus. This will be further discussed below (Figure 2.13). Some lamina I neurons project to the hypothalamus, most of them via the posterior thalamus (Dado *et al.* 1994).

We do not know whether lamina I neurons are specialized to project to particular nuclei only. However, as it is likely that every tract neuron projects to several sites (Spike *et al.* 2003; Zhang *et al.* 1995b; Kostarczyk *et al.* 1997), there may exist subgroups of ascending tract

Figure 2.11 Supraspinal and spinal projections of neurons in lamina I (a) or laminae V to VII (b) of the dorsal horn. (a) Projection of lamina I neurons. (b) Projection of neurons in laminae V to VII. i.l., internal lateral nucleus in the parabrachial complex; NTS, nucleus tractus solitarii; for thalamic nuclei see legend of Figure 2.13. For details see text. Modified after Villanueva and Nathan (2000) with permission.

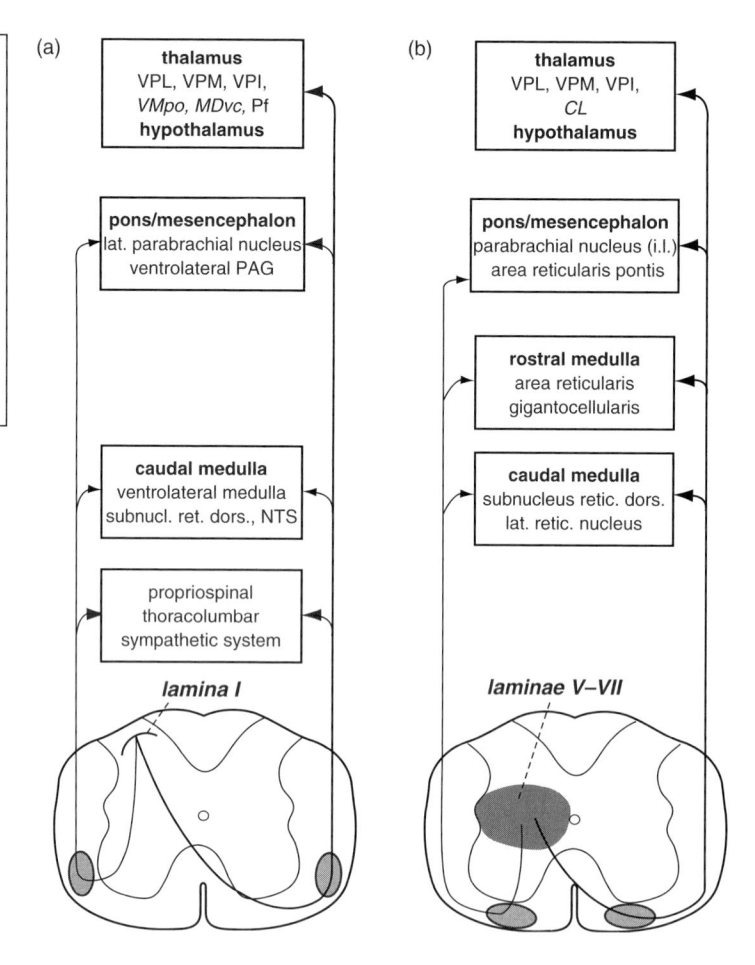

neurons that project to specific groups of target nuclei (e.g., nuclei of the posterior thalamus; Zhang and Giesler 2005).

Functions and projections of deep dorsal horn neurons

Other ascending projection neurons that transmit information in afferent neurons with small-diameter (Aδ, C) fibers from skin, deep somatic tissues and viscera are located in lamina V (a few in lamina IV) and deeper laminae of the spinal grey matter (laminae VII, VIII, X). The projection neurons in lamina V have been most extensively studied, those in the other deep laminae not; therefore the latter projection neurons are functionally poorly characterized.[6] Most lamina V projection neurons are activated by physiological stimulation of large- and small-diameter myelinated (Aβ, Aδ) and unmyelinated (C) afferent fibers from skin, deep somatic tissues and viscera, the latter synaptic connections being polysynaptic. Most of these neurons are therefore multireceptive (MR; see note 4) neurons; a few are nociceptive specific (NS) neurons and a few low-threshold neurons (Figure 2.10c) (Craig 2003a; Dostrovsky and Craig 2005). Projection neurons in deeper laminae probably have similar functional properties.

Most ascending tract neurons in lamina V and deeper laminae have large receptive fields and relatively high ongoing discharge rates. Some have no ongoing activity. They encode in their activity the intensity of stimuli applied to the peripheral tissue spanning the innocuous and noxious range. They are somatotopically organized; however, this organization is relatively poor. They integrate the activity in many afferent inputs to the dorsal horn. They are modality-ambiguous, i.e. they poorly differentiate the modality of innocuous or noxious stimulation or the tissue of origin. However, they may respond preferentially to one type of physiological stimulus and weaker to other physiological stimuli. Thus, they may have their strongest activation to noxious stimuli and weak activation to innocuous stimuli or strong activation to visceral stimuli and weak activation to cutaneous or deep somatic stimuli. Whether they are involved in the generation of pain and visceral sensations is a matter of controversial discussions as far as primates are concerned (Craig and Blomqvist 2002; Willis *et al.* 2002; Craig 2003a). To clarify this important question it is necessary to look at the physiology of *populations* of deep spinal tract neurons. Such studies have not been done yet. These studies may show that spinothalamic neurons in lamina V and deeper laminae are able to discriminate modality (e.g., noxious, visceral and other events) and location of peripheral stimuli.

Ascending tract neurons in spinal lamina V and deeper laminae preferentially project through the ventromedial spinal tract (Craig 1996; Villanueva and Nathan 2000; Gauriau and Bernard 2002; Dostrovsky and Craig 2005; Figure 2.11b):

- bilaterally to the caudal medulla oblongata (subnucleus reticularis dorsalis, lateral reticular nucleus);
- bilaterally to the rostral medulla (gigantocellular reticular area), bilaterally to pons and mesencephalon (internal lateral nucleus of the parabrachial area [which does not receive synaptic input from lamina I neurons], pontine reticular area); and
- contralaterally to various nuclei of the thalamus (see below). This projection to the thalamus is strong in primates (about 25% each of the spinothalamic neurons are located in lamina V and deeper laminae in primates; the remaining 50% are located in lamina I) (Willis *et al.* 2001, 2002; Dostrovsky and Craig 2005) but weak in the rat (Gauriau and Bernard 2004b) although this point is also a matter of controversial discussion (Dado *et al.* 1994; Zhang and Giesler 2005).

Projections through the postsynaptic dorsal column pathways

Based on clinical investigations and experimental investigations conducted on rats or monkeys it is discussed whether an ascending tract in the dorsal columns is conveying information about visceral nociception in the pelvic, abdominal and thoracic viscera to the

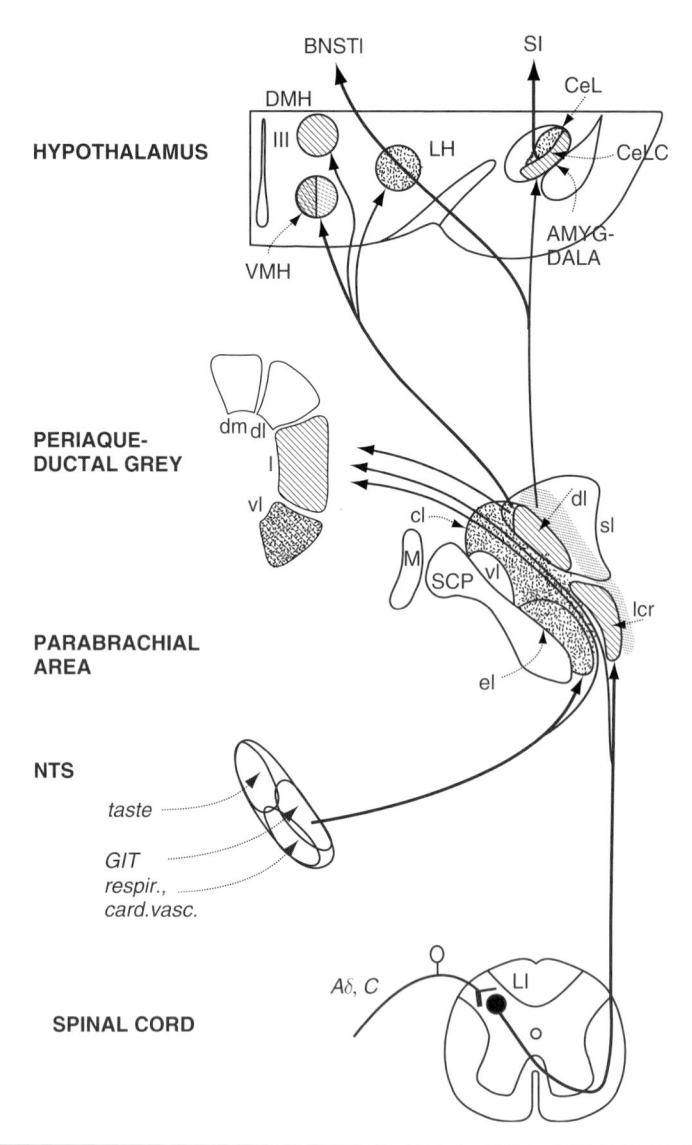

Figure 2.12 Projection of lamina I neurons and second-order neurons of the nucleus tractus solitarii (NTS) to the parabrachial area and projection of parabrachial nuclei to nuclei involved in autonomic and endocrine regulation. Spinal lamina I neurons project to the dorsolateral (dl) and the lateral crescent (lcr) nuclei (hatched) and to sites lateral to these nuclei (shaded) in the parabrachial area. Second-order neurons in the NTS receiving synaptic input from cardiovascular, respiratory or gastrointestinal vagal afferents project to the central lateral (cl) and external lateral (el) nuclei (dotted) of the lateral parabrachial area (for details see Herbert et al. 1990). The parabrachial nuclei project to the lateral (l) and ventrolateral (vl) periaqueductal grey (PAG), to the dorsomedial, ventromedial and lateral hypothalamus (DMH, VMH, LH), to the central amygdala (lateral capsular division [CeLC] and lateral division [CeL]) and to the lateral part of the bed nucleus of the stria terminalis (BNSTl). Parabrachial nuclei receiving synaptic input from lamina I neurons or NTS neurons exhibit differential projections (see hatched [lamina I], shaded [lamina I], dotted [NTS] areas). Not included is the NTS projection to the external medial parabrachial nucleus that projects to the VMb of the thalamus (see Figure 2.14). dl, dm, dorsolateral, dorsomedial PAG; GIT, gastrointestinal tract; M, mesencephalic trigeminal tract; SCP, superior cerebellar peduncle; SI, substantia innominata; sl, vl, superior lateral nucleus, ventrolateral nucleus of the parabrachial area. Modified from Bernard and Bandler (1998) and Gauriau and Bernard (2002) with permission.

thalamocortical system. Myelotomy in the midline of the dorsal columns relieves or significantly attenuates chronic visceral pain in some patients (e.g., of the colon due to cancer [Hirshberg et al. 1996]). Dorsal horn neurons in the sacral spinal cord, which are located close to the central canal and can be activated by colon distension or inflammation, project through the dorsal columns close to the septum to the gracile nucleus (postsynaptic dorsal column [PSDC] tract). In the gracile nucleus these neurons synapse with neurons that project to the ventral posterior lateral nucleus of the thalamus (VPL). Activation of neurons in the VPL by colon distension is largely reduced after interruption of the PSDC, but not when the anterolateral tracts are interrupted bilaterally. Activation of these thalamic neurons by cutaneous noxious stimuli is not affected when the PSDCs are lesioned. For the thoracic and abdominal visceral organs a corresponding PSDC pathway seems to exist that has its origin in the thoracolumbar spinal segments and projects through the dorsal columns to the cuneate nuclei (Al-Chaer et al. 1996a, b, 1997, 1999; Hirshberg et al. 1996; Willis et al. 1999).

Projection to the parabrachial nuclei
Spinal and trigeminal lamina I neurons strongly project to the lateral parabrachial area. Spinal lamina I neurons project heavily to the dorsolateral (dl) and lateral crescent (lcr) parabrachial nuclei and less intensely to areas lateral to these nuclei (shaded; e.g., the superior lateral [sl] nucleus) (Figure 2.12) and trigeminal lamina I neurons to the external medial parabrachial nucleus (not shown in Figure 2.12). The spinal projections to the parabrachial nuclei are largely separate from the projection of the second-order neurons in the NTS related to taste, cardiovascular, respiratory or gastrointestinal afferent neurons. Figure 2.12 (dotted area) demonstrates the NTS projections related to the gastrointestinal tract and probably cardiovascular system that include the central lateral (cl) and external lateral (el) nuclei.

The neurons in the dorsolateral, lateral crescent and external medial parabrachial nuclei have the same functional characteristics as nociceptive lamina I neurons, although their receptive fields are significantly larger than those in lamina I. In the rat, these neurons are excited by either cutaneous noxious mechanical and/or heat stimuli or by cutaneous noxious cold stimuli or by both. About two-thirds of these neurons are additionally excited by noxious visceral stimuli (colorectal distension), i.e. they are viscero-somatic convergent neurons. Neurons excited only by visceral stimuli are absent. Furthermore, neurons specifically excited by innocuous cold or innocuous mechanical stimuli are absent (Bernard and Besson 1990; Bernard et al. 1994; Menendez et al. 1996).

Neurons in the parabrachial nuclei receiving synaptic input from lamina I neurons project to the lateral and ventrolateral periaqueductal grey, heavily to the ventromedial nucleus (dorsomedial part) of the hypothalamus and to the dorsomedial hypothalamus (and to a lesser extent to other hypothalamic nuclei), to the amygdala (central nucleus), to the lateral division of the bed nucleus of the stria terminalis (Figure 2.12), and (via the medial thalamus) to the anterior cingulate cortex (the *limbic motor cortex* representing homeostatic autonomic and endocrine regulations). It is hypothesized that the spino(trigemino)-parabrachial system and its projection areas are involved in pain (nociception)-related autonomic and emotional reactions (see Subchapter 11.3).

Projection to the thalamocortical system
Figure 2.13 summarizes the organization of the thalamocortical systems that receive their synaptic input from spinothalamic (and trigeminothalamic) neurons in lamina I, lamina V and deeper laminae and are involved in different types of body sensation including pain. Traditionally the thalamic nuclei involved in nociception (and probably other body sensations) are divided in nuclei located in the lateral thalamus and nuclei located in the medial thalamus (both groups of nuclei are divided in Figure 2.13 by a vertical dotted line). This division into lateral and medial thalamic system probably also applies to the cortical areas to which these thalamic nuclei project: the primary (SI) and secondary somatosensory (SII) cortex belong to the lateral system, the anterior and middle cingulate cortex (ACC, MCC) to the medial system and the insula would take an intermediate position. The lateral system would represent somatosensory exteroreception and the medial and intermediate system interoception (Treede *et al.* 1999, 2000; Craig 2003a; Gauriau and Bernard 2004b; Vogt 2005):

- In *primates*, lamina I neurons project to the ventral posterior nuclei (to the VPL [from the spinal cord], to the ventroposterior medial nucleus [VPM; from the spinal trigeminus nucleus]), the ventral posterior inferior nucleus (VPI), to the ventral portion of the medial dorsal nucleus (MDvc) and to the parafascicular nucleus (Pf). Furthermore, a dense projection has been proposed to an area in the postero-lateral thalamus named by Craig as posterior part of the ventromedial nucleus (VMpo) (Craig *et al.* 1994; Craig 2004b). In the *rat*, the lamina I neurons project particularly to the triangular posterior nucleus (which may correspond to the VMpo in primates [Gauriau and Bernard 2004a]), to the VPL/VPM, to the posterior nucleus (PO; which is large in rat but small in monkey) and to the MDvc and Pf nuclei.
- In *primates*, lamina V neurons project to the VPL/VPM (Zhang *et al.* 2000a, b) and VPI and heavily to the centrolateral nucleus. In the *rat*, the projections of lamina V neurons to the VPL/VPM and VPI are weak.
- Spinothalamic neurons in deeper laminae project to the centrolateral nucleus (CL).
- Neurons in VPL/VPM project heavily to SI and less strongly to SII. Neurons in VPI project to SI and SII. These cortical areas are

involved in exteroception and likely in discriminative aspects of body sensations including pain, thermal sensations and visceral sensations (although this is very much doubted by Craig; see below).
- Neurons in the region called VMpo project preferentially to the dorsal insula (Figure 2.14), possibly including parts of SII and the adjacent orbitofrontal cortex. A small projection from VMpo exists to area 3a in SI.
- Neurons in the MDvc, Pf and CL project mainly to the ACC and to the MCC. The neurons in MDvc project to area 24c in the ACC that is located anterior to the visceral cingulate motor areas (Craig and Blomqvist 2002; Dostrovsky and Craig 2005; Vogt 2005).

It has been proposed by Bud Craig that the layer of spinal and trigeminal lamina I neurons represents an interface between body tissues and thalamocortical systems that consists of various functionally distinct neural channels related to skin, deep somatic tissues and viscera and that is topographically organized (mediolaterally, craniocaudally) (Figure 2.13; Craig 2003a, b, c):

- The projections of the lamina I neurons to the posterior part of the ventromedial nucleus of the thalamus (named VMpo) are considered to be essential for the generation of sensations related to the states of the body tissues (including viscera), such as mechanical, thermal, chemical, metabolic states. Craig argues that this nucleus only exists in primates and is particularly large in humans. Its size is about half the size of the ventral posterior nuclei (VPL, VPM) that receive their main ascending synaptic input from the medial lemniscal system. The VMpo is located caudally to the VMb that receives viscerotopic synaptic input from vagal afferents via the NTS and the external medial parabrachial nucleus (see below). In the rat the posterior triangular nucleus may correspond to the VMpo in primates (Gauriau and Bernard 2004a, b); however, it is unclear whether this nucleus is just a part of the posterior thalamic nuclei.
- The lamina I neurons project topographically to the VMpo, the lumbar segments being represented posteriorly in the VMpo and the cervical segments and trigeminal part anteriorly. The anterior part of the VMpo is continuous with the representation of taste and internal organs in the VMb related to the NTS, taste afferents and vagal afferents (see below). This would imply (1) that *interoception of the body* (including pain and thermal sensations) is represented in the anterior-posterior axis of the VMb/VMpo and (2) that this thalamic representation of interoception is perpendicularly organized to the mediolateral axis of the VPM/VPL system representing *exteroception and proprioception of the body* (Craig 2004b).
- Studies in humans (microstimulation, functional imaging studies) suggest that the VMpo is important to mediate body sensations related to the tissue states (pain, temperature, itch, muscle sensations, visceral sensations).

76 | VISCERAL AFFERENT NEURONS AND AUTONOMIC REGULATIONS

EXTEROCEPTION
LATERAL SYSTEM

INTEROCEPTION
MEDIAL SYSTEM

CORTEX: SI[6] (3a), SII[6], insula, MCC ACC[7] (24c)

THALAMUS: VPL, VPM[1] | VPI[2] | PO[3] | VMpo[4] | MDvc, Pf | CL

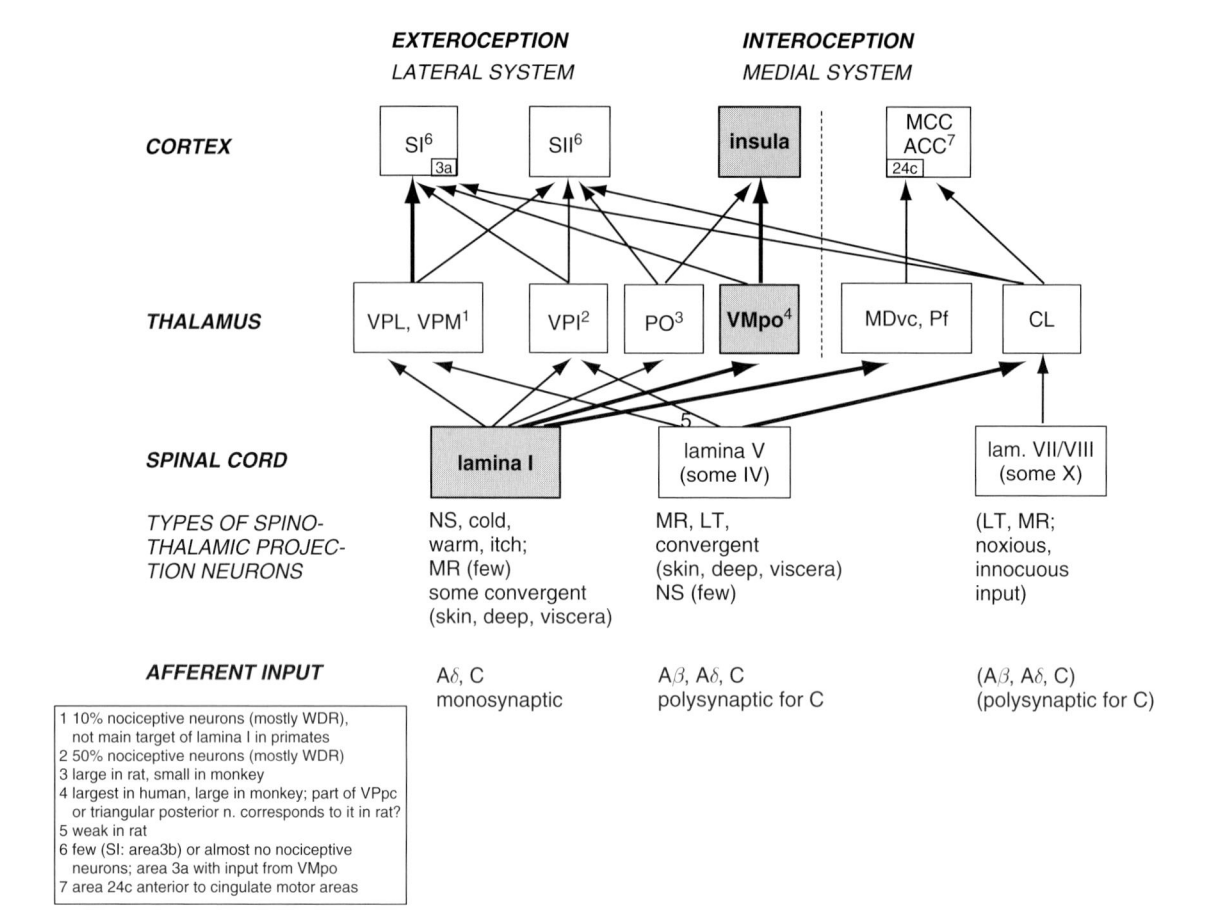

SPINAL CORD: lamina I | lamina V (some IV) | lam. VII/VIII (some X)

TYPES OF SPINO-THALAMIC PROJECTION NEURONS:
- NS, cold, warm, itch; MR (few) some convergent (skin, deep, viscera)
- MR, LT, convergent (skin, deep, viscera) NS (few)
- (LT, MR; noxious, innocuous input)

AFFERENT INPUT:
- $A\delta$, C monosynaptic
- $A\beta$, $A\delta$, C polysynaptic for C
- ($A\beta$, $A\delta$, C) (polysynaptic for C)

1 10% nociceptive neurons (mostly WDR), not main target of lamina I in primates
2 50% nociceptive neurons (mostly WDR)
3 large in rat, small in monkey
4 largest in human, large in monkey; part of VPpc or triangular posterior n. corresponds to it in rat?
5 weak in rat
6 few (SI: area3b) or almost no nociceptive neurons; area 3a with input from VMpo
7 area 24c anterior to cingulate motor areas

Figure 2.13 The spino-thalamocortical systems being involved in body sensations including pain. Systems connected to the primary (SI) and secondary (SII) somatosensory cortex are involved in exteroception (and proprioception) and systems connected to the insula cortex and the anterior and middle cingulated cortex (ACC, MCC) are involved in interoception. The pathway from lamina I neurons to the posterior part of the ventromedial nucleus of the thalamus (VMpo) and the dorsal insula, emphasized in bold and grey, is mainly involved in the generation of pain in skin, deep somatic tissues or viscera, cold, warm and itch sensations, and non-painful sensations from deep somatic tissues and viscera in primates. Pathways from lamina I to SI and SII via the thalamic nuclei VPL, VPM (trigeminal) and VPI may be involved in discriminative aspects of these sensations. Pathways from lamina V to VPL/VPM and VPI may also be involved in discriminative aspects of these sensations, however this is a controversial issue. In the rat these pathways are weak. Pathways to the ACC via the thalamic nuclei MDvc, Pf and CL are involved in the affective aspects of these sensations. The pathway via the thalamic nucleus PO is large in rat but not in primates. The vertical dotted line divides lateral and medial thalamic nuclei. The lower part of the figure shows the afferent inputs to the spinothalamic neurons (afferent neurons with $A\beta$ axons from low-threshold [LT] receptors; with $A\delta$-axons from nociceptors and cold receptors; with C-fibers from nociceptors, cold receptors, warm receptors or itch receptors) and the functional types of spinothalamic neurons (NS [nociceptive-specific], cold [non-noxious], MR [multireceptive] neurons). For details see text. ACC, anterior cingulate cortex; CL, centrolateral nucleus; Pf, parafascicular nucleus; MDvc, ventrocaudal part of medial dorsal nucleus; PO, posterior nucleus; VMpo, posterior part of ventromedial nucleus; VPI, ventral posterior inferior nucleus; VPL, ventral posterior lateral nucleus (spinal input); VPM, ventral posterior medial nucleus (trigeminal input); WDR, wide dynamic range. Based on Vogt et al. (1979), Willis (1985), Apkarian and Shi (1994), Craig (1996, 2003a). References for notes in box lower left: 1. Dostrovsky and Craig (2005); 2. Apkarian and Shi (1994), Dostrovsky and Craig (2005); 3. Craig and Blomqvist (2002), Gauriau and Bernard (2004b); 4. Gauriau and Bernard (2004a); 5. Gauriau and Bernard (2004b), Willis et al. (2001, 2002); 6. Dostrovsky and Craig (2005). Changed from Treede et al. (1999).

- The thalamic nucleus VMpo projects to the posterior part of the dorsal insula. This projection is topographically organized (Hua et al. 2005). The posterior part of the dorsal insula is, together with the more rostrally located part, which receives the NTS projections via the thalamic nucleus VMb, the primary sensory representation of the state of the body tissues and therefore the *primary interoceptive representation* in the cortex *(limbic sensory cortex)*.
- Craig proposes that the system consisting of lamina I, VMpo and dorsal insula is responsible for the generation of body sensations in primates such as sharp and burning pain, itch, sensual touch, muscle burn and pain and different types of visceral sensation including visceral pain. This includes the sensory-discriminative component (intensity coding, spatial resolution) as well as the affective component.
- The organization of the pathways underlying visceral sensations (and also muscle sensations) has not been as well explored compared to those underlying the cutaneous sensations.

The idea that the lamina I-VMpo-posterior insula is the primary representation of painful, thermal, deep somatic and visceral body sensations in primates is controversially debated and needs further independent verification. Based on anatomical tracing and immune-histochemical studies, Jones very much doubts that the VMpo, as described by Craig, is a thalamic nucleus in its own right. He argues that the VMpo has no existence as a specific thalamic nucleus and that the thalamus zone described by Craig as VMpo (Craig et al. 1994; Craig 2004b) is an integral part of the VPM, that the high projection density of lamina I neurons to this part of the thalamus is apparent, and that lamina I neurons also have a considerable projection to the neurons in the ventrobasal complex (VPL, VPM) that project to the primary and secondary somatosensory cortices (Jones 2002, 2006; Graziano and Jones 2004; Zhang et al. 2000a, b). To my understanding the idea of Craig does not necessarily collide with the more traditional view showing that, *first*, the VPL/VPM–SI/SII system and, *second*, the population of spinothalamic neurons in lamina V and deeper laminae of the dorsal horn are involved in discriminative aspects of cutaneous painful, thermal and other body sensations generated by activation of $A\delta$- and C-afferents (see Craig and Blomqvist 2002; Willis et al. 2002; Craig 2003a; Dostrovsky and Craig 2005). In fact, in primates, lamina I neurons and deep dorsal horn neurons also project to the ventrobasal complex (VPL, VPM). However, in the rat, one major projection area of lamina I neurons to the thalamus are the posterior thalamic nuclei, which includes the posterior triangular nucleus, and the ventrobasal complex (Gauriau and Bernard 2002; but see Dado et al. [1994]; Zhang and Giesler [2005]). Deep dorsal horn neurons in the rat exhibit weak projections to the ventrobasal complex and to the posterior thalamus (Gauriau and Bernard 2002; but see Zhang and Giesler [2005]). Area 3a in the postcentral sulcus may be involved in discriminative and motor aspects of pain. However, this area receives synaptic input from VMpo too. There is agreement that the ACC and the MCC are involved in affective aspects of pain and non-nociceptive body sensations (Dostrovsky and Craig 2005; Vogt 2005).

2.6.2 Vagal afferents and the thalamocortical system

As will be described in Chapter 8 in more detail, taste afferents and vagal afferents from the gastrointestinal tract, from the respiratory system and from the cardiovascular system project viscerotopically to the NTS. The rostrocaudal and mediolateral projection pattern of vagal afferents is transformed into the viscerotopic organization of the second-order neurons in the NTS and preserved for the ascending pathways to the dorsal insular cortex. The middle and posterior dorsal insular cortex represent the sensations related to the organ systems and mediate the information from the viscera to the *autonomic motor cortices* (e.g., in the *anterior cingulate cortex*) and to other cortices that are responsible for the emotional feelings accompanying the organ sensations. Therefore middle and posterior dorsal insular cortex are called *primary visceral sensory cortex*.

As described in Subchapter 2.3, excitation of vagal afferents does not generate, with a few exceptions, visceral pain sensations but is related to hunger, satiety, thirst, nausea, fullness, desire to cough etc. (see Table 2.1). The pathway believed to be essential to elicit non-painful visceral sensations associated with the excitation of vagal afferents innervating thoracic or abdominal organs includes the VMb. This nucleus is called the most medial part of the parvicellular part of the ventroposterior nucleus (VPpc or VPMpc) in rats (Cechetto 1995; Saper 1995, 2002).

In primates (but not in rats) second-order neurons in the taste area of the NTS (rostral part of the NTS) project directly to the contralateral VMb; these projections are stronger than those to the ipsilateral VMb (Beckstead *et al.* 1980). Whether the caudal part of the NTS that represents the afferent (vagal) input areas from the respiratory tract, the cardiovascular system and the gastrointestinal tract project directly to the contralateral VMb in primates is unexplored. The main projection from the NTS to the VMb occurs via the medial external nucleus of the parabrachial area in rats and probably in primates too. This projection via the parabrachial external nucleus is suggested to be important for sensations generated by activation of vagal afferents. Neurons in the VMb/VPpc project viscerotopically to the middle portion of the dorsal insula. The pathways transmitting information in taste, gastrointestinal, respiratory or cardiovascular afferents are viscerotopically organized. This viscerotopy is preserved in the parabrachial external medial nucleus (although this has not been shown yet), in the VMb/VPpc of the thalamus and in the middorsal insula, taste being represented rostrally, sensations related to the respiratory and cardiovascular systems caudally and sensations related to the gastrointestinal tract in between (Figure 2.14).

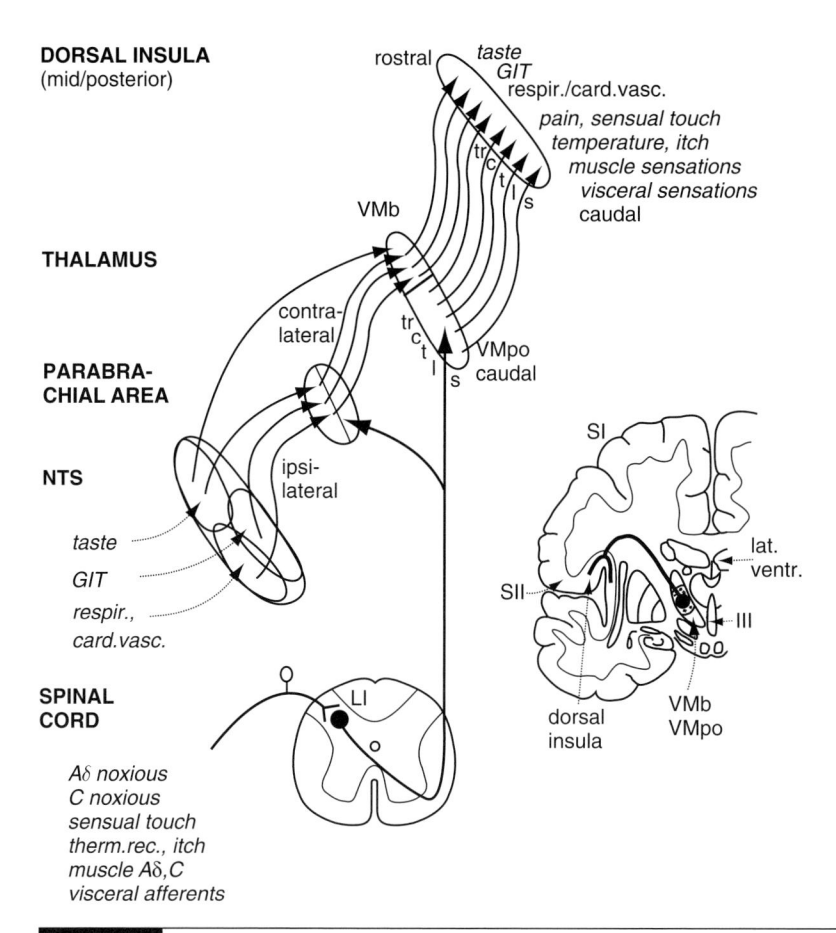

Figure 2.14 Neural pathways responsible for eliciting body and visceral sensations (including taste sensations) to stimulation of vagal and spinal visceral afferents in primates. Vagal and taste afferents project viscerotopically to the nucleus tractus solitarii (NTS) (see Figures 8.11, 8.12). Second-order neurons in the NTS project viscerotopically, via the external medial nucleus in the parabrachial area and the basal part of the ventromedial nucleus of the thalamus (VMb; the VPpc in rats) to the middle part of the posterior dorsal insula (primary visceral sensory cortex), taste being represented rostrally, respiratory and cardiovascular system caudally and the gastrointestinal tract (GIT) in between. In primates neurons in the rostral NTS (taste) may also project directly to the contralateral VMb (Beckstead et al. 1980). In the parabrachial area NTS neurons and lamina I neurons project to separate subnuclei (see Figure 2.12). Lamina I (LI) neurons encode activity from various afferents in somatic tissues (thermoreceptors, nociceptors, ergoreceptors etc.) and visceral afferents (see Figure 2.10); they project via the posterior part of the ventromedial nucleus of the thalamus (VMpo located caudal to the VMb) to the dorsal insula caudal to the projection from the VMb. The inset on the right illustrates on a frontal section the location of the dorsal insula and the VMb/VMpo. tr, c, t, l, s, trigeminal, cervical, thoracic, lumbar, sacral; SI, SII, primary and secondary somatosensory cortex; III, third ventricle; lat. ventr., lateral ventricle. For details see text. Figure designed after Saper (2002) and Craig (2003a). Inset on the right from Craig (2003a).

Conclusions

1. Visceral organs are innervated by vagal and spinal visceral afferent neurons. Most of these afferent neurons have unmyelinated axons and some thinly myelinated ones. The visceral afferent

neurons encode in their activity mechanical and chemical events. They are involved in peripheral ("efferent" or effector) functions and central (afferent) functions. These functions are closely associated with autonomic and neuroendocrine regulation of body tissues.

2. About 80% to 85% of the axons in the vagal nerves are afferents innervating the respiratory tract, the cardiovascular system and the gastrointestinal tract (down to the transverse colon). They have their cell bodies in the inferior (nodose) ganglion of the vagus nerve (a few in the superior (jugular) and petrosal ganglia). Almost all inferior ganglion cells are peptide-negative.

3. Vagal afferents innervating each visceral organ system are subdifferentiated into several functional types. They are involved in autonomic reflexes and various types of regulation (see Chapter 8 and 11) and in visceral sensations such as hunger, satiety, thirst, nausea, respiratory sensations. They are (with a few exceptions related to the proximal esophagus and trachea) not involved in visceral pain. Vagal afferents innervating the mucosa of the gastroduodenal section are involved in nociceptive (protective) reflexes.

4. Subgroups of vagal afferents (in particular those innervating the gastrointestinal tract and projecting through the hepatic vagal branch) are involved in general body protection. They are activated by inflammatory processes and processes of the immune system and are sensitized or activated by the proinflammatory cytokines tumor necrosis factor α and interleukin 1β. Their activation generates illness responses, one component being hyperalgesia. Vagal afferents also modulate the inhibitory control of transmission of nociceptive impulses in the spinal dorsal horn.

5. Vagal afferents project viscerotopically to the nucleus tractus solitarii (NTS) in the medulla oblongata (see Subchapter 8.3). Sensations generated by activation of vagal afferents (and taste afferents) are represented in the middle part of the dorsal insula (limbic sensory cortex). In primates the sensations elicited by vagal afferent neurons are mediated by a viscerotopically organized pathway consisting of NTS, the external medial nucleus of the parabrachial area and the basal part of the ventromedial thalamus (VMb; in rats the parvicellular part of the ventroposterior nucleus [VPpc]) and from here to the insula. Taste is represented rostrally, respiratory and cardiovascular system caudally and the gastrointestinal tract in between. In primates second-order neurons in the NTS also project directly to the VMb.

6. Spinal visceral afferent neurons have their cell bodies in the thoracic, upper lumbar and sacral dorsal root ganglia. They project with pre- and postganglionic sympathetic or parasympathetic (sacral) neurons through splanchnic nerves to visceral organs. The number of spinal visceral afferent neurons is low compared to the total number of spinal afferent neurons, being in the range of 1.5% to 2.5% (in some dorsal root ganglia up to 8%). Many spinal visceral

afferent neurons are peptidergic, containing substance P and/or calcitonin gene-related peptide (CGRP).

7. Spinal visceral afferent neurons are involved in multiple organ reflexes, organ regulation (pelvic organs), extraspinal "peripheral" reflexes (mediated by prevertebral ganglia), protective "axon reflex"-mediated effector reactions, non-painful visceral sensations (particularly sacral visceral afferents) and visceral pain. An individual spinal visceral afferent neuron innervates one organ or organ section only.

8. Many spinal visceral afferent neurons (in particular thoracolumbar ones) have peripheral "efferent" effector functions. Their activation generates local increase in blood flow, change of motility, secretion, absorption and other changes by release of peptides. In addition, these afferents seem to have trophic functions (e.g., maintenance of the barrier function of the urothelium). Spinal visceral afferents may be specialized with respect to the peripheral "efferent" (effector) functions or to the central (afferent) functions (reflexes, sensations).

9. Most thoracolumbar spinal visceral afferent neurons are polymodal and activated by mechanical (distension, contraction) and chemical stimuli (occurring during ischemia and inflammation). They do not seem to signal specific organ events to the spinal cord. The distinction of nociceptive and non-nociceptive thoracolumbar spinal visceral afferents is difficult.

10. Sacral visceral afferent neurons are specialized with respect to the bladder and bowel and probably the internal reproductive organs. They are mechano- (responding to distension and contraction) and chemosensitive. Several types can be distinguished: low-threshold, a few high-threshold and some mechanoinsensitive (extremely high-threshold or silent) afferents. All types of sacral visceral afferents can be sensitized. The distinction of nociceptive and non-nociceptive sacral afferents by different criteria is problematic. Sacral visceral afferents are involved in specific organ regulations, specific sacro-lumbar reflexes, non-painful sensations and pain.

11. Spinal visceral afferents project to lamina I, lamina V and deeper laminae of the spinal grey matter (sparing laminae II to IV). Single thoracolumbar visceral afferent neurons project over the whole mediolateral width of lamina I and V and over several segments rostral and caudal to their spinal cord entry.

12. Sensations related to spinal visceral afferent neurons are probably triggered by ascending tract neurons in lamina I, together with thermal sensations, pain and other sensations that are related to the state of somatic body tissues. All spinal neurons receiving synaptic input from spinal visceral afferents are convergent viscero-somatic neurons. The decoding of activity in spinal visceral afferent neurons by second-order neurons in the dorsal horn leading to distinct organ regulations, organ reflexes and sensations (including visceral pain) is poorly understood.

13. The layer of spinal lamina I neurons represents an interface between body tissues and supraspinal autonomic centers in the brain stem, hypothalamus and limbic system on one side and thalamocortical systems on the other. It is composed of functionally distinct neural channels that are activated by afferent neurons with small-diameter myelinated and unmyelinated (Aδ, C) axons innervating skin (noxious, temperature, itch), deep somatic tissues or viscera.
14. Lamina I neurons project via various nuclei in the medulla oblongata (ventrolateral medulla, NTS, lateral reticular nucleus, subnucleus reticularis dorsalis) and the lateral parabrachial area to autonomic and neuroendocrine centers in brain stem, hypothalamus and limbic system.
15. In primates, lamina I neurons project topographically to the posterior part of the ventromedial nucleus of the thalamus (VMpo) that is situated caudal to the VMb. The VMpo projects topographically to the dorsal insula caudal to the projection field of the VMb.
16. Neurons in laminae V and deeper that can be activated by spinal visceral afferents (convergent viscero-somatic tract neurons) project to various nuclei in the brain stem, hypothalamus and thalamus. They are involved in autonomic regulation and somatomotor functions. They may also be involved in discriminative aspects of pain, however, this issue is controversial.
17. The dorsal insula is the primary sensory cortex of interoception and represents sensations related to the states of the tissues of the body, including the visceral organs. It is synaptically connected to limbic centers that are involved in autonomic regulation, emotional feelings and conscious experience.

Notes

1. This applies in principle also to primary afferent neurons with small diameter myelinated and unmyelinated fibers that have nociceptive function or monitor the metabolic or thermal state of the tissues. The cell bodies of primary afferent neurons consist by histological criteria of large light neurons (A-afferent neurons) and small dark neurons (B-afferent neurons). Both groups of afferent neurons are distinguished by several criteria (ontogeny, cell phenotype, functional characteristics). Thus nociceptors, thermoreceptors and metaboreceptors of skeletal muscle would belong to the class of B-afferent neurons across the three body domains (skin, deep somatic tissues, viscera). Prechtl and Powley (1990b) have put forward the idea that B-afferent neurons innervating skin, deep somatic tissues or viscera are particularly associated with the autonomic nervous system. They proposed to group these primary afferent neurons together as part of a common reflex system involved in homeostatic regulations and to label these afferents "autonomic". The idea of Prechtl and Powley met quite some criticism (see Open Peer Commentary in Prechtl and Powley [1990b]) and was overall not accepted. However, this idea resembles the concept of a general body sense of interoception propagated by Craig (2002, 2003c). The opinion, to label afferents

innervating viscera as neutrally spinal visceral or vagal visceral does not collide with the view that these afferents, together with small-diameter afferents innervating deep somatic tissues and skin, are important in the generation of body feelings that are related to the states of body tissues and to homeostatic body regulations. These body feelings include pain elicited from all tissues, all visceral sensations, thermal sensations, sensations from skeletal muscle during vigorous exercise etc. and belong to interoception "as the sense of the physiological conditions of the entire body." They have distinct cortical representations in the insular cortex (Craig 2002, 2003a, b, c; see Subchapter 2.6).

2. Axon reflex, neurogenic inflammation: Some primary afferent neurons with unmyelinated nerve fibers (and a few with small-diameter myelinated ones) have dual functions: (1) *Afferent function*: They encode physical and chemical stimuli in their activity by specific transduction mechanisms in their peripheral terminals. The activity is transmitted by action potentials to the central terminals and then synaptically transmitted to the second-order neurons. The transmitter(s) involved is glutamate (and/or aspartate) and possibly one or two neuropeptides (such as substance P and calcitonin gene-related peptide CGRP). (2) *Efferent function*: During excitation peptidergic afferents release one or two neuropeptides (such as neurokinins [substance P and neurokinin A] and CGRP) at their peripheral terminals and induce precapillary arteriolar vasodilation and postcapillary venular plasma extravasation. The afferent-induced vasodilation was originally called "axon reflex" (Bayliss 1901; Bruce 1910, 1913). This is a misnomer since no synapse is involved and since the afferent terminal excited by physical or chemical stimuli releases the neuropeptide(s). The afferent-induced plasma extravasation is called "neurogenic inflammation", a concept introduced by the Hungarian Jancsó (Jancsó 1960; Jancsó et al. 1967, 1968). Sometimes the afferent-induced vasodilation and plasma extravasation (together with other afferent-induced peripheral changes, see Figures 2.2 and 2.3) are called collectively neurogenic inflammation. The peptides released act either directly on the effector cells (smooth muscle cells, endothelial cells) or via other cells (e.g., mast cells). Arteriolar vasodilation is primarily generated by release of CGRP and enhanced in some tissues by neurokinins (Häbler et al. 1999b). Plasma extravasation is induced by release of neurokinins. Afferent-induced vasodilation and plasma extravasation are present in many tissues such as: skin (human skin does not exhibit afferent-induced plasma extravasation); mucosal tissues of the oronasal cavities, of the proximal esophagus, of trachea and bronchi, of the urinary tract (urinary bladder, urethra) and of the anus; the serosa and mesenteries of intestinal organs; deep somatic tissues (fascia, joint capsule, synovia); dura and pia mater. Only subpopulations of small-diameter primary afferent fibers are involved in afferent-induced vasodilation, most of them probably having nociceptive function. In the rat a small subpopulation of polymodal nociceptors with C-fibers is involved in afferent-induced vasodilation, but most of them in plasma extravasation. In the pig only heat nociceptors with C-fibers but no polymodal nociceptors are involved in vasodilation (Lynn 1996a, b; Lynn et al. 1996). In humans chemonociceptors with C-fibers are involved in vasodilation but no polymodal nociceptors are (Schmelz et al. 2000; Schmidt et al. 2000). Cutaneous small-diameter myelinated (Aδ) fibers in the rat are involved in vasodilation but not in plasma extravasation (Jänig and Lisney 1989) (for review see McDonald [1997]; Holzer [1998b]).

3. Figures of spinal visceral afferents are normally drawn as if the same afferent fiber innervates all three structures (see Figures 2.3 and 2.4). However, there is no experimental basis supporting this.
4. Dorsal horn neurons that process nociceptive afferent information are functionally divided as follows: (1) *Nociceptive-specific neurons* that can only be excited by noxious stimuli. They are synaptically excited by stimulation of afferent neurons with Aδ- or C-axons. (2) *Multireceptive neurons* are excited by noxious and non-noxious stimuli. They are synaptically activated by stimulation of afferent neurons with Aβ-, Aδ- or C-axons. These neurons are also called wide-dynamic range (WDR) neurons. (3) *Convergent nociceptive neurons* are activated from skin, deep somatic tissues and/or viscera. These neurons can be nociceptive-specific or multireceptive (Willis and Coggeshall 2004a).
5. In view of the rather small number of spinothalamic neurons in lamina I and deeper laminae that are involved in painful and non-painful somatic and visceral sensations in relation to the number of spinal interneurons and propriospinal neurons, it is theoretically possible that spinal tract neurons receiving selective or relatively selective synaptic input from visceral afferent neurons, are difficult to detect and have been overlooked. For example, it is conceivable to assume that there exist ascending tract neurons in the sacral grey matter that are functionally relatively specific for the urinary tract, the hindgut or the sexual organs. Otherwise it would be difficult to understand the distinct regulation of these organ systems and the distinct sensations elicited from them.
6. Many lamina V projection neurons have dendrites that project into laminae I and II. The functional significance of this projection is unknown.

Part II

Functional organization of the peripheral autonomic nervous system

Chapter 3

The final autonomic pathway and its analysis

3.1 The final autonomic pathway   *page* 88
3.2 Functions of the autonomic nervous system and levels of integration   90
3.3 Activity in peripheral autonomic neurons reflects the central organization   92
3.4 Reflexes in autonomic neurons as functional markers   95
3.5 Some methodological details about recording from peripheral autonomic neurons in vivo   96
3.6 Confounding effects of anesthesia in animal experiments   104

In Chapter 1 I described the anatomical and physiological characteristics of the peripheral autonomic nervous system on the macroscopic level. The overall conclusion from this conservative approach is that the peripheral autonomic neurons are integrated in the neural regulation of many target cells of the body (see Table 1.2). In other words, peripheral autonomic pathways that transmit signals from the spinal cord and brain stem to the peripheral effector cells must have some functional specificity with respect to these effector cells, in the sympathetic and the parasympathetic nervous system. Otherwise it would be impossible to understand how the precise autonomic regulation that is the basis for the continuous adaptation of the body during various demands occurs. Implicit in this idea is that these peripheral autonomic pathways are connected to distinct neuronal circuits in the spinal cord, brain stem, hypothalamus and telencephalon.

    In this and the next chapter I will give arguments, and describe in some detail, that principally each type of target cell that is innervated by autonomic neurons is influenced by one or two autonomic pathways and that these pathways transmit distinct messages to the periphery and are connected to distinct central circuits. This chapter describes the final autonomic pathway and its analysis. It concentrates particularly on the neurophysiological analysis. Morphological analysis of autonomic circuits will be described in Chapter 8 and elsewhere in the book.

3.1 | The final autonomic pathway

The concept that the motoneurons are the "final common motor paths" (Sherrington 1906), which are shared by segmental and propriospinal reflex pathways and descending systems from the brain stem and cortex, has dominated the analytical approach to the somatomotor system for a long time. This concept can also be applied to the autonomic nervous system as will be emphasized in the description of the functional properties of autonomic neurons (Chapters 4, 9 and 10). The peripheral sympathetic and parasympathetic autonomic systems consist of several functionally distinct subsystems, each associated with a different type of target tissue (Table 1.2). Each autonomic system is based on a set of preganglionic and postganglionic neurons that are synaptically connected in autonomic ganglia and constitute a pathway that transmits the central message to its target tissue. I will therefore call this pathway the "*final autonomic pathway*" (Jänig 1986).

The *final autonomic pathways* are the building blocks of the peripheral autonomic nervous system and the concept described by this term probably applies to all sympathetic and parasympathetic pathways. It seems to be generally accepted that, except for the vagal pathways to the gastrointestinal tract (see Furness and Costa [1987]; Furness *et al.* [2003a]; Furness [2005]; Subchapters 5.6 and 10.7), the parasympathetic pathways appear to be more distinct and simpler in organization than the sympathetic ones. This may be true for the parasympathetic pathways to some target organs, such as the sphincter pupillae and ciliary muscle (concerned with the pupillary light reflex and accommodation via the ciliary ganglion), the salivary and lacrimal glands (controlling fluid secretion via the pterygopalatine, otic and submandibular ganglia) or the helical arteries of the penis (responsible for erection via neurons in the pelvic ganglia). However, it seems unlikely to be the case for the parasympathetic pathways to other target organs, such as to the urinary bladder and colon (that produce evacuation via neurons in the pelvic ganglia), to the heart (decreasing heart rate and atrial contractility via neurons in the cardiac plexus) or to smooth muscle of the trachea and bronchi (producing constriction via neurons of the paratracheal ganglia). This suggests that the simplicity results more from the relatively small size and simple anatomy of the former compared to the latter group of target organs. The sympathetic pathways to organs like the pineal gland and dilator pupillae are probably just as simple in their organization.

As far as transmission of the central message to the target organs is concerned, the concept of the "final autonomic pathway" is for most autonomic systems similar to that of the "final common motor path" in the somatomotor system, in the sense that it corresponds to the innervation of a skeletal muscle or group of

3.1 THE FINAL AUTONOMIC PATHWAY

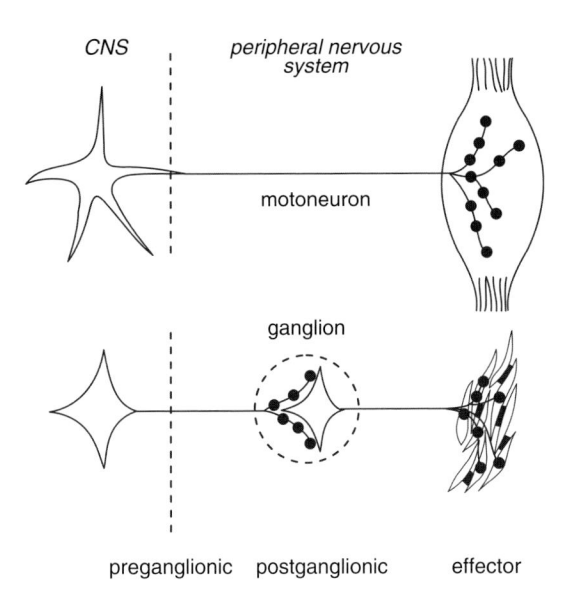

Fig. 3.1 The "final autonomic pathway" (lower part) in comparison to the "final common motor path" (upper part) of the somatomotor system. CNS, central nervous system.

muscles with the same function by a pool of α-motoneurons (Figure 3.1). The main differences between the somatomotor and autonomic systems are:

- The same autonomic target organ can be innervated by more than one "final autonomic pathway". The component tissues are either independently innervated (e.g., in the eye) or the same tissue cells are innervated by two final autonomic pathways (e.g., pacemaker cells and atria of the heart). However, the latter is more the exception than the rule (see Chapter 1 and Table 1.2).
- The central message may undergo quantitative changes within autonomic ganglia because of convergence and divergence and the variable effectiveness of different preganglionic inputs (see Subchapters 6.2 and 6.5).
- In prevertebral ganglia and in some other ganglia, synaptic inputs from the periphery may summate with those from preganglionic neurons. Intestinofugal neurons (of the enteric nervous system) and collateral branches of spinal peptidergic visceral afferent neurons may establish peripheral autonomic circuits with postganglionic neurons, which are integrated in the final autonomic pathways (see Subchapter 6.5; Jänig 1995a).
- The neurally derived signals may interact with other parameters in the target organ or in the ganglion. These include local neural influences (e.g., peptide released from activated afferent terminals), remote and local hormones, local metabolites and the endogenous activity of the target organ (e.g., myogenic activity) and substances released from local cells (e.g., from the endothelium). Such factors vary for the neuroeffector transmission in different functional pathways (see Subchapter 7.4).

3.2 | Functions of the autonomic nervous system and levels of integration

The autonomic nervous system regulates body functions in order to enable the body to act in a coordinated way under various challenging conditions. It is a nervous system in its own right, like the somatomotor system or the sensory systems. This does not mean that the functioning of the autonomic nervous system can be understood and described independently of the sensory systems (including the visceral afferent neurons). However, this applies to the somatomotor system as well.

The autonomic nervous subsystems are hierarchically organized and represented in the peripheral and in the central nervous system. Figure 3.2 and Table 3.1 show in a schematic form the different levels of function of the autonomic nervous system:

- The lowest level occurs at the *target tissue*. The effector responses of the target cells may depend on several classes of signals, which potentially impinge on them. The target cells are under the control of autonomic neurons and some individual tissues are supplied by more than one type of efferent innervation

Figure 3.2 Levels of integration of the autonomic nervous system. IN, interneuron; PA, primary afferent neuron.

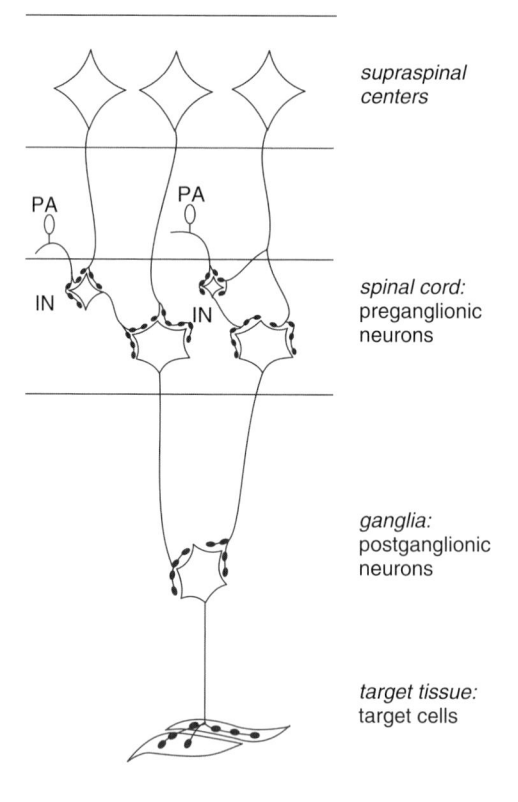

(e.g., pacemaker cells of the heart, some blood vessels; see Table 1.2). The effectiveness of the signals of postganglionic autonomic neurons in generating an effector response may also depend on signals arising from other sources (e.g., spontaneous myogenic activity of some smooth muscle cells; local physical factors, e.g., P_{CO_2}, pH, temperature; interstitial cells of Cajal associated with the enteric nervous system [see Subchapter 5.4]), other neurons (e.g., nociceptive primary afferent terminals; see Subchapter 2.2), remotely derived hormones (e.g., circulating angiotensin, adrenaline, vasopressin), endothelium-derived factors in blood vessels (e.g., nitric oxide) or local paracrine signals (e.g., cytokines) (see Subchapter 7.4).
- The next level of integration occurs in some *autonomic ganglia*. Postganglionic neurons may integrate signals derived from several convergent preganglionic neurons. Some postganglionic neurons, particularly in prevertebral and possibly in some parasympathetic ganglia (e.g., in the heart), also integrate signals from other peripheral neurons. These might be interneurons of the enteric nervous system (intestinofugal neurons) or collateral branches of spinal visceral primary afferent neurons. These neurons establish reflex pathways that are completely organized outside of the spinal cord (see Subchapter 6.5).
- *Preganglionic neurons* are situated in the spinal cord and brain stem. These neurons integrate a diverse range of synaptic activity from interneurons, primary afferent neurons and systems in higher levels in the central nervous system. This level of organization is associated with reflexes and regulating systems that are represented in the spinal cord and lower brain stem

Table 3.1 | *Functional levels of integration in the autonomic nervous system*

Level of integration	Component parts	Functions
Autonomic nervous system	Sympathetic, parasympathetic nervous system	Adaptation of body functions (e.g. arterial blood pressure, body temperature, micturition)
Single autonomic pathways	Functionally defined by groups of pre- and postganglionic neurons, interneurons, etc.	Control of specific body functions (e.g., sweating, blood flow through skin, resistance in skeletal muscle, secretion of saliva)
Sets of neurons	Groups of interneurons belonging to specific autonomic pathways	Baroreceptor and chemoreceptor reflexes in vasoconstrictor neurons, spinal reflexes in sudomotor neurons
Single neurons	Neuroeffector junction, post-, preganglionic neurons	Neuroeffector transmission, signaling patterns
Subcellular, molecular composition of neurons	Specialized receptors, ion channels, protein synthesis, second messenger systems	Maintenance and plasticity of cellular networks

(e.g., cardiovascular system, gastrointestinal tract, evacuative systems) (see Chapters 9 and 10).
- *Central pathways* in the upper brain stem and hypothalamus are antecedent to the spinal and lower brain stem reflex centers. They integrate activity from several sources concerned with autonomic homeostatic regulation, neuroendocrine regulation and regulation of the somatomotor system (see Chapter 11).
- *Pathways in the telencephalon* (limbic system and neocortex) adapt the complex homeostatic regulations to the needs of the organism according to the environmental conditions and through memory processes according to previous experiences.

While generally applicable, this type of functional hierarchical organization is of course a simplification since the afferent and efferent communication between levels of central integration does not only occur between adjacent levels but also across levels. For example, some neurons from the hypothalamus project directly to preganglionic neurons or autonomic interneurons in the spinal cord (Figure 8.15) and second-order neurons in lamina I of the spinal cord may project directly to the hypothalamus. Second-order neurons in the nucleus tractus solitarii (NTS) do project to nuclei in the brain stem, hypothalamus and limbic system (Figure 8.13). Furthermore, the functional hierarchical organization is not uniform across the different autonomic pathways. In some systems the hypothalamus is the main integrative structure that determines the firing characteristics of the peripheral autonomic neurons, while in others the lower brain stem or even the spinal cord serves this function (see Chapters 4, 9 and 10 for extensive description).

3.3 | Activity in peripheral autonomic neurons reflects the central organization

The discharge pattern recorded from the peripheral autonomic neurons is the result of integrative processes in the central representations of the respective autonomic system. The circuits of these central representations are located in the spinal cord, brain stem, hypothalamus and telencephalon. Although details about many of these circuits are still missing, we can forward the idea that the organization of these central circuits is specific for each autonomic system. The evidence comes from the following sets of observation:

1. Neural regulation of all autonomic effector organs is amazingly precise. This observation, though apparently trivial and a universally accepted fact, has to be remembered. The mechanisms behind most of these precise regulations under various behavioral conditions are still unknown and are to be found in the central integration of autonomic systems. Walter

Bradford Cannon had this in mind when he wrote his book *The Wisdom of the Body* (Cannon 1939).

2. Recordings from single peripheral autonomic neurons show a bewildering variety of distinct discharge patterns related to peripheral afferent and centrally generated events. This aspect is extensively discussed in Chapter 4.

3. At least some neurophysiological studies of neurons of central circuits that are related to peripheral autonomic pathways demonstrate rather specific reflexes that are typical for the neurons of the peripheral autonomic pathways. Most of these studies concentrated on central autonomic circuits related to peripheral autonomic pathways innervating resistance vessels and heart (such as muscle and visceral vasoconstrictor neurons, sympathetic and parasympathetic cardiomotor neurons) (see Spyer 1981, 1994; Guyenet 1990, 2000; Dampney 1994; Guyenet *et al.* 1996; Blessing 1997). But recently some studies concentrated on central circuits that are involved in regulation of activity in cutaneous vasoconstrictor neurons (Kanosue *et al.* 1998; Rathner *et al.* 2001), neurons innervating the adrenal medulla (Morrison and Cao 2000; Morrison 2001a) or neurons innervating the brown adipose tissue (lipomotor neurons; Morrison 1999; 2001a, b). These aspects will be discussed extensively in Chapters 9 and 10.

4. A breakthrough in unraveling aspects of the microanatomy of the central organization of autonomic systems came (1) with the introduction of axon tracer methods in the 1970s and (2) with the introduction of transneuronal labeling of neuron populations using neurotropic viruses at the end of the 1980s by Arthur Loewy's group. The first method allows labeling of cell bodies of autonomic pre- and postganglionic neurons and of central autonomic neurons (e.g., sympathetic and parasympathetic premotor neurons) as well as the field of termination of their axons. The second method principally allows labeling of whole networks of neurons in the neuraxis that are connected with a specific autonomic output system. Using these morphological techniques new ideas about the various neuronal networks in spinal cord, brain stem, hypothalamus and telencephalon, which are involved in the regulation of the final autonomic pathways, were obtained. In this way the groundwork for future physiological studies on the central organization of the different autonomic systems was laid down (Jansen *et al.* 1993, 1995a; Strack *et al.* 1989b; Saper 1995). Results obtained with these techniques will be discussed in Chapter 8.

5. Further progress was made in the last 10 to 20 years in the research on the mechanisms of integrative action of autonomic systems by combining various techniques. Some examples are:
    - The development of a working-heart-brain-stem (WHBS) preparation in the mouse and in the rat in which the isolated

lower brain stem is continuously perfused. In this preparation the efferent (parasympathetic) connections to the heart and the afferent (vagal) connections from heart, arterial baroreceptors and arterial chemoreceptors are intact. This in vitro preparation allows detailed intracellular studies in neurons *in situ* related to cardiovascular regulation (e.g., in the NTS, caudal and rostral ventrolateral medulla etc.) (see Chapter 10; Paton 1996a, b, 1999; Paton and Kasparov 2000).

- The WHBS preparation can be combined with an attached (and perfused) spinal cord, allowing the study of sympathetic neurons *in situ* and with intact cardiovascular centers (Chizh et al. 1998).
- Labeling of functionally (neurophysiologically) identified neurons. Markers are injected into the neurons, using intracellular or juxtacellular injection techniques (see note 3 in Chapter 10). In this way neurons with distinct function can be visualized.
- Neurophysiological recording from central neurons that have been labeled by a marker before (e.g., preganglionic neurons projecting to distinct ganglia; sympathetic premotor neurons projecting to the spinal cord).
- Cellular markers of neurons that are activated when the neurons are activated can be used to label populations of central neurons following physiological stimulation of afferent neurons (e.g., arterial baroreceptors or chemoreceptors; distinct classes of afferent neurons from the gastrointestinal tract or pelvic organs). The cellular marker used is the immediate early gene *c-fos* and its protein (see note 1 in Chapter 10).

These points strongly support the methodological approach as outlined in Figure 3.3 in the analysis of the organization of the sympathetic (and principally also the parasympathetic) systems in the periphery and in the central nervous system (see Subchapter 3.5).

Three further points strengthen the idea of the functional differentiation of the autonomic nervous system:

1. Histochemical investigations of sympathetic pre- and postganglionic neurons show that functionally different types of neurons (as defined by their target tissue) are characterized by neuropeptides colocalized with the "classical" transmitter (*neurochemical coding*). It is irrelevant in this context that we do not know the function of most of these peptides in the autonomic neurons and that there are species differences (see Gibbins [1995]) (see Subchapters 1.4 and 5.1).
2. Activity is transmitted function specifically from preganglionic to postganglionic neurons in the autonomic ganglia (McLachlan 1995; see Chapter 6).
3. Where investigated with morphological, neurophysiological and pharmacological techniques, it has been shown that the

activity in sympathetic and parasympathetic postganglionic axons is transmitted by distinct neuroeffector junctions to the target cells (see Chapter 7).

3.4 | Reflexes in autonomic neurons as functional markers

Physiologists have always known that autonomic involvement in the regulation of organ function is marked by the precision with which this occurs in relation to the overall behavior of the organism. This is the basis of homeostasis and the ability to adapt to various external and internal perturbations (allostasis; McEwen 1998, 2000, 2001b; see Chapter 11). Such precise control implies that there are subgroups of pre- and postganglionic autonomic neurons that are discrete with respect to the function they control in their target organs. Differences in the subgroups of pre- and postganglionic neurons are reflected in the discharge patterns elicited by afferent stimuli and can be measured in neurophysiological experiments.

This addresses the question whether individual autonomic neurons in vivo can be recognized to belong to one of the autonomic subsystems listed in Table 1.2 by particular reflex discharge patterns. If this is possible, we would have *functional markers* for different systems of neurons that are independent of recorded responses from target organs. Such functional markers should be sufficiently characteristic to allow them to be recognized wherever they are recorded, for example in the preganglionic cervical or lumbar trunks. The functional markers might be correlated with other characteristics of the neurons such as passive and active biophysical properties (determining the firing of action potentials in response to synaptic inputs), projection and geometry of dendrites and axons, histochemical characteristics (neuropeptide content), and synaptic transmission in the autonomic ganglia and at the neuroeffector junctions. The strategy used to define such functional markers in experiments on animals and human beings has been:

1. to select appropriate nerves in which the autonomic pre- or postganglionic axons to be recorded from project only to known targets,
2. to use natural (adequate) stimuli to excite afferent receptors, which are appropriate to elicit reflex changes of activity in the pathways to those targets and consequently in the autonomic neurons to be analyzed, and,
3. to record the target organ/cell responses.

The types of reflexes elicited in autonomic neurons by afferent stimuli and the correlation between this activity and other centrally generated parameters (e.g., the centrally generated respiratory cycle and the cycle of sleep and wakefulness) depend on the organization of the different control systems in the spinal cord, brain stem, hypothalamus or higher structures and therefore reflect this organization. Thus, this approach may be applied in future studies of the central organization of the autonomic nervous system by recording autonomic reflexes under appropriate conditions (e.g., experiments on animals or human subjects with transected spinal cords or central disruptions, such as focal lesions or focal activation of known populations of neurons). However, knowing the function of the stimulated afferents, the function of the efferent neurons from which recordings are made, and even the organization of the central reflex pathways, does not necessarily reveal how the autonomic neurons behave under closed-loop conditions[1] (i.e., during ongoing regulation of the target organs). For example, the neural elements of the arterial baroreceptor and chemoreceptor reflexes, their projections and the transmitters involved at each synapse are essentially known (Guyenet 1990, 2000; Guyenet et al. 1996; see Chapter 10). However, the functions of these reflex pathways during normal regulation can only be studied under closed-loop conditions. Therefore experiments in the intact, non-anesthetized (awake) organism are required in which autonomic regulation is investigated under closed-loop conditions and in which the activity is recorded from autonomic neurons (see Kirchheim et al. [1998]).

3.5 | Some methodological details about recording from peripheral autonomic neurons in vivo

3.5.1 Neurophysiological recordings in animals

Recording from autonomically innervated target organs under closed-loop conditions in anesthetized and awake animals gives valuable insight into the overall capacity and efficiency of the autonomic systems and their target organs in the maintenance of homeostasis and in modifying the inner milieu during various physiological behaviors (e.g., regulation of arterial blood pressure, organ and tissue blood flow, micturition and defecation, body temperature, sexual organ function, etc.) (Folkow and Neil 1971; Korner 1979; Randall 1984; Eckberg and Sleight 1992; Rowell 1993; Greger and Windhorst 1996). However, in such studies, the nervous system is treated like a black box. This approach has given only limited insight into the organization of the peripheral autonomic nervous system and its central control mechanisms. Therefore, it is necessary to record from single autonomic neurons or small groups of neurons in vivo or in vitro and to combine these (mostly neurophysiological) recordings with other techniques of recording target organ function (see Loewy and Spyer 1990b).

The neurophysiological approach in vivo has been applied extensively in studies of the activity of some groups of autonomic neurons in cats, rats and some other species, using extracellular recording from peripheral autonomic axons. This approach has concentrated

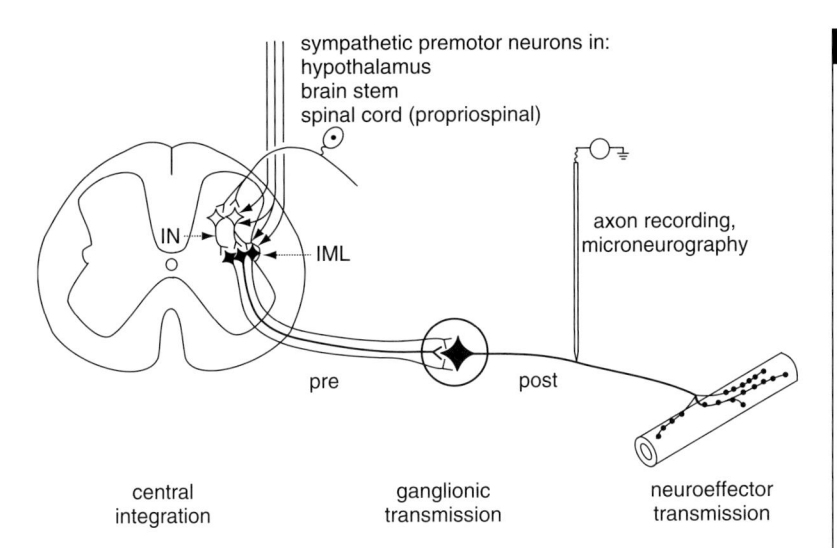

Figure 3.3 Arrangement of recording situation using microneurography from bundles with postganglionic axons in human subjects or from fiber strands isolated from peripheral nerves in anesthetized animals in vivo. Note that signals recorded from the postganglionic axons in both experimental situations reflect the result of central integrative processes in the respective sympathetic system; the impulses in the sympathetic pathways are specifically transmitted through the sympathetic ganglia and to the target cells by the neuroeffector junctions (see Chapters 6 and 7). IML, intermediolateral nucleus; IN, interneuron; pre, preganglionic neuron; post, postganglionic neuron. Modified from Jänig and Häbler 2003 with permission.

on pre- and postganglionic neurons of the lumbar sympathetic system supplying skeletal muscle, skin and pelvic organs and on preganglionic neurons of the thoracic sympathetic system innervating postganglionic neurons in the superior cervical ganglion, which are destined for target organs in the head and upper neck (Jänig, 1985a, 1986, 1988a, 1996a; Jänig and McLachlan 1987; Jänig et al. 1991; Boczek-Funcke et al. 1992a, b, c, 1993; Häbler et al. 1992, 1993c, 1994a, b, 1996, 1999; Grewe et al. 1995; Bartsch et al. 1996, 1999). The advantage of this experimental approach is that the signals in the neurons of the final sympathetic pathways reflect the central organization of the respective system and can be correlated with the effector responses under controlled experimental conditions. The limitations are clearly to be seen in the technical difficulties to isolate signals from single neurons routinely.

A few in vivo experiments of a similar type have been performed on parasympathetic systems in which spontaneous and reflex activity was recorded in vivo from preganglionic neurons (e.g., to the heart, the respiratory tract and the upper gastrointestinal tract) or postganglionic neurons (e.g., in the ciliary ganglion; see Jänig 1995a). There are limitations to the application of the method if exposure of the nerve of interest involves intrusive or extensive surgery that may compromise the responses of the autonomic neurons.

Figure 3.3 illustrates the arrangement of the recording from peripheral sympathetic neurons in both animals and humans. As mentioned, the discharge patterns recorded from the peripheral neurons in this way are the result of integrative processes in the central representations of the respective sympathetic system.

The methodological approach as outlined in Figure 3.3 in the analysis of the functional organization of the sympathetic (and principally also the parasympathetic) systems in the periphery and in the central nervous system is to record from bundles with few autonomic axons and, if possible, single autonomic axons. In view of the thousands of postganglionic neurons projecting into individual peripheral somatic or visceral nerves (McLachlan and Jänig 1983; Baron *et al.* 1985a, b) and in view of the fact that tens of thousands of postganglionic neurons may innervate the same type of target organ and have therefore the same function (e.g., muscle, visceral or cutaneous vasoconstrictor neurons) it is sometimes criticized that conclusions made from single unit recordings were not representative for the whole population of sympathetic neurons of the same function. This turned out not to be true and has clearly been refuted in experiments in which activity from two bundles containing sympathetic postganglionic axons with the same function was recorded simultaneously (e.g., muscle vasoconstrictor axons in humans, see Sundlöf and Wallin 1977) or from functionally different types of sympathetic postganglionic or preganglionic axons in the same bundle (see Chapter 4). In fact experiments using this type of recording from sympathetic neurons have enabled us to give the ultimate underpinning for the differential organization of the sympathetic nervous system, in the periphery and in the central nervous system.

Recording of multiunit activity in autonomic axons from whole nerves or large bundles isolated from peripheral nerves may turn out to be favorable in certain experimental situations. For example, spontaneous activity and reflex patterns elicited to stimulation of arterial baroreceptors, arterial chemoreceptors, nociceptors etc. in sympathetic axons of the major splanchnic nerve probably occurs exclusively in visceral vasoconstrictor neurons. Activity in other types of sympathetic neuron that project in the major splanchnic nerve and that is related to regulation of the motility or secretion of the gastrointestinal tract or to other non-vasoconstrictor functions cannot be recognized in the recordings of multiunit activity from the major splanchnic nerve. Therefore, splanchnic nerve recording can be used as reference recording for vasoconstrictor neurons innervating resistance blood vessels of the viscera in the analysis of central circuits that are connected with this sympathetic pathway (see Figures 10.12, 10.15). The same applies in the rat to the nerve innervating the adrenal medulla (in rats and humans most preganglionic neurons innervate cells that synthesize and release adrenaline), to the nerves innervating the interscapular brown adipose tissue in rats (most postganglionic axons innervate the adipocytes of the brown adipose tissue; see Figure 10.15), or to nerves innervating the rat tail (most sympathetic postganglionic axons in these nerves innervate blood vessels involved in thermoregulation).

As already mentioned, recordings of multiunit activity from functionally homogeneous groups of pre- or postganglionic neurons are used in the neurophysiological analysis of neuronal circuits in the neuraxis (see e.g. Figures 10.12, 10.15). These recordings from identified central neurons are then combined with other methods to characterize the neurons morphologically (size and location of the neurons, projection of axons and dendrites), histochemically (neurotransmitters and their receptors; neuropeptides and their receptors; intracellular signaling pathways), by their synaptic inputs, etc.

3.5.2 Representative examples of recordings from sympathetic neurons in vivo in animals

Most data shown in Chapter 4 were obtained in recordings from post- or preganglionic axons made with metal electrodes in anesthetized animals; some were done with microelectrode recordings from the cell bodies of the autonomic neurons. Recordings from the axons are stable (once one succeeds in isolating the axons from the respective nerve and obtaining a sufficient signal-to-noise ratio between the extracellularly recorded action potentials and the recording noise) and can last for hours from the same axon or axon bundle.

Figures 3.4 and 3.5 illustrate three examples showing how post- and preganglionic neurons are identified and how activity is recorded from the axons of these neurons that were isolated in bundles under a microscope from the respective nerves. The bundles were positioned on recording platinum electrodes and the activity in the axons was recorded and amplified by a high-impedance amplifier. The size of the signals recorded from unmyelinated and small-diameter myelinated nerve fibers is normally in the range of 20 to 100 µV (sometimes larger) and the recording noise is in the range of 10 to 20 µV.

- In the first example (Figure 3.4b) the activity was recorded from a bundle isolated from a nerve innervating hairy skin of the cat hindlimb. The bundle contained three postganglionic fibers conducting at less than 1 m/s (i.e., they were unmyelinated), as shown by the responses to single pulse stimulation of the peripheral nerve (stim. nerve in Figure 3.4a). The responses in axons 1 to 3 appeared at latencies of about 75 to 90 ms (Figure 3.4b, lower trace), showing that the action potentials traveled at 0.66 to 0.8 m/s over a distance of 60 mm between stimulation and recording electrodes. Axons 1 and 2 were also activated by electrical stimulation of the preganglionic axons in the lumbar sympathetic trunk (LST) with single pulses (Figure 3.4b, upper trace). Axon 1 responded with three action potentials to stimulation of the LST because there is a synapse between pre- and postganglionic neurons and there is convergence of several preganglionic axons on one postganglionic cell body. This leads in some postganglionic neurons to repetitive responses upon preganglionic single pulse stimulation (for details see Chapter 6). The latencies of the responses elicited from the LST were long (here about 200 to 250 ms) because the conduction distance between the stimulation site at the LST and the peripheral recording site was long and the fibers were unmyelinated. The lower trace of Figure 3.4b shows what happens when both the LST and the peripheral nerve are stimulated *simultaneously* with single pulses. Now the first action potential in postganglionic axon 1 and the action potential in postganglionic axon 2 cannot be elicited from the LST because these action potentials collided with antidromically traveling action potentials generated in the same axons by

Figure 3.4 Identification of pre- and postganglionic sympathetic neurons in the cat. (a) Arrangement of stimulation (stim. WR, LST, nerve) and recording electrodes at the pre- and postganglionic sides (rec. pre, rec. post). (b) Recording from a bundle isolated from the superficial peroneal nerve (skin nerve) with two (unmyelinated) postganglionic fibers (1, 2) and one unmyelinated afferent fiber (3). Simultaneous stimulation of the LST (10 V) and of the peripheral nerve subthreshold for unmyelinated fibers (upper trace) and suprathreshold for unmyelinated fibers (lower trace). Postganglionic fiber 1 was activated with three spikes by electrical stimulation of the LST; the first of these three action potentials collided with the antidromic signal when the postganglionic axon was excited. The LST-evoked action potential in postganglionic axon 2 collided with the antidromic signal after suprathreshold peripheral stimulation. Note that one afferent Aδ-fiber [i] was activated. (c) Recording from a bundle with two axons dissected from the LST. Electrical stimulation of the WRL2 (stim. WR) excited the two axons. Axon 5 had ongoing activity and was therefore preganglionic (see spont. lower trace). Axon 4 was silent (it was either a visceral afferent axon or a preganglionic axon). LST, lumbar sympathetic trunk; WR, white ramus. Modified from Jänig and Szulczyk (1981) and Blumberg and Jänig (1982) with permission.

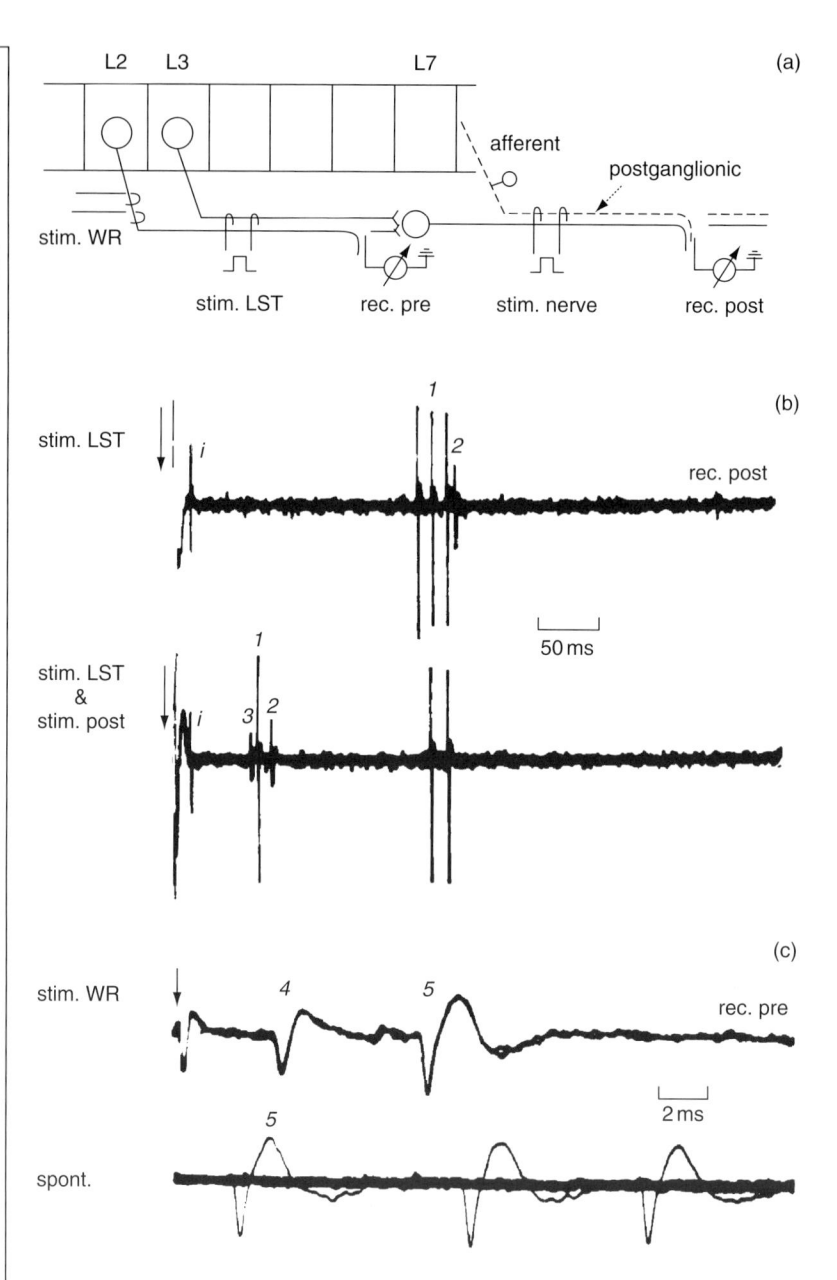

peripheral nerve stimulation. This example clearly demonstrates that axons 1 and 2 were postganglionic and axon 3 was (by exclusion) afferent (because it could not be activated from the LST and because afferent neurons that project to the extremities do not do so through the LST [McLachlan and Jänig 1983]; see Chapter 2).

- In the second example in Figure 3.4, activity was recorded from a bundle that was isolated from the lumbar sympathetic trunk (Figure 3.4a) and that contained two axons (axons 4 and 5 in Figure 3.4c). Electrical stimulation of the lumbar white ramus L2 with single pulses elicited responses in these axons at short

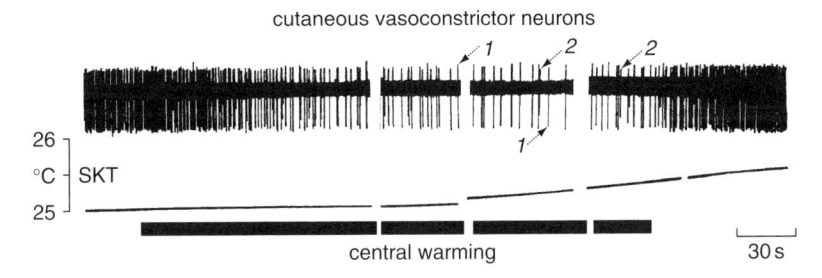

Figure 3.5 Decrease of activity in postganglionic cutaneous vasoconstrictor axons innervating the plantar skin of the cat hindpaw after central warming in relation to change in skin temperature. Activity was recorded from a fine nerve fiber bundle, isolated from the medial plantar nerve of the cat paw, which contained about three to four postganglionic vasoconstrictor fibers with resting activity at about 38 °C body core temperature. Two fibers (fiber *1* and *2*) could be discriminated by the size of their action potential. Skin temperature (SKT) was recorded from the surface of the central pad. The black bar indicates simultaneous warming of the spinal cord (by warm water flowing through a U-shaped tubing positioned epidurally dorsal to the spinal cord) and hypothalamus (by warm water perfused through a thermode positioned into the anterior hypothalamus). Note the strong inhibitory effect on the vasoconstrictor activity during central warming and its reversal after termination of warming. As a consequence of this decrease of activity in the cutaneous vasoconstrictor neurons the cutaneous blood vessels dilated, the blood flow through skin increased and the SKT increased. Increase and decrease in SKT were delayed with respect to neural activity because of the thermal capacity of the skin. Modified from Grewe et al. (1995) with permission.

latencies (upper trace in Figure 3.4c) demonstrating that these axons were myelinated. Axon 5 had spontaneous activity that was centrally generated (lower trace in Figure 3.4c; note the characteristic shape of the extracellularly recorded action potentials recorded from axon 5). Axon 4 had no spontaneous activity (and could not be reflexly activated). Thus this axon was either a preganglionic axon being silent under the experimental conditions or an axon of an afferent neuron that projected through the white ramus L2 to the viscera.

- The third example demonstrates the recording of spontaneous activity from two postganglionic axons in one bundle innervating hairless skin of the cat hindpaw (the bundle was isolated from the medial plantar nerve of the hindpaw) and of the temperature on the surface of the hairless skin (Figure 3.5). Warming of the hypothalamus and the spinal cord (which occurs in vivo when the body is overheated) decreases the activity in the postganglionic neurons, which is then followed by dilatation of the cutaneous blood vessels, increase in blood flow and heat transfer through the skin and a subsequent increase in skin temperature (SKT). Thus the postganglionic neurons recorded from in Figure 3.5 were most likely cutaneous vasoconstrictor neurons (for details see Chapter 4).

Size and shape of extracellularly recorded signals from pre- or postganglionic axons are the basis to discriminate signals from different axons using window discriminators and template (shape) analysis. In this way simultaneously recorded signals in the same microbundle can be analyzed separately.

3.5.3 Neurophysiological recordings in humans

The introduction of microneurography at the Department of Clinical Neurophysiology in Uppsala under the guidance of the Swedish neurologist Karl-Erik Hagbarth in the 1960s (Vallbo et al. 2004) has made it possible to study activity of sympathetic postganglionic axons in peripheral nerves of human beings. Insulated tungsten microelectrodes, with fine uninsulated tips having diameters of 1 to 5 µm, are inserted manually through the intact skin into fascicles of underlying skin and muscle nerves (see Figure 3.6a). This technique allows activity in peripheral afferent and efferent axons traveling in skin and muscle nerves to be studied in conscious subjects who can communicate freely with the experimenters. Activity in these neurons can be correlated with sensory perceptions, somatomotor responses, autonomic effector responses (e.g., blood pressure, heart rate, electrocardiogram [ECG], respiration, blood flows, galvanic skin responses [sweat gland activity]) and central commands (Vallbo et al. 1979; Wallin and Fagius 1988).

The advantages of this technique are obvious. However, the signals recorded from unmyelinated axons have limited resolution when recording from postganglionic axons with relatively low-impedance electrodes. The signals usually consist of multiunit ongoing activity (because several unmyelinated fibers usually run together in one Schwann cell bundle and the bare tip of a tungsten electrode is large when compared to the diameter of the unmyelinated axons [Fig. 3.6b]).

Recently, using metal electrodes with a rather fine bare tip for recording, i.e. electrodes with high impedance, recording from single postganglionic axons is possible, although this single unit recording cannot be used routinely. To find single postganglionic axons generating signals with sufficiently large signal-to-noise ratio for discrimination (see single unit marked by asterisk and the shape of its action potential in Figure 3.6c, upper two records) is much more difficult since the "seeing distance" of the fine-tipped high-impedance electrodes is short when compared to the "seeing distance" of the low-impedance electrodes (see Vallbo et al. [1979]; Macefield and Wallin [1996]; Macefield et al. [2002]). Finally, microneurography can only be applied to nerves in human beings that are located relatively superficially but not to visceral nerves and other deeply located nerves.

The situation is conceptually and methodologically comparable to the analysis of the neuronal control of skeletal muscle during movement. This field initially obtained its impetus (starting with Sherrington in 1906, see Sherrington [1947]) from the application of the reflex concept and the subsequent formulation of hypotheses that could be tested during ongoing movements in animals and human beings, such as locomotion, target reaching, ballistic movements and manipulation (Baldissera et al. 1981; Granit 1981; Jankowska and Lundberg 1981).

Figure 3.6c shows a microneurographic recording from a bundle containing muscle vasoconstrictor axons in a human subject. To

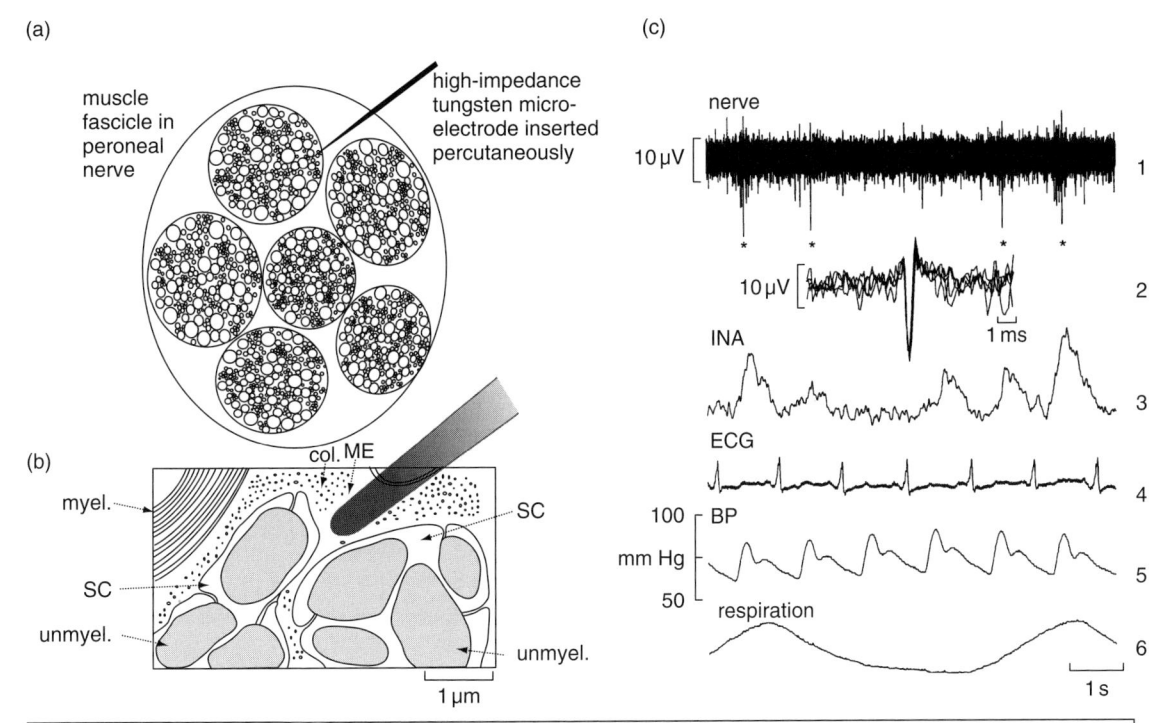

Figure 3.6 Microneurographic recording of activity from postganglionic muscle vasoconstrictor axons in an awake human being. (a) Schematic representation of a tungsten microelectrode inserted percutaneously into a human nerve. (b) The bundles of unmyelinated fibers recorded from in (c) can only be seen in large magnification. col., collagen bundles; ME, tip of tungsten microelectrode; myel., myelinated nerve fiber; SC, Schwann cell cytoplasm; unmyel., profiles of unmyelinated nerve fibers. (c) Recording from a bundle with several postganglionic axons (multiunit activity) in the deep peroneal nerve. Traces 1 and 2: one single unit was identified by the size and shape of the signal (see asterisk and superimposed spikes on expanded time scale in second trace to demonstrate that these spikes have the same shape). Trace 3: integrated nerve activity (INA) representing the activity in several postganglionic axons (see large and small spikes in trace 1). Traces 4 to 6: electrocardiogram (ECG), arterial blood pressure (BP) and respiration parameter. This record was made from a patient with heart failure. Modified from Macefield et al. (2002) with permission.

obtain this recording, a tungsten microelectrode was inserted percutaneously into the deep peroneal nerve of a human being. The original record (upper trace) shows multiunit activity, which could be discriminated from the recording noise and the activity in a single postganglionic axon (marked by asterisks). The electronic discrimination of the single unit is demonstrated in the second trace. This type of "clean" discrimination of activity in single postganglionic axons is more the exception than the rule (Macefield et al. 2002). The multiunit activity has a relatively poor signal-to-noise ratio. The third to sixth recording traces show the integrated neural activity,[2] the ECG, the arterial blood pressure and a parameter of inspiration and expiration, respectively. These traces show that the activity in the postganglionic axons is correlated with the ECG and the pulse pressure wave. This type of microneurographic recording is the basis of all data obtained in human beings from postganglionic neurons innervating skin and skeletal muscle (see Subchapters 4.1 and 4.2).

3.6 | Confounding effects of anesthesia in animal experiments

Other than anesthesia (see Jänig and Räth 1980; Häbler et al. 1994b), the conditions under which discharges in axons are measured are probably close to normal for the autonomic systems under investigation. This can be judged from the degree of spontaneous activity in the autonomic neurons and the reactions of the effector organs, such as for example skin temperature, systemic blood pressure and blood flow through skeletal muscle, skin and viscera. All vital parameters, such as rate of ventilation, end-tidal CO_2, acid–base balance, body core temperature, can be kept as close to normal as possible. Consequently, it is not surprising that both the level and the pattern of discharge recorded from single sympathetic axons in anesthetized animals are comparable to those observed in equivalent sympathetic systems in conscious human beings (Wallin and Fagius 1988; Wallin and Elam 1994; Wallin 2002; Jänig and Häbler 2003). It seems likely then that they are also similar to those in the awake animal. For example, the cardiac and respiratory pattern of discharge in muscle vasoconstrictor neurons of the anesthetized cat and that of human subjects resting in a prone position are almost indistinguishable except that cats have a higher heart rate and a higher frequency of respiration (Häbler et al. 1994a; Jänig and Häbler 2003; see Figure 10.22).

The central effect of the anesthesia used, of course, produces quantitative and sometimes even qualitative distortions of some reflexes and regulatory outflow. This has been demonstrated for thermoregulation and baro- and chemoreceptor regulation of cardiovascular parameters (Hensel 1981, 1982; Eckberg and Sleight 1992; Kirchheim et al. 1998). Any anesthetic will affect brain functions that are dependent on the cortex and limbic system and many also affect glutamatergic synapses at the spinal level. These confounding effects on the discharges of autonomic neurons have to be taken into account when comparing results obtained in different species (including human beings) under different conditions (see Jänig and Räth [1980]). Thus, anesthesia is not a major problem in the analysis of the neural organization of autonomic systems in the periphery and in the neuraxis; but it should always be taken into account in the interpretation of the data. It certainly is a major problem in the analysis of the regulation of autonomic circuits by the forebrain.

Conclusions

1. Spinal cord and brain stem are connected to the autonomic target cells by two-neuron chains of the peripheral sympathetic and

parasympathetic nervous systems. These chains consist of populations of preganglionic neurons and postganglionic neurons that are synaptically connected in the autonomic ganglia. They transmit messages from the central nervous system to the target cells. By analogy to the motoneurons these pathways are called "final autonomic pathways".
2. The final autonomic pathways are the building blocks of the peripheral autonomic nervous system. The main difference between the final somatomotor pathways and the final autonomic pathways is that the central messages may undergo quantitative changes in the autonomic ganglia, and that some effector cells are innervated by more than one functional pathway.
3. The impulse pattern transmitted by these peripheral autonomic pathways to the target cells is the result of central integration in the spinal cord, brain stem, hypothalamus and telencephalon. Reflex patterns that are generated by afferent stimuli in peripheral autonomic neurons serve as physiological markers to analyze the functional structure of the autonomic circuits in the neuraxis.
4. Using this approach by neurophysiological recording from single autonomic neurons in vivo, detailed knowledge has accumulated about the organization of the autonomic nervous system in animals and the human being.

Notes
1. In *closed-loop conditions* neurons are studied when all afferent and efferent systems including the afferent feedback from these effector systems are intact. An example is given in Figure 4.6, which demonstrates the activity in muscle vasoconstrictor neurons in a conscious human subject under resting conditions. In *open-loop conditions* the responses of neurons to experimental afferent or other stimuli are studied in vivo or in vitro. Examples are demonstrated in Chapter 4, showing the responses of neurons to well-defined afferent stimuli (e.g., applied to skin, arterial chemoreceptors, urinary bladder etc.).
2. The original signal is recorded at a 700 to 2000 Hz bandpass filter, amplified and fed through an amplitude discriminator (in order to improve the signal-to-noise ratio). A resistance-capacitance (RC) integrator network with a time constant of 0.1 seconds is used to obtain the integrated nerve activity of the multiunit neural activity (INA [mean voltage display] in Figure 3.6).

Chapter 4

The peripheral sympathetic and parasympathetic pathways

4.1 Sympathetic vasoconstrictor pathways	*page* 107
4.2 Sympathetic non-vasoconstrictor pathways innervating somatic tissues	123
4.3 Sympathetic non-vasoconstrictor neurons innervating pelvic viscera and colon	136
4.4 Other types of sympathetic neuron	141
4.5 Adrenal medulla	143
4.6 Sympathetic neurons innervating the immune tissue	148
4.7 Proportions of preganglionic neurons in major sympathetic nerves	151
4.8 Parasympathetic systems	153

In this chapter I describe the reflex patterns for different groups of autonomic neurons, in particular sympathetic neurons. For other groups of autonomic neurons that have not yet been investigated in this way, I will draw indirect conclusions by analogy to those that have been investigated. Using this approach we gather information about the functional specificity of different neurons, about the relation between activity in certain types of neurons and the responses of the target tissue as well as information about the principal organization of the central circuits that determine the discharge pattern of these neurons (see Chapters 8 to 11). Thus, the experimental data described in this chapter are an important cornerstone of this book: they show that each autonomic pathway exhibits a characteristic pattern of discharge and that this is dependent on the structure of the central circuits in the neuraxis and the synaptic connections of these circuits with the different groups of afferent input to the neuraxis. This type of analysis gives the ultimate underpinning for the concept that the autonomic nervous system consists of functionally distinct building blocks (Jänig and McLachlan 1992a, b). As I have emphasized in Chapter 3, this description does not show how these autonomic systems function during ongoing regulation of autonomic function. This will be discussed in Chapters 5 to 10.

A similar approach has been used for the analysis of the somatomotor system. Here too a detailed analysis of the spinal and supraspinal reflex loops linked to the Ia, Ib, joint, cutaneous and other primary afferent neurons gave valuable insight into the system, which then became the basis for further analysis in order to understand how length and strength of muscle and the coordination of muscles are regulated by the brain so as to understand how movements are brought about by the brain and their underlying mechanisms. This systematic analysis of the somatomotor system used functionally distinct afferent inputs and efferent outputs (motoneurons) in order to unravel the central circuits (see Subchapter 9.4 and Chapter 3). Similarly, the autonomic reflex loops are analyzed with respect to the various afferent feedback signals from the target organs, from the extracellular fluid matrix of the body and from somatic body structures. These afferent feedbacks include neural, hormonal and humoral (e.g., glucose concentration) signals (see Figure 1).

4.1 | Sympathetic vasoconstrictor pathways

Large and small arteries, arterioles and most veins are innervated by sympathetic noradrenergic neurons. Capillaries (exchange vessels) and most venules are not innervated. The density of the anatomical innervation varies considerably between vascular beds in different tissues, between different sections of the same vascular tree and between different functional types of blood vessel and, to some extent, between species. Activation of noradrenergic neurons generates vasoconstriction leading to increased resistance to flow and decreased capacitance of veins. These neurons are therefore called *vasoconstrictor neurons*. Large numbers of sympathetic neurons controlled from all spinal levels innervate blood vessels in tissues throughout the body. These pathways have been studied in detail in the anesthetized cat at both pre- and postganglionic levels, in conscious human beings at the postganglionic level and to a lesser extent in the rabbit and in the rat at the postganglionic level. Some functional properties for different types of vasoconstrictor neurons in the *cat* (and some in the rabbit) are listed in Table 4.1.

4.1.1 Vasoconstrictor neurons in animals

Neurophysiological studies have been conducted mainly in the cat and the rat. Reflex patterns in vasoconstrictor neurons in the two species are similar. However, those in the rat appear to be less differentiated. Furthermore, the respiratory patterns in the activity of the vasoconstrictor neurons of the two species are different (Häbler *et al.* 1993c, 1994a, b, 1996, 1999a, 2000; Bartsch *et al.* 1996, 1999, 2000).

Muscle vasoconstrictor neurons
About 90% of the postganglionic axons in muscle nerves of the cat hindlimb consist of vasoconstrictor axons; these noradrenergic

Table 4.1 | *Functional classification of sympathetic vasoconstrictor neurons in the cat based on reflex behavior in vivo*

Likely function	Location	Target organ	Likely target tissue	Major identifying stimulus	Ongoing activity[a]
Muscle vasoconstrictor	Lumbar	Hindlimb muscles	Resistance vessels	Baro-inhibition	Yes
	Cervical	Head and neck muscles	Resistance vessels	Baro-inhibition	Yes
Cutaneous vasoconstrictor	Lumbar	Hindlimb skin	Thermoregulatory blood vessels	Inhibited by CNS warming	Yes
	Cervical	Head and neck skin	Thermoregulatory blood vessels	Inhibited by CNS warming	Yes
Visceral vasoconstrictor	Lumbar splanchnic	Pelvic viscera	Resistance vessels	Baro-inhibition	Yes
Renal vasoconstrictor	Thoracic splanchnic	Kidney	Resistance vessels	Baro-inhibition	Yes

Notes:
For details about rates of ongoing activity in pre- and postganglionic neurons, reflexes to various afferent stimuli, spinal and supraspinal reflex pathways, coupling to regulation of respiration and conduction velocities of pre- and postganglionic axons see Table 6.2 and Jänig (1985a, 1988a), Jänig and McLachlan (1987), Jänig et al. (1991), Boczek-Funcke et al. (1992b, c, 1993), Häbler et al. (1994b), Grewe et al. (1995).
[a] Some neurons do not have spontaneous activity and are recruited under special functional conditions.
CNS, central nervous system.
Modified from Jänig and Häbler (1999).

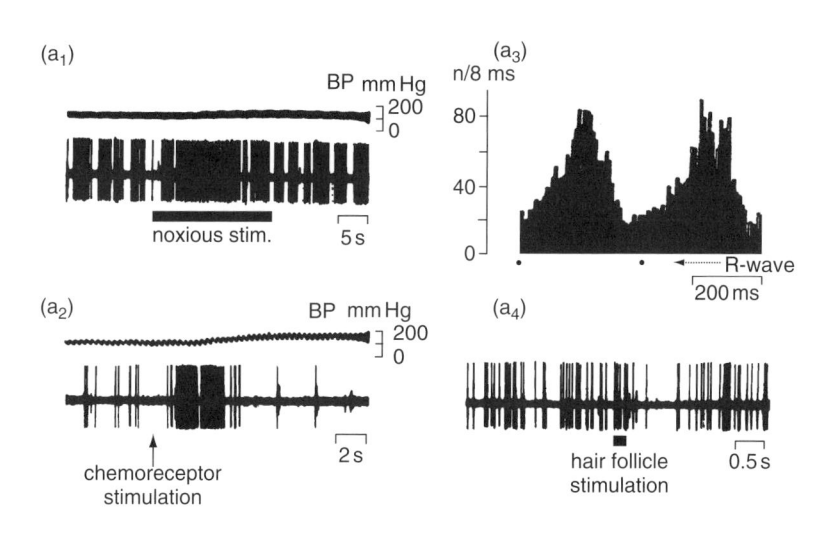

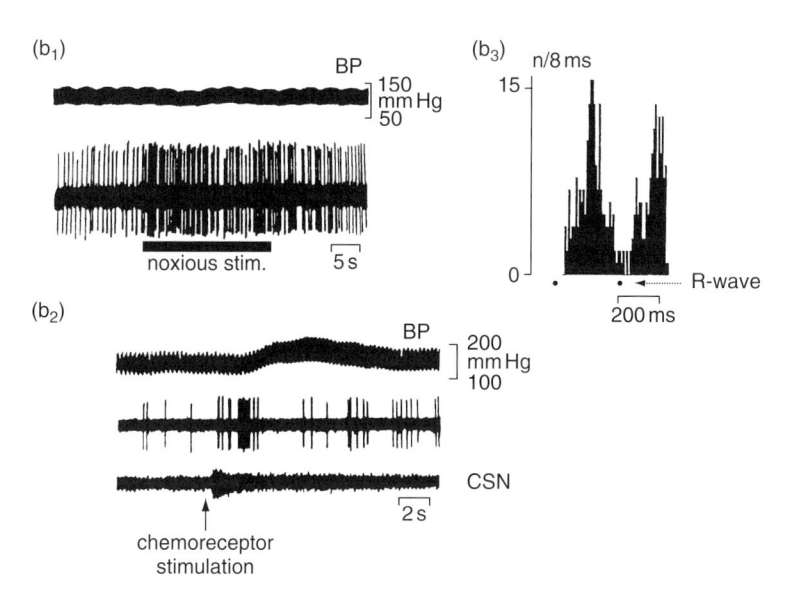

Figure 4.1 Reflexes in sympathetic neurons with putative muscle vasoconstrictor function in the anesthetized cat. (a) Postganglionic neuron projecting to the peroneal muscle. Original neural activity in (a_1), (a_2) and (a_4). (a_1) Excitation in response to mechanical noxious stimulation of a toe of the ipsilateral hindpaw. (a_2) Excitation in response to stimulation of arterial chemoreceptors projecting through the carotid sinus nerve by retrograde bolus injection of 0.8 ml CO_2-enriched Ringer solution into the left lingual artery. (a_3) Strong rhythmic changes in activity with respect to phasic stimulation of arterial baroreceptors by pulsatile blood pressure (and therefore phasic inhibition of activity; "cardiac rhythmicity" of the activity, 2000 periods of activity superimposed, triggered by the R-wave of the electrocardiogram [indicated by dots]). (a_4) Inhibition in response to short-lasting stimulation of hair follicle receptors on the trunk of the cat by air jets (ten trials superimposed). (b) Thoracic preganglionic neuron projecting in the cervical sympathetic trunk. Original neural activity in lower trace of (b_1) and in middle trace of (b_2). (b_1) Excitation in response to mechanical noxious stimulation of the ear. (b_2) Excitation in response to stimulation of arterial chemoreceptors (injection of 0.2 ml CO_2-enriched Ringer solution into the left lingual artery). Activity of chemoreceptor afferents in carotid sinus nerve (CSN) shown. (b_3) Same as (a_3) (500 periods of activity superimposed). BP, blood pressure; (a), from Blumberg et al. (1980) and unpublished; (b), modified from Boczek-Funcke et al. (1992a).

axons are associated with small and large arterial blood vessels but not with veins. A few may have vasodilator and other functions (see Subchapter 4.2.2). The vast majority of muscle sympathetic neurons are spontaneously active; a few appear to be silent under experimental conditions. The rate of spontaneous activity is in the range of 0.5 to 3 imp/s in the anesthetized cat (Jänig 1985a, 1988a) and 1.4 ± 0.5 imp/s (range 0.3 to 2.4 imp/s) in the anesthetized rat (Häbler et al. 1994a) (see Table 6.2). The proportion of spontaneously active and silent neurons may vary between individual animals. However, it is safe to say that the ongoing activity recorded from

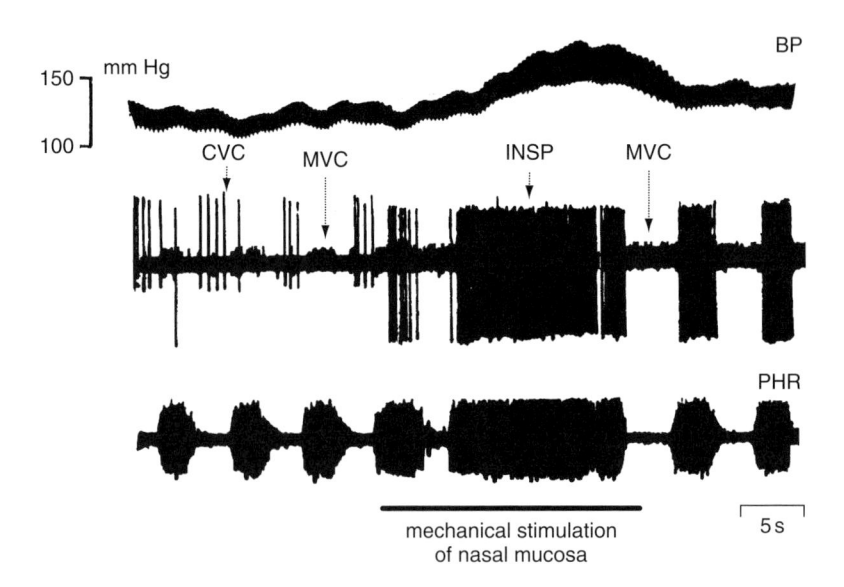

Figure 4.2 Reflexes in sympathetic neurons elicited by mechanical stimulation of the nasal mucosa. Simultaneous recording of the activity in a preganglionic cutaneous vasoconstrictor (CVC) neuron, a preganglionic inspiratory (INSP) neuron and a preganglionic muscle vasoconstrictor (MVC) neuron in an anesthetized cat. The activity was recorded from a strand of nerve fibers isolated from the cervical sympathetic trunk and from the phrenic nerve (PHR). Before stimulation, the CVC neuron was active in expiration, the MVC neuron in inspiration and expiration and the INSP neuron was almost silent. Mechanical (probably noxious) stimulation of the nasal mucosa inhibited the CVC neuron, activated the MVC neuron in inspiration as well as expiration and activated the INS neuron, but only in inspiration. Note that the reflexes in the neurons outlasted the stimulus and that the increase in blood pressure (BP) was correlated with the continuous MVC discharge. Modified from Boczek-Funcke et al. (1992b).

bundles containing postganglionic axons isolated from muscle nerves arises only from muscle vasoconstrictor axons.

Muscle vasoconstrictor neurons have the following key functional properties, some of which are demonstrated in Figure 4.1:

1. They are under powerful inhibitory control by the arterial baroreceptors. This generates rhythmic firing due to periods of inhibition resulting from the rhythmic activation of the arterial baroreceptors by the increase of blood pressure during each cardiac cycle (Figure 4.1a_3, b_3).
2. They are excited by most inputs from the body surface (e.g., nociceptors, Figure 4.1a_1, b_1), from the viscera (e.g., distension-sensitive receptors in the urinary bladder and colon), from arterial chemoreceptors (Figure 4.1a_2, b_2) and from trigeminal receptors (Figure 4.2). However, stimulation of low-threshold mechanoreceptive afferents from the skin leads to inhibition of the activity in muscle vasoconstrictor neurons (Figure 4.1a_4).
3. Stimulation of central thermoreceptors (by hypothalamic and spinal cord warming) does not influence muscle vasoconstrictor neurons (Grewe et al. 1995).
4. The activity of muscle vasoconstrictor neurons is modulated during the respiratory cycle in a characteristic way. They are excited during central inspiration (particularly when the respiratory drive

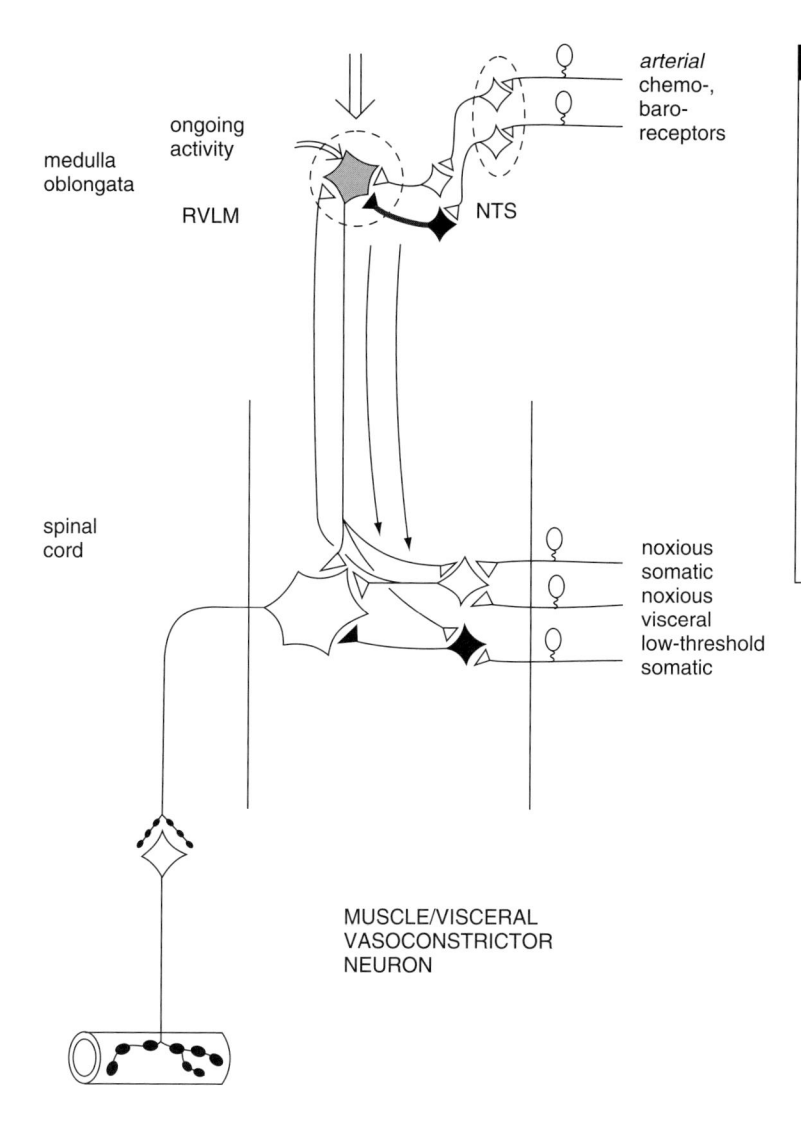

Figure 4.3 Simplified schematic diagram of the central pathways involved in the reflexes in muscle and visceral vasoconstrictor neurons elicited by physiological stimulation of spinal afferents, arterial baroreceptor afferents or arterial chemoreceptor afferents. The reflex pattern in the neurons is generated by spinal and supraspinal circuits. Excitatory interneurons not filled. Inhibitory interneurons filled black. Sympathetic premotor neuron in medulla oblongata shaded. NTS, nucleus tractus solitarii; RVLM, rostral ventrolateral medulla. See Subchapters 9.2, 10.2 and 10.3.

is high), with an ensuing period of decreased activity and sometimes with a period of decreased activity in early inspiration. This profile of respiratory modulation of activity in muscle vasoconstrictor neurons interacts with the inhibitory effects of arterial baroreceptors and possibly other cardiovascular afferents, which are stimulated rhythmically when the arterial blood pressure rises and falls with inspiration and expiration (see Subchapter 10.6 and MVC in Figure 4.2; Häbler et al. 1993c, 1994b, 1999a).

The distinctive functional properties of muscle vasoconstrictor neurons are determined by reflex pathways in the spinal cord (Jänig 1996a) and by supraspinal reflex pathways through the medulla oblongata and higher central structures. Figure 4.3 demonstrates schematically some of these pathways which can explain the discharge pattern observed in muscle vasoconstrictor neurons. These reflex pathways will be discussed in more detail in Chapters 9 and 10. The respiratory modulation of activity in muscle vasoconstrictor neurons is determined by the coupling between cardiovascular sympathetic premotor neurons in the medulla oblongata (e.g., in the rostral ventrolateral medulla [RVLM], see Figure 4.3) and neurons of the ponto-medullary respiratory network (Richter and Spyer 1990; Richter et al. 1991; St.-John 1998; see also Häbler et al. [1994b]). This has been worked out in experiments on anesthetized animals in which respiratory parameters (activity in phrenic nerve) were recorded in parallel with the activity from the muscle vasoconstrictor neurons. Its underlying mechanism will be discussed in more detail in Subchapter 10.6.

The observed responses in *postganglionic* muscle vasoconstrictor neurons are typical of what would be expected for vasoconstrictor neurons that determine peripheral vascular resistance. As expected these neurons terminate virtually exclusively on resistance vessels (arterioles and small arteries 20 to 250 μm in diameter). Activation of these neurons is followed by an increase in arterial blood pressure. An identical discharge pattern has been observed in lumbar *preganglionic* neurons in pathways that project to the hindlimb (Jänig and Szulczyk 1980, 1981) and in thoracic *preganglionic* neurons that project to the head and neck via the superior cervical ganglion (Boczek-Funcke et al. 1992a, b; compare a and b in Figure 4.1). In the cat, about 20% of the preganglionic neurons projecting through the cervical sympathetic trunk to the superior cervical ganglion (targets in head and neck; Boczek-Funcke et al. 1992a, 1993) and 10% of the preganglionic neurons projecting to paravertebral ganglia distal to the lumbar ganglion L5 (targets in hindlimb and tail; Jänig and Szulczyk 1980) have functional properties of muscle vasoconstrictor neurons (see Table 4.5 and Subchapter 4.7).

The pattern of activity present in muscle vasoconstrictor neurons is found in peripheral pathways throughout the body. Because of the ubiquity of small resistance vessels, the constriction of which is the primary determinant of the level of arterial blood pressure, discharge patterns in sympathetic neurons with predominantly cardiac and respiratory rhythms are the archetypal patterns that dominate almost all recordings from whole "*sympathetic*" nerve trunks, including the cervical and lumbar sympathetic trunks, the renal nerves and the splanchnic nerves (Häbler et al. 1994b).

Visceral vasoconstrictor neurons

A discharge pattern that is very similar to that of muscle vasoconstrictor neurons has been found in three separate groups of sympathetic neurons: (1) in a subpopulation of sympathetic *preganglionic* neurons in the lumbar segments that project in the lumbar splanchnic nerves to the inferior mesenteric ganglion; (2) in a subpopulation of sympathetic *postganglionic* neurons in the inferior mesenteric ganglion that project to colon and pelvic viscera; and (3) in a subpopulation of spontaneously active *postganglionic* neurons in the renal nerves. Thus, these axons show spontaneous activity, strong inhibitory control of activity by arterial baroreceptors, increased activation during stimulation of arterial chemoreceptors and rhythmic changes of activity with central respiration, as do muscle vasoconstrictor neurons. These features suggest that they are vasoconstrictor neurons innervating arterial blood vessels involved in the regulation of peripheral vascular resistance, i.e. in the regulation of arterial blood pressure. It is likely that pre- and postganglionic visceral vasoconstrictor neurons, projecting through the major and minor splanchnic nerves to the celiac and superior mesenteric ganglia and from here through mesenteric and other nerves to blood vessels in the viscera, also have the same functional discharge pattern. In the cat, about 15% of the preganglionic neurons projecting in the lumbar splanchnic nerves (target vessels in the pelvic organs and colon; Bahr *et al.* 1986c) have functional properties of visceral vasoconstrictor neurons (see Table 4.5 and Subchapter 4.7).

There are quantitative differences between the reflex patterns in visceral and muscle vasoconstrictor neurons indicating that the central outputs to these two sets of vasoconstrictor pathways are not identical. Furthermore, renal vasoconstrictor neurons may differ functionally from vasoconstrictor neurons innervating other viscera and from sympathetic neurons controlling non-vascular functions in the kidney (see Subchapter 4.4; Bahr *et al.* 1986b; Dorward *et al.* 1987; Jänig 1988a; Jänig *et al.* 1991; Kopp and DiBona 1992; DiBona and Kopp 1997).

Cutaneous vasoconstrictor neurons

Most postganglionic axons in skin nerves supply blood vessels, sweat glands and piloerector muscles. The sympathetic innervation of the blood vessels is largely vasoconstrictor but in some parts of the skin there may also be sympathetic vasodilator axons (see Subchapter 4.2). The cutaneous vasoconstrictor axons probably differ according to the section of the vascular bed they innervate (small muscular artery, arteriole, arteriovenous anastomosis, vein) and according to the type of skin (hairless [glabrous] skin of the distal extremities, hairy skin, skin of the trunk and proximal extremities, skin of the face). This is reflected in a differential colocalization of neuropeptides in noradrenergic neurons to the cutaneous vasculature of the guinea pig where different levels of the arterial tree are supplied by postganglionic axons with different peptide content (Gibbins and Morris 1990; Morris 1995; see Subchapter 1.4). This suggests different functions for different types of cutaneous vasoconstrictor neurons. Thus, it would not be unexpected that the activity pattern in cutaneous vasoconstrictor neurons is not as uniform as that in muscle vasoconstrictor neurons, as has proved to be the case. However, the functional identification of cutaneous vasoconstrictor neurons innervating different sections of the cutaneous vascular bed remains to be elucidated.

Many cutaneous vasoconstrictor neurons are spontaneously active in the anesthetized cat and rat. However, the exact percentage

Figure 4.4 Reflexes in sympathetic neurons with putative cutaneous vasoconstrictor function in the anesthetized cat. (a) Postganglionic neuron (large action potential 1) innervating hairy skin of the hindlimb (innervation territory of the superficial peroneal nerve). (a_1) Inhibition in response to mechanical noxious stimulation of a toe of the ipsilateral hindpaw. (a_2) Inhibition in response to stimulation of arterial chemoreceptors projecting through the carotid sinus nerve (see legend Figure 4.1a_2). (a_3) No rhythmic changes in activity with respect to phasic stimulation of arterial baroreceptors by pulsatile blood pressure (1000 periods of activity superimposed, triggered by the R-wave of the electrocardiogram [indicated by dots]). See Figure 4.1a_3. (a_4) Excitation in response to short-lasting stimulation of hair follicle receptors on the trunk by air jets (10 trials superimposed). Note that the bundle recorded from contained two other postganglionic axons, one behaving like a cutaneous vasoconstrictor neuron (axon 2) and the other like a muscle vasoconstrictor neuron (axon 3). (b) Recordings from a single thoracic preganglionic neuron (large signal) projecting in the cervical sympathetic trunk. (b_1) Inhibition in response to mechanical noxious stimulation of the ear. (b_2) Inhibition in response to stimulation of arterial chemoreceptors (injection of 0.2 ml CO_2-enriched Ringer solution into the left lingual artery). Activity of chemoreceptor afferents in carotid sinus nerve (CSN) shown. (b_3) No modulation of the activity with respect to phasic stimulation of arterial baroreceptors by the pulsatile blood pressure (500 periods of activity superimposed with respect to R-wave of the electrocardiogram). BP, blood pressure; (a), from Blumberg et al. (1980) and unpublished; (b), modified from Boczek-Funcke et al. (1992a) with permission.

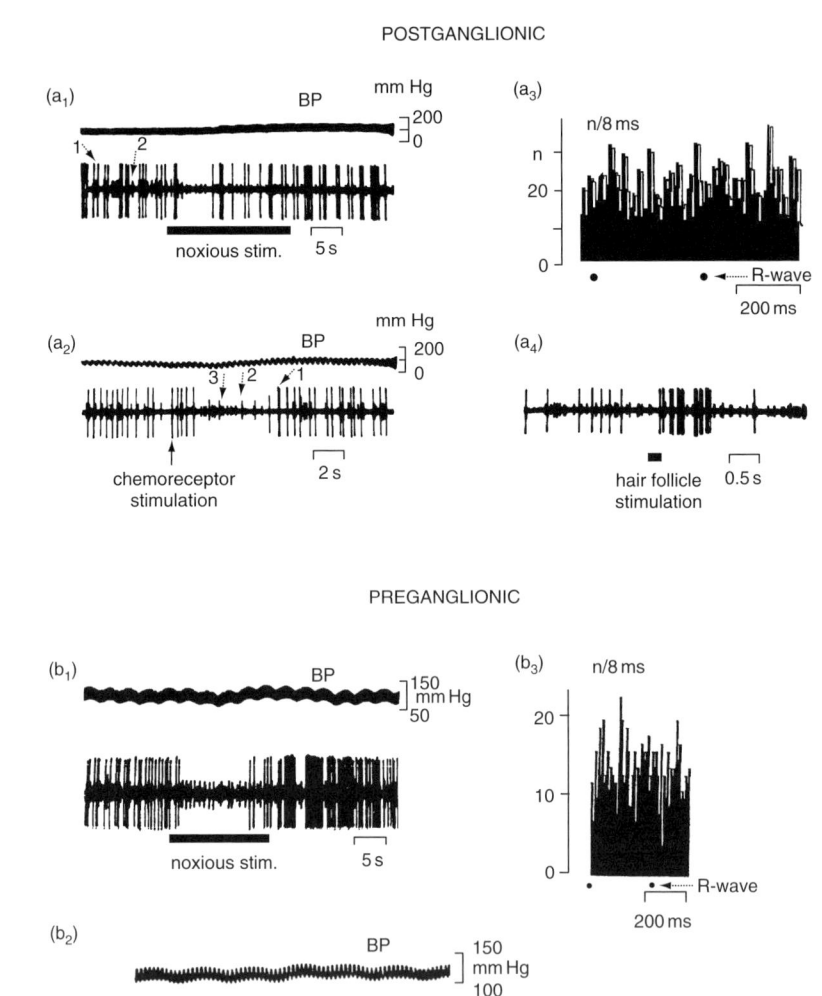

of spontaneously active postganglionic cutaneous vasoconstrictor neurons under thermoneutral conditions is unknown. The rate of spontaneous activity in these postganglionic neurons is 1.2 ± 0.7 imp/s (mean $\pm$ SD; range 0.1 to 4 imp/s) in the cat (hairy and hairless skin; Jänig 1985a, 1988a), 1.2 ± 0.6 imp/s (range 0.3 to 2.4 imp/s) in cutaneous vasoconstrictor neurons innervating the rat hairy skin of the hindlimb (Häbler et al., 1994a) and 1.1 ± 0.7 imp/s (range 0.23 to 2.6 imp/s) in those innervating the rat hairy skin of the tail (Häbler et al. 1999a) (see Table 6.2).

Most cutaneous vasoconstrictor neurons that innervate the hairy and hairless skin of the cat and rat hindpaw or tail (i.e., the distal skin of the extremities or tail) have complex discharge patterns that differ from those of muscle vasoconstrictor neurons (Figure 4.4) in the following ways (data taken from Jänig 1985a, 1988a; Häbler et al. 1992, 1994a; Grewe et al. 1995 [cat hindpaw, tail]):

1. In about 80% of cutaneous vasoconstrictor neurons innervating hairy skin and probably in all cutaneous vasoconstrictor neurons innervating hairless skin of the paw, inhibition of activity elicited by stimulation of arterial baroreceptors is usually either weak or absent. In these neurons the phasic bursts of discharge in parallel with the pulse pressure wave, which are typical of muscle vasoconstrictor activity, are absent or weak (Figure 4.4a_3, b_3).
2. Warming the hypothalamus and/or the spinal cord inhibits activity in almost all cutaneous vasoconstrictor neurons (both hairy and hairless skin). This also applies to neurons that are responsive to arterial baroreceptor stimulation. Decreased activity in the cutaneous vasoconstrictor neurons is followed by increased skin temperature.
3. Stimulation of cutaneous nociceptors in the ipsilateral paw inhibits most cutaneous vasoconstrictor neurons innervating the stimulated paw, whereas stimulation of cutaneous nociceptors at more remote body sites has weaker inhibitory effects or is even excitatory (Horeyseck and Jänig 1974b). This type of reflex pattern is also very pronounced for cutaneous vasoconstrictor neurons innervating the cat tail (Grosse and Jänig 1976). This unique inhibitory nociceptor reflex is typical for cutaneous vasoconstrictor neurons in animals. It was first described in the rabbit by the Swede Otto Christian Lovén (1866) in Carl Ludwig's laboratory in Leipzig and became known as the Lovén reflex[1] (Figure 4.4a_1, b_1).
4. Stimulation of low-threshold mechanoreceptive afferents from the skin (e.g., hair follicle receptors) leads to excitation of the activity in cutaneous vasoconstrictor neurons (Figure 4.4a_4).
5. Stimulation of other afferent inputs (from pelvic organs, from arterial chemoreceptors, from trigeminal nasal afferents) is also inhibitory in the cat (Figures 4.2, 4.4a_2, b_2).

6. In the cutaneous vasoconstrictor neurons there is either no respiratory rhythm or activity decreases during inspiration and increases during expiration or, alternatively, only increases during inspiration. These distinct respiratory patterns in the activity of cutaneous vasoconstrictor neurons show that the ponto-medullary respiratory network in the lower brain stem and the circuits regulating activity in cutaneous vasoconstrictor neurons are coupled and that this coupling is different from that to the vasoconstrictor pathways innervating resistance vessels in skeletal muscle and viscera (see Subchapter 10.6; see CVC in Figure 4.2).

Cutaneous vasoconstrictor neurons innervating the hairy skin of the rat hindpaw (Häbler *et al.* 1994b) or the rat tail (Häbler *et al.* 1999a, 2000; Owens *et al.* 2002) have discharge patterns similar to those in the cat; however, the reflexes are not as pronounced as in the cat.

There are quantitative differences in functional properties between the populations of postganglionic cutaneous vasoconstrictor neurons innervating hairy skin and those supplying hairless skin. A small class of cutaneous vasoconstrictor neurons innervating hairy skin (<20%) behaves at least qualitatively, if not quantitatively, like muscle and visceral vasoconstrictor neurons, i.e. they are under strong inhibitory reflex control from arterial baroreceptors and are excited by stimulation of arterial chemoreceptors and of nociceptors (Blumberg *et al.* 1980). This class of cutaneous vasoconstrictor neurons has not been found among vasoconstrictor neurons to hairless skin of the cat paw (Jänig and Kümmel 1977; Figures 4.10, 4.11). This is consistent with the existence of functional subgroups of cutaneous vasoconstrictor neurons innervating different sections of the vascular bed (nutritive blood vessels, arteriovenous anastomoses, capacitance vessels). All or most cutaneous vasoconstrictor neurons innervating skin of the distal extremity are likely to be involved in thermoregulation (Grewe *et al.* 1995).

The typical reflex pattern seen in most *postganglionic* cutaneous vasoconstrictor neurons has also been identified in lumbar *preganglionic* neurons projecting to the hindlimb and in thoracic preganglionic neurons projecting to the superior cervical ganglion (compare [a] and [b] in Figure 4.4). About 12% to 13% of the preganglionic neurons projecting in the lumbar sympathetic trunk distal to the paravertebral ganglion L_5 (targets vessels in hindlimb and tail; Jänig and Szulczyk 1980) or in the cervical sympathetic trunk (target vessels in head and neck; Boczek-Funcke *et al.* 1992a, 1993) exhibit functional properties of cutaneous vasoconstrictor neurons (see Table 4.5 and Subchapter 4.7).

The complex reflex patterns in cutaneous vasoconstrictor neurons are most likely not the result of integration of functionally different synaptic inputs in sympathetic paravertebral ganglia although we have no absolute experimental proof that preganglionic neurons with muscle vasoconstrictor-like discharge pattern and preganglionic neurons with cutaneous vasoconstrictor-like discharge pattern

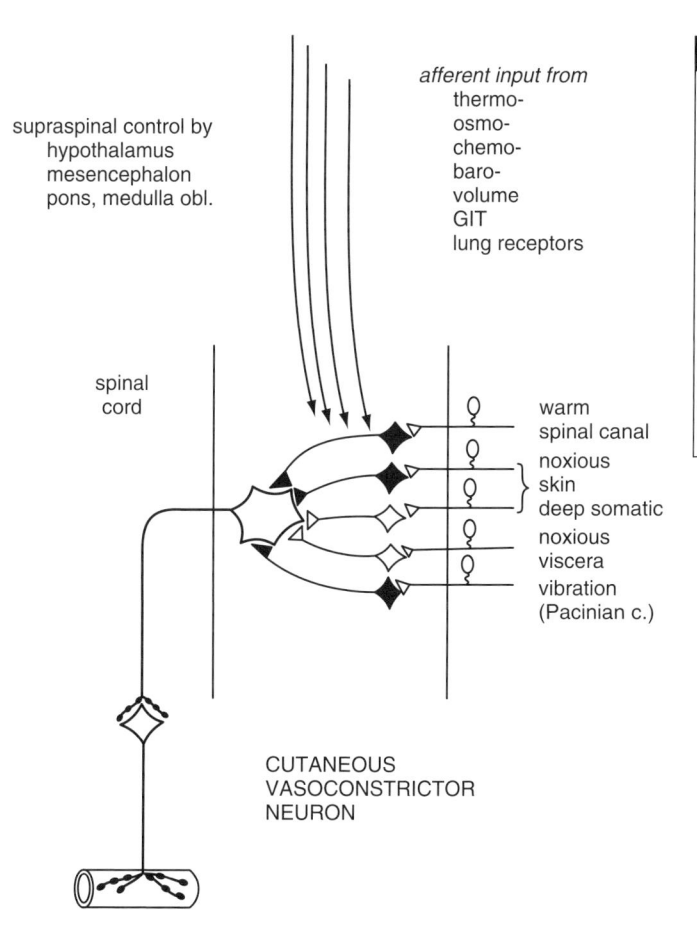

4.1 SYMPATHETIC VASOCONSTRICTOR PATHWAYS | 117

Figure 4.5 Schematic diagram of the central pathways involved in the reflexes in cutaneous vasoconstrictor neurons elicited by physiological stimulation of spinal afferents, vagal afferents or thermosensitive neurons. The reflex pattern in the neurons is generated by spinal and supraspinal circuits. The supraspinal components are related to the hypothalamus, upper and lower brain. Inhibitory neurons filled. See Subchapters 9.2 and 10.5.

do converge on the same postganglionic neurons (see Chapter 6). The functional properties of cutaneous vasoconstrictor neurons are determined by reflex pathways in the spinal cord, lower brain stem and hypothalamus. Experiments in cats with chronically transected spinal cord show that there are several spinal reflex pathways that are specific for the cutaneous vasoconstrictor system (Figure 4.5; see Subchapter 9.2). Generally these reflex pathways seem to be more complex than those of muscle and visceral vasoconstrictor neurons (Jänig 1985a, 1996a). They are under complex control of the lower brain stem and hypothalamus; but available information is limited (see Chapters 9 and 10).

4.1.2 Vasoconstrictor neurons in human beings

Muscle vasoconstrictor neurons

Healthy humans and anesthetized cats and rats (with intact vagus and baroreceptor nerves) have very similar discharge patterns in muscle vasoconstrictor neurons but in resting recumbent humans the rate of spontaneous activity is significantly lower than in the anesthetized cat and rat (0.33 ± 0.04 imp/s, mean $\pm$ SD, range 0.09

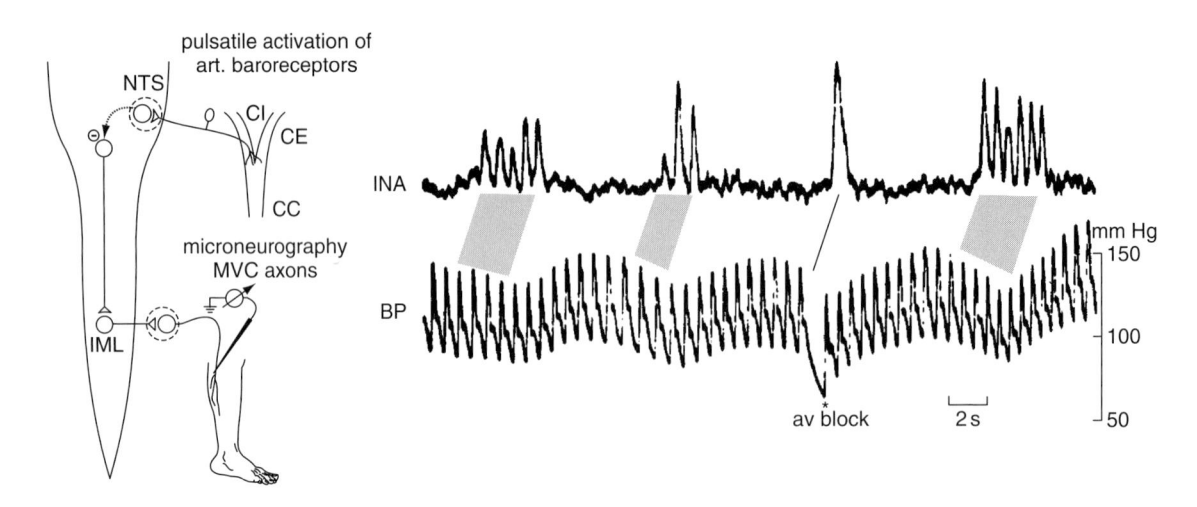

Figure 4.6 Pulsatile unloading of arterial baroreceptors leads to pulsatile inhibition and excitation of muscle vasoconstrictor (MVC) neurons in a human being. Integrated multiunit neural activity (INA) in postganglionic MVC axons recorded microneurographically (see Figure 3.6) from the deep peroneal nerve in an awake human being. Lower trace, arterial blood pressure (BP). Respiration-related increase in BP is followed by inhibition of MVC activity. Note that during respiration-related decrease of BP the MVC neurons are activated in a pulsatile manner. This is generated by pulsatile unloading of arterial baroreceptors. An atrioventricular (av) block is followed by a large decrease of BP (asterisk) and subsequently by strong activation of the MVC neurons. CC, common carotid artery; CE, external carotid artery; CI, internal carotid artery; IML, intermediolateral nucleus. Modified from Wallin and Fagius (1986) with permission.

to 0.69 imp/s, n = 33; Macefield and Wallin 1999a). In humans there are considerable interindividual differences in the rate of resting activity in muscle vasoconstrictor neurons, which are reproducible over weeks, months and years (Fagius and Wallin 1993; Sundlöf and Wallin 1977); the reason for these interindividual differences is unclear. Overall, the resting activity increases with age (Mano 1999); however, it is unclear if the percentage of active muscle vasoconstrictor neurons is higher or if individual postganglionic neurons have a higher activity or both.

The activity in muscle vasoconstrictor neurons exhibits rhythmic changes that are correlated with the pulse pressure wave (Figure 4.7b) and with respiration (Figure 4.6). The pulse synchronous bursts of activity are triggered by rhythmic unloading of the arterial baroreceptors and are clearly equivalent to the cardiac rhythmicity of the activity in muscle vasoconstrictor neurons of animals (see Fig. 4.1a$_3$, b$_3$; Hagbarth and Vallbo 1968; Delius et al. 1972; Sundlöf and Wallin 1978; Eckberg et al. 1985; Wallin and Fagius 1988). The "respiratory" rhythmicity of the activity in the muscle vasoconstrictor neurons is mostly closely linked to the respiratory fluctuations of the arterial blood pressure, indicating that both rhythmic changes are generated by activation of arterial baroreceptors during each increase in arterial blood pressure. However, there are also respiratory rhythmicities that are independent of the arterial baroreceptors (for details see Subchapter 10.6).

Figure 4.7 Microneurographic recordings in human subjects from bundles containing muscle vasoconstrictor axons in the deep peroneal nerve. (a) Activation by mechanical noxious stimulation (black bar; open bar, non-noxious mechanical stimulation). Note pulsatile modulation of neural bursts. Multiunit activity. (b) Activation during apnea (cessation of respiration). Note rhythmic multiunit and single unit (*) activity with respect to electrocardiogram (ECG). BP, blood pressure; INA, integrated neural activity. Modified from Nordin and Fagius (1995) and Macefield and Wallin (1999a) with permission.

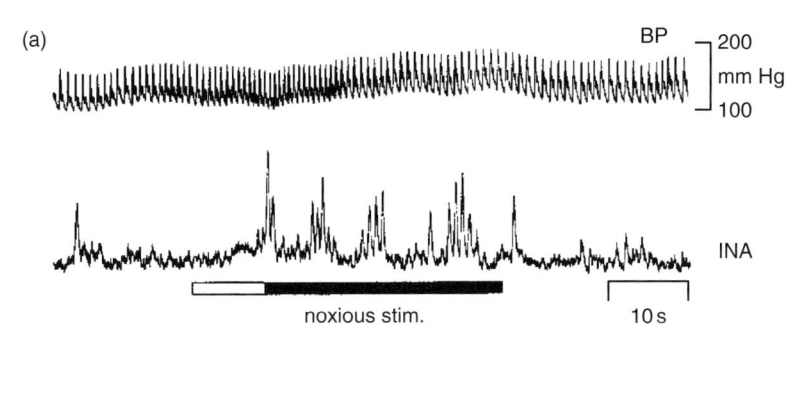

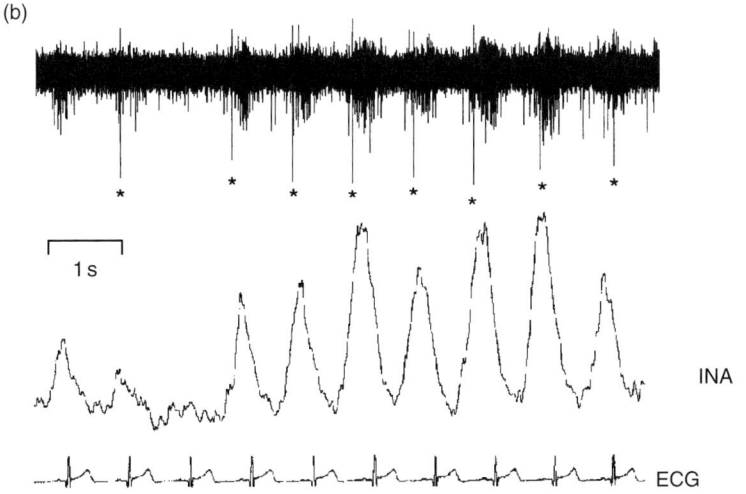

Any procedure which leads to changes of arterial blood pressure or changes of volume in the venous capacitance system generates reflex changes of activity in the muscle vasoconstrictor neurons ("sympathetic bursts") via stimulation or unloading of arterial baroreceptors or of intrathoracic low-pressure receptors in the right atrium and large veins (Sundlöf and Wallin 1978; Vissing et al. 1994). For example, a short transient block of atrioventricular conduction in the pacemaker system of the heart is followed by a transient drop in arterial blood pressure (because the heart stops to pump one stroke volume into the arterial system); this is then followed by a reflex activation of muscle vasoconstrictor neurons (see asterisk in Figure 4.6). Noxious stimuli (e.g., excitation of cutaneous nociceptors by mechanical stimulation [Figure 4.7a] or by radiant heat or with iced water [the cold pressor test] [Victor et al. 1987]), stimulation of trigeminal receptors (e.g., by immersion of the face in water [Fagius and Sundlöf 1986]) or apnea (arrest of respiration) activate muscle vasoconstrictor neurons in human subjects, as they do in the cat.

Cutaneous vasoconstrictor neurons

In humans, cutaneous vasoconstrictor neurons projecting to the distal parts of the extremities have ongoing activity at neutral

Table 4.2 | *Functional classification of sympathetic neurons innervating skeletal muscle and skin in human beings based on microneurographic studies*

Likely function	Target organ	Likely target tissue	Major identifying stimulus	Ongoing activity[a]
Muscle vaso-constrictor	Leg, arm	Resistance vessels	Baro-inhibition	Yes
Cutaneous vasoconstr.	Leg, arm	Thermoregul. blood vessels	Excited by general cooling	Yes
Sudomotor	Leg, arm	Sweat glands	Excited by general warming	Yes
Cutaneous vasodilator	Leg, arm, face	Blood vessels	?	No (?)
Pilomotor	Hairy skin	Piloerector muscles	?	No (?)

Notes:
All data are obtained from postganglionic neurons in awake humans. For details about rates of ongoing activity, reflexes to various stimuli and maneuvers, coupling to respiration and conduction velocity of postganglionic axons see Hagbarth *et al.* (1972), Bini *et al.* (1980a, b), Nordin (1990), Macefield *et al.* (1994), Noll *et al.* (1994), Nordin and Fagius (1995), Macefield and Wallin (1996, 1999a, b), Wallin *et al.* (1998).
[a] Some neurons do not have spontaneous activity and are recruited under special functional conditions.
Modified from Jänig and Häbler (1999).

ambient temperatures. They discharge at rates that are lower than those in anesthetized cats and rats. The rates (measured in cooled subjects) are 0.53 ± 0.11 imp/s (mean $\pm$ S.E.M.; range 0.08 to 2.04, n = 17; Macefield and Wallin 1999b). Most cutaneous vasoconstrictor neurons that supply more proximal skin areas in human beings seem to be silent at neutral ambient temperatures (Bini *et al.* 1980b). The same may be true for cutaneous vasoconstrictor neurons innervating the trunk. Thermal stimuli are the most specific stimuli that change the activity in these cutaneous vasoconstrictor neurons: exposure to warm and cold environments is followed by a decrease or increase, respectively, of activity in vasoconstrictor neurons innervating the distal skin of the extremities (Figure 4.8a; Bini *et al.* 1980a). The ongoing activity in groups of cutaneous vasoconstrictor neurons innervating hairless skin of the palm in human subjects is not, or only weakly, under the control of either arterial baroreceptors (Bini *et al.* 1981; Fagius *et al.* 1985) or low-pressure baroreceptors in the right atrium (Vissing *et al.* 1994). Therefore, in human beings changes of activity in cutaneous vasoconstrictor neurons with the pulse pressure wave are absent or small. However, human subjects undergoing whole-body cooling exhibit cardiac rhythmicity in the activity recorded microneurographically from bundles in the superficial peroneal nerve which innervate hairy skin (Macefield and Wallin 1999b). Thus, as in the cat, there is a difference between cutaneous vasoconstrictor neurons innervating hairy skin and cutaneous vasoconstrictor neurons innervating hairless skin: those innervating hairless (glabrous) skin are under weak or no control of the baroreceptor reflexes, whereas those innervating hairy skin are under some baroreceptor control.

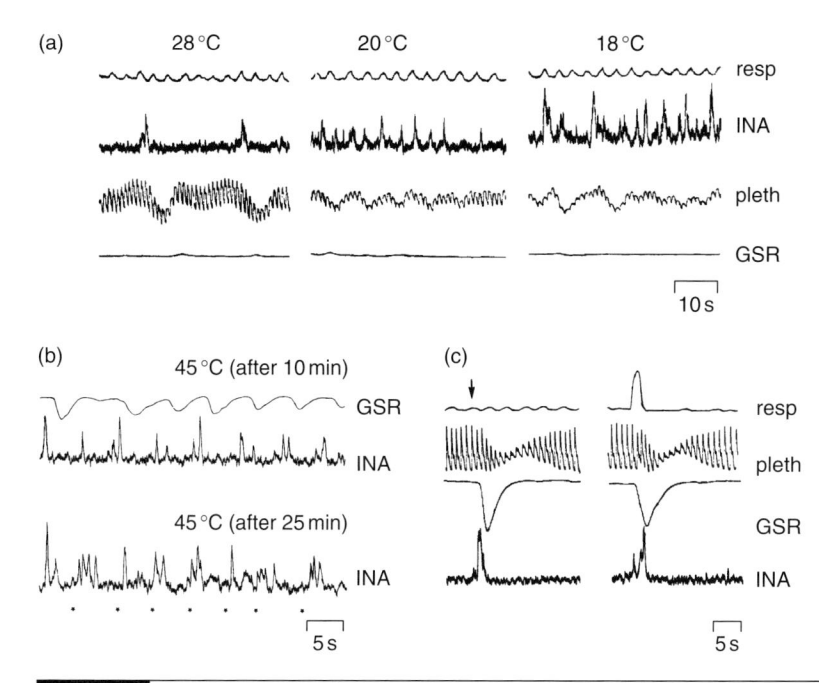

Figure 4.8 Microneurographic recordings of activity in cutaneous vasoconstrictor neurons and sudomotor neurons in human subjects. (a) Integrated cutaneous vasoconstrictor activity (median nerve) at three different ambient temperatures in relation to finger pulse plethysmogram (pleth, vasoconstriction downward and reduction of amplitude) and palmar skin resistance (GSR, galvanic skin response; reduction of skin resistance downward). Note increase of neural activity and decrease of pulse amplitude with decrease of ambient temperature and relation of neural bursts to phasic vasoconstriction. Upper record, respiration (resp).
(b) Integrated sudomotor activity at high central (and ambient) temperature in relation to skin resistance change (GSR). Activation of sudomotor neurons is accompanied by perception of heat waves (indicated by *). (c) Activation of both cutaneous vasoconstrictor neurons and sudomotor neurons by arousal stimulus (sudden shout, arrow left) or by deep inspiration (right). INA, integrated neural activity (mean voltage neurogram). Modified from Wallin and Fagius (1986) and Bini et al. (1980a) with permission.

It is unclear whether cutaneous vasoconstrictor neurons in humans are reflexly inhibited during noxious stimulation of the skin. This may be due to the fact that the excitation of cutaneous vasoconstrictor neurons in conscious human is mediated by the telencephalon, which overrides the inhibition mediated by circuits lower in the neuraxis (see below). However, painful intraneural electrical microstimulation in the superficial peroneal nerve at the ankle, which probably excites thinly myelinated nociceptive afferents, elicits reflex dilation (increased blood flow) in skin areas lying adjacent to, as well as in, the territory of the stimulated nerve (Blumberg and Wallin 1987).[2] The dilation is larger in the skin of the stimulated hindlimb than in the skin of the hindlimb contralateral to the noxious stimulus. The dilation in the ipsi- and contralateral skin is abolished by local anesthesia of the nerve proximal to the stimulation site and the dilation is enhanced by body cooling (i.e., when the activity in cutaneous vasoconstrictor neurons is high). Thus, this reflex in human beings appears to be very similar to the inhibitory reflex in cutaneous vasoconstrictor neurons elicited by cutaneous noxious stimuli in anesthetized cats and rats (Figure 4.9; see above and also Figure 4.4a_1, b_1).

Figure 4.9 Reflex vasodilation in hairy skin generated by intraneural microstimulation of afferent Aδ-fibers in the superficial peroneal (SP) nerve in an awake human being. The stimulation parameters were 1.1 V, 0.2 ms pulse duration and 5 Hz for 1 min. Blood flow was measured by laser Doppler flowmetry in the innervation territory of the stimulated SP nerve (territory a), in the innervation territory of the contralateral SP nerve (territory c), and outside of the SP nerve innervation territory of the contralateral foot (territory b). Electrical intraneural stimulation elicited vasodilation in the three skin territories (b, d). Local anesthesia (l.a.) of the contralateral SP nerve prevented the vasodilation in territory c but not in territory b (c, d). Local anesthesia of the ipsilateral SP nerve proximal to the stimulation electrode prevented the contra- and ipsilateral vasodilation (e). Possible mechanisms underlying this reflex vasodilation are inhibition of activity in cutaneous vasoconstrictor neurons (a) or activation of cutaneous vasodilator neurons. DH, dorsal horn; IML, intermediolateral nucleus. Modified from Blumberg and Wallin (1987) with permission.

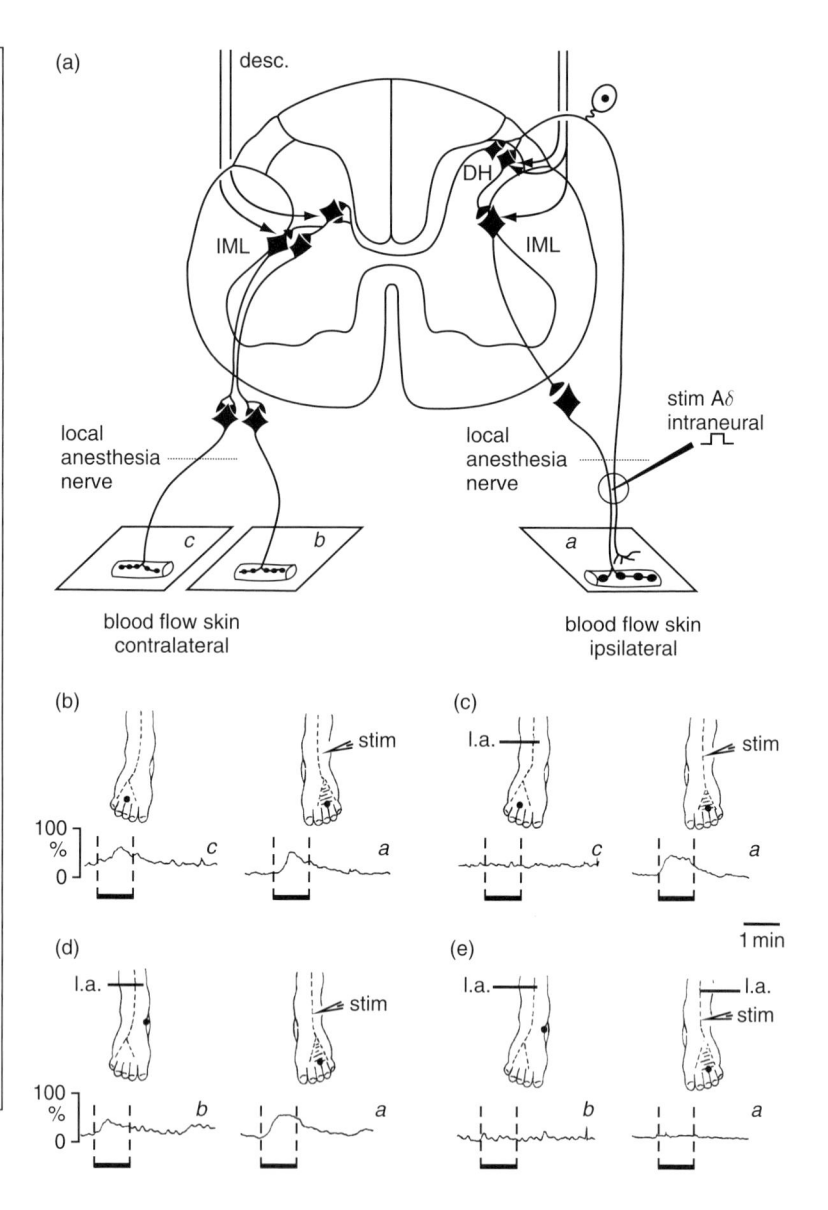

Cutaneous vasoconstrictor neurons in human beings are inhibited during stimulation of trigeminal receptors by immersion of the face in water (Fagius and Sundlöf 1986) and during hypoglycemia (Berne and Fagius 1986). The same stimuli simultaneously excite muscle vasoconstrictor neurons. Arousal, emotional stimuli, deep breaths and hyperventilation do activate cutaneous vasoconstrictor neurons in human beings (Figure 4.8c).[3] This is typical in the conscious subject and largely depends on activity in the cerebral cortex and limbic system. Thus, excitatory and inhibitory reflexes to somatic and visceral stimuli evoked, in the anesthetized cat, in cutaneous vasoconstrictor neurons that are organized at the spinal cord level are probably masked in conscious human beings.

The data from human subjects show that cutaneous vasoconstrictor neurons are very similar in their discharge patterns to those in the anesthetized cat. The differences probably result because any change of mental state in human subjects (which certainly occurs during noxious or other uncomfortable stimuli) is reflected in their cutaneous vasoconstrictor activity as it is in the sudomotor system in human beings (see Subchapter 4.2).

In *summary*, the behavior of muscle and cutaneous vasoconstrictor neurons is at least as distinctive in human beings as in anesthetized animals, indicating that the activity in the two populations of neurons is regulated by different central mechanisms. It is also important to remember that cutaneous vasoconstrictor neurons are likely to include functionally heterogeneous subtypes that are related to the type of cutaneous blood vessel and to the type of skin area (e.g. distal hairless [acral] skin and proximal hairy skin) they innervate.

4.2 | Sympathetic non-vasoconstrictor pathways innervating somatic tissues

Skin and deep somatic tissues are also innervated by sympathetic neurons that have other functions than vasoconstrictor neurons. So far, sudomotor neurons, pilomotor neurons and vasodilator neurons supplying target organs in skin or skeletal muscle have been studied. Some functional properties of these types of sympathetic neurons in the cat are listed in Table 4.3. Functional properties of neurons innervating the skin of human beings are listed in Table 4.2. From physiological studies, it is evident that other types of neurons must also exist, for example pupillodilator neurons, neurons supplying the pineal gland, sympathetic neurons supplying fat cells, etc. These neurons are likely to be activated by quite distinct central control mechanisms (see Subchapter 4.4).

4.2.1 Sympathetic postganglionic neurons innervating skin

Sudomotor neurons

STUDIES ON ANIMALS

Postganglionic cholinergic sudomotor neurons innervate eccrine sweat glands in the hairless skin of the paw pads of the *cat* and other mammals (for review of autonomic regulation of eccrine sweat glands see Gibbins [1997]). In the cat, sweat glands are confined to the hairless skin and do not appear to be involved in thermoregulation (Grewe *et al.* 1995).

Sudomotor neurons have a low rate of spontaneous activity (up to 0.4 Hz under chloralose anesthesia; Figure 4.10c), or they are silent. Activity in sudomotor neurons generates changes of potential on the surface of the hairless skin (Figures 4.10, 4.11). The spontaneous and

Table 4.3 | *Functional classification of sympathetic non-vasoconstrictor neurons in the cat based on reflex behavior in vivo*

Likely function	Location	Target organ	Likely target tissue	Major identifying stimulus	Ongoing activity
Muscle vasodilator	Lumbar	Hindlimb muscles	Muscle arteries, feeding vessels	Hypothalamic stim., emotional stim.	No
Cutaneous vasodilator	Lumbar	Hindlimb skin	Skin blood vessels	Excited by CNS warming	No
Sudomotor	Lumbar	Paw pads	Eccrine sweat glands	Vibration (in cat)	Yes
Pilomotor	Lumbar	Skin tail, back	Piloerector muscles	Hypothalamic stim., emotional stim.	No
Inspiratory	Cervical	Airways?	Nasal mucosal vasculature	Inspiration	Yes
Pupillo-motor	Cervical	Iris	Dilator pupillae muscle	Inhibition by light	Yes
Motility-regulating					
Type 1	Lumbar splanchnic	Hindgut, urinary tract	Visceral smooth muscle	Bladder distension	Yes
Type 2	Lumbar splanchnic	Hindgut, urinary tract	Visceral smooth muscle	Inhibited by bladder distension	Yes
Reproduction	Lumbar splanchnic	Reproductive organs	Visceral smooth muscle	Central stim (?)	No

Notes:
For details see Footnote of Table 4.1.
Modified from Jänig and Häbler (1999).

evoked skin potential changes can be as large as −30 mV in amplitude and are blocked by atropine, indicating that they are generated by muscarinic action of acetylcholine released by the sudomotor axons. The potential changes are generated by synaptic activation of secretory cells and most likely extracellularly recorded secretory potentials. They are negative with respect to the extraglandular tissue, because synaptic activation of the glandular cells by the cholinergic sudomotor fibers increases the potassium conductance of the extraluminal membrane, leading to hyperpolarization of the extraluminal membrane.[4] Discharges in single sudomotor axons innervating the cat paw are followed by transient skin potentials, at constant latencies of about 600 to 800 ms (see large extracellularly recorded action potentials SM in Figure 4.10b). Simultaneously recorded activity in cutaneous vasoconstrictor neurons innervating the same hairless skin area is not followed by transient skin potential changes (see small extracellularly recorded action potentials CVC in Figure 4.10c). The observation that single action potentials in individual postganglionic sudomotor neurons are followed by fast transient skin potential changes (Figure 4.10b, c) indicates that many sudomotor neurons must discharge synchronously. This can be explained by a large divergence of preganglionic sudomotor axons in paravertebral ganglia (Chapter 6) and/or by central synchronizing processes.

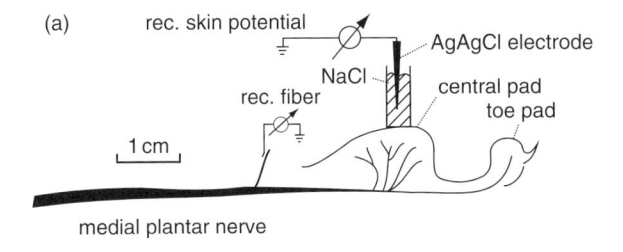

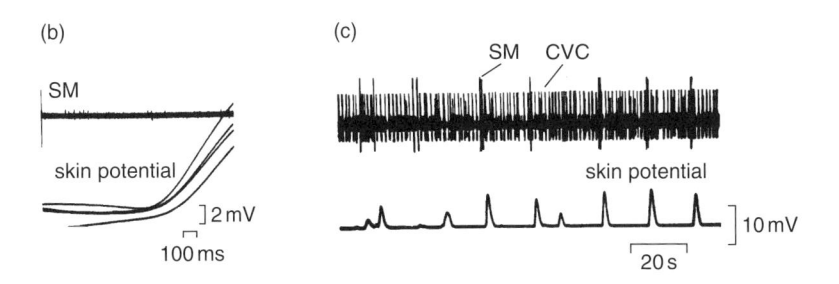

Figure 4.10 Activity in postganglionic sudomotor (SM) and cutaneous vasoconstrictor (CVC) axons innervating hairless skin in the cat. Recording of activity in filaments isolated from fascicles of the medial plantar nerve innervating the hairless skin of the hindpaw; lower records, skin potential (negative transient potential changes). (a) Experimental setup. Skin potential (direct current) was recorded from the surface of the central pad with AgAgCl electrodes via a fluid bridge of physiological saline. (b) Ongoing discharges in an SM axon are followed by deflections of the skin potential at about 600 ms latency (4 traces superimposed). (c) Ongoing activity in a single SM axon and several CVC axons and skin potential. Most SM discharges are followed by transient skin potential changes; activity in CVC neurons is not correlated with the skin potential. Modified from Jänig and Kümmel (1977) with permission.

The most unique stimulus that leads to reflex activation of sudomotor neurons in the cat hindpaw is vibration that excites Pacinian corpuscles (largely in the hindpaw) (Figure 4.11). Individual deflections of the skin potential elicited during this type of afferent stimulation may be preceded by individual sudomotor discharges. This excitatory reflex is a definitive functional marker for sudomotor neurons in the *cat*: no other type of sympathetic neuron can be activated in this way and stimulation of other cutaneous mechanoreceptors with large-diameter Aβ-fibers (e.g., from hair follicles) does not lead to reflex activation of sudomotor neurons. The discharge rate of sudomotor neurons during vibration stimuli, which elicit large skin potentials of 10 to 20 mV and ranges from 0.2 to 2.1 Hz, also implies that the sudomotor neurons work in a low-frequency range under physiological conditions. Interestingly, in some experiments, cutaneous vasoconstrictor neurons innervating the same hairless skin were inhibited during the vibration stimulation (Figure 4.11a; Jänig and Kümmel 1977; Jänig and Räth 1977; see Figure 9.5b$_2$).

Excitatory reflexes are also elicited in sudomotor neurons by noxious cutaneous stimuli (Figure 4.11b), by stimulation of visceral receptors in the pelvic organs (e.g., by distension and contraction of the urinary bladder [Häbler *et al.* 1992]) and by stimulation of arterial chemoreceptors (Figure 4.11c; Jänig and Kümmel 1981). Stimulation

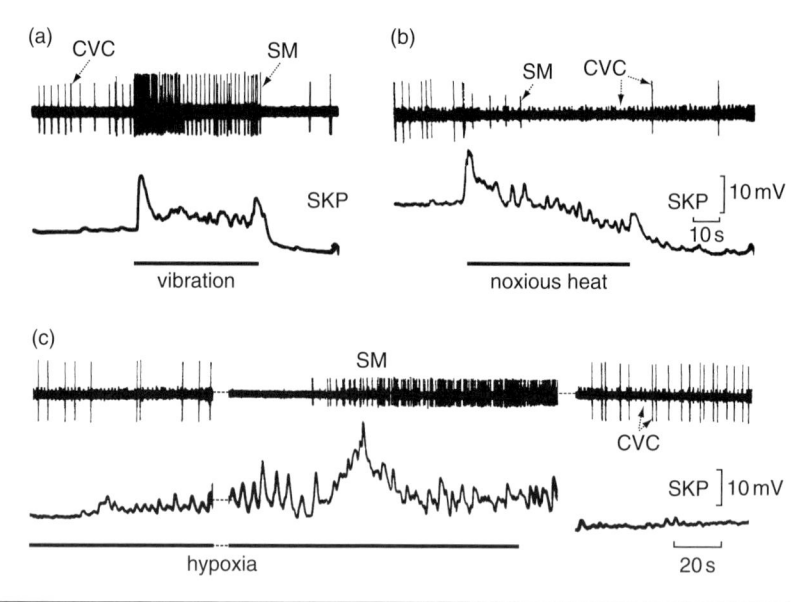

Figure 4.11 Reflex patterns, recorded simultaneously in the cat, in sudomotor (SM) neurons and cutaneous vasoconstrictor (CVC) neurons innervating hairless skin. Upper records, postganglionic activity; lower records, skin potential (SKP; negativity up). (a) Reflex activation of a single SM neuron and reflex inhibition of a single CVC neuron during a vibration stimulus (stimulation of Pacinian corpuscles in the paws). (b) Excitation of a single SM neuron and inhibition of CVC neurons (large signal: single CVC axon; small signals: several CVC axons) to noxious stimulation of a toe of the ipsilateral hindpaw by radiant heat (55 °C). For the slow positivation of skin potential after activation of the SM neurons see note 4. (c) Excitation of a single SM neuron and inhibition of CVC neurons during stimulation of arterial chemoreceptors by ventilating the cat with a gas mixture of 8% O$_2$ in N$_2$ (hypoxic ventilation started about 1 min before the record). Same preparation as in (b). Modified from Jänig and Kümmel (1977) with permission.

of arterial baroreceptors has no effect on the activity of sudomotor neurons. Warming the hypothalamus or the spinal cord does not activate sudomotor neurons in the cat (Grewe et al. 1995). Activity in sudomotor neurons is modulated by central respiration (Boczek-Funcke et al. 1992c; Häbler et al. 1994b). The pattern of this respiratory modulation is different from that in muscle and most cutaneous vasoconstrictor neurons, indicating a distinct coupling between the respiratory network in the medulla oblongata and presympathetic sudomotor neurons (see Subchapter 10.6).

The reflex pattern in sudomotor neurons is organized reciprocally to that in the cutaneous vasoconstrictor neurons innervating the cat paw: sudomotor neurons are excited by afferent stimuli that inhibit cutaneous vasoconstrictor neurons. For example, the nerve fiber bundle recorded from in Figure 4.11b, c contained one sudomotor axon (*SM*) and two cutaneous vasoconstrictor axons (*CVC*) innervating the hairless skin of the cat hindpaw. Noxious heat stimulation of skin and stimulation of arterial chemoreceptors by hypoxia inhibited the activity in the cutaneous vasoconstrictor neurons and excited the sudomotor neuron. This reciprocal reflex pattern is largely based on spinal circuits (Jänig and Kümmel 1981; Jänig 1996a; see Subchapter 9.2). The consequence of the reciprocal organization is that neural activation of sweat glands is accompanied by increased blood flow through the skin and probably subcutaneous tissue.

The function of the "vibration reflex" in the sudomotor neurons is to hydrate the epithelium of the skin in order to keep the surface of the hairless skin soft and flexible, by secretion of sweat, which diffuses into the epithelium. This increases friction during contact and creates optimal conditions for somatosensory discrimination since any contact of the paws with surfaces during movements and manipulations always excites the Pacinian corpuscles in the paw skin. I presume that this reflex activation of sweat glands could be an ideal way to mark territory in felines and it has been speculated that the secretion is important for scent marking of walking trails (Matthews 1969).

STUDIES ON HUMAN BEINGS

In contrast to the cat and many other hairy mammals (not horse), sweat glands are present in skin all over the body surface in the *human being*. Sudomotor neurons in human subjects are active at high ambient temperatures (Figure 4.8b) and silent at low ambient temperatures (Figure 4.8a). Some sudomotor neurons show rhythmic discharges with the arterial pressure wave at high temperatures (Bini et al. 1981). This may be because of rhythmic activation of low-pressure cardiovascular receptors: it has been shown that low body negative pressure and head tilt up leads to a reduction of activity in sudomotor neurons (Dodt et al. 1995). Sudomotor neurons supplying sweat glands in proximal hairy skin of the extremities have a lower threshold for activation in response to warming of the body surface than sudomotor neurons supplying hairless skin, which are only activated at relatively high and unpleasant ambient temperatures. This suggests that sudomotor activity dominates the sympathetic activity in the skin nerves to, for example, the proximal forearm skin, because there is almost no activity in cutaneous vasoconstrictor neurons at ambient temperatures whereas cutaneous vasoconstrictor activity dominates in nerves to the glabrous skin of the hand (Bini et al. 1980b). This may indicate that the primary function of sudomotor neurons destined to human glabrous skin[5] is again to keep the skin flexible for optimal sensory discrimination. This function is fully consistent with the function of the vibration reflex in sudomotor neurons in the cat.

Like cutaneous vasoconstrictor neurons, subsets of sudomotor neurons are activated by arousal and mental (emotional) stimuli as well as by deep breathing (Figure 4.8c; Bini et al. 1980a); these include sudomotor pathways to the glabrous skin of hands and feet, to the armpits and to some parts of the face. The simultaneous activation of both sympathetic pathways to skin of the extremities is initiated from the forebrain and shows that the neural circuits in the neuraxis that are responsible for the reciprocal activation of both systems are inhibited by the central signals initiated in the forebrain.

Activity of the sudomotor system is monitored by measuring electrodermal activity (either change of skin potential or change of skin resistance upon activation of the sudomotor neurons). Electrodermal activity has been used for more than 100 years by psychophysiologists to monitor intrapsychic processes (Boucsein 1992) and is the basis of the lie detector test (Lykken 1998). Any change in mental state in human subjects (e.g., generated by arousal or emotional stimuli) will activate these subsets of the sudomotor system,[6] depending on the thermal condition of the body (Bini et al. 1980a).

Pilomotor neurons

Piloerector muscles exist in the hairy skin of the *cat* all over the body surface. These smooth muscles are strongly developed in the hairy skin of the tail, back, dorsal part of the upper hindleg and dorsal part of the head, but not in other parts of hairy skin (Strickland and Calhoun 1963). Repetitive electrical stimulation of the sympathetic chain is followed by piloerection on the tail, back and dorsal parts of the head and weak piloerection on the dorsal part of the hindlimb, but no piloerection in other parts of the hairy skin (Langley and Sherrington 1891; Langley 1894a). Piloerection with this distribution can be observed in the cat and other mammals during different behavioral states (e.g., defense behavior). It is unclear whether only piloerector muscles in those parts of the hairy skin that exhibit piloerection on nerve stimulation are innervated or whether the putative innervation of the small piloerector muscles is functionally ineffective to elicit piloerection.

The pilomotor neurons supplying the tail are silent in anesthetized and probably also in awake cats under thermoneutral and emotionally neutral conditions. In the anesthetized cat the central nervous pilomotor circuits cannot be activated by any natural stimulus except asphyxia (Grosse and Jänig 1976). It is unknown whether pilomotor neurons are activated by body cooling. It could well be that piloerection observed in cats in cold environments is produced by an increase in sensitivity of erector pili muscles to circulating adrenaline and noradrenaline (Hellmann 1963). It is likely that pilomotor neurons are activated specifically during certain species-specific behavior. They may be controlled predominantly by the hypothalamus and the limbic system and normally not influenced by stimuli applied to the body surface (Figure 4.12). However in cats, in which cortex and large parts of the limbic system have been removed ("hypothalamic cats"), noxious and non-noxious stimuli may lead to piloerection (Bard 1928; Bard and Rioch 1937; Ectors 1941; Bard and Macht 1957).

Pilomotor neurons also innervate piloerector muscles in *human beings*. They are activated during exposure to a cold environment during central cooling, fever, strong emotional stimuli and paradoxical cold sensations (e.g., during a hot bath after strong exercise). All these stimuli may lead to piloerection and to "goose bumps" of the skin.

Vasodilator neurons supplying skin

STUDIES ON ANIMALS
When noradrenergic neuroeffector transmission to blood vessels is blocked by guanethidine,[7] electrical stimulation of preganglionic axons in the lumbar sympathetic trunk leads to vasodilation in both hairless and hairy skin of the cat paw. This vasodilation is resistant to atropine (an antagonist of muscarinic cholinergic transmission, which blocks sweat gland activation) but abolished by blockade of impulse transmission in autonomic ganglia with hexamethonium[8] (Bell *et al.* 1985). Thus, blood vessels of the cat paw are supplied by cutaneous vasodilator neurons that are neither noradrenergic nor cholinergic.[9]

Using a thermoregulatory paradigm (warming of the spinal cord or hypothalamus) and assuming that sympathetic cutaneous vasodilator neurons are involved in thermoregulation, an attempt was made to recognize putative cutaneous vasodilator neurons directly which project to the skin of the cat hindlimb. Warming of spinal cord and/or hypothalamus inhibits activity in cutaneous vasoconstrictor neurons and increases blood flow through skin and therefore increases skin temperature. Under the same conditions, some normally silent unmyelinated fibers are activated in nerves to hairless and hairy skin (Gregor *et al.* 1976; Jänig and Kümmel 1981; Grewe *et al.* 1995). The responses elicited by spinal cord warming are graded. Interestingly, the neurons exhibiting these responses to spinal cord warming did not project through the sympathetic chain. They could neither be activated by electrical stimulation of the sympathetic chain nor could their activity, generated by spinal cord warming, be prevented by blockade of synaptic transmission in sympathetic ganglia using hexamethonium. This indicates that these neurons activated by spinal cord warming are either not sympathetic postganglionic or that the impulse transmission through the sympathetic ganglia is not nicotinic for this system (Jänig and Kümmel 1981). I hypothesize that we have recorded antidromically conducted impulses in unmyelinated axons of spinal afferent neurons and that their cell bodies in the dorsal root ganglia or the axons projecting in the dorsal roots were excited by central warming. If these afferent neurons were peptidergic they could generate arteriolar vasodilation and increased blood flow in the skin (see Häbler et al. 1997b). Recently it has been suggested that this neurogenic mechanism may contribute to inflammation (Willis 1999). An afferent-induced vasodilation of this type could be a protective mechanism against central overheating.

STUDIES ON HUMAN BEINGS

In human subjects, body heating increases blood flow through the skin, which is closely paralleled by increased sweating. Blocking conduction in the nerves to the forearm and hand abolishes sweating at the peak of the vasodilator response in the corresponding skin areas. It also blocks most of the vasodilation in the forearm skin but it does *not* block the vasodilation in the hand (Roddie et al. 1957). This observation suggests that increased blood flow through acral skin (hand and foot) occurs due to release from activity in cutaneous vasoconstrictor neurons. In contrast, increased blood flow through the forearm (and other) hairy skin is elicited largely by *active vasodilation*. Several arguments support the idea that increased blood flow in human hairy skin during heat load is generated by sympathetically mediated vasodilation mechanisms (Joyner and Halliwill 2000): (1) In patients with a surgically sympathectomized extremity vasodilation can no longer be produced by whole-body heating (Roddie et al. 1957). (2) In patients with autonomic neuropathy[10] whole-body-heating-induced vasodilation is absent. (3) The magnitude of vasodilation in skin during increase of body core temperature is far greater than that achieved by removal of vasoconstrictor activity (Roddie et al. 1957; Kellogg et al. 1989). (4) Atropine (which abolishes sweat production by blocking muscarinic cholinergic transmission) does not abolish the vasodilation during heat load but only weakly attenuates it (Roddie et al. 1957; Kellogg et al. 1995). (5) Local injection of botulinum toxin into the skin (which prejunctionally blocks the fusion of the secretory vesicles in the cholinergic terminals with the plasma membrane and therefore prevents the release of acetylcholine) prevents active vasodilation in skin as well as sweating during heat load (Kellogg et al. 1995). These results would argue that the vasodilation is dependent on the activation of the sweat glands. (6) Cutaneous vasoconstrictor neurons supplying forearm skin are nearly silent at neutral ambient temperatures (Bini et al. 1980b).

These points argue that active vasodilation in human hairy skin is not directly generated by a cholinergic mechanism but that cholinergic (possibly sudomotor) neurons may be indirectly involved. Thus, whether the active vasodilation seen in proximal skin of human extremities during heat load is generated by a distinct population of cutaneous vasodilator neurons or is generated in association with the activation of sweat glands by sudomotor neurons is unclear. The transmitters released by the postganglionic neurons or by nonneural cells responsible for the vasodilation are still a matter of debate (see Rowell [1993]; for discussion and references see Jänig [1985a, 1990a]; Joyner and Halliwill [2000]).

Evidence that human skin is innervated by cutaneous vasodilator neurons is largely indirect because direct recordings from putative postganglionic vasodilator axons with microneurography in skin nerves of human subjects have not been published (Bini et al. 1980a; Wallin 1990; Wallin et al. 1998). Specifically during non-REM (rapid eye movement) sleep in human subjects single skin sympathetic activity bursts and short periods of increased skin sympathetic activity are sometimes followed by increased skin blood flow with and without skin resistance changes (which indicates activity in sudomotor fibers) (Noll et al. 1994). During mild body heating bursts of skin sympathetic activity in the superficial peroneal nerve is mostly followed by sweat expulsion and vasodilation in skin and very rarely by vasodilation only (Sugenoya et al. 1998). Furthermore, Nordin (1990) has shown that sympathetic multiunit activity, occurring in bursts in the supraorbital nerve, elicited by mental stress, arousal or whole-body warming are followed by vasodilation in the forehead skin and decreased skin resistance (activation of sweat glands). It is unclear whether the cutaneous vasodilation is generated by activation of cutaneous vasodilator neurons or mediated via activation of sweat glands. The results of both groups indicate indirectly that cutaneous blood vessels of the extremity and forehead in humans may be innervated by sympathetic vasodilator neurons and that these neurons may be activated under various circumstances.

There is indirect evidence that parasympathetic vasodilator neurons innervate the skin of face and mucosa of humans (Drummond 1995) and rats and cats (Izumi 1999). However, this issue is rather controversial and unsolved (for discussion see Jänig 1985a, 1990a; Rowell [1993]).

Sympathetic postganglionic neurons supplying sensory receptors in skin

Some cutaneous mechanoreceptors with myelinated afferents seem to be supplied by unmyelinated fibers. It is believed that this innervation is sympathetic postganglionic. Studies using catecholamine fluorescence have shown that catecholaminergic axons are present but, in most cases, direct proof is missing (see Akoev [1981]). By the same token, electrical stimulation of the sympathetic supply at unphysiological frequencies may excite primary afferent fibers in normal somatic tissues. However, there is no functional evidence that a distinct class of sympathetic neuron innervates cutaneous sensory receptors and regulates their sensitivity under physiological conditions (for extensive discussion see Jänig and Koltzenburg [1991a]).[11]

The situation may be different under pathophysiological conditions after nerve lesions, in which case stimulation of sympathetic neurons may lead to excitation and sensitization of somatic afferent terminals (see Jänig and McLachlan [1994]; Jänig *et al.* [1996]; Jänig and Häbler [2000b]; Jänig and Baron [2001, 2002, 2003]; Jänig [2002]).

4.2.2 Sympathetic postganglionic neurons innervating deep somatic structures: skeletal muscle, joints, bone, etc

All deep somatic structures are innervated by sympathetic noradrenergic postganglionic neurons. It is commonly believed that these are vasoconstrictor in function and that their reflex patterns are similar to those of muscle vasoconstrictor neurons. However (1) there is no evidence that the reflex patterns in these neurons are identical to those of muscle vasoconstrictor neurons and (2) it is unclear whether all sympathetic postganglionic neurons projecting to these tissues have vasoconstrictor function. Furthermore, there is evidence that the bone is innervated by sympathetic neurons that contain vasoactive intestinal peptide (VIP) and are presumably not noradrenergic. This sympathetic innervation has no vascular function and is probably involved in bone mineralization (Hohmann *et al.* 1986). Here I will only concentrate on muscle vasodilator neurons and on sympathetic postganglionic neurons that appear to supply skeletal muscle cells and muscle spindles.

Vasodilator neurons supplying skeletal muscle

STUDIES ON ANIMALS

Integral components of the defense reaction in animals are increased blood flow through skeletal muscle with concomitant decreased blood flow through viscera and skin, increased blood pressure and heart rate, pupil dilation, sweat gland activation, activation of piloerector muscles and activation of the adrenal medulla and adrenal cortex (see Subchapter 11.3). This type of behavior is organized at the hypothalamic and mesencephalic levels and under the control of the limbic system and cortex. The defense reaction has puzzled and fascinated physiologists and psychologists since Darwin published his famous book *The Expression of the Emotions in Man and Animals* (Darwin 1872). Figure 4.12 illustrates a cat in rage, as originally shown in Darwin's book, which should have all the autonomic responses mentioned above. Today we know that this behavior consists of the behaviors confrontational defense, flight and quiescence, which are differentiated according to their motor, autonomic, neuroendocrine and endogenously generated analgesic responses. They are organized in rostrocaudal cell columns of the dorsolateral and ventrolateral mesencephalon. These defensive behaviors are displayed by animals (and probably human beings) under stress and

Figure 4.12 A cat in rage. Note the piloerection and dilation of the pupil. Muscle vasodilator neurons innervating the left extremity that is going to strike are activated during this behavioral state. From Darwin (1872) with permission.

pain and are archetypal integrated protective responses of the body (Subchapter 11.3; Bandler *et al.* 1991, Bandler and Shipley 1994; Bandler and Keay 1996; see also Akert [1981]; Jänig [1995b]).

When the defense reaction is elicited by "emotional stimuli" in an awake cat (e.g., by confrontation with a dog or another cat), an initial rapid increase in blood flow occurs in the skeletal muscle of the limb that strikes or is going to strike (Mancia *et al.* 1972; Ellison and Zanchetti 1973). This vasodilation is blocked by atropine, suggesting that it is mediated by cholinergic sympathetic fibers. Repetitive electrical stimulation of the hypothalamic defense area in anesthetized cats also elicits an atropine-sensitive vasodilation in skeletal muscles of the hindlimbs in conjunction with the other characteristic cardiovascular responses (Eliasson *et al.* 1951; Uvnäs 1954, 1960; for references see Jänig [1985a]).

During atropine-sensitive vasodilation in the hindlimb of anesthetized cats, elicited by local electrical stimulation of the hypothalamic defense area (perifornical area), about 10% of the postganglionic neurons supplying skeletal muscle which probably have vasodilator functions are activated (Figure 4.13b right side). Electrical stimulation

Figure 4.13 Reaction of a postganglionic muscle vasodilator neuron to electrical stimulation of the hypothalamus inducing atropine-sensitive vasodilation (right) or vasoconstriction (left) in the hindlimbs of a cat. Postganglionic axon isolated from the gastrocnemius-soleus nerve. Blood flow was measured 1 cm proximal to the aortic bifurcation using an electromagnetic flow probe. (a) Stimulation of the preganglionic axons in the lumbar sympathetic trunk (LST) between paravertebral ganglia L4 and L5 with single pulses (5 V, 0.2 ms pulse duration) excited the postganglionic neuron with three action potentials because of synaptic impulse transmission in a lumbar paravertebral ganglion (see Figure 3.4). (b) Upper: Neural activity before, during (middle traces) and after bipolar stimulation at two hypothalamic sites close-by generating vasoconstriction (left) or vasodilation (right, hypothalamic defense area). Stimulation parameters, 8 to 10 V (pulse duration 1 ms) at 100 Hz for 5 s. (b) Lower: Measurement of arterial blood flow (flow, left ordinate scale) and blood pressure (BP, right ordinate scale). Change of blood flow resistance indicated by numbers. Modified from Horeyseck et al. (1976) with permission.

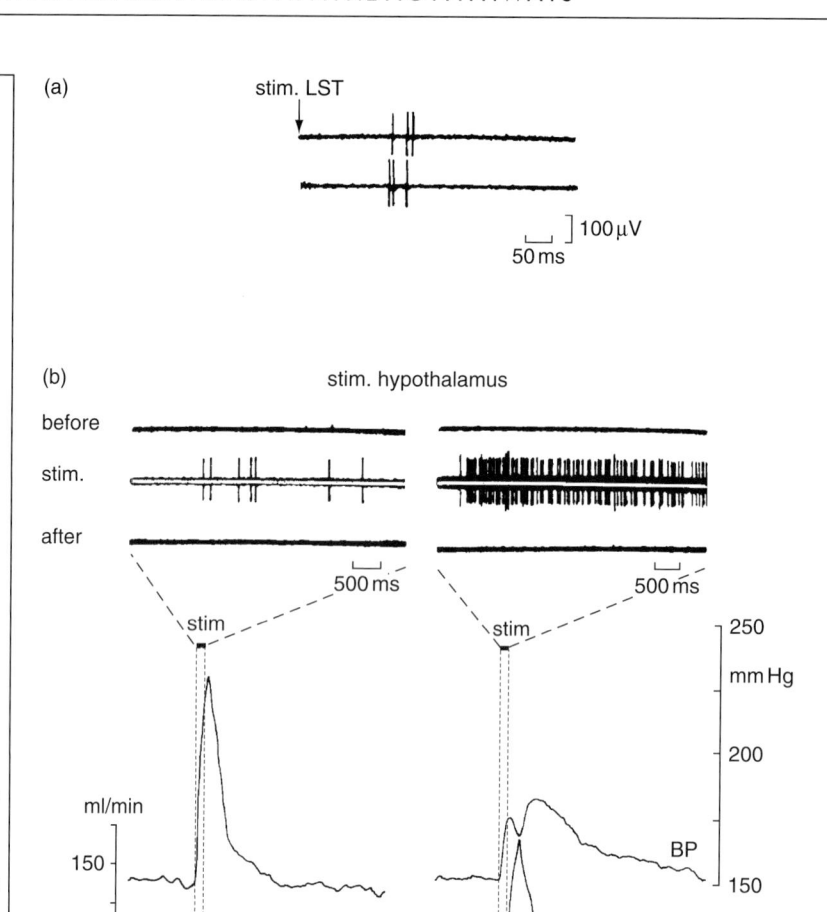

of a nearby hypothalamic site that leads to vasoconstriction in the skeletal muscle does not or only weakly activates the putative vasodilator neurons (Figure 4.13b left side) but activates vasoconstrictor neurons. These vasodilator neurons are not spontaneously active and do not exhibit reflex activation in response to stimulation of somatic and cardiovascular afferents. They innervate only large resistance vessels in skeletal muscle upstream of the small resistance vessels, which are densely innervated by vasoconstrictor fibers (Bolme and Fuxe 1970; Anderson et al. 1996; Grasby et al. 1997). Thus, by acting at this strategically favorable vascular site, activation of only relatively few sympathetic cholinergic postganglionic vasodilator neurons will generate a rapid increase in blood flow through the vascular bed. Cutaneous vasoconstrictor neurons are activated (leading to decreased blood flow through skin) and muscle vasoconstrictor neurons are inhibited during active vasodilation in skeletal muscle (Horeyseck et al. 1976). Therefore it is believed that the cholinergically mediated vasodilation in skeletal muscle of the cat is brought about by activation of a specific sympathetic muscle vasodilator pathway. Although all animal species display typical defense behaviors, only some of them seem to have a muscle vasodilator system (e.g., cat and dog, Bolme et al. 1970).

STUDIES ON HUMAN BEINGS

It is controversial whether the muscle vasodilator system exists in human beings. During mental stress, blood flow through skeletal muscle of the forearms increases rapidly. This increase in blood flow is absent after block of the stellate ganglion with a local anesthetic or after sympathectomy. Thus, it is neurogenically mediated. However, it is difficult to explain the blood flow increase by withdrawal of activity in muscle vasoconstrictor neurons because it remains when adrenoceptors are blocked; furthermore cholinoceptor-blocking drugs have ambiguous effects (Blair et al. 1959; Barcroft et al. 1960; Anderson et al. 1987; see Greenfield [1966]; Roddie [1977]). It has been shown that increased blood flow through skeletal muscle of the human forearm during mental stress is reduced by atropine as well as by a nitric oxide (NO) synthase inhibitor (Dietz et al. 1994). However, it remains entirely unclear whether a neuronally mediated active mechanism exists in human skeletal muscle, which is responsible for increase of blood flow through skeletal muscle during mental stress. Recent evidence suggests that this vascular dilation is generated by circulating adrenaline released by the adrenal medulla acting via β_2-adrenoceptors on vascular smooth muscles and endothelial cells (here leading to release of NO) (see Joyner and Halliwill [2000]; Joyner and Dietz [2003]).

Sympathetic postganglionic neurons supplying muscle spindles

The number of postganglionic neurons innervating non-vascular structures in skeletal muscle is very small when compared to the number of vasomotor neurons. Some muscle spindles (but no tendon organs) are anatomically associated with postganglionic axons (Barker and Saito 1981; see Akoev [1981] for literature). By the same token, occasional postganglionic fibers seem to have terminals near skeletal muscle fibers.

Electrical stimulation of the sympathetic supply of skeletal muscle may activate some Ia-fibers from muscle spindles,[12] but not the Ib-fibers from tendon organs. This activation is usually small and can only be elicited at stimulation frequencies that are supraphysiological (for discussion see Jänig and Koltzenburg [1991a]). In humans, resting discharge in muscle spindles is not modulated by increase of activity in sympathetic neurons to skeletal muscle (Macefield et al. 2003). Therefore it is debatable whether these few sympathetic fibers have any function or whether the sensitivity of muscle spindles and therefore muscle afferents can be changed by activity in sympathetic neurons.

4.3 | Sympathetic non-vasoconstrictor neurons innervating pelvic viscera and colon

In Subchapters 4.1 and 4.2 I have discussed the functional properties of sympathetic neurons innervating target cells in the skin and skeletal muscle and of visceral vasoconstrictor neurons. Here I will discuss the functional properties of sympathetic neurons that innervate viscera and have no vasoconstrictor function.

Viscera receive an abundant sympathetic innervation and most of these neurons are not involved in regulation of vascular resistance, but rather in regulation of motility and secretion/reabsorption of the gastrointestinal tract or regulation of motility and secretion in pelvic organs. Others are probably involved in regulation of immune function (e.g., spleen and Peyer's patches; see Subchapter 4.6) and possibly in other functions (Figures 5.13 and 5.15). The cell bodies of the postganglionic neurons of the visceral sympathetic pathways are largely located in the prevertebral ganglia. Some that are associated with pelvic organs are located in the pelvic ganglia (Keast 1995a, b, 1999). Using neurophysiological and morphological techniques, the sympathetic innervation of pelvic organs and distal colon is the most systematically studied part of this innervation in the cat. All the organ systems in the pelvic region have storage and evacuation as major functions. The lower urinary tract (urinary bladder and urethra) retains and evacuates urine (continence and micturition; Jänig 1996b; de Groat 2002). The distal part of the large bowel (distal colon, rectum, anus) stores and evacuates feces and also reabsorbs water and sodium ions. The internal reproductive organs and erectile tissues of the external reproductive organs are more specialized but have basically similar functions related to retention and transport of semen and ova, together with, in the female, implantation of fertilized ova, storage and development of the fetus, and subsequently birth of the new individual (Jänig and McLachlan 1987; Jänig 1996c see Subchapter 9.3). Finally a sympathetic vasodilator pathway innervates the erectile tissue of the reproductive organs. This pathway may duplicate the sacral parasympathetic vasodilator pathway. Its functioning under physiological conditions is unknown (see Subchapter 9.3; Jänig 1996c).

In the cat, the sympathetic innervation of pelvic organs and the colon has its main preganglionic representation in the lumbar spinal segments L3 to L5 (see Chapter 8, Figures 8.6 and 8.8). Even if there are vasoconstrictor axons, the sympathetic supply to the pelvic organs is primarily non-vascular. Activity in these neurons is involved in inhibition of motility and mucosal secretion, excitation of sphincteric muscles and excitation of smooth muscles of male reproductive organs. The instances in which these neurons are excited or inhibited at the same time as the vasoconstrictor neurons are rare. Many sympathetic neurons (both pre- and postganglionic) in nerves projecting to the pelvic viscera (e.g., through the lumbar splanchnic and hypogastric nerves) (Bahr *et al.* 1986a, b, c; Bartel *et al.* 1986; Jänig and McLachlan 1987; Jänig *et al.* 1991) have the following functional characteristics:

- They do not respond to stimulation of typical cardiovascular afferents (i.e., of arterial baroreceptors or arterial chemoreceptors).
- Most of them do not display respiratory rhythmicity in their activity (Boczek-Funcke *et al.* 1992c).
- They exhibit distinct reflexes in response to activation of *sacral* visceral afferents from the lower urinary tract, colon or anal canal by innocuous stimuli, which do not affect the visceral vasoconstrictor neurons.[13] These reflexes fall into the following categories: excitation or inhibition from the urinary bladder, from the colon and from the anal canal. Most of these neurons are spontaneously active (0.8 ± 0.7 imp/s for preganglionic neurons and 0.7 ± 0.5 imp/s for postganglionic neurons; Table 6.2; Bahr *et al.* 1986c; Jänig *et al.* 1991).
- At least three functional types of sympathetic non-vasoconstrictor neuron that innervate pelvic organs have been identified. These neurons have been called motility-regulating (MR) neurons although they may also be involved in secretory processes (see Table 4.2). Two types of MR neurons exhibit reciprocal reflex patterns to physiological stimulation of sacral afferents of the urinary bladder and colon. Type 1 MR neurons are excited during contraction and/or distension of the urinary bladder, and inhibited (or not affected) during contraction and/or distension of the colon (Figure 4.14a). Type 2 MR neurons are inhibited during contraction or distension of the urinary bladder and excited (or not affected) by contraction or distension of the colon (Figure 4.14b). There are other types of MR neuron, which are either only excited by stimulation of the mechanoreceptors in the anal canal, or not excited by any stimulus under standardized experimental conditions but have resting activity, or are entirely silent. These MR neurons have not been further characterized. A small group (about 3% to 4%) of pre- and postganglionic sympathetic visceral neurons projecting into the lumbar splanchnic and hypogastric nerves has functional properties of both visceral vasoconstrictor neurons and of MR neurons (Bahr *et al.* 1986a, b; Jänig *et al.* 1991).
- One of the most prominent and unique reflexes in sympathetic neurons are those generated during excitation of visceral afferents from the anal canal by mechanical shearing stimuli. These stimuli excite a specific class of myelinated low-threshold sacral afferent neuron (Bahns *et al.* 1987; Jänig and Koltzenburg 1991c). Physiological stimulation of these afferents for 10 to 20 seconds by mechanical shearing stimuli generates excitatory reflexes in

Figure 4.14 Reflexes in motility-regulating (MR) neurons in response to stimulation of the urinary bladder or colon in the cat. Recording from postganglionic axons isolated from the hypogastric nerve. (a) Two MR neurons type 1: activation during distension of the urinary bladder ([a_1], filling with 20 ml saline) and inhibition during distension/contraction of the colon ([a_2], filling with 20 ml saline); during filling of the colon (black bar) and afterwards the colon contracted. (b) Motility-regulating neuron type 2: activation during contraction of the colon (b_1) and inhibition during contraction of the urinary bladder (b_1, b_2). In (b_2) 20 contractions are superimposed. Ordinate scales, intraluminal pressure or impulse rate (upper records in [b_1] and [b_2]) (a), Jänig, Schmidt, Schnitzler and Wesselmann, unpublished; (b), modified from Jänig et al. (1991).

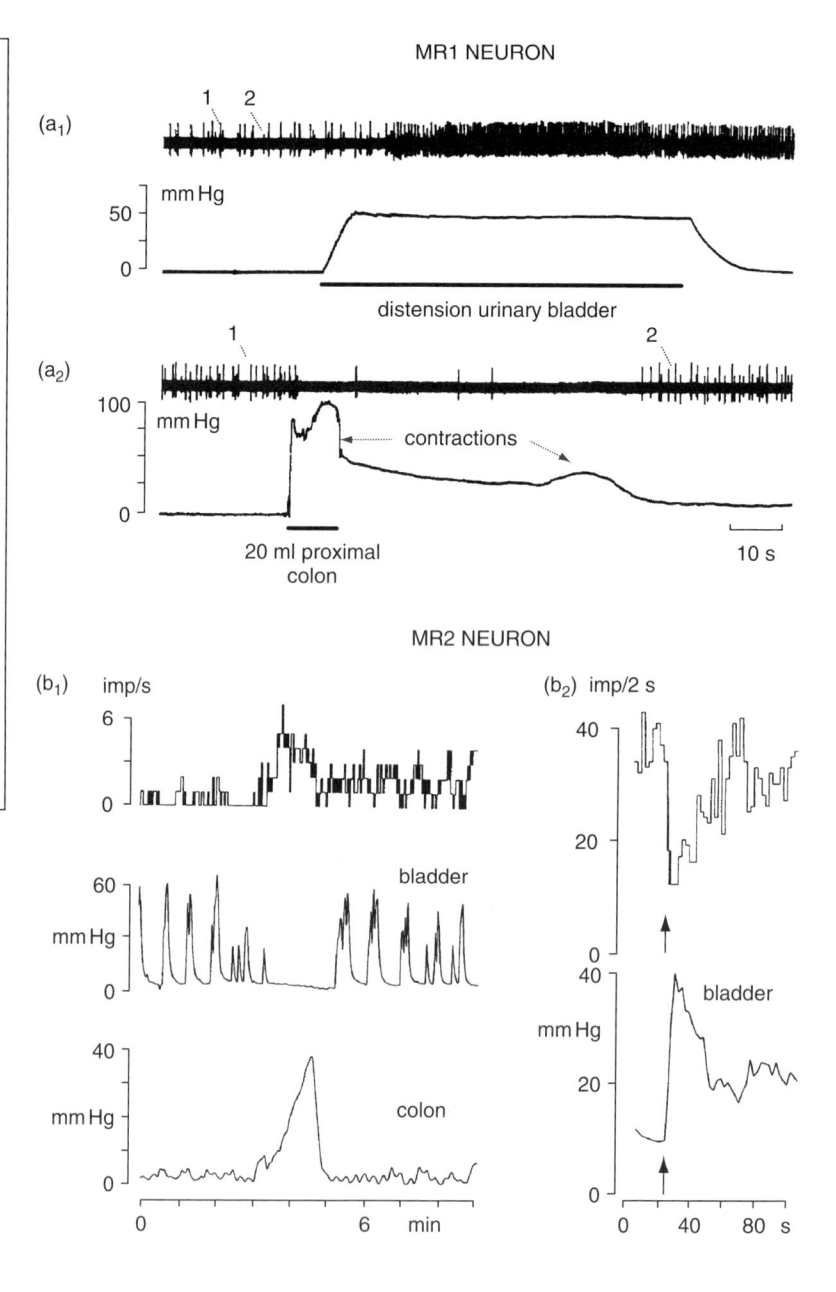

about 80% of the identifiable MR neurons (Figure 4.15) and inhibition in 8% of them. In 60% of these neurons the excitation is followed by afterdischarges lasting for 1 to 12 minutes (mean 4.8 minutes). This unique reflex activity can practically only be elicited from the anal canal, not from the colon-rectum and only rarely or not at all from the perianal hairy skin. The long-lasting afterdischarge is a property of the spinal reflex pathway and is not due to afterdischarges in the sacral visceral afferent neurons (Bahns et al. 1987; Jänig and Koltzenburg 1991c).

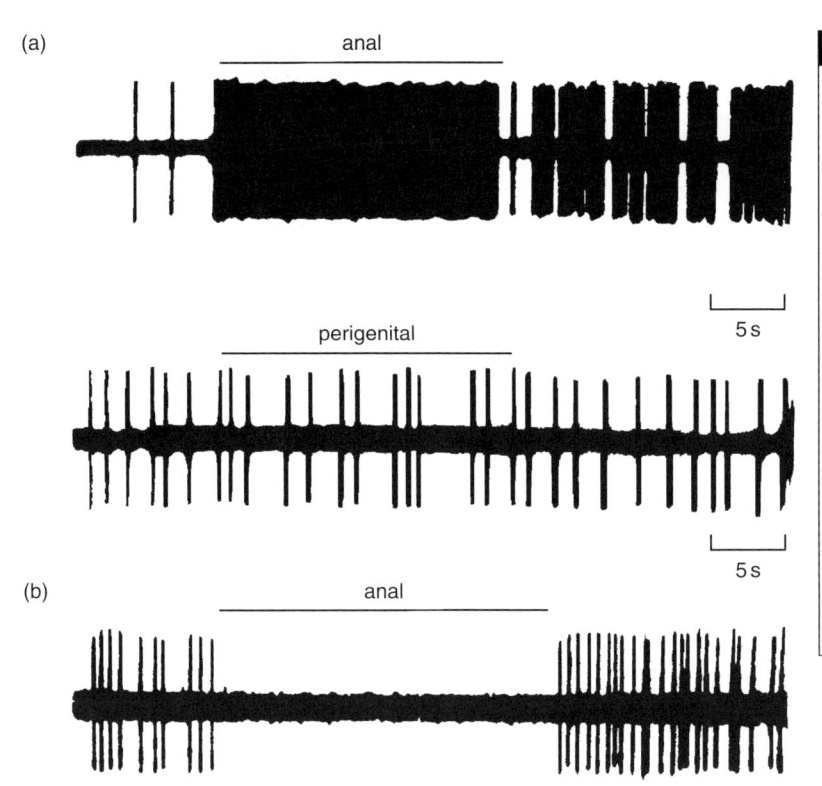

Figure 4.15 Reactions of preganglionic motility-regulating neurons in response to mechanical stimulation of the mucosa of the anal canal. Preganglionic axons were isolated from a lumbar splanchnic nerve in the cat. Anal stimulation consisted of a light shearing stimulus applied at a moving frequency of about 0.5 to 1 Hz with a spatula to the anal mucosa for 20 s. (a) Activation and afterdischarge in response to anal stimulation. No activation in response to perigenital mechanical stimulation. (b) Inhibition in response to anal stimulation. (c) Long-lasting afterdischarge after reflex activation in response to anal stimulation. Modified from Bahr et al. (1986a) with permission.

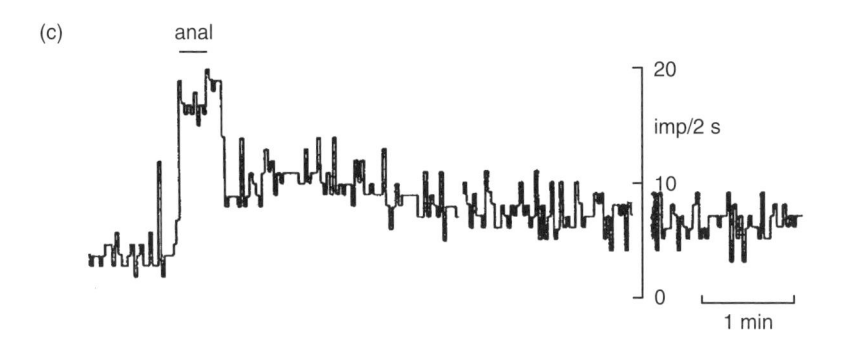

- Most excitatory and inhibitory reflexes elicited in the non-vasoconstrictor (MR) neurons from pelvic organs are preserved after *acute* thoracic spinalization, but those in the visceral vasoconstrictor neurons are abolished. Similar to the other vasoconstrictor systems, the visceral vasoconstrictor system is dependent on supraspinal centers for its resting activity and cardiovascular, respiratory and other reflexes (Figure 4.3). In contrast, the MR sympathetic systems are less (or not at all) dependent on supraspinal centers. The vasoconstrictor systems (and the sudomotor system) therefore exhibit "spinal shock" (total areflexia) after transection of the spinal cord, but the sympathetic systems regulating motility and probably other non-

Figure 4.16 Schematic diagram of the pathways involved in the reflexes elicited in motility-regulating neurons type I by stimulation of sacral visceral afferents from pelvic organs. The reflex pattern in the neurons can be fully explained by sacro-lumbar spinal reflex pathways. The supraspinal pathways are almost unknown and are probably related to the hypothalamus, upper and lower brain stem. Inhibitory neurons are filled. See Subchapter 9.2.

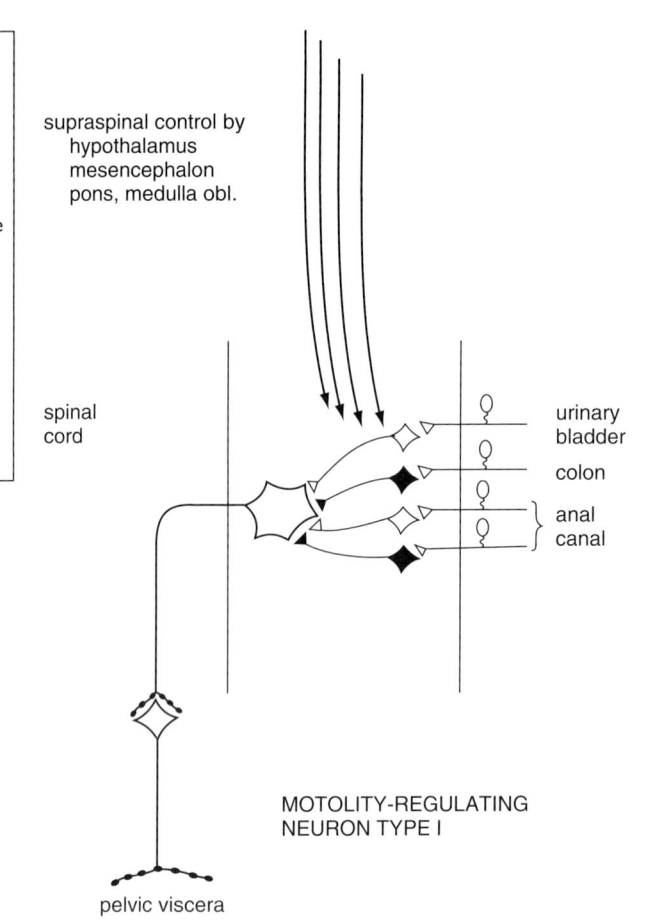

vascular parameters do not (for details see Subchapter 9.2). This shows (1) that multiple sacro-lumbar reflex circuits connected to the sympathetic non-vasoconstrictor systems do not require the descending influence from supraspinal centers in order to function (Figure 4.16), and (2) that spinal reflex circuits to vasoconstrictor neurons require the descending excitatory influence from supraspinal centers (Figure 4.3; see Subchapter 9.2).

- The functional distinction between vasoconstrictor neurons and non-vasoconstrictor neurons of the lumbar sympathetic visceral outflow is supported by differences in some other characteristics including: conduction velocity of the preganglionic axons (vasoconstrictor vs. motility-regulating 2.8 vs. 8.1 m/s), segmental distribution of preganglionic neurons (L1 to L4 vs. L3/L4), ongoing activity (1.6 vs. 0.8 imp/s), segmental and suprasegmental reflexes upon electrical stimulation of somatic or visceral afferents (see Figure 9.2) and neurochemistry of the neurons (see Subchapter 1.4).

It is still unclear how individual sympathetic non-vasoconstrictor pathways are linked to the different target cells in the pelvic viscera and colon. However, the results of the analysis reported above demonstrate beyond any doubt that this system is highly differentiated. In the cat, about 40% of the preganglionic neurons projecting in the lumbar splanchnic nerves have functional properties of MR neurons (Bahr *et al.* 1986c). The functional differentiation between different sympathetic visceral pathways is further detailed in Subchapter 1.4 on neuropeptides and "neurochemical coding" of autonomic neurons and in Subchapter 6.5 on transmission of impulses in autonomic ganglia.

The lumbar sympathetic outflow to pelvic organs and the colon is entirely separate from the lumbar sympathetic outflow to the hindlimbs and the tail of the cat. Otherwise all sympathetic systems to skin, skeletal muscle and pelvic viscera are represented in lumbar segment L3 in the cat. I will discuss in Subchapter 8.2 viscerotopic aspects of the lumbar preganglionic neurons being involved in regulation of pelvic organs, colon or autonomic targets in somatic tissues.

The sympathetic non-vasoconstrictor outflow projecting via the celiac and superior mesenteric ganglia to the gastrointestinal tract has not been studied at the single neuron level using neurophysiological methods. However, based on functional, morphological and other studies it can be inferred that this sympathetic outflow is also differentiated into the categories of MR and secretomotor neurons, both major pathways probably consisting of several subtypes.

4.4 | Other types of sympathetic neuron

So far functional properties of sympathetic neurons have been described, which have been explicitly investigated using neurophysiological and morphological methods. There are other target organs that are most likely also innervated by functionally distinct groups of sympathetic neuron; however, they have not been studied systematically or not at all on the single neuron level (see Table 1.2):

- The heart is innervated by a large group of sympathetic cardiomotor neurons, which are probably similar to muscle and visceral vasoconstrictor neurons in their reflex discharge pattern. These cardiomotor neurons may consist of functional subgroups as far as the innervation of ventricles, atria and pacemaker system is concerned.
- In the guinea pig, indirect experimental evidence shows that the airway smooth muscle cells are innervated by a sympathetic pathway that is functionally distinct from the sympathetic innervation of the vasculature of the airways. Noxious stimulation of the airway mucosa (e.g., by capsaicin) activated this sympathetic bronchomotor pathway leading to relaxation of the airways mediated by β_1- and β_2-adrenoceptors (Oh *et al.* 2006). It is unknown whether this type of sympathetic pathway exists in other species, including primates. It is generally believed that

circulating catecholamines are the main component influencing airway resistance via β_2-adrenoceptors. However, this may turn out not to be true.
- The sympathetic innervation of the kidney modifies glomerular filtration, renin release from the juxtaglomerular cells, tubular Na^+ reabsorption, as well as vascular resistance (DiBona and Kopp 1997; Kopp and DiBona 2000). Activation of these sympathetic neurons leads to vasoconstriction, renin secretion and tubular sodium (and water) reabsorption. Physiological studies indicate that the glomerular filtration rate is differentially controlled by separate sympathetic vasoconstrictor neurons (Denton et al. 2004; Eppel et al. 2004). However, whether renin release or tubular Na^+ reabsorption are controlled by functionally different types of sympathetic neuron and by anatomically separate sympathetic pathways is unclear (Luff et al. 1992).
- Lipocytes of the brown adipose tissues in rodents are innervated by a group of functionally distinct sympathetic neurons. These neurons are probably similar, but very unlikely identical, to cutaneous vasoconstrictor neurons in their functional properties. Like cutaneous vasoconstrictor neurons, they are involved in thermoregulation and their activity is dependent on the integrative processes in the hypothalamus, medulla oblongata (e.g., caudal raphe nuclei) and spinal cord (Himms-Hagen 1991; Morrison 1999, 2001a, b). These neurons may be called lipomotor neurons.
- Sympathetic neurons may be directly involved in glycogenolysis and gluconeogenesis of liver cells and in lipolysis of fat cells of the white adipose tissue (Rosell and Belfrage 1979; Rosell 1980; Frayn and MacDonald 1996; Bartness and Bamshad 1998; Dodt et al. 1999).
- The dilator muscle of the pupil is innervated by a distinct sympathetic pathway (see Boczek-Funcke et al. [1992a]). This corresponds to a dense innervation of the iris dilator muscle and a sparse innervation of the iris sphincter muscle and ciliary muscle by noradrenergic fibers.
- Salivary glands are innervated by sympathetic neurons that are involved in secretion. Stimulation of the sympathetic supply to salivary glands leads to mucous secretion. In the rat and mouse, 15% to 25% of the neurons in the superior cervical ganglion are estimated to innervate salivary glands (Gibbins 1991). These sympathetic neurons are involved in regulation of blood flow (vasoconstrictor neurons), secretion of mucous saliva and activation of myoepithelial cells (Bartsch et al. 1996). The main function of the sympathetic innervation of the salivary glands seems to be to modify the composition of the saliva as to the protein and mucous content (Garrett et al. 1999).
- Some 10% of the thoracic preganglionic neurons projecting through the cervical sympathetic trunk to the superior cervical ganglion are

only active during inspiration ("inspiratory" neurons). Most of these neurons are not under the control of arterial baroreceptors. They decrease their activity during hyperventilation and increase their activity during hypoventilation (increase in arterial P_{CO2}), and are activated by stimulation of nociceptors, arterial chemoreceptors, nasopharyngeal and other receptors (Figure 4.2). Most of their cell bodies are located in the thoracic segments T1 or T2 and their axons conduct significantly faster than those of preganglionic muscle and cutaneous vasoconstrictor neurons (Boczek-Funcke et al. 1992a, 1993). Thus this class of sympathetic preganglionic neuron is functionally distinct from the other main classes of preganglionic neurons. Their target cells are unknown, but may be related to the respiratory tract.
- The pineal gland is innervated by a group of sympathetic neurons, which are dependent in their activity on the suprachiasmatic nucleus in the hypothalamus. These neurons are involved in the circadian induction of melatonin synthesis (Moore 1996, 2003).
- Cells of the adrenal medulla synthesizing adrenaline or noradrenaline are innervated by two separate groups of sympathetic preganglionic neurons (Morrison and Cao 2000; Morrison 2001b; see Subchapter 4.5).
- The immune tissue may be innervated by a functionally distinct sympathetic pathway. This will be discussed in Subchapter 4.6.

4.5 | Adrenal medulla

4.5.1 Circulating adrenaline and noradrenaline

The adrenal medulla when activated releases both adrenaline and noradrenaline (in addition to other substances, which, in the present context, are functionally unimportant [Winkler 1988]). There are considerable differences between species in the proportion of each catecholamine that is released. For example, the rabbit releases 100% adrenaline, the whale 100% noradrenaline, the cat 50% of each and the human subject as well as the rat 80% adrenaline and 20% noradrenaline. The reason for these differences between species remains unclear (von Euler 1956).

The release of the two catecholamines from the adrenal medulla is regulated by activity in preganglionic sympathetic neurons that project from the thoracic spinal cord through the splanchnic nerves to the adrenal medulla (Holman et al. 1994). These neurons form synapses with adrenal medulla cells (Coupland 1965). Histological and immunohistochemical studies show that adrenaline and noradrenaline are synthesized by different cells, which are also innervated by different groups of preganglionic neurons (Edwards et al. 1996). These data suggest that there are two separate sympathetic pathways to the adrenal medulla, one mediating the release of adrenaline and the other noradrenaline (see Vollmer [1996]). This idea is supported by functional studies which show that the release of adrenaline and noradrenaline from the adrenal medulla can be regulated differentially in animals (Folkow and von Euler 1954; see Folkow [1955] [for early studies]; Vollmer [1996]).

Morrison and Cao (2000) have shown in the rat that preganglionic neurons that innervate adrenergic cells of the adrenal medulla are not affected by arterial baroreceptor reflexes and respiratory reflexes but are strongly excited by hypoglycemia. In contrast preganglionic neurons that innervate noradrenergic cells of the adrenal medulla behave like muscle vasoconstrictor neurons (i.e., they are not excited by hypoglycemia and they are under strong arterial baroreceptor reflex control). These results are consistent with the idea that the two groups of adrenal medullary cells are differentially regulated by the brain. However, the context in which this occurs, its functional meaning as well as the central pathways involved, are only incompletely understood (Morrison 2001b).

In humans under resting conditions, the concentration of noradrenaline in the blood plasma is 1 to 1.5 pmol/ml (0.17 to 0.25 µg/l) and of adrenaline about 0.25 pmol/ml (0.05 µg/l; see Kopin [1989]). Furthermore, under almost all physiological conditions, the concentration of circulating noradrenaline is three to five times higher than the concentration of adrenaline (Figure 4.17; Cryer 1980; Kopin 1989). In humans only about 2% to 8% of the circulating noradrenaline is released by the adrenal medulla and the rest is released by sympathetic nerve endings (Cryer 1980; Esler et al. 1990). Systematic studies of various effector organs in the cat (Celander 1954) have shown that adrenaline, released from the adrenal medulla by electrical stimulation of the preganglionic axons innervating the adrenal medulla, has almost no effect on peripheral effector organs, which are under the control of sympathetic noradrenergic neurons. The only important exception, which is supported by experimental investigations, is that circulating adrenaline may generate a vasodilation in skeletal muscle by acting on β_2-adrenoceptors, e.g. during defense behavior in animals or emotional excitation in humans (see Subchapter 4.2).[14] Steady state intravenous (i.v.) infusion of adrenaline moderately increases heart rate and systolic blood pressure and decreases diastolic blood pressure at plasma concentrations of ≥ 100 ng/l and mobilization of glucose and lactate at plasma concentrations of ≥ 200 ng/l (Figure 4.17a; Clutter et al. 1980).

In essence, adrenaline in the plasma released from the adrenal medulla under physiological conditions is primarily a *metabolic hormone*, which chiefly serves to catalyze the mobilization of glucose and lactic acid from glycogen in liver and skeletal muscle, and of free fatty acids from adipose tissue. In addition, it has moderate cardiovascular effects in humans (Kopin 1989). Overall, it seems that adrenaline secretion by the adrenal medulla is unimportant in the maintenance of circulatory homeostasis in everyday life (Young and Landsberg 2001).

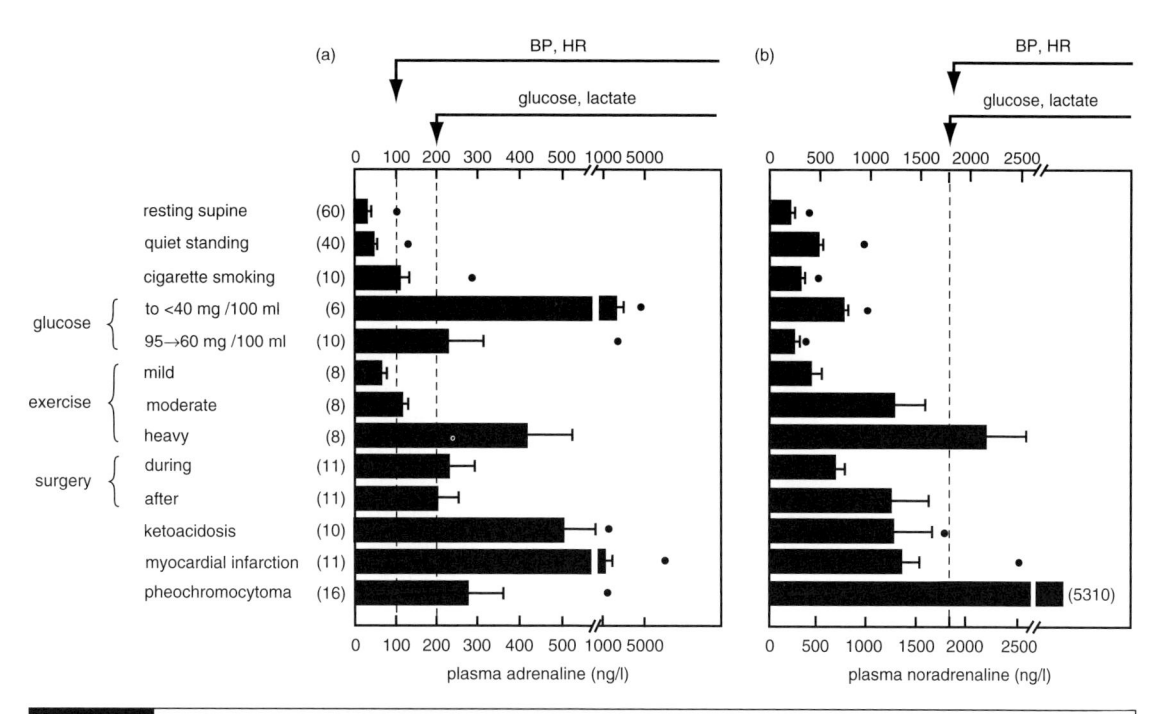

Figure 4.17 Relative changes of plasma concentration of adrenaline and noradrenaline in various physiological and some pathophysiological conditions in humans. The horizontal bars represent mean (+1 S.E.M.) of adrenaline (a) or noradrenaline (b) in ng per liter. The numbers of subjects tested are given in brackets. The highest values measured are indicated by the solid circles. The vertical dashed lines indicate the steady state plasma concentrations of adrenaline or noradrenaline that elicit cardiovascular changes (BP, blood pressure; HR, heart rate) or metabolic changes (glucose, lactate) when adrenaline or noradrenaline are continuously intravenously infused. Adrenaline is released by the adrenal medulla. Most noradrenaline in the blood is released by nerve endings of sympathetic postganglionic noradrenergic neurons. About 2% to 8% are released by the adrenal medullae. Note that hypoglycemia (low glucose concentration in plasma) stimulates relatively selectively the release of adrenaline and that during the other conditions both catecholamines increase, the concentration of noradrenaline being three to five times higher than the concentration of adrenaline. From Cryer (1980) with permission. Infusion experiments (see vertical lines) from Clutter et al. (1980) and Silverberg et al. (1978) with permission.

Noradrenaline infused i.v. at concentrations of ≤1800 ng/l has neither hemodynamic nor metabolic effects in human subjects. These plasma concentrations may occur in human subjects only under extreme conditions (Silverberg et al. 1978; Cryer 1980). Noradrenaline, infused in physiological concentrations i.v. in the cat, has only negligible effects on peripheral effector organs when compared with the effects of sympathetic nerve stimulation (Celander 1954). Thus, plasma noradrenaline, spilled over from the varicosities of activated sympathetic fibers and released to a minor degree by the adrenal medulla, has no known function in practically all physiological conditions; it primarily reflects activity in sympathetic postganglionic neurons (Kopin 1989; Esler et al. 1990).

Taken together, the available data challenge Cannon's concept that the sympathico-adrenal system is a unitary system (see Folkow [1955]). In fact, the data clearly support the general idea that the sympathetic nervous system consists of functionally distinct pathways. One of these pathways regulates the release of adrenaline from the adrenal medulla. However, the situation may change dramatically under *pathophysiological conditions*, for example when peripheral organs (blood vessels, heart) are denervated and become sensitive to circulating catecholamines (Kopin 1989).[15]

4.5.2 Previously unrecognized functions of circulating adrenaline

Role of adrenaline in mechanical hyperalgesia and inflammation

Inflammation and hyperalgesia following tissue trauma are *protective body reactions* observed in all tissues, which further *healing*. Mechanisms of inflammation are commonly considered to be confined to the periphery involving immune-competent and related inflammatory cells as well as vascular cells. The main mechanism of hyperalgesia during inflammation in this view is confined to the sensitization of nociceptors by inflammatory mediators leading to central changes (central sensitization) and appropriate protective behavior. However, using animal (rat) models of experimental inflammation of the knee joint synovia and mechanical hyperalgesic behavior to intradermal injection of the inflammatory mediator bradykinin it can be shown that both of these types of protective body reactions,[16] namely inflammation and nociceptor sensitization are potentially under the control of the sympatho-adrenal system. More specifically, adrenaline released by the adrenal medulla may have a previously unknown function. Although both animal models (bradykinin-induced plasma extravasation in the synovia of the knee joint; bradykinin-induced mechanical hyperalgesic behavior) are rather artificial and specific and although there exist up to now no human models and no clinical situation that correspond to these animal models, it is of interest to learn from these animal models in which way the sympatho-adrenal system – and therefore the brain – can principally influence peripheral mechanisms of inflammation and nociceptor sensitization:

- Noxious stimulation (of skin or viscera) depresses bradykinin-induced venular plasma extravasation in the synovia of the rat knee. This depression is mediated by reflex activation of the adrenal medulla and most likely by the release of adrenaline. The released adrenaline does not inhibit plasma extravasation by vasoconstriction in the synovia but by preventing the increased permeability of the endothelium for plasma proteins generated by the inflammatory mediator bradykinin (probably by acting on β_2-adrenoceptors). The cellular mechanisms of this effect of adrenaline on the experimental inflammatory process are unknown (Miao *et al.* 2000, 2001).
- In rats injection of the inflammatory mediator bradykinin into the dermis of the dorsal skin of the hindpaw reduces paw-withdrawal

threshold to mechanical stimulation of the dorsum of the rat hindpaw. This bradykinin-induced reduction of paw-withdrawal threshold is enhanced and baseline paw-withdrawal threshold (to saline injected into the skin) is reduced after subdiaphragmatic vagotomy. The vagotomy-induced decrease of paw-withdrawal threshold is prevented by prior denervation or removal of the adrenal medulla. The interpretation of these results is that adrenaline released by the adrenal medulla sensitizes nociceptors in the skin for mechanical stimulation. The time course of the development and reversal of this sensitization is slow; it takes from days to about two weeks. The mechanism underlying this sensitization of nociceptors to mechanical stimulation is unknown, but may involve inflammatory cells in the skin (Khasar *et al.* 1998a, b, 2003; Jänig *et al.* 2000).

Role of adrenaline in memory consolidation
A third example of an unprecedented function of adrenaline released by the adrenal medulla is that it influences the consolidation of memory in animals and humans under specific behavioral conditions. It is hypothesized that adrenaline, e.g. released by the adrenal medulla during stressful behavior, activates vagal afferents, which in turn leads to direct or indirect activation of central ascending noradrenergic neurons, via the nucleus tractus solitarii (NTS). Activation of the central noradrenergic neurons in the NTS and locus coeruleus activates (via α_1- and β-adrenoceptors) the cyclic adenosine monophosphate-protein kinase-A (cAMP-PKA) pathway of pyramidal cells of the basolateral amygdala. The noradrenergic activation of the cAMP-PKA pathway leads to N-methyl-D-aspartate (NMDA)-dependent plasticity in the amygdala neurons, resulting in modulation of memory consolidation in the hippocampus and other brain regions (Williams *et al.* 1998; Clayton and Williams 2000; McGaugh 2000; see McGaugh and Roozendaal [2002] for review). This interesting idea would assign a novel function to adrenaline and to the central pathways that regulate the release of adrenaline by the adrenal medulla. However, to prove this hypothesis it has to be shown whether a distinct class of vagal afferent neurons is excited specifically by adrenaline released by the adrenal medulla. Alternatively, circulating adrenaline might act on neurons in the NTS via the area postrema, which lacks a blood–brain barrier (see Subchapter 8.3).

In *conclusion* (Figure 4.18), the release of adrenaline by the adrenal medulla is controlled by central pathways that are distinct from those linked to other sympathetic pathways. Under physiological conditions adrenaline may be involved in the following functions:

1. regulation of metabolism such as catalyzing the mobilization of glucose and lactic acid from glycogen and probably of fatty acids from adipose tissue;

Figure 4.18 Functions of adrenaline released by the adrenal medulla. For details see text. NTS, nucleus tractus solitarii; vag aff, vagal afferents.

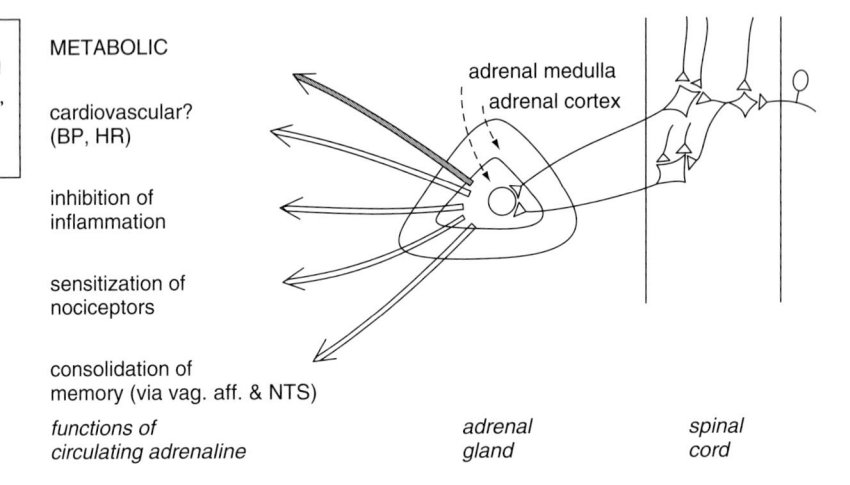

METABOLIC

cardiovascular?
(BP, HR)

inhibition of inflammation

sensitization of nociceptors

consolidation of memory (via vag. aff. & NTS)

functions of circulating adrenaline

adrenal medulla
adrenal cortex

adrenal gland

spinal cord

2. modulation of sensitivity of nociceptors;
3. modulation of inflammation;
4. consolidation of memory (via vagal afferents and the NTS); and
5. *probably not* to support responses of target organs elicited by activation of other sympathetic pathways (with the possible exception of β_2-adrenoceptor-mediated vasodilation in skeletal muscle) except under strong emergency conditions.

Functions 2 to 4 are still hypothetical and require direct experimental proof. Furthermore, in view of the fact that (1) the concentration of circulating noradrenaline is three to five times higher than the concentration of circulating adrenaline in humans and (2) there exist considerable differences between species in the proportions of the two catecholamines released by the adrenal medulla, it is debated whether the adrenal medulla is important in humans.

4.6 | Sympathetic neurons innervating the immune tissue

Anatomical, physiological, pharmacological and behavioral experiments on animals support the notion that the sympathetic nervous system can influence the immune system and therefore control protective mechanisms of the body at the cellular level (Hori *et al.* 1995; Madden and Felten 1995; Madden *et al.* 1995). Control of the immune system by the sympathetic nervous system would mean that the brain is principally able to influence, probably via the hypothalamus, immune responses. This is an attractive and important idea and based on clinical observations (Jänig and Häbler 2000a). The mechanisms of this influence remain largely unsolved (Besedovsky and del Rey 1992; Ader and Cohen 1993; Saphier 1993). This has conceptual and methodical reasons. In view of the functional specificity of the sympathetic pathways as outlined in this chapter the key question to be asked is:

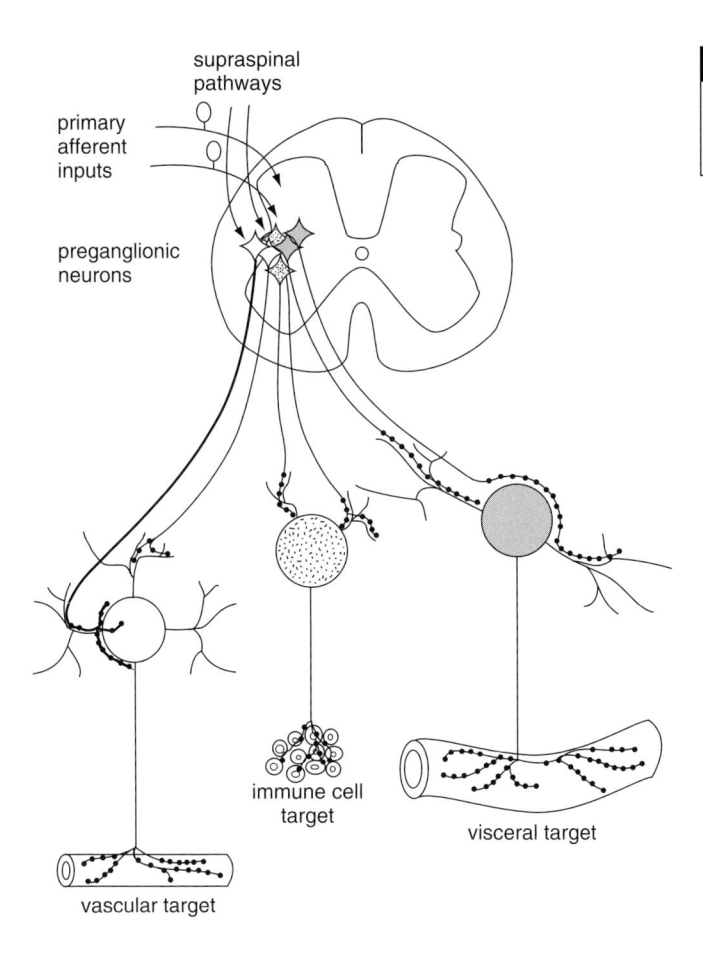

Figure 4.19 A separate sympathetic pathway to the immune tissue: a hypothesis. For details see text.

Does a specific sympathetic subsystem exist which communicates signals from the brain to the immune system or is this efferent communication a general function of the sympathetic system? In other words, is the immune system supplied by a sympathetic pathway which is distinct from other sympathetic pathways (see above) and mediates only an immunomodulatory effect? (Figure 4.19)

Several observations support the idea that an important channel of efferent communication from the brain to the immune system occurs via the sympathetic nervous system and that this channel *is functionally distinct from all other sympathetic systems*:

1. Primary and secondary lymphoid tissues are innervated by postganglionic noradrenergic sympathetic neurons. Varicosities of the sympathetic terminals can be found in close proximity with T lymphocytes and macrophages (Besedovsky and del Rey 1995; Madden *et al.* 1995) as described for other sympathetic target cells.
2. The spleen of the cat is innervated by sympathetic postganglionic neurons that are numerically, relative to the weight of the organs, three times as large as the sympathetic innervation of the kidneys (Baron and Jänig 1988). Many sympathetic neurons innervating

the spleen are not under control of the arterial baroreceptors and show distinct reflexes to stimulation of afferents from the spleen and the gastrointestinal tract, which are different from those in vasoconstrictor neurons, e.g. innervating the kidney (Dorward et al. 1987; Meckler and Weaver 1988; Stein and Weaver 1988), suggesting that many sympathetic neurons innervating the spleen have a function other than to elicit vasoconstriction.

3. Functional studies performed on the spleen of rodents have shown that (for review see Hori et al. [1995]): (1) Surgical and chemical sympathectomy alters the splenic immune responses (e.g., increase of natural killer cell cytotoxicity, lymphocyte proliferation responses to mitogen stimulation and production of interleukin-1β). (2) Stimulation of the splenic nerve reduces the splenic immune responses mediated by β-adrenoceptors. (3) Lesion or stimulation of distinct hypothalamic sites or microinjection of cytokines at distinct hypothalamic sites (in particular the ventromedial nucleus of the hypothalamus) activate some splenic immune responses. These changes are no longer present after denervation of the spleen. (4) Activity in the splenic nerve is affected by these central manipulations and changes in neural activity are correlated with the changes of the splenic immune responses. A *hypothalamo-sympathetic neural system that controls the immune system* has been postulated by Hori et al. (1995).

4. Many sympathetic postganglionic neurons projecting in skin nerves do not exhibit spontaneous and reflex activity. Do some of these postganglionic neurons not innervate the "classical" sympathetic targets (blood vessels, sweat glands, arrector pili muscles) but are associated with the skin immune system (Bos and Kapsenberg 1986; Bos 1989; Williams and Kupper 1996)?

Using classical neurophysiological recordings in vivo from sympathetic neurons it should be possible to discriminate neurons innervating the immune tissue from other functional types of sympathetic neuron (e.g. vasoconstrictor neurons to skin, skeletal muscle, kidney or spleen). This is exemplified in Table 4.4 showing the target cells in particular organs which are innervated and possibly controlled by sympathetic neurons and the functions of these neurons. One sympathetic channel in three of these organs projects potentially to the immune tissue and possibly regulates the immune response.

Thus, it should theoretically be possible to characterize the sympathetic neurons innervating lymphoid tissues by using stimuli that are adequate to elicit immune responses and to assign to these sympathetic neurons characteristic functional markers (see Hori et al. [1995]). This idea leads to the formulation of *two alternative testable hypotheses*:

1. Neurons of sympathetic pathways are functionally specific for the immune tissues and can be characterized by way of distinct reflex patterns elicited in these neurons by adequate stimuli that are

Table 4.4 | *Examples of function-specific sympathetic pathways to different targets within some organs with emphasis on immune tissue*

Organ	Target cells	Function
Spleen	BV, *IC*	VC, *IR*
Hairy skin	BV_{skin}, *IC*	VC (VD?), *IR*
Hairless skin	BV_{skin}, SG, *IC*	VC (VD?), SM, *IR*
Kidney	BV, JGA	VC, renin release
Skel. muscle	BV_{muscle}	VC (VD?)

Notes:
Abbreviations: BV, blood vessel; VC, vasoconstriction; VD, vasodilation; *IC*, immune cells; *IR*, immune response; SG, sweat gland; SM, sudomotor response; JGA, juxtaglomerular cells.

related to the immune system and therefore related to defense and protection of the organism.

2. The alternative hypothesis would be that reflex responses in sympathetic neurons that modulate immune responses are found indiscriminately in all sympathetic neurons; these responses would therefore not be functionally specific for the lymphoid tissue. This could mean that more or less *all* sympathetic noradrenergic pathways have, in addition to their specific target-organ related functions, a general function, which is related to defense and protection of the tissues. Verification of this hypothesis would render an argument that might *unify* both Cannon's concept about the general function of the sympathetic nervous system (Cannon 1939) and the concept of the functional differentiation of the sympathetic nervous system.

4.7 | Proportions of preganglionic neurons in major sympathetic nerves

The discussion in this chapter and Table 1.2 show that the peripheral sympathetic system is composed of many functionally different types of pathways that are defined by their target cells. For some of these pathways it has clearly been shown that the pre- and postganglionic neurons exhibit distinct reflex patterns ("functional fingerprints"), indicating that these pathways are functionally distinct and implying that they are linked with distinct reflex pathways in the neuraxis. For other pathways this has not been shown yet or it has been demonstrated that they exhibit no reflexes and no spontaneous activity under the experimental conditions in which they were studied. This applies, for example, to pilomotor neurons and to vasodilator neurons. In view of this situation it is of interest to look at some numerical data about *preganglionic* neurons, projecting in major

Table 4.5 | *Numbers and proportions of major classes of preganglionic neurons in the cat[a]*

	CST[b]	LST[c]	LSN[d]
Location of neurons	T1–T5	L1–L4	L1–L5
Target tissues	Head, upper neck	Hindlimb, tail[e]	Pelvic organs, colon
Number of neurons	6200	4500	2300
MVC/VVC neurons	20%	10%	15%
CVC neurons[f]	12%	13%	–
MR neurons[g]	–	–	41%
INSP neurons[h]	10%	–	–
Silent neurons[i]	45%	70%	40%

Notes:

[a] The percentages are estimates and do not add up to 100% since some preganglionic neurons have spontaneous activity but exhibit no known reflexes or have characteristics of sudomotor neurons (LST).
[b] Cervical sympathetic trunk (CST); Boczek-Funcke et al. (1992a, 1993), Wesselmann and McLachlan (1984).
[c] Lumbar sympathetic trunk (LST) distal to lumbar paravertebral ganglion L5; Jänig and Szulczyk (1980), Jänig and McLachlan (1986a, b).
[d] Lumbar splanchnic nerves (LSN); Bahr et al. (1986c), Baron et al. (1985c).
[e] A few preganglionic fibers in the distal LST synapse with postganglionic neurons in the sacral paravertebral ganglia that project to pelvic organs and probably innervate blood vessels (see Jänig and McLachlan 1987).
[f] CVC neurons innervating the distal parts of the extremities and the tail. Some 20% of the postganglionic CVC neurons innervating hairy skin have a reflex pattern similar to that in MVC neurons (excitation to the stimulation of arterial chemoreceptors, strong baroreceptor control, excitation during stimulation of nociceptors; see Subchapter 4.1). These CVC neurons cannot be recognized by way of their reflex pattern in the population of preganglionic neurons. Thus, these CVC neurons are hidden in the population of preganglionic MVC neurons.
[g] MR neurons project in the LSN and are activated or inhibited by at least one of the following stimuli: distension/contraction of the urinary bladder or colon, mechanical stimulation of anal or perianal skin. A few MR neurons have no ongoing activity.
[h] INSP neurons are only active during inspiration (see Subchapter 4.4).
[i] Silent neurons have no ongoing activity and no reflex activity under the experimental conditions tested. A few silent neurons can be recruited (e.g. by hypercapnia or hypoxia).
Abbreviations: CVC, cutaneous vasoconstrictor; INSP, inspiratory-type; MR, motility-regulating; MVC, muscle vasoconstrictor; VVC, visceral vasoconstrictor.

"sympathetic" nerves of the cat, that have been investigated neurophysiologically and morphologically. These nerves are the lumbar sympathetic trunk (LST) distal to the paravertebral ganglion lumbar L5 (innervation territory mainly hindlimb and tail), the cervical sympathetic trunk (CST; innervation territory head and upper neck) and the lumbar splanchnic nerves (LSN; innervation territory pelvic organs and colon). These nerves contain about 4500 (LST), 6200 (CST) and 2300 (LSN) preganglionic axons respectively (Table 4.5).

Based on the neurophysiological investigations (Table 4.5) it can be estimated that:

- About 10% to 20% of the preganglionic neurons in the three nerves have vasoconstrictor function and are involved in blood pressure regulation (muscle and visceral vasoconstrictor neurons).

- About 12% to 13% of the preganglionic neurons projecting in the CST and LST participate in regulation of cutaneous blood flow (cutaneous vasoconstrictor neurons). This percentage may be somewhat higher since some 20% of the postganglionic cutaneous vasoconstrictor neurons exhibit a reflex pattern similar to that in muscle vasoconstrictor neurons.
- About 10% of the preganglionic neurons projecting in the CST are neurons with the discharge pattern of inspiratory neurons (INSP). The target of these neurons may be the respiratory tract (see Figure 4.2 and Subchapter 4.4).
- About 40% of the preganglionic neurons projecting in the LSN are involved in regulation of visceral organ functions (MR neurons).
- Varying between the "sympathetic" nerves (CST, LST, LSN), about 40% to 70% of the preganglionic neurons exhibit no spontaneous activity and no reflexes under the experimental conditions. It is unlikely that this high proportion of silent preganglionic neurons is mainly related to the anesthesia under which the experiments were conducted:
  - Some of the silent preganglionic neurons have distinct functions and are only activated during specific behavioral conditions (e.g., pilomotor neurons, vasodilator neurons supplying skin or skeletal muscle, neurons supplying internal reproductive organs).
  - Other silent neurons may have vasoconstrictor function and are only recruited under extreme conditions (e.g., emergency such as during severe blood loss, extreme exposure to cold).
  - Silent preganglionic neurons may be involved in the regulation of metabolism (e.g., white adipose tissue) or the immune tissue (see Subchapter 4.6).

The numerical estimates of the four major groups of functionally identified sympathetic preganglionic neurons clearly show that only small fractions of the total number of preganglionic neurons are involved in cardiovascular regulation or regulation of cutaneous blood flow. Recordings of multiunit activity from "sympathetic" nerves (e.g., lumbar or major splanchnic nerves, renal nerve, sympathetic trunk) show that the discharge patterns are rather uniform and always dominated by the discharge pattern in neurons that are involved in cardiovascular regulation. This is important to know once the central pathways in the neuraxis that are linked to the preganglionic neurons are worked out (see Chapters 8 to 10).

4.8 Parasympathetic systems

Pre- and postganglionic parasympathetic neurons have always been presumed to constitute distinct peripheral pathways which transmit centrally derived signals to their target organs and exhibit distinct reflexes to afferent stimuli which are appropriate for their function. Compared to the amount of experimentation on the reflex discharge in single sympathetic pre- and postganglionic neurons, corresponding studies of the functional properties of parasympathetic neurons are relatively few (Table 4.6). The lack of systematic studies is probably because:

- Many parasympathetic ganglia are located close to or on the wall of their target organs and are therefore less well defined than the sympathetic ganglia. The postganglionic axons are therefore very short and often cannot be visualized for dissection.
- Parasympathetic preganglionic neurons are relatively difficult to record from and to separate with respect to their target organs. This applies to recordings from cell bodies of the neurons (e.g., in the dorsal motor nucleus of the vagus [DMNX], in the nucleus ambiguus, in the sacral spinal cord) as well as to recordings from their axons.
- Functionally and anatomically, peripheral parasympathetic neurons constitute more discrete pathways to the target organs than sympathetic neurons. It is widely accepted that most parasympathetic ganglia act as simple relay stations (i.e. without integration; see Subchapter 6.5). However, this is certainly not true for some parasympathetic ganglia (e.g., cardiac parasympathetic ganglia [see Subchapter 6.5; Figure 6.12] and some pelvic ganglia [see Keast [1995a, 1999]).

The main difference between sympathetic and parasympathetic systems is that each type of target organ of the parasympathetic system is spatially restricted (such as sphincter pupillae and ciliary body, salivary and other exocrine glands of the head, the heart, the blood vessels of the erectile tissue and the smooth muscle cells of the urinary tract and hindgut) whereas some types of target organs of the sympathetic nervous system are widely distributed (e.g., blood vessels in skeletal muscle, in skin, in viscera, sweat glands, erector pili muscles, etc.). This has consequences for the organization of the autonomic ganglia in each of the systems. However, some target organs of the sympathetic nervous system are as restricted as those of the parasympathetic pathways (e.g., dilator pupillae muscle, pineal gland). Furthermore, the sympathetic innervation of vascular beds of skeletal muscle or skin may be subdivided according to the anatomical region (e.g., distal or proximal skin of the extremities, skin of trunk; skeletal muscle of extremities or of trunk).

4.8.1 Parasympathetic pathways to pelvic organs

Pelvic organs (which include the rectum-sigmoid of the descending colon) are innervated by and regulated via sacral (in the rat sacral and caudal lumbar [L_6]) spinal autonomic (parasympathetic) systems (Figure 4.20). Cranial parasympathetic (vagal) systems are not involved

Table 4.6 Functional classification of parasympathetic neurons in animals based on reflex behavior in vivo

Likely function	Location	Target organ	Likely target tissue	Major identifying stimulus	Ongoing activity[a]
Pupillo-constrictor	Ciliary ganglion, postggl.[b]	Iris	Constrictor pupillae muscle	Excitation/inhibition by light	Yes
	EW lateral, preggl.[c]	Iris	Constrictor pupillae muscle	Excitation by light	Yes
Accommodation[d]	EW lateral, preggl.	Ciliary body	Ciliary muscle	Excitation by target moving	Yes
Cardiomotor[e]	N. ambig. preggl.	Heart	Pacemaker cells, atrial muscle cells	Excited by stim. of baroreceptors	Yes
Bronchomotor[f]	N. ambig, preggl., tracheobronch. ggl.	Trachea, bronchi	Smooth muscle cells	Stim. of tracheal mucosa, excitation in inspiration	Yes
Bronchosecreto-motor[f]	Tracheobronch. ggl.	Trachea, bronchi	Glands	Excited in expiration	Yes
Gastromotor excitatory[g]	DMNX preggl.	Stomach	Smooth muscle	Inhibition in resp. to duodenal distension	Yes
Gastromotor inhibitory[g]	DMNX preggl.	Stomach	Smooth muscle	Excitation in resp. to duodenal distension	Yes
Urinary bladder[h]	Sacral spinal cord, preggl.	Urinary bladder	Smooth muscle	Excitation in resp. to bladder distension	No
Colon[h]	Sacral spinal cord, preggl.	Colon	Smooth muscle	Excitation in resp. to colon distension	Yes

Notes:
[a] Not all neurons have spontaneous activity
For details about rates of ongoing activity in pre- and postganglionic neurons and reflexes to various afferent stimuli see
[b] Inoue (1980), Johnson and Purves (1983), Melnitchenko and Skok (1970), Nisida and Okada (1960)
[c] Sillito and Zbrozyna (1970a, b), Gamlin and Clarke (1995), Gamlin (2000)
[d] Gamlin et al. (1994), Gamlin (2000)
[e] Jewett (1964), Katona et al. (1970), Kunze (1972), McAllen and Spyer (1978a)
[f] Tomori and Widdicombe (1969), Widdicombe (1966), McAllen and Spyer (1978a), Mitchell et al. (1987)
[g] Grundy et al. (1981), Roman and Gonella (1994), see Subchapter 10.7
[h] de Groat et al. (1982)

Abbreviations: preggl., preganglionic; postggl., postganglionic; N. ambig., Nucleus ambiguus; EW, Edinger–Westphal nucleus; DMNX, dorsal motor nucleus of the vagus; tracheobronch. ggl., tracheobronchial ganglia
Modified from Jänig and Häbler (1999)

Figure 4.20 Parasympathetic pathways from the sacral spinal cord to the pelvic organs. In the rat the parasympathetic preganglionic neurons are also located in the lumbar segment L_6. Excitatory pathway to the body of the urinary bladder (detrusor vesicae), inhibitory pathway to the urethra (transmitter nitric oxide [NO] and vasoactive intestinal peptide [VIP]), inhibitory pathway to the erectile tissue (transmitter NO and VIP) and excitatory pathway to the distal colon/rectum (this pathway possibly consists of three neurons, i.e. the postganglionic neurons form synapses with enteric neurons [myenteric and serosal plexus]). There may exist further functionally distinct spinal parasympathetic pathways to the pelvic organs.

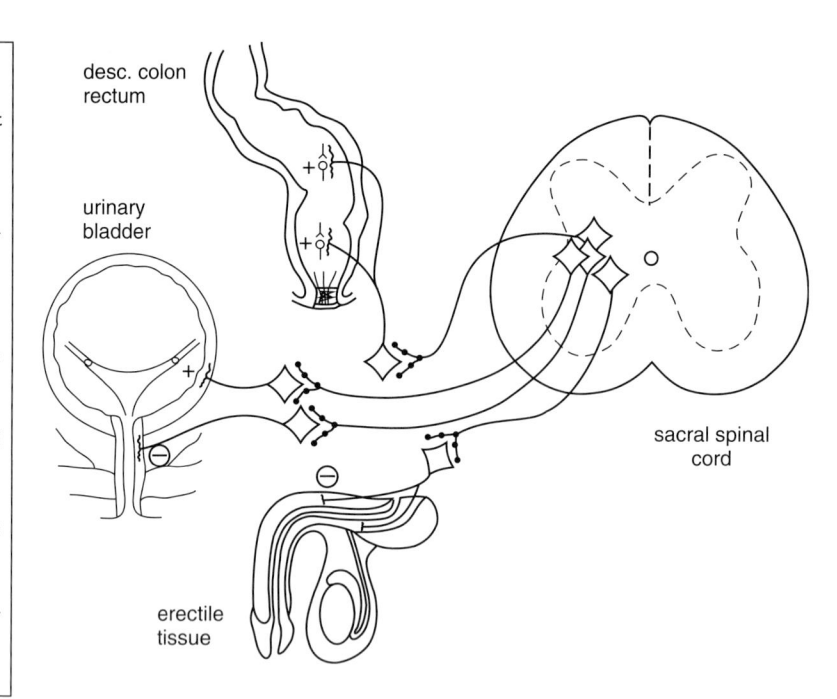

in their regulation. Some details about the anatomy and neural regulation of these systems have been described in Chapter 8 and Subchapter 9.3 and in the literature (see below).

Hindgut

The sacral parasympathetic pathways to the hindgut are rather complex and little understood in their physiology. They largely consist of (pre-postganglionic) neuron chains that innervate enteric neurons in the myenteric plexus (but not in the submucosal plexus), or neurons in the serosal plexus, similar to the sympathetic non-vasoconstrictor pathways to the gastrointestinal tract. Thus, the neural connection between the sacral spinal cord and target cells in the rectum-sigmoid consists of chains of at least three neurons connected synaptically. Only a few postganglionic neurons in pelvic splanchnic ganglia innervate targets (e.g., circular smooth muscle) directly. No projections exist to mucosa, submucosa, longitudinal muscle and submucosal blood vessels (Fukai and Fukuda 1985; Luckensmeyer and Keast 1998a, b).

It appears as if the sacral parasympathetic innervation of the hindgut is only involved in the regulation of its motility and above all in the control of its evacuation (defecation) and continence (the latter together with the upper lumbar sympathetic supply; see Subchapter 9.3). The sacral parasympathetic neurons to the hindgut are excited by stimulation of sacral afferents from the hindgut (most of them being unmyelinated) generated by distension or contraction, and possibly inhibited by stimulation of afferents from the lower urinary tract (urinary bladder and urethra) (see Subchapter 9.3 and Figure 9.10). Stimulation of afferents from the anal canal (which constitute a distinct population of visceral afferents [Bahns *et al.*

1987; Jänig and Koltzenburg 1991c]) probably inhibit the sacral efferent neurons innervating the hindgut.

Lower urinary tract
The lower urinary tract (urinary bladder and urethra) is supplied by at least two sacral parasympathetic pathways. One pathway innervates the bladder body (detrusor vesicae). Its activation leads to contraction of the urinary bladder. Neurons of this pathway are activated by stimulation of sacral bladder afferents and probably inhibited by stimulation of sacral afferents from the colon or from the anal canal. The other sacral parasympathetic pathway innervates the urethra. Its activation (e.g. by stimulation of sacral vesical afferents during micturition) relaxes the urethra, probably by release of NO and VIP. This parasympathetic pathway to the lower urinary tract is anatomically and functionally less well characterized than the pathway to the bladder body (Jänig 1996b; de Groat 2002; see Subchapter 9.3).

Reproductive organs
The sacral spinal autonomic pathways to the reproductive organs have a complex organization, which is related to the target cells and to the integration of lumbar (sympathetic) spinal systems and sacral spinal autonomic systems at the level of the target cells and of the splanchnic pelvic ganglia (McKenna 1999, 2000, 2001; see Jänig and McLachlan [1987], Jänig [1996c]; see Subchapter 9.3).

The spinal parasympathetic innervation of the reproductive organs consists of at least one pathway innervating the erectile tissues (sinusoids and trabecular tissues of the corpus cavernosum and corpus spongiosum of the penis and the helicine arteries that feed the erectile tissues in males). In males the postganglionic neurons innervating the erectile tissues project through the cavernous nerve. These neurons are cholinergic and use NO and VIP as transmitters (Jänig 1996c; de Groat 2002; see Figure 9.12). It is unclear whether postganglionic neurons projecting through the cavernous nerve also receive convergent synaptic input from lumbar sympathetic preganglionic neurons or whether the sympathetic pathway responsible for penile erection (e.g., in patients or animals with destroyed sacral spinal cord) is separate from the parasympathetic pathway (see Figure 9.12 and Subchapter 9.3). Neurons of the parasympathetic pathways of the erectile tissue are reflexly activated by stimulation of pudendal afferents from the penis and surrounding tissue (Figure 9.11). Further characterization of these neurons has not been done. Parasympathetic spinal neurons may also be involved in activation of glands of the reproductive organs.

The role of spinal parasympathetic pathways in the regulation of female external and internal reproductive organs (during sexual arousal) has been little explored. However, it is generally believed that the spinal parasympathetic pathway(s) and the underlying mechanism involved are similar to those in the male. Thus, the erectile tissue of females is innervated by a parasympathetic cholinergic vasodilator pathway as in the male. Furthermore, engorgement of the vaginal wall with blood and transudation of mucoid fluid through the epithelium of the vagina seems to be generated by reflex activation of spinal parasympathetic neurons, which generate vasodilation and increase of permeability of the capillaries (McKenna 2000, 2002; Giuliano et al. 2001).

4.8.2 Parasympathetic pathways from the brain stem

Heart

The heart is innervated by at least two parasympathetic pathways. One pathway is involved in the regulation of frequency of heart beat, by acting on the pacemaker cells of the sinoatrial node and on the atrioventricular node, and in the regulation of force of contraction of the atria. The axons of most preganglionic neurons of this pathway are myelinated (in the cat, rat and human). The cells bodies of the preganglionic neurons are situated in the external formation of the nucleus ambiguus and the cell bodies of the postganglionic neurons are in the parasympathetic cardiac ganglia associated with the right atrium. Some neurons of this cardiomotor pathway exhibit spontaneous activity. The neurons are excited by stimulation of arterial baroreceptors and exhibit cardiac rhythmicity in their activity (Figure 4.21c). The activity in most cardiomotor neurons of this pathway exhibit respiratory rhythmicity: (in the cat) the neurons are inhibited during inspiration and activated in expiration (mostly postinspiration). This coupling to the central respiratory network is the basis of respiratory sinus arrhythmia of the heart beat. Stimulation of arterial chemoreceptors, nasopharyngeal receptors, or other receptors excites these cardiomotor neurons, leading to bradycardia (McAllen and Spyer 1978a, b; Izzo and Spyer 1997; Cheng and Powley 2000). Chemical stimulation of the cell bodies of the cardiomotor neurons in the nucleus ambiguus leads to decrease of heart rate (Figure 4.21d). In fact excitation of a single parasympathetic preganglionic cardiomotor neuron already slows heart frequency, showing how powerful this parasympathetic pathway is to generate cardiac bradycardia (McAllen and Spyer 1978a).

A second parasympathetic cardiomotor pathway, which is entirely separate from the first one, originates in the DMNX (Figure 4.21a). Preganglionic neurons of this pathway have unmyelinated axons (in the rat and probably in the cat). Some of these cardiomotor neurons exhibit spontaneous activity. This spontaneous activity is neither modulated by activity in arterial baroreceptors nor by the central respiratory drive. The function of this cardiomotor pathway is unknown. Its activation generates weak bradycardia. However, it may primarily generate coronary vasodilation when activated (Izzo and Spyer 1997; Feigl 1998; Jones et al. 1998; Cheng et al. 1999).

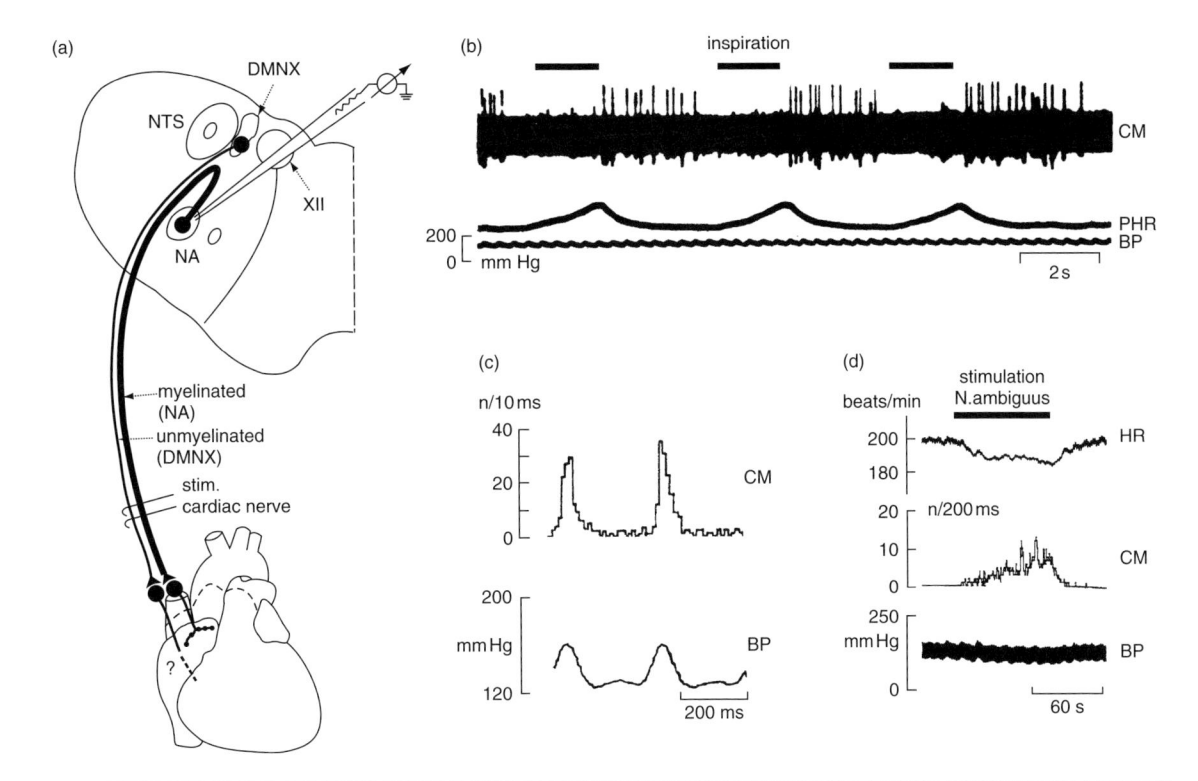

Figure 4.21 Discharge pattern of parasympathetic cardiomotor neurons (CM) in the cat. (a) Experimental set-up: recording from the cell bodies with a microelectrode. The cell bodies of the neurons are located in the external formation of the nucleus ambiguus (NA). The neurons project through the cardiac nerves to the heart. The neurons were identified by electrical stimulation of their axons in the cardiac branch of the vagus nerve. Cardiomotor neurons in the NA have myelinated axons. Cardiomotor neurons in the dorsal motor nucleus of the vagus (DMNX) have unmyelinated axons. Their function is unknown (see text). The CM neurons are normally silent under the experimental conditions (probably due to the anesthesia). The ongoing activity was generated by continuous ionophoretic application of an excitatory amino acid (DL-homocysteic acid) through a second barrel of the microelectrode. NTS, nucleus tractus solitarii; XII, hypoglossus nucleus. (b) Respiratory modulation of the activity. The neuron was active in expiration (mostly postinspiration) and inhibited during inspiration. BP, blood pressure; PHR, integrated phrenic nerve activity. (c) Cardiac rhythmicity of the activity in a cardiomotor neuron in the NA. Blood pressure pulse-triggered histogram of the activity in the neuron. Activity in 256 cardiac cycles superimposed. Lower record: averaged pulse wave of the arterial blood pressure recorded in the femoral artery. (d) Stimulation of cell bodies of the CM neurons in the NA by ionophoretic application of an excitatory amino acid through a microelectrode decreased heart rate (HR) and BP. In this experiment only a few (possibly only one!) CM neurons were stimulated. Modified from McAllen and Spyer (1978a) with permission.

Airways

Tracheobronchial airways of the respiratory tract seem to be innervated by three parasympathetic pathways. Two pathways supplying the airway smooth muscle (parasympathetic bronchomotor neurons) and one pathway to the glands of the airways (bronchosecretomotor neurons; Canning and Mazzone 2005). One bronchomotor pathway contracts the airway smooth muscle cells by release of acetylcholine. Neurons of this pathway are activated by stimulation of airway nociceptors, arterial chemoreceptors, upper airway mechanoreceptors

Figure 4.22 Reflex discharge pattern of postganglionic bronchomotor neurons. Intracellular recordings from the cell bodies of neurons in tracheal ganglia in anesthetized cats (upper records). Lower or middle records, activity in the phrenic nerve (PHR). (a) Neuron 1 active in inspiration and postinspiration. The activity of this neuron was depressed by increase of transpulmonary pressure (TP). This type of neuron may innervate the smooth musculature of airways. (b) Neuron 2 largely active during expiration. This neuron was activated by increase of TP. It may innervate tracheobronchial glands. Modified from Mitchell et al. (1987) with permission.

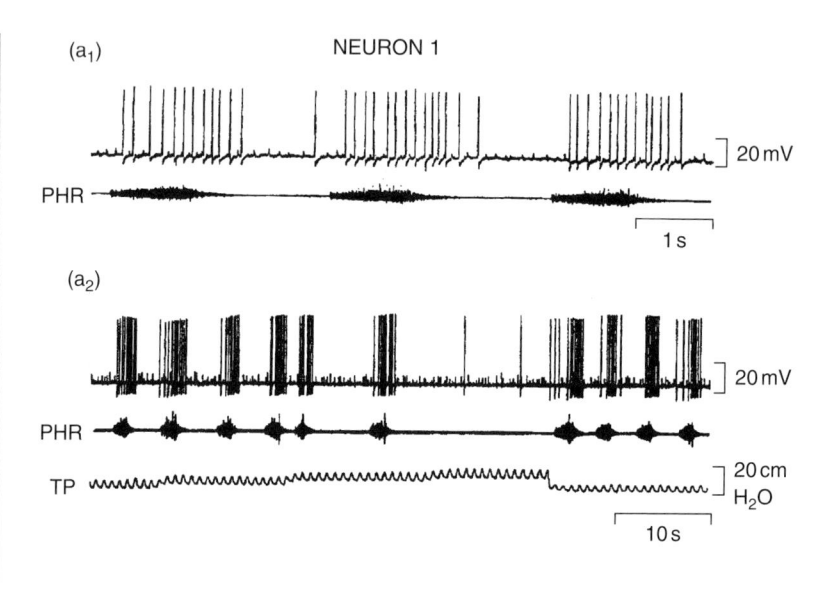

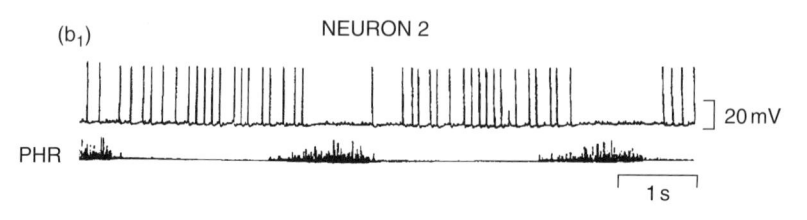

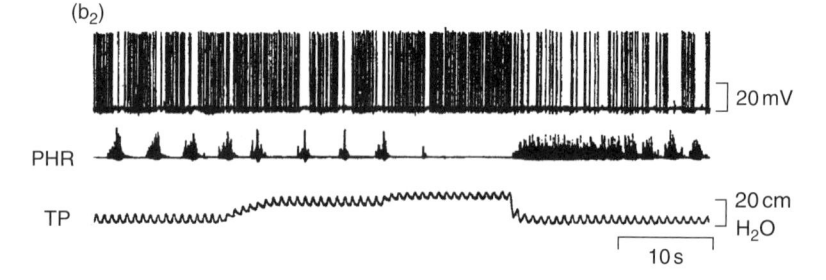

or esophageal afferents. The second parasympathetic bronchomotor pathway relaxes the airway smooth musculature by release of NO and possibly a peptide (e.g., vasoactive intestinal peptide). This pathway may also be activated by noxious airway stimulation or stimulation of rapidly adapting stretch afferents of the lung. Bronchosecretomotor neurons may be reflex activated by airway nociceptors or by lung stretch receptors.

Most neurons of these bronchomotor pathways seem to be spontaneously active. One type of bronchomotor pathway is active during inspiration and postinspiration and not under control of arterial baroreceptors. It is inhibited by hyperinflation of the lung (generated by increase in transpulmonary pressure [Figure 4.22a]) and excited by stimulation of arterial chemoreceptors and other stimuli that lead to reflex bronchoconstriction. The postganglionic neurons of this bronchomotor pathway may innervate the bronchial smooth muscle cells. The second functional type of bronchomotor neuron is active during expiration and excited by lung hyperinflation (Figure 4.22b). These bronchomotor neurons may innervate secretory glands of the airways, and thus are probably bronchosecretomotor neurons (Mitchell et al. 1987; see Jordan [1997a]). The discharge pattern of the type of parasympathetic postganglionic bronchomotor neuron that relaxes airway smooth muscle cells is unknown; however, these neurons may exhibit spontaneous activity too (Kesler et al. 2002). The cell bodies of the preganglionic neurons of the first type of bronchomotor neuron are located in the external formation of the nucleus ambiguus, mostly rostral to the cell bodies of the preganglionic cardiomotor neurons with myelinated fibers (McAllen and Spyer 1978a). The location of the second type of bronchomotor neuron is unknown, but probably also in the external formation of the nucleus ambiguus.

Gastrointestinal tract

Cranial parasympathetic pathways to the gastrointestinal tract are complex. The preganglionic neurons are located in the DMNX; the postganglionic neurons are neurons of the enteric nervous system. Experimental studies of preganglionic neurons clearly show that these neurons are functionally differentiated with respect to regulation of various gastrointestinal functions (motility, secretion of exocrine glands, secretion of endocrine glands) (details in Subchapters 5.7 and 10.7; see Subchapter 10.7 for recording from preganglionic neurons in the DMNX). For some functions related to regulation of motility, the reflex pathways have been worked out. For neurons of parasympathetic pathways to exocrine glands or endocrine cells (e.g., G-cells secreting gastrin or the insular cells of the pancreas secreting insulin or glucagon) the reflex patterns and pathways are unknown.

Salivary glands

Preganglionic neurons projecting to parasympathetic ganglia innervating salivary glands (and possibly nasopharyngeal glands) are located in the superior salivary nucleus (for the submandibular and sublingual salivary glands; for the nasopharyngeal glands) and in the inferior salivary nucleus (for the parotid gland and the lingual glands) (see Figure 1.4). Studies of the biophysical properties and the reflex pattern show that these neurons may consist functionally of two types, secretomotor neurons and vasodilator neurons. The main afferent physiological stimuli activating these neurons are taste stimuli and mechanical stimuli in the naso-oro-pharyngeal cavity (Matsuo and Yamamoto 1989; Matsuo and Kang 1998; Matsuo et al. 1998; Kim et al. 2004; Fukami and Bradley 2005).

Eye

Preganglionic neurons of the pupillomotor pathway (pupilloconstriction) are activated (some inhibited) by light. Preganglionic neurons of the accommodation pathway (contraction of ciliary body) are specifically activated during target tracking. Preganglionic neurons of the vasodilator pathway (vasodilation of the chorioid blood vessels) are activated by light. In *birds* and *monkeys* the preganglionic neurons are organized topically according to their function in the Edinger–Westphal nucleus: the preganglionic pupilloconstrictor neurons are located in the caudal part of the lateral Edinger–Westphal nucleus (only about 3% of the preganglionic neurons in this lateral part are involved in pupilloconstriction). Most neurons in the lateral Edinger–Westphal nucleus are involved in accommodation. Neurons in the medial Edinger–Westphal nucleus are involved in vasodilation of chorioid blood vessels. Each parasympathetic pathway in the nucleus Edinger–Westphal seems to be connected to distinct central pathways, including the pretectal olivary nucleus (pupillary light reflex), the suprachiasmatic nucleus (chorioid vasodilation reflex), the cerebellum and their nuclei (fastigial and interpositus nucleus; accommodation reflex), which determine the distinct discharge patterns (see Reiner *et al.* [1983] and Gamlin [2000] for review and references).

Conclusions

1. The peripheral autonomic nervous system supplies each group of target cells by one (sometimes two) pathway(s) each consisting of sets of pre- and postganglionic neurons with distinct patterns of reflex activity. This has been established for the lumbar sympathetic outflow to skin, skeletal muscle and viscera, for the thoracic sympathetic outflow to the head and neck and for some parasympathetic pathways. It is likely that this type of organization applies to *all* autonomic pathways, (i.e., also to those that have not been analyzed in detail).
2. There are good reasons to assume that the principle of organization into functionally discrete pathways is the same in both the sympathetic and the parasympathetic nervous systems, the only difference being that some functional targets of the sympathetic system are widely distributed throughout the body (e.g., muscle blood vessels, skin blood vessels, sweat glands, erector pili muscles, fat tissue). However, other target organs are spatially restricted in both sympathetic and parasympathetic systems.
3. Experimental investigations of sympathetic systems innervating skin or skeletal muscle in humans, using microneurography, fully support the idea of functionally discrete sympathetic pathways developed in animal studies.
4. Figure 4.23 shows the peripheral sympathetic pathways to somatic tissues (skin and skeletal muscle of hindlimb and tail) and to pelvic organs (including the distal colon), which have their origin in the upper lumbar spinal cord and which have been

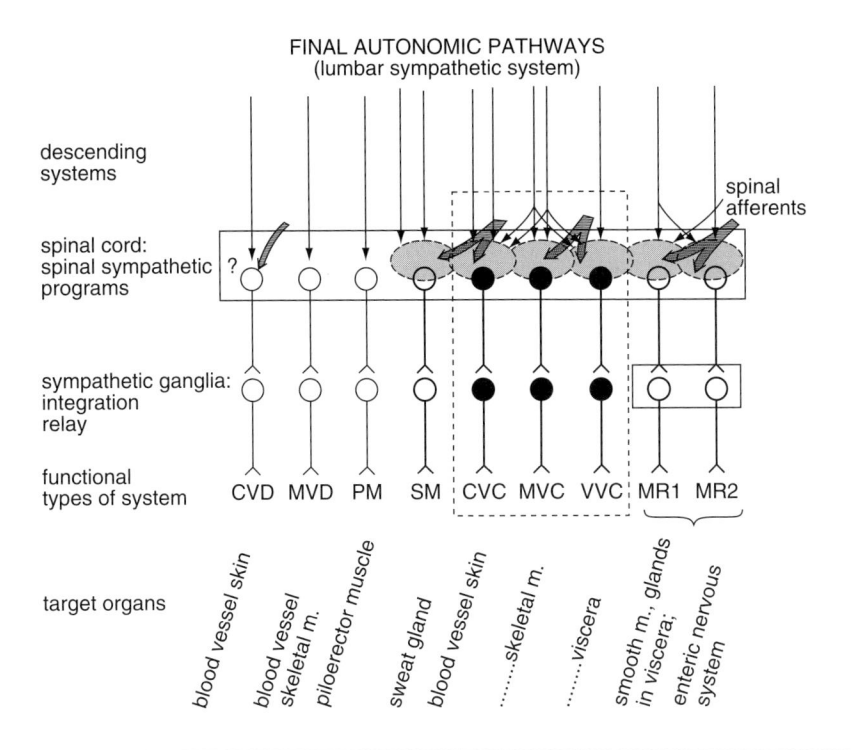

Figure 4.23 Lumbar sympathetic systems supplying skeletal muscle and skin of the hindlimb and pelvic organs (including distal colon) in the cat. The preganglionic neurons of these systems are largely located in the lumbar segments L1 to L5 (see Figure 8.8) and project in the lumbar sympathetic trunk distal to paravertebral ganglion L5 or in the lumbar splanchnic nerves and the hypogastric nerves. The first group projects (with a few exceptions related to pelvic organs) to postganglionic neurons innervating somatic tissues and the second group to postganglionic neurons innervating pelvic organs and colon. The second group includes more than two pathways as shown here (related to smooth musculature or gland in the hindgut, urinary bladder, urethra, erectile tissue, vas deferens). Some integration does occur in the sympathetic prevertebral ganglia in non-vasoconstrictor pathways to the gastrointestinal tract (ganglia in boxes; see Subchapter 6.5). Preganglionic neurons are associated with spinal circuits (shaded areas) and are under multiple supraspinal control via descending systems (see Chapters 10 and 11). CVC, cutaneous vasoconstrictor; CVD, cutaneous vasodilator; MR, motility-regulating; MVC, muscle vasoconstrictor; MVD, muscle vasodilator; PM, pilomotor; SM, sudomotor; VVC, visceral vasoconstrictor. Modified from Jänig (1986).

studied extensively in the anesthetized cat, many of them at both preganglionic and postganglionic level:

a. The reflex patterns observed in each group of autonomic neurons are the result of integrative processes in the spinal cord, brain stem and hypothalamus (all of them probably being under the control of the forebrain).

b. The neurons in many of these pathways have ongoing activity (outlined in bold in Figure 4.23), but neurons in some pathways are silent normally and/or under anesthesia and are activated only under special behavioral conditions (thin lines in Figure 4.23).

c. Functionally similar preganglionic and postganglionic neurons are synaptically connected in the autonomic ganglia, probably with little or *no* "cross-talk" between different peripheral functional pathways (see Chapter 6). This implies that little integration occurs in ganglia.

d. Vasoconstrictor neurons (black in Figure 4.23) consist of several subtypes and have a wide spatial distribution. Cutaneous vasoconstrictor neurons are functionally subdifferentiated with respect to different sections of the vascular tree and different types of skin (hairy, hairless; distal, proximal). This probably applies to muscle and visceral vasoconstrictor pathways too.

e. Neurons of the other types of sympathetic pathways supplying somatic tissues are either silent or have a very low rate of spontaneous activity (e.g., sudomotor neurons).

f. The description of motility-regulating neurons that control the pelvic organs is certainly not complete as there probably exist other types of motility-regulating neurons. These neurons innervate spatially restricted groups of target cells in the viscera (e.g., erectile tissue).

g. Release of adrenaline by the adrenal medulla is mediated by a distinct sympathetic pathway. Adrenaline is primarily a metabolic hormone, but also has some other functions.

h. Based on the functional specificity of the reflex patterns and the centrally evoked discharges in the sympathetic neurons, it is postulated that each group of preganglionic neuron receives connections from specific spinal reflex pathways and specific groups of sympathetic premotor neurons that are located in the brain stem and hypothalamus (Chapters 8 to 10).

5. Pre- or postganglionic neurons of only a few parasympathetic pathways have been studied systematically for their reflex pattern. However, where studied (e.g., those to the urinary bladder, gastroduodenal region, heart, airways) the parasympathetic neurons exhibited discharge patterns as predicted from their target cell responses. Neurons of parasympathetic pathways to blood vessels (pelvic erectile tissue, cranial blood vessels) and to glands of the head (salivary, lacrimal, nasopharyngeal) and of pelvic organs have not been studied systematically using neurophysiological methods.

6. These peripheral autonomic building blocks are the basis for the neural regulation of body functions during internal and external challenges. They are also the basis for the adaptive responses of the body during different types of behavior including basic emotions and an important basis for all protective reactions of the body.

7. In view of this organization, the concept that the sympathetic nervous system operates in an "all-or-none" fashion, without distinction between different effector organs, is misleading. The same applies to the idea of a simple functional antagonism between the sympathetic and parasympathetic nervous systems. This is corroborated by the finding that circulating noradrenaline

has no detectable function under physiological conditions. Circulating adrenaline, the release of which from the adrenal medulla is regulated by a distinct sympathetic pathway, is a metabolic hormone and does not otherwise elicit autonomic effector responses under physiological conditions. Thus, these findings refute the idea of a generally functioning sympatho-adrenal system.

Notes

1. Lovén performed his experiments on unanesthetized and immobilized (curarized) rabbits. He observed the saphenous artery (constriction, vasodilation/increase in pulsation) and the blood flow through the ear. He found that electrical stimulation of the central stump of the dorsal nerve of the hindpaw, a branch of the superficial peroneal nerve (Lovén called it the Nervus dorsalis pedis), leads to a dilation of the saphenous artery, no vasodilation but sometimes vasoconstriction in the ear and increase of blood pressure. Stimulation of the central stump of the auricular nerve generates vasodilation in the ear, no vasodilation but sometimes vasoconstriction of the saphenous artery and increase in blood pressure. He concluded (1) that vasoconstrictor fibers innervating saphenous artery and ear blood vessels must be different from vasoconstrictor fibers responsible for increase in blood pressure and (2) that vasodilation of skin vessels is generated reflexly by stimulation of afferents which innervate the same territory that is innervated by the cutaneous vasomotor fibers or a territory close by (Lovén 1866). Lovén became known as the discoverer of the taste buds at the Department of Anatomy of the Karolinska Institute in Stockholm.
2. This study was conducted on subjects with skin temperature on the big toe adjusted to 20 to 22 °C by mild whole-body cooling. Thus, the subjects were in a thermoregulatory state in which the activity in the cutaneous vasoconstrictor neurons was present (see note 3).
3. Responses of blood flow through glabrous skin in humans (monitored by laser Doppler flowmetry or photoelectrical pulse plethysmography) to arousal, mental stress and deep breath very much depend on the thermoregulatory state of the subjects (generated experimentally by whole-body cooling or warming). At skin temperatures of ≥ 30 °C (when activity in cutaneous vasoconstrictor neurons is low or absent) these stimuli generate vasoconstriction in skin. At skin temperatures of ≤ 25 °C (when activity in cutaneous vasoconstrictor neurons is high) these stimuli generate vasodilation in skin. The vasodilation is generated by decrease of activity in cutaneous vasoconstrictor neurons, the vasoconstriction in the warm state by activation of cutaneous vasoconstrictor neurons (Oberle et al. 1988; Wallin et al. 1998).
4. In the cat, the potential recorded from the surface of the hairless skin via a fluid bridge as shown in Figure 4.10a using non-polarizable silver-silver chloride electrodes (surface versus subcutaneous [extraglandular] electrode) consists of three components: (1) A direct current (DC) potential, which is negative on the surface of the skin when the sweat glands are *not* activated by sudomotor neurons. (2) This DC potential decreases during activation of sweat glands by the sudomotor neurons. Reflex activation of the sudomotor neurons leads to a slow shift of the skin potential into positive direction (see Figures 4.11a and b). (3) Activity in the sudomotor

neurons generates fast transient negative skin potentials, which are preceded by the impulses in the postganglionic sudomotor neurons (Figure 4.10b, c, Figure 4.11a, b; Jänig and Kümmel, unpublished observations).
5. Glabrous skin is the palmar and plantar skin of primates, which is different in its organization from hairless skin of non-primate mammals.
6. It has also been observed that sudomotor neurons supplying glabrous skin of the hands of hyperhydrotic patients (who suffer from bouts of profuse sweating) can be activated by stimulation of low-threshold mechanoreceptors (probably Pacinian corpuscles of the hand; Marchettini, Torebjörk, Culp & Ochoa, unpublished observation). This reflex appears to be similar to the vibration reflex in cats (Figure 4.11a) and may normally be inhibited by supraspinal mechanisms. It may function in healthy human subjects during motor commands which control movements of the hand during manipulation. I hypothesize that cortically initiated motor commands have access to sudomotor circuits which regulate the activity in the final sudomotor pathway innervating sweat glands in hairless (acral) skin.
7. Guanethidine is a compound that is actively taken up by the noradrenergic nerve terminals and concentrated in the neurosecretory vesicles where it replaces noradrenaline. As a consequence of this replacement, the noradrenergic terminals are no longer able to release noradrenaline when excited, which results in a block of noradrenergic transmission.
8. Hexamethonium is a quaternary ammonium agent that blocks cholinergic nicotinic transmission in autonomic ganglia.
9. It has been proposed that this vasodilation is generated by vasoactive intestinal peptide (VIP), which coexists with acetylcholine in sudomotor neurons, and is released by sudomotor axon terminals when activated (Lundberg 1981). Evidence against this possibility is (1) that hexamethonium differentially blocks vasodilation and sweat gland activation (i.e., a low concentration of hexamethonium blocks the vasodilation in skin but not the activation of sweat glands, whereas a high concentration of hexamethonium blocks both), (2) that intense activation of sudomotor neurons by vibration does not generate a measurable vasodilation in the skin, and (3) that vasodilation induced by released VIP is probably restricted to the tissue around the sweat glands deep below the surface of the skin (see Bell et al. 1985).
10. Neuropathy describes generically a disease involving peripheral nerves, i.e., primary afferent neurons, autonomic neurons or motoneurons. For example, during severe diabetes mellitus afferent and autonomic neurons may be affected (sensory and autonomic neuropathy).
11. Sensations elicited in the skin (e.g. upper arm, back or head) during emotional arousal are most likely due to piloerection produced by activation of pilomotor neurons and subsequent contraction of piloerector muscles. This leads to activation of low-threshold mechanoreceptors innervating the hair follicles and probably also of low-threshold slowly adapting mechanoreceptors that respond to stretch of skin (Grosse and Jänig unpublished) (see Willis and Coggeshall [2004a] for cutaneous sensory receptors).
12. Muscle spindles and Golgi tendon organs are encapsulated structures in skeletal muscle. Muscle spindles are arranged in parallel with extrafusal muscle fibers and are innervated by fast-conducting (afferent) Ia-fibers (in

addition to afferent group II fibers and efferent γ-fibers), which measure length of muscle and its derivatives. Golgi tendon organs are located at the junction between extrafusal muscle fibers and tendon. They are innervated by fast-conducting Ib-fibers, which measure force and its derivatives of extrafusal muscle fibers.

13. The distinct reflexes in these sympathetic neurons are mediated by sacral visceral afferents and not lumbar visceral afferents because they are unchanged after interruption of the lumbar visceral afferents.

14. In the cat, dilation in skeletal muscle is generated by electrical low-frequency (<2 Hz) stimulation of the innervation of the adrenal medulla or by intravenous infusion of adrenaline (<0.5 μg/[kg min]) (Celander 1954). In humans intra-arterial infusion of 0.05 to 0.1 μg/min adrenaline generates vasodilation in skeletal muscle but vasoconstriction in skin (Golenhofen et al. 1962). This adrenaline-induced vasodilation in skeletal muscle is most likely not due to the metabolic effect of adrenaline on the skeletal muscle fibers (release of lactic acid) but due to its direct effect on the vascular smooth muscle (Schmidt-Vanderheyden and Koepchen 1967).

15. Some autonomic effector organs, such as smooth muscle, cardiac muscle, exocrine glands and pineal gland, develop adaptive supersensitivity after denervation and to a certain degree also after decentralization (interruption of the preganglionic axons). This supersensitivity may be interpreted as a compensatory mechanism by which the excitable tissue increases its excitability to extrinsic signals such as neurotransmitters and other compounds. The supersensitivity may be relatively specific (e.g., increased sensitivity of the cardiac muscle for catecholamines) or non-specific. The underlying cellular changes are multiple and include change of postreceptor pathways, membrane potential, electrogenic Na^+–K^+ transport, ion channels, intracellular Ca^{2+} sequestration and other changes. There is no evidence for an increase of density and affinity of receptors for a neurotransmitter in autonomic target cells after denervation. These cellular mechanisms vary between effector tissues and have been poorly investigated (Fleming and Westfall 1988). Denervation in vivo may result in responses of heart, pupil and vasculature to circulating catecholamines. This supersensitivity may even be maintained over a long time when the target tissue is reinnervated (Jobling et al. 1992; Koltzenburg et al. 1995).

16. *Venular plasma extravasation* is a component of inflammation in all tissues. It is under neural (not in the skin of humans) and neuroendocrine control (see note 2 in Chapter 2). Sensitization of nociceptor for mechanical stimulation leads to *mechanical hyperalgesic behavior*. Both venular plasma extravasation and mechanical hyperalgesic behavior serve the protection of body tissue, leading to healing after noxious (tissue-damaging) stimuli. These are both called *protective body reactions*.

Chapter 5

The enteric nervous system

5.1 Anatomy, components and global functions of the enteric nervous system	page 169
5.2 The enteric nervous system is an autonomic system in its own right	178
5.3 Regulation of motility and intraluminal transport in the small and large intestine: the neural basis of peristalsis	181
5.4 Integration of enteric neural, pacemaker and myogenic mechanisms in generation of motility patterns	188
5.5 Regulation of secretion and transmural transport	194
5.6 Defense of the gastrointestinal tract and enteric nervous system	196
5.7 Control of the enteric nervous system by sympathetic and parasympathetic pathways	200

In his chapter "The sympathetic and related systems of nerves" in *Schäfer's Textbook of Physiology* (1900) Langley defined for the first time the idea that the gastrointestinal tract has a nervous system of its own. He called this system the enteric nervous system (Langley 1900). He repeated this idea in his short monograph in 1921 where he clearly described the division of the autonomic nervous system into three parts: *sympathetic*, *parasympathetic* and *enteric nervous system*. This classification is still used today (Langley 1921). The existence of the plexuses of Auerbach (plexus myentericus) and Meissner (plexus submucosus) has been known since the second half of the nineteenth century. Langley recognized that this system can act to a large extent independently of the central nervous system. He separated the myenteric and submucosal ganglia from the sympathetic (and parasympathetic) nervous system and classified them as a *third autonomic nervous system* for the following reasons: (1) they have a distinct histology compared to the histology of the paravertebral and prevertebral sympathetic ganglia; (2) it was unclear at that time whether they are connected with the central nervous system by sympathetic and parasympathetic neurons and (3) the sympathetic postganglionic fibers either send collaterals to or form synapses with the neurons of the enteric nervous system (Langley 1900). Up to about 1970 rather little or no attention was given to the enteric nervous system of the gastrointestinal tract; in fact this system was practically ignored.

Our present knowledge about the neurobiology of the enteric nervous system is chiefly based on systematic experimental studies of this system in the guinea pig and recently in the mouse (including various mutant mice) combining various methods (e.g., neurophysiology, histology, immunohistochemistry, pharmacology, etc.). In the center of these experimental investigations are different types of in vitro preparations of the small intestine and distal colon of the guinea pig, which allow one to study the enteric neurons under visual control and with respect to their function. This type of systematic research is correlated with research on the intrinsic neural control of the gastrointestinal tract in other species including the human. Two research groups represented by John Furness and Marcello Costa have performed pioneering experimental work in the last 30 years. Mainly based on this research, I will describe the anatomy, the general concepts and some general functional characteristics of the enteric nervous system. For further details, some comprehensive reviews are available (see Furness and Costa [1987]; Gershon [1994]; Wood [1994]; Kunze and Furness [1999]; Furness *et al.* [2003a, b, 2004]; Brookes and Costa [2002]; Furness [2005]).

5.1 | Anatomy, components and global functions of the enteric nervous system

5.1.1 Anatomy and global functions

The gastrointestinal tract serves various global functions: ingestion of energy-rich compounds, water, electrolytes and some other vital substances (e.g., vitamins); transport of its content; enzymatic breakdown of the energy-rich and other vital substances; resorption of the breakdown products, electrolytes, water and vitamins; and evacuation of waste products. To accomplish these functions the gastrointestinal tract is endowed with a nervous system of its own, the enteric nervous system. Under normal conditions the central nervous system has delegated many basic functions to the enteric nervous system and does not interfere with the individual functions of the gastrointestinal tract. It has special commands over the regulation of the intake and of the evacuation of the gastrointestinal tract. Furthermore it adapts the functioning of the gastrointestinal tract during special behaviors of the organism.

Anatomically the enteric nervous system consists of the myenteric plexus and the submucosal plexus and the axons these supply to various effectors (Figure 5.1). The plexuses contain the neurons of the enteric nervous system. The nerve cell bodies are grouped in

Figure 5.1 Transverse section through the small intestine and its mesentery (a) and representation of the enteric plexuses in relation to the layers of the intestine (b). Indicated are smooth muscle layers, neural plexuses, submucosa and mucosa viewed macroscopically. Modified from Furness and Costa (1980) with permission.

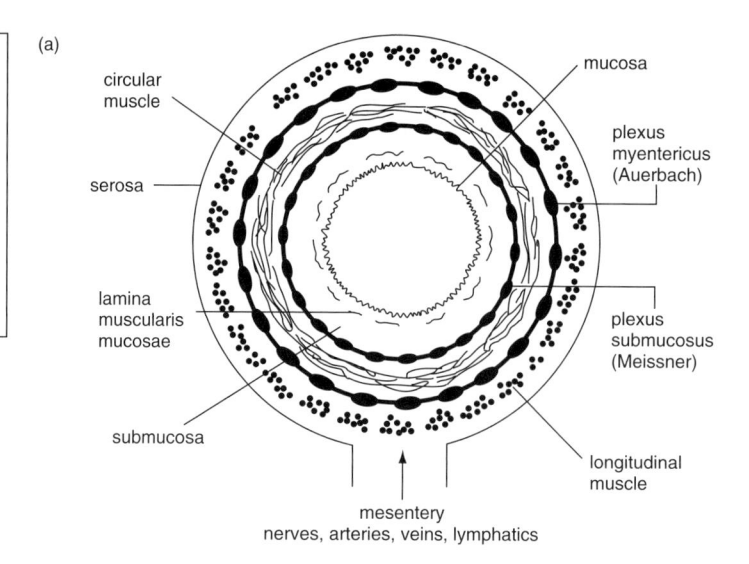

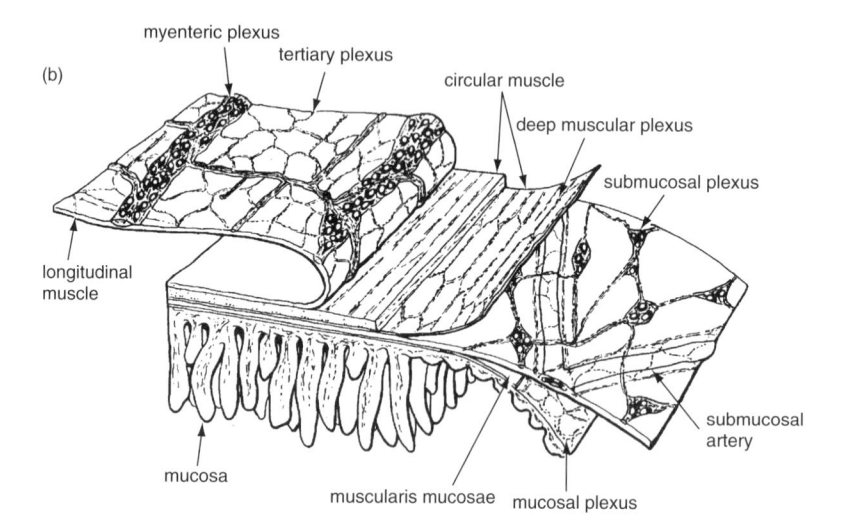

small enteric ganglia, which are connected by bundles of nerve cell processes (axons and dendrites). The myenteric plexus is located between the circular and longitudinal smooth musculature and the submucosal plexus is arranged between the circular musculature and the submucosa. In many species the submucosal plexus consists of an inner plexus (closer to the mucosa) and an outer plexus (closer to the circular muscle). This plexus is prominent in the small intestines and colon and almost absent in the esophagus and stomach (although there are species differences). The myenteric plexus is present throughout the entire gastrointestinal tract. The enteric nervous system is also found in the gallbladder wall and pancreas, including the bile ducts, all three having developed from the small intestine. Individual ganglia in both plexuses contain up to 200 nerve cell bodies that are arranged in the quasi two-dimensional planes of the plexuses.

The enteric nervous system contains approximately 10^7 to 10^8 neurons (varying according to species and counting procedure; e.g., in humans 2 to 6 times 10^8 neurons [Furness 2005]). This number is in the same order of magnitude as the number of neurons in the spinal cord. The myenteric plexus is mainly (but not exclusively) involved in the regulation of different motility patterns whereas the submucosal plexus is mainly (but not exclusively) involved in the regulation of secretion of water and electrolytes. The motor neurons of the submucosal plexus project preferentially to target cells in the mucosa, but also to cells of the circular muscles. The motor neurons of the myenteric ganglia project preferentially to the smooth musculature, but also to the mucosa and to the submucosal plexus.

The effector cells of the enteric nervous system are smooth muscle cells of the longitudinal and circular musculature, mucosal muscle cells, secretory mucosal cells (which include secretory glands), hormone-secreting cells in the mucosa, and mucosal and submucosal arterioles. Smooth muscle cells and cells of secretory epithelia are electrically coupled by gap junctions. Thus these cells form functional and electrical syncytia that comprise the effector tissues of the enteric nervous system. Interstitial cells of Cajal (ICC), which form several plexuses of electrically coupled cell in the gastrointestinal tract that are arranged in parallel with the smooth muscle cell layers, are also effector cells of the enteric nervous system (see Subchapter 5.4). It is debated whether the gut-associated lymphoid tissue (GALT), which is by cell count the largest immune organ in the body, is also an effector tissue of the enteric nervous system (Shanahan 1994; Wood 1994, 2002).

5.1.2 Components of the enteric nervous system

The enteric nervous system contains three classes of neurons in the myenteric plexus and submucosal plexus: intrinsic primary afferent neurons, motor neurons and interneurons. Each class of neurons consists of several types, which are characterized by their anatomy (location and projection of axons), neurochemical coding (content of neuropeptides, primary transmitters as well as cotransmitter(s) (Figure 5.2, Table 5.1).[1]

Intrinsic primary afferent neurons (IPANs)
About 20% of the enteric neurons in the small intestine are IPANs (Furness *et al.* 1998, 2003b, 2004).[2] The cell bodies of these neurons lie in the myenteric and submucosal plexuses. They are activated by various physiological stimuli, such as mechanical stimuli (e.g., shearing and pressure stimuli applied to the mucosa; distension and contraction of the wall of the gut), or intraluminal chemical stimuli (e.g., inorganic acids, bases, short-chain fatty acids or glucose). Excitation of the enteric afferent neurons by chemical or mechanical stimulation of the mucosa is partly mediated by enterochromaffin cells in the mucosa releasing 5-hydroxytryptamine

Figure 5.2 Types of neurons of the enteric nervous system in the small intestine of the guinea pig. The neurons have been defined by their functions, cell body morphology, projection to targets, neurochemistry and primary transmitters. Intrinsic primary afferent neurons (IPANs) hatched, motor neurons in black, interneurons shaded. Motor neurons innervating the muscularis mucosae or endocrine cells are not shown. The numbers adjacent to the neurons correspond to the numbers in Table 5.1, which lists each of the neuron types by their function, neurochemistry, primary transmitter and percentage of cell bodies in the myenteric or submucosal plexus. BV, blood vessel; CM, circular muscle; LM, longitudinal muscle; MP, myenteric plexus; Muc, mucosa; SMP, submucosal plexus. Modified from Furness *et al.* (2004) with permission.

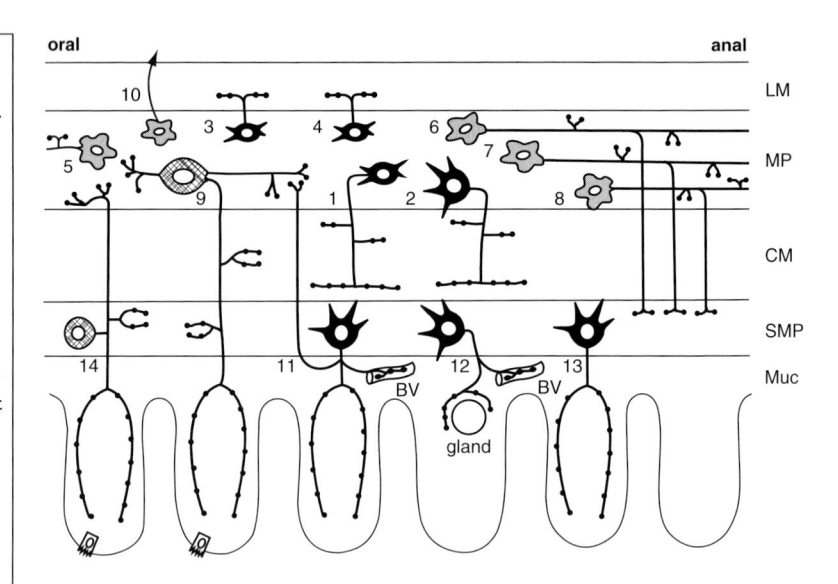

(5-HT), and possibly by other enteric hormone cells in the mucosa (e.g., cells releasing cholecystokinin [CCK] or motilin). Thus, the mechanosensitive enteric afferent neurons would also be in strict sense chemosensitive.

Intrinsic primary afferent neurons whose cell bodies are located in the submucosal plexus have their receptive endings in the mucosa. They respond to mechanical distortion of the mucosa (e.g., shearing stimuli exerted by the luminal content) and to intraluminal chemical stimuli. Intrinsic primary afferent neurons located with their cell bodies in the myenteric plexus have their receptive endings in the external muscle layer and in the mucosa. They respond to contraction and distension of the enteric musculature and to intraluminal chemical stimuli. In the *small intestine* of the *guinea pig*, most IPANs are polymodal, i.e. they respond to various mechanical as well as chemical stimuli. Some IPANs may have a relative stimulus specificity, i.e. they respond preferentially to one type of adequate stimulus and weaker to other stimuli. It is unknown whether this relative specificity is functionally important (Furness *et al.* 2004). In the *colon* of the *guinea pig*, the mechanosensitive IPANs consist functionally of two types. One type is stretch-sensitive but does not respond to contraction. The second type responds preferentially to contraction of the smooth musculature but less to distension of the colonic wall (Spencer *et al.* 2002, 2003a).

Most IPANs are afterhyperpolarizing (AH) neurons (i.e. they exhibit long-lasting afterhyperpolarization after excitation, which is dependent on two types of potassium channels; see note 1). These neurons use acetylcholine and tachykinins as transmitters. Enteric afferent neurons connect synaptically to other enteric afferent neurons to form *networks of enteric afferent neurons* in the myenteric plexus. These networks connect to enteric interneurons and to enteric motor neurons (see Figure 5.6).

Motor neurons

Enteric neurons innervating target cells are called enteric motor neurons. These neurons are specialized with respect to the different groups of target cells they innervate:

- *Excitatory muscle motor neurons* supplying smooth muscle cells innervate the longitudinal or circular smooth musculature and are involved in descending and ascending reflexes. The cell bodies of these motor neurons are located in the myenteric plexus, representing about 40% of all neurons in this plexus. They are cholinergic and use acetylcholine as the primary transmitter, which acts via cholinergic muscarinic receptors. Tachykinins (substance P, neurokinin A and K) are colocalized with acetylcholine in these neurons and may serve as secondary transmitters. Some excitatory muscle motor neurons also innervate muscle fibers in the lamina muscularis mucosae.
- Motor neurons that induce relaxation of circular smooth muscles are involved in descending inhibitory reflexes and therefore in accommodation of the stomach. They are located in the myenteric plexus and represent about 16% of all neurons in this plexus. These *inhibitory muscle motor neurons* use nitric oxide (NO), adenosine triphosphate (ATP), "pituitary adenylyl cyclase activating peptide" (PACAP) and vasoactive intestinal peptide (VIP) as transmitters. In most inhibitory muscle motoneurons NO may serve as the primary transmitter and the other compounds as secondary transmitter(s). The specific combination of these substances, which potentially serve as inhibitory transmitters, varies between groups of inhibitory motoneurons depending on the section of the gastrointestinal tract and on the species. Both excitation and inhibition generated by enteric muscle motor neurons in the smooth muscles are mediated, at least in part, by intramuscular interstitial cells of Cajal (ICC) (Burns *et al.* 1996; Ward *et al.* 1998; Ward and Sanders 2001; see Subchapter 5.4).
- The lamina muscularis mucosae is also innervated by inhibitory and excitatory motor neurons. The cell bodies of these motor neurons are located in the myenteric plexus.
- Neurons that innervate the mucosal epithelium in the small and large intestines and gallbladder (causing water, electrolyte and bicarbonate secretion) are called *secretomotor neurons*. Almost all secretomotor neurons are located in the submucosal plexus and represent about 90% of all neurons in this plexus. Neural control of secretion by the enteric nervous system is closely linked to the control of local blood flow by enteric *vasodilator neurons*. Both appear to be exerted by the same populations of secretomotor neurons. Thus some secretomotor neurons innervate both gland cells and the initial segments of arterioles in the submucosa and in the outer part of the mucosa. The submucosal blood vessels are the distributing vessels for the blood supply of the mucosa and even in part of the external musculature. About 30% of these secretomotor

neurons have only secretory function and it is not excluded that the mucosal blood vessels are also controlled by a distinct class of enteric vasodilator neurons (see Figure 5.12). The secretomotor neurons are either cholinergic and use acetylcholine as primary transmitter, acting on muscarinic cholinergic receptors, or they are non-cholinergic and use VIP as primary transmitter. Neurons with dual (vasodilator and secretomotor) function seem to be preferentially non-cholinergic. In the stomach neural control of parietal cells secreting proton ions is exerted by cholinergic secretomotor neurons. This secretion is accompanied by local vasodilation, which is either also mediated by enteric cholinergic neurons or is reactive consequent to the activation of the parietal cells (reactive or functional hyperemia).

- Several types of enteric endocrine cells are located in the densely innervated mucosa of the gastrointestinal tract. There is evidence that these endocrine cells are innervated too and therefore are under neural control. The best studied example is the control of gastrin release by G-cells in the antral mucosa by enteric motor neurons. It is likely that release of other hormones of the gastrointestinal tract is under neural control as well (e.g., secretin released by S-cells in the duodenal mucosa, hormones [insulin, glucagon, pancreatic polypeptide] released by endocrine cells in the pancreas). The G-cells are innervated by cholinergic neurons in the myenteric plexus. These neurons express gastrin-releasing peptide (GRP). Location and type of motor neurons associated with other enteric endocrine cells are unknown.

Motor neurons of like type are not synaptically connected with each other and therefore do not form motoneuron networks as do intrinsic primary afferent neurons and interneurons.

Interneurons

The enteric nervous system contains various groups of neurons that have neither afferent nor motor neuron function and that are interposed between these two groups of neuron. So far, in the myenteric plexus of the guinea pig small intestine, three classes of interneuron projecting anally with their axons and one class of interneuron projecting orally with its axons have been identified (Figure 5.2); they form about 16% of all neurons in this plexus. Interneurons seem to be absent in the submucosal plexus. Thus, IPANs are synaptically connected to secretomotor neurons in the submucosal plexus without intervening interneurons (Figure 5.12).

Probably all interneurons are cholinergic, but synaptic transmission from them to motor neurons and between interneurons is not purely cholinergic. Interneurons of like type are synaptically connected with each other and form chains of like neurons that run either orally or anally (Figure 5.6). The ascending interneurons are involved in regulation of motility and transmit their activity by nicotinic action of acetylcholine (ACh). The descending interneurons

Table 5.1 | Types of neuron of the enteric nervous system in the guinea pig small intestine

Myenteric neurons[a]	Proportion	Chemical coding	Functions/comments
Excitatory circular muscle motor neurons (1)	12%	Short: ChAT/TK/ENK/GABA Long: ChAT/TK/ENK/NFP	To all regions; PT **ACh**, CT **TK**
Inhibitory circular muscle motor neurons (2)	16%	Short: NOS/VIP/PACAP/ENK/NPY/GABA Long: NOS/VIP/PACAP/Dynorphin/GRP/NFP	Several CT with varying prominence: NO, ATP, VIP, PACAP
Excitatory longitudinal muscle motor neurons (3)	25%	ChAT/Calretinin/TK	PT **ACh**, CT **TK**
Inhibitory longitudinal muscle motor neurons (4)	about 2%	NOS/VIP/GABA	Several CT with varying prominence: NO, ATP, VIP, PACAP
Ascending interneurons (local reflex) (5)	5%	ChAT/Calretinin/TK/ENK	PT **ACh**, CT **TK**
Descending interneurons (local reflex) (6)	5%	ChAT/NOS/VIP ± GRP ± NPY	PT **ACh**, CT **ATP**?
Descending interneurons (secretomotor reflex) (7)	2%	ChAT/5-HT	PT **ACh**, **5-HT** (at 5-HT$_3$ receptors)
Descending interneurons (migrating myoelectric complex) (8)	4%	ChAT/SOM	PT **ACh**
Myenteric intrinsic primary afferent (primary sensory) neurons (9)	26%	ChAT/calbindin/TK/NK$_3$ receptor	PT **TK** and **ACh** (CGRP and ACh in other species)
Intestinofugal neurons (10)	<1%	ChAT/GRP/VIP/NOS/CCK/ENK	PT **ACh**, mainly from colon
*Motor neurons to gut endocrine cells	N/A	N/A	E.g., myenteric neurons innervating gastrin cells; also in submucosal ganglia
*Secretomotor neurons to gastric glands	N/A	ChAT	Stimulate gastric acid secretion

Table 5.1 (cont.)

Submucosal neurons[a,b]	Proportion	Chemical coding	Functions/comments
Non-cholinergic secretomotor/vasodilator neurons (11)	45%	VIP/PACAP/GAL	PT **VIP**; a few with cell bodies in myenteric ganglia
Cholinergic secretomotor/vasodilator neurons (12)	15%	ChAT/calretinin/dynorphin	PT **ACh**
Cholinergic secretomotor (non-vasodilator) neurons (13)	29%	ChAT/NPY/CCK/SOM/CGRP/dynorphin	PT **ACh**; a few with cell bodies in myenteric ganglia
Submucosal intrinsic primary afferent (primary sensory) neurons (14)	11%	ChAT/TK/Calbindin	PT **ACh** (?TK) (CGRP and ACh in other species)
*Excitatory motor neurons to musculares mucosae	N/A	ChAT/TK	PT **ACh**
*Inhibitory motor neurons to musculares mucosae	N/A	NOS/VIP	PT?, similar to myenteric inhibitory motor neurons

Notes:
Types of neurons in the small intestine, some of their defining characteristics and percentage of occurrence found in the enteric nervous system of the guinea pig.
[a] see Figure 5.2 for numbers.
[b] see also Figure 5.12 for numbers.
*Types of motor neuron found in other parts of the digestive tract.
Abbreviations: ACh, acetylcholine; ATP, adenosine triphosphate; ChAT, choline acetyltransferase; CCK, cholecystokinin; CGRP, calcitonin gene-related peptide; CT, cotransmitter; ENK, enkephalin; GABA, γ-aminobutyric acid; GAL, galanin; GRP, gastric-releasing peptide; 5-HT, 5-hydroxytryptamine (serotonin); NA, not applicable; NFP, neurofilament protein; NK, neurokinin; NO, nitric oxide; NOS, nitric oxide synthase; NPY, neuropeptide Y; PACAP, pituitary adenylyl cyclase activating peptide; PT, primary transmitter; SOM, somatostatin; TK, tachykinin; VIP, vasoactive intestinal peptide. From Furness *et al.* (2004) with permission and Brookes (2001).

probably use 5-HT (ACh/5-HT descending interneurons that are involved in secretory reflexes) or ATP (ACh/ATP/NOS [nitric oxide synthase] descending interneurons that are involved in inhibitory motor reflexes), in addition to acetylcholine, as cotransmitter. Whether NO is used as a transmitter by these interneurons is unclear. If it is, its effect is minor. Descending cholinergic interneurons containing somatostatin seem to be involved in the transmission of the migrating myoelectrical complex (MMC, see Subchapter 5.4).

The interneurons vary considerably in type between different sections of the gut and are difficult to identify anatomically, histochemically and physiologically. This is not surprising in view of the difficulties in functionally identifying spinal interneurons related to the somatomotor system, to spinal autonomic systems (see Chapter 9) or to sensory systems, or in identifying interneurons in the lower brain stem that are related to the arterial baro- and chemoreceptor reflexes (see Subchapters 10.3 and 10.4).

Finally, intestinofugal neurons that project through mesenteric nerves to sympathetic prevertebral ganglia constitute a further type of interneuron. These neurons form cholinergic synapses with noradrenergic neurons projecting to the gastrointestinal tract. Intestinofugal neurons are particularly numerous in the colon and rectum. They are involved in extraspinal intestinointestinal reflexes that relay signals from more distal to more proximal regions of the gastrointestinal tract and are involved in regulation of motility and possibly also of secretion (see Figures 5.15 and 6.11). Intestinofugal neurons are synaptically activated by IPANs responding to, e.g., distension. Some intestinofugal neurons may be a type of intrinsic primary afferent neuron activated by intestinal distension.

5.1.3 Neurochemical coding of enteric neurons

Enteric neurons contain many (more than 30) substances that could potentially be neurotransmitters. However, only a few substances have been established as neurotransmitters. It is very possible that most substances contained in these neurons are not used as neurotransmitters or neuromodulators. There is consensus that a few substances serve as primary transmitters and that these primary transmitters are more or less uniformly found in the same groups of enteric neurons across species (e.g., acetylcholine, 5-HT, tachykinins [substance P, neurokinin A and K], VIP). Substances that are colocalized with the primary transmitters may serve as secondary or subsidiary transmitters or neuromodulators (see Table 5.1). These vary in distinct groups of enteric neurons and between species. The presence of these different substances, in variable combinations, in enteric neurons is a neurochemical code for these neurons. This code can be used to recognize and identify the neurons and is a valuable tool in the functional experimental studies of enteric neurons (Costa *et al.* 1996).

5.2 | The enteric nervous system is an autonomic system in its own right

The global functions of the gastrointestinal tract are intraluminal transport and mixing of its content, enzymatic and chemical processing of ingested food, absorption of nutrients, water, electrolytes and some other vital substances, evacuation of waste products and vomiting, and protection of the body against poisonous substances and pathogens. These functions are adjusted and coordinated at any moment by the enteric nervous system, independent of the brain, but in concert with gastrointestinal hormones (e.g., in gastric acid secretion). The enteric nervous system contains a limited repertoire of neural programs to organize these functions according to the state of the gastrointestinal tract (e.g., fasting or digestion). The component parts of the enteric nervous system that are important to accomplish these functions have been discussed in the last subchapter. Synaptic connections between intrinsic afferent neurons, interneurons and motoneurons are organized in such a way that excitation of intrinsic primary afferent neurons by physiological stimuli leads to many distinct reflexes mediated by the enteric nervous system; the underlying mechanisms of some of them have been worked out in detail (see Subchapter 5.3). Under physiological conditions these reflexes are spatially and temporally coordinated and lead to typical behaviors of the gastrointestinal tract. Single reflexes, as they are experimentally analyzed for their underlying mechanisms, are isolated parts of these integrated behaviors and valuable tools in this experimental analysis. This situation is conceptually very similar to that in the experimental analysis of the integrative motor or autonomic processes represented in the spinal cord (see Chapter 9) or in the lower brain stem (e.g., the dorsal vagal complex that consists of the nucleus tractus solitarii, the dorsal motor nucleus of the vagus [DMNX] and the area postrema; see Subchapter 10.7)

The diagram in Figure 5.3 graphically outlines the integrative functioning of the enteric nervous system and its control by the brain. Afferent neurons, interneurons and motor neurons form various reflex circuits, which are defined by the reacting target tissue of the motor neurons and therefore by the function(s) of the motor neurons. Some reflex circuits are monosynaptic between afferent and motor neurons, others are polysynaptic. These reflex circuits include those in the same part as well as those between different parts of the gastrointestinal tract (e.g., gastro-gastric reflexes, duodeno-gastric reflexes [response to glucose], ileo-gastric reflexes [response to lipids], esophageal-gastric reflexes [response to distension] etc.). The interneurons, which are defined by their synaptic input, their synaptic output and projection of their axons, probably play an important role in integrating the various reflex circuits. The

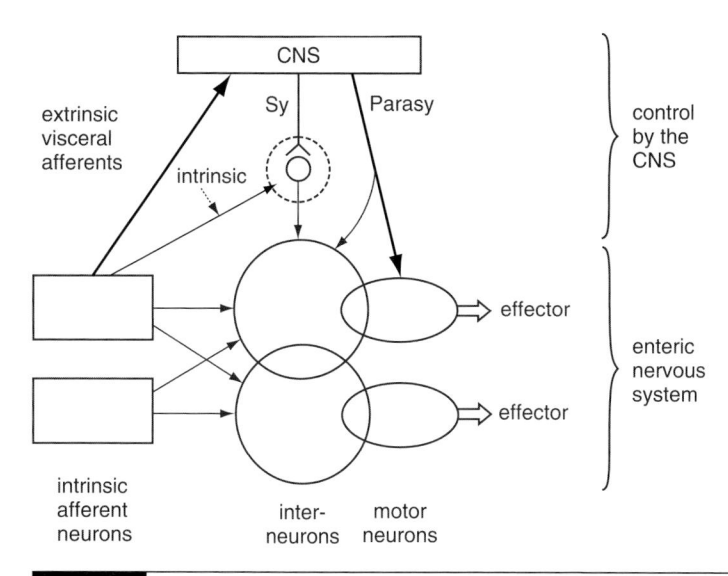

Figure 5.3 Concept of functioning of the enteric nervous system and its control by the central nervous system (CNS). The enteric nervous system consists of intrinsic afferent neurons, interneurons and motoneurons, which form various reflex circuits. Effector cells can be smooth muscle cells, interstitial cells of Cajal (ICC), secretory epithelia, hormone-secreting cells, arterioles and possibly immune cells. Each category of neuron is subdivided into several types according to its physiological response properties, morphology (location of cell body, projection of axon and dendrites), immunohistochemistry and transmitter(s) (Table 5.1). The circuits and their reciprocal synaptic connections represent the "sensorimotor programs" of the enteric nervous system, which regulate movement patterns and secretory processes. Interneurons are important in the coordination of the different processes. Extrinsic visceral afferent neurons projecting to the spinal cord or to the nucleus tractus solitarii provide detailed information about events in the gastrointestinal tract to the CNS. The CNS acts on the enteric nervous system via sympathetic (Sy) and parasympathetic (Parasy) pathways. With some exceptions, it does not regulate single motor functions of the gastrointestinal tract but interferes with the sensorimotor programs of the enteric nervous system. For example, activation of sympathetic motility-regulating neurons may interrupt peristalsis by interfering with enteric neurons and close sphincters directly. The CNS has direct control over intake of nutrients, disposal of waste products and resistance vessels. Intrinsic neurons (intestinofugal interneurons) project to prevertebral sympathetic ganglia giving afferent feedback to postganglionic secretomotor and motility-regulating neurons. Modified from Wood (1994) with permission.

population of interneurons is an important component of the various *sensorimotor programs* that are represented in the enteric nervous system. This organization is the basis for the autonomous functioning of the enteric nervous system, independent of the brain. Therefore the enteric nervous system is sometimes also called "*brain of the gut.*" During normal regulation of the gastrointestinal tract some of these enteric reflex components may not be readily visible (as is the case with the functioning of spinal autonomic circuits; see Chapter 9). However, this does not mean that they are unimportant and secondary.

The central nervous system is continuously informed about the state of the gastrointestinal tract by impulse activity in extrinsic vagal and spinal visceral afferent neurons and directly and indirectly by non-neural signals (e.g., hormones of the gastrointestinal tract that act via the area postrema of the medulla oblongata or the arcuate nucleus of the hypothalamus on central circuits, such as gastrin, CCK, ghrelin, glucagon-like peptide 1, peptide YY [see Subchapter 10.7]; glucose concentration in the blood; cytokines of the immune system of the gut). The neural afferent feedback signals from the gut are highly differentiated (see Chapter 2; Table 2.1); they include information about the mechanical and chemical state of the gastrointestinal tract, about inflammatory and other noxious processes and about processes in the immune system of the gut (the GALT) that forms the most powerful defense barrier of the body to the outside world (Shanahan 1994). The importance of the afferent feedback signals is reflected in the high number of visceral afferent neurons projecting in the vagus nerves to the lower brain stem (nucleus tractus solitarii; see Chapter 2 and Subchapter 8.3); about 85% of the axons in the subdiaphragmatic vagus nerves are afferent. An important component in the coordination of motility and secretion of different segments of the gut are the sympathetic prevertebral ganglia. Sympathetic secretomotor and motility-regulating neurons in these ganglia receive synaptic input from enteric intestinofugal interneurons, which are activated by intrinsic afferent neurons (Figure 5.3; see Subchapter 6.5). Signals from the central nervous system and the enteric nervous system are thus integrated in prevertebral ganglia (Figure 6.10b; Subchapter 6.5).

The central nervous system influences the functioning of the enteric nervous system via parasympathetic pathways projecting from the DMNX through the vagus nerves and from the sacral spinal cord through pelvic splanchnic nerves and via sympathetic systems in the thoracolumbar spinal cord (see Subchapters 4.3 and 4.8). The numbers of preganglionic parasympathetic neurons influencing the gastrointestinal tract are rather small when compared to the total number of neurons of the enteric nervous system (e.g., in the rat approximately 5000 preganglionic parasympathetic neurons project through the vagus nerve to the gut). The central nervous system does not interfere directly with most individual effector tissues in the gut but indirectly by giving commands to the enteric neural circuits. Most noradrenergic non-vasoconstrictor neurons in the prevertebral ganglia innervate enteric neurons; the smooth musculature of sphincters are directly innervated by noradrenergic postganglionic neurons. Some sympathetic postganglionic secretomotor neurons may directly innervate the mucosa. Even in the distal colon of rats most postganglionic parasympathetic neurons of the pelvic splanchnic ganglion that project to the colon do innervate the myenteric plexus but not the smooth musculature (Luckensmeyer and Keast 1998a). Thus, here also there are chains of three neurons connected serially between the sacral spinal cord and the effector cells in the distal gastrointestinal tract.

However, at the beginning and at the end of the gut, the central nervous system seems to have more or less direct command over some gastrointestinal functions, such as regulation of defecation and storage of the colon, and in the control of some functions of the stomach. This is also reflected in a high innervation density of both parts of the gut by extrinsic afferent neurons. It is consistent with the idea that these parts of the gut are precisely and rapidly adapted to the behavior of the organism. The central nervous system also has direct command over the resistance of arterial blood flow in the gut exerted via visceral vasoconstrictor neurons. This centrally regulated function is important for the regulation of arterial blood pressure and regional blood flow. It appears to be entirely separate from regulation of the fluid balance in which the mucosal and submucosal arterioles are also involved.

5.3 | Regulation of motility and intraluminal transport in the small and large intestine: the neural basis of peristalsis

5.3.1 Motility patterns and slow waves

Stomach and small and large intestine show two basic patterns of motility in all mammalian species: the interdigestive motility pattern, characterized by the migration myoelectric complex (MMC), and the fed pattern of activity.

- The *interdigestive motility pattern* passes along the intestine at regular intervals (in the human at intervals of 80 to 110 minutes and somewhat faster in smaller mammals). It occurs between the feeding and digesting periods. In continuously feeding mammals it also passes along the intestine during feeding.
- The *fed pattern of motility* consists of ongoing irregular phasic contractions. Some 30% to 40% of the contractions propagate like peristaltic waves over distances of up to 10 cm to the anal site. Thus, this pattern includes what is called segmentation, cycling as well as peristaltic movements of the gut.

Figure 5.4 demonstrates electrical activity and mechanical myogenic activity (contractions) of a proximal segment of the mouse small intestine in vitro. Compound electrical activity was recorded extracellularly from the serosal surface of the intestinal segment at three sites with electrodes separated by about 2 cm and pressure was recorded intraluminally at the corresponding sites. At a preset intraluminal pressure gradient of about 1 cm water or more the electrical activity shows rhythmic waves that are called *slow waves*. Some groups of slow waves are followed by local rhythmic contractions of the intestinal segment in a 1:1 manner leading to rhythmic transient increase of the intraluminal pressure and pulsatile outflow from the intestinal tube. These contractions only occur if the slow electrical waves show action potentials on their plateau phases. Slow waves without action potentials are not followed by contractions. The action potentials (see arrows) are not seen very well in this type of extracellular recording. However, simultaneous recording of potential changes in the smooth muscle cells and of contraction clearly shows that the slow waves without action potentials are not followed by contractions whereas those with action potentials are

Figure 5.4 Slow waves and distension-induced periodic occurrence of action potentials superimposed on slow waves in an isolated proximal segment of the mouse small intestine (a) associated with propagating bursts of transient increases in intraluminal pressure (b). Slow waves were recorded extracellularly from the serosal surface at three different sites, 10 mm apart (i, ii, iii). Intraluminal pressure was recorded at the same sites. The intestinal segment was distended setting the intraluminal pressure at 40 mm H$_2$O. (a) Action potentials occurred in five to ten slow waves, followed by two to six slow waves without action potentials. The extracellularly recorded compound action potentials are indicted by arrows (indicated in the third record on the right). (b) Periodic increase in intraluminal pressure in a 1:1 correlation with the slow waves carrying action potentials. Pulsatile outflow from the intestine is indicated by black boxes. Lines indicate propagating slow waves and contraction waves (10 mm/s). Modified from Huizinga et al. (1998) with permission.

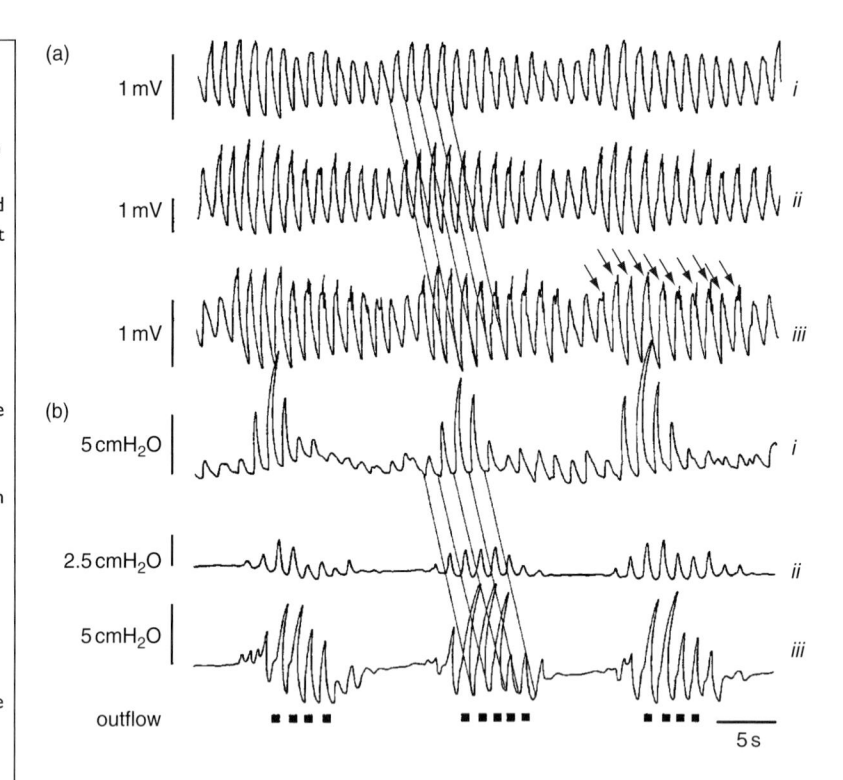

(Figure 5.5). Both slow electrical waves and contractile activity are generated by the smooth musculature; they are synchronized and propagate in the anal direction at a characteristic velocity, the speed of the latter depending on the section of the gastrointestinal tract and the animal species. The action potentials are generated by inflow of Ca^{2+} through L-type calcium channels and the contractions are mediated by an increase of intracellular Ca^{2+} in the cytosol. This coordinated pattern of electrical and contractile activity is the basis of all active motility patterns generated by the gastrointestinal tract (which does not include reactive relaxation and accommodation).

The distinct motility patterns as they occur in vivo under *physiological conditions* are dependent on the coordinated activity of the following components:

1. myogenic mechanisms;
2. a functioning enteric nervous system;
3. functioning networks of the interstitial cells of Cajal (ICC), which are important in the generation and propagation of slow waves of the gastrointestinal tract and therefore of the MMC and in the neuroeffector transmission from enteric muscle motor neurons to the smooth musculature in some parts (e.g., the stomach) of the gastrointestinal tract (see Subchapter 5.4).

The extrinsic autonomic (parasympathetic and sympathetic) innervation including the innervation by the prevertebral ganglia and gastrointestinal hormones is not necessary in the generation of the

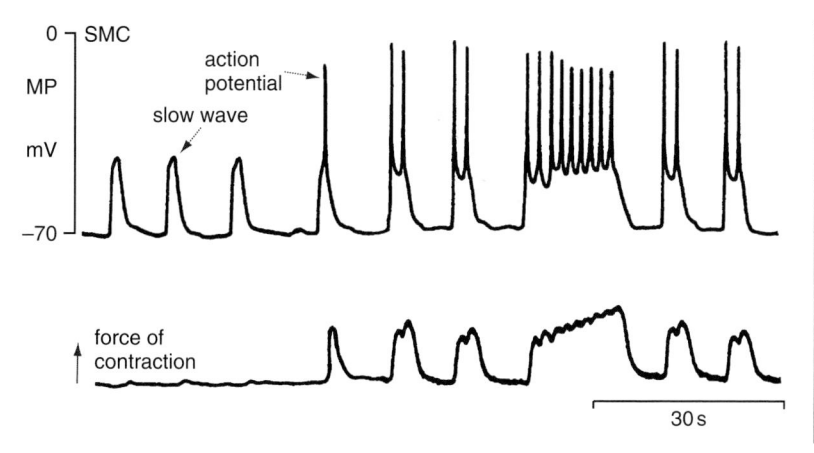

Figure 5.5 Slow waves, action potentials and contractions of smooth musculature. Slow wave potentials recorded intracellularly from a smooth muscle cell (SMC) syncytium with and without action potentials. The action potentials are generated by opening of L-type calcium channels, which trigger intracellularly the excitation–contraction mechanism, leading to contraction of the smooth musculature. Modified from Huizinga et al. (1997).

motility pattern. However, the motility pattern is modulated by these extrinsic neural and hormonal influences.

5.3.2 Peristalsis

The term *peristalsis* describes the propulsion of intestinal contents from the oral site to the anal site by circular constriction of the intestinal musculature. Bayliss and Starling (1899, 1900) defined peristalsis as consisting of a contraction of the smooth muscles oral to an intraluminal bolus and the relaxation of the smooth muscles anal to the bolus. Both the circular and the longitudinal muscle contract and relax simultaneously, whereby contraction and relaxation are usually stronger in the circular muscle than in the longitudinal muscle. Bayliss and Starling (1899) called these polarized excitatory and inhibitory reflexes "The law of the intestine." Figure 5.7a illustrates one of the earliest in vivo experiments from Bayliss and Starling (1899) on the dog small intestine in which they recorded contraction and relaxation of the circular and longitudinal muscles separately. It is interesting to read their own description:

> In the experiment from which the curves in Fig. [5.7a] were obtained, two enterographs were placed at right angles to one another at a point 130 cm from the pylorus. The position of the levers is shown in Fig. [5.7a, upper], *a* and *b* being the levers of the longitudinal enterograph (*a* being the movable lever), *c* and *d* the levers of the enterograph recording the contractions of the circular muscle. At the beginning of the observation the intestinal wall was contracting rhythmically, the contractions affecting both coats, and being synchronous in both. At *A* a bolus made of cotton-wool coated with vaseline was inserted by an opening into the intestine 4½ inches above the enterographs. It will be seen that the contractions of the circular coat cease instantly, and this inhibition is accompanied by a gradually increasing relaxation. There is some relaxation of the longitudinal coat, but the rhythmic contractions do not altogether cease. On inspection of that intestine it was seen that the introduction of the bolus caused appearance of a strong constriction above it. This constriction passed downwards, driving the bolus in front of it. The numbers above the tracing of the circular fibers indicate the distance of

Figure 5.6 Assemblies of intrinsic primary afferent neurons (IPANs), ascending interneurons and descending interneurons with cell bodies in the myenteric plexus. These assemblies of neurons are important in generation of peristalsis (see Figure 5.8). (a) The projections of the IPANs to the mucosa are mechano- and/or chemosensitive. IPANs are reciprocally connected by slow synaptic excitatory transmission mediated by tachykinin. Their output connects to assemblies of interneurons and to motoneurons. (b, c) The assemblies of interneurons are reciprocally connected by excitatory synapses. They form either ascending (b) or descending chains (c), which connect synaptically to motor neurons (excitatory and inhibitory; not shown). Modified from Kunze and Furness (1999) with permission.

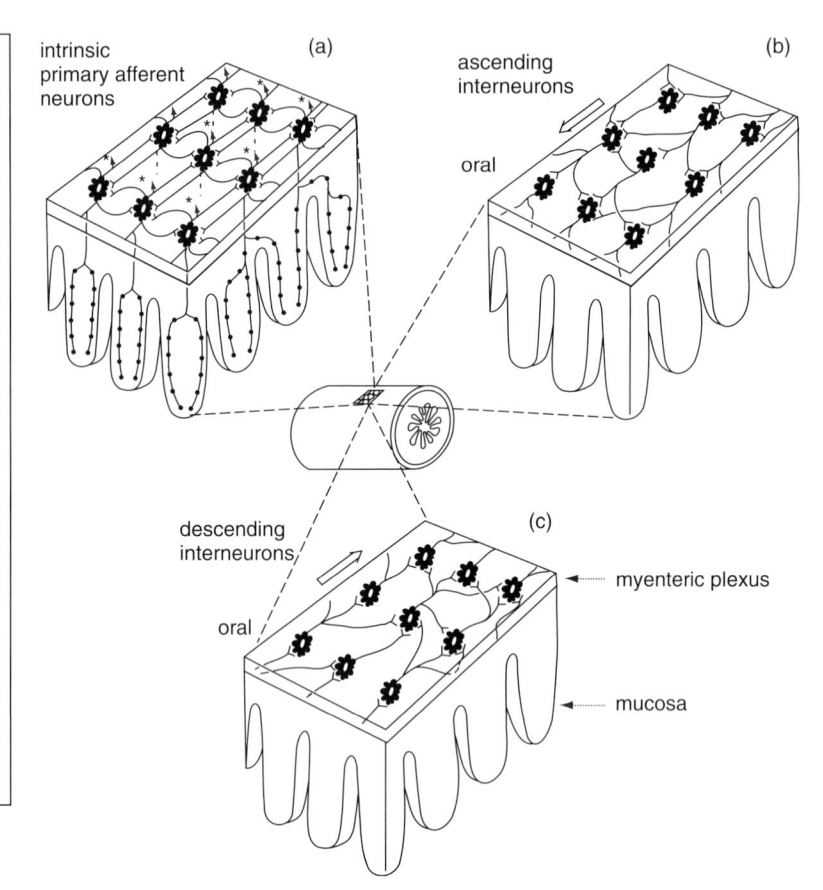

the bolus in inches from the uppermost enterograph lever. At B the bolus had arrived at the upper longitudinal lever and at C had passed this and was directly under the transverse enterograph, or a little below it. At this point a strong tonic contraction of both coats occurs, expelling the bolus beyond the levers. This strong contraction passes off to be succeeded by another, which like the first is moving down the intestine …

On the basis of their observations Bayliss and Starling inferred that mechanical stimulation of the mucosa and/or distension of the intestinal wall by the bolus excite mechanoreceptive afferent neurons, which then generate polarized excitatory (oral) and inhibitory (anal) reflexes. Today we know that mechanical as well as intraluminal chemical stimuli in the small intestine elicit peristalsis, both stimuli occurring together under physiological conditions. The underlying neural mechanisms of this propulsive peristalsis can now be almost fully described.

The IPANs are mediators of peristalsis. Their cell bodies lie in the myenteric plexus and project to the mucosa. In the small intestine these neurons are excited by chemical and/or mechanical mucosal stimuli, by distension and particularly by contraction of the circular smooth musculature. In the colon the IPANs involved in peristalsis consist of two separate types, those that are only stretch-sensitive and

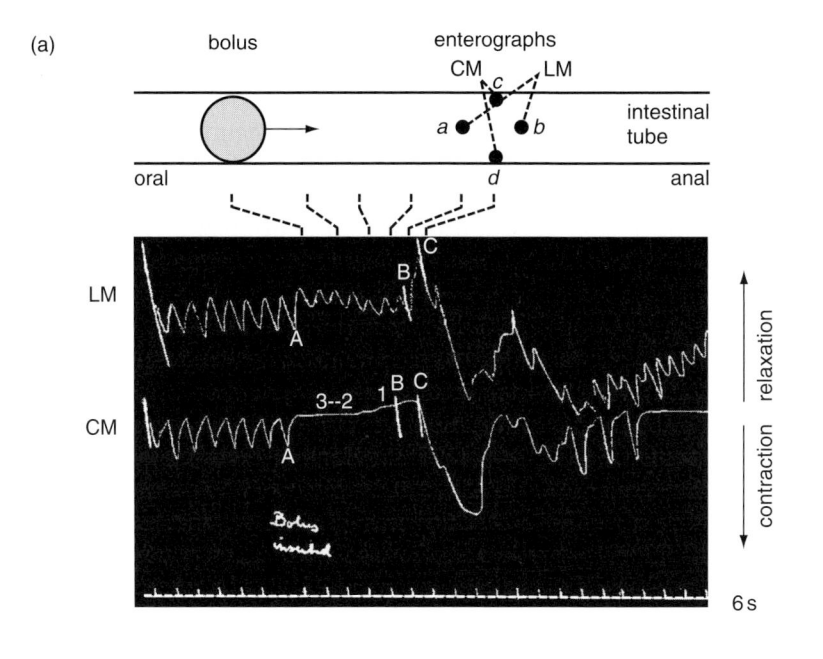

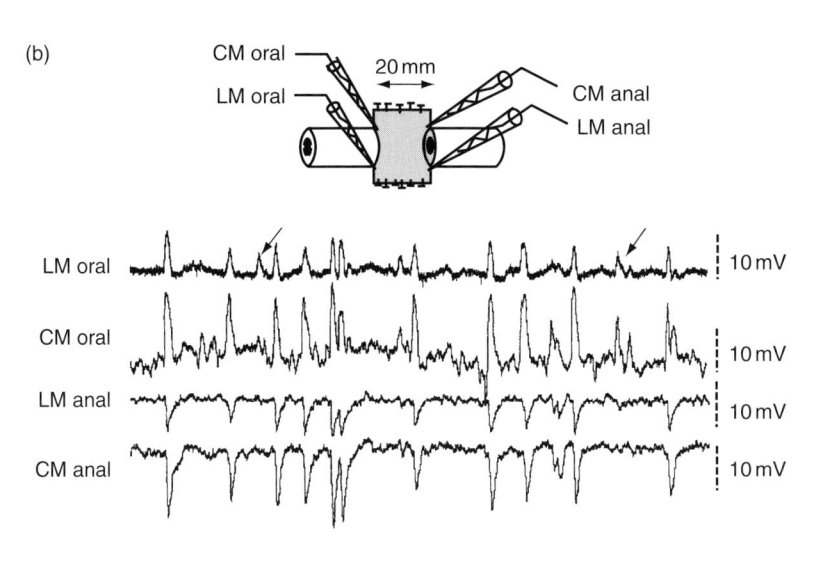

5.3 THE NEURAL BASIS OF PERISTALSIS | 185

Figure 5.7 Neuronal mechanisms of peristalsis. (a) Contraction and relaxation of the circular (CM) and longitudinal musculature (LM) of the small intestine in a dog in vivo before and during propulsive transport of a bolus from oral to anal. When the bolus was inserted into the intestinal tube about 11 cm (4.5 inch) oral to the recording sites the CM and LM relaxed. When the bolus reached the recording sites both muscles contracted. Contraction and relaxation of the CM and LM were recorded mechanographically across the transverse axis and the longitudinal axis of the intestine, respectively. See text for details. Modified from Bayliss and Starling (1899) with permission. (b) Simultaneous intracellular recordings from two cells of the CM and two cells of the LM in the guinea pig distal colon in vitro. The colon was opened on a length of 20 mm, fixed in a two-dimensional plane and left connected to the intact oral and anal tubes. Intracellular records were made from two cells at the oral site of the flattened preparation (LM oral, CM oral) and from two cells at its anal end (LM anal, CM anal) during circumferential stretch of the preparation (stimulating in this way stretch-sensitive intrinsic primary afferent neurons). Excitatory junction potentials (EJPs) in both the CM and LM at the oral end of the colon as well as inhibitory junction potentials (IJPs) in both the CM and the LM 20 mm anally mostly occur synchronously; a few EJPs are not accompanied by IJPs (arrows). From Spencer et al. (2003a) with permission.

those that are preferentially contraction-sensitive, similar to the mechanosensitive IPANs in the small intestine. The neurons that are only stretch-sensitive are interneurons in the myenteric plexus of the colon (Spencer and Smith 2004). As mentioned before, IPANs of like type are synaptically connected with each other and form assemblies of neurons (Figure 5.6a). These assemblies of IPANs are synaptically connected to at least three assemblies of synaptically connected interneurons in the myenteric plexus (Figure 5.6b, c). One assembly of interneurons projects orally and is synaptically connected to excitatory muscle motor neurons. The other assembly of interneurons projects in the anal direction and is synaptically connected to inhibitory muscle motor neurons. Both groups of muscle motor neurons are *not* synaptically connected with each other, thus they do not form motor neuron assemblies.

This arrangement of afferent neurons, interneurons and muscle motor neurons, together with the functional properties of the individual classes of involved enteric neuron largely (but not entirely) explains the neural mechanism of oral–anal (aboral) propulsive peristalsis induced by distension (Figure 5.8) with some differences between the small intestine and the colon. These differences are probably related to differences in transport: in the small intestine the content is transported in liquid state and in the distal hindgut in a more solid state (in guinea pigs in solid fecal pellets).

1. In the *small intestine* the polarized reflex organization applies to the circular muscle (CM). Stimulation of IPANs leads to simultaneous excitation of the CM orally as well as anally. Simultaneously, and preceding the reflex excitation of the CM, the CM is anally inhibited. This inhibition modulates the anal contraction of the CM leading to a decrease of amplitude and rate of rise of its contraction. This inhibition is possibly responsible for the anal propulsion of the liquid content of the small intestine. The longitudinal muscle (LM) is excited orally as well as anally to the site of activation of the IPANs (Figure 5.8a). Thus at least three reflex pathways are involved in peristalsis of the small intestine (Hirst et al. 1975; Spencer et al. 1999).
2. The IPANs involved in the reflexes in the small intestine are polymodal afferent neurons responding to stretch, contraction as well as mucosal chemical stimuli. The neurons are cholinergic and release a tachykinin when excited that generates slow excitatory postsynaptic potentials (slow EPSPs). In this way, the IPANs form a self-reinforcing afferent neuronal network (Figure 5.6a).
3. In the *colon* the polarized reflex organization applies to the CM as well as the LM (Figure 5.8b). Both the CM and LM located anally to the stimulation are also excited; this excitation leads to rebound contraction of the colonic musculature following its relaxation (not shown in Figure 5.8b).
4. The IPANs involved in these reflexes in the distal colon consist of two types: (1) Stretch-sensitive IPANs that do not respond to contraction. These IPANs are probably myenteric interneurons. They are cholinergic and do not release a tachykinin during activation. (2) Polymodal IPANs, preferentially responding to contraction, which are functionally similar to those in the small intestine.
5. In the colon a descending assembly of cholinergic interneurons connects synaptically to two populations of inhibitory motor neurons innervating the CM and the LM, respectively (Figure 5.6c). The motor neurons to the CM use NO, ATP and VIP as transmitter (increase of potassium conductance). The motoneurons to the LM

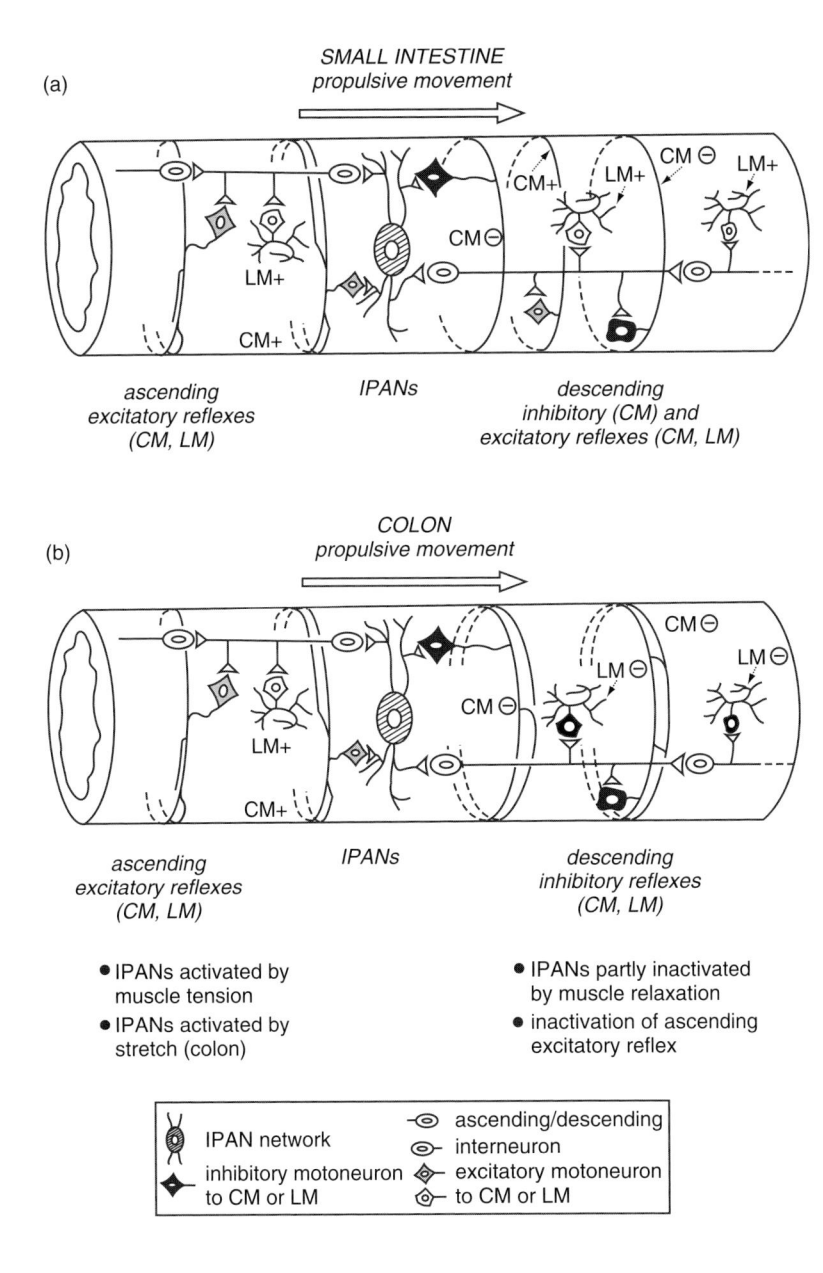

Figure 5.8 Organization of reflex circuits responsible for peristalsis in the small and large intestine. (a) Small intestine: intrinsic primary afferent neurons (IPANs), ascending interneurons and descending interneurons consist of synaptically connected assemblies of neurons (see Figure 5.6). Three reflex circuits to the circular muscles (CM; ascending excitatory, descending inhibitory and descending excitatory) and two reflex circuits to the longitudinal muscles (LM; ascending excitatory and descending excitatory) are indicated. (b) Colon: the ascending reflex pathways are the same as in the small intestine. The two descending reflex circuits are inhibitory both to the CM and the LM. In (a) and (b) only *one* ascending and descending network of interneurons, respectively has been indicated. Each may consist of two (ascending) or three (descending) networks. +, excitation; −, inhibition. Modified from Furness et al. (2003a), Kunze and Furness (1999) and Sanders and Smith (2003) with permission.

use NO as transmitter. Activation of the inhibitory motor neurons by this assembly of interneurons generates synchronous inhibitory junction potentials (IJPs) in the CM and LM followed by relaxation of both muscles (Figure 5.7b; in this experiment IPANs that are only stretch-sensitive have been activated) (Spencer and Smith 2001; Smith et al. 2003).

6. An ascending assembly of cholinergic interneurons (Figure 5.6b) connects to two populations of excitatory motor neurons innervating the CM and the LM, respectively. Both groups of motor neurons use acetylcholine as transmitter. Activation of the excitatory motor neurons by this assembly of interneurons

generates synchronous EJPs in the CM and LM followed by contraction of both muscles. Excitatory junction potentials in the muscles oral to the stimulation site and IJPs in the muscles anal to the stimulation site are also synchronous (Figure 5.7b; Spencer and Smith 2001). The intestinal tube is pulled over its content by the reflex activation of the LM (this is particularly important in the colon).

In *summary*, stimulation of the assembly of IPANs leads to synaptic activation of both assemblies of interneuron (the ascending one and the descending one). This synaptic activation triggers the following events (Figure 5.8):

- Reflex contraction of the orally located muscles by activation of excitatory muscle motor neurons and further activation of the IPANs by muscle contraction.
- Relaxation of the anally located muscles by activation of inhibitory muscle motor neurons and decrease of activity in IPANs due to decreased muscle tension.
- Additional activation of orally located IPANs enhances the polarized reflex activity.
- In the colon, decrease of activity in anally located IPANs, due to relaxation of the circular muscle, probably reduces ascending reflex activation.

5.4 | Integration of enteric neural, pacemaker and myogenic mechanisms in generation of motility patterns

The enteric reflex pathways and their coordination, as described in the preceding section do not sufficiently explain the oral–aboral propulsive (and local non-propulsive, pendular or segmental) movement patterns of the gastrointestinal tract as they occur in vivo. These enteric reflex pathways are part of a complex interrelationship between enteric neurons, smooth muscle cells and a third group of cells that are called interstitial cells of Cajal (ICC). Interaction between these three categories of cells will ultimately explain and clarify the role of the enteric neurons in the generation of the motility patterns of the gastrointestinal tract.

5.4.1 The interstitial cells of Cajal (ICC) and slow waves

Cajal (1995) described a group of cells in the wall of the gastrointestinal tract that are distinct in their morphology from smooth muscle cells and from enteric neurons, but closely associated with both of them. He believed that these cells are neurons, being intercalated between the terminals of enteric neurons and smooth muscle cells, and that they are important in generating pacemaker activity in the gut. Investigation of the development of the ICC shows that they are of mesenchymal origin, as smooth muscle cells are. For their development and differentiation they need the protooncogene *c-kit*, which encodes the receptor tyrosine kinase protein (by which these cells can be recognized histochemically). Loss of *c-kit* leads to loss of development of ICC from progenitor cells which then transform into smooth muscle cells (Sanders *et al.* 1999). Systematic morphological investigations have shown that ICC form several networks that are oriented along the longitudinal axis of smooth muscle cells or of the enteric nerve plexuses (Figure 5.9):

- A network of ICC in the same plane as the myenteric plexus between the circular and longitudinal muscle layers (myenteric ICC, ICC-MY) in most regions of the gastrointestinal tract.
- A network of ICC in the smooth musculature (intramuscular ICC, ICC-IM).
- A network of ICC at the inner surface of the circular muscular layer of the small intestine (ICC at the deep muscular plexus, ICC-DMP).
- Groups of ICC that are located in the connective tissue septa between blocks of circular muscle in the stomach of larger animals (ICC-SEP).

The ICC of these networks are coupled electrically with each other by close appositions and/or gap junctions forming electrical and functional syncytia. The ICC also form close appositions and/or gap junctions with the smooth muscle cells. Finally, ICC within the smooth muscles of the stomach and possibly elsewhere (but not in the

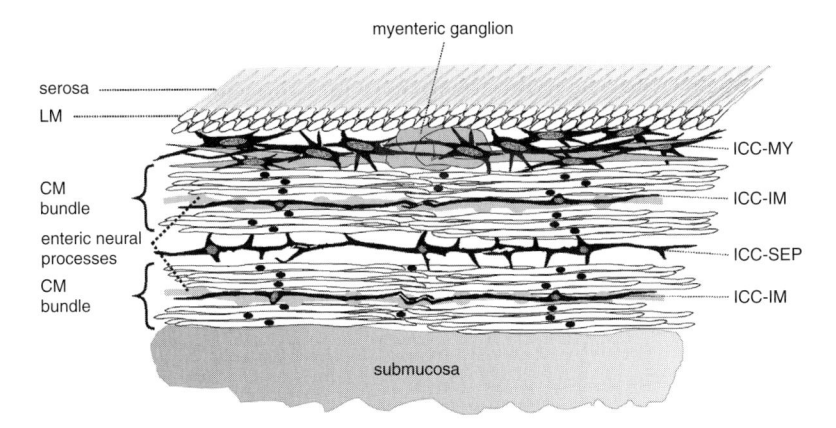

Figure 5.9 Functional organization of the interstitial cells of Cajal (ICC) in the canine gastric antrum. The antral wall contains two layers of smooth muscle cells: the outer longitudinal muscle layer (LM) lacking ICC and the inner circular layer (CM) in which individual smooth muscle cells are organized into bundles. In this layer, intramuscular ICC (ICC-IM) are distributed through individual CM bundles. A network of ICC lies between the LM and CM close to the myenteric plexus (ICC-MY). In CM bundles, ICC-IM function both to augment depolarizations reaching them from ICC-MY and as essential intermediaries in the transmission of information from enteric neural processes to nearby smooth muscle cells. ICC-SEP form functional cables that transmit information from pacemaker ICC-MY to deep CM bundles present in the stomach of larger animals. It may well be that ICC-SEP and ICC-MY are directly connected; alternatively ICC-SEP may be electrically excited by activity in CM bundles lying closer to the ICC-MY pacemaker network. From Hirst (2001) with permission.

longitudinal muscle of the small intestine and colon) are intercalated between nerve terminals of enteric neurons and smooth muscle cells (Thuneberg 1982; Huizinga *et al.* 1997; Sanders *et al.* 1999, 2000).[3]

Functional investigations of various parts of the gastrointestinal tract (antrum, small and large intestines), using intracellular recordings from smooth muscle cells and from ICC, have shown that the electrical slow waves in the smooth muscle of the gastrointestinal tract are generated by the ICC.[4] Interstitial cells of Cajal of the myenteric plexus normally produce the pacemaker potentials. In mutant mice without myenteric ICC, pacemaker activity and slow waves are absent. In the stomach, the pacemaker potentials in the myenteric ICC are amplified by the ICC in the musculature and electrically transmitted to the smooth muscle cells. In the small intestine, the pacemaker currents of myenteric ICC seem to be sufficiently strong to depolarize passively the musculature and to generate slow waves.

Figure 5.10b shows simultaneous recordings from a pacemaker cell of the myenteric plexus (ICC-MY) and from a cell of the circular smooth muscle closeby in the guinea pig stomach in vitro. The (driving) potential generated in the syncytium of myenteric ICC cells is electrically transmitted to the syncytium of smooth muscle cells. The pacemaker potential exhibits an initial fast component and a plateau component. The intracellular record from the circular musculature has two components: a passive component that is transmitted via the gap junctions from the myenteric ICC to the smooth muscle cells and a secondary regenerative component (Figure 5.10b). In animals with absent intramuscular ICC the secondary regenerative components seen in the smooth muscle cells is absent, but the passive component is present. Thus the secondary regenerative component is generated in the intramuscular ICC. In this way the transmission of the pacemaker potentials from the myenteric ICC to the smooth muscle cells is augmented (for discussion and literature see Hirst and Ward [2003]). The driving potential is generated by voltage-independent opening of a non-specific cation channel for sodium and calcium, which in turn is regulated by regenerative intracellular release of calcium from the sarcoplasmic reticulum (Horowitz *et al.* 1999).[5]

The induced changes in membrane potential in the smooth muscle cells that follow the pacemaker potential are called slow waves (Figure 5.5). These slow wave potentials may generate action potentials in smooth muscle cells by opening L-type calcium channels leading to calcium inflow and contraction of the smooth musculature (Figure 5.5). The slow waves in the smooth muscle syncytium are propagated from oral to aboral since the frequency of pacemaker activity in the ICC is higher at the oral site than at the aboral site. The propagation does not occur in the smooth musculature but in the networks of electrically coupled ICC. Thus, the function of the ICC is not only to generate and augment pacemaker activity but also to propagate the slow waves (Sanders *et al.* 2000; Ward and Sanders 2001).

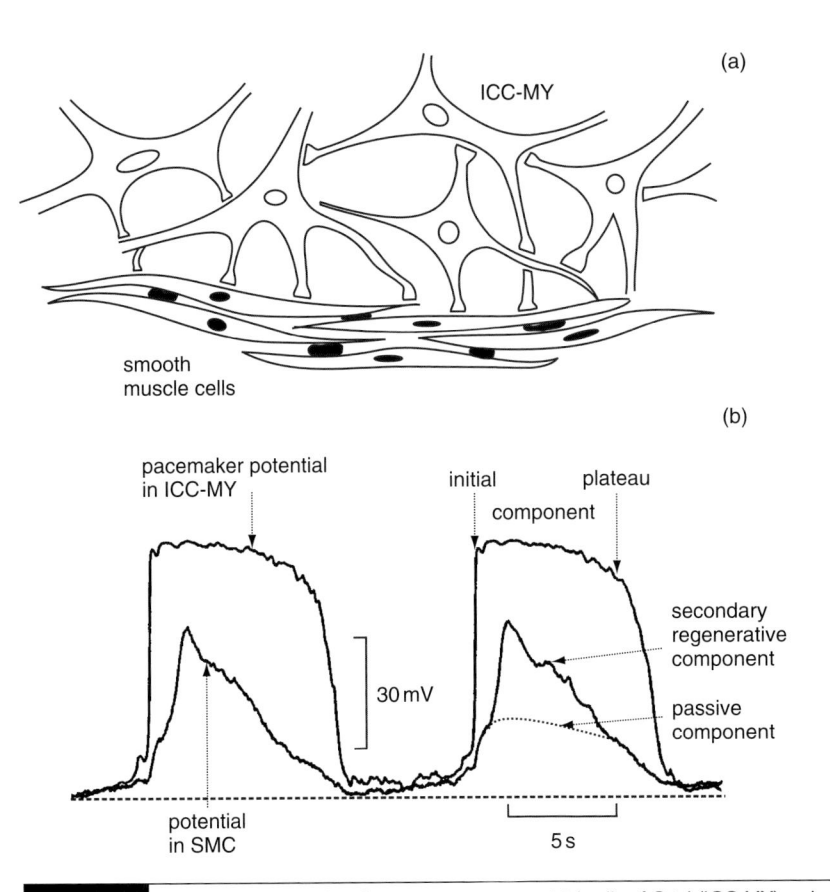

Figure 5.10 Simultaneous recording from myenteric interstitial cells of Cajal (ICC-MY) and smooth muscle cells (SMC) in the guinea pig stomach. (a) Network of ICC and SMC. The ICC are coupled electrically by gap junctions with each other and with the SMC, which are also electrically coupled by gap junctions. Thus both form electrical and functional syncytia. (b) Simultaneous intracellular recording from ICC-MY and SMC of the circular smooth muscle. The driving potential in the ICC syncytium (pacemaker potential) produces a secondary potential (slow wave potential) in the SMC syncytium by current flowing through the gap junctions. The driving potential in the ICC has a primary rapid component, which is followed by a long-lasting plateau component. In the SMC, a passive initial component of the slow wave is followed by a secondary (regenerative) component, which is related to the amplifying effects of the intramuscular ICC. The passive electrotonically transmitted component seen in the SMC in the absence of intramuscular ICC (i.e., when the amplifying effect of the ICC is absent) is indicated by the dotted-line curve. In vitro experiment under nifedipine blocking L-type calcium channels in the SMC in order to prevent the generation of action potentials and contractions. Modified from Dickens et al. (1999) with permission.

5.4.2 ICC-myogenic activity and enteric neurons

As mentioned above, intramuscular ICC in the stomach, which form gap junctions and close appositions with smooth muscle cells, are also closely associated with enteric nerve fibers (Figure 5.9). Morphologically some of these nerve fibers form intimate synaptic contacts with the ICC in the musculature. Excitatory and inhibitory postsynaptic potentials in circular smooth muscle cells generated by electrical stimulation of enteric neurons are absent or reduced in W/W^V mutant mice, which lack the ICC in the musculature (Burns et al. 1996; Ward et al. 2000a; Iono et al. 2004). This shows that the effects of excitatory and inhibitory enteric motor muscle neurons are mediated, at least in part, by the intramuscular ICC. This is the third main function of the ICC, in addition to those of generating pacemaker activity and propagating slow waves.

How do the three components, ICC, smooth muscle cells and enteric neurons, act together to generate the different motility patterns of the gut under normal biological conditions? It has been shown in isolated segments of mouse small intestine that distension-induced peristalsis with pulsatile release of the intraluminal content can be generated experimentally under the following conditions (Huizinga et al. 1998):

- In mutant W/W^V-mice lacking the ICC of the myenteric plexus, which cannot generate pacemaker activity and slow waves, distension-induced peristalsis is present and fully dependent on the enteric neural motor program as described in the preceding section (see Figure 5.5).
- In normal mice that have been treated with tetrodotoxin, blocking active conduction of the enteric neurons, the slow waves are present and the distension-induced rhythmic contractions must be initiated by a myogenic mechanism (possibly a stretch-sensitive ionic mechanism in the smooth muscle cells).
- Even in W/W^V-mutant mice lacking the ICC of the myenteric plexus distension-induced peristalsis can be present under tetrodotoxin when action potential conduction in enteric neurons is blocked. In this condition the distension-induced contraction is fully dependent on myogenic mechanisms.
- However, under these three experimental conditions, in which the myenteric ICC or the enteric neurons or both have been eliminated, *distension-induced contractions of an isolated intestinal segment are not normal.* They lack precise oral–aboral coordination, have a high threshold, lack full pulsatile relaxation (which is generated by the inhibitory muscle motor neurons) or lack the burst-like pattern of contraction. Thus, under biological conditions, a cooperation between the motor programs of the enteric nervous system, the slow wave mechanism linked to the ICC and the myogenic mechanism is necessary to produce a well-coordinated peristalsis for a proper transit of the intestinal content.

The way the other local movement patterns (pendular, segmental, cycling) are generated is poorly understood. It is not far-fetched to assume that they require the cooperative action of the three mechanisms too. It has, for example, been shown that local patchy contractions of the small intestine are dependent on slow waves generated by the ICC, which then lead to generation of action potentials in the plateau phase of the slow waves (see Figure 5.5) and local contractions, which may propagate over short but not long distances (Lammers 2000).

The MMC is phase III of the interdigestive movement pattern of the gastrointestinal tract. It is prevented by blockade of cholinergic nicotinic synaptic transmission between neurons with hexamethonium and by blockade of muscarinic synaptic transmission to the smooth musculature with atropine. Thus, the MMC is fully dependent on functioning enteric nerve circuits. However, contrary to expectations, the MMC is still present in the absence of the myenteric ICC network that is responsible for the intrinsic pacemaker activity, as shown in transgenic mice lacking this network (Spencer et al. 2003b). In the MMC up to 100% of the slow waves are accompanied by action potentials and contractions of the smooth muscle. In the silent phase of the interdigestive movement pattern the slow waves are present but do not generate action potentials and no contractions occur. The MMC is not triggered by distension of the gut (the gut is relatively empty!) and is independent of signals in parasympathetic and sympathetic neurons although it can be suppressed by commands from the brain (e.g., during stress behavior). It is hypothesized that the MMC is linked to an assembly of cholinergic interneurons in

◀—— PROPAGATION OF SLOW WAVES IN ICC NETWORK ——▶

ICC network in pacemaker region

smooth muscle cells

intramuscular ICC

enteric motoneuron

① spontaneous generation of pacemaker current

② electrotonic conduction of slow waves

③ depolarisation and activation of L-type Ca^{2+} channels

④ neural input to smooth muscle mediated by ICC, regulates responses to slow waves

⋏ gap junction

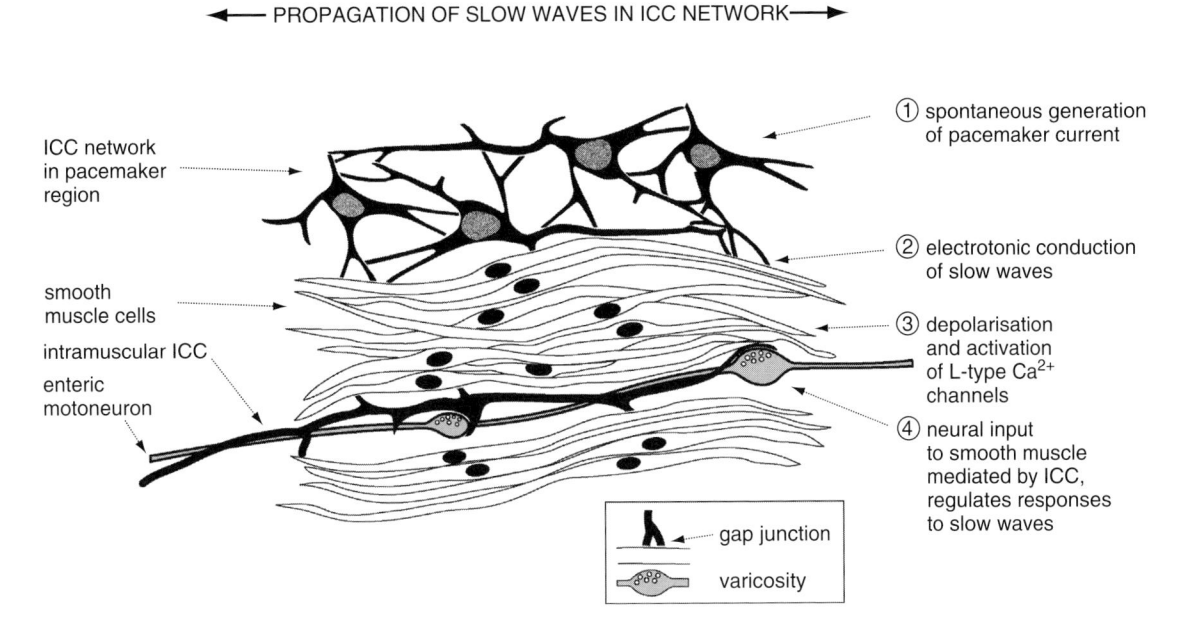

varicosity

Figure 5.11 The rhythmoneuromuscular apparatus to generate gastrointestinal motility. Interstitial cells of Cajal (ICC; specifically of the myenteric plexus of stomach, small and large intestines) generate the pacemaker potentials. ICC form electrically coupled networks (syncytia), which in turn are electrically coupled to the smooth muscle syncytium by gap junctions. The pacemaker current flows through the gap junctions and induces in the smooth muscle cells slow wave potentials, which are amplified by the intramuscular ICC (see Figures 5.4 and 5.5). The slow wave potentials lead to activation of voltage-dependent L-type calcium channels with massive inflow of calcium and triggering of smooth muscle contractions. Intramuscular ICC amplify the pacemaker potentials in the transmission to the smooth muscle cells and mediate inhibitory and excitatory effects of enteric muscle motor neurons. Varicose nerve terminals of these neurons form close (20 nm) specialized junctions with these ICC. Neural influence modulates the slow wave potentials via the ICC. This in turn modulates the generation of action potentials and therefore contraction of the smooth muscle cells. Modified from Horowitz et al. (1999) with permission.

the myenteric plexus, which project with their axons aborally and form synapses with inhibitory muscle motor neurons that innervate the circular musculature (see Figures 5.6, 5.8). In the guinea pig these interneurons contain somatostatin. This assembly of interneurons receives no or only sparse synaptic input from intrinsic primary afferent neurons, which are important for the initiation of peristaltic reflexes. It is unclear what triggers and maintains the MMC. This triggering may be linked to a local hormone (e.g., motilin).

In *summary*, ICC, smooth muscle cells and enteric motoneurons form a "*rhythmoneuromuscular apparatus,*" which generates the pattern of gastrointestinal motility. Figure 5.11 depicts the elements and connections of this somewhat hypothetical model that is propagated by Kent Sanders and his coworkers (Horowitz *et al.* 1999; Sanders *et al.* 2000):

1. Pacemaker currents are generated in the network of ICC of the myenteric plexus.
2. These currents are electrotonically transmitted to smooth muscle cells and to intramuscular ICC. The intramuscular ICC amplify the pacemaker potentials. The syncytium of smooth muscle cells generates the slow electrical wave potentials, which are propagated from oral to aboral sites. This propagation is also dependent on the ICC networks.
3. During the slow wave depolarization voltage-dependent L-type calcium channels are opened in the smooth muscle cells, which leads to generation of calcium action potentials if threshold is achieved. Inflow of calcium triggers the contraction of the smooth muscle cells and leads to the mechanical slow waves.
4. In some parts of the gastrointestinal tract (e.g., circular muscles of small intestine, colon and stomach) excitatory and inhibitory enteric muscle motor neurons form synaptic contacts with the ICC within the muscles and in this way (at least in part) influence the smooth muscle cells, increasing or decreasing the effectiveness of slow waves in generating action potentials and contractions.

5.5 | Regulation of secretion and transmural transport

Reflex pathways involved in regulation of secretory cells and/or local blood vessels (arterioles) have been much less explored than enteric reflex pathways that are involved in regulation of motility. This applies to the organization of the reflex pathways connected to the mucosa and mucosal blood vessels as well as to their coordination with the reflex pathways that are integrated in regulation of motility. Activation of secretomotor neurons stimulates epithelial cells to secrete chloride ions into the intestinal lumen, taking with them sodium ions and water. The physiological stimuli that induce a reflex

5.5 REGULATION OF SECRETION AND TRANSMURAL TRANSPORT

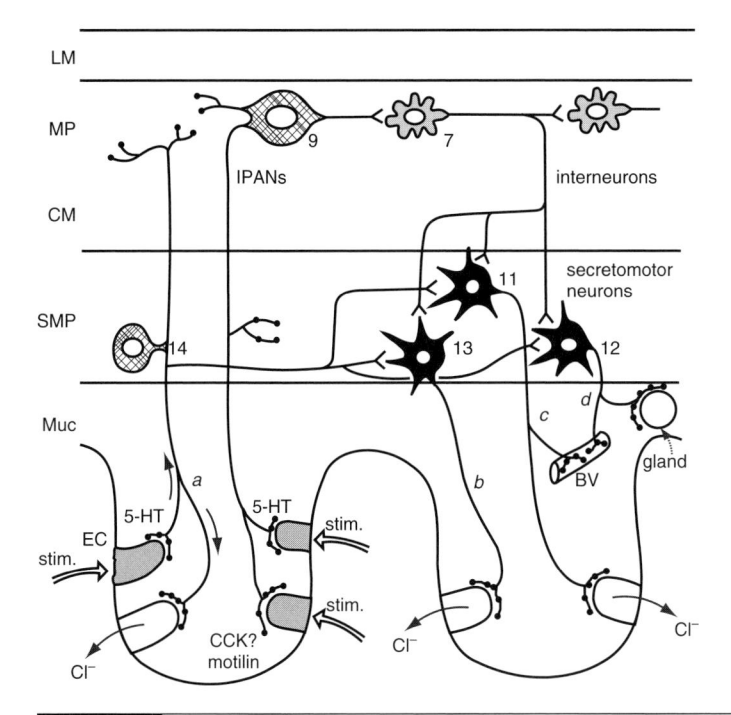

Figure 5.12 Enteric reflex circuits related with secretion and local vasodilation. Intrinsic primary afferent neurons (IPANs) in the submucosal plexus (SMP) or myenteric plexus (MP) are activated by mechanical (distension, contraction of the musculature; intraluminal shearing stimuli) and/or intraluminal chemical stimuli. Intraluminal stimuli may be mediated by enterochromaffin cells (EC) releasing 5-hydroxytryptamine (5-HT) or possibly by other cells (e.g., releasing motilin or CCK). Secretomotor neurons in the SMP are either activated monosynaptically or di- or polysynaptically via interneurons in the myenteric plexus (e.g., by cholera toxin). This induces either secretion of chloride ions (reflex b) or secretion of chloride ions as well as dilation of submucosal arterioles (reflex pathways c and d). The first is generated by release of acetylcholine, the second by release of vasoactive intestinal peptide. IPANs themselves may activate mucosal secretory cells (and possibly inhibit arterioles [not shown]) via an axon-reflex-like mechanism (pathway a). BV, blood vessel; CCK, cholecystokinin; CM, circular muscle; LM, longitudinal muscle. The numbers adjacent to the neurons correspond to the numbers in Table 5.1. IPANs, hatched; interneurons, shaded; motor neurons, black. From Cooke (1994, 1998) and Lundgren (1988, 2000). Modified from Furness et al. (2004) with permission.

activation of the secretomotor neurons are mechanical stimulation of the mucosa (shearing stimuli) and chemical stimuli (e.g., glucose or proton ions). The intraluminal stimuli triggering secretion appear to be the same as those causing enteric motility reflexes in the intestine. The stimuli are mediated to the IPANs via enterochromaffin cells (EC) in the mucosa, which release 5-HT and excite the IPANs (Figure 5.12). The IPANs excite enteric neurons synaptically by release of acetylcholine and substance P. Intraluminal toxins, such as cholera toxin, also lead to strong reflex activation of secretomotor neurons. Secretomotor reflexes elicited by intraluminal physiological stimuli are always paralleled by enteric vasodilator reflexes mediated via the enteric neurons.

Figure 5.12 illustrates the enteric reflex pathways that lead to activation of the secretomotor neurons resulting in secretion and vasodilation in the mucosa. Based on experimental studies (Lundgren

1988; Cooke 1994, 1998; Cooke and Reddix 1994) there are four configurations by which stimulation of IPANs activate secretory cells of the mucosa via enteric interneurons and motor neurons coordinated with local vasodilation of submucosal arterioles and therefore local increase in blood flow. The dilation of the blood vessels may be generated by the same enteric secretomotor neurons that activate the secretory cells or possibly also by separate ones (not shown in Figure 5.12):

1. Intrinsic primary afferent neurons themselves may directly innervate secretory cells and/or possibly blood vessels although structural evidence does not support the latter (not shown in Figure 5.12). In this way both groups of effector cells are influenced by the afferent neurons directly in an axon-reflex-like manner (pathway *a* in Figure 5.12).
2. Reflex secretion triggered by intraluminal chemical or mechanical stimulation is mediated through the submucosal plexus. No interneurons seem to be intercalated between the intrinsic primary afferent neuron and the secretomotor neuron. The secretomotor neurons projecting only to secretory cells use acetylcholine as transmitter (pathway *b* in Figure 5.12).
3. Reflex secretion and vasodilation triggered by intraluminal chemical or mechanical stimulation is mediated through the submucosal plexus and secretomotor neurons that innervate both secretory cells and blood vessels. These motor neurons use VIP as transmitter (pathways *c* and *d* in Figure 5.12).
4. Reflex secretion generated by cholera toxin is mediated by an interneuron in the myenteric plexus that projects to secretomotor neurons in the submucosal plexus (pathways *c* and *d* in Figure 5.12). It is possible that the reflex pathway via the myenteric plexus also contributes to normal reflexes, e.g., those evoked by mechanoreceptor activation (Reed and Vanner 2003).

5.6 Defense of the gastrointestinal tract and enteric nervous system

The gut has developed powerful mechanisms to defend the body against invading antigens derived from food, bacteria and parasites, toxins and other compounds. Its epithelial lining constitutes the inner surface of the body that is highly permeable in most parts of the gastrointestinal tract, in particular in the small intestine. The gut contains the largest immune system of the body, collectively called "gut-associated lymphoid tissue" (GALT) that continuously surveys antigens in the lumen of the gut. This immune system of the gut includes the immune cells of Peyer's patches, M-cells of the epithelial lining (modified intestinal epithelial cells [enterocytes]), lymphocytes in the lamina propria, macrophages and mast cells. The GALT is integrated with the enteric nervous system, the gut endocrine system and spinal as well as vagal primary afferent neurons and constitutes a powerful local defense system that is continuously working and protecting the body. A histological and histochemical expression of the interaction between GALT and nervous system is the finding that Peyer's patches are innervated by enteric neurons, postganglionic sympathetic neurons and extrinsic primary afferent neurons (Heel et al. 1997).

The central nervous system is informed about the actions of this peripheral defense system by activity in vagal and spinal primary afferent neurons, by endocrine signals (hormones in the blood) and by messages from the GALT (cytokines in the blood, such as interleukin-1β [IL-1β], IL-6, tumor necrosis factor α [TNFα]). Activity in both types of extrinsic primary afferent neurons initiates a host of protective reflexes and reactions, which may act back on the local defense mechanisms of the gut via parasympathetic and sympathetic pathways and possibly via endocrine systems. Activity in these afferent neurons trigger discomfort, pain and hyperalgesia of the gut, the latter being elicited by activity in spinal afferents (see Subchapter 2.4). Ideas as to how the enteric nervous system might be involved in this defense system have been discussed in the literature (Downing and Miyan 2000; Furness and Clerc 2000; Holzer 2002a, b; Sharkey and Mawe 2002; Wood 2002).

The immune system of the intestine is involved in every process of the gastrointestinal tract that potentially leads to its damage (inflammation, infections, injury, erosions by HCl, allergy etc.). This involves M-cells of the intestinal epithelium, which transport antigens across the epithelium to macrophages and dendritic cells. The antigens are processed by these cells and presented to T lymphocytes that stimulate B lymphocytes. The B lymphocytes proliferate in the lamina propria of the mucosa and produce antibody, preferentially IgA. Some B lymphocytes migrate to mesenteric lymph nodes and continue to mature and to proliferate. They enter the general circulation and localize to the mucosa-associated lymphoid tissue in the intestine and elsewhere.

Signaling molecules of the immune system, such as IL-1β, IL-6 and TNFα, are synthesized and released during injuries by immune cells in the mucosa and muscle layers or cells associated with the immune system (e.g., enterocytes, monocytes, macrophages, fibroblasts, enteric glia cells). Interleukin-1β and IL-6 increase the excitability of most neurons of the submucosal plexus and of some neurons of the myenteric plexus. This enhances mucosal fluid secretion and motility of the gastrointestinal tract. Cytokines also sensitize and probably activate IPANs of the enteric nervous system and extrinsic spinal and vagal visceral afferent neurons innervating the gastrointestinal tract.

Figure 5.13 summarizes some aspects of the role of the enteric nervous system and other peripheral systems in the protection of the gastrointestinal tract. This diagram demonstrates various hypothetical

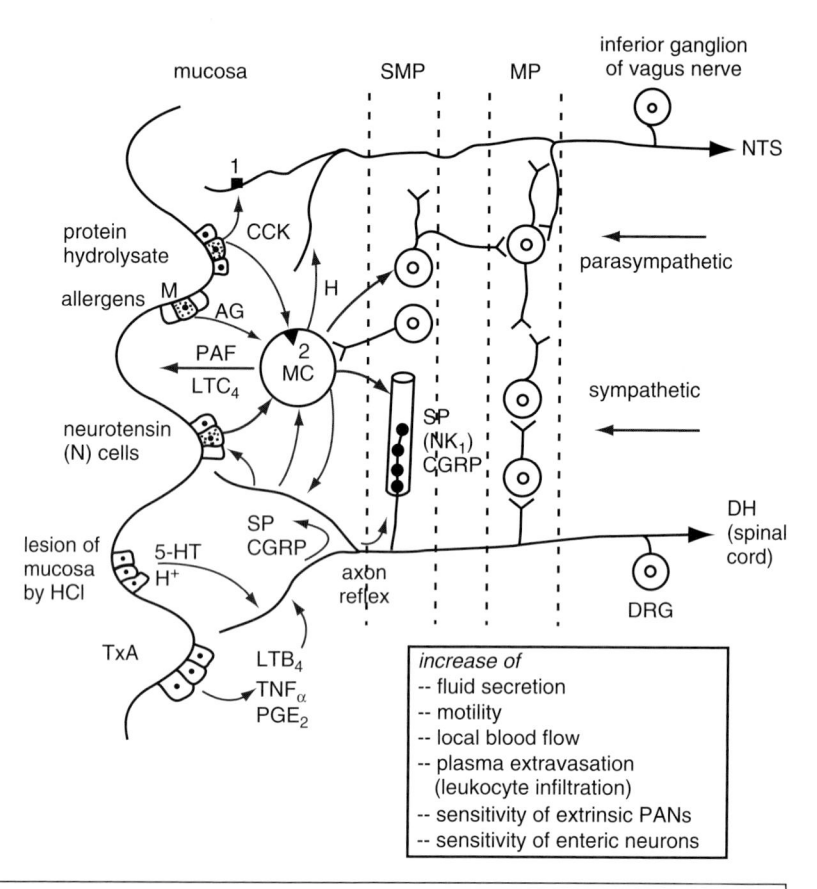

Figure 5.13 Enteric nervous system and protection of the gastrointestinal tract: a summary figure. Emphasized are mast cells (MC), enteric neurons and extrinsic (spinal) and vagal visceral afferent neurons. Intraluminal processes (protein hydrolysates, allergens, toxins [e.g., TxA, toxin A of *Clostridium difficile*], HCl etc.) that may generate injury activate MC and intrinsic and extrinsic primary afferent nerve neurons (in addition to the activation of the gut-associated lymphoid tissue [GALT]; not shown here). This activation occurs via endocrine cells (cholecystokinin [CCK]; neurotensin [N]), antigens (AG), prostanoids (prostaglandin E_2 [PGE_2]), leukotrienes [e.g., LTB_4], cytokines [e.g., tumor necrosis factor α [$TNF\alpha$], interleukin-1β [IL-1β]] etc.) and is in part mediated by substance P (SP) and calcitonin gene-related peptide (CGRP) released by terminals of spinal primary afferent neurons. Mast cells activate enteric neurons, activate and sensitize extrinsic afferent nerve fibers, increase local blood flow, increase plasma extravasation (leading to leukocyte infiltration) and act directly on the enterocytes of the mucosal lining (leading to fluid secretion). They are under control of enteric neurons and possibly extrinsic autonomic neurons. Extrinsic afferent terminals generate neurogenic inflammation (arteriolar vasodilation and venular plasma extravasation) via the "axon reflex" by release of SP and CGRP. They furthermore activate neurons of the myenteric plexus. The global protective effects of these interactions are listed in the lower right. 1, 2: CCK_1 and CCK_2 receptors; NK_1, neurokinin 1 receptor for SP; DH, dorsal horn of the spinal cord; DRG, dorsal roof ganglion; MP, myenteric plexus; NTS, nucleus tractus solitarii; PAF, platelet-activating factor; SMP, submucosal plexus. Modified from Downing and Miyan (2000) and Sharkey and Mawe (2002).

pathways by which the protective reactions are initiated and maintained, involving enteric neurons, extrinsic afferent neurons, cells associated with the immune system, endocrine cells, enterocytes and vascular cells. Details about the functioning of these pathways and their activation during various harmful or potentially harmful stimuli acting at the epithelial lining have to be worked out. Some of these potentially harmful stimuli are shown on the left side in Figure 5.13:

- Macroscopically the protective reactions of the gastrointestinal tract consist of the following components: increase of mucosal secretion, motility, local blood flow, leukocyte infiltration, sensitivity of enteric neurons and sensitivity of the terminals of extrinsic primary afferent neurons.
- Mast cells can be activated by several agents including antigens (AG), CCK (via the CCK_2 receptor), neurotensin, substance P (SP) and neurokinin A (NKA) released by the terminals of spinal afferent neurons. Activated mast cells can release many signaling molecules, such as leukotrienes, interleukins, prostaglandins, TNFα, tryptase, histamine and 5-HT. They are under the control of enteric neurons of the submucosal plexus, probably influenced by noradrenergic sympathetic postganglionic neurons and possibly under neuroendocrine control. Their activation sensitizes extrinsic afferent nerve fibers, e.g. via release of histamine and other compounds, stimulates enteric neurons, stimulates leukocyte extravasation and generates directly fluid secretion by the mucosa (e.g., by release of platelet-activating factor [PAF] and leukotriene C_4 [LTC_4]). Mast cells are considered to mediate stress-induced barrier defects generated by "psychological" stimuli (Söderholm and Perdue 2001; Yu and Perdue 2001).
- Spinal visceral peptidergic afferent neurons innervating the gut have dual functions: in the periphery they mediate various protective functions; centripetal impulse activity elicits protective reflexes, mediated by spinal cord and brain stem, and pain, discomfort and other sensations (see Subchapters 2.2 to 2.4).
- Some vagal visceral afferent neurons innervating the gastrointestinal tract are involved in protective reflexes, central modulation of nociceptive impulse transmission and so-called sickness behavior that may develop during harmful affections of the gastrointestinal tract (Holzer 1998a, 2002a, b). They are not directly involved in the generation of visceral pain (see Subchapters 2.3 and 2.4).
- Lesion of the mucosa by HCl leads to activation of enterocytes as well as acidification of the mucosal interstitial tissue. Peptidergic afferent capsaicin-sensitive fibers are activated. This leads locally, mediated by calcitonin gene-related peptide (CGRP) and NO (probably released by enteric neurons), to increase of blood flow through the mucosa, bicarbonate secretion and mucus secretion. With time release of SP and NKA triggers trophic processes (proliferation of endothelial cells, vascular muscle cells and fibroblasts with angiogenesis) and healing (Holzer 2002a, b).
- Toxin A of *Clostridium difficile* activates enterocytes. This leads to release of prostaglandin E_2 (PGE_2), TNFα and leukotriene B_4 (LTB_4) by the enterocytes. These substances activate and sensitize the terminals of spinal afferent neurons that mediate various peripheral effects involving mast cells, enteric neurons and blood vessels.

- Allergens reacting with M-cells in the mucosal lining activate mast cells. This activation is enhanced by cytokines released by cells related to the immune system.
- Protein hydrolysates (and other chemical stimuli) activate entero-endocrine cells in the duodenum releasing CCK. Cholecystokinin activates mast cells (via CCK_2 receptors) and vagal afferents (via CCK_1 receptors). Both together lead to changes of motility patterns of the small intestine.
- Cholera toxin activates enterochromaffin cells of the mucosa, which release 5-HT and PGE_2. This activates enteric secretory reflex pathways generating active fluid secretion and vasodilation (see reflex pathways *c* and *d* in Figure 5.12).

5.7 | Control of the enteric nervous system by sympathetic and parasympathetic pathways

As mentioned before, the parasympathetic systems have their dominant effects and overall access to target cells at the oral and anal sites of the gastrointestinal tract. Otherwise they do not interfere with single groups of effector cells in the gastrointestinal tract but influence enteric sensorimotor programs. The sympathetic non-vasoconstrictor systems act throughout the gastrointestinal tract and also largely do not interfere directly with single groups of effector cells but modulate the enteric sensorimotor programs too. The visceral vasoconstrictor pathways influence visceral blood vessels directly. The numbers of preganglionic parasympathetic and sympathetic neurons involved in gastrointestinal functions are low in comparison to the numbers of enteric neurons, being in the range of $\leq 1\textperthousand$!

5.7.1 Parasympathetic (vagal) pathways

Rather little is known about parasympathetic preganglionic neurons in the dorsal vagus motor nucleus that project to the gastrointestinal tract. These neurons are probably highly specialized with respect to various motility, secretomotor and enteric endocrine functions. They innervate motor neurons and interneurons of the enteric nervous system (see Subchapter 4.8 and Table 4.6). Some functional properties of parasympathetic preganglionic neurons that are involved in regulation of gastrointestinal functions will be described in Subchapter 10.7. Figure 5.14 summarizes the way in which parasympathetic preganglionic neurons in the DMNX (see Figures 8.10 and 10.25) may be involved in the regulation of different motility patterns of the gastrointestinal tract. This scheme is hypothetical in several aspects. However, it shows that the brain does interfere, via the preganglionic parasympathetic neurons, with many different functions in which the enteric nervous system is involved. The diagram may aid understanding of the complex peripheral mechanisms that are activated by the different classes of preganglionic parasympathetic neurons.

Parasympathetic preganglionic neurons in the DMNX may influence the motility of the gastrointestinal tract (above all stomach, duodenum and ileum) via

- enteric neurons that excite the smooth muscle directly,
- enteric neurons that excite the smooth muscle via ICC,
- enteric neurons that inhibit smooth muscle cells, and
- enteric neurons that activate endocrine cells (ENDC) releasing a hormone (e.g., gastrin), which in turn influences smooth muscle cells and enteric neurons that are involved in regulation of motility.

In addition to vagal preganglionic neurons involved in regulation of motility there exist other types of parasympathetic preganglionic neurons that innervate enteric secretomotor neurons in the submucosal plexus and enteric neurons supplying endocrine cells that are not involved in direct regulation of motility. The cell bodies of these preganglionic neurons are possibly also situated in the DMNX.

To reiterate, *first*, the number of preganglionic parasympathetic neurons is very small when compared to the number of enteric neurons and, *second*, the small number of preganglionic neurons is subdifferentiated into several functional types.

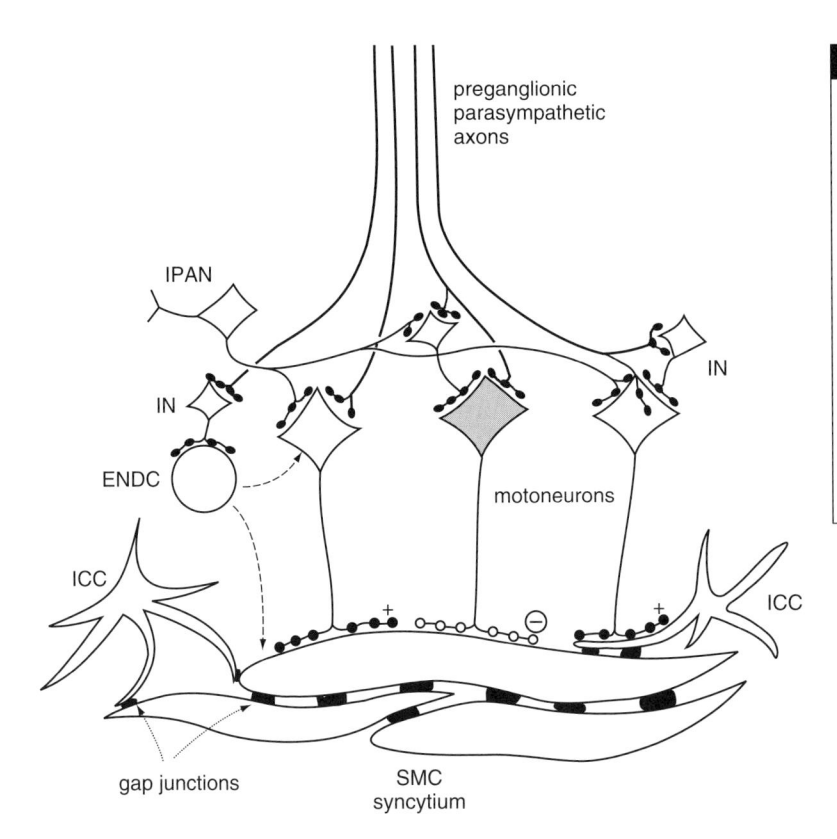

Figure 5.14 Schema showing relation between vagal parasympathetic preganglionic neurons (cell bodies in the dorsal vagus motor nucleus) and neurons of the enteric nervous system in regulation of motility. Various peripheral pathways are involved in regulation of the motility of the smooth muscle syncytia. For details see text. ENDC, endocrine cell; ICC, interstitial cell of Cajal; IN, interneuron; IPAN, intrinsic primary afferent neuron; SMC, smooth muscle cell. +, excitation; −, inhibition.

5.7.2 Sympathetic pathways

The sympathetic innervation of the gastrointestinal tract is also insufficiently explored. The postganglionic sympathetic neurons are located in the prevertebral and some in the paravertebral ganglia. Most postganglionic visceral vasoconstrictor neurons innervating the vasculature (mainly arterioles) of the gastrointestinal tract seem to be located in the paravertebral ganglia. They receive synapses from preganglionic neurons in the spinal cord but not from peripheral intestinofugal neurons. Furthermore, the postganglionic visceral vasoconstrictor neurons do not receive synaptic input from collaterals of spinal peptidergic primary afferent fibers (Messenger et al. 1999; Gibbins et al. 2003c; see Subchapter 2.2).

The population of sympathetic non-vasoconstrictor neurons consists of motility-regulating neurons and secretomotor neurons (Figure 5.15). These groups of neurons are most likely further subdifferentiated with respect to various functions of the gastrointestinal tract. They innervate neurons of the myenteric or submucosal plexus, respectively. Activation of these sympathetic neurons inhibits enteric neurons; this inhibition occurs either presynaptically by decrease of release of excitatory transmitter or postsynaptically. A few postganglionic sympathetic secretomotor neurons innervate the mucosa directly and a few sympathetic motility-regulating neurons innervate the non-sphincteric smooth muscles, leading in both cases to inhibition when activated. Sphincter muscles are directly innervated by sympathetic postganglionic fibers and contract when these fibers are excited. Non-vasoconstrictor neurons receive synaptic input from enteric intestinofugal neurons and form extraspinal reflex circuits (see Subchapter 6.5). These intestinofugal neurons are cholinergic and may have VIP colocalized. In the guinea pig, collaterals of spinal peptidergic afferents containing substance P appear to influence only postganglionic neurons in prevertebral ganglia that express somatostatin (in addition to noradrenaline). Only these postganglionic neurons that have secretomotor function express the tachykinin NK_1 receptor for substance P but not the other postganglionic neurons (Messenger et al. 1999; see Subchapter 6.5).

Activity of sympathetic secretomotor neurons is related to the balance of fluid volume and electrolytes in the body. Decrease of both leads to activation of these neurons and consequently to inhibition of chloride and bicarbonate secretion. This results in decreased transport of sodium and water into the intestinal lumen and conservation of body water. The central circuits of this regulation are little known; it is unlikely that they are identical to those linked to the visceral vasoconstrictor pathway. However, it may be speculated that the central pathways linked to sympathetic secretomotor neurons innervating the gastrointestinal tract receive input from vagal atrial stretch receptors and osmoreceptors and that these pathways project to the preganglionic neurons or associated interneurons in

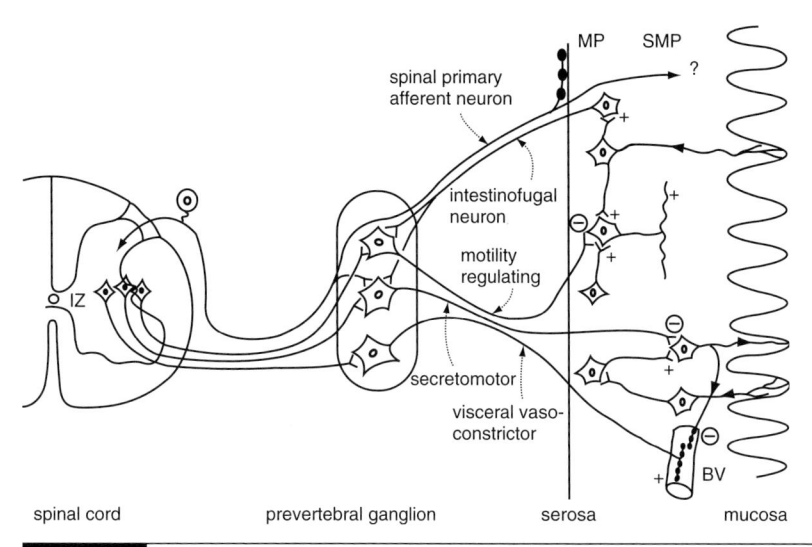

Figure 5.15 Sympathetic outflow to the small intestine. Most sympathetic postganglionic neurons projecting to the intestine are situated in the prevertebral ganglia. These neurons consist of three classes: (1) Vasoconstrictor neurons innervating blood vessels (BV). These neurons do not get synaptic input from collaterals of spinal primary afferent neurons (peptidergic) or from enteric intestinofugal neurons (cholinergic). The cell bodies of many visceral vasoconstrictor neurons are located in paravertebral ganglia. (2) Secretomotor neurons innervating neurons of the submucosal plexus (SMP). (3) Motility-regulating neurons innervating neurons of the myenteric plexus (MP). Both secretomotor and motility-regulating neurons exert (pre- and postsynaptically) inhibitory effects on enteric neurons; they receive peptidergic synaptic input from spinal primary afferent neurons and intestinofugal neurons (cholinergic nicotinic) and show integration of preganglionic and peripheral synaptic inputs (but see Subchapter 6.5). IZ, intermediate zone; +, excitation; −, inhibition. Modified from Furness et al. (2003a) with permission.

the thoracolumbar spinal cord. The central circuits linked to the sympathetic motility-regulating and secretomotor neurons are largely unknown (see Subchapter 4.3).

Conclusions

The enteric nervous system is an autonomic nervous system in its own right and functions in many aspects independently of the central nervous system. It has about as many neurons as the spinal cord and is metaphorically called the "brain of the gut."

1. The enteric nervous system has neurons with cell bodies in the myenteric and submucosal plexus of ganglia and consists of intrinsic primary afferent neurons, interneurons and motor neurons to various effectors. Each group of neurons is subdifferentiated according to the physiological stimuli that activate the afferent neurons, the target cells of the motor neurons, their synaptic connections with other neurons, their primary and secondary transmitters, the projection of dendrites and axons and other criteria.
2. Intrinsic primary afferent neurons of functionally similar types are reciprocally synaptically connected and form in this way

afferent networks. Interneurons of functionally similar types are also synaptically connected and form several cell assemblies. Motoneurons are not synaptically connected with each other.

3. The primary transmitter in most excitatory enteric neurons is acetylcholine (muscarinic action on effector cells; nicotinic action on neurons). Some enteric neurons use ATP as primary transmitter. Intrinsic primary afferent neurons use tachykinins as additional primary transmitter. Inhibitory motor neurons use several cotransmitters to varying degrees (NO, VIP, PACAP, ATP). Some secretory motor neurons use VIP as transmitter. Enteric neurons are neurochemically characterized by colocalization of various neuropeptides. The functions of these colocalized peptides are largely unknown.

4. Afferent neurons, interneurons and motor neurons form reflex circuits. Different reflex circuits related to regulation of motility or secretion are coordinated and represent the sensorimotor programs of the enteric nervous system. These programs underlie the neural regulation of motility, secretion, local blood flow and probably other functions such as hormone release and gut-immune function.

5. The regulation of motility patterns are mainly directed from the myenteric plexus. The neural basis of *peristalsis*, which has been most extensively studied, consists of the coordinated activation of ascending and descending reflex pathways. In the *small intestine*, the circular and longitudinal muscle layers are influenced by ascending and descending excitatory reflex pathways. The circular muscles are additionally influenced by a descending inhibitory reflex pathway that modulates the contraction of the anally located circular muscle and leads in this way to anal propulsion of the intestinal content. In the *colon*, the circular and longitudinal muscle layers are powerfully inhibited by descending inhibitory reflex pathways and excited by an ascending reflex pathway, supporting expulsion.

6. Inhibitory and excitatory reflex circuits are organized and coordinated with pacemaker activity of the interstitial cells of Cajal (ICC) and myogenic activity to generate the different movement patterns. The ICC are electrically coupled and form several networks. These networks are electrically coupled to the smooth musculature. The networks of ICC generate the pacemaker activity responsible for the slow waves and mediate some excitatory and inhibitory synaptic activity of enteric neurons to the circular smooth musculature.

7. Neural regulation of fluid and electrolyte transport, which is less well explored, is largely controlled through the submucosal plexus. Reflex activation of secretomotor neurons is always paralleled by local reflex vasodilation in the mucosa. Secretion and vasodilation are triggered by the same type of secretomotor neuron. Several distinct reflex pathways can be described.

The coordination and integration of these reflex pathways with the enteric neural motor programs regulating motility are unexplored.

8. The enteric nervous system is involved, in interaction with the gut-associated lymphoid tissue (GALT) and the gut endocrine system, in protective reactions of the gastrointestinal tract. Intraluminal antigenic and toxic stimuli are sensed by cells of the endothelial lining and lead to concerted actions of these systems resulting in an increase of secretion, motility and mucosal blood flow. Local protective processes are referred by endocrine, immune and afferent neural signals to the brain. They are under neural and hormonal control of the brain.

9. The brain modulates the functions of the enteric nervous system via the parasympathetic and sympathetic nervous system. It receives detailed information from the gastrointestinal tract by impulse activity in visceral vagal and spinal afferent neurons as well as by gastrointestinal hormones. In this way functions of the enteric nervous system are adapted to the behavior of the organism.

10. The efferent extrinsic autonomic systems are differentiated with respect to various functions of the enteric nervous system. Except for the direct control of the oral and anal sites of the gastrointestinal tract and of blood vessels, the pathways of both extrinsic autonomic nervous systems do not interfere directly with the final effectors of the enteric nervous system. They give command signals to enteric neural circuits and in this way influence the enteric reflex programs and their coordination.

Notes

1. The enteric neurons were originally divided into two electrophysiological classes: S neurons and AH neurons (Hirst *et al.* 1974; Furness and Costa 1987). (1) S neurons receive fast excitatory synaptic inputs (therefore S, synaptic). They have a distinct morphology and are uniaxonal, most of them being Dogiel type I neurons. The S neurons are interneurons and motoneurons. (2) AH neurons exhibit prolonged afterhyperpolarization following an action potential (therefore AH). They receive no fast excitatory synaptic inputs. They have relatively large cell bodies and multipolar processes (Dogiel type II neurons). These neurons are enteric afferent neurons.

    Some enteric interneurons can have afferent function (e.g., in the myenteric plexus of the guinea pig colon [Spencer and Smith 2004]). Thus the term intrinsic primary afferent neuron (IPAN) is functionally defined and does not only include AH neurons but also some S neurons.

2. The percentages given are for the guinea pig small intestine.

3. Cajal stained the cells that later became known as the ICC by methylene blue and by Golgi impregnation. To him these cells were similar in their shape to intraganglionic neurons and resembled an end formation of the sympathetic nervous system. Therefore he originally called these cells sympathetic interstitial neurons. He believed that these interstitial cells

form networks or accessory plexus that are closely associated with intrinsic nerves and target cells and that mediate neural signals to the autonomic target cells (smooth muscle cells, epithelial cells) and are important in the regulation of autonomic effector cells (Cajal 1995). Research starting in the 1980s and being intensified after 2000 clearly shows that these cells are of mesenchymal origin and that they are not restricted to the gastrointestinal tract but also present in other extradigestive cavity organs of various species (such as corpus spongiosum, prostate, urethra, urinary bladder, uterus, fallopian tube, vas deferens, mesenteric lymphatic vessels, mesenteric artery, portal vein) and even in the pancreas (as already described by Cajal). As in the gastrointestinal tract, they are probably universally involved in pacemaker activity, peristalsis, regulation of neurotransmission and secretion. The modern criteria to identify ICC are immunohistochemical staining for the c-kit protein; formation of networks of c-kit positive cells with ≥ 2 processes that are connected to target cells (smooth muscle cells, epithelial cells) and intrinsic nerve fibers; gap junctions between ICC (forming in this way a functional syncytium) and between ICC and smooth muscle cells; close contacts to intrinsic nerve fibers without interposition of basal lamina (position of ICC between nerve fibers and effector cells) (Thuneberg 1999; Harhun et al. 2005; Huizinga and Faussone-Pellegrini 2005; Popescu et al. 2005a, b).

4. Originally it was thought that pacemaker activity and slow waves of depolarization leading to rhythmic contractions of different sections of the gastrointestinal tract are initiated by the smooth musculature. Thus, it was believed that certain regions of the smooth muscular syncytium have pacemaker properties and that the regenerative process, generated by opening of calcium channels, is propagated aborally through the smooth musculature, the frequency of the pacemaker potentials being highest in the oral regions of the different sections of the gastrointestinal tract. However, smooth muscle cells isolated from the gastrointestinal tract do not show pacemaker activity and do not generate slow electrical waves.

5. The cellular basis for generation of pacemaker potentials and of secondary regenerative potentials in the ICCs is complex and not entirely understood.

(1) *The rhythmic electrical (pacemaker) activity* of the myenteric ICC depends on several cellular processes: (a) Pulsatile (rhythmic) release of Ca^{2+} into the cytoplasm from inositol 1,4,5-triphosphate (IP_3)-receptor operated intracellular Ca^{2+} stores. (b) Ca^{2+} release from the IP_3-receptor operated intracellular Ca^{2+} stores triggers Ca^{2+} uptake by mitochondria. (c) Mitochondrial Ca^{2+} uptake is linked to activation of pacemaker currents by an unknown mechanism; it is probably related to opening of non-selective cation channels or of chloride selective channels in the membrane of the ICC (probably triggered by Ca^{2+} released from the IP_3 stores). (d) The pacemaker cycle probably is completed and reset by uptake of cytoplasmic Ca^{2+} into the intracellular store.

This molecular process occurs in an intracellular compartment of the ICC near the cell membrane that includes the sarcoplasmic reticulum (calcium store), the mitochondria and the cluster of non-selective cation-selective ion channels and chloride channels (see Ward et al. [2000b], Hirst and Ward [2003] and Koh et al. [2003] for discussion and literature).

(2) Slow wave regeneration depends on the *active (regenerative) propagation of the pacemaker potentials* in the network of electrically coupled ICC (not

between smooth muscle cells). This propagation also depends on the intrinsic property of the ICC to generate spontaneous activity. Regenerative propagation is believed to consist of an entrainment of the spontaneous pacemaker activity of the different ICC within the network. It is hypothesized that this entrainment is brought about either by voltage-dependent activation of the IP_3 synthesis or by voltage-dependent inflow of Ca^{2+} or by both (see Koh *et al.* [2003] for discussion and literature).

Part III

Transmission of signals in the peripheral autonomic nervous system

In Chapters 3 and 4 the idea of the functional organization of the "final autonomic pathways" that transmit the central information to the peripheral targets was described. This idea sets the limits for the ensuing discussion of transmission of impulses along the final autonomic pathways (i.e., in the autonomic ganglia and to the target cells) as it occurs in vivo. In this section I want to emphasize that the centrally generated signals are faithfully transmitted from the preganglionic neurons through autonomic ganglia to the postganglionic neurons and from the postganglionic axons to the effector cells at the neuroeffector junctions. I will describe the principles of this signal transmission and how it can be modulated by peripheral reflexes and humoral mechanisms. I will not, however, extensively discuss the details of synaptic transmission in autonomic pathways including the different types of receptors for the neurotransmitters, their pharmacology and postreceptor pathways and chemical neuroanatomy (see articles in Burnstock and Hoyle [1992]; Elfvin *et al.* [1993]; McLachlan [1995]; Skok [2002]).

Chapter 6

Impulse transmission through autonomic ganglia

6.1 Morphology, divergence and convergence in
    autonomic ganglia   *page* 213
6.2 Strong and weak synaptic inputs from
    preganglionic neurons   217
6.3 The autonomic neural unit: structural and functional
    aspects   223
6.4 Electrophysiological classification, ionic channels,
    functions and morphology of sympathetic
    postganglionic neurons   225
6.5 Different types of autonomic ganglia and their
    functions in vivo   230
6.6 Non-nicotinic transmission and potentiation resulting
    from preganglionic stimulation in sympathetic ganglia   240

Postganglionic neurons are the final autonomic motoneurons. Their cell bodies are aggregated in peripheral autonomic ganglia. They receive synaptic input from preganglionic neurons and in some ganglia (notably sympathetic prevertebral ganglia) from peripheral neurons of the enteric nervous system and from peptidergic spinal afferent fibers. As already mentioned in Chapter 1, most sympathetic ganglia are located at distance from their target cells and parasympathetic ganglia are located close to their targets.

Sympathetic ganglia have fascinated investigators since ancient times. It was believed that these structures are "little brains," which integrate, carry and distribute the "animal spirits" from the brain to the periphery, leading to coordinated actions of the peripheral target organs (the "sympathies") in association with the activity of the brain (Fulton 1949; Pick 1970; Spillane 1981; Karczmar *et al.* 1986).

However, it turns out that the primary function of most peripheral sympathetic and parasympathetic pathways is to distribute messages to the periphery from relatively small pools of preganglionic neurons to relatively large pools of postganglionic neurons. This particularly applies to the neural regulation of autonomic body functions, which are chiefly under central control, e.g., regulation of systemic blood pressure, thermoregulation, gastrointestinal functions, evacuative functions (micturition, defecation), erection, salivation, pupil diameter etc.

Acetylcholine is released by all preganglionic axon terminals at their synapses in ganglia and the effects of nerve activity are antagonized by blockade of nicotinic acetylcholine receptors. Under some experimental and possibly physiological conditions, repetitive activation of preganglionic axons can release enough acetylcholine to activate extrasynaptic muscarinic receptors and may release peptides to activate peptidergic receptors.

Experimental work in the last 50 years has shown that many if not most autonomic ganglia have rather complex neurophysiological, morphological, neurochemical and pharmacological organizations (see Karczmar *et al.* [1986]; McLachlan [1995]; Skok and Ivanov [1989]). In the last 35 years, most experiments on autonomic ganglia have been conducted in vitro in order to study synaptic transmission and its modification under controlled conditions. These preparations are very stable and enable long-term studies of the behavior of intact synapses. This type of research has accelerated in recent times with the introduction of molecular methods and the possibility of conducting experiments in which the structure and function of receptors, postreceptor pathways and cellular effectors can be analyzed at the molecular level. Most of this work is conducted on isolated cells (with or without culture in vitro) so that synapses cannot be studied. Research on autonomic ganglia from mutant animals in which components of receptors, ionic channels or intracellular pathways have been changed genetically promises to reveal more.

The functional relevance of fast and slow synaptic events, of divergence and convergence of preganglionic neurons onto postganglionic neurons and of the functional specificity of synaptic connections in autonomic ganglia for neural regulation of autonomic target organs have mostly not been addressed in these experiments. One may intuitively assume that the complexities worked out in the in vitro experiments have implications for impulse transmission through autonomic ganglia during the autonomic regulation of organ function. However, this assumption may not be true at all. Only when this regulation is studied in vivo can the extent to which the in vitro work is applicable to normal function be clarified. In vivo experiments of this type have so far been conducted by McLachlan, Purves and Skok and their coworkers (Johnson and Purves 1983; Skok and Ivanov 1983, 1989; Ivanov and Purves 1989; McLachlan *et al.* 1997, 1998).

The term "autonomic ganglia" includes sympathetic, para- and prevertebral and parasympathetic ganglia. Here it does not include the enteric nervous system (Chapter 5). The transmission of impulses is mainly dealt with here in terms of the division of sympathetic and parasympathetic nervous systems into functional subunits as defined by their target organs (see Chapter 4). The diverse physiological, pharmacological and molecular research in which autonomic ganglia are used as models will not be addressed in this book (see McLachlan [1995]). The main question addressed in this chapter is: how is the centrally generated impulse activity of preganglionic neurons transmitted across autonomic ganglia to postganglionic neurons under physiological conditions?

6.1 Morphology, divergence and convergence in autonomic ganglia

6.1.1 Morphology of postganglionic neurons and their synaptic inputs

The size of the cell body and the number of dendrites of sympathetic postganglionic neurons is dependent on the number of target cells innervated by the neurons and the function of the neurons. In guinea pig (and probably other mammalian species) vasoconstrictor neurons are smaller than all other functional types of sympathetic neuron; pilomotor neurons are smaller than secretomotor neurons (innervation of sweat gland or salivary glands). This size corresponds to the conduction velocity of the axons of these neurons in the cat. Parasympathetic postganglionic neurons are morphologically comparatively simple, having either no or only a few dendrites.

Only about 1% to 2% of the surface of postganglionic neurons is covered by synapses, most of the neuron surface being occupied by Schwann cells (Figure 6.1). About 50% of the surface of soma and proximal dendrites of motoneurons is covered by synapses. Less than 50% of the boutons formed by preganglionic axons or (in prevertebral ganglia) by axons of intestinofugal neurons of the enteric nervous system form synapses with the postganglionic neurons. These synapses are characterized by close contacts between boutons

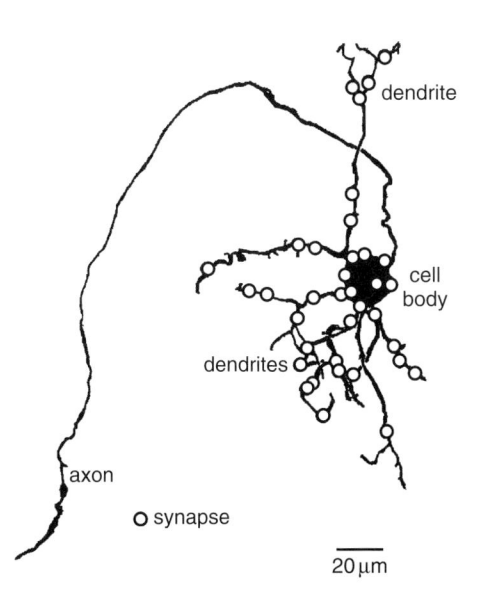

Figure 6.1 Sympathetic postganglionic neuron in a thoracic paravertebral ganglion of the mouse. The neuron has been filled with an intracellular injection of neurobiotin and reconstructed from a through-focus series of confocal sections. Note the dendrites and the axon. Hypothetical distribution of synapses formed by boutons of preganglionic axons are indicated by the circles. As based on systematic investigations using electron microscopy in the guinea pig (Gibbins et al. 1998) this distribution of synapses is random. Approximately 1% to 2% of the surface of the postganglionic neurons are covered by synapses. The density and distribution of synapses probably are representative across species. With permission from Gibbins et al. (2000).

and postganglionic membrane, by electron-dense membrane specializations at the point of contact and by clustering of vesicles at the presynaptic specialization. Boutons not forming synapses are entirely ensheathed by Schwann cells. It is likely, although not proven, that only boutons forming synapses release acetylcholine on excitation. However, colocalized peptide (e.g., vasoactive intestinal peptide [VIP] in intestinofugal axons; substance P or calcitonin gene-related peptide [CGRP] in some groups of preganglionic axons) may also be released by boutons not forming synapses.[1]

The number of preganglionic fibers innervating one postganglionic neuron, and therefore also the total number of boutons forming synapses, is related to the total surface of soma and dendrites of the postganglionic neuron. For example, in guinea pig the postganglionic neurons in the lateral celiac ganglion, most of them being vasoconstrictor in function, have a small surface of about $1300\,m\mu^2$. Each neuron receives about 80 synaptic boutons formed by 2 to 10 preganglionic axons. The postganglionic neurons in the medial celiac ganglion, most of them being motility-regulating or secretomotor in function, have a large surface of about $3500\,m\mu^2$. They receive about 300 to 400 synaptic boutons formed by preganglionic axons or axons of intestinofugal neurons (Gibbins *et al.* 2003c).

6.1.2 Divergence and convergence

Pre- and postganglionic neurons are synaptically connected in the ganglia by divergence and convergence of preganglionic axons (Figure 6.2). Divergence is a mechanism by which a relatively small number of preganglionic neurons connects with a large number of postganglionic neurons. Thus the functional role of divergence is to amplify and distribute the central signal with minimal central representation. Convergence is a mechanism by which several preganglionic neurons form synapses with a single postganglionic neuron. The functional role of convergence is not clear for postganglionic neurons that receive one (or rarely two) strong (suprathreshold) synaptic input from preganglionic axons (see Subchapter 6.2); is it vestigial or is it "designed" for some special functional purpose? The degree of divergence and convergence of preganglionic axons with postganglionic neurons in autonomic ganglia varies between ganglia, between species and, within the same ganglion, between populations of neurons with different functions (i.e. in different final autonomic pathways) (Wang *et al.* 1995).

Studies of the superior cervical ganglion in mammals have shown that the number of preganglionic and postganglionic neurons increases with body size, i.e. with increasing target organ size. The increase in preganglionic cell number is smaller than the rate of increase in postganglionic cell number and both are much smaller than the rate of increase in target size (Ebbeson 1968a, b; Purves *et al.* 1986). For example the ratios of body weights of mouse, guinea pig, cat and human are about 1:10:100:2000 (30 g, 300 g, 3 kg and 60 kg), yet the ratios of the numbers of preganglionic neurons projecting to

Figure 6.2 Convergence and divergence in autonomic ganglia. The degree of divergence of individual preganglionic neurons on postganglionic neurons probably is a function of the size of target organ (and therefore body size) and of the type of autonomic final pathway (e.g., low divergence in the pupillomotor pathway and high degree of divergence in vasoconstrictor pathways). The degree of convergence varies between different autonomic pathways. Here preganglionic axon b (bold) diverges to contact all postganglionic neurons and postganglionic neuron 3 receives convergent input from preganglionic axons a to c (boxed).

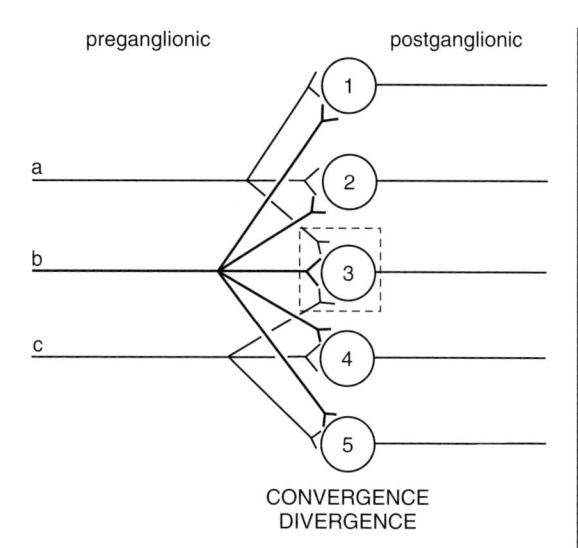

the superior cervical ganglion are about 1:4:9:14 (700, 2800, 6000 and 10 000 neurons) and the ratios of numbers of the postganglionic neurons in the superior cervical ganglion are about 1:3.5:10:100 (10 000, 35 000, 100 000 and 1 000 000 neurons) (Ebbeson 1968a, b; Wesselmann and McLachlan 1984; Purves et al. 1986; see Gabella [1976]). Thus, the average numbers of postganglionic neurons innervated by the same number of preganglionic neurons is about 7 times larger in the human than in the mouse (from 14:1 in small animals to 100:1 in humans), whereas the average size of the target tissue innervated by the same the number of postganglionic neurons increases by a factor of 20 from mouse to human. Furthermore, with increasing target size, the mean number of preganglionic neurons converging on each postganglionic neuron in the superior cervical ganglion increases, the numbers in mouse, guinea pig and rabbit being about 5, 12 and 16 preganglionic neurons (Purves et al. 1986; Ivanov and Purves 1989) and possibly higher in larger animals.

Thus, the degree of convergence of sympathetic preganglionic neurons on postganglionic neurons is related to the total surface of the postganglionic neurons and the total surface is related to the size of the target tissue innervated by the postganglionic neurons (Voyvodic 1989). This is an approximation and future investigations will show that there are differences between functionally different types of sympathetic pathways (e.g., vasoconstrictor pathways vs. vasodilator or sudomotor pathways). Originally it was assumed that the number of converging preganglionic axons matches the number of dendrites of a postganglionic neuron and that each preganglionic axon innervates one dendrite of a postganglionic neuron ("domain theory"; Purves and Hume 1981; Forehand and Purves 1984; Purves and Lichtman 1985; Purves et al. 1988). However, this view is too simplistic since it has been shown that one dendrite may be innervated by more than one preganglionic axon (Forehand 1987; Murphy et al. 1998).

The detailed mechanisms involved in this matching between pre- and postganglionic neurons are unknown, although it is believed that retrograde transport of trophic signals from the target tissue to the postganglionic neurons (e.g., nerve growth factor for sympathetic neurons, neurturin for parasympathetic neurons; Heuckeroth et al. 1999) and from the postganglionic neurons to the preganglionic neurons (brain-derived neurotrophic factor [BDNF]; Causing et al. 1997) is involved (Purves and Lichtman 1978; Purves 1988). Thus, the degree of convergence and divergence of preganglionic axons in autonomic ganglia is dependent on the size and type of target organ. This principle is probably universally valid for autonomic ganglia and there are no principal differences between parasympathetic and sympathetic ganglia (Wang et al. 1995).

Increasing the degree of convergence and divergence in autonomic ganglia with body size may be the best strategy to cope with a larger amount of target tissue. This would avoid increasing the number of neurons in order to meet the distribution function of autonomic ganglia under various functional conditions (Purves 1988). Convergence and divergence vary between functionally different autonomic pathways. It is likely, although unproven, that convergence and divergence occur from preganglionic axons to postganglionic neurons of the same autonomic pathway, i.e. within the muscle vasoconstrictor, cutaneous vasoconstrictor, pilomotor pathway etc. Alternatively, it is possible that this applies only to the strong (suprathreshold) synaptic inputs but not the weak (subthreshold) synaptic inputs (see Subchapter 6.2). Presumably, like in somatic motor units, this is related to the precision of functional control and target size.

6.1.3 Segmental origin of converging sympathetic preganglionic neurons

Preganglionic sympathetic axons converging on one postganglionic neuron originate from several contiguous spinal segments. In the guinea pig, postganglionic neurons in paravertebral ganglia are innervated by preganglionic neurons from four to five (range one to seven) contiguous spinal segments (Nja and Purves 1977a, b; Lichtman et al. 1979; 1980). The general principle of this organization is that the synaptic input to the postganglionic neurons is dominated by one spinal segment (usually the one where the paravertebral ganglion lies for targets of the trunk), which contributes the highest number of preganglionic axons per segment and elicits on average larger excitatory postsynaptic potentials per preganglionic axon. This spinal segment probably supplies the preganglionic axon forming the strong (suprathreshold) input to the postganglionic neurons (see Subchapter 6.2). This principle of segmental organization also applies to the thoracic sympathetic outflow of other species such as the cat, rabbit and hamster (Langley 1892, 1894a, 1903a) and to the lumbar sympathetic outflow to viscera (Krier et al. 1982).

Different types of pathways are represented in different groups of contiguous segments in relation to the position of target organs in the body: some are distributed over the entire thoracolumbar preganglionic cell column (e.g., the vasoconstrictor and pilomotor neurons); others are restricted to groups of contiguous thoracic or lumbar segments (e.g., preganglionic pupillomotor [T1 to T2], sudomotor [in the cat] T2 to T3, L1 to L2, motility-regulating neurons [pelvic organs L3 to L4]; parasympathetic preganglionic neurons to pelvic organs S1 to S2). This was quite obvious from experimental studies of the reaction of target organs (vasoconstriction in skin, piloerection, dilation of pupil and palpebral retraction) to electrical stimulation of the segmental preganglionic outflow (Langley 1892, 1894b, 1895, 1897; Nja and Purves 1977a, b; Figure 8.8). It is also evident from neurophysiological experiments in which the segmental origin of functionally identified sympathetic lumbar preganglionic neurons projecting to the inferior mesenteric ganglion of the cat (Bahr et al. 1986c) and of functionally identified sympathetic thoracic preganglionic neurons projecting to the superior cervical ganglion (Boczek-Funcke et al. 1993) were studied. Finally, it is evident from morphological experiments in which populations of preganglionic neurons have been labeled using tracers that are injected into distinct targets tissue and transported retrogradely transsynaptically by the postganglionic neurons and preganglionic axons to the preganglionic cell bodies (Strack et al. 1988; see Chapter 8 and Table 8.2).

6.2 | Strong and weak synaptic inputs from preganglionic neurons

When stimulated, preganglionic axons converging on a postganglionic neuron elicit excitatory postsynaptic potentials (EPSPs) of different size. This depends on several factors (see Jänig and McLachlan [1987]):

1. The number of quanta of acetylcholine that are released. This is related to the number of varicosities of the preganglionic axon that synapse on the postganglionic neuron and to the average probability of release from these varicosities (McLachlan 1975).
2. The membrane properties of the postganglionic neuron. It is clear that an EPSP occurring for a given synaptic current decreases or increases in duration and less so in amplitude when the input resistance of a neuron decreases or increases as a consequence of a conductance change generated by, e.g., the muscarinic action of synaptically released acetylcholine, or by peptides and hormones released presynaptically or by other processes. While the membrane conductance may vary along the soma-dendritic arbor, the dendritic impedance is such that the properties of the soma and proximal dendrites dominate in determining the size of the potential change.

3. A possible third factor is the geometry of the synapse, i.e. the location of the synaptic contacts of the preganglionic axon on soma and dendrites of the postganglionic neuron. However, this may turn out to be unimportant for electrical reasons since soma and proximal dendrites determine the output of the neurons (see Jamieson et al. [2003]).

Investigations on sympathetic ganglia in vitro in the rat and guinea pig (Hirst and McLachlan 1984, 1986; Cassell and McLachlan 1986) have shown that most postganglionic neurons in paravertebral and some postganglionic neurons in prevertebral ganglia are normally innervated by one (sometimes two, rarely three) preganglionic axons that, when stimulated, always evoke a suprathreshold EPSP in the postganglionic neuron. Other preganglionic axons (up to 15 depending on the functional type of neuron, the ganglion and the species; Purves et al. 1986; Ivanov and Purves 1989), when stimulated, evoke subthreshold EPSPs (Figure 6.3). Preganglionic axons that elicit suprathreshold EPSPs are called "strong" inputs. The underlying EPSP elicited by these axons can be up to 100 mV in size, which is seen if the postganglionic neuron is hyperpolarized to prevent generation of action potentials (Figure 6.5c). These EPSPs are like the endplate potential of the skeletal neuromuscular junction. The number of quanta of acetylcholine released is >20 generating currents up to several nA in amplitude. Neurons receiving one strong synaptic input develop only one discrete current on depolarization; postganglionic neurons receiving two strong preganglionic synaptic inputs develop two distinct currents upon depolarization. This suggests that the Purves' "domain" principle applies to the strong inputs. Hirst and McLachlan (1986) proposed that the synaptic contacts of a strong preganglionic axon are associated with a single dendrite and particularly with a region of the postsynaptic membrane containing a cluster of voltage-dependent calcium channels. Action potentials elicited by these large unitary EPSPs have no inflection on their rising phase; the fast sodium current is activated before the calcium current.

Preganglionic axons eliciting subthreshold EPSPs are called "weak" inputs. The number of quanta released is much smaller than for strong inputs. The synaptic contacts are not associated with voltage-sensitive calcium channels and so they are likely to be in other "domains" of the surface of the postganglionic neurons.

From the studies of Hirst and McLachlan (1986), it is an open question whether "one particular preganglionic axon is predetermined to make synaptic contacts on the dendrites that bear a high density of calcium channels" or whether the formation of calcium channels in the postganglionic neurons is induced by synaptic contacts of particular preganglionic axons. It seems likely that strong inputs have specificity related to the function and, therefore, the target cells of the postganglionic neurons. In the rat lumbar chain ganglia, such contacts develop 7 to 14 days postnatally and at 21 days

6.2 STRONG AND WEAK SYNAPTIC INPUTS FROM PREGANGLIONIC NEURONS | 219

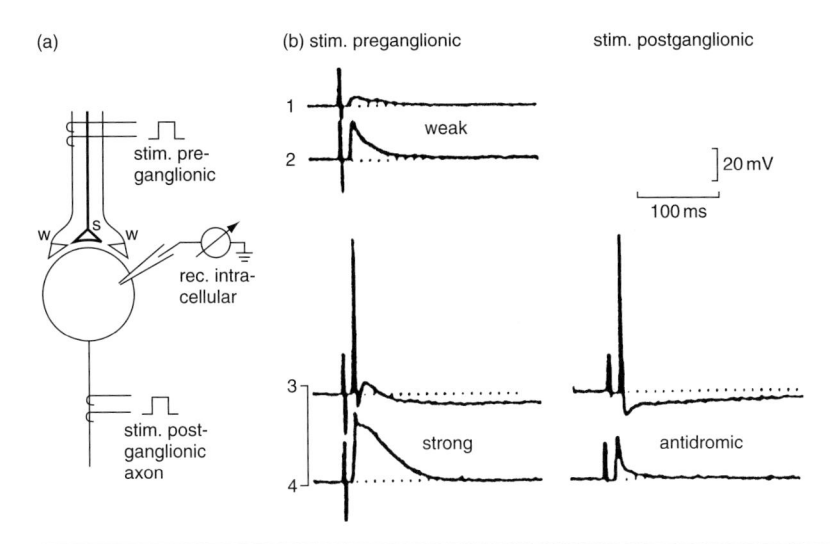

Figure 6.3 Relay function of synaptic transmission in autonomic ganglia. Intracellular recording from a sympathetic postganglionic neuron in a paravertebral ganglion. In vitro experiment. (a) Experimental set-up. (b) Electrical stimulation of converging preganglionic axons with single pulses (the stimulus strength increased from 1 to 3) generates several small subthreshold (weak, w) postsynaptic potentials (see 1 and 2) and one large suprathreshold (strong, s) postsynaptic potential (see 3 left). Strong synaptic responses can be identified by their "all-or-none" onset and the characteristic form of the postsynaptic potential when the action potential is blocked by hyperpolarization (see record 4 left: brief peak and humped decay phase). Electrical stimulation of the peripheral nerve elicits an antidromically generated action potential. After hyperpolarization, the antidromic response is characterized by its brief decay time course relative to the postsynaptic potential. The relay function, in which only discharge of strong inputs is transmitted across the ganglion, occurs in practically all paravertebral sympathetic neurons, in many prevertebral sympathetic neurons and in parasympathetic neurons. Modified from Jänig and McLachlan (1992a) with permission.

after birth the majority of the postganglionic neurons in the lumbar paravertebral ganglia receive one or two strong synaptic inputs typical of the adult animal (Hirst and McLachlan 1984, 1986).

In the rat, all postganglionic sympathetic neurons in paravertebral ganglia, some 50% of sympathetic neurons in prevertebral ganglia and probably all parasympathetic postganglionic neurons receive at least one strong synaptic input from preganglionic neurons. Activity in these postganglionic neurons is entirely dependent on activity in these preganglionic neurons. This activity is generated within the central nervous system.

Some sympathetic postganglionic neurons in prevertebral ganglia projecting to the gastrointestinal tract receive mainly weak or just suprathreshold synaptic inputs from preganglionic neurons. These neurons are non-vasoconstrictor in function. They are involved in regulation of motility and possibly in secretion/absorption of fluid across the gut mucosa (motility-regulating neurons, secretomotor neurons [see Subchapter 4.3]). These neurons receive abundant synaptic inputs from peripheral afferent neurons (intestinofugal enteric neurons, see Subchapter 6.5). Almost all of these latter are subthreshold (weak). Some postganglionic neurons also receive peptidergic inputs from collateral branches of spinal primary afferent neurons (Figure 6.4; see also Subchapters 2.2 and 5.7). These

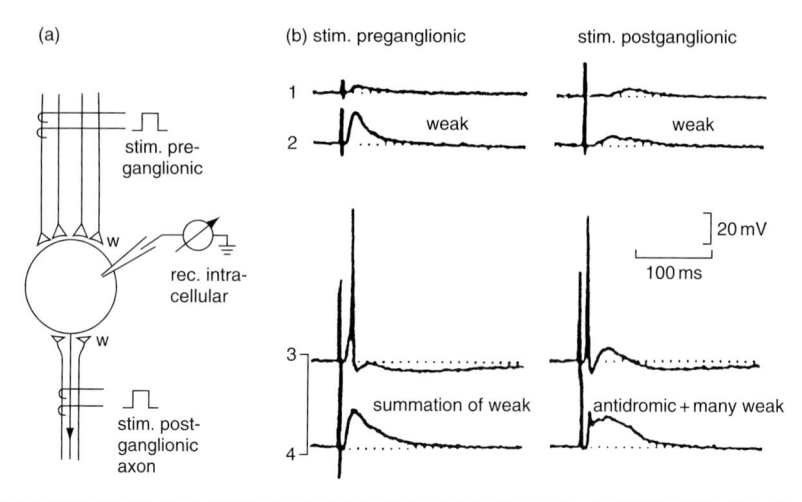

Figure 6.4 Integrative function of synaptic transmission in autonomic ganglia. Intracellular recording from a sympathetic postganglionic neuron in a prevertebral ganglion in vitro. (a) Experimental set-up. (b) Graded electrical stimulation of the converging preganglionic axons elicits several small (weak) excitatory synaptic potentials with similar latencies. These summate at resting membrane potential to initiate an action potential (see *1* to *3* left). Graded electrical stimulation of a mesenteric (peripheral) nerve containing the postganglionic axon also produces many weak excitatory synaptic potentials with varying latencies and an antidromic action potential (see records *1* to *3* right), which can be identified as occurring earlier than the summed synaptic potentials when the membrane is hyperpolarized (record *4* right). This integration by summation of multiple subthreshold (weak) inputs occurs in sympathetic prevertebral motility-regulating neurons and possibly secretomotor neurons. Modified from Jänig and McLachlan (1992a) with permission.

postganglionic neurons function by integrating the many synaptic inputs and are activated mainly by summation of weak EPSPs (for correlation between types of synapses in sympathetic ganglia, biophysical membrane properties, "neurochemical coding" and function see Chapter 6.4).

These findings are conceptually very important. They show that EPSPs elicited by "strong" preganglionic axons do not represent the extreme of EPSPs elicited by "weak" preganglionic converging axons, e.g., simply by a preganglionic axon with a few more synaptic contacts. Rather the large EPSPs represent a class of EPSP that is distinct from those elicited by excitation of "weak" preganglionic axons. Release of acetylcholine at "strong" synapses is dependent on N-type calcium channels (35%) and calcium channels resistant to known blockers (65%); acetylcholine release at "weak" synapses is dependent on N- and P-type calcium channels (40% each) and on calcium channels resistant to blockers (20%) (Ireland et al. 1999). Though not yet explicitly investigated, it is likely that this generally applies to *all* sympathetic pathways through the paravertebral ganglia that innervate targets in somatic tissues, to some sympathetic pathways projecting to the viscera (e.g., the visceral vasoconstrictor pathway), and possibly to all parasympathetic pathways although these pathways have not been systematically investigated in this respect.

Intracellular in vivo recordings show that most neurons in the ciliary ganglion (Melnitchenko and Skok 1970; Johnson and Purves

6.2 STRONG AND WEAK SYNAPTIC INPUTS FROM PREGANGLIONIC NEURONS | 221

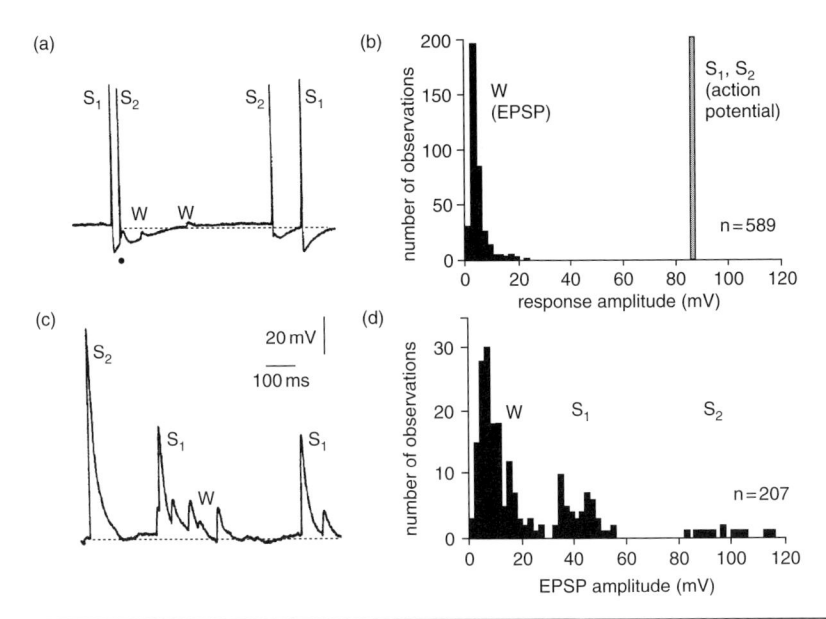

Figure 6.5 Ongoing postsynaptic activity recorded intracellularly in a postganglionic neuron of the rat superior cervical ganglion in vivo. Intracellular recordings were made from a postganglionic neuron of the superior cervical ganglion with its preganglionic innervation intact in a pentobarbital-anesthetized rat. Recordings were made at −50 mV (a, b) and with the neuron hyperpolarized to −130 mV (c, d) to block action potential generation. (a) Inputs generating EPSPs with amplitudes >20 mV at resting membrane potential triggered action potentials of two configurations (S_1, S_2). S_1 had a larger afterhyperpolarization than S_2 whereas S_2 had a depolarization following the spike, more readily seen during the afterhyperpolarization (dot). (c) At the hyperpolarized membrane potential, all EPSPs were increased in amplitude due to the greater driving potential but the largest EPSPs (underlying S_1 and S_2) had distinct amplitudes of ∼40 mV and ∼100 mV respectively. (b, d) Distribution of peak amplitudes of the responses (n = 589 responses and n = 207 responses in (b) and (d) respectively). Note the difference in amplitude between weak (W) and strong (S) responses. From McLachlan et al. (1997) with permission.

1983) and most postganglionic neurons of the superior cervical ganglion of the rabbit and rat (Skok and Ivanov 1983, 1987; Tatarchenko et al. 1990; McLachlan et al. 1997) exhibit synaptically evoked action potentials with no inflection on their rising phase. Intracellular recording of spontaneous activity and reflex activity from postganglionic sympathetic neurons in the superior cervical ganglion in vivo show that spontaneously active neurons (McLachlan et al. 1997; Figure 6.5):

1. receive one or two preganglionic synaptic inputs that are always suprathreshold;
2. also receive several subthreshold preganglionic synaptic inputs; and
3. are not normally activated physiologically by summation of subthreshold preganglionic synaptic inputs because the firing rates of individual convergent inputs are too low.

Thus, the output of these neurons that is "seen" by the target cells is *only* generated by suprathreshold (strong) preganglionic synaptic inputs and *not* by summation of converging subthreshold synaptic inputs. This concept of weak and strong synaptic contacts in autonomic ganglia is fully supported by the following observations:

1. Electrical stimulation of the preganglionic axons in the lumbar sympathetic trunk of the cat with single pulses at graded stimulus strengths evokes discharges in postganglionic neurons to skin and skeletal muscle (e.g., vasoconstrictor, sudomotor and pilomotor neurons) at constant latencies with a small scatter of mostly less than 1 ms and at a well-defined electrical threshold.
2. The scatter of the latency does not change when the strength of the electrical stimulus applied to the preganglionic axons is raised (recruiting in this way more, mostly weak preganglionic axons) nor was there a difference in scatter of latency in most postganglionic neurons whether the preganglionic axons were stimulated close to the ganglion cell body or some 30 to 60 mm proximally, which would have been expected to disperse the EPSPs due to the difference in conduction velocity of the stimulated preganglionic axons (Figure 6.6) (Grosse and Jänig 1976; Jänig and Szulczyk 1980, 1981).

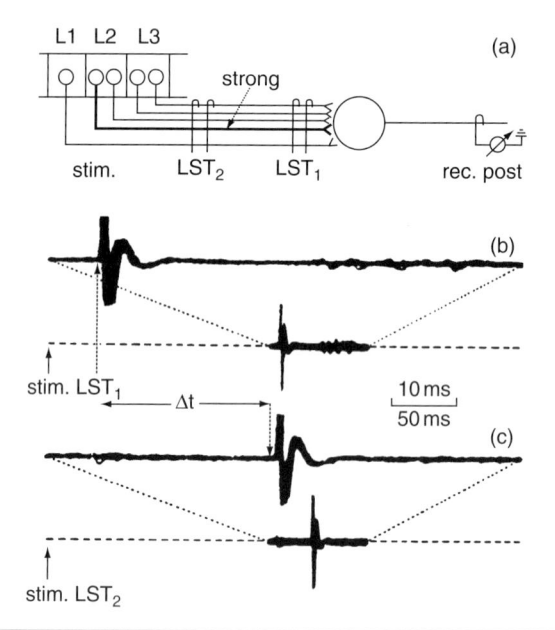

Figure 6.6 Electrical stimulation of preganglionic axons in the sympathetic chain elicits action potentials at distinct latencies and thresholds in postganglionic neurons. Recording from a vasoconstrictor axon innervating the cat hindlimb and electrical stimulation of the lumbar sympathetic trunk (LST) at two sites that were about 60 mm apart. (a) Experimental set-up. Indicated are one strong preganglionic input (bold axon) and several weak ones (thin axons). L1 to L3, lumbar segments 1 to 3. (b) Stimulation at the distal site (LST_1). (c) Stimulation at the proximal site (LST_2). Recordings 10 to 20 times superimposed. Upper traces represent an expanded time scale of intensified part of lower traces. Note that the variability of the latency of evoked responses was small and the same for both responses. This and the distinct threshold voltages at which both responses had been elicited argue that *one* strong preganglionic axon was stimulated at both sites of the LST and that it generated a large suprathreshold excitatory postsynaptic potential. Δt ($= 27$ ms) is the latency difference between the two responses; thus the conduction velocity of the strong preganglionic axon was 2.2 m/s (60 mm/27 ms). Time base: 10 ms applies to the upper records and 50 ms to the lower ones. Modified from Jänig and Szulczyk (1980, 1981) with permission.

These results can only be explained in the following way: the discharges evoked in postganglionic neurons are generated by activation of *one* of the converging preganglionic axons, which forms a strong synapse with the postganglionic neuron and generates a large suprathreshold postsynaptic potential. These results are fully consistent with the concept that summation of postsynaptic potentials generated by activation of weak preganglionic axons rarely initiates action potentials in the postganglionic neurons as revealed by intracellular recording (Figure 6.5). They support the idea that individual postganglionic vasoconstrictor, sudomotor and pilomotor neurons supplying the skin and skeletal muscle of the cat are innervated by one or a few "strong" preganglionic axons.

6.3 | The autonomic neural unit: structural and functional aspects

The "motor unit" of the final common motor pathway of the skeletomotor system consists of the motoneuron and the muscle cells innervated by it. The size of the motor unit varies from a few muscle cells to several hundred muscle cells depending on the function of the skeletal muscle. By analogy, Purves defined the "neural unit" of the final autonomic pathways (see Johnson and Purves [1981]; Purves and Wigston [1983]; Purves *et al.* [1986]). It consists of the number of postganglionic neurons innervated anatomically by one preganglionic neuron. Its mean size for a given autonomic pathway is determined by the ratio of the number of postganglionic to preganglionic neurons and by the mean degree of convergence of preganglionic axons on a postganglionic neuron. The size of an autonomic neural unit is therefore the number of postganglionic neurons divided by the number of preganglionic neurons multiplied by the degree of convergence. For example in Figure 6.7a, five preganglionic neurons converge on one postganglionic neuron and the ratio between pre- and postganglionic neurons is 1:3; the autonomic neural unit consists therefore of one preganglionic neuron and 15 postganglionic neurons. The size of the autonomic neural unit varies with the size of the animal, the size of the target organ and probably the function of the postganglionic neuron (i.e. the type of autonomic final pathway). For the superior cervical ganglion, the mean size of the autonomic neural unit increases from 60 postganglionic neurons in the mouse to 420 postganglionic neurons in the rabbit (Purves and Wigston 1983; Purves *et al.* 1986; Ivanov and Purves 1989).

The size of the autonomic neural unit is in this way defined *structurally*. For autonomic systems in which each postganglionic neuron is innervated by *only one* preganglionic axon (for example, neurons in the submandibular ganglion of the rat and rabbit [Lichtman 1977] or in many ganglia of lower vertebrates [Blackman *et al.* 1963]), the synaptic input from that preganglionic axon is

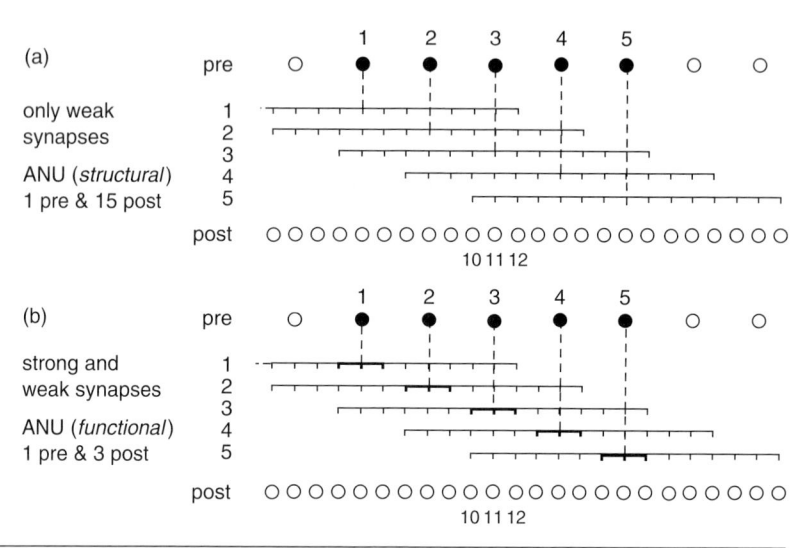

Figure 6.7 The autonomic neural unit (ANU). Schematic representation for a convergence of five preganglionic neurons on one postganglionic neuron. The ratio between post- and preganglionic neurons is assumed to be 3:1. (a) The *structural autonomic neural unit* ("neural unit" of Purves [see Johnson and Purves [1981]; Purves and Wigston [1983]; Purves et al. [1986]]) consists of one preganglionic neuron and 15 postganglionic neurons (which are innervated by each preganglionic neuron). The ANUs exhibit considerable overlap (see units 1 to 5, solid line brackets). Only postganglionic neurons 10, 11 and 12 receive convergent inputs from the five preganglionic neurons. The postganglionic neurons would discharge only with closely coincident firing of several converging preganglionic neurons if all synaptic connections were subthreshold (weak). (b) Examples of *functional autonomic neural units*, which consist of one preganglionic neuron and three postganglionic neurons when each postganglionic neuron receives only one strong (suprathreshold) synaptic input (in addition to subthreshold synaptic inputs). Now the postganglionic neuron would discharge when the preganglionic neuron forming a strong synapse with it discharges. Functional autonomic units do not overlap when each postganglionic neuron receives one strong synaptic input only. Modified from Jänig (1995a) with permission.

generally suprathreshold (i.e. strong) and it is relatively easy to understand how the pattern of discharge is transmitted in these autonomic systems from the central nervous system to the target tissue. The size of the autonomic neural unit in these systems is therefore structurally and functionally identical. For this case the analogy between skeletomotor and peripheral autonomic pathways is obvious. The main difference from the motor unit of the skeletomotor system is that a structurally defined autonomic neural unit shares the same target cells with other autonomic neural units of the same type. The latter is an assumption since it is unknown whether all target cells of a structurally defined autonomic neural unit are functionally of the same type. As in the skeletomotor unit, the response of the autonomic target organ to neural stimulation would depend on the mean frequency of the preganglionic neurons and the successive recruitment of different autonomic neural units.

In sympathetic final pathways in which there is considerable convergence of preganglionic axons onto postganglionic neurons, the analogy between the "motor unit" of the final common skeletomotor pathway and the functional autonomic neural unit is less easy to define because several autonomic neural units share subpopulations of postganglionic neurons (Figure 6.7). The size of the autonomic neural unit is usually smaller functionally than structurally because most preganglionic synaptic inputs are subthreshold and only a few are suprathreshold (strong; bold in Figure 6.7b) (see Subchapter 6.2). This applies to most autonomic ganglia in mammals, in particular all sympathetic ones. The size of the functional autonomic neural unit probably varies depending on the type of final common autonomic pathway. However, the overlap between functional autonomic neural units is relatively small. This follows from the finding that most postganglionic neurons of sympathetic final pathways to somatic tissues, postganglionic neurons of some sympathetic pathways to viscera and postganglionic neurons of parasympathetic pathways receive only *one strong synaptic* preganglionic input and their output depends only on this and not on summation of weak synaptic inputs. These systems therefore behave relatively simply as relay stations in the pathway to the target. The situation differs for some sympathetic motility-regulating and possibly secretomotor pathways innervating the gastrointestinal tract that receive many peripheral synaptic inputs and few strong preganglionic synaptic inputs and are probably activated by summation of synaptic potentials (see Subchapter 6.5).

In conclusion, the autonomic neural unit was originally structurally defined by the number of postganglionic neurons innervated by one preganglionic neuron. Its size increases with increasing degrees of divergence of preganglionic axons. The functional autonomic neural unit is usually much smaller because most synaptic inputs from preganglionic neurons are subthreshold and do not elicit discharges in the postganglionic neurons.

6.4 | Electrophysiological classification, ionic channels, functions and morphology of sympathetic postganglionic neurons

General electrophysiological properties of postganglionic neurons as well as mechanisms underlying fast and slow synaptic potentials have been extensively reviewed in the literature and will not be discussed here (Adams and Harper 1995; Tokimasa and Akasu 1995). Experimental investigations in vitro of postganglionic neurons in paravertebral and prevertebral ganglia of the guinea pig (but also other species such as rat, mouse and cat) show that these sympathetic neurons can be classified by way of their responses to suprathreshold depolarizing current pulses passed through an intracellular microelectrode into three types (Figure 6.8): phasic neurons (rapidly adapting), tonic neurons (slowly adapting) and neurons with a long afterhyperpolarization following the action potential (LAH neurons). Phasic and tonic refers to the pattern of discharge of action potentials. Phasic neurons respond with a burst of action potentials at the beginning of the suprathreshold current pulse. Tonic neurons discharge

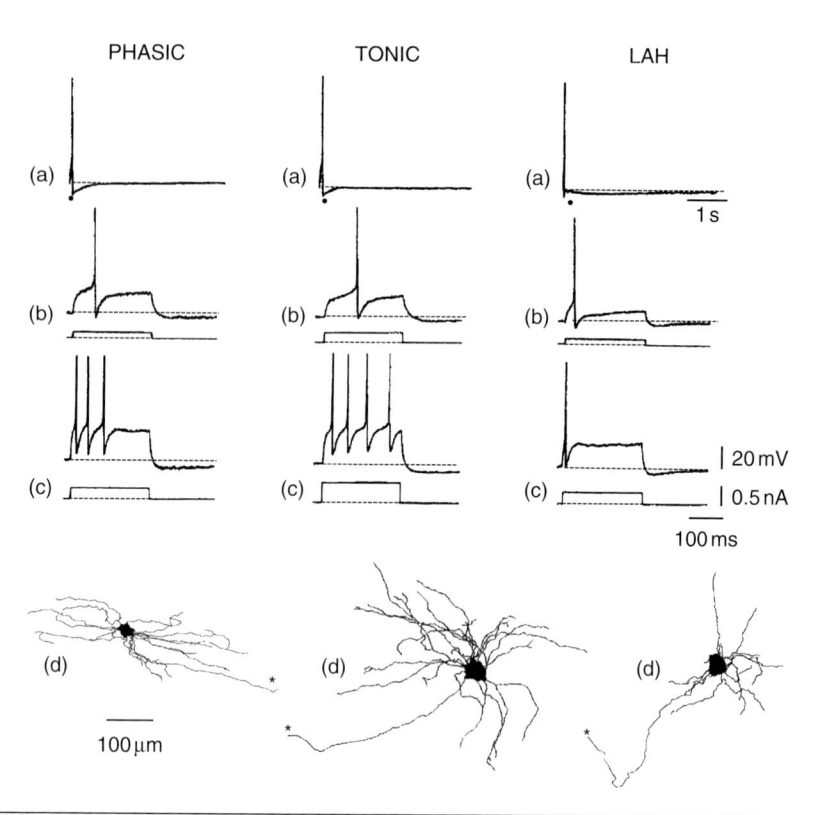

Figure 6.8 Three major classes of sympathetic postganglionic neuron in the guinea pig celiac ganglion defined by discharge characteristics and morphology. Individual neurons were characterized neurophysiologically. (a) Action potentials elicited by 50 ms current steps (dots) showing relative duration of afterhyperpolarization in each class of neuron. (b, c) Voltage responses to long intracellular depolarizing current pulses (lower traces: in (b) just threshold; in (c) approximately twice threshold). (d) Neurophysiologically characterized neurons were filled with biocytin. The cell bodies, dendrites and axons (indicated by *) of the biocytin-filled neurons were reconstructed and analyzed quantitatively. Furthermore, the biocytin-filled neurons were investigated for their neuropeptide content (here somatostatin [SOM] and neuropeptide Y [NPY]) using immunohistochemistry (see Table 6.1 for details). *Phasic neurons* respond with a burst of action potentials to long-lasting depolarizing current pulses. Many of them contain NPY. They do not contain SOM. They have relatively small cell bodies and small dendritic trees. *Tonic neurons* respond repetitively to depolarizing current pulses. They contain SOM but not NPY. They have large cell bodies and an extensive dendritic tree. *Long afterhyperpolarizing* neurons respond with one spike to long-lasting depolarizing currents. Many of these neurons contain NPY. They have large cell bodies and small dendritic trees. LAH, long afterhyperpolarizing neuron. Modified from McLachlan and Meckler (1989) (neurophysiology) and Boyd et al. (1996) (morphology) with permission.

rhythmically throughout the current pulse. The LAH neurons discharge only one action potential at the beginning of the current pulse.

These three classes of electrophysiologically classified sympathetic postganglionic neurons exhibit differential distributions in different sympathetic ganglia (Figure 6.9) and have phenomenologically the following characteristics (Figures 6.8, 6.9; Table 6.1):

1. Almost all postganglionic neurons in the paravertebral ganglia and some 15% to 25% of the postganglionic neurons in the prevertebral ganglia are phasic neurons. Most of these neurons contain, in addition to noradrenaline, neuropeptide Y (NPY). They receive one or two strong and several weak preganglionic synaptic inputs

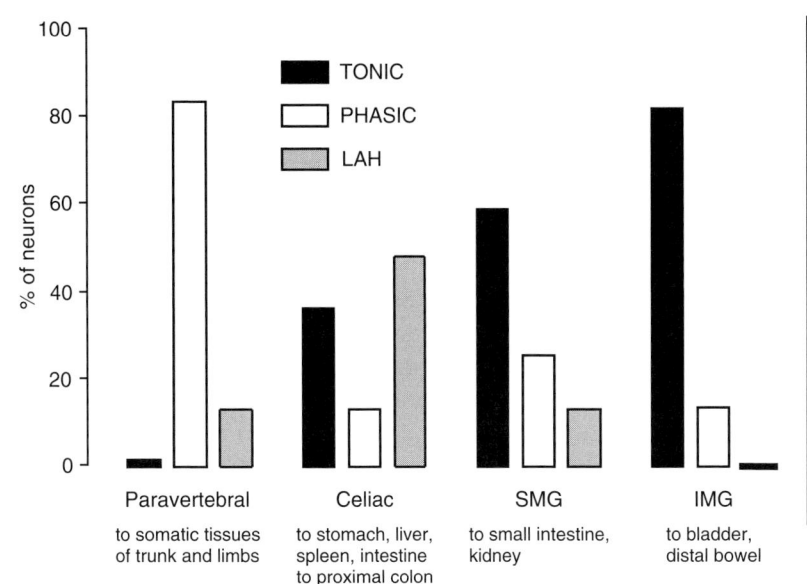

Figure 6.9 Proportions of neurons of different classes in various sympathetic ganglia of the guinea pig. The target tissues of the neurons are indicated at the bottom. Data for celiac ganglia (n = 175), superior mesenteric ganglia (SMG; n = 80) and inferior mesenteric ganglia (IMG; n = 156) from Keast et al. (1993). Data for paravertebral ganglia in the lumbar sympathetic chain (n = 100) from Ireland (unpublished), Davies (unpublished), Christian and Weinreich (1988) and Hamblin et al. (1995). From Elspeth M. McLachlan.

and rarely peripheral synaptic input. In prevertebral ganglia their cell bodies are the smallest of the three classes of neurons and their total dendritic length lies between that of the tonic and the LAH neurons. Phasic neurons in the paravertebral ganglia are involved in regulation of target organs in skin and deep somatic tissues (blood vessels, arrector pili muscles, sweat glands). Most of the phasic neurons in the prevertebral ganglia are probably involved in regulation of vascular resistance; some located in the inferior mesenteric ganglion are involved in the regulation of the vas deferens and other functions related to the pelvic organs. All of these neurons are fully under the control of the central nervous system.

2. Tonic postganglionic neurons are numerous in prevertebral ganglia and are almost absent from paravertebral ganglia. In guinea pigs, they contain, in addition to noradrenaline, either no known peptide or somatostatin. They receive weak preganglionic inputs and rarely a strong one, which is not as large as those in the paravertebral ganglia. Many other weak inputs originate from peripheral neurons (e.g., from intestinofugal neurons of the enteric nervous system, see Chapter 5). However, it is unclear whether this also applies to tonic neurons in the inferior mesenteric ganglion that project through the hypogastric nerve and innervate pelvic organs. Tonic neurons have large cell bodies and many dendrites giving a large total dendritic length. In the gastrointestinal tract, tonic neurons containing somatostatin innervate largely the myenteric plexus and probably are involved mainly in regulation of motility of the gastrointestinal tract. Tonic neurons containing no known peptide project to the submucous plexus and probably are involved mainly in regulation of secretion

Table 6.1 *Biophysical properties of sympathetic postganglionic neurons and their correlation with content of peptides and function*

	Phasic	Tonic	LAH
Discharge during depolarization	Transient burst	Rhythmic	Single
Morphology			
Relative soma size	0.75	1.25	1.0
No of primary dendrites	16	20	13
Total dendritic length	2336 µm	3574 µm	1325 µm
Anatomical distribution			
SCG	85%	<1%	15%
LSC	95%	<1%	<5%
CG	15%	35%	50%
IMG	18%	80%	2%
Resting membrane potential	−51 mV	−63 mV	−54 mV
Resting conductance	27 KΩ.cm^2	40 KΩ.cm^2	15 KΩ.cm^2
Action potential			
Repolarization	Ca^{2+} through N- and L-type channels activates BK channels	few BK channels	few BK channels
Ca^{2+}-dependent afterhyperpolarization	Ca^{2+} through N-type channels activates SK channels		Ca^{2+} through L-type channels activates BK/SK/other channels
Prominent K$^+$ channel			
M	yes	few	few
SK (medium AHP)	yes	yes	yes
multiple Ca^{2+}-activated K$^+$	no	no	yes
A (half-inactivation)	−85 mV	−73 mV	−80 mV
D	rare	common	none
Synaptic input			
from CNS (preganglionic)	1–2 strong, several weak	rarely strong, several weak	1 strong, several weak
from periphery (visceral afferent)	rare weak	numerous weak	very rare
Peptide content	many NPY	SOM	many NPY
Likely functional role	vasoconstriction, pilomotor	motility, secretion	motility, secretion

Notes:

Potassium channels (see Adams and Harper [1995]; Alexander *et al.* [2004]): *A channel*: voltage-sensitive, rapid activation and inactivation. *BK channel*: large-conductance Ca^{2+}-sensitive. *D channel*: voltage-sensitive slowly activating and inactivating slow A-type channel. *M channel*: Muscarine-sensitive, voltage-sensitive (slow activating, non-inactivating). *SK channel*: small-conductance Ca^{2+}-sensitive. CG, celiac ganglion; IMG, inferior mesenteric ganglion; LSC, lumbar sympathetic chain; NPY, neuropeptide Y; SCG, superior cervical ganglion; SOM, somatostatin. Modified from Jänig and McLachlan (1992a) by Elspeth McLachlan. Data for guinea pig ganglia from Cassell *et al.* (1986), McLachlan and Meckler (1989), Keast *et al.* (1993), Boyd *et al.* (1996), Davies *et al.* (1996, 1999), Parr and Sharkey (1996), Martinez-Pinna *et al.* (2000), Jamieson *et al.* (2003).

(see Subchapter 5.7 and Figure 5.15; see Furness and Costa [1987]; Furness [2005]). Tonic postganglionic neurons appear normally to be mainly under peripheral control and the centrally generated signals in preganglionic neurons may require facilitatory synaptic input from peripheral neurons in order to be effective (see Subchapter 6.5).

3. Most LAH neurons are also found in prevertebral ganglia and are almost exclusively located in the celiac ganglion and superior mesenteric ganglion; a few are located in the paravertebral ganglia, in particular the superior cervical ganglion. Many of them contain, in addition to noradrenaline, NPY. They receive one strong and several weak preganglionic inputs and very rarely receive synaptic inputs from peripheral neurons. They have medium-sized cell bodies and a small dendritic arbor. In the gastrointestinal tract, they may innervate the submucosal plexus and mucosal cells and are mainly (but not exclusively) involved in secretion and absorption. These neurons are under central control and are important for regulation of fluid balance (see Sjövall et al. 1987; Lundgren 1988). The LAH neurons in the celiac ganglion also project to the spleen. Their function may be vasoconstrictor or they may modulate immune cells (Jobling, unpublished observations).

This description shows that there are no straightforward correlations between the functions of these postganglionic neurons as defined by their target cells and their neurophysiological properties. Indeed this description covers probably the most well-characterized group of neurons. Furthermore, the peptide content of these neurons can only be used as a functional label to a limited degree. There are variations between species in the peptide content of different neurophysiologically characterized postganglionic neurons.

The mechanisms underlying the neurophysiological characteristics of the postganglionic neurons (action potential, afterhyperpolarization, responses to depolarizing current pulses) have been extensively studied, mainly in the guinea pig and rat. These characteristics depend on the presence (or absence) of different voltage- and calcium-dependent potassium channels, which are differentially expressed in the membranes of these neurons. The main potassium channels that are involved are listed in Table 6.1. For example:

- The phasic responses of sympathetic neurons to current steps (Figure 6.8) depend on the presence of the M-type potassium channel (muscarine-sensitive potassium channel). Opening of this channel by depolarization increases the K^+ current and limits the discharge of phasic neurons. After blockade of this channel, by a muscarinic agonist, phasic neurons exhibit a tonic response to suprathreshold depolarizing current pulses.
- The rhythmic firing of tonic neurons is determined by A-type and D-type potassium channels with distinct voltage characteristics

that make it dominate membrane behavior in the subthreshold voltage range.
- The long-lasting afterhyperpolarization in LAH neurons is generated by various calcium-dependent potassium channels (BK, SK and other potassium channels; see Table 6.1).

6.5 Different types of autonomic ganglia and their functions in vivo

6.5.1 Paravertebral sympathetic ganglia

The function of sympathetic paravertebral ganglia is to transmit discharges from preganglionic neurons to postganglionic neurons without integrating additional synaptic inputs from the periphery. Experimental investigations of sympathetic neurons in vivo in the cat show that the frequency of ongoing discharge is remarkably similar in pre- and postganglionic neurons of the same functional type under the same experimental conditions (Table 6.2).[2] Thus, convergence of many preganglionic axons on one postganglionic neuron, which presumably occurs in all paravertebral sympathetic systems, does not lead to a higher suprathreshold activity in the postganglionic neurons. The simplest explanation for this finding is that each postganglionic neuron receives strong synaptic inputs from one (or two) preganglionic axons (Figure 6.10a). Evidence for this has been presented for paravertebral ganglia in the rat, guinea pig and rabbit as described in Subchapter 6.2 (Skok and Ivanov 1983; Cassell et al. 1986; Hirst and McLachlan 1986; McLachlan et al. 1997).

Traditionally it was assumed that summation of subthreshold EPSPs leads to firing of the postganglionic neurons (Blackman 1974). However, as discussed above, this mechanism probably cannot work unless there are synchronizing mechanisms for the discharges of the converging preganglionic neurons (see Skok and Ivanov 1987). Synchronization of discharges in preganglionic neurons converging on one postganglionic neuron under physiological conditions could be generated by rhythmic firing with the phasic activation of the arterial baroreceptor reflex by the pulse pressure wave (see Figures 4.1, 4.6, 4.7) or with respiration or by central endogenous mechanisms (see Subchapters 10.2 and 10.6). Thus, there is evidence of the tendency for sympathetic neurons of particular types to discharge more or less in rhythmic bursts. However, (1) only synaptic events that occur at time intervals of less than 50 ms lead to effective temporal and spatial summation of EPSPs (see McLachlan et al. 1997), and (2) this type of physiological synchronization of activity in preganglionic neurons only occurs in some final autonomic pathways (e.g., muscle and visceral vasoconstrictor neurons, sympathetic "inspiratory" neurons [see Table 4.1]) but not in most other

Table 6.2 | Discharge rates of sympathetic neurons in vivo in the cat and rat

Neuron type	Resting activity		Maximal activity		Refs
	Preganglionic imp/s	Postganglionic imp/s	Preganglionic imp/s	Postganglionic imp/s	
MVC$_L$	1.8 ± 1.3 (26) (0.1–4.6)	(0.5–3)	approx. 12	approx. 10	8,9
		1.4 ± 0.5 (44) (0.3–2.4)			6
MVC$_C$	1.1 ± 0.8 (54)				
	1.1 ± 0.8 (36)^a (0.2–3.5)	0.7 ± 0.4 (21)^b (0.2–1.5)			6,3,4
VVC$_L$	1.6 ± 0.9 (46) (0.3–4)	1.1 ± 1.1 (14) (0.2–2.5)	<5		2,9
CVC$_L$	0.9 ± 0.6 (47) (0.1–3.4)	1.2 ± 0.7 (44) (0.4–4)			8,9
		1.2 ± 0.6 (65)^c (0.3–2.4)			6,7
CVC$_C$	1.5 ± 1.1 (30)				5
SM$_L$	low	0.2 ± 0.1 (15)		approx. 5	8,9
MR$_L$	0.8 ± 0.7 (91) (0.1–3.8)	0.7 ± 0.5 (67) (0.1–2.6)	5.3 ± 1.5 (54) (max 8)	2.7 ± 1.5 (22) (max 5)	1,2,10
INSP$_C$	1.4 ± 1.4 (24)				5

Notes:
The measurements were done under chloralose anesthesia in artificially ventilated cats or in pentobarbital-anesthetized and artificially ventilated rats (rat data in italic). Resting activity was recorded extracellularly from the axons of the neurons at mean arterial blood pressure of ≥ 100 mmHg and end-tidal CO_2 of 4%. Rates of maximal activity were estimated from reflex responses to systemic hypoxia (MVC$_L$, SM$_L$, VVC$_L$), to mechanical shearing stimuli applied to the mucosa of the anal canal (MR$_L$) and to mechanical stimulation of the nasopharyngeal mucosa (MVC$_C$, INSP$_C$). CVC, cutaneous vasoconstrictor; INSP, inspiratory type; MR, motility-regulating; MVC, muscle vasoconstrictor; SM, sudomotor; VVC, visceral vasoconstrictor; C, thoracic sympathetic outflow to the superior cervical ganglion; L, lumbar sympathetic outflow to skin, skeletal muscle or pelvic organs; means ± SD (n), range of activity in parentheses.
^{a} Cervical sympathetic trunk in rat: most neurons behave like MVC neurons
^{b} Postganglionic neurons to submandibular gland in rat with spontaneous activity and high cardiac rhythmicity
^{c} Rat tail data for CVC 1.1 ± 0.7 imp/s (51), range 0.2–2.6 imp/s (Häbler et al. 1999)
References: 1 Bahr et al. (1986a); 2 Bahr et al. (1986c); 3 Bartsch et al. (1996); 4 Bartsch et al. (2000); 5 Boczek-Funcke et al. (1993); 6 Häbler et al. (1994a); 7 Häbler et al. (1999a); 8 Jänig (1985a); 9 Jänig (1988a); 10 Jänig et al. (1991)
From Jänig (1995a).

autonomic pathways. I am skeptical whether generation of action potentials in postganglionic neurons by spatial and temporal summation of EPSPs is important at all.

What is the function of the subthreshold EPSPs, generated by weak converging preganglionic axons, that do not summate to suprathreshold events but occur more often than the frequency of action potentials in most neurons? Is the existence of these subthreshold potentials left over from the development of the connectivity

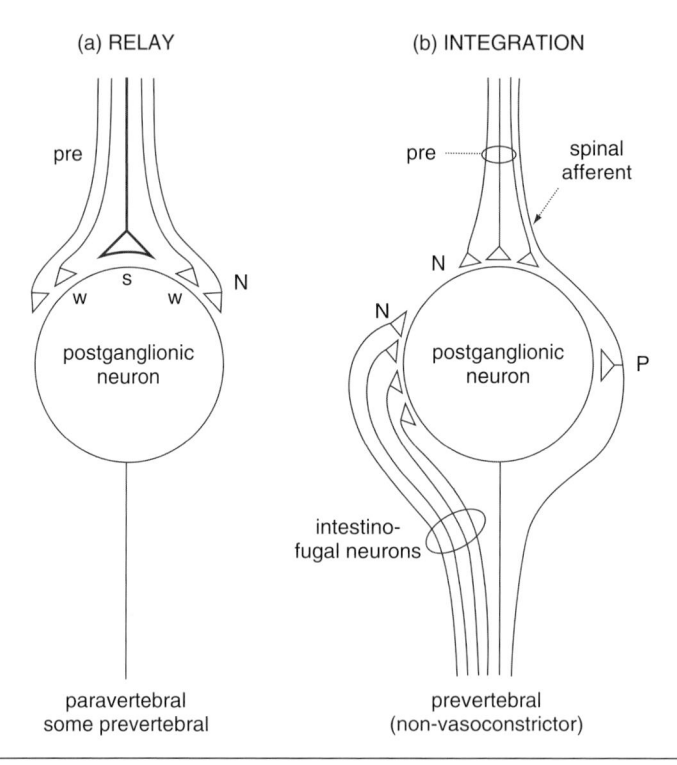

Figure 6.10 Relay and integrative functions of autonomic ganglion cells. (a) Postganglionic neuron with one (and sometimes two) strong (s; suprathreshold) preganglionic synaptic input and several weak (w; subthreshold) synaptic inputs. This connectivity occurs in most paravertebral sympathetic neurons, some prevertebral sympathetic neurons and most parasympathetic neurons. These connections mainly function to transmit the activity from specific preganglionic neurons to postganglionic neurons, i.e. they determine the discharge pattern of the postganglionic neuron. (b) Postganglionic neuron with synaptic inputs from both preganglionic neurons and intestinofugal neurons of the enteric nervous system and also from collaterals of spinal visceral afferents. The first two are cholinergic and nicotinic (N) but subthreshold (weak); the intestinofugal afferents may also contain neuropeptides, which may function as transmitters (e.g., vasoactive intestinal peptide in guinea pig). The collaterals of spinal visceral afferent neurons release substance P (P), which produces a slow depolarization of the postganglionic neuron (calcitonin gene-related peptide that is colocalized with substance P in the afferent neuron is probably also released but has no known function). These postganglionic neurons innervate neurons of the enteric nervous system and possibly other target organs in the viscera. They fire only after integration of several subthreshold inputs with or without the peptide-induced slow depolarization.

between preganglionic axons and postganglionic neurons (Hirst and McLachlan 1984), i.e. do they have any function relevant to the ongoing regulation of the target organs? Skok (1986) has expressed the idea that these subthreshold EPSPs may be used to fire particular groups of postganglionic neurons under special functional conditions in which the discharge in preganglionic neurons that converge on these groups of neurons becomes synchronized. This mechanism would require a considerable degree of central organization in order to shape the temporal firing of selective groups of preganglionic neurons converging on individual postganglionic neurons. This becomes even more difficult to imagine as a successful strategy for the central nervous system when the dispersion of conduction velocity of preganglionic axons is taken into account (Jänig and Szulczyk 1980; Jänig 1985a; Boczek-Funcke et al. 1993).

Do all converging preganglionic neurons generating strong suprathreshold or subthreshold synaptic events in an individual postganglionic neuron have the same functional properties (i.e. have discharge characteristics of either muscle vasoconstrictor, or cutaneous vasoconstrictor or sudomotor etc. neurons)? This has not been tested systematically, although experiments performed so far indicate that this is often the case (Skok and Ivanov 1987; McLachlan et al. 1997). In this context I want to emphasize: (1) this idea can only be tested in vivo in neurophysiological experiments, i.e. when the central circuits are intact and the afferent inputs are stimulated physiologically to generate reflexes; (2) these experiments require intracellular measurements in postganglionic neurons; and (3) this type of experiment is extremely difficult to conduct (see McLachlan et al. 1997, 1998).

It is still possible that supra- and subthreshold non-summating synaptic events measured in the same postganglionic neuron have different reflex patterns. If this were the case, then the suprathreshold events generated in a postganglionic neuron by strong preganglionic synaptic inputs may be related to the development and maintenance of the functional specificity of the postganglionic neurons with respect to their target organs and therefore may be responsible for the separation of different types of final autonomic pathways. This would mean that these pathways are functionally but not anatomically separate. This idea is consistent with regeneration experiments. After partial denervation of a paravertebral sympathetic ganglion, surviving preganglionic axons sprout and innervate the denervated postganglionic neurons. This sprouting of intact axons leads preferentially to the formation of strong synapses with the same characteristics as those in normal ganglia (Ireland 1999). The pattern of functional synaptic transmission is restored but not the number of synaptic connections. The strong synaptic inputs are newly formed, in addition to the existing weak synapses. These newly formed strong synaptic inputs derive either from preganglionic neurons that also form strong synapses on other preganglionic neurons or are induced by signals generated by the postganglionic neurons. Thus, this type of regeneration study raises the possibility that preganglionic neurons are specialized to form either strong or weak synapses only.

The following questions remain to be addressed:

1. Are all synaptic events in individual postganglionic neurons of the same functional type, i.e. do all converging preganglionic neurons have the same functional properties? Evidence so far available is positive (McLachlan et al. 1997).
2. What is the degree of convergence of preganglionic neurons in different types of final autonomic pathways? In vasoconstrictor neurons the degree of convergence is high and in secretomotor neurons to salivary glands it is low.
3. How many of the preganglionic neurons converging on postganglionic neurons with spontaneous activity are silent, at least under anesthesia?

4. Do central synchronizing mechanisms contribute to spatial and temporal summation of subthreshold EPSPs and in this way to synaptic activation of postganglionic neurons that receive a strong (suprathreshold) input? This idea is not supported by experiments so far conducted (McLachlan et al. 1997, 1998).
5. How do strong suprathreshold synaptic inputs develop postnatally with respect to the target cells? Strong synaptic inputs form at about the same time as the connections of postganglionic axons with peripheral targets (Hirst and McLachlan 1984). There must exist functional matching between preganglionic neurons and postganglionic neurons, otherwise it would be impossible to get correct restoration of function after denervation following complete or partial lesion of preganglionic axons (Langley 1895; Murray and Thompson 1957; Guth and Bernstein 1961; Nja and Purves 1977a).

6.5.2 Prevertebral sympathetic ganglia

The sympathetic prevertebral ganglia have a number of different functions during ongoing regulation of the abdominal organs, pelvic organs and blood vessels. They contain different types of neurons as judged by morphological, electrophysiological, neurochemical and other criteria (Table 6.1). Many neurons in these ganglia receive multiple synaptic inputs from the spinal cord, from the periphery and from collaterals of visceral primary afferent axons (Figure 6.10b; see Chapter 2). Distinct types of postganglionic neurons are involved in different functions such as regulation of vascular resistance, motility of the gastrointestinal tract and pelvic organs and secretion/reabsorption across the mucosa of the gastrointestinal tract etc. Results obtained in in vivo experiments show, at least for the cat inferior mesenteric ganglion and the projection of preganglionic neurons in the hypogastric nerve to pelvic organs, that the central message is faithfully transmitted to the periphery in different pathways to distal colon and pelvic organs (see Subchapter 4.3).[3] In this context it is worth noting that many postganglionic vasoconstrictor neurons projecting to abdominal and pelvic organs are situated in the paravertebral ganglia (Kuo et al. 1984; Hill et al. 1987; see Jänig and McLachlan [1987]).

Mediation of peripheral reflexes

Peripheral extraspinal reflexes have been well established for the gastrointestinal tract in the guinea pig; distension of one part of this tract leads to inhibition of the motility of other parts (e.g., colo-colonic, gastro-colic reflexes) mediated via the inferior mesenteric ganglion and the celiac plexus (Kreulen and Szurszewski 1979a, b; see Szurszewski and King [1989]). These reflexes may also exist in other mammals, such as cat and dog as well as humans. These were first described by Kuntz in the dog (inset (a) in Figure 6.11; Kuntz 1940; Kuntz and Saccomanno 1944). The neurons involved have their cell

6.5 DIFFERENT TYPES OF AUTONOMIC GANGLIA AND THEIR FUNCTIONS IN VIVO

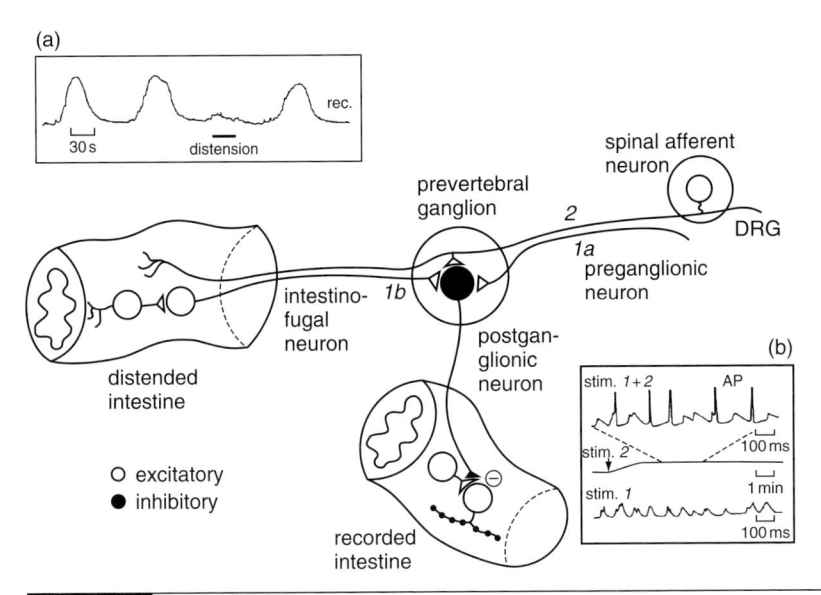

Figure 6.11 Peripheral reflexes involving the enteric nervous system and sympathetic prevertebral ganglia. Distension of the bowel activates afferent terminals of local enteric neurons, which project to the prevertebral ganglia or are linked to intestinofugal neurons (neuron 1b). It also activates visceral primary afferent neurons (neuron 2) with cell bodies in the dorsal root ganglia (DRG). The intestinofugal neurons project only to prevertebral ganglia where (1) they excite postganglionic neurons projecting to another part of the bowel. This leads to inhibition (−) of enteric excitatory muscle motor neurons and relaxation. (2) Collateral branches of primary afferent neurons (neuron 2) release substance P around the same postganglionic neurons, depolarizing it and summing with discharge generated by inputs from intestinofugal and preganglionic neurons. Inset (a): Recording of the intraluminal pressure in an isolated proximal segment of the colon in the cat and distension of the rectum. The inferior mesenteric ganglion (IMG) was decentralized by section of the preganglionic axons in the lumbar splanchnic nerves. The regular contraction waves of the proximal colon were inhibited during distension of the rectum. This inhibition was mediated by the IMG. From Kuntz (1940). Inset (b): lower trace, ongoing subthreshold cholinergic postsynaptic potentials elicited in a prevertebral postganglionic neuron during activity in intestinofugal (neuron 1b) and preganglionic neurons (neuron 1a); middle trace, long-lasting postsynaptic potential generated by activity in peptidergic spinal visceral afferent neurons (decrease in K^+ conductance); upper trace, enhancement of nicotinic postsynaptic potentials when peptidergic synapses are activated, leading to an increase in size and duration of postsynaptic potentials with generation of action potentials (AP). Idealized in vivo behavior derived from in vitro experiments. Modified from Jänig and McLachlan (2002).

bodies in the wall of the gut (intestinofugal neurons) and can be synaptically activated by intrinsic primary afferent neurons (IPANs; see Subchapter 5.1) in the enteric nervous system (Crowcroft et al. 1971). Synaptic transmission to the postganglionic neurons is cholinergic and nicotinic, but may also be non-cholinergic (i.e., peptidergic) (Figure 6.11).

These peripheral reflex pathways may also exist, at least anatomically, for other target tissues of the sympathetic prevertebral neurons in the gastrointestinal tract and its appendages (such as the submucosal tissue, pancreas, gallbladder) or sympathetic neurons in the stellate ganglion projecting to the heart and possibly for sympathetic pathways to other organs. For example, it has been proposed that neurons in the stellate ganglion that project to the heart and lung receive synaptic input from peripherally located afferent neurons (Bosnjak and Kampine 1982, 1984, 1985). These extraspinal neural circuits may be of considerable importance for the understanding of neural regulation of the internal organs in health and disease, yet we need more rigorously controlled physiological experiments to demonstrate their recruitment and relative importance in organ regulation.

Integration of impulse activity from spinal cord and periphery

Most peripheral synaptic and central synaptic inputs to tonic prevertebral sympathetic postganglionic neurons that innervate non-vascular target organs, in particular the gastrointestinal tract, are subthreshold (see Szurszewski [1981]; Jänig and McLachlan [1987]):

- Spatial summation of synaptic inputs is necessary in order to excite the neurons. It is proposed that ongoing afferent synaptic activity from the periphery sets the firing threshold of the prevertebral postganglionic neurons for the preganglionic synaptic input. At low levels of synaptic activity from the periphery (e.g., during small distensions of the gastrointestinal tract) the central messages may not reach the peripheral targets at all. At high levels of synaptic activity induced from the periphery, the central message from several preganglionic inputs will reach the peripheral target tissue. Thus, by lowering or increasing afferent activity from the periphery, the state of the periphery would determine the conditions under which the central messages reach the target organs.
- Alternatively, one could argue that the preganglionic impulse activity sets the firing threshold of the postganglionic neurons for the peripheral afferent synaptic input (Figure 6.11).
- Experimental data showing that the prevertebral abdominal ganglia function in this integrative mode in vivo are lacking. However, peripheral integration of this type is probably important for pathways to some non-vascular target cells of the gastrointestinal tract, unimportant for regulation of vascular resistance and unimportant for regulation of motility and secretion of pelvic organs.
- As already discussed in Chapter 5, sympathetic pathways to the gastrointestinal tract by which the central nervous system directly controls target tissue have strong preganglionic synapses and are not controlled by peripheral reflexes. This applies to the neural control of vascular resistance.
- The sympathetic pathways involved in regulation of motility (and possibly other functions) are not under dominant control by the central nervous system and are modifiable by peripheral reflexes.
- The sympathetic pathway regulating secretion and absorption may be under both dominant central control and peripheral control.

In vivo experiments on animals need to be done to test the idea that there is reciprocal integration of central and peripheral cholinergic synaptic input in prevertebral postganglionic neurons. The main question to be tested is: can central reflex activation of the postganglionic neurons be gated by the peripheral input from the gastrointestinal tract and vice versa?

Putative function of peptidergic synapses made by visceral afferents

Postganglionic neurons in the prevertebral *abdominal* ganglia may receive synaptic input from collaterals of spinal visceral afferent neurons that have their cell bodies in the thoracolumbar dorsal root ganglia. Anterograde labeling of spinal thoracolumbar visceral afferents in guinea pig shows that these afferents may project to postganglionic neurons in the inferior mesenteric ganglion (Dalsgaard *et al.* 1982; Matthews and Cuello 1984); some of them forming synapses with postganglionic neurons (Matthews *et al.* 1987). Similar results have also been obtained for the stellate ganglion (Quigg *et al.* 1990). These collaterals form boutons associated with non-vascular postganglionic neurons. The peptides present in these collaterals are substance P and CGRP (for review see Elfvin *et al.* 1993). Postganglionic neurons in the celiac and inferior mesenteric ganglion of guinea pig expressing the neurokinin receptor 1 (NK_1) for substance P are not closely surrounded by pericellular baskets of substance-P-immunoreactive boutons formed by the afferent collaterals (Messenger *et al.* 1999). The expression of the NK_1-receptor is almost restricted to noradrenergic neurons containing somatostatin. Thus, there is no correlation between the release site for substance P from the afferent terminals and its receptor in the postganglionic neurons.

In the guinea pig, repetitive electrical stimulation of the visceral afferents in the upper lumbar dorsal roots, which have been sectioned between the spinal cord and the stimulation electrodes (so as to prevent spinal reflex activation of preganglionic neurons), elicits long-lasting postsynaptic potentials in postganglionic neurons of the prevertebral ganglia (produced by an increase of membrane resistance due to closure of potassium channels). Electrical stimulation of the mesenteric nerves elicits slow EPSPs in tonic postganglionic neurons and prolonged inhibition of the afterhyperpolarization in postganglionic LAH neurons (Zhao *et al.* 1996). These changes in the postganglionic neurons are generated by impulses in unmyelinated visceral afferents. They can also be generated by substance P and be partially blocked by an NK_1-receptor antagonist. Thus, the transmitter involved in this is probably substance P. Calcitonin gene-related peptide, which is probably also released by the afferent fibers, has no known function on the postganglionic neurons. The anatomical data argue that substance P acts by volume conduction on the postganglionic neurons in the prevertebral ganglia and not by direct synaptic action. The biophysical and pharmacological properties of this effect of substance P released by afferent axons have been described in the literature (Tsunoo et al. 1982; Dun 1983; Katayama and Nishi 1986; Tokimasa and Akasu 1995).

What is the function of this peptidergic transmission from collaterals of spinal visceral afferents on prevertebral sympathetic neurons? These visceral afferents respond to various mechanical and chemical stimuli, most of them being polymodal and many of them having low mechanical threshold. They are involved in spinal and supraspinal reflex discharges (including not only visceral sympathetic systems, but also sympathetic systems supplying skin and skeletal muscle and skeletomotor systems) and various forms of visceral sensations, the most important one being visceral pain and discomfort (see Chapter 2). Reflexes and sensations elicited by stimulation of these afferents may be integrative components of a protective behavior, which is elicited when, for example, the gastrointestinal tract or other visceral organs are abnormally affected (e.g., by peritonitis, by inflammation of the organs, by overdistension, by obstruction). Activation of the spinal visceral afferents would induce slow non-cholinergic depolarizations of the prevertebral neurons, which would enable the preganglionic and peripheral afferent cholinergic synaptic potentials to reach threshold (see Figure 6.11, inset [b]). This peripheral mechanism would enhance inhibitory intestino-intestinal reflexes mediated by sympathetic prevertebral ganglia to the gastrointestinal tract and contributes to the protection of the organ. Evidence for this hypothesis comes from experiments conducted on postganglionic neurons in the guinea pig inferior mesenteric ganglion with an attached segment of the colon in vitro: distension of the colon elicits non-cholinergic depolarizations in many neurons, leading to an increased excitability (Kreulen and Peters 1986; see also Krier and Szurszewski [1982]). Similar slow depolarizations can be elicited in neurons of the inferior mesenteric ganglion by distension of a mesenteric vein attached to the ganglion (Keef and Kreulen 1986) and by distension of the ureter (Amann et al. 1988). The ureter and the mesenteric vein do not contain peripheral afferent neurons that project to the inferior mesenteric ganglion, therefore these slow depolarizations are most likely generated by excitation of collaterals of spinal visceral afferents.

6.5.3 Parasympathetic ganglia

Parasympathetic ganglia are different from sympathetic ones in some anatomical and functional ways. As already discussed in Chapter 1 the postganglionic neurons are mostly located in small ganglia close to their targets. Most parasympathetic postganglionic neurons are structurally relatively homogeneous, having few or no dendrites, but this varies between species. They typically receive a small number of preganglionic synaptic inputs, many of them being strong, but also this is not a discriminating criterion against sympathetic neurons. Parasympathetic ganglia are commonly considered to have only relay function. This may not be entirely true and may also be related to the fact that parasympathetic postganglionic neurons have not been studied as extensively as sympathetic ones (Akasu and Nishimura 1995).

No evidence for interneurons or peripheral synaptic inputs has been detected in cranial parasympathetic ganglia (e.g., the ciliary, otic, submandibular and pterygopalatine ganglia). Ganglia in pancreas or gallbladder, both having embryologically the same origin as the enteric neurons, are similar to enteric ganglia but are otherwise also typical of parasympathetic ganglia. The neurons in these ganglia are involved in regulation of exocrine and endocrine functions of the pancreas or of motility and secretion of the gallbladder. They receive synaptic input from enteric neurons as well as from preganglionic parasympathetic neurons and probably have integrative function. These ganglia are considered to be accessory enteric ganglia (Mawe 1995, 1998).

Intracardiac parasympathetic ganglia have been studied in more detail using neurophysiological methods (Edwards et al. 1995). In these ganglia only a proportion of the ganglion cells receive input from preganglionic axons and project to cardiac muscle. A subpopulation of neurons that cannot be activated synaptically may be afferent neurons (P neurons in Figure 6.12). These cells have intrinsic activity, which might be responsible for ongoing synaptic activity recorded in other neurons within the ganglia. A third group of smaller neurons receives local synaptic inputs and may also terminate on cardiac muscle (S neurons in Figure 6.12). Both the location of the endings of the putative afferent neurons and the adequate stimuli that excite them remain unclear. However, these peripheral connections are obviously a feature of parasympathetic cardiac ganglia.

Pelvic ganglia contain postganglionic neurons supplying pelvic viscera and are often considered to be parasympathetic. However, this is misleading. Many postganglionic neurons in pelvic ganglia are noradrenergic. Most cholinergic neurons receive synaptic input from preganglionic neurons in the sacral spinal cord (and in rats the sixth lumbar spinal segment) projecting through the pelvic nerve. Most noradrenergic neurons receive synaptic input from preganglionic neurons in the upper lumbar spinal cord projecting through the hypogastric nerves (see note 3). Very few postganglionic neurons in the pelvic ganglia receive convergent synaptic input from both the sacral spinal cord and the upper lumbar spinal cord. Interneurons are absent or rare in the pelvic ganglia. Most cholinergic vasodilator neurons in pelvic ganglia of guinea pigs are activated either by sacral or upper lumbar preganglionic neurons, very few by both (Jobling et al. 2003). Pelvic neurons supplying distal colon and rectum (and innervating myenteric plexus or circular musculature) may receive additional synaptic input from intestinofugal neurons located in the myenteric plexus or the serosal ganglia. Overall there are

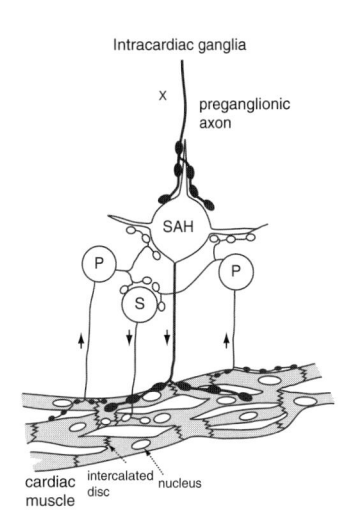

Figure 6.12 Integration in parasympathetic cardiac ganglia. Diagram of the component neurons of intracardiac ganglia of guinea pig. Only one neuron type (SAH, synaptic with after-hyperpolarization) receives preganglionic synaptic inputs via the vagus nerve (X). The SAH cells and another neuron type (S, synaptic) receive synaptic input arising from the third neuron type, which is spontaneously active (P, pacemaker neurons). The P cells may be sensory neurons with afferent terminals within the heart. The physiological stimuli to activate them are unknown. Modified from Jänig and McLachlan (2002).

considerable species differences in organization of the pelvic ganglia and the postganglionic neurons of these ganglia have been less well studied than those of other ganglia (Jänig and McLachlan 1987; Keast 1995a, 1999).

6.6 | Non-nicotinic transmission and potentiation resulting from preganglionic stimulation in sympathetic ganglia

Studies of sympathetic ganglia of amphibians and some mammals have shown that electrical stimulation of sympathetic preganglionic axons elicits not only fast nicotinic EPSPs, but also slow muscarinic and slow non-cholinergic EPSPs (see Elfvin [1983]; Karczmar et al. [1986]; North [1986]). This has led to the speculation that fast nicotinic synaptic transmission in autonomic ganglia is modulated by long-lasting (muscarinic and non-cholinergic) conductance changes and that peptides may be acting as transmitters (Akasu and Koketsu 1986; Katayama and Nishi 1986). In fact the only neuropeptide in autonomic ganglia that has been thoroughly studied so far and shown to have postsynaptic effects as a putative transmitter is a luteinizing hormone-releasing hormone (LHRH)-like peptide in the bullfrog. This peptide coexists with acetylcholine in sympathetic preganglionic neurons with unmyelinated axons that project to vasoconstrictor neurons in paravertebral ganglia (C neurons) regulating blood vessels. It is absent in sympathetic preganglionic neurons that synapse with postganglionic secretomotor neurons (B neurons) in paravertebral ganglia that regulate cutaneous glands. The peptide is released when the preganglionic neurons innervating the C neurons are excited repetitively and leads to long-lasting postsynaptic potentials in the B neurons (after diffusing to them). Furthermore, acetylcholine, which is released after repetitive stimulation of the preganglionic axons, generates slow EPSPs by muscarinic action. Both conductance changes are generated by closing of potassium channels of the M-type (see Table 6.1) (Jan et al. 1979, 1980a, b; Jan and Jan 1982). It has not been shown up to now, even in this well-established example, whether non-nicotinic transmission is relevant in vivo during neural regulation of skin glands and blood vessels in the bullfrog.

There are some reports in the literature indicating that postganglionic sympathetic neurons projecting to the heart (cardiomotor neurons) and postganglionic muscle vasoconstrictor neurons can be activated in vivo from the central nervous system or through a central reflex pathway via non-nicotinic synaptic mechanisms, i.e. when the nicotinic transmission is blocked (Brown 1967, 1969; Henderson and Ungar 1978). Furthermore, blood pressure responses in humans to asphyxia (Freyburger et al. 1950a, b) and to tilting (Fielden et al. 1980) are not completely abolished by blockade of nicotinic transmission in autonomic ganglia.

The important questions to be addressed are:

1. Can postganglionic neurons supplying skeletal muscle and skin be activated via non-nicotinic mechanisms in paravertebral ganglia after electrical stimulation of the preganglionic axons?
2. Can the postganglionic neurons be activated or their physiological discharges be modulated by these non-nicotinic mechanisms?

To answer these questions, activity was recorded from postganglionic axons isolated from skin and muscle nerves of the hindlimb and tail of the anesthetized cat. The preganglionic neurons innervating the postganglionic neurons were stimulated electrically or physiologically (by a reflex) (Figure 6.13a). The postganglionic neurons were vasoconstrictor (skin, skeletal muscle), sudomotor or pilomotor in function.

6.6.1 Responses to electrical stimulation of preganglionic axons

Decentralized preparations
In decentralized preparations (preganglionic axons cut, see Figure 6.14a), repetitive supramaximal electrical stimulation of preganglionic axons in the lumbar sympathetic trunk with short trains of impulses (50 stimuli at 25 Hz) elicits (early) high-frequency responses during the train in all postganglionic neurons projecting to the hindlimb and (late) low-frequency responses after the train in many postganglionic neurons. The early high-frequency response is generated by nicotinic action of acetylcholine (i.e., it is blocked by an antagonist of nicotinic receptors like hexamethonium). The late response is resistant to blockade of nicotinic transmission and is therefore non-nicotinic. These non-nicotinic responses have the following characteristics (Hoffmeister et al. 1978; Jänig et al. 1982, 1983, 1984; Figure 6.13a):

1. They start at a latency of 2 to 10 seconds following the burst of fast (nicotinic) responses and last up to 2 minutes.
2. Sometimes the early component of the slow response is blocked by atropine, but the late part is not, even at high concentrations of atropine (which blocks muscarinic cholinergic receptors).
3. The responses in the postganglionic neurons are elicited with trains of 50 stimuli applied to the preganglionic axons at at least 3 to 4 Hz or by 3 to 10 stimuli at 25 Hz.
4. The responses can *only* be elicited when slowly conducting, mostly unmyelinated, preganglionic axons are stimulated repetitively: i.e. repetitive electrical stimulation of myelinated preganglionic axons at 25 to 50 Hz for 2 to 5 seconds does not elicit the late responses. Nicotinic responses in the same postganglionic neurons are also elicited when low-threshold faster conducting preganglionic axons are stimulated. Preganglionic neurons with unmyelinated axons that have functional properties of muscle and cutaneous vasoconstrictor neurons have been identified in

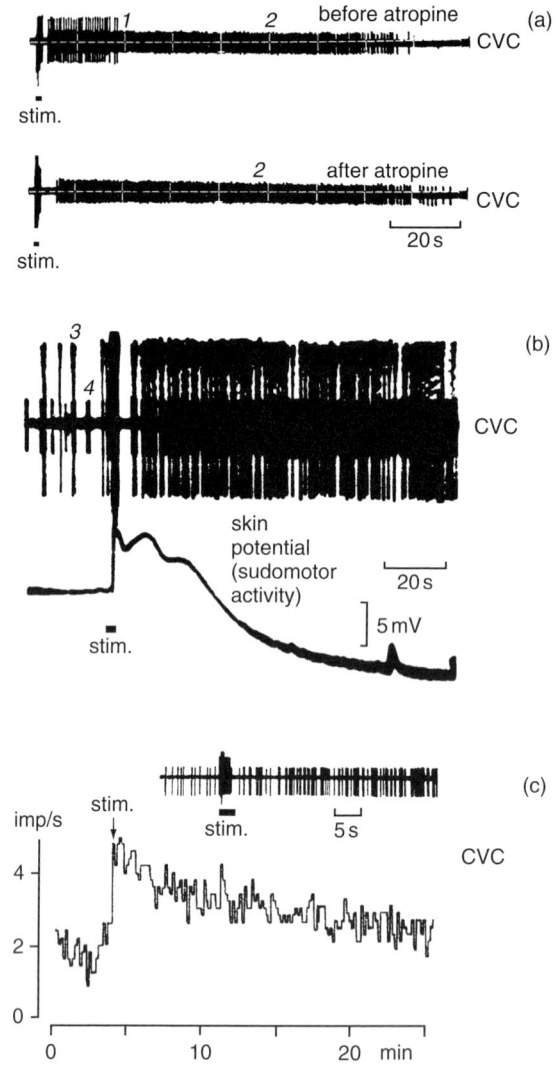

Figure 6.13 Long-lasting responses in postganglionic neurons induced by repetitive electrical stimulation of preganglionic axons. Recording from postganglionic neurons projecting to the cat hindlimb in response to electrical stimulation of the lumbar sympathetic trunk (stim.). Fifty stimuli at 25 Hz, stimulus strength supramaximal for all preganglionic axons (10 V in [a], 8 V in [b] and [c]), pulse duration 0.2 ms (experimental set-up see right part of Figure 6.14a). The first high frequency response is mediated by nicotinic transmission and can be blocked by hexamethonium. (a) Decentralized preparation (preganglionic axons cut central to the stimulation electrode). Response of two postganglionic neurons supplying hairy skin (neuron 1 and 2 identified by amplitude, both probably being vasoconstrictor neurons). The short-lasting high-frequency response was masked by the stimulus artifact (bar). The long-lasting response of neuron 1 and the first few seconds of the long-lasting response in neuron 2 were blocked by atropine (0.6 mg/kg i.v.). (b) Preganglionic axons intact. Responses of two cutaneous vasoconstrictor neurons (neurons 3 and 4) supplying hairless skin of the cat hindpaw and of the skin potential recorded from the central pad of the hindpaw (indicating activation of sudomotor neurons; negativity up [see Subchapter 4.2 and Figures 4.10 and 4.11]). Note the long-lasting activation of the vasoconstrictor neurons and the absence of fast negative deflections in the skin potential record indicating that sudomotor neurons did not exhibit long-lasting responses to repetitive preganglionic stimulation. For the long-lasting positive change of skin potential see note 4 in Chapter 4. (c) Preganglionic axons intact. Long-lasting response of a cutaneous vasoconstrictor neuron supplying hairless skin of the cat hindpaw. Note the duration of the enhanced activity. Upper trace: activity before, during and after stimulation. CVC, cutaneous vasoconstrictor; (a) from Hoffmeister et al. (1978); (b) and (c) from Blumberg and Jänig (1983a) with permission.

the lumbar sympathetic outflow and in the sympathetic outflow to the head and neck (Jänig and Szulczyk 1980, 1981; Boczek-Funcke *et al.* 1993). The physiological firing frequencies of these preganglionic neurons may reach 10 Hz, e.g. by chemoreceptor stimulation during hypoxia (see Table 6.2).

5. The responses occur in most postganglionic neurons supplying skeletal muscle and in about 30% of the postganglionic neurons supplying hairy skin of the cat hindlimb, both being most likely vasoconstrictor in function.
6. Postganglionic pilomotor neurons supplying the tail skin and sudomotor neurons supplying sweat glands in the paw pads do not exhibit late, long-lasting discharges following repetitive electrical stimulation of the preganglionic axons (Fig. 6.13b).

These results clearly demonstrate that impulses can be elicited in sympathetic postganglionic neurons in vivo via non-nicotinic transmission from the preganglionic site in the same way as in vitro; that this synaptic transmission appears to be specific for vasoconstrictor neurons; and that preganglionic neurons with slowly conducting, unmyelinated axons have to be recruited to activate the postganglionic neurons via the non-cholinergic synaptic mechanisms. These responses have been demonstrated to exist in cats but not in rodents and rabbits. This effect probably involves a maintained depolarization (with decreased conductance) of the postganglionic neuron.

Intact preparations
Can the rate of ongoing discharge present in postganglionic vasoconstrictor neurons in vivo (when the preganglionic axons are not interrupted) be enhanced by non-nicotinic mechanisms following repetitive electrical stimulation of the preganglionic axons? To answer this question, the same experimental design was used as in the experiments on the decentralized preparation but the preganglionic axons were left intact and spontaneous activity (which is fully dependent on the synaptic input from preganglionic axons) was recorded from the postganglionic neurons (Blumberg and Jänig 1983a; Figure 6.13b, 6.13c):

1. Ongoing discharge in most preparations with cutaneous and muscle vasoconstrictor neurons was enhanced for 4 to 40 minutes or longer following repetitive electrical stimulation of preganglionic axons for 2 seconds at 25 Hz. This was not the case for sudomotor neurons.
2. The enhancement reached peak values of 120% to 600% of the control rate of ongoing discharge 1 to 2 minutes after the train.
3. The enhancement could only be elicited by stimulus strengths at which the nicotinic response was maximal, i.e. when slowly conducting preganglionic axons in the sympathetic trunk were recruited.

4. The enhancement could be elicited heterosynaptically by stimulation of a cut white ramus.
5. The enhancement was abolished or partially abolished by blockade of muscarinic transmission (by atropine) in some preparations but not in others.
6. The enhancement was associated with a long-lasting reduction of blood flow through the skin (Jänig and Koltzenburg 1991b).

These results warrant speculation that the enhancement of ongoing discharge occurs in postganglionic neurons that receive many weak convergent preganglionic axons, which generate a high rate of subthreshold ongoing synaptic activity in vasoconstrictor neurons. After repetitive activation of the small-diameter preganglionic axons, some of these small ineffective synaptic events could be enhanced by an increase in membrane resistance generated by closure of potassium channels (e.g., of the M-type). Alternatively, weak preganglionic terminals could exhibit post-tetanic facilitation. Evidence against this possibility is that the enhancement of ongoing activity can also be generated heterosynaptically. Finally, it is possible that there exists a special class of preganglionic neurons with unmyelinated fibers that release peptide selectively.

6.6.2 Reflexes in vasoconstrictor neurons mediated by non-nicotinic mechanisms

Stimulation of arterial chemoreceptors (e.g., by systemic hypoxia: ventilation of the animal with a gas mixture of 8% O_2 in N_2) elicits reflex activation of muscle vasoconstrictor neurons and reflex inhibition of cutaneous vasoconstrictor neurons that are among the strongest reflex responses known in these neurons. This reflex is mediated by neuronal circuits through the medulla oblongata for the muscle vasoconstrictor neurons and probably also through the hypothalamus for the cutaneous vasoconstrictor neurons (Jänig 1975; Gregor and Jänig, 1977; see Subchapter 10.4) (Figure 6.14b, left). After blockade of nicotinic transmission (with hexamethonium) in sympathetic ganglia, which leads to disappearance of the centrally generated spontaneous activity in the vasoconstrictor neurons and to disappearance of fast responses generated by electrical single pulse stimulation of preganglionic axons, muscle vasoconstrictor neurons could still be activated reflexly, as in the control (before blockade of nicotinic transmission). This response was accompanied by an increased blood pressure and was either partially or completely blocked by atropine or was not affected (Figure 6.14b, 6.14c, left side). Corresponding to the reflex activation of the muscle vasoconstrictor neurons via a non-nicotinic mechanism, the postganglionic neurons exhibited afterdischarges to repetitive electrical stimulation of the preganglionic axons (Figure 6.14b, 6.14c right side). Cutaneous vasoconstrictor neurons that were normally reflexly inhibited during arterial chemoreceptor stimulation were not activated after blockade of nicotinic transmission and after additional blockade of muscarinic

6.6 NON-NICOTINIC TRANSMISSION AND POTENTIATION | 245

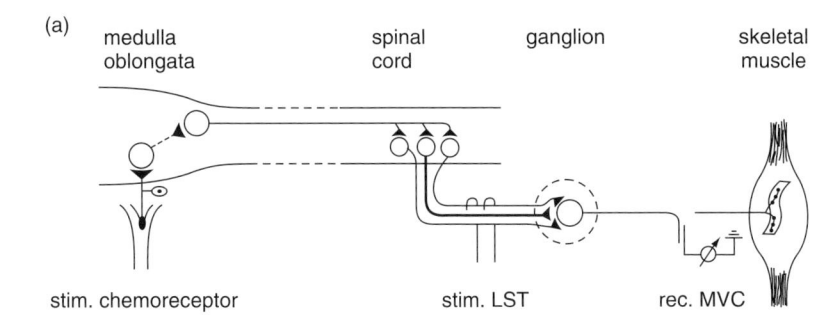

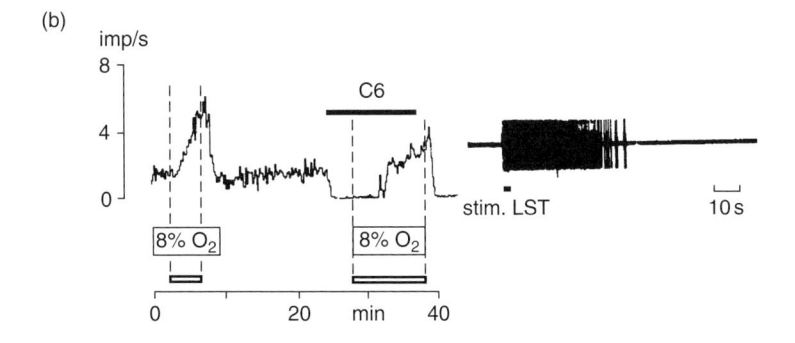

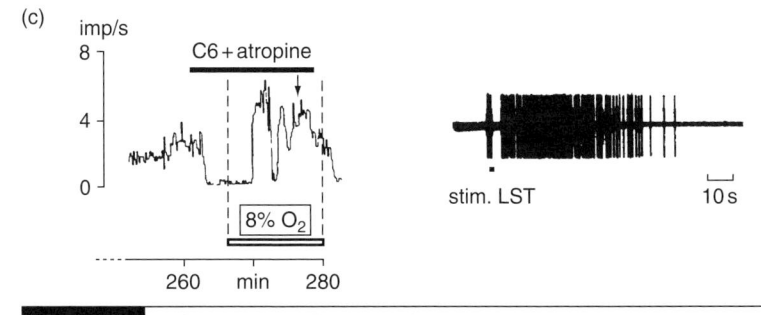

Figure 6.14 Reflex activation of a postganglionic muscle vasoconstrictor neuron projecting to the cat hindlimb via non-nicotinic transmission in a paravertebral ganglion. Activity was recorded in a single muscle vasoconstrictor neuron (MVC) (a). Stimulation of arterial chemoreceptors (ventilation of the animal with a gas mixture of 8% O_2 in N_2 [lower open bars]) activated the MVC neuron (b) before and after i.a. injection of hexamethonium (C6, 12.1 mg/kg infused over 13 min) and (c) after intra-arterial (i.a.) injection of C6 (26 mg/kg infused over 19 min) in the presence of atropine (1 mg/kg i.a.). Drugs were injected retrogradely into the internal iliac artery close to the caudal lumbar paravertebral ganglia. At 276 min in (b) an additional dose of 0.4 mg atropine was injected i.a. (arrow). (Activity in cutaneous vasoconstrictor neurons (CVCs) recorded simultaneously was inhibited during stimulation of the arterial chemoreceptors [not shown; see Subchapter 4.1]. The CVCs were not activated after i.a. injection of hexamethonium and atropine during stimulation of arterial chemoreceptors.) The records on the right show the responses of the MVC neuron to repetitive electrical stimulation of the preganglionic axons in the lumbar sympathetic trunk (stim. LST; 50 stimuli at 25 Hz, 8 V/0.2 ms) during block of nicotinic transmission by hexamethonium before (b) and after (c) atropine. Note that the early part of the long-lasting response in the MVC neuron was blocked by atropine. Nicotinic blockade generated by hexamethonium was regularly confirmed by the absence of responses of the postganglionic neurons to single suprathreshold preganglionic stimuli presented at 1 per 5 s. Time scale in (b) and (c) (left) indicates time after start of the experiment. Modified from Jänig et al. (1983).

Figure 6.15 Transmission of impulses from pre- to postganglionic neurons in paravertebral ganglia in the cat. Stimulation of myelinated preganglionic fibers (conduction velocity >3 m/s) with single pulses or repetitively (at 25 to 50 Hz for 2 to 5 s) excites the postganglionic neurons only via nicotinic synaptic transmission. Stimulation of unmyelinated and some small-diameter myelinated preganglionic axons (conduction velocity <3 m/s) may excite the postganglionic vasoconstrictor neurons via nicotinic and muscarinic mechanisms and via a non-cholinergic (possibly peptidergic) mechanism. The receptors for the muscarinic and non-cholinergic mechanisms are probably (or in particular) also located extrasynaptically. Thus, acetylcholine and the putative peptide reach these receptors by volume conduction. Postganglionic sudomotor and pilomotor neurons cannot be excited via these non-nicotinic mechanisms.

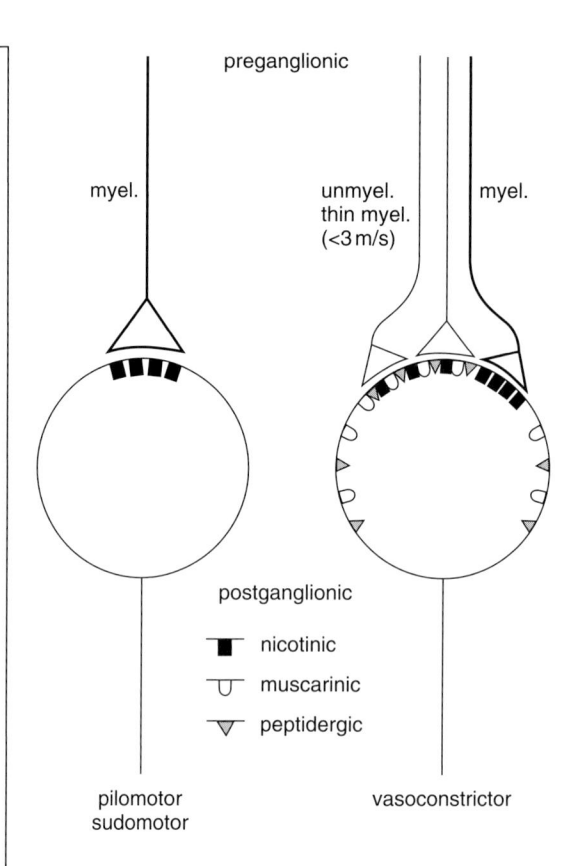

transmission. However, they showed late discharges in response to repetitive electrical stimulation of the preganglionic axons (not shown in Figure 6.14).

These results clearly show that

1. postganglionic vasoconstrictor neurons can be physiologically activated after complete blockade of nicotinic transmission or, in some cases, blockade of both nicotinic and muscarinic transmission in the ganglia, and
2. the artificially elicited synchronous discharges in the preganglionic neurons, as they occur during electrical stimulation, are not necessary for driving the postganglionic neurons via non-nicotinic mechanisms in the paravertebral ganglia.

In *summary* (Figure 6.15), nicotinic transmission in postganglionic vasoconstrictor neurons can undergo a long-lasting enhancement when preganglionic neurons with slowly conducting axons are activated. This enhancement is probably heterosynaptic and involves either muscarinic and/or non-cholinergic mechanisms in different neurons. It is specific for vasoconstrictor pathways. The membrane events that mediate long-term synaptic processes and the transmitter(s) (peptides?) that mediates the noncholinergic discharges in the vasoconstrictor neurons are unknown. The long-lasting enhancement of ongoing discharge elicited in the vasoconstrictor neurons in vivo may be related to the long-term potentiation of weak synaptic events observed in sympathetic ganglia in vitro after repetitive preganglionic and pharmacological stimuli. The cholinergic muscarinic and putative peptidergic receptors in the postganglionic cell membrane are probably mainly localized extrasynaptically. Thus, acetylcholine and peptide are reaching these receptors by volume conduction. However, the finding that high-frequency stimulation of myelinated preganglionic axons (conduction velocity >3 m/s) does not generate long-lasting discharges in the vasoconstrictor neurons whereas stimulation of unmyelinated preganglionic axons does strongly argues that the preganglionic neurons eliciting the long-lasting discharge are distinct. Thus, weak synapses formed by myelinated axons can theoretically be enhanced but this enhancement must occur heterosynaptically.

Conceivably, the central nervous system could use this mechanism of long-lasting enhancement of synaptic transmission in sympathetic ganglia to modulate the level of ongoing firing of postganglionic vasoconstrictor neurons by means of short, high-frequency bursts of preganglionic discharge at long intervals.

6.6.3 Non-cholinergic ganglionic transmission to postganglionic vasodilator neurons supplying uterine artery

Arteries of the female internal sexual organs of guinea pigs are innervated by vasodilator neurons located in the paracervical ganglia that receive synaptic input from sympathetic preganglionic neurons projecting through the lumbar splanchnic and hypogastric nerves. Each postganglionic vasodilator neuron receives synaptic input from about two preganglionic axons, one being strong (Jobling *et al.* 2003). Activation of the preganglionic neurons generates vasodilation by release of nitric oxide (NO) and VIP from the postganglionic vasodilator fibers. Repetitive electrical stimulation of the sympathetic preganglionic axons at 10 Hz generates continuous discharge of the postganglionic vasodilator neurons, produced by a slow excitatory postsynaptic potential, and a vasodilation, both outlasting the train of stimulation considerably. Continuous activation of postganglionic neurons and vasodilation are only slightly reduced by blockade of nicotinic transmission in the paracervical ganglia. The ganglionic transmitter involved is unknown, but unlikely to be substance P, adenosine triphosphate, 5-hydroxytryptamine, glutamate or acetylcholine (muscarinic action) (Morris *et al.* 2005). The continuous discharge of the postganglionic neurons cannot be a long-term potentiation of cholinergic transmission since it also occurs after complete block of cholinergic transmission and since synaptic transmission normally occurs via a strong cholinergic synapse.

This mechanism to generate and maintain activity in pelvic vasodilator neurons (and possibly secretomotor neurons) ties up with observations, made in the cat, that preganglionic non-vasoconstrictor neurons (here collectively called motility-regulating neurons) projecting in the lumbar splanchnic nerves can be activated reflexly for up to 12 minutes by short-lasting mechanical stimulation of sacral afferents (e.g., from the anal canal) as described in Subchapter 4.3 (see Figure 4.15). The mechanism underlying this long-lasting reflex activation may lie in the spinal pathways. Thus, a centrally potentiated reflex in preganglionic neurons would be further amplified by non-cholinergic synaptic transmission in paracervical ganglia. This sequential cascade of amplification of a central signal could be important in the neural regulation of reproductive organs.

Conclusions

1. Most autonomic ganglia transmit the central message with high fidelity to the postganglionic neurons. This concurs with the idea that the target organs innervated by the neurons in these ganglia are predominantly under central control.
2. A relatively small number of preganglionic neurons connect with a large number of postganglionic neurons by divergence of the preganglionic axons. This divergence has primarily a distribution function. In addition, postganglionic neurons receive convergent synaptic input from several preganglionic neurons. The degree of divergence and convergence varies between different final autonomic pathways and different species.
3. Almost all neurons in sympathetic paravertebral and parasympathetic ganglia and some neurons in sympathetic prevertebral ganglia with convergent synaptic input probably receive one (sometimes two and rarely three) strong (suprathreshold) synaptic inputs, the other synaptic inputs being weak (subthreshold). Discharges in postganglionic neurons are generated by these strong synaptic inputs but usually not by summation of several weak synaptic inputs.
4. Based on neurophysiological properties and mainly the presence of voltage- and calcium-dependent potassium-channels, sympathetic postganglionic neurons consist electrophysiologically of three broad classes:
   a. Phasic neurons respond with a burst of action potentials at the beginning of suprathreshold current steps. Almost all paravertebral and 15% to 25% of the prevertebral neurons are phasic. These neurons receive one or two strong synaptic inputs from preganglionic axons and are fully under the control of the central nervous system. They do not receive peripheral synaptic input.
   b. Tonic neurons discharge rhythmically throughout a current step. These neurons are numerous in prevertebral ganglia and practically absent in paravertebral ganglia. They receive weak preganglionic synaptic inputs and weak synaptic inputs from the periphery (mainly from the gastrointestinal tract) and have integrative function. They are non-vasoconstrictor in

function and mainly involved in regulation of motility of the gastrointestinal tract.
   c. Neurons with long afterhyperpolarization following an action potential (LAH neurons) discharge one action potential at the beginning of a current step. Most LAH neurons are located in the celiac ganglion and a few in paravertebral ganglia. They receive one strong synaptic input and rarely peripheral synaptic inputs. They are fully under the control of the central nervous system and have relay function. They are non-vasoconstrictor in function and mainly involved in secretion and absorption.
5. The number of postganglionic neurons innervated by one preganglionic neuron is called autonomic neural unit (ANU). Functionally defined ANUs are smaller than anatomically defined ANUs since only postganglionic neurons that receive a strong synaptic input from a preganglionic neuron are activated. The size of ANUs varies according to function; e.g., ANUs in a vasoconstrictor pathway are larger than ANUs of the pupillomotor pathway.
6. Many neurons in *sympathetic prevertebral ganglia* and in some parasympathetic ganglia receive, in addition to several preganglionic inputs, cholinergic synaptic inputs from peripheral intestinofugal neurons and/or peptidergic inputs from collaterals of spinal visceral afferents. These postganglionic neurons are probably involved in the regulation of non-vascular functions.
7. In many of the sympathetic neurons in prevertebral ganglia, the central synaptic input is weak or even may play a subordinate role. The way in which central synaptic input and peripheral synaptic inputs are integrated in vivo by these prevertebral postganglionic neurons is unclear. In cardiac parasympathetic ganglia, postganglionic neurons receive, in addition to strong preganglionic inputs, peripheral afferent synaptic inputs. Their function is unknown.
8. Ongoing and reflex activity are transmitted through sympathetic ganglia by nicotinic synaptic transmission. This activity can be enhanced by muscarinic and non-cholinergic (peptidergic) mechanisms. For vasoconstrictor systems, it has been shown in vivo that ongoing activity can be enhanced for up to tens of minutes by these non-nicotinic mechanisms following short bursts of synaptic input from preganglionic neurons.
9. Reflex activity can be transmitted to postganglionic vasoconstrictor neurons by non-nicotinic mechanisms. This leads to amplification of the efferent signal in the time domain. Few functional contexts in which these events may occur in vivo are known.

Notes
1. Release of acetylcholine requires the presence of SNARE (**s**oluble **N**SF [N-ethylmaleimide-sensitive factor] **a**ttachment protein **re**ceptors) proteins in the presynaptic boutons, which are necessary for the fusion of

the synaptic vesicle membrane with the presynaptic plasma membrane. Release of peptides does not seem to require SNARE proteins and is therefore insensitive to botulinum toxin. The SNARE proteins are present in presynaptic specializations, but absent or reduced in boutons not forming synapses (Gibbins *et al.* 2003b).

2. The ongoing discharge in sympathetic postganglionic neurons is physiologically entirely of central origin. Decentralization of paravertebral ganglia (transection of the preganglionic axons innervating the postganglionic neurons) in cat or rat always abolishes the ongoing activity in the postganglionic neurons supplying skin or skeletal muscle (Häbler and Jänig, unpublished observation). Furthermore, less than 5% of acutely denervated postganglionic neurons of the inferior mesenteric ganglion that project in the hypogastric nerve exhibit resting discharge at a rate of about 0.1 imp/s in vivo in the cat. Taking the rate of ongoing activity and the fraction of postganglionic neurons with ongoing activity in the inferior mesenteric ganglion before denervation into account, the total ongoing impulse traffic from the inferior mesenteric ganglion to the pelvic organs decreases to about $\leq 1\%$ after denervation (Jänig *et al.* 1991). Also, this experiment shows that the ongoing activity of the postganglionic neurons is of central origin. These data are seemingly at variance with data obtained on isolated ganglia in vitro showing that postganglionic neurons may generate spontaneous activity without synaptic input from preganglionic neurons and from peripheral neurons (celiac ganglion [Gola and Niel 1993]; colonic ganglia [Krier and Hartman 1984]; intracardiac ganglia [Xi *et al.* 1991; Selyanko 1992]; inferior mesenteric ganglion [Julé and Szurszewski 1983]; vesical pelvic ganglia [Griffith *et al.* 1980]). It was postulated that the ganglia contain pacemaker cells, which may synaptically transmit their activity to other ganglion cells (Julé and Szurszewski 1983; Gola and Niel 1993). However, other investigations performed in vitro did not find spontaneous activity and pacemaker activity in postganglionic sympathetic neurons (e.g., in the celiac ganglion; McLachlan and Meckler 1989).

3. In the cat about 90% of the efferent axons in the hypogastric nerve are postganglionic and 10% preganglionic (Baron *et al.* 1985b). In the male rat less than 30% of the efferent axons in the hypogastric nerve are postganglionic and 70% preganglionic (Baron and Jänig 1991). The same applies to the hypogastric nerve of the male guinea pig; in the female guinea pig about 50% of the efferent axons are preganglionic (McLachlan 1985). These histological data imply that many sympathetic postganglionic neurons innervating pelvic organs are located in the caudal part of the inferior mesenteric ganglion in cat and in the pelvic ganglia of rat and guinea pig.

Chapter 7

Mechanisms of neuroeffector transmission

7.1 Transmitter substances in postganglionic neurons   *page* 252
7.2 Principles of neuroeffector transmission in the
       autonomic nervous system   255
7.3 Specific neuroeffector transmissions   264
7.4 Integration of neural and non-neural signals
       influencing blood vessels   273
7.5 Unconventional functions of sympathetic
       noradrenergic neurons   276

Impulse activity in the preganglionic neurons is the result of integrative processes in the spinal cord and brain stem (see Chapters 9 and 10). Such top down signaling (i.e. central message) is not usually distorted in the ganglia of the final autonomic pathways. It is modified by synaptic input from peripheral neurons in a few autonomic pathways to the viscera (see Subchapters 5.7 and 6.5). The central message is distributed to a large population of postganglionic neurons but confined to the respective autonomic pathways. Thus, transmission of impulses from preganglionic neurons to postganglionic neurons occurs within the same final autonomic pathway, but not between autonomic pathways. In this chapter I will discuss some mechanisms by which centrally generated messages are transmitted from postganglionic neurons to the target cells (effectors).

The mechanism of neuroeffector transmission varies considerably between target tissues. This is a consequence of there being many different types of cells innervated by postganglionic neurons (see Table 1.2). However, neuroeffector transmission has been studied in only a few cases. Such studies clearly demonstrate that the neural impulse activity is transmitted to the effector cells in a rather specific way and this may apply to *all* autonomically innervated effector cells. This view does not preclude the idea that the activity in some autonomic effector cells is also modified by multiple non-neural influences and that these non-neural components may operate during ongoing regulation of the target tissues.

However, in the context of this book, the important point in the neural regulation of autonomic effector cells is that (1) the central message reaches the target tissue via differentiated structures and (2) these specific peripheral mechanisms are an important basis for the precise regulation of autonomic target organs by the brain.

In the last part of this chapter I will (1) summarize the interaction of the centrally generated neural signal with other neural and non-neural signals in the regulation of target cells and (2) describe some unconventional functions of sympathetic postganglionic fibers that are not dependent on generation of action potentials or on release of noradrenaline.

7.1 | Transmitter substances in postganglionic neurons

The principles of chemical transmission were originally defined in the peripheral autonomic nervous system based on the release of the "conventional" neurotransmitters, acetylcholine and noradrenaline. However, it is now clear that, in addition to acetylcholine and noradrenaline, other putative neurotransmitters may be contained within individual autonomic neurons. These have multiple actions on effector tissues (Furness *et al.* 1989; Morris and Gibbins 1992; Hoyle *et al.* 2002):

- Most sympathetic postganglionic axons release noradrenaline, but sympathetic sudomotor and, in some species, muscle vasodilator axons are generally cholinergic. Whether these cholinergic vasodilator axons exist in human beings remains controversial (see Subchapter 4.2). Significantly, not all effects generated by stimulation of postganglionic sympathetic nerve fibers can be abolished by antagonists of either adrenoceptors or muscarinic receptors.
- All parasympathetic postganglionic neurons are cholinergic, defined by the presence of choline acetyltransferase (ChAT) in these neurons (Keast 1995b; Keast *et al.* 1995; McLachlan 1995). However, not all effects of stimulating parasympathetic nerve fibers are blocked by muscarinic antagonists; these must therefore be mediated by another transmitter or other transmitters. Nicotinic cholinergic receptors are absent in effector cells.
- Postganglionic autonomic neurons whose effects of excitation are not in any way modified by antagonists to the conventional transmitters acetylcholine or noradrenaline have been called non-adrenergic non-cholinergic (NANC) neurons and by analogy the transmission is referred to as NANC transmission.
- Non-adrenergic non-cholinergic transmission has been studied extensively in the enteric nervous system, where this transmission is thought to mediate inhibition of gastrointestinal smooth muscle

by enteric neurons. The compounds that may be involved in NANC transmission are nitric oxide (NO), adenosine-triphosphate (ATP) or a neuropeptide (e.g., vasoactive intestinal peptide [VIP], neuropeptide Y [NPY], galanin [GAL], etc.). "Nitrergic" nerves (containing the enzyme neuronal nitric oxide synthase [NOS]) are vasodilator or relaxant in several tissues. Neuropeptides (tachykinins, calcitonin gene-related peptide [CGRP], etc.) released upon activation from sensory endings also take part in NANC transmission (see Subchapter 6.5).

Responses of tissues (contraction, secretion, etc.) to nerve-released noradrenaline and acetylcholine usually only occur following repetitive activation of many axons. In contrast, high-frequency stimuli, particularly in bursts, may produce effector responses due to the concomitant release of a neuropeptide. Immunohistochemistry has revealed the presence of many peptides in the postganglionic neurons (cell bodies and nerve terminals). This led to the interesting concept of "neurochemical coding" of autonomic neurons (Subchapter 1.4). However, only a few of these peptides have been demonstrated to modify effector function after release from nerve terminals in vivo and the functional role of these peptides during ongoing regulation of the target cells still awaits clarification. Some examples follow.

- Neuropeptide Y in sympathetic vasoconstrictor axons evokes vasoconstriction in some arteries, whereas in others it potentiates the effects of noradrenaline on the contractile apparatus (Lundberg 1996). In the heart, NPY released by sympathetic cardiomotor neurons attenuates decrease of heart frequency produced by activation of parasympathetic cardiomotor neurons (Potter 1987).
- Vasoactive intestinal peptide is thought to be the primary *vasodilator* transmitter released from cholinergic vasodilator, sudomotor and secretomotor axons (Gibbins 1994; Lundberg 1996).

Vasoconstriction in some vascular beds, such as those in skin and mesentery, may be in part regulated by interactions between sympathetic axons and the terminals of primary afferent nociceptive neurons that express substance P and CGRP. However, it is a matter of debate whether this interaction only occurs at the level of the vascular smooth muscle or also between both types of neuronal terminals (Häbler et al. 1997c). Calcitonin gene-related peptide released by axon reflex activation[1] of afferent terminals during inflammation is a potent vasodilator (Holzer 1992, 2002a, b). Substance P released by axon reflex activation also contributes to precapillary vasodilation, but mainly to venular plasma extravasation (Häbler et al. 1997c, 1999).

Table 7.1 lists transmitter substances (primary transmitters) or putative transmitter substances in autonomic postganglionic neurons. In most neurons the primary transmitter is either

Table 7.1 | *Transmitter substances in autonomic neurons*

System	Neuron	Transmitter	Cotransmitters
Parasympathetic	Preganglionic	ACh	
	Postganglionic	ACh	VIP and/or NO
Sympathetic	Preganglionic	ACh	
	Postganglionic	NAd	ATP and/or NPY
		some ACh	VIP and/or NO
Enteric	[Vagal & pelvic inputs ACh]		
	[Sympathetic inputs NAd]		
	Intrinsic afferent neurons	Substance P (tachykinins)	
	Interneurons	ACh	some ATP
	Motor Excitatory	ACh	Substance P (tachykinins)
	Inhibitory	NO/VIP/PACAP/?ATP	
	Secretomotor	ACh	VIP

Notes:
ACh   acetylcholine
NAd   noradrenaline
ATP   adenosine triphosphate
NO   nitric oxide
NPY   neuropeptide Y
VIP   vasoactive intestinal polypeptide
PACAP   pituitary adenylate cyclase-activating peptide
For further data about cotransmitters and neuropeptides in enteric neurons see Table 5.1.
Modified from Jänig and McLachlan (2002).

acetylcholine or noradrenaline; in a few neurons another compound has been established as a transmitter. However, in most neurons with colocalized neuropeptides the functions of the peptides as transmitters are entirely unclear. The presence of a neuropeptide in an autonomic neuron, the release of that neuropeptide during nerve stimulation and the presence of the receptors for the same neuropeptide on the target cells (neurons or effector cells) do not reveal whether the neuropeptide is normally, or even under pathophysiological conditions, used during neural regulation of that target tissue. For example, no entirely convincing experiment has been performed that illustrates a functional context for neurally released NPY on autonomically innervated target tissues, although it has been shown that this peptide is released into the blood during activity of muscle vasoconstrictor neurons and that its concentration increases during exercise (Morris *et al.* 1986; Pernow 1988). The same is true for VIP, which is colocalized with acetylcholine in sympathetic postganglionic neurons innervating sweat glands (Lundberg 1981) and in parasympathetic postganglionic neurons innervating the erectile tissue of sexual organs (de Groat and Booth 1993; de Groat 2002).

7.2 | Principles of neuroeffector transmission in the autonomic nervous system

The membranes of the autonomic effector cells contain a large number of membrane receptors that mediate the effect of signaling molecules, to which the cells are potentially exposed, leading to responses of the cellular effectors (ionic channels, contractile proteins, transcription, metabolic processes, secretion). The membrane receptors are connected to the cellular effectors in three ways: ligand-coupled receptors, via G-protein-coupled receptors and catalytic receptors. The high number of membrane receptors, the limited number of intracellular pathways to the cellular effectors, the interaction of these intracellular pathways and the quantitative changes of receptors and intracellular pathways leads to an enormous variety of responses of autonomic effector organs. However, postganglionic autonomic neurons use under physiological conditions only a limited number of the membrane receptors and cellular pathways connected to these channels in order to generate the cellular effector responses. This is the basis for the diversity of the autonomic effector responses generated by neural signals (see Table 1.2).

In peripheral tissues, the effects of activity in autonomic nerve terminals can be due to the release of several different compounds. However, failure to block effector responses by either adrenoceptor or muscarinic receptor antagonists at concentrations that entirely abolish the response to exogenous transmitter is not proof for the existence of NANC transmitters. This is so because exogenous transmitter acts dominantly on extrajunctional receptors whereas neuronally released transmitter acts dominantly on postjunctional receptors and both types of receptors may be different. The effects of exogenous transmitter substances on cellular functions are known for many tissues but the mechanisms of activation of these tissues by neurally released transmitters (see Table 7.1) have been little investigated. This latter type of activation is biologically the most important, as it transmits the centrally generated neural signal from the postganglionic neurons to the target cells. Where these studies have been performed, the mechanisms of neuroeffector transmission have been found to be diverse. The important concept that has emerged from these studies is that the mechanism utilized by endogenously released transmitter is often not the same as that activated by exogenous transmitter substances or their analogues (Hirst *et al.* 1992, 1996).

Excitation or inhibition may involve a range of cellular events:[2]

- Brief openings of ligand-gated channels (as at many neuronal synapses) or slower conductance changes mediated by second messenger systems.
- The ensuing depolarization may open voltage-dependent calcium channels leading to Ca^{2+} influx.

- Receptor activation may cause G-protein activation and release of Ca^{2+} from intracellular Ca^{2+} stores or modulation of the Ca^{2+} sensitivity of the contractile/secretory mechanism.
- Receptor activation (e.g., β-adrenoceptors) may be linked to adenylate cyclase and modify cell function by changing intracellular levels of cyclic adenosine monophosphate (cAMP).
- Inhibition (relaxation) often involves the activation of cyclic guanosine monophosphate (cGMP)-dependent protein kinases.

7.2.1 The autonomic effector cells and their innervation

The autonomic effector cells are very diverse, the most common type being smooth muscle cells, cardiac muscle cells and secretory epithelia (Table 1.2). The cytoplasm of the adjacent cells in these autonomic effector tissues is connected by gap junction channels that allow electrical and chemical signals to pass directly between neighboring cells. Thus, the cells in these tissues form *functional syncytia*, with neurally evoked signals spreading directly from cell-to-cell. In the cases of electrical signals, current injected by transmitter action on an individual cell flows most readily through the low-resistance gap junction channels into neighboring cells and potential change at the point source is brief. When nerves are stimulated, transmitter is released at many points (e.g., from many varicosities) throughout the tissue and current spreads through the gap junction channels from each point source to change the membrane potential of the whole tissue. In this case, where the tissue is uniformly polarized, the time course of the potential change will be determined by a relatively slow process of charge dissipation across the cell membrane; i.e., the time course is dependent on the resistance and capacitance of the cell membranes. This behavior of the electrically coupled cells makes the study of synaptic transmission from postganglionic autonomic fibers to the effector cells technically difficult and difficult to interpret.[3]

The axons of a postganglionic autonomic neuron branch near the cells they innervate and form many fine varicosities. These varicosities are about 1 μm or less in diameter. They contain numerous vesicles that store transmitter substances and contain the biochemical machinery for synthesis, release, reuptake and metabolism of the transmitter substances. Each postganglionic axon forms hundreds to thousands of varicosities in its terminal branches depending on the size of the target tissue and its function. Axons of sympathetic postganglionic vasoconstrictor neurons have up to 10 000 varicosities, while parasympathetic postganglionic axons projecting to the eye or axons of enteric motoneurons acting locally on the circular or longitudinal smooth muscles have several hundred varicosities.

At the skeletal neuromuscular endplate, acetylcholine is released in quanta from the stores of vesicles in the nerve terminals into the synaptic cleft at the neuromuscular endplate. After release the transmitter reaches a high concentration in the synaptic cleft for a short time and reacts with acetylcholine receptors in the postsynaptic membrane close to the site of release. This causes transient opening of specific non-selective cation channels resulting in a brief excitatory synaptic current and a consequent endplate potential. The enzyme choline esterase, present in the synaptic cleft, breaks down acetylcholine to form choline and acetic acid in about 0.1 ms and thereby limits the action of acetylcholine and prevents diffusion of acetylcholine out of the synaptic cleft. This process of chemical transmission of an electrical signal from the terminal of motor axons to the skeletomotor fibers is restricted to the neuromuscular endplate and shows a high safety factor for transmission. Chemical transmission at neuroeffector junctions is also spatially restricted but does not exhibit the high safety factor.

A widely held misconception that arose from early electron microscopic studies is that tissues innervated by postganglionic autonomic neurons do not have a specialized neuroeffector apparatus. Rather, varicosities were found to be located at variable distances from the effector cells with few varicosities in close apposition to the effector cell. The consequences of this arrangement would be that the transmitter is released at variable distances with neurotransmitters acting like a local hormone to activate widely distributed receptors on the effector cells. This type of neuroeffector communication has been assumed to be the *status quo* particularly for the sympathetic postganglionic fibers. It has fostered the belief that the effect of postganglionic noradrenergic neurons on the effector cells can be equated with the effect of noradrenaline or a cotransmitter superfused over the preparation, i.e. that endogenous and exogenous neurotransmitters act in the same way. However, experimental, morphological and functional studies performed recently radically contradict this conventional concept:

1. In many autonomically innervated target tissues so far investigated quantitative structural analysis of the neuroeffector apparatus at the electron microscopic level clearly shows that a large proportion of the varicosities form organized neuroeffector contacts (Table 7.2).
2. Transmitter released from these varicosities often acts quite differently from transmitter applied exogenously to the preparation. Neurally released transmitter interacts with specialized junctional receptors, which have different properties from extrajunctional receptors activated by the same transmitter (Hirst *et al.* 1992, 1996).

7.2.2 Morphology of the neuroeffector junction

Precise transmission of the signals in postganglionic neurons to the effector cells requires a spatially close association between the varicosities and the effector cells. However, random histological thin sections give the false impression that most varicosities do not make close contact with the effector cells, that the gaps between varicosities and effector cells are very variable and that close contacts are

Table 7.2 | *Properties of neuroeffector junctions on autonomic effector cells determined by successive sections of tissues using the electron microscope*

Tissue	Percentage of varicosities forming junctions	Area occupied by junctions (μm^2) mean ± S.E.M.	Varicosity volume (μm^3) mean ± S.E.M.	Ref.
(a) Sympathetic neuroeffector junctions in blood vessels				
Rabbit juxtaglomerular arterioles				
afferent	70	0.48 ± 0.32 (n = 4)	0.52 ± 0.25 (n = 8)	1
efferent	69	0.38 ± 0.05 (n = 6)	0.36 ± 0.13 (n = 8)	1
Guinea pig submucosal arterioles*	90	1.26 ± 0.29 (n = 11)	1.2 ± 0.25 (n = 11)	2
Guinea pig mesenteric veins*	98	0.64 ± 0.08 (n = 47)	0.81 ± 0.15 (n = 17)	4
(b) Sympathetic neuroeffector junctions on cardiac muscle and in structures associated with the rat iris dilator muscle				
Toad sinus venosus*	88	0.44 ± 0.07 (n = 25)	0.68 ± 0.39 (n = 4)	3
Guinea pig sinoatrial node*	90	0.15 ± 0.03 (n = 9)	0.31 ± 0.06 (n = 9)	5
Rat iris dilator myoepithelium*	26	0.23 ± 0.07 (n = 6)	0.21 ± 0.07 (n = 4)	6
Rat iris dilator melanophores	35	0.21 ± 0.05 (n = 8)		6
(c) Parasympathetic neuroeffector junctions on cardiac muscle and in structures associated with the rat iris dilator muscle				
Toad sinus venosus*	96	0.45 ± 0.06 (n = 28)	0.60 ± 0.16 (n = 9)	3
Guinea pig sinoatrial node*	85	0.21 ± 0.04 (n = 13)	0.48 ± 0.07 (n = 13)	5
Rat iris dilator myoepithelium*	36	0.50 ± 0.12 (n = 14)	0.36 ± 0.09 (n = 14)	6
Rat iris dilator melanophores	71	0.44 ± 0.08 (n = 27)		6

Notes:
*Autonomic effector structures in which neuroeffector transmission has been studied
References: 1, Luff et al. (1992); 2, Luff et al. (1987); 3, Klemm et al. (1992); 4, Klemm et al. (1993); 5, Choate et al. (1993b); 6, Hill et al. (1993). From Hirst et al. (1996).

rare. The shortcoming of such measurement is that the relationship between a varicosity and the target tissue is assumed to be retained throughout the entire surface of the varicosity. This turns out *not* to be the case when the varicosities are investigated, using electron microscopical serial sections, and three-dimensionally reconstructed, and the relationship between pre- and postjunctional membranes is determined. Such an approach has been used for various autonomically innervated tissues (references see Table 7.2). The following parameters were studied quantitatively: percentage of varicosities forming junctions, size of the area on the effector cell surface occupied by the junction and volume of the varicosity (for detailed data see Table 7.2).

The sympathetic and parasympathetic neuroeffector junctions investigated so far have the following characteristics (Figure 7.1, Table 7.2):

- Most varicosities that lose part of their Schwann cell sheath form organized neuroeffector junctions with nearby target

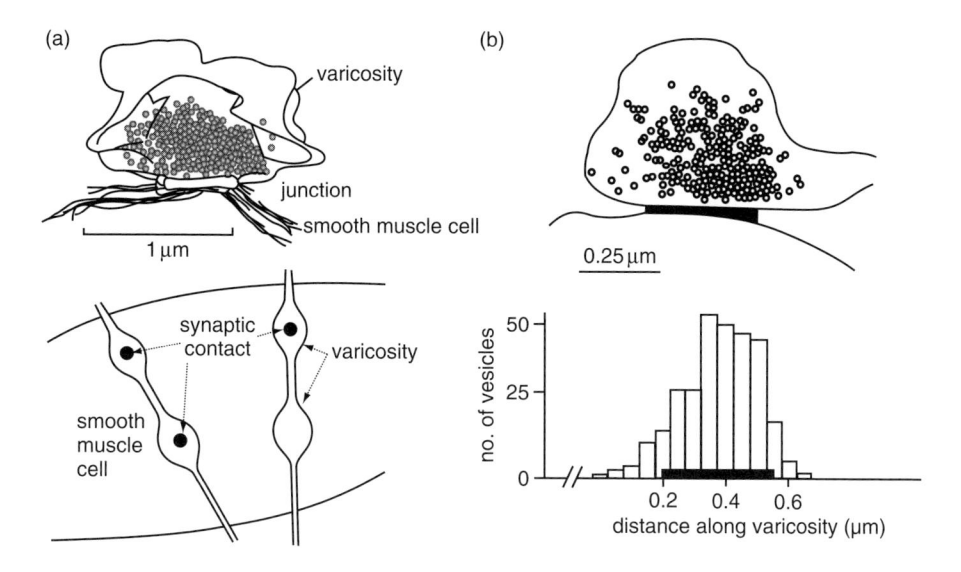

Figure 7.1 Morphology of a neuroeffector synapse to smooth muscle cells. (a) Three-dimensional reconstruction. The varicosity and its postjunctional cell was serially sectioned and reconstructed. The varicosity forms a close contact with a smooth muscle cell (on a small part of its surface). The neuroeffector cleft is 20 to 80 nm wide. The varicosity loses part of its Schwann cell sheath. The basal laminae fuse. Pre- and postjunctional specializations are mostly absent. The lower panel shows schematically that some varicosities form a synaptic contact with the effector cell via a small part of their surface area. (b) Outline of a varicosity and the position of individual vesicles (small open circles) close to the neuroeffector cleft (black bar). Lower diagram: distribution of vesicles along the varicosity. The vesicles are distributed along the neuroeffector cleft. Histological data supplied by Susan Luff.

cells. These junctions cover about 0.1% to 1% of the effector cell surface.
- At the junction, the basal laminae covering the effector cell and the varicosity fuse.
- The neuroeffector gap is 20 to 80 nm wide.
- Small vesicles that contain the transmitter(s) accumulate at the junctional cleft. Large vesicles that contain a neuropeptide, in addition to a classical transmitter, do not accumulate at the junctional cleft.
- In varicosities, which are exposed through a gap in the Schwann cell covering but do not form neuroeffector contacts (no fused basal laminae), the synaptic vesicles are homogeneously distributed.
- Some prejunctional specialization may exist but no postjunctional one.

These morphological characteristics of the neuroeffector junctions clearly indicate that chemical transmission may occur at these sites. The physiological function of non-contacting varicosities remains unclear. There appears to be a difference in neuroeffector junctions between arterioles and large arteries. In the rat tail artery ≤50% of the varicosities form close neuroeffector junctions and in the rat mesenteric artery about 15% (Luff et al. 1995, 2000; Luff personal communication).

7.2.3 Electrophysiology of the neuroeffector transmission

Experimental investigations of a few effector tissues show that the neuroeffector transmission is tissue specific. There is now full agreement that the transmitter or combination of a "classical" (primary) transmitter and one (or more) peptide transmitter(s) are localized in the synaptic vesicles. Importantly, as for other synapses, release of transmitter at these junctions occurs in quanta with a low probability of release from individual varicosities. In regard to the former, it has long been held in contention whether the release of transmitter from individual varicosities of autonomic nerve fibers occurs quantally through emptying of the varicosity or through fractional release, i.e. whether during the brief transient fusion of a vesicle with the plasma membrane its total content is released or only a fraction of it. The majority of experimental evidence supports quantal release although some scientists still believe that this problem is not solved and that under physiological conditions of activity in postganglionic neurons fractional release of transmitter does occur (Folkow and Nilsson 1997; Folkow 2000).

Evidence for quantal release can be gleaned from electrophysiological experiments when the following conditions are fulfilled (Hirst et al. 1996):

1. The transmitter released must act at *ligand-gated* ion channels. This leads to a *rapid* opening of the ionic channels and to a postsynaptic potential at *short latency* and with a brief duration. If the transmitter were to act on the ionic channels through a G-protein and second messenger system it would be impossible to detect the unitary postsynaptic events upon release of single quanta because the kinetics of the intracellular multistage reaction pathways between membrane receptors and ionic channel are too slow.
2. The transmitter must be released at a close neuroeffector junction that leads, upon the release of the content of a vesicle, to a short-lasting high concentration of the transmitter in the cleft. In wide clefts the transmitter released prejunctionally would be diluted with the second power of the distance between release site and membrane receptor.
3. The recording of the postjunctional potential must be made either from a neuroeffector junction involving a varicosity on a single electrically isolated cell or from a small syncytial conglomerate of electrically coupled cells that behaves approximately like a single, spherical cell. Recording from single cells that are not electrically coupled to other cells and that are innervated has not been described for autonomic neuroeffector transmission.

With these limitations it has been shown for some blood vessels and for the vas deferens that transmitter is released from the varicosities in quanta. Figure 7.2b shows an intracellular recording from a short segment of an arteriole of about 200 µm length, which is innervated by two to four noradrenergic postganglionic axons with about

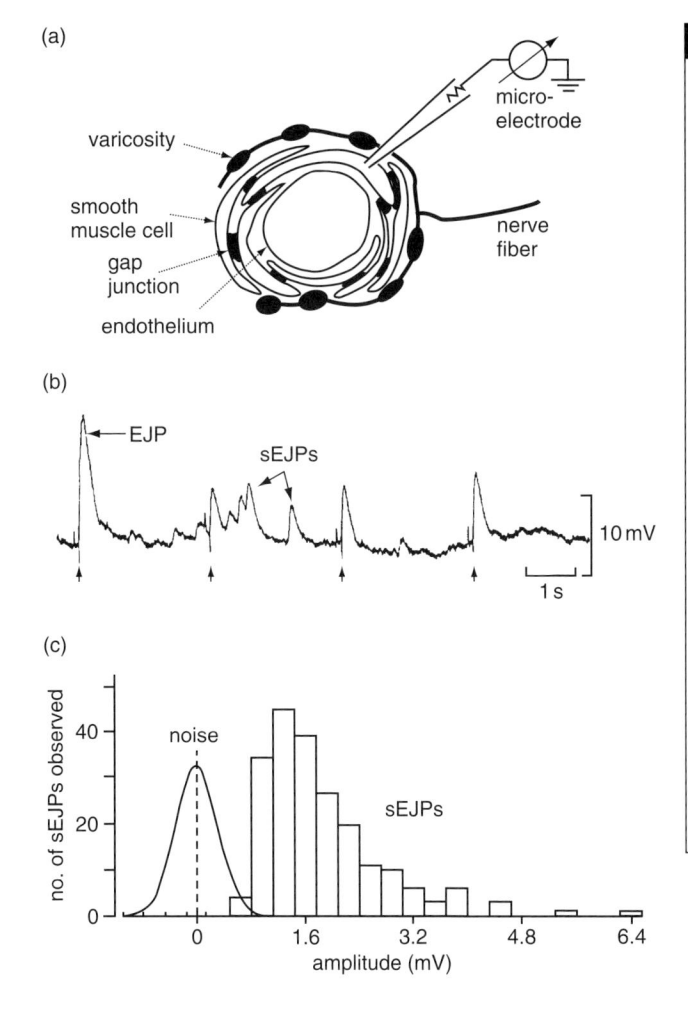

Figure 7.2 Neuroeffector transmission in arterioles. (a) Experimental set-up. Intracellular recording from a 200 μm segment of an arteriole isolated from the guinea pig submucosa. The innervation was electrically stimulated transmurally with single impulses of 0.5 ms duration at a rate of 0.5 Hz. (b) Recording of excitatory junction potentials (EJPs) evoked by nerve stimulation (arrows) and of spontaneous excitatory junction potentials (sEJPs) occurring randomly. The size of the EJPs is close to that of the sEJPs. This implies that the quantal content of each EJP is low. (c) Comparison of the amplitude frequency histogram of the sEJPs with the normal distribution of the recording noise. Spontaneous EJPs have a separate unimodal amplitude distribution. All responses are the result of transmitter release from sympathetic varicosities. With permission from Hirst and Neild (1980).

200 varicosities that make neuromuscular junctions in this vascular segment. This short segment of arteriole contains about 100 to 200 electrically coupled smooth muscle cells and current injected into a single cell uniformly polarizes the whole segment, so that its electrical behavior approximates that of a single spherical cell. This preparation exhibits typical postsynaptic events. *First*, electrical activation of the postganglionic axons by single pulses evokes stimulus-locked excitatory junction potentials (EJPs). *Second*, in the absence of stimulation, spontaneous excitatory junction potentials (sEJPs) are recorded that are thought to be produced by spontaneous release of the transmitter content of a single vesicle (quantum). These junction potentials are generated by excitatory junction currents (EJCs), which have a latency of 1–2 ms, a rise time of about 10 ms and a decay time of 100 to 150 ms. The duration of the EJPs is in the range of 1 second, which is related to the capacitance of the syncytium of the electrically coupled cells.

The time course of the sEJP and EJP in these short arteriolar segments is similar and hence it is possible to determine directly the relationship between these two events. The amplitude distribution of the sEJPs is unimodal and their amplitudes correspond to the smallest EJPs recorded. The unimodal distribution of sEJP amplitudes indicates that transmission must occur at a population of varicosities that make close neuroeffector contacts with the smooth muscle. Assuming that the EJP is composed of quantal units having the same size as the sEJP, the amplitude distribution of EJPs indicates that the majority of EJPs in this preparation had a quantal content of one to four. Furthermore, since all varicosities of this preparation are excited by the electrical stimulation of the postganglionic axons with single pulses it follows that only about 1% of the varicosities release the content of a single vesicle when the innervation is activated by a single electrical stimulus. These findings indicate that transmitter release from an individual varicosity occurs in quanta (i.e., is not fractional).

These results have been fully confirmed by extracellular recording of electrical activity at sympathetic neuromuscular junctions in both vas deferens and muscular arteries by using a suction microelectrode (Figure 7.3a). The technique allows recording of both evoked and spontaneous junction currents (EJCs) from small populations of varicosities and demonstrated that both events have a similar amplitude and time course (Figure 7.3b, c). This finding indicates that both events result from release of the transmitter content of a single vesicle. The extracellular recording technique also allows recording of the nerve terminal action potential (NTAP in Figure 7.3b, c) and demonstrated that electrical excitation of the postganglionic axon leads to action potential invasion into *all* the varicosities. Therefore, the probability of releasing single quanta of transmitter from the invaded varicosities is very low (range of probability of 0.01 to 0.1 depending on the frequency of excitation) (Cunnane and Stjärne 1982; Brock and Cunnane 1988, 1992; 1993; Lavidis and Bennett 1992).

In conclusion, individual EJCs, generated by the release of one vesicle from a distinct varicosity, are statistically evoked about once during 100 electrical stimuli applied to the innervation of the preparation. These individual EJCs are identical in shape, size and duration to the sEJCs generated by spontaneous release from the same vesicle.

In the sympathetically innervated tissues neither the sEJPs nor the evoked EJPs could be blocked by α- or β-adrenoceptor antagonists. This shows that these junction potentials are not mediated by noradrenaline. The compound that is believed to elicit the junction potentials is ATP. This is colocalized with noradrenaline in the small vesicles and released together with noradrenaline from the varicosity, either spontaneously or when the postganglionic axon is

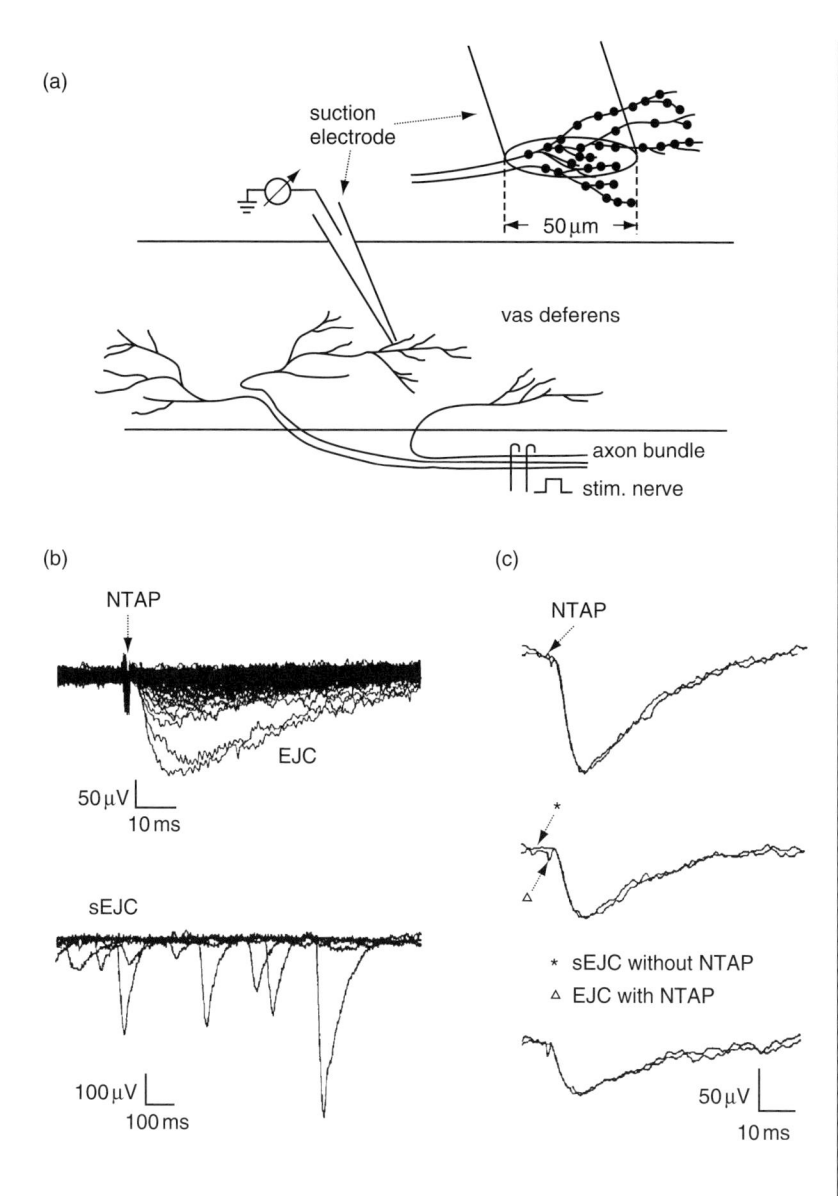

Figure 7.3 Extracellular recordings of action potentials from varicosities and junction potentials generated by varicosities in smooth muscle cells using a suction microelectrode. (a) Experimental set-up. The suction microelectrode is set on the surface of the vas deferens with a high resistance between the spaces inside and outside of the microelectrode. The inner diameter of the microelectrode is, at its tip, about 50 μm; a small number of varicosities are on the membrane patch covered by the microelectrode (inset upper right). This arrangement allows extracellular recording of the action potentials of the nerve fibers and varicosities (nerve terminal action potential, NTAP) and the junction potentials in the effector cells. (b) Spontaneous excitatory junction currents (sEJCs) and excitatory junction currents (EJCs) evoked by nerve stimulation. Transmitter release is quantal (in vesicles). Nerve stimulation with single pulses at 1 Hz leads to action potentials in every varicosity but to only intermittent release of the transmitter (in one vesicle) from an individual varicosity with a probability of about ≤ 0.01. (c) The shape of sEJCs and EJCs generated by release from an individual varicosity are similar. Three pairs of similar sEJC (without NTAP) and evoked EJC (with NTAP) in the same preparation are shown. These three pairs of junction current were most likely generated by the release of transmitter from three individual varicosities. Modified from Brock and Cunnane (1988) with permission.

excited. The evidence that the junction potentials are generated by release of ATP are as follows:

- They are blocked by suramine, a purinoreceptor blocker.
- They are blocked by α-β-methylene-ATP, an agonist that rapidly desensitizes the purinoceptors.
- They are absent in tissues of transgenic mice in which P_{2X1}-purinoceptors have been knocked out (vas deferens; Mulryan et al. 2000). This is so far the best evidence for ATP as transmitter to date.

Ironically, we know more about the molecular mechanisms of ATP release than about those of noradrenaline release (see further discussion below in the context of neurovascular transmission).

7.3 | Specific neuroeffector transmissions

In the following Subchapter I will describe the neuroeffector transmission for three groups of effector cells: sinoatrial node and the sinus venosus of the heart, small blood vessels and the longitudinal muscle of the gastrointestinal tract. In these three cases stimulation of the autonomic (sympathetic, parasympathetic or enteric) nerves and superfusion of the preparation with the respective transmitter, acetylcholine or noradrenaline, causes the same effector responses. However, detailed examination of the changes in membrane potential of the effector cells following neurally released transmitter and following transmitter added to the preparation reveals that the mechanisms leading to the effector responses are quite different. This, together with the morphology of the neuroeffector junctions, supports the notion that the transmission of neural signals to the effector cells occurs through these junctions.

7.3.1 Innervation of the heart

The neuroeffector transmission in the heart has been studied in the toad and in some mammals (in particular guinea pig). For the parasympathetic control there are parallels between toad and mammals. For the sympathetic control there are some differences. These differences do not change the overall message to be made that the transmission of the neural signals to the heart is functionally specific.

Parasympathetic cardiomotor neurons

Stimulation of the preganglionic cardiomotor axons in the vagus nerve slows action potential firing in atrial pacemaker cells. During nerve stimulation, the frequency of the action potentials decreases without any change in their configuration and without membrane hyperpolarization. The pacemaker potential exhibits a slower depolarization and therefore the time to reach threshold for initiation of the action potential is increased. This neurally induced slowing of the pacemaker depolarization is generated by inhibition of sodium channels (that are distinct from non-selective calcium channels [Edwards et al. 1993]), which is accompanied by an increase in membrane resistance. These sodium channels are normally open, tending to depolarize the pacemaker cells (Figure 7.4b, d). Stronger stimulation of the vagus nerve would lead to further inhibition of these sodium channels and complete abolition of pacemaking.

Exogenous acetylcholine superfused over the preparation also slows firing of the pacemaker cells. However, the membrane is hyperpolarized and the amplitude and duration of the action potentials are reduced. Exogenous acetylcholine opens potassium channels causing hyperpolarization (Figure 7.4c, d; Bolter et al. 2001). Thus, the action potentials are decreased in size and shortened (due to an increase in the membrane conductance).

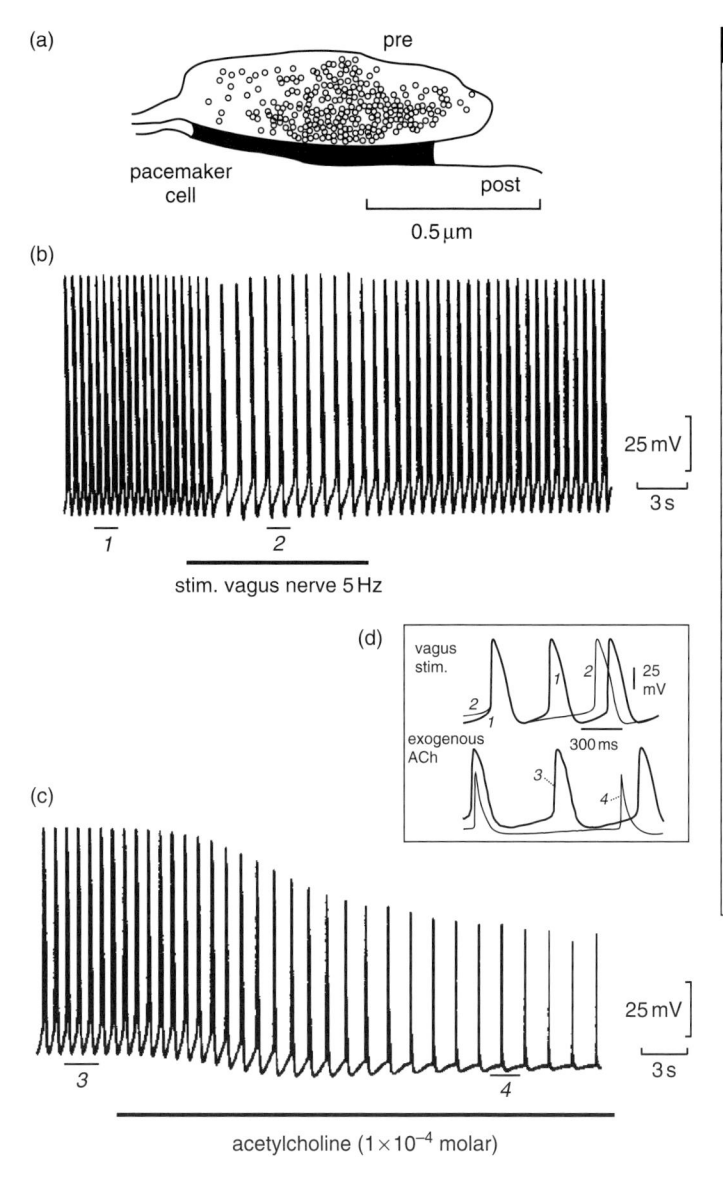

Figure 7.4 Effect of vagus nerve stimulation and acetylcholine on cardiac pacemaker cells. (a) Schematic outline of a cholinergic neuromuscular junction on a cell of the guinea pig sinoatrial node. (b, c) Intracellular record from a pacemaker cell in an in vitro preparation with attached vagus nerve. Expanded records of time periods marked with numbers 1 to 4 are shown in the inset (d). The cell shows pacemaker potentials with action potentials. Electrical stimulation of the vagus nerve at 5 Hz (b) shows no change in action potential configuration but a reduced slope of the pacemaker potential (upper records in inset). Exogenous acetylcholine (ACh) (c) hyperpolarizes the membrane and shunts the action potentials (due to opening of potassium channels and decrease in membrane resistance), reducing their amplitude and duration (lower records in inset). (a) According to Choate et al. (1993a). (b, c) Modified from Campbell et al. (1989) with permission.

When pacemaking is abolished by blocking the calcium channels (e.g., by nicardipine) the membrane potential of the pacemaker cells settles to values of about −35 to −40 mV because of the pacemaker-associated open sodium channels. Both electrical stimulation of the vagus nerve and superfusion of the preparation with acetylcholine now hyperpolarize the pacemaker cells. However, the onset of hyperpolarization following nerve stimulation is much faster and shorter than that following acetylcholine superfusion (Figure 7.5). Importantly, these mechanisms are different as while the response to acetylcholine superfusion is prevented after blocking potassium channels with barium ions (Ba^{2+}) in the arrested preparation, the hyperpolarization to nerve stimulation is not affected (Figure 7.5;

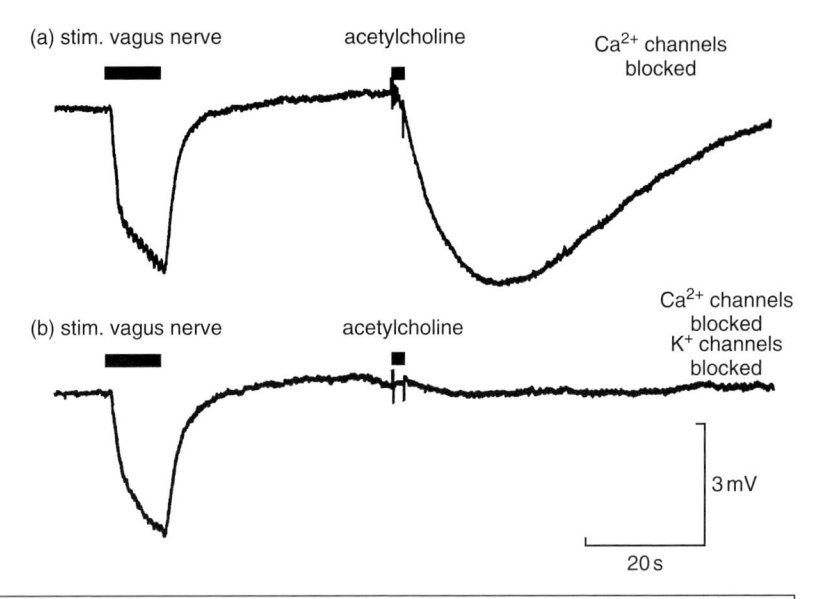

Figure 7.5 Inhibition of pacemaker activity generated by electrical stimulation of the vagal nerve and by applied acetylcholine (ACh) is mediated by different cellular mechanisms. Intracellular recording in the sinus venosus of toad *Bufo marinus* in vitro with attached vagus nerve. The pacemaker cells are arrested by blocking the calcium channels with nicardipine (10 μM). Under this condition the membrane potential settles at about −35 mV because of the high Na^+ permeability. (a) Vagal stimulation for 10 s at 1 Hz hyperpolarizes the pacemaker cells. Ionophoretic application of ACh applied through a micropipette for 2 s (10^{-4} M, 50 nA) also hyperpolarizes the cells. Onset and decay of the hyperpolarization are longer when generated by exogenously applied ACh than during vagus nerve stimulation. (b) Recording from the same cell about 15 min later after adding Ba^{++} (1 mM), which blocks potassium channels, to the physiological saline superfusate. Response to vagus nerve stimulation is unaffected but the response to ionophoretically applied ACh is abolished. Acetylcholine released by vagus nerve stimulation leads to closing of sodium channels coupled to subjunctionally located muscarinic receptors. Superfused ACh leads to opening of potassium channels coupled indirectly via a G-protein and cAMP to extrajunctionally located muscarinic receptors. Modified from Hirst et al. (1991) with permission.

Bramich et al. 1994). Both responses are blocked by atropine, an antagonist for muscarinic cholinergic receptors, in the beating and in the arrested preparation. As the postreceptor conductance changes clearly differ, the muscarinic receptors activated by acetylcholine released by the vagal fibers cannot be the same as those to which exogenous acetylcholine binds (Campbell et al. 1989).

Further investigations have shown that the post-receptor pathways connecting muscarinic receptors to ionic channels are different for the junctional and the extrajunctional acetylcholine receptors. The junctional receptors are connected via a yet to be defined intracellular pathway or even directly to the sodium channels and the extrajunctional acetylcholine receptors are connected via a G-protein, adenylate cyclase and cAMP to potassium channels and other channels (see Demir et al. [1999]). Since it is unlikely that significant acetylcholine escapes from the neuroeffector junction during neural stimulation (due to choline esterase near the synaptic cleft) it is likely that the extrajunctional acetylcholine receptors play no role in the control of heart rate under physiological conditions. As mentioned before (see Subchapter 7.2), only $\leq 1\%$ of the surface of the pacemaker cells are covered by neuroeffector junctions. Thus, the biologically important transmission of neural signals from parasympathetic cardiomotor neurons to these pacemaker cells occurs at these junctions only.

Sympathetic cardiomotor neurons
Similar investigations have been conducted on the sympathetic innervation of the pacemaker cells, which increases the rate of pacemaking when activated. In *mammals*, both neurally released and exogenous noradrenaline activate β-adrenoceptors. In the *toad*, sympathetic postganglionic cardiomotor neurons use adrenaline as transmitter. Nerve-released adrenaline activates junctional adrenoceptors that are neither α- nor β-adrenoceptors; applied adrenaline activates β-adrenoceptor.[4] Noradrenaline or adrenaline released from the nerve terminals of the sympathetic cardiomotor neurons modifies the pacemaker current whereas application of exogenous noradrenaline or adrenaline produces its effect by increasing cAMP in the pacemaker cells leading to phosphorylation of voltage-dependent ionic channels, which are activated during pacemaker activity (Choate *et al.* 1993a).

Thus, as for vagal innervation of the heart, it is likely that the signals arising from the sympathetic innervation of the heart are mediated by junctional adrenoceptors. At least in the toad, the intracellular postreceptor pathways to the ionic channels are different for the junctional adrenoceptors and the extrajunctional adrenoceptors: the extrajunctional adrenoceptors involve cAMP and the junctional adrenoceptors an intracellular pathway still not identified; in amphibians it may involve inositol 1, 4, 5-trisphosphate (IP_3), which releases Ca^{2+} from intracellular stores (Bramich *et al.* 2001).

The function of the extrajunctional adrenoceptors in neuroeffector transmission in the heart is unknown. However, they may play a role in mediating the effects of circulating adrenaline; whether this is functionally important under physiological conditions remains to be shown (see Chapter 4.5 and Figure 4.17; see here that circulating noradrenaline, in physiological concentration, neither increases heart rate nor blood pressure). This may become particularly important in the sympathetically denervated heart (e.g., in the transplanted heart or in patients with severe diabetic neuropathy in whom the sympathetic postganglionic neurons to the heart are metabolically destroyed); however, under these conditions the adrenoceptors may have dramatically changed in number, affinity and coupling to the post-receptor pathways.

7.3.2 Sympathetic innervation of blood vessels

Small arteries and arterioles
Postganglionic vasoconstrictor neurons have long axons and form many branches in the adventitia of blood vessels with thousands of varicosities (Figure 7.6). Many of the varicosities on small muscular blood vessels with a diameter of <0.5 mm form close neuromuscular junctions as described above. The proportion that does not form

junctions varies between vessels at different sites throughout the vascular bed (Luff and McLachlan 1989, Luff et al. 1995, 2000; see above). The neural signal is transmitted from these varicosities to the vascular smooth muscle cells, which form a functional syncytium. The small vesicles in the varicosities contain noradrenaline and ATP; the large vesicles (which are much less frequent than the small ones) contain noradrenaline, ATP and NPY. In some vascular beds (e.g., of the skin) different vascular sections (e.g., large arteries, small arteries, arterioles, veins) are innervated by neurochemically different types of vasoconstrictor neurons (see Subchapter 1.4).

Most blood vessels constrict when exogenous noradrenaline is applied to them. This pharmacological action is largely mediated by α_1-adrenoceptors, but also by α_2-adrenoceptors located extrajunctionally (note that only about 1% of the surface of autonomic effector cells [e.g., arterioles, heart] are occupied by synapses). Most of the Ca^{2+} triggering vasoconstriction is released from intracellular stores by second messengers. Nerve stimulation evokes brief depolarizations in the vascular smooth muscle cells (Figure 7.2b), which are followed in some blood vessels by long-lasting depolarizations. The brief depolarizations have durations of about one second and a time-constant of decay of 200 to 300 ms and cannot be blocked by adrenoceptor antagonists. Since they can be blocked by an antagonist to purinoceptors (e.g., by suramin) or by desensitization of the purinoceptors by α-β-methylene-ATP they are suggested to be mediated by ATP. The receptor involved is the P_{2X1} purinoceptor (Mulryan et al. 2000). These brief potentials are generated by activation of ligand-gated cation channels, which under appropriate levels of activity can summate to depolarize the membrane potential to threshold for opening of voltage-sensitive calcium channels. This generates an action potential and can cause a transient constriction depending on the type of blood vessel. The slow depolarization reflects the action of noradrenaline on α_2-adrenoceptors, which causes maintained constriction (Jobling et al. 1992; Hirst et al. 1996).

Constriction of some small arteries during sympathetic nerve activity is abolished by α-adrenoceptor antagonists but is only partly dependent on membrane depolarization (Brock et al. 1997). In contrast, in some small arterioles (e.g., the arterioles of the submucosa of the gastrointestinal tract), vasoconstriction produced by nerve stimulation is unaffected by α-adrenoceptor blockade and is entirely dependent on opening of voltage-dependent calcium channels (Evans and Surprenant 1992). However, in other small arterioles (e.g., those in the rat iris) vasoconstriction produced by nerve stimulation is due solely to release of noradrenaline and activation of α_1-adrenoceptors. Activation of these receptors leads to the intracellular formation of IP_3, which in turn triggers the release of calcium from intracellular stores. Cytosolic calcium activates the contractile apparatus (vasoconstriction) and opening of chloride channels in the cell membrane with chloride outflow (along its electrochemical gradient) resulting in depolarization (Gould and Hill 1996; see note 5).

Thus, neural control of constriction in the small arteries and arterioles involves multiple transmitters (such as ATP and noradrenaline) and both voltage-dependent and voltage-independent mechanisms in different parts of the arterial vascular tree. In addition, there is evidence that the relative role of the cotransmitters changes depending on the pattern of nerve stimulation (e.g., Evans and Cunnane 1992). However, the way in which signals generated by the cotransmitters and the membrane receptors that they activate are integrated by the vascular smooth muscle (Figure 7.7) so as to regulate vascular tone and control of blood flow through tissues remains a puzzle.

This apparent paradox may resolve in future if one looks at comparable densely innervated arteries in the same species, such as fine branches of the superior mesenteric artery and the tail artery of the rat. The fine branches of the superior mesenteric artery have about four layers of smooth muscle cells and a lumen diameter of approximately 200 μm. The rat tail artery has several layers of smooth muscle cells (up to 15) and a lumen diameter of 200 to 400 μm:

- In the *rat mesenteric artery*, electrical nerve stimulation at 10 Hz produces ATP-mediated junction potentials but no α-adrenoceptor-mediated slow depolarizations (Brock and van Helden 1995). However, a part of the contraction produced by electrical stimulation of the vascular innervation can be prevented by an α-adrenoceptor blocker, another part by blockade of purinergic transmission (Luo et al. 2003). Contraction produced by exogenous noradrenaline is closely correlated with the depolarization produced by this agent (Nilsson et al. 1994). However, concentrations greater than 1 μM are required to produce significant depolarization (Mulvany et al. 1982). This shows that exogenously applied noradrenaline does *not* mimic the effect of nerve stimulation. Electrochemical studies show that the concentration of noradrenaline at the adventitia during electrical nerve stimulation at 10 Hz can reach levels in the *micromolar range* (Dunn et al. 1999). These concentrations can presumably activate adrenoceptors located at a distance from the release sites (i.e., extrajunctionally). These micromolar concentrations cannot be produced by circulating noradrenaline under physiological conditions, which is in the picomolar range in the rat (see Subchapter 4.5 and Figure 4.17).
- In the *rat tail artery*, electrical nerve stimulation with trains of impulses at 10 Hz produces junction potentials that are most likely mediated by ATP and purinoceptors and slow depolarizations mediated by noradrenaline acting at α-adrenoceptors (Jobling and McLachlan 1992). Nerve-evoked contractions of this artery are primarily generated by released noradrenaline with only a minor role of ATP (depending on the frequency of stimulation) (Bao et al. 1993). The electrical and contractile effects of nerve-released noradrenaline are mimicked by exogenously applied α-adrenoceptor agonists (Brock et al. 1997). However, also

Figure 7.6 Morphology of the postganglionic vasoconstrictor neuron and its neuroeffector synapse. (a) The postganglionic neuron and its blood vessels. The neuron has a long axon and forms many 1000 varicosities close to the blood vessel. (b, c) The terminal branches of the postganglionic axon form a noradrenergic plexus at the adventitial site around the blood vessel. Varicosities are arranged close to the smooth muscle cells. Smooth muscle cells are electrically and chemically coupled by gap junction channels, forming a functional syncytium. (d) The neuroeffector synapse formed by a varicosity. The varicosity forms close contact with the smooth muscle cells without intervening Schwann cell cytoplasm. The basal laminae of smooth muscle cell and varicosity fuse. The vesicles are arranged close to the synaptic cleft.

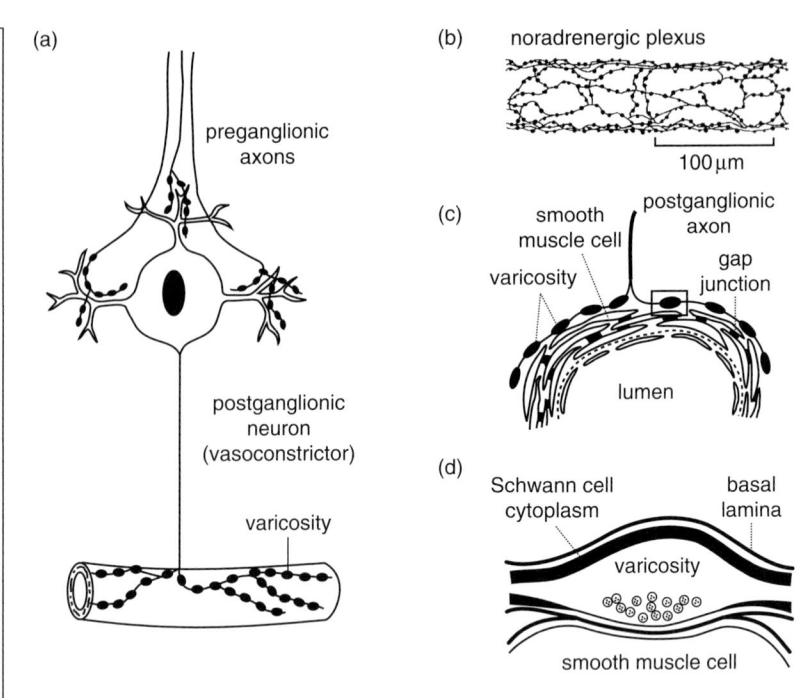

for this artery the concentration of exogenously applied noradrenaline required to induce constriction never occurs in in vivo conditions. Thus, the nerve-induced constriction of the rat tail artery is mainly mediated by noradrenaline that is released by the varicosities and acts on extrajunctional α-adrenoceptors probably reaching here also concentrations in the micromolar range.

These examples and the example of the submucosal arterioles, which are fully dependent on a non-adrenergic (purinergic) mechanism for their contraction, demonstrate that the mechanisms of nerve-induced constriction are different for different vascular sections.

Neuropeptide Y can play a neuromodulatory role by potentiating the contractile response to nerve stimulation in some blood vessels. However, as high bursts of activity are necessary to elicit NPY action and such activity is normally absent in postganglionic vasoconstrictor neurons it remains unclear as to the importance of this mechanism.

Veins and pulmonary artery

Veins (portal, mesenteric) and the pulmonary artery contain sympathetic neuroeffector junctions. Noradrenaline applied to these veins mimics the effect of noradrenaline released by the sympathetic nerve fibers, both causing contraction of the blood vessels via similar mechanisms. Both effects are abolished by α-adrenoceptor blockers. It is likely that the adrenoceptors in the venous smooth muscle cells are linked to an intracellular pathway, which involves IP_3. This leads to intracellular release of calcium and contraction as well as opening

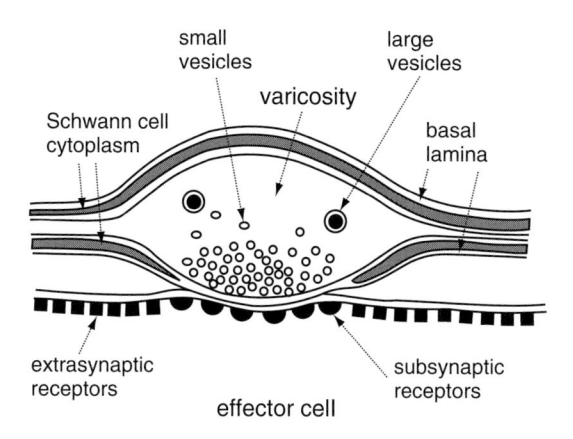

Figure 7.7 Simplified scheme of neuroeffector transmission to autonomic target cells (arterioles, heart). Subsynaptic receptors mediate the effect of transmitter released by the nerve terminals during excitation under physiological conditions. The cell surface at which this nerve–effector communication occurs is ≤1% of the total effector cell surface. These subsynaptic receptors are ligand gated or second-messenger coupled to cellular effectors (e.g., ionic channels). Extrasynaptic receptors for the transmitter are either different from the subsynaptic ones and/or are coupled by different intracellular second-messenger pathways to the cellular effectors. The function of the extrajunctionally located receptors are unclear for most innervated effector organs. Nerve-induced constriction of small arteries is entirely or partially mediated by noradrenaline acting on extrajunctionally located adrenoceptors. Small vesicles containing the transmitter are located close to the synaptic cleft. Large vesicles are also present in many varicosities. In these large vesicles neuropeptides are colocalized with "classical" transmitter. Large vesicles are not located close to the synaptic cleft. The physiological role of the neuropeptides is, in most cases, unclear.

of calcium-activated chloride channels, which are responsible for the fast EJPs[5] but not the long-lasting EJPs. Purinoceptor-mediated mechanisms do not seem to play a role in sympathetic neuron-induced constriction of veins although these vessels are very reactive to applied ATP (van Helden 1988a, b, 1991; Hirst et al. 1996).

7.3.3 Transmission to smooth muscle of the ileum

The longitudinal muscle of the ileum is innervated by cholinergic neurons of the myenteric plexus. The varicose terminals of these neurons form close junctions with the smooth muscle cells, which are the sites of signal transmission between enteric neurons and effector cells (Klemm 1995). Nerve-evoked contractions and contractions induced by exogenous acetylcholine are both mediated by muscarinic receptors and therefore abolished by atropine. Both elicit membrane potential changes (depolarizations) at relatively long latencies of 500 to 600 ms, Ca^{2+} action potentials and muscle contractions. The long latency of the depolarizations shows that both activate complex intracellular pathways and are not generated by activation of ligand-gated ionic channels. However, the mechanisms underlying both are entirely different (for details see Cousins et al. [1993, 1995]; Hirst et al. [1996]; Figure 7.8):

- Nerve-released acetylcholine (ACh) generates EJPs that trigger Ca^{2+} muscle action potentials and muscle contraction. However, when the action potentials are prevented (by calcium channel

MECHANISMS OF NEUROEFFECTOR TRANSMISSION

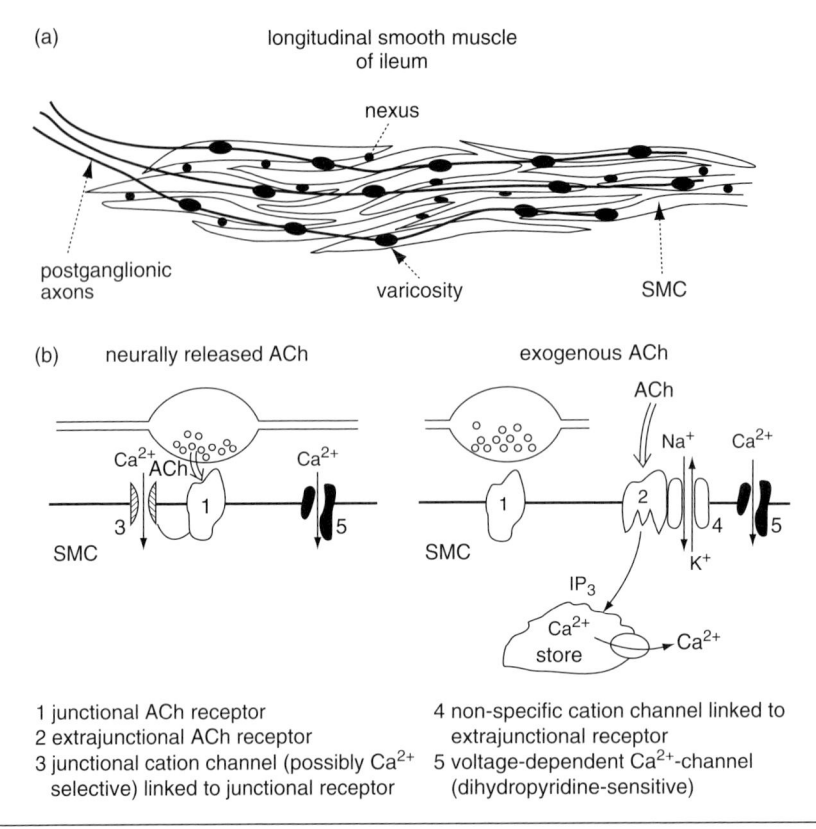

Figure 7.8 Neuroeffector transmission to the longitudinal muscle of the ileum. (a) Cholinergic innervation of the longitudinal smooth muscle forms close synaptic contacts with the muscle cells with all characteristics of synapses (see Figure 7.1). In the circular smooth musculature enteric neurons also form synapses with interstitial cells of Cajal (ICC), which seem to mediate many neurally induced excitatory and inhibitory effects to the smooth muscle cells (see Chapter 5.4). Acetylcholine (ACh) released from the nerve terminal is prevented from leaving the synaptic cleft by acetylcholine esterase. (b) Acetylcholine released by nerve impulses from the varicosities activates junctional muscarinic receptors leading to depolarization by activation of Ca^{2+}-selective ion channels. This depolarization opens voltage-activated calcium channels. In contrast, ACh applied by superfusion to the preparation activates extrajunctional muscarinic receptors linked to non-selective cation channels, leading to depolarization and activation of voltage-activated calcium channels. In addition, exogenous ACh releases Ca^{2+} from intracellular stores via the inositol 1,4,5-triphosphate (IP_3) pathway. Increase in intracellular Ca^{2+} from both sources contributes to muscle contraction. SMC, smooth muscle cell. After Cousins et al. (1995).

blockers) EJPs still evoke contractions. Excitatory junction potentials probably open channels with high Ca^{2+} selectivity. This is supported by the observation that EJPs and transient increase of intracellular Ca^{2+} in the smooth muscle cells occur simultaneously.

- Ionophoretically applied ACh opens non-selective cation channels (for Na^+, Ca^{2+} and K^+) yet the Ca^{2+} entering the muscle cells is negligible. The depolarization following opening of these channels opens voltage-dependent calcium channels generating action potentials and contraction (through inflow of Ca^{2+}). Additionally, ionophoretically applied ACh causes increase of intracellular

Ca^{2+} from an intracellular store. This release of Ca^{2+} is pulsatile and involves the IP_3 pathway.[6]

These results suggest that increase in intracellular Ca^{2+}, and subsequent contraction, caused by neurally released ACh and by ionophoretically applied ACh involves different intracellular pathways, the latter involving IP_3 and pulsatile release of Ca^{2+} from an intracellular store and the former not. This difference is supported by different sensitivities to organic Ca^{2+} antagonists of the responses to nerve-released ACh and to applied ACh.

These findings suggest that muscarinic receptors involved in transmission from cholinergic nerves to the longitudinal smooth muscle of the ileum are restricted to the junctional membrane and differ from those activated by exogenous acetylcholine. The findings also suggest that the intracellular pathways leading to contraction differ (Cousins *et al.* 1993, 1995). Thus, neural transmission in this smooth muscle occurs at the organized neuroeffector junctions. The function of the extrajunctionally located ACh receptors is unknown. Acetylcholine esterase in the junctional cleft prevents ACh from leaking out of the cleft and hence ACh is unlikely to activate extra-junctionally located ACh receptors.

Interstitial cells of Cajal (ICC) in the smooth musculature of the gastrointestinal tract mediate, at least in part, the excitatory and inhibitory effect of enteric motor neurons to circular smooth muscle cells (see Subchapter 5.4; Burns *et al.* 1996; Ward *et al.* 2000a). It appears that the nerve fibers of the motor neurons also form synapses with the ICC (see Burns *et al.* [1996], Iino *et al.* [2004], Ward *et al.* [2000a]; Chapter 5.4).

7.4 | Integration of neural and non-neural signals influencing blood vessels

The reactions of many autonomic effector cells do not only depend on impulses in the autonomic postganglionic neurons. They also depend on impulse activity in peptidergic afferent neurons, on circulating compounds (hormones), on intrinsic properties of the effector cells, on locally released compounds and directly on physical stimuli (mechanical, thermal). These remote and local neural and non-neural influences vary between effector cells and in different functional conditions of the micromilieu of the effector cells. For example, contraction of the smooth muscles of the eye is only dependent on impulses in the postganglionic neurons innervating the pupil and iris. In contrast, contraction and relaxation of arterioles and small arteries in skeletal muscle, while *dependent on their innervation by vasoconstrictor neurons*, are also dependent on many other factors. There are differences between different types of blood vessels in regard to these neural and non-neural influences. Potential

REMOTE CONTROL

LOCAL CONTROL

vasoconstrictor
(noradrenergic)

vasodilator
(non-adrenergic)

blood borne
influences
(hormones
e.g., adrenaline,
vasopressin)

myogenic activity
(autoregulation)

stretch by intraluminal
pressure

vasodilator action
of tissue metabolites

endothelium derived
factors (e.g., NO)

non-neural cells
(mast cells, PMNL)

afferent
(peptidergic)

environmental
(e.g. temperature)

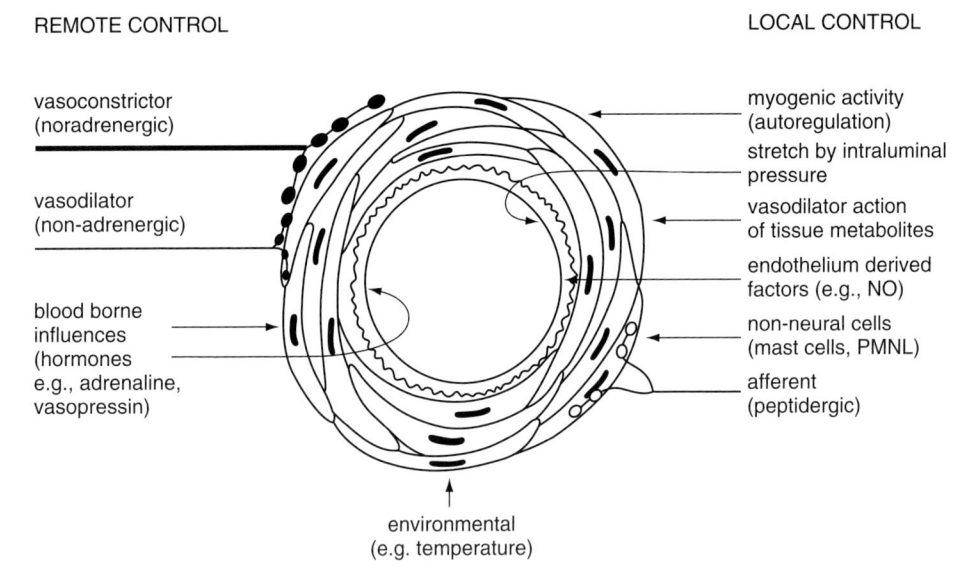

Figure 7.9 The blood vessels are under multiple remote and local controls. *Left side*: Remote controls. Vasoconstrictor neurons (ongoing activity). Vasodilator neurons (normally silent; only in some blood vessels). Blood-borne influences: vasopressin (VP; e.g., during severe blood loss); angiotensin II (e.g., during severe blood loss); adrenaline (vasodilation; only blood vessels in skeletal muscle and during special behavioral conditions; see Chapter 4.5). *Right side*: Local control. Spontaneous myogenic activity. Stretch-evoked contractions. Endothelium-induced vasodilation (e.g., by nitric oxide [NO] released by the endothelial cells by shearing forces exerted by the blood column). Vasodilation generated by tissue metabolites (e.g., in skeletal muscle during exercise; in brain vessels by increase in P_{CO2}; in coronary arteries by decrease in P_{O2}). Vasodilation during inflammation involving mast cells, polymorphonuclear leukocytes (PMNL) and other cells. Vasodilation by peptides (calcitonin gene-related peptide, substance P) released by unmyelinated and small-diameter myelinated afferent nerve fibers. Environmental temperature (e.g., enhancement of neuroeffector transmission from vasoconstrictor fibers to skin blood vessels during cold). Modified from Folkow and Neil (1971).

long-range and local influences acting on blood vessels are exemplified in Figure 7.9:

- Some vascular beds are innervated by sympathetic or parasympathetic vasodilator neurons that are not integrated in the homeostatic regulation of blood flow and resistance. For example, small arteries in skeletal muscle are innervated by sympathetic cholinergic vasodilator neurons in some species (see Subchapter 4.2). Sinusoids and helical arteries of the erectile tissue of sexual organs are innervated by sacral parasympathetic vasodilator neurons, and cranial arteries by parasympathetic cholinergic vasodilator neurons (see Subchapter 4.8).
- Remote non-neural influences on blood vessels are blood borne (e.g., circulating catecholamines, angiotensin, vasopressin). These blood-borne substances act on vascular muscle cells or on the varicosities of the postganglionic axons, modulating transmitter release. For many of these long-range influences it is unclear how important they are in generating or in attenuating vasoconstriction under physiological conditions. For example, it is unlikely that circulating noradrenaline is important in the regulation of

densely innervated blood vessels. Circulating adrenaline may generate β_2-adrenoceptor-mediated vasodilation in skeletal muscle (see Subchapter 4.5). The endothelium is relatively impermeable for these catecholamines in physiological concentrations (Lew et al. 1989).

- Many blood vessels are innervated by peptidergic afferent nerve fibers (e.g., superficial cutaneous, visceral, deep somatic and intracranial blood vessels). These nerve fibers have nociceptive function. Their excitation elicits arteriolar vasodilation and venular plasma extravasation, both responses being peripheral protective reactions independent of the spinal cord. Vasodilation is mediated by CGRP and to a lesser extent by substance P; plasma extravasation is mediated by substance P (Häbler et al. 1999b; Holzer 1992). It is likely that subpopulations of peptidergic afferents are specialized with respect to these functions (see Subchapter 2.2, note 2). Impulses in the peptidergic afferents at frequencies of ≤ 1 Hz or higher overrides the constriction generated by impulse activity in vasoconstrictor neurons in the physiological range (≤ 5 Hz) (Häbler et al. 1997c, 1999b).
- Some blood vessels (e.g., in skeletal muscle and kidney) generate spontaneous myogenic activity and react to stretch (increase in transmural pressure). These properties are the basis for autoregulation of vascular beds. The mechanisms underlying the spontaneous myogenic activity, myogenic responses to distension of the blood vessels and the interaction between myogenic activity and nerve-induced vasoconstriction (e.g., in resistance vessels), however, are poorly understood (see note 3 in Chapter 5). Rhythmic spontaneous contractions of blood vessels that are independent of its innervation may involve cyclic Ca^{2+} release from intracellular stores without contribution of voltage-dependent calcium channels in the cell membranes. This rhythmic intracellular release of Ca^{2+} is entrained between different muscle cells of the syncytium by the oscillating membrane potential via gap junction channels (Hill et al. 1999; Peng et al. 2001). Stretch-activated myogenic responses involve mechanosensitive ionic channels and complex intracellular pathways leading to intracellular increase of Ca^{2+} and contractions (for review see Davis and Hill [1999]; Hill et al. [2001]).
- Mechanical shearing forces exerted by the pulsating blood flow at the luminal endothelial surface trigger synthesis and release of NO by endothelial cells. Nitric oxide leads to relaxation of the vascular smooth muscle cells and to vasodilation. This local mechanism seems to operate continuously in various arterioles and small arteries. It is unclear whether this endothelium-derived production of NO and therefore the vasodilation are also under some neural control (see Häbler et al. 1997b), and whether there are differences between the endothelia of different vascular beds in regard to this mechanism of vasodilation. Furthermore, there are several other compounds released by the vascular

endothelium that influence the vascular musculature. The physiological relevance of most of these paracrine activators and neural influences on these agents or their actions is unclear (Busse *et al.* 2002).
- Skeletal muscles undergo dramatic changes during exercise resulting in the accumulation of several metabolites (e.g., lactate, phosphate), decrease in pH and P_{O2}, increase in P_{CO2}, increase in osmolality, increase in adenosine etc. It is believed that during exercise these changes directly influence the vascular muscles to generate a vasodilation that attenuates neurally evoked vasoconstriction. However, the mechanisms of local vasodilation are still poorly understood and no single factor that dilates blood vessels under physiological conditions has so far been identified. Furthermore, the vasoconstrictor influence on blood vessels in skeletal muscle during exercise does not seem to be diminished by the metabolic changes (for discussion see Rowell [1993]).
- Local mechanisms related to inflammation influence blood vessels, generating precapillary vasodilation and postcapillary plasma extravasation. These changes are brought about by compounds released by inflammatory cells (e.g., mast cells, leukocytes), which either act directly on the muscle cells, via the endothelium or via peptidergic afferents.
- Environmental temperature changes may influence cutaneous blood vessels directly or indirectly by changing neurovascular transmission from vasoconstrictor fibers. Warming dilates these vessels and attenuates neurovascular transmission; cooling constricts these blood vessels and enhances neurovascular transmission (Cassell *et al.* 1988).

This summary of the different components that may influence the reactivity of the vascular smooth musculature shows that this autonomic effector tissue is potentially under multiple local and remote controls. These interact with impulses from vasoconstrictor neurons. Local and remote non-neural influences may also influence other autonomic effector tissues (e.g., sweat glands) and interact with their innervation. However, here I want to emphasize that (1) for most of these processes the mechanisms of interaction with the neural influence are not very well explored, and (2) the neural signal in the vasoconstrictor neurons seems to have preference over most other signals under many physiological conditions.

7.5 | Unconventional functions of sympathetic noradrenergic neurons

New ideas emerging are whether noradrenergic sympathetic terminals have functions that are not dependent on centrally generated impulse activity, not on their excitability and even not on release of noradrenaline. Arguments will be given showing that sympathetic terminals may serve in bradykinin-induced plasma extravasation and as a mediator element in the sensitization of nociceptors by the inflammatory mediator bradykinin (and possibly by other inflammatory mediators). Thus, terminals of sympathetic noradrenergic neurons may have functions that are entirely different from their conventional function of transmitting centrally generated impulses to peripheral target tissues.

7.5.1 Involvement of the sympathetic postganglionic fibers in neurogenic inflammation

The synovium of joints continuously secretes fluid into the joint cavity. This secretion occurs at the venular site of the synovial vascular bed; it is essential for adequate functioning of joints and is finely adjusted to the joint activity. Decrease of secretion of synovial fluid leads to damage of joints and finally to arthrosis; increase of synovial secretion occurs during synovial inflammation and also leads to a restriction of joint function. Here I describe experiments demonstrating that synovial secretion is dependent on the sympathetic innervation of the synovia and that the role of the sympathetic postganglionic fibers is, first, to control blood flow (as expected) and, second, to serve as a mediator element in the control of venular plasma extravasation.

Bradykinin[7] is a potent inflammatory mediator and generates plasma extravasation by increasing the permeability of the postcapillary venules. Perfusion of the knee joint cavity of the rat hindlimb with bradykinin solution at a concentration of 160 ng/ml increases the synovial plasma extravasation about fivefold (open circles in Figure 7.10c):

- Resting and bradykinin-evoked plasma extravasation are not dependent on the innervation of the joint capsule by unmyelinated afferent fibers (Coderre et al. 1989) but are to some extent dependent on the innervation of the synovia by sympathetic postganglionic nerve fibers. Both are significantly reduced 7 to 14 days after surgical sympathectomy (Figures 7.10c and 7.11; Miao et al. 1996b).
- Cutting the preganglionic axons to chronically decentralize the lumbar sympathetic ganglia that contain the postganglionic neurons to the rat hindlimb does not significantly change the bradykinin-induced plasma extravasation in the knee joint capsule (Figure 7.10). Similarly, acute interruption of the lumbar sympathetic chains during ongoing bradykinin-induced plasma extravasation does not reduce this plasma extravasation (Miao et al. 1996b). Both interventions leave the postganglionic axons intact.
- However, as might be expected, activation of the sympathetic postganglionic neurons by electrical stimulation of the

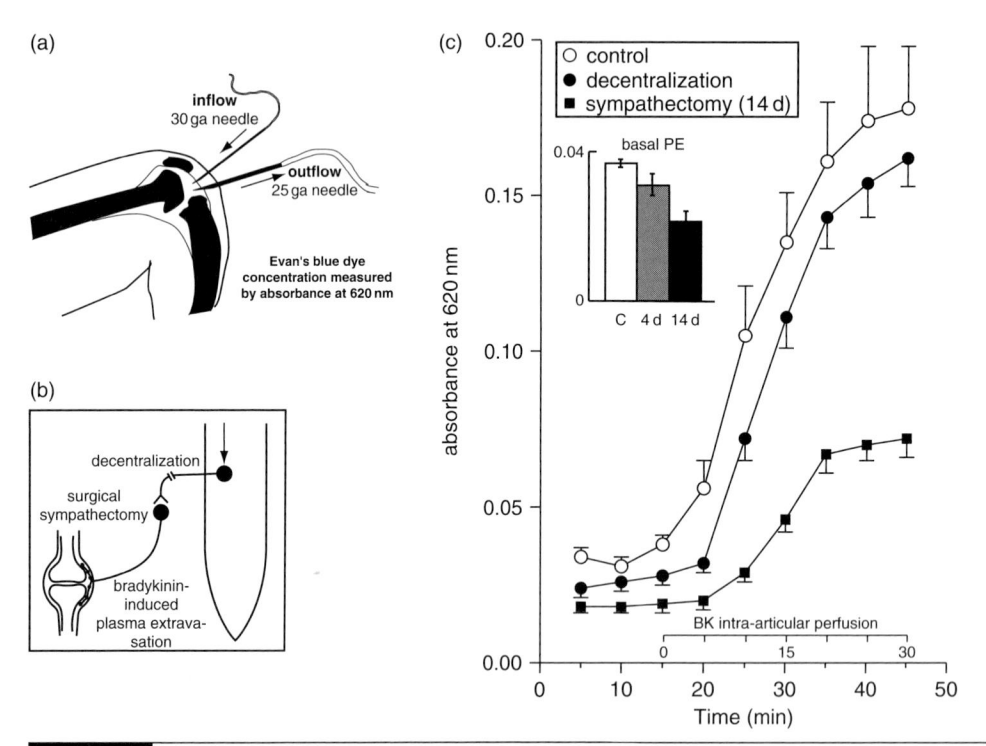

Figure 7.10 Bradykinin-induced plasma extravasation in the synovia of the knee joint is largely dependent on the sympathetic innervation but not on activity in the sympathetic neurons. (a) The rat knee joint was perfused with saline at a constant rate of 250 μl/min in the anesthetized rat. Rats were pretreated with Evan's blue dye given i.v. (Evan's blue binds to albumin and does not normally leave the vascular space). Plasma extravasation of the synovia into the knee joint cavity was determined by measuring Evan's blue dye extravasation into the perfusate in 5 min samples. The concentration of Evan's blue, determined spectrophotometrically at a wave length of 620 nm, is proportional to the degree of extravasation (ordinate scale in [c]). Bradykinin (BK, 160 ng/ml, 1.5×10^{-7} M), an inflammatory mediator, was added to the perfusate 15 min after beginning of the perfusion and the perfusion lasted for another 30 min. (b) Three preparations were used: intact lumbar sympathetic system; decentralized lumbar sympathetic system (preganglionic axons interrupted 7 days before the experiment by sectioning the white rami); surgical sympathectomy 14 days before the experiment (removal of the paravertebral ganglia [see Baron et al. [1988]]). (c) Increase of plasma extravasation induced by BK in control rats (open circles, n = 12 knees). This increase was significantly smaller 14 days after sympathectomy (closed squares, n = 12 knees) but was not significantly different from control after decentralization (closed circles, n = 12 knees). Inset: baseline plasma extravasation (PE) over 15 min preceding infusion of BK was also lower in animals 4 days and 14 days after sympathectomy than in control animals. (c) Modified from Miao et al. (1996b) with permission; (a) from Jänig et al. (2006).

sympathetic chain reduces both resting and bradykinin-evoked plasma extravasation because of a reduction of blood flow through the synovia.

- Quantitative analysis of the synovial plasma extravasation generated by different concentrations of bradykinin in the perfusate shows that the sympathetically mediated component is particularly large at bradykinin concentrations that have been measured in inflamed tissues (between 10^{-8} and 10^{-7} M; Hargreaves et al. 1993; Swift et al. 1993) and almost undetectable at higher (pharmacological) concentrations, probably because bradykinin also acts directly on the endothelial cells and because this direct endothelial

7.5 UNCONVENTIONAL FUNCTIONS OF SYMPATHETIC NORADRENERGIC NEURONS | 279

Figure 7.11 Concentration-dependent bradykinin-induced plasma extravasation in control rats and in sympathectomized rats. For experimental procedure, see legend to Figure 7.10. Bradykinin was added cumulatively from 10^{-8} to 10^{-5} M to the perfusate. The bradykinin concentrations that have been measured in inflamed tissue (Hargreaves et al. 1993; Swift et al. 1993) are indicated in bold on the abscissa scale. In sympathectomized rats, the extravasation is significantly reduced at 3×10^{-8} to 3×10^{-7} M bradykinin. At higher (pharmacological) concentrations sympathectomy has no effect because the effect of bradykinin is generated via other than the sympathetic-mediated pathways. Modified from Miao et al. (1996a) with permission.

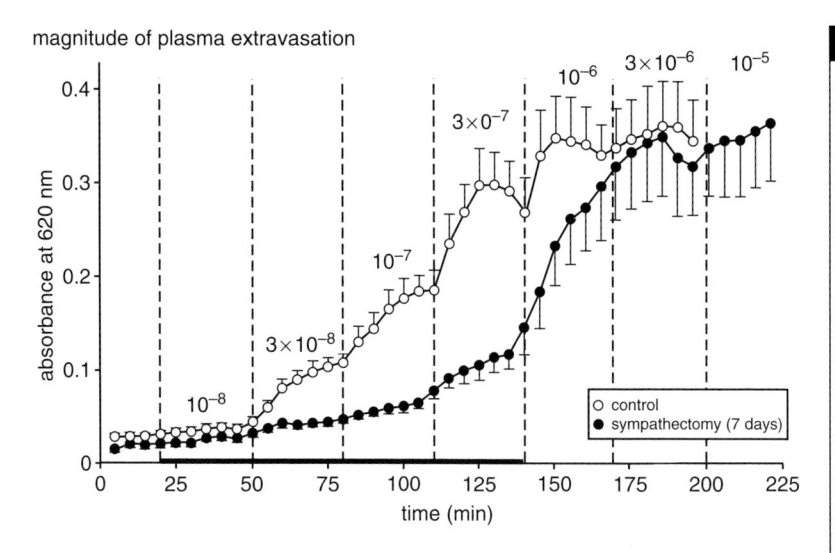

effect is maximal at pharmacological bradykinin concentrations (Figure 7.11; Miao et al. 1996a).

These experiments suggest that bradykinin-induced plasma extravasation is dependent on the presence of the terminals of the sympathetic postganglionic fibers in the synovia but not on action potentials in these fibers.

These types of experiments suggest that sympathetic postganglionic neurons innervating the joint capsule and its synovium have two functions (Figure 7.12), to regulate blood flow (vasoconstrictor function) and to mediate vascular permeability. The latter function is a newly described component of neurogenic inflammation. The first occurs at the precapillary resistance vessels by vesicular release of transmitter(s), which induces vasoconstriction and which is regulated by action potentials in the sympathetic vasoconstrictor neurons. The second function occurs at the postcapillary venules by non-vesicular release of a chemical substance(s) (possibly prostaglandin E_2 and/or related substances) (Sherbourne et al. 1992), which is independent of the electrical activity in the sympathetic neurons. Whether this substance is released by the sympathetic terminals or by other cells in association with these terminals is unknown. This sympathetically mediated component of neurogenic inflammation is an entirely peripheral function of the sympathetic terminals. Whether both functions of the sympathetic postganglionic neurons are represented in the same class of neuron or in distinct ones remains to be studied.

The results obtained from experiments investigating bradykinin-induced plasma extravasation in the rat knee joint raises questions as to whether:

- the noradrenergic sympathetic neurons are involved in the maintenance of inflammatory processes that normally occur in various

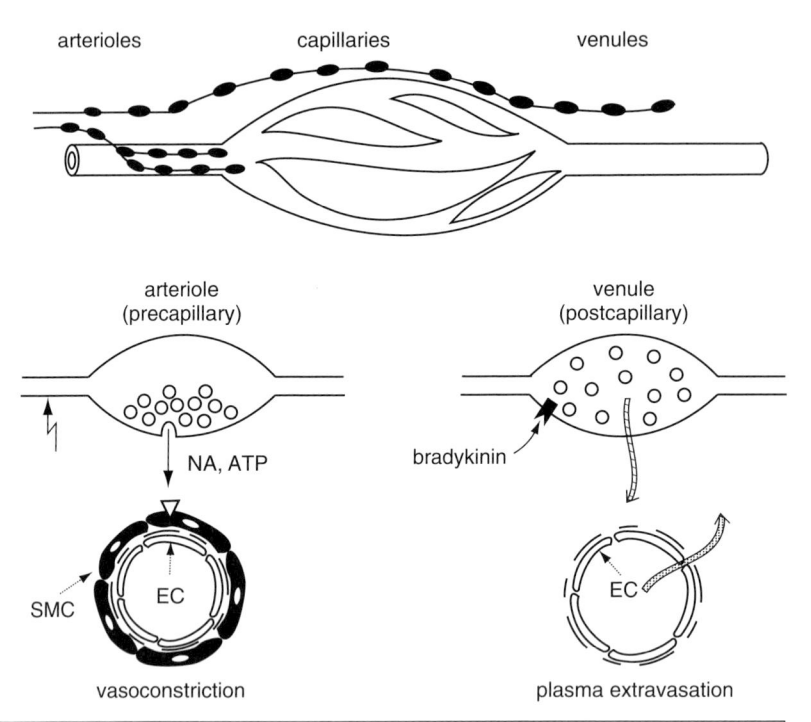

Figure 7.12 Schematic diagram illustrating that sympathetic fibers innervating the synovia have two functions. *Left side*: varicosities of vasoconstrictor axons form close synaptic contacts with the smooth muscle cells of the precapillary resistance vessels (arterioles). Impulse activity in these postganglionic neurons leads to vesicular release of transmitter(s) and to constriction of the resistance vessels. *Right side*: reaction of bradykinin with bradykinin receptors in the varicosities leads to activation of the cyclooxygenase pathway of arachidonic acid metabolism and to synthesis and release of a prostaglandin (probably PGE). This release occurs by *de novo* synthesis and not by vesicular release, and leads to opening of the endothelial gaps of venules and subsequent increase in plasma extravasation. This process is independent of activity in the sympathetic neurons. It is unknown whether synthesis and release of prostaglandin occurs from the varicosities or from other cells in association with the varicosities. It is furthermore unknown whether vasoconstriction and plasma extravasation are generated by the *same* group of sympathetic postganglionic neurons or by *different* groups of sympathetic postganglionic neurons. ATP, adenosine triphosphate; EC, endothelial cell; NA, noradrenaline; SMC, smooth muscle cells. Modified from Miao et al. (1996b) with permission.

compartments of the body (e.g., the skin, the viscera, the deep somatic tissues);
- inflammatory mediators other than bradykinin generate increases in permeability of postcapillary venules via the sympathetic noradrenergic varicose terminals (Pierce et al. 1995); and
- an interaction between sympathetically mediated neurogenic inflammation and afferent-mediated neurogenic inflammation occurs and what the nature of this interaction is (see note 2, Subchapter 2.2).

Bradykinin-induced plasma extravasation in the synovia can be modulated by the hypothalamo–pituitary–adrenal axis and by the sympatho-adrenal axis. Activation of both neuroendocrine axes (e.g., by noxious cutaneous or visceral stimulation) depresses this experimental inflammation. This depression is mediated by corticosterone and adrenaline, which are released by the adrenal cortex and the adrenal medulla, respectively. The effects of activation of the hypothalamo–pituitary–adrenal axis and of intravenously administered corticosterone on the bradykinin-induced plasma extravasation in the synovia are dependent on the presence of the sympathetic terminals in the synovia. This shows that the sympathetic fibers are in a strategic, yet unexpected, position in the control of synovial plasma extravasation by the brain (Green *et al.* 1995, 1997). The depression of bradykinin-induced plasma extravasation in the synovia by adrenaline released by the adrenal medulla is not mediated by vasoconstriction in the synovia; its mechanism is unknown (Jänig *et al.* 2000; Miao *et al.* 2000, 2001).

7.5.2 Involvement of the sympathetic postganglionic fibers in sensitization of nociceptors and mechanical hyperalgesia

Sensitization of nociceptors during peripheral inflammation leading to hyperalgesia (increased sensitivity to painful stimuli) may depend on the sympathetic innervation of the inflamed tissue.[8] Two animal models of hyperalgesic behavior in which the sympathetic postganglionic nerve fibers are supposed to be involved have been proposed.

Cutaneous mechanical hyperalgesia elicited by the inflammatory mediator bradykinin

The paw-withdrawal threshold to mechanical stimulation is a behavioral measure of pain elicited by this stimulus. This, as studied in rats, is dose-dependently decreased by intracutaneous injection of bradykinin at the site of stimulation. Following a single intracutaneous injection into the dorsal skin of the hindpaw the decrease in threshold to mechanical stimulation lasts for more than one hour (Taiwo and Levine 1988). This mechanical hyperalgesic behavior is mediated by the B_2 bradykinin-receptor (Khasar *et al.* 1995, 1998a) and does not develop when bradykinin is injected subcutaneously, i.e. away from the cutaneous nociceptors (Khasar *et al.* 1993). The decrease in paw-withdrawal threshold to mechanical stimulation generated by bradykinin is prostaglandin (prostaglandin E_2), which is an arachidonic acid metabolite and sensitizes cutaneous nociceptors. Blockade of arachidonic acid metabolism by indomethacin (injected intraperitoneally 30 minutes before the measurements and injected together with bradykinin intracutaneously), prevents the release of prostaglandin and therefore also the mechanical hyperalgesia.

The cutaneous mechanical hyperalgesic behavior can no longer be generated by intracutaneous injection of bradykinin after *surgical sympathectomy* of the rat hindlimb (removal of the sympathetic paravertebral ganglia; performed ≥ 8 days before the behavioral measurements). However, *decentralization* of the lumbar sympathetic trunk (cutting the preganglionic axons and leaving the postganglionic neurons in the paravertebral ganglia intact) does not change the mechanical hyperalgesic effect of bradykinin. It is hypothesized that the sensitization of the cutaneous nociceptors for mechanical stimulation by a prostaglandin is dependent on the sympathetic terminals but not on activity in the sympathetic neurons. The release of prostaglandin, either from the sympathetic terminals or from cells in association with the sympathetic innervation of the skin, is triggered by bradykinin via B_2-receptors presumably in the sympathetic terminals.

This role of the sympathetic fiber as mediator in sensitization of cutaneous nociceptors independent of activity in the fiber is an interesting and new idea, which assigns to the sympathetic fiber a function that is entirely different from its usual one. However, this interpretation raises several open and critical questions (see Jänig and Häbler 2000b). The mechanisms of bradykinin-induced sensitization of cutaneous nociceptors for mechanical stimulation and of sympathetic fibers in facilitating this effect are rather unclear and have to be more closely investigated. It is necessary to perform neurophysiological experiments on cutaneous nociceptive afferents in vivo and/or in vitro in order to show that the nociceptors or subpopulations of them are sensitized by bradykinin for mechanical stimulation and that this effect is mediated by the sympathetic terminals.

Cutaneous hyperalgesia generated by nerve growth factor (NGF)

Systemic injection of NGF is followed by a transient thermal and mechanical hyperalgesia in rats (Lewin et al. 1993, 1994) and in humans (Petty et al. 1994). During experimental inflammation (evoked by Freund's complete adjuvant[9]) in the rat hindlimb thermal and mechanical hyperalgesia is paralleled by a significant increase of NGF in the inflamed tissue (Donnerer et al. 1992; Woolf et al. 1994). Anti-NGF antibodies prevent the increase of NGF in the inflamed tissue and the development of hyperalgesia (Lewin et al. 1994; Woolf et al. 1994). These results argue that NGF may be responsible for the development of hyperalgesia during inflammation. The mechanisms responsible are direct sensitization of nociceptors via high-affinity NGF-receptors (trkA receptors) and increased synthesis of the peptides CGRP and substance P in the afferent cell bodies induced by NGF taken up by the afferent terminals and transported to the cell body (McMahon 1996; Woolf 1996). Furthermore, the NGF-induced sensitization of nociceptors also seems to be mediated indirectly by the sympathetic terminals. Behavioral experiments on rats show that local injection of NGF (200 ng) into the plantar skin is followed by heat and mechanical hyperalgesic behavior lasting up to 24 hours (i.e., decreased paw-withdrawal latency to stimulation of polymodal nociceptors by heat, decreased paw-withdrawal threshold to stimulation of nociceptors with von Frey hairs). Thermal and mechanical hyperalgesic behavior were no longer present or significantly reduced in rats that were chemically sympathectomized (treatment of the neonatal rats with daily intraperitoneal injections of 50 mg/kg guanethidine[10] for 2 weeks; Woolf et al. 1996); thermal hyperalgesia was slightly reduced in rats that were surgically sympathectomized (Andreev et al. 1995). These data suggest that NGF released during inflammation by inflammatory cells acts on the sympathetic terminals via the high-affinity trkA receptors, which could lead to release of inflammatory mediators and subsequently to sensitization of nociceptors for mechanical and heat stimuli. The trkA receptors have been shown to exist on sympathetic terminals (Smeyne et al. 1994). These experiments indicate too, that sympathetic fibers may have a function which is entirely different from their classical one, namely to transmit impulses to effector cells. However, as interesting as these experiments may be, they *are not definitive* and raise a number of questions:

- How do cutaneous nociceptors behave after intraplantar injection of NGF and during inflammation after surgical sympathectomy? Are all nociceptors sensitized or only a subpopulation?
- What is the effect of controlled surgical sympathectomy in adult animals on inflammation-induced mechanical hyperalgesia?
- Is activity in the sympathetic neurons necessary to mediate the sensitizing effect of NGF, i.e. how do animals behave when the sympathetic outflow to the affected extremity is decentralized (e.g., interruption of preganglionic axons to the paravertebral ganglia)?
- Which inflammatory mediators are released by the sympathetic terminals or by other cells in association with the sympathetic terminals?

Conclusions

Axons of postganglionic neurons form many branches close to their effector cells with hundreds to thousands of varicosities, which contain transmitter(s) packed in vesicles. Excitation of the postganglionic neurons spreads over all branches and normally invades all varicosities.

1. Signal transmission from postganglionic neurons to most effector cells occurs through neuroeffector synapses, which are characterized by a close junction between varicosities and membranes of the effector cells, by fusion of the basal laminae and by accumulation of synaptic vesicles close to the synaptic junction.
2. Release of transmitter during excitation of the postganglionic neuron occurs in quanta (i.e. release of the entire content of a vesicle). Individual varicosities release the transmitter of one vesicle with low probability of about $p = 0.01$ when the discharge frequency is low. This probability of transmitter release may

increase up to ten times with increasing discharge rate of the postganglionic neurons.
3. The release of the content of a vesicle from a sympathetic varicosity leads to a short-lasting increase in the concentration of the transmitter in the junctional cleft, which may reach into the millimolar range, and the subsequent interaction of the transmitter with junctional receptors in the effector cells.
4. In the heart, neurally released acetylcholine reacts only with junctional muscarinic receptors that are coupled via a distinct intracellular second-messenger pathway to the cellular effectors (e.g., ionic channels). Extrajunctional cholinergic muscarinic receptors are not activated by neuronally released acetylcholine. These extrajunctional muscarinic receptors are coupled by another intercellular second-messenger pathway to the cellular effectors. Their function during neural organ regulation is unclear.
5. In the heart, noradrenaline released by the varicosities of the sympathetic cardiomotor axons reacts with junctional β_1-adrenoceptors, which are coupled via a specific but as yet unknown intracellular pathway to ionic channels modifying the pacemaker current. Activation of extrajunctional β_1-adrenoceptors increases cAMP and leads to phosphorylation of ionic channels. It is unclear whether this pathway is used during ongoing regulation of the heart by the sympathetic neurons.
6. Arterioles and small arteries are influenced by neural release of noradrenaline and ATP from the varicosities of the vasoconstrictor axons. The ATP reacts with junctional purinoceptors and opens ligand-gated cation channels, which depolarize the membrane causing activation of voltage-sensitive calcium channels and an action potential. Noradrenaline released from the postganglionic vasoconstrictor terminals reacts with extrajunctionally located α-adrenoceptors leading to slow depolarization in some blood vessels.

- Nerve-induced constriction of small arterioles (e.g., in the submucosa of the gut) is only mediated by ATP and purinoceptors.
- Nerve-induced constriction of some larger arteries (e.g., the rat tail artery) is mediated by noradrenaline and α-adrenoceptors.
- Nerve-induced constriction in some arteries is mediated by both the purinergic and adrenergic mechanism. The integration between these cellular processes during ongoing regulation of small arteries and arterioles is unclear.
- Circulating catecholamines cannot be involved in the regulation of these innervated blood vessels in the concentrations as they occur in vivo.

7. Contractility of veins and the pulmonary artery is regulated by neuronal release of noradrenaline via adrenoceptors coupled to an intracellular pathway involving IP_3.

8. Cholinergic neurons of the enteric nervous system that innervate the longitudinal musculature of the ileum activate the smooth muscle cells via junctional muscarinic receptors. This activation leads, via an intracellular second-messenger pathway, to opening of *calcium channels* and subsequently to contraction. The cellular mechanisms mediating contraction evoked by activation of junctional and extrajunctional muscarinic receptors are different.
9. The functions of neuropeptides in neuroeffector transmission have been little explored and are, in most cases, unknown.
10. The influence exerted by autonomic neurons on their effector tissues may interact with and may be modulated by other local and remote non-neural signals. The influence of these local and long-range signals varies between different effector tissues.
11. Experimental bradykinin-induced inflammation of the knee joint synovia and experimental bradykinin-induced mechanical hyperalgesic behavior in rats indicate that sympathetic postganglionic nerve fibers may have entirely unexpected functions that are independent of their function to transmit impulses to effector cells. These functions may include the mediation of venular plasma extravasation and the sensitization of nociceptors generated during inflammation.
12. The examples of neuroeffector transmission investigated so far indicate the diversity of mechanisms for neuroeffector transmission in different targets of the autonomic nervous system. They show that nerve-generated signals are transmitted to the autonomic effector cells by distinct cellular pathways linked to the neuroeffector junctions. As the neural responses of more tissues are studied, the traditional ideas about the ability of conventional transmitters to mimic effects of nerve activity are being reconsidered.
13. As exceptions have not yet been found, it is not far-fetched to assume that neuroeffector transmission is specific for all target cells innervated by postganglionic neurons. To interfere with ongoing nerve-mediated effects in vivo, it is more useful to determine which specific antagonists reduce or abolish the effects of nerve activity than to know which receptor subtypes are present in the tissue.
14. In conclusion, activity in postganglionic sympathetic and parasympathetic neurons and in enteric motoneurons, generated by integrative action of central autonomic circuits or/and circuits of the enteric nervous system, is reliably transmitted by various neuroeffector mechanisms to the target cells. These mechanisms lead to the precise regulation of autonomic effector organs.

Notes
1. See note 2 on the axon reflex concept in Chapter 2.
2. For intracellular signaling involving receptor-activated G-proteins and second-messenger systems see Howard Schulman and James Roberts'

"Intracellular signaling" in Squire et al. *Fundamental Neuroscience* (2003), pp. 259–298.

3. The passive electrical properties of functional syncytia of electrically coupled smooth muscle cells can be treated as one-, two- or three-dimensional systems depending on the type of syncytium. In small arterioles, current spreads approximately in one dimension; in thick-walled arteries or gastrointestinal smooth muscles (e.g., of the stomach or the colon), current spreads in three dimensions. This electrical geometry influences the time course of potential changes and the distance of electrotonic spread of a potential generated by current injection in the syncytia. For resting arterioles that can be treated approximately as one-dimensional cables, the time constant is in the range of 300 to 700 ms (time of the potential to decrease to 1/e [37%]; product of total membrane resistance and capacitance) and the length constant in the range of 1.5 mm (distance taken for electrotonic potentials to decrease to 1/e [37%]; square root of ratio of membrane resistance to axial resistance). The electrical behavior of the syncytia of electrically coupled cells is described in the literature (Jack et al. 1975; Tomita 1975; Hirst and Edwards 1989).

4. The argument that nerve-released adrenaline in the toad activates adrenoceptors that are neither α nor β is based on pharmacological experiments. These show that the postsynaptic effects (generation of junction potentials) on nerve stimulation (1) are not abolished by α- or β-adrenoceptor blockers, (2) are abolished after depletion of adrenaline from the postganglionic terminals by bretylium, and (3) are blocked by dihydroergotamine (which blocks the effect of applied adrenaline but not of applied ATP by acting on the adrenoceptors). The experiments show furthermore that the effect of applied adrenaline on the sinus venosus is blocked by a β_2-adrenoceptor antagonist, arguing that applied adrenaline reacts with extrajunctionally located β_2-adrenoceptors (Morris et al. 1981; Bramich et al. 1990, 1993).

5. Vascular smooth muscle cells have a high intracellular chloride concentration due to transport of chloride from extracellular to intracellular. This results in a chloride equilibrium potential that is more positive than the membrane resting potential (in the range of -30 mV). Opening of the chloride channels by elevated intracellular calcium generates an outflow of chloride along its electrochemical gradient producing a depolarization (i.e. excitatory junction potentials; van Helden 1988a, b). This mechanism applies to arterioles in the rat iris too (see above; Gould and Hill 1996).

6. The pulsatile Ca^{2+} release from the intracellular store does not only activate the contractile machinery but also potassium channels in the muscle cell membrane. This leads to transient membrane hyperpolarizations superimposed on the membrane depolarizations generated by the opening of unspecific cation channels.

7. Bradykinin is a peptide and is released in inflamed tissues. It mediates multiple functions, one of them being the generation of plasma extravasation through postcapillary venules.

8. With the exception of the parenchyma of liver and brain, all tissues are innervated by primary afferent neurons that are activated by tissue damaging stimuli or by stimuli indicating impending damage (thermal, mechanical, chemical stimuli). These afferent neurons are called nociceptive neurons. They have unmyelinated or small-diameter myelinated axons. *Sensitization of nociceptors* occurs during inflammation;

it is characterized (1) by a decrease in threshold to heat and mechanical stimulation, (2) by the development of spontaneous activity, (3) by increased responses to noxious stimuli and (4) by the recruitment of nociceptive primary afferent neurons that cannot normally be activated (or have extremely high threshold for activation) (see Subchapter 2.4). *Hyperalgesia* denotes increased pain generated by a stimulus that is normally painful and excites nociceptors. The mechanisms of hyperalgesia are peripheral (sensitization of nociceptors) and central (sensitization of central neurons, e.g., in the dorsal horn of the spinal cord). *Animal models of hyperalgesic behavior* are used to investigate the peripheral and central mechanisms of hyperalgesia. Standardized reactions of the animals to mechanical or thermal stimuli are measured and quantified (e.g., paw-withdrawal threshold to mechanical stimulation of the hindpaw; latency of paw withdrawal to a heat stimulus applied to the hindpaw; latency of tail flick to heat stimulation etc.) (Belmonte and Cervero 1996; Meyer *et al.* 2005).

9. *Freund's adjuvant* consists of a water-in-oil emulsion incorporating antigen, in the aqueous phase, into paraffin oil with the aid of an emulsifying agent. Injection of this mixture induces strong persistent antibody formation. In *Freund's complete adjuvant* killed, dried mycobacteria (e.g., *Mycobacterium butyricum* or *M. tuberculosis*) are added to this mixture. This elicits cell-mediated immunity as well as humoral antibody formation.

10. Guanethidine is a compound that is actively taken up by the noradrenergic nerve terminals and concentrated in the neurosecretory vesicles where it replaces noradrenaline. As a consequence of this replacement, the noradrenergic terminal is no longer able to release noradrenaline when excited, which results in a block of noradrenergic transmission.

Part IV

Central representation of the autonomic nervous system in spinal cord, brain stem and hypothalamus

In the preceding chapters I described that sympathetic and parasympathetic systems consist of many functionally *separate* pathways that supply the peripheral target organs. The neurons in these pathways have characteristic reflex discharge patterns, which are centrally generated. The final central output neurons are the preganglionic neurons in the spinal cord and in the brain stem. Integrative processes in general do not seem to occur *between* functionally distinct autonomic pathways but only *within* some non-vasoconstrictor pathways to viscera (see Chapter 4). The discharge patterns measured in postganglionic neurons are unlikely to be generated by qualitatively different discharge patterns in subpopulations of preganglionic neurons converging on the same postganglionic neuron, although we have no direct experimental proof for this. This does not collide with the finding that neurochemically different preganglionic neurons may converge on the same postganglionic neuron (Murphy *et al.* 1998). Thus, in the autonomic ganglia, the discharge patterns are not changed qualitatively, but may be modified quantitatively, i.e. enhanced by the converging synaptic input. At the neuroeffector junctions, temporal and spatial aspects of the neural signals transmitted to the effector cells contribute to transmission; furthermore, various non-neural signals may modulate the neuroeffector transmission pre- and postjunctionally.

This section will focus on some principles of the organization of central regulation of peripheral autonomic pathways integrated in the spinal cord, brain stem and hypothalamus. I will not describe the central regulation of specific autonomic functions that are related to the regulation of the cardiovascular system, pelvic organs, gastrointestinal tract, body core temperature, metabolism, etc. in detail (see Loewy and Spyer [1990b]; Blessing [1997]; Appenzeller [1999]; Mathias and Bannister [2002]) but only use some of these data to exemplify what is known about the organization of autonomic systems in the neuraxis.

The reference point for this description is the *functional specificity* of the peripheral autonomic pathways with respect to the target organs derived from the measurements of the reflexes in post- and preganglionic neurons (see Chapter 4). It should be kept in mind that these reflexes are mostly isolated fragments of neural regulating systems; they are as such experimental artifacts (such as the monosynaptic stretch reflex as it is used in the research on motor physiology). Some reflexes may easily be interpreted as functionally meaningful (e.g., baro- and chemoreceptor reflexes in vasoconstrictor neurons [to regulate cardiovascular functions; see Chapter 10]; locally restricted inhibitory nociceptive reflexes in cutaneous vasoconstrictor neurons [to increase blood flow through damaged tissue; see Subchapter 4.1]; thermoregulatory spinal and supraspinal reflexes in cutaneous vasoconstrictor neurons; vibration reflexes generated by stimulation of Pacinian corpuscles in sudomotor neurons [to keep the stratum corneum flexible during sensory manipulation; see Subchapter 4.2]; excitatory and inhibitory sacro-lumbar reflexes in sympathetic motility-regulating neurons innervating pelvic organs [to regulate continence; see Subchapter 4.3]), whereas others may not. The application of the reflex concept provides an insight into the neuronal structure underlying regulation of autonomic target organs. The advantage of this approach is obvious as it was in the analysis of the somatomotor system (Granit 1981):

1. the types of efferent neurons being controlled are known;
2. the afferent neurons stimulated are known;
3. the experimental conditions can be defined;
4. the reflexes can be studied in various types of preparation (e.g., in vivo brain-intact, decerebrate or spinal animals; in vitro whole-heart-brain-stem preparation with attached heart in which the afferent and efferent pathways are intact; Paton 1996a, b);
5. the mechanisms of synaptic transmission in these well-defined reflex pathways can be studied;
6. once the reflexes, their pathways (including interneurons) and the underlying mechanisms are well defined, how these pathways function and interact during ongoing regulation of the autonomic target organs can be elucidated.

The central representations of the autonomic systems are organized caudorostrally in spinal cord, brain stem, hypothalamus and telencephalon. This organization exhibits some hierarchy in the degree of functional complexity, simple functions being represented at the level of the spinal cord and most complex autonomic functions at the forebrain level (hypothalamus and telencephalon). The description of the central representations of autonomic functions will follow this general idea:

1. The first level of integration occurs in the spinal cord for the sympathetic systems and the sacral parasympathetic systems and is represented in the various autonomic reflex pathways and their mutual synaptic connections via autonomic interneurons (see Subchapter 8.3). Here I will use the term *spinal autonomic systems* (Nilsson 1983). An analogous situation exists for the parasympathetic systems in the brain stem (see Subchapter 10.7).
2. Integration is more complex in the lower brain stem (medulla oblongata and pons). It mainly involves (1) regulation of arterial blood pressure to maintain perfusion of the body's tissues with blood, (2) regulation of the transport of oxygen and carbon dioxide (regulation of respiration) and (3) regulation of gastrointestinal function (ingestion, digestion and absorption of nutrients and fluid). These three global *homeostatic regulations* are closely integrated. Included in them are the spinal autonomic systems and equivalent parasympathetic systems in the brain stem. Additionally, the lower brain stem represents regulation of evacuative functions (urinary bladder, hindgut) and reproductive organs. The underlying supraspinal mechanisms of these regulations are largely unknown and will not be further discussed (see Jänig [1996b, c]; de Groat [1999]; McKenna [2000, 2001]).
3. Integration in the hypothalamus and mesencephalon is related to thermoregulation, regulation of energy balance and metabolism, regulation of body protection, etc. These regulations include autonomic, somatomotor and neuroendocrine systems. Integrated in these regulations are the homeostatic regulations represented in the lower brain stem and the spinal autonomic systems.
4. The most rostral level of autonomic integration occurs in the telencephalon. The homeostatic autonomic regulations are adapted to the needs of the organism according to environmental challenges (e.g., when the organism is physically threatened, is potentially in a state of starvation, etc.).[1]

This concept of hierarchical organization of central autonomic systems is not absolute and must not be taken too literally. It is not new as can be seen from previous discussions of control of cardiovascular system by the brain (see Rushmer and Smith [1959]; Bard [1960]). However, it helps to understand the structural and physiological basis of the regulation of peripheral autonomic pathways by the brain. Here I would like to put a word of caution: in the last 10 to 20 years we have made considerable progress in the research on the central autonomic nervous systems; however, we are far from understanding how the multiple homeostatic regulations are generated, integrated with each other and integrated with the behavior of the organism.

In Chapters 9 to 11, spinal cord, brain stem and hypothalamus are separated. This separation is helpful because it reflects to a certain degree a functional hierarchy in the organization of the central autonomic regulations mentioned above. Chapter 11 describes some overall aspects of homeostatic regulation of autonomic functions that are represented in the upper brain stem and hypothalamus and associated structures without going into details in describing autonomic hypothalamic and telencephalic functions.

This section will be introduced by Chapter 8 on the anatomy of the central autonomic systems, which concentrates mainly on the preganglionic neurons, the sympathetic and parasympathetic premotor neurons and the neurons that are antecedent to the autonomic premotor neurons. Further aspects of the anatomy of central autonomic systems are described in Chapter 10, in relation to the description of the physiology of central autonomic systems in the lower brain stem, and by Saper (1995, 2002).

Notes

1. The adaptation of homeostatic regulations represented in brain stem and hypothalamus in response to internal and environmental challenges (during exercise, temperature load, lack of fluid, lack of nutrients, physical threat, mental stress) is sometimes called *allostasis* (Sterling and Eyer 1988; McEwen 2001b). This concept will be discussed in Chapter 11.

Chapter 8

Anatomy of central autonomic systems

8.1 Tools to investigate the anatomy of the central
   autonomic systems   page 293
8.2 Morphology and location of preganglionic neurons   297
8.3 Nucleus tractus solitarii   311
8.4 Sympathetic and parasympathetic premotor neurons
   in brain stem and hypothalamus   317

8.1 Tools to investigate the anatomy of the central autonomic systems

The functional specificity of the autonomic regulation of target organs and the neurophysiological recordings from peripheral autonomic neurons (see Chapter 4) argue that central autonomic systems must be differentiated. It seems likely then that the central organization is reflected in the micro- and macroanatomy of the central autonomic systems. Originally, central stimulation and lesion studies, involving recordings from autonomic nerves or autonomic effector responses (e.g., blood pressure, heart rate, gastrointestinal motility), gave clues about the location of the autonomic centers in the neuraxis. However, the results from these studies were imprecise so that it was not possible to identify anatomically distinct populations of central neurons as being associated with distinct autonomic output systems. Furthermore, focal electrolytical lesioning and focal electrical stimulation did not discriminate between destruction or excitation of cell bodies and passing axons. With the introduction of chemical stimulation of cell bodies (by microionophoretic application of excitatory amino acids, such as glutamate, aspartate or DL-homocystic acid, or inhibitory amino acids, such as γ-aminobutyric acid [GABA] or glycine) it was possible to topically excite or inhibit small populations of central neurons selectively.

This technique turned out to be a valuable tool in physiological experimentation.

The breakthrough in unraveling aspects of the microanatomy of the central organization of autonomic systems came (1) with the introduction of axon tracer methods in the 1970s and (2) with the introduction of retrograde transneuronal labeling of neuron populations using neurotropic viruses at the end of the 1980s by Arthur Loewy's group (Strack et al. 1989a, b):

- The first method allows the localization of the cell bodies of autonomic pre- and postganglionic neurons as well as of central autonomic neurons (e.g., sympathetic and parasympathetic premotor neurons) with substances that were applied to their cut axons or axon terminals, taken up by the axons and transported retrogradely by axoplasmic flow to their cell bodies (e.g., horseradish peroxidase [HRP], wheat-germ agglutinin horseradish peroxidase, cholera toxin subunit B [CTb], C-fragment of tetanus toxin, Fluoro-Gold [FG], Fast Blue [FB]; Figure 8.1, left side). Other markers applied close to the cell bodies are taken up and transported orthodromically to the axon terminals (e.g., the lectin *Phaseolus vulgaris* leuco-agglutinin [PHA-L]; Gerfen and Sawchenko 1984) or tracers are transported bidirectionally (e.g., biotinylated dextran amines). The markers can be histochemically visualized, allowing quantitative analyses of location, number and structure of the labeled cells or terminals.
- The second method principally allows the localization of whole networks of neurons that are connected with a specific autonomic output system (Loewy 1998). A suspension of a live neurotropic virus is injected into a peripheral tissue (e.g., heart, a gland, urinary bladder wall etc.) or autonomic ganglion (e.g., stellate ganglion, celiac ganglion, pterygopalatine ganglion etc.). The virus is taken up by nerve terminals and transported to the cell bodies. Within the cell bodies, the virus undergoes replication in the nucleus. The virions exit the nucleus and acquire an envelop derived from the nuclear membrane. The newly formed viruses are transported intracellularly by the endoplasmic reticulum and Golgi apparatus. In order to trace functionally related neurons of a neuronal network it is critical to use viruses that are transported by *transsynaptic mechanisms* and not by unspecific release through the cell membranes of the neurons, which would infect neighboring neurons via the extracellular space, glia cells and other cells in an unspecific way (Figure 8.1, right side). This has been achieved using weak suspensions of herpes viruses and the Bartha strain of the pseudorabies virus (PRV), which turned out to be particularly useful as a highly specific retrograde transneuronal synaptic marker in the autonomic nervous system (Card et al. 1990; Strack and Loewy 1990; Jansen et al. 1993; Enquist and Card 2003; references see Tables 8.2 and 8.3). With this strain of PRV (or genetically engineered substrains) and careful monitoring, transneuronal synaptic

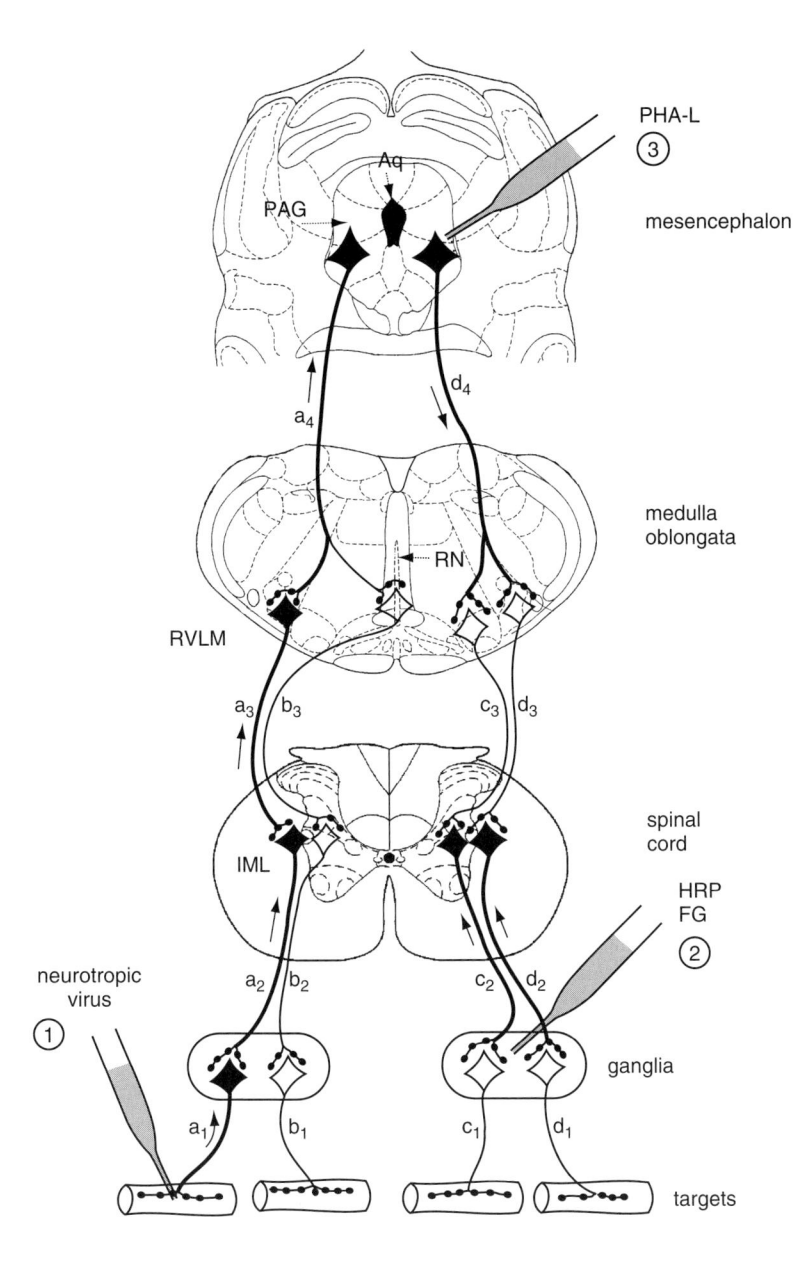

Figure 8.1 Schematic drawings to illustrate the design of experiments in which autonomic neurons are labeled by a neurotropic virus (left side) or by markers not transported transsynaptically (right side). (1) A neurotropic virus (e.g., Bartha strain of the pseudorabies virus) is injected in the target tissue. The virus is transported retrogradely and replicates in the synaptically connected neurons a_1 to a_4 (black neurons) but not in the neurons b_2 and b_3. (2) A marker, that is taken up by the axon terminals and transported to the cell bodies but not transported transsynaptically, is injected into an autonomic ganglion or into the target tissue (e.g., horseradish peroxidase, HRP, or Fluoro-Gold, FG). This marker is only found in the cell bodies of neurons c_2 and d_2 but not in the cell bodies of the other neurons. (3) A marker that is taken up by the cell bodies and transported orthogradely to the synaptic terminals is applied close to the cell bodies of the neurons d_4 (e.g., *Phaseolus vulgaris* leuco-agglutinin, PHA-L). This marker is found in the synaptic terminals on neurons c_3 and d_3 but not in the other synaptic terminals. These marker techniques can be combined. Aq, cerebral aqueduct; IML, intermediolateral cell column; PAG, periaqueductal grey matter; RN, raphe nuclei; RVLM, rostral ventrolateral medulla.

transport of the virus and labeling of first-, second-, third- or fourth-order neurons depends on time after virus application. Taking these aspects and some other limiting factors into account only neurons that are synaptically connected to the cell bodies of the infected neurons are labeled (Figure 8.1, left side).

- Combining the retrograde transsynaptic labeling technique with "classical" tracing methods, and combining two different sub-strains of a neurotropic virus that can be differentiated histochemically are powerful anatomical tools to label central groups of neurons that form direct or indirect connections with the preganglionic neurons of the peripheral autonomic pathways.

Various neuronal networks in spinal cord, brain stem, hypothalamus and cerebral hemispheres involved in the regulation of the final autonomic pathways have been described.
- The axonal tracing techniques can be combined with other techniques labeling neuropeptides, enzymes etc. in the neurons or their axon terminals, using immunohistochemistry. They can be combined with neurophysiological experiments in which reflex activity is recorded from functionally identified neurons. Examples are:
  - Single neurons that are functionally characterized with neurophysiological techniques can be labeled by tracers injected intracellularly or applied juxtacellularly (Pinault 1996; Pilowsky and Makeham 2001; see note 3 in Chapter 10; Figures 8.3, 8.9, 10.10).
  - Experiments can be performed on animals in which distinct populations of neurons have been eliminated by a toxin that is conjugated with an agonist for a specific membrane receptor or with a specific enzyme. The toxin–agonist–receptor or toxin–enzyme complex is taken up and internalized by the neurons (e.g., by the cell bodies or axon terminals of catecholaminergic neurons or serotonergic neurons projecting to the spinal cord; by neurons expressing specific receptors, e.g. the neurokinin 1 receptor in the respiratory rhythm generator in the medulla oblongata). The internalized toxin leads then to a selective suicide killing of the neurons (see note 9 in Chapter 10).
  - Expression of *c-fos* messenger RNA (mRNA) and its protein in neurons during their activation can be used as a marker for activation of populations of functionally related neurons during physiological stimuli (e.g., stimulation of arterial baro- or chemoreceptors) (see note 1 in Chapter 10 and Schulman and Hyman in Squire *et al.* [2003]).

Knowledge about the functional anatomy of central autonomic systems that was obtained with older tracing and degeneration techniques has been confirmed and considerably extended. In this way the basis for future physiological studies on the central organization of the different autonomic systems has been laid down.

In this chapter I will summarize the data about the anatomy of central autonomic systems, including the location of preganglionic neurons in spinal cord and brain stem and the nucleus tractus solitarii (NTS). The point of reference of all experimental investigations about central autonomic neurons and their functions are the different groups of preganglionic neurons. Details about the anatomy of central autonomic systems (mostly in rats; some in pigeons, pig and cat) are described and discussed by Cabot (1990), Cechetto and Saper (1990), Loewy (1990a), Loewy and Spyer (1990b), Saper (1995) and Blessing (1997). In future the mouse will be used in this anatomical experimental work. This shift will enable the use of genetically engineered animals. Tables 8.2 and 8.3 list the groups of neurons in the different nuclei of spinal cord, brain, hypothalamus and cerebral hemispheres that have been transneuronally labeled from various target tissues using a neurotropic virus (in most cases the Bartha strain of the PRV).

8.2 | Morphology and location of preganglionic neurons

8.2.1 Sympathetic preganglionic neurons and interneurons

Preganglionic neurons

Sympathetic preganglionic neurons lie in the intermediate zone of the thoracolumbar spinal cord, extending from the white matter to the central canal (Figure 1.2). This zone is identical to the dorsal part of lamina VII of the spinal grey matter according to Rexed (1952, 1954). The intermediate zone has been subdivided into four sub-nuclei (Petras and Cummings 1972; Oldfield and McLachlan 1981): the funicular part of the intermediolateral nucleus (ILf) for neurons lying in the white matter, the principal part of the intermediolateral nucleus (ILp) for densely packed neurons lying just medial to the border between white and grey matter, the intercalate spinal nucleus (IC) for neurons lying medial to the ILp across the spinal cord in the intermediate zone and the central autonomic nucleus (CA) for neurons lying lateral and dorsal to the central canal, which is located in lamina X of Rexed (Figure 8.2). In rodents the CA is large in the lumbar segments and mainly located dorsal to the central canal and therefore has been called dorsal commissural nucleus. This reflects a similarly named nucleus in the sacral segments that does not contain preganglionic neurons.

Sympathetic preganglionic neurons have cell bodies that are larger than most dorsal horn neurons but significantly smaller than those of motoneurons. They have various groups of dendrites extending mainly rostrocaudally in the lateral intermediate zone and some medially across the intermediate zone or laterally into the white matter. The dendritic field of the preganglionic neurons is almost strictly confined to the intermediate zone. A few dendrites may project ventrally or dorsally along the border between white and grey matter (Figure 8.3). The distances over which the dendrites project along the rostrocaudal axis are considerable, reaching 1.5 to 2.5 mm in the cat (Dembowsky et al. 1985). The surface of the cell bodies is much smaller than the surface of the dendrites, thus most synaptic inputs occur at the dendrites (Cabot et al. 1994; Cabot 1996). The preganglionic neurons project their axons around the lateral edge of the ventral horn and into the ventral root. The majority of preganglionic axons do not have recurrent axon collaterals in the spinal cord (in analogy to the collaterals of the axons of motoneurons that form synapses with Renshaw cells). Some examples have been

Figure 8.2 Location and morphology of sympathetic preganglionic neurons in the spinal cord in the rat. (a) Diagram of transverse section. Subnuclei in the intermediate zone (IZ) in the cat: ILf, ILp, funicular and principal part of the intermediolateral nucleus; IC, intercalate nucleus; CA, central autonomic nucleus. (b) Photomicrographs showing sympathetic preganglionic neurons labeled with cholera toxin subunit B in transverse sections (A, B) and in a horizontal section (C) of the rat. A. Spinal level C8. Preganglionic neurons are widely distributed in the ILp and ILf. B. Spinal level T2. Labeled preganglionic neurons are located in the IML (here named IMf and IMp) and some in the CA. C. Preganglionic neurons form periodic clusters at intercluster distances of 300 to 600 μm connected by longitudinal dendritic bundles (LDB). Most dendrites are oriented rostrocaudally and some dendrites are oriented mediolaterally in bundles across the intermediate zone. A few dendrites are oriented dorsally or ventrally. AF, LF, PF, anterior, lateral and posterior funiculus; cc, central canal; DC, dorsal columns; DH, dorsal horn; TDB, transversal dendritic bundle; VH, ventral horn. From Hosoya et al. (1991) with permission.

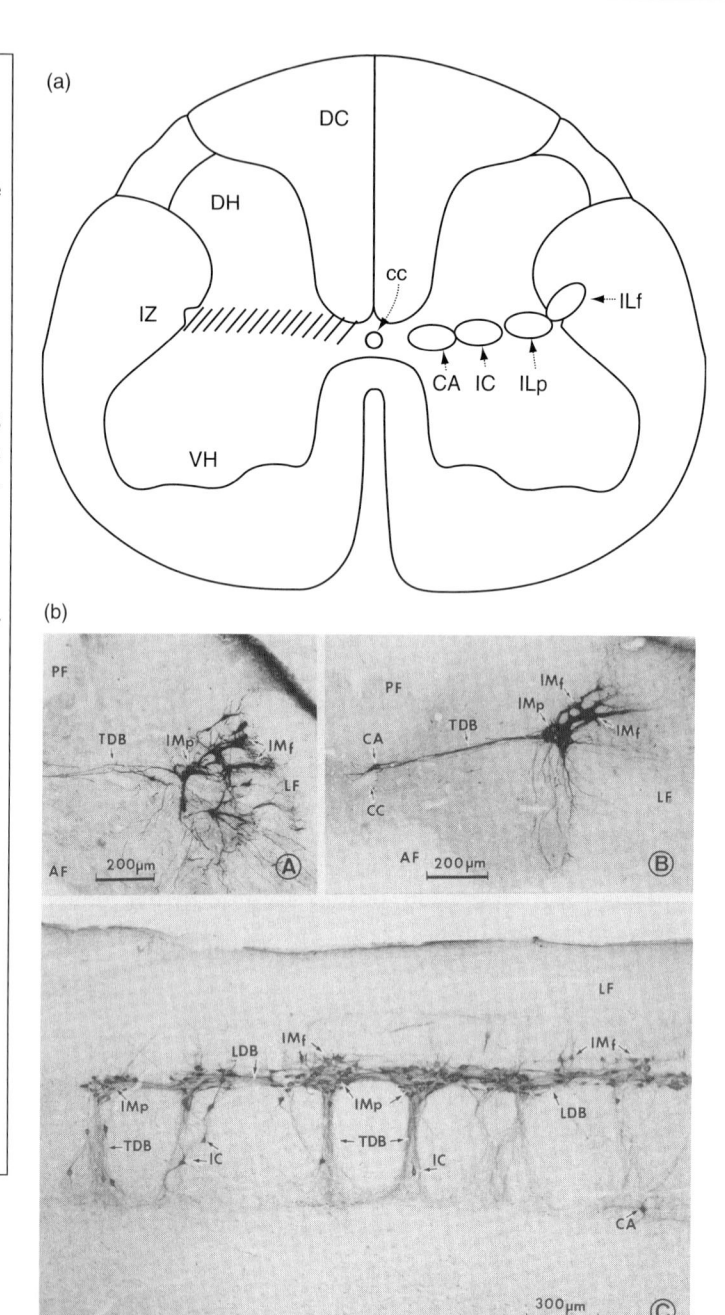

demonstrated but the function of these collaterals is unknown (see Weaver and Polosa [1997]).

The preganglionic cell bodies are arranged in clusters with intercluster distances varying from 100 to 500 μm (Figure 8.2b C). There is no indication that preganglionic neurons of individual clusters are homogeneous in function; thus, different preganglionic neurons of the same cluster may project into different nerves and to postganglionic neurons with different functions (e.g., to the adrenal medulla

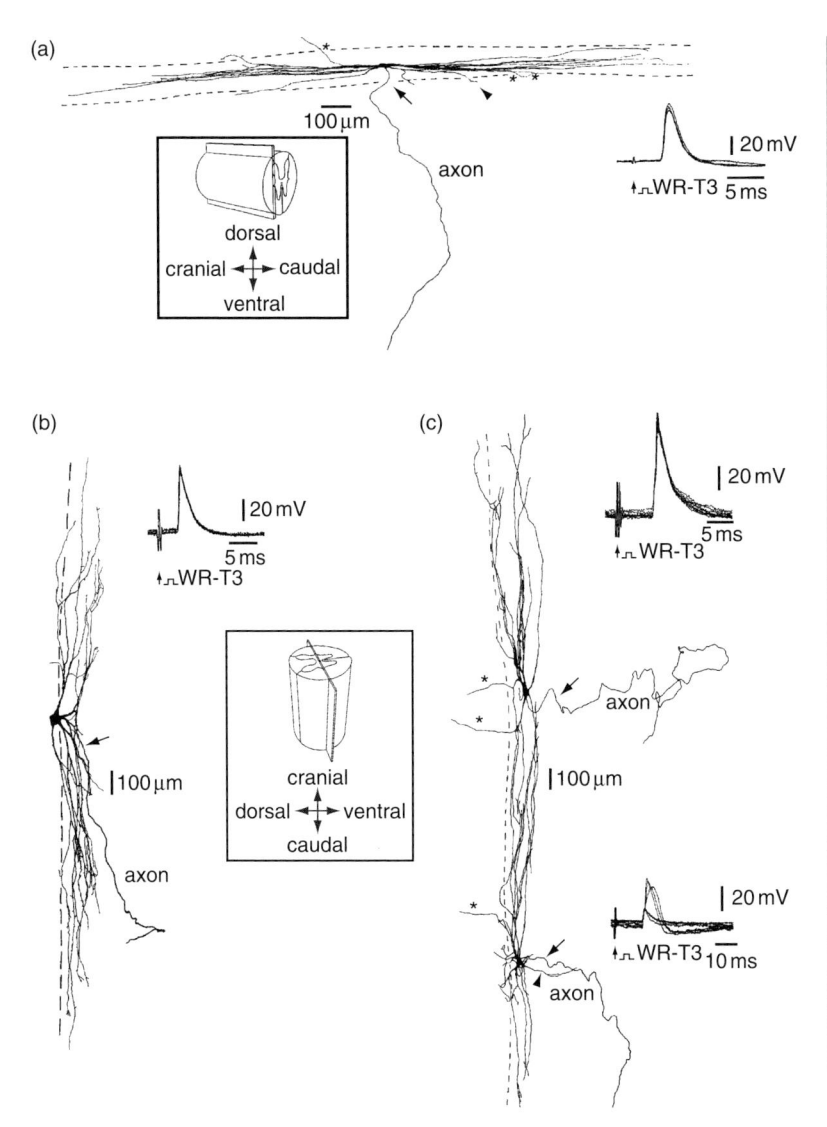

Figure 8.3 Morphology of single preganglionic neurons located in the intermediolateral cell column in the thoracic segments T3 in the cat. Activity was recorded from the preganglionic neurons using an intracellular microelectrode filled with horseradish peroxidase (HRP). The neurons were identified by stimulating their axons in the white ramus T3 (WR-T3) with single pulses (see the superimposed action potentials). After identification the neurons were filled with HRP. After filling the neurons the animals were perfused, the spinal cord was removed and cut longitudinally in serial sections (30 to 50 μm thick) either parasagittally or horizontally. The neurons were reconstructed from the serial sections. (a) Neuron in a parasagittal plane. The dashed lines mark the dorsal and the ventral border of the intermediate zone. (b, c) Neurons in a horizontal plane. In (c) two neurons labeled in the same animal. The dashed lines mark the border between grey and white matter. Asterisks, dendrites projecting laterally or dorsally; arrow head, dendrites projecting ventrally; arrows, origin of axon. Modified from Dembowsky et al. (1985) with permission.

or to the superior cervical ganglion or to the stellate ganglion; Jansen et al. 1993). Furthermore, there is no evidence that axons of individual preganglionic neurons branch and project into different nerves (e.g., both in a lumbar splanchnic nerve and in the lumbar sympathetic trunk distal to the most caudal lumbar splanchnic nerve) or up and down in the sympathetic trunk. Thus, axons of individual preganglionic neurons do not branch and diverge before reaching their target ganglia. This does not preclude that an individual preganglionic axon may form synapses with postganglionic neurons in adjacent prevertebral ganglia. Finally, sympathetic preganglionic neurons projecting to distinct sympathetic ganglia or the adrenal medulla are segmentally organized. For example, in the cat, preganglionic neurons projecting to the superior cervical ganglion are located in the segments T1 to T7, those projecting in the lumbar sympathetic trunk distal to the paravertebral ganglion L4 in the segments T12 to L5, and those projecting in the lumbar splanchnic nerves in the segments L2 to L5 (Figure 8.8b) (see Jänig [1985a], Jänig and McLachlan [1987]). Similar distinct segmental distributions of different groups of preganglionic neurons have been shown to exist in the rat (Strack *et al.* 1988). Table 8.2 (column spinal cord) lists the segmental location of sympathetic preganglionic neurons that are associated with different target tissues in the rat as has been studied using transsynaptic transport of a neurotropic virus injected into the target tissue. These data correspond to the results of experiments in which autonomic effector responses elicited by electrical stimulation of preganglionic axons in ventral roots were recorded (see Figure 8.8a).

Since Langley, preganglionic sympathetic neurons in the lumbar spinal cord have been most extensively studied, in several species, using morphological, immunohistochemical and neurophysiological methods. These neurons project in the lumbar sympathetic trunk, lumbar splanchnic nerves and hypogastric nerves; they are involved in various functions (see Tables 4.1 and 4.3) in somatic tissues, pelvic organs and colon (Figure 8.6a). Therefore, I will describe these sympathetic preganglionic neurons more extensively. The results of these experiments show (Baron *et al.* 1985a, b, c; Jänig and McLachlan 1986a, b):

- Almost all preganglionic neurons that project in the lumbar sympathetic trunk distal to the most caudal lumbar splanchnic nerve (Figure 8.6a), which are involved in regulating target cells in skin and skeletal muscle, lie in the ILf or ILp.[1] Very few neurons are located medial to the ILp (Figure 8.4). Most preganglionic neurons are situated in the segments L1 to L4 with a few in the segments T12, T13 and L5 (Figure 8.8b). Practically all preganglionic neurons are ipsilateral (Figure 8.4a).
- All preganglionic neurons that project in the lumbar splanchnic nerves lie in the intermediate zone medial to the border between white and grey matter. Many of these neurons are located in the ILp, but lie somewhat medial to those that project caudally in the lumbar sympathetic trunk and innervate postganglionic neurons projecting to somatic tissues. Many preganglionic neurons projecting in the lumbar splanchnic nerves are located medial to the ILp extending across to the central canal. These preganglionic neurons are situated in the segments L2 to L5, most of them in segments L3 and L4 (Figure 8.8b). Practically all preganglionic neurons are located ipsilaterally (Figure 8.5).
- All preganglionic neurons that project in the hypogastric nerves lie in the intermediate zone medial to the border between white and grey matter. However, most of these neurons are located in the ILp and only very few medial to the ILp. Again these preganglionic neurons are situated medial to those projecting in the lumbar

8.2 MORPHOLOGY AND LOCATION OF PREGANGLIONIC NEURONS | 301

Figure 8.4 Spatial distribution of preganglionic cell bodies in lumbar segments L1 to L4 innervating postganglionic neurons that project to somatic tissues in the cat. Horseradish peroxidase (HRP) was applied to the cut lumbar sympathetic trunk distal to the most caudal lumbar splanchnic nerve (see Figure 8.6a). The enzyme is transported retrogradely to the preganglionic cell bodies two days after application and the reaction product is demonstrated histochemically in the cell bodies. Horizontal sections of the spinal segments are cut serially and the location of the HRP-labeled preganglionic cell bodies reconstructed by superimposing their position in sections through the entire intermediate region. (a) Rostrocaudal distribution of cells. The labeled cell column stops in the caudal half of segment L4. Note that all labeled preganglionic cell bodies are located ipsilaterally. (b) The lengths of L1 to L4 marked by broken lines in (a) (a–e, each about 4 mm long) have been rotated by 90 degrees and displayed transversely. The broken lines in the transverse sections indicate the limits of the horizontal sections that were included in this analysis. Note that many cell bodies are located in the ILf and very few medial to the ILp (see Figure 8.2). For abbreviations see legend of Figure 8.2. From Jänig and McLachlan (1986a) with permission.

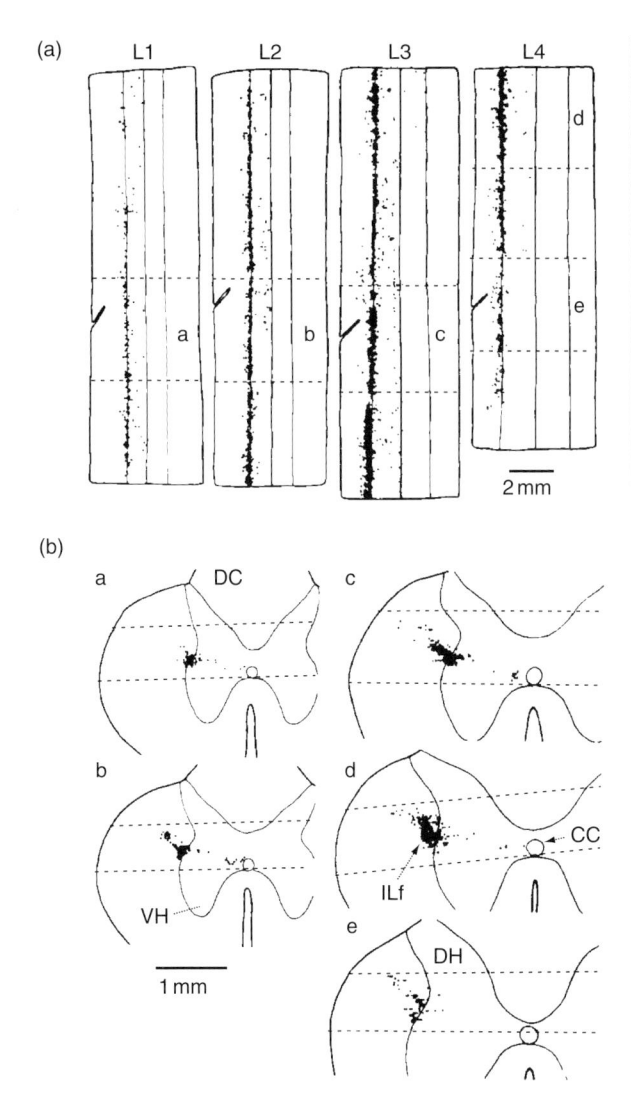

sympathetic trunk. About 20% of these preganglionic neurons are located in the contralateral intermediate zone.

- When subtracting the position of the preganglionic neurons that project through the hypogastric nerves to the pelvic organs from those projecting through the lumbar splanchnic nerves (Figure 8.5) one obtains the position of preganglionic neurons that synapse with postganglionic neurons in the inferior mesenteric ganglion, many of them being associated with target organs in the hindgut. These preganglionic neurons are also located medial to those projecting in the lumbar sympathetic trunk, many of them lying in the grey matter extending up to the central canal.
- The approximate positions of the three groups of preganglionic neurons (associated with somatic tissues, pelvic organs and colon, respectively) are schematically outlined in Figure 8.6b, c. In the cat,

Figure 8.5 Spatial distribution of preganglionic cell bodies in lumbar segments L3 to L5 innervating postganglionic neurons that project to pelvic organs and colon in the cat. Horse radish peroxidase was applied to the cut lumbar splanchnic nerves (see Figure 8.6a). (a) Superimposed horizontal sections showing the rostrocaudal distribution of cells in the intermediate zone. All labeled cell bodies are located ipsilaterally. (b) The length of L3 (1.6 mm) and L4 (1.4 mm) marked by the broken lines in (a) have been rotated by 90 degrees and displayed transversely. Broken lines indicate the limits of the horizontal sections that were included in the analysis. Note that no neurons were located in the ILf and that many neurons were located medial to the ILp. For abbreviations see legend of Fig. 8.2. From Baron et al. (1985c) with permission.

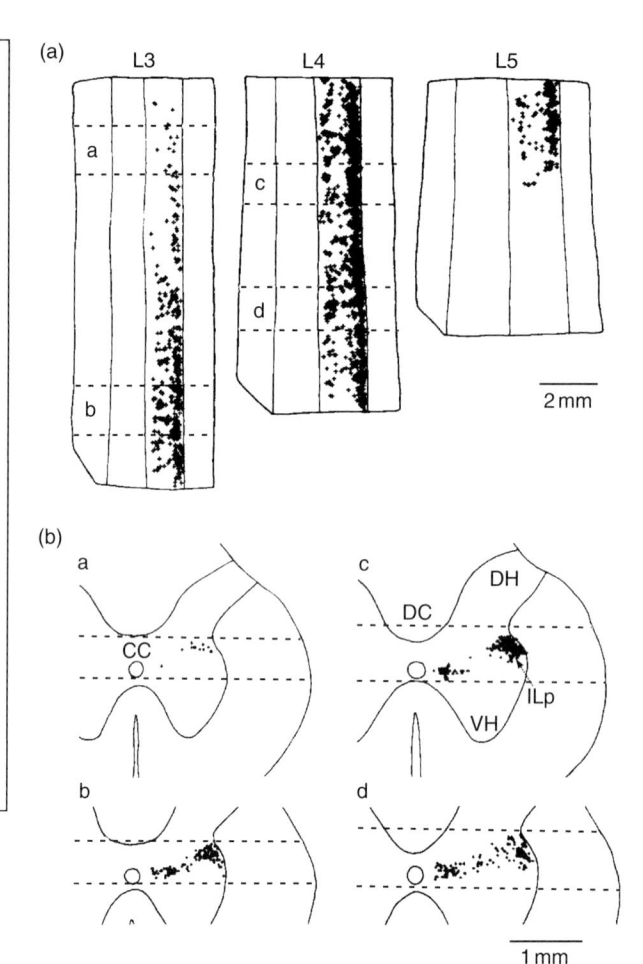

these lumbar preganglionic neurons exhibit some viscerotopic organization:

- Almost all neurons projecting in the lumbar sympathetic trunk are located in the "classical" intermediolateral cell column and most of them in the white matter (ILf). Most of these neurons are vasoconstrictor in function (1 in Figure 8.6b, c).
- Almost all neurons projecting in the hypogastric nerves are located medial to those projecting in the lumbar sympathetic trunk, extending in clusters up to the central canal (3 in Figure 8.6b, c).
- Most preganglionic neurons that synapse in the inferior mesenteric ganglion are also located in the grey matter in the ILp, but medial to those projecting in the lumbar sympathetic trunk and ventral to those projecting in the hypogastric nerves (2 in Figure 8.6b, c). Most of these neurons and those projecting in the hypogastric nerve are probably not vasoconstrictor in function but involved in regulation of visceral organs (Bahr et al. 1986c).

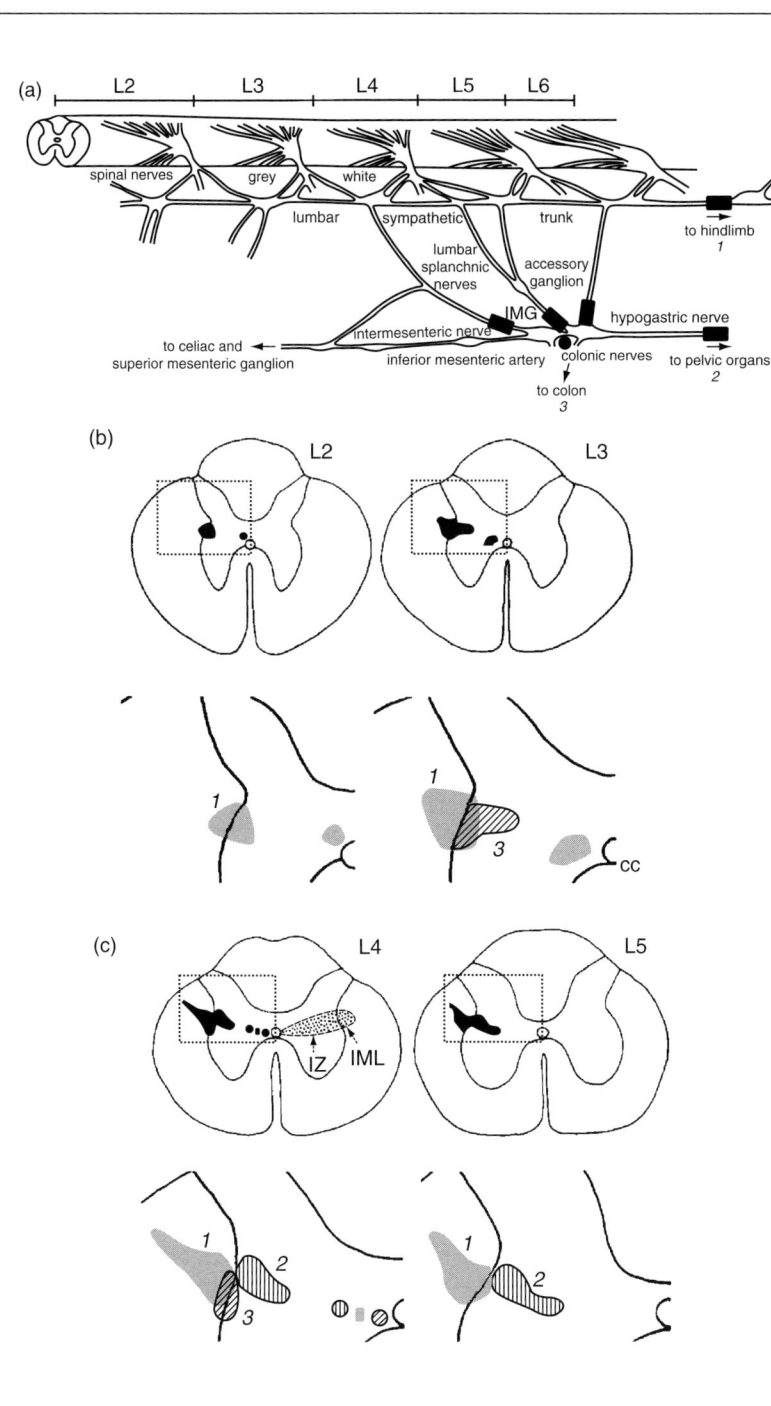

Figure 8.6 Topographic organization of sympathetic preganglionic neurons in the lumbar spinal cord projecting in the different nerve trunks of the cat. (a) Diagram of the anatomy of the lumbar nerve pathways in the cat, indicating various sites at which horseradish peroxidase was applied. (b, c) Transverse sections illustrate the regions containing the highest density of preganglionic neurons in the most rostral 2 mm of each segment L2 to L5. The lower parts in (c) and (b) show enlargements of the intermediate zone (see boxes) of cells labeled from the lumbar sympathetic trunk (grey, 1) and from the hypogastric nerve (right hatching, 3), together with those projecting in the lumbar splanchnic but not hypogastric nerves (vertical hatching, 2). The latter group has been derived from the first two and is likely to be involved with innervation of the colon. IMG, inferior mesenteric ganglion; IML, intermediolateral nucleus; IZ, intermediate zone. From Jänig and McLachlan (1986b) with permission.

Similar results on the location of lumbar sympathetic preganglionic neurons projecting to viscera or somatic target tissues have been obtained in the guinea pig (Dalsgaard and Elfvin 1982; McLachlan 1985) and in the rat (Baron and Jänig 1991; see Table 8.2 and 8.3 for references). However, in these species, the medially located preganglionic neurons that project to the viscera are situated in the dorsal commissural nucleus.

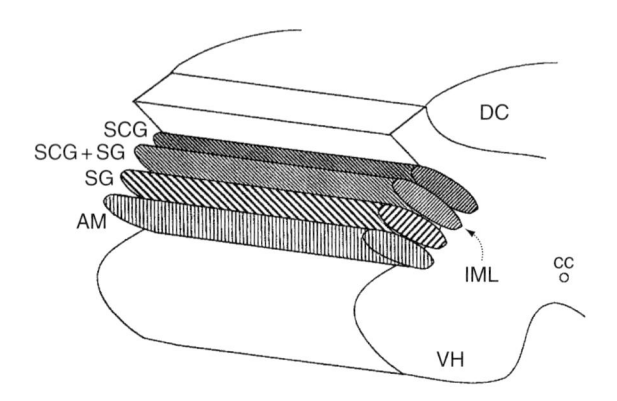

Figure 8.7 Arrangement of functionally distinct sympathetic preganglionic neurons in horizontal columns in the rat thoracic spinal cord. Diagram of the grey matter of the right side of the spinal cord indicating the location in the intermediolateral nucleus of different groups of sympathetic preganglionic neurons in the thoracic segment T5. Note the arrangement of the preganglionic neurons in horizontal rostrocaudal columns. Preganglionic neurons projecting to the superior cervical ganglion (SCG) are located dorsally and neurons projecting to the adrenal medulla (AM) ventrally. Neurons projecting to the stellate ganglion are located in between and close to the neurons projecting to the AM. SCG + SG, region containing both groups of preganglionic neuron. In more rostral thoracic segments there is little AM representation whereas at more caudal segmental levels (such as T10) there is little representation of SG and no SCG. Sympathetic preganglionic neurons to each of the three targets were simultaneously labeled with fluorescent dyes, either Fluoro-Gold, Fast Blue, or Diamidino Yellow. cc, central canal; DC, dorsal column; IML, intermediolateral nucleus; VH, ventral horn. From Pyner and Coote (1994) with permission.

This principle of organization also applies to the sympathetic outflow of the thoracic spinal cord in the rat: populations of preganglionic neurons projecting to the superior cervical ganglion, stellate ganglion, or adrenal medulla, respectively, are arranged in horizontal columns in the intermediolateral nucleus (mainly principal part; Figure 8.7) (Pyner and Coote 1994).

Figure 8.8 projects the topographical anatomical situation on the functional situation for the preganglionic neurons of the lumbar sympathetic outflow, which project in the lumbar sympathetic trunk and the lumbar splanchnic nerves of the cat. From one side of the spinal cord about 4500 preganglionic neurons project in the lumbar sympathetic trunk and 2300 neurons in the lumbar splanchnic nerves. Not included in these numbers are preganglionic neurons that innervate only postganglionic neurons in paravertebral ganglia projecting into segmental nerves (ventral and dorsal rami) that supply the skin and deep somatic tissues of the trunk since these preganglionic neurons were not labeled in the tracing experiments (see Figure 8.6a) (see Baron *et al.* 1995). These are probably numerically small.

Based on the stimulation studies performed by Langley and co-workers and some other investigators (see legend of Figure 8.8), most preganglionic neurons involved in regulation of target cells in somatic tissues, pelvic viscera and colon are situated in the segments L1 to L5. This fully corresponds to the neurophysiological studies on pre- and postganglionic neurons of the lumbar sympathetic outflow as reported in Chapter 4. Segments L3 and L4 contain the highest density of preganglionic neurons; they consist of at least ten

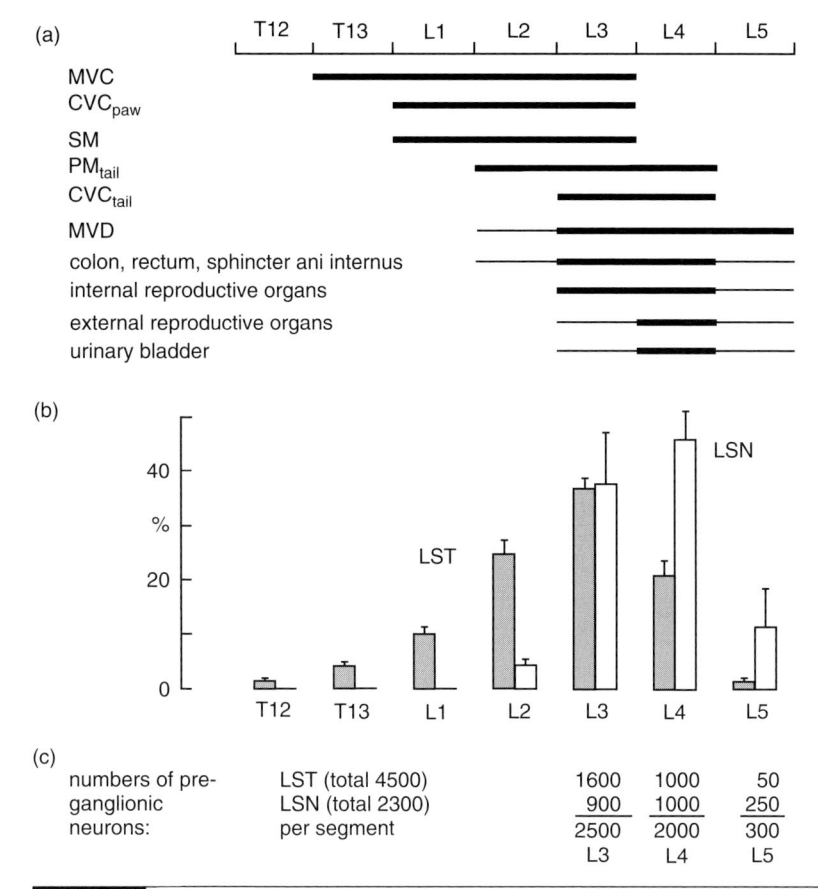

Figure 8.8 The lumbar spinal sympathetic outflow: functional systems, segmental distribution and numbers in the cat. (a) Segmental distribution of sympathetic spinal systems. Autonomic effects generated by electrical stimulation of the ventral roots in the cat. MVC, MVD: muscle vasoconstriction and vasodilation in the hindlimb (Sonnenschein and Weissman, 1978). CVC: cutaneous vasoconstriction in paw and tail skin (Langley 1894b). SM: sudomotor activity and sweat secretion on foot pads (Langley 1891, 1894b). PM: pilomotor activity and piloerection on the tail (Langley 1894a; Langley and Sherrington 1891). *Colon, rectum, sphincter ani internus*: decrease of motility, contraction, pallor (Langley and Anderson 1895a). *Internal reproductive organs*: contraction, pallor (Langley and Anderson 1895d). *External reproductive organs*: pallor, contraction, retractor penis and cutaneous smooth muscles (Langley and Anderson 1895c). *Urinary bladder*: short contraction (Langley and Anderson 1895b). (b) Segmental distribution of preganglionic neurons that project in the lumbar sympathetic trunk (LST) distal to paravertebral ganglion L5 (distal to the most caudal lumbar splanchnic nerve, shaded columns) and in the lumbar splanchnic nerves (LSN, open columns) of the cat. These experimental distributions were determined from experiments as described in Figures 8.2, 8.4 and 8.5. Mean ± S.E.M. LST, n = 9 (Jänig and McLachlan 1986a). LSN, n = 4 (Baron et al. 1985c). (c) Numbers of preganglionic neurons. These numbers were derived from the animals with highest numbers of horseradish peroxidase-labeled cells (LST, n = 6; LSN, n = 3) (Baron et al. 1985c; Jänig and McLachlan 1986a, b). Modified from Jänig (1986) with permission.

functionally different types of preganglionic neurons (see Tables 4.1 and 4.3). There is no indication that the preganglionic neurons are arranged in the different subnuclei of the intermediate zone with respect to their function although up to now it has not been possible to test this explicitly.

Spinal autonomic interneurons and propriospinal neurons
Sympathetic preganglionic neurons labeled transsynaptically are commonly associated with labeled neurons in the spinal cord that

cannot be labeled by non-viral not transsynaptically transported tracers applied to the preganglionic axons (e.g., HRP or FG). Thus, these neurons do not project with their axons through the ventral roots. It is hypothesized that these neurons are interneurons, which form synapses with the preganglionic neurons. Three major groups of putative autonomic interneurons detected after labeling of preganglionic or postganglionic neurons with a transsynaptically transported virus have been identified in this way:

1. Interneurons that are situated in the same or neighboring spinal segments as the preganglionic neurons. These interneurons are found in laminae I, II, V, VI, IX and X (close to the central canal)[2] of the grey matter as well as close to the preganglionic neurons in the intermediate zone (Table 8.2, Figure 8.15). Experimental evidence suggests that all interneurons within the region of the intermediolateral nucleus have activity related to the activity in sympathetic preganglionic neurons, suggesting that these neurons are antecedent to the sympathetic preganglionic neurons (Deuchars et al. 2001b; Brooke et al. 2002). It has been shown in spinal cord slices that interneurons in the intermediolateral region exhibit a fast firing pattern, which is dependent on the presence of specific voltage-gated potassium channels. Interneurons labeled transneurally from the adrenal gland contain the channel subunit Kv3.1 of these potassium channels (Deuchars et al. 2001b; Brooke et al. 2002). Figure 8.9 illustrates the morphology and electrophysiology of such an interneuron.
2. Propriospinal neurons located in spinal thoracolumbar segments remote from the labeled preganglionic neurons in laminae I, V, VII and X (e.g., in segments C1 to C6; Jansen and Loewy 1997; Smith et al. 1998).
3. Propriospinal neurons lying within the white matter of the cervical spinal cord. These spinal interneurons are found in the lateral funiculus and in the lateral spinal nucleus of spinal segments C1 to C4 (Jansen et al. 1995a, b; Jansen and Loewy 1997; see Table 8.2, Figure 8.15).

Although no spinal sympathetic interneuron has been functionally identified and characterized so far, it is not far-fetched to assume that these interneurons are particularly important in the generation of the differentiated reflex patterns in sympathetic pre- and postganglionic neurons that have been worked out in neurophysiological experiments (see Chapters 4 and 9).

8.2.2 Parasympathetic preganglionic neurons and interneurons

Sacral parasympathetic systems
The sacral spinal cord contains at least three groups of parasympathetic preganglionic neurons, which are involved in the regulation of

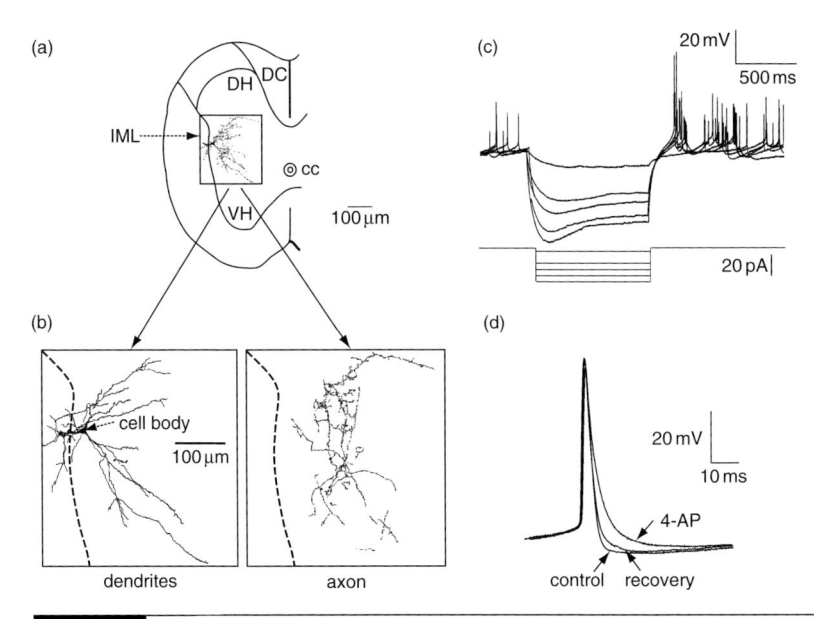

Figure 8.9 Morphology and electrophysiology of an autonomic interneuron in the thoracic segment T9. The neuron was recorded intracellularly in a transverse spinal cord slice (300 μm thick). The neuron showed characteristic high-frequency discharges and short action potentials to depolarizing pulses (c, following hyperpolarizing pulses), both being due to a specific type of voltage-gated potassium channel. The neurons are sensitive to the potassium channel blockers 4-aminopyridine (4-AP; d) and tetraethylammonium suggestive for the presence of the Kv3 potassium channel. It has been shown that autonomic interneurons labeled transneuronally from the adrenal medulla contain the Kv3.1 subunit of these channels (Deuchars et al. 2001a; Brooke et al. 2002). After recording the neuron was filled through the microelectrode with biotinamide. The spinal cord slice was serially sectioned and the neuron was reconstructed (axon arborization, dendrites and cell body). (a) shows the reconstructed neuron in a transverse section. In (b) cell body and dendrites (left) and axon ramification (right) of the neurons are shown. cc, central canal; DC, dorsal columns; DH, dorsal horn; IML, intermediolateral nucleus; VH, ventral horn. From Deuchars et al. (2001a) with permission.

lower urinary tract, hindgut and sexual organs. In the cat, some morphological differentiation and viscerotopic organization are present.

- Preganglionic neurons associated with the hindgut are small, lie in the dorsal band of the intermediate zone and have primarily a mediolateral orientation of their dendrites.
- Preganglionic neurons associated with the urogenital tract lie in the lateral band of the intermediate zone. These neurons have a complex dendritic organization, one type projecting its dendrites mainly into the (superficial) lamina I of the dorsal horn and into the ventral funiculus and the second type having its major dendritic projections into the dorsolateral and lateral funiculi of the white matter and medially. These two types of sacral preganglionic neurons may innervate the urinary tract and reproductive organs, respectively. Axons of the preganglionic neurons in the lateral band exhibit extensive intraspinal collaterals, which project bilaterally to various regions of the ventral and dorsal horn (Morgan et al. 1993; de Groat et al. 1996).

In the rat, the preganglionic neurons projecting to pelvic organs are located in the sacral preganglionic nucleus lying across the intermediate zone in the segments S2, S1 and L6. Also here several groups of putative interneurons have been found that are labeled by a transsynaptically transported neurotropic virus after application to the sacral parasympathetic preganglionic axons or when injected into the walls of the pelvic organs. These do not project through the ventral roots and are not labeled by a classical (not transsynaptically transported) tracer applied to the preganglionic axons. These interneurons are located in lamina I of the dorsal horn, in lamina X, in the dorsal commissural nucleus and in the intermediate zone just dorsal to the preganglionic neurons (see Table 8.3). Some of the interneurons have been characterized using neurophysiological techniques (see Chapter 9 and Figure 9.8).

Cranial parasympathetic systems

The location of the preganglionic neurons of the cranial parasympathetic systems have been extensively studied, mainly in the rat but also in other species, using retrograde labeling and other techniques. Table 8.1 and Figure 1.4 show, for different functional groups, the location of the cranial preganglionic neurons, the cranial nerves through which the preganglionic neurons project and the location of the postganglionic neurons (see also Table 8.3).

- Almost all parasympathetic preganglionic neurons with myelinated axons innervating the heart (cardiomotor neurons) and parasympathetic neurons innervating airways and lung (bronchomotor neurons, probably also secretomotor neurons) are located in the external formation of the nucleus ambiguus (NA). A few are located in the intermediate region between NA and dorsal motor nucleus of the vagus (DMNX). However, this issue is controversial. Those parasympathetic cardiomotor neurons that are located in the DMNX have unmyelinated axons. They are probably located in the most lateral part of the DMNX (lateral to those neurons projecting to the cecum [Figure 8.10]). In this region, there seem to exist some species differences (Bieger and Hopkins 1987; Altschuler et al. 1989, 1992; Izzo et al. 1993; Hopkins et al. 1996). The functions of the cardiomotor neurons in the DMNX are unknown (Jones et al. 1998; see Subchapter 4.8).
- Preganglionic parasympathetic neurons innervating the proximal part of the gastrointestinal tract (stomach, duodenum, small intestine including pancreas and liver) are located in the DMNX. This nucleus is in practice the motor nucleus of the foregut in which the stomach has the highest representation. In the rat it contains about 10 000 neurons (5000 neurons on each side), 7500 neurons being preganglionic and projecting to the gastrointestinal tract, the remaining neurons being interneurons or

Table 8.1 | Classification of parasympathetic motoneurons

Peripheral target	Brain stem nucleus[a]	Cranial nerve	Final motor neuron
Iris muscle (constriction of pupil)	Edinger–Westphal nucleus, lateral division	III	Ciliary ganglion
Ciliary muscles (accommodation)	Edinger–Westphal nucleus, lateral division	III	Ciliary ganglion
Chorioid blood vessels (vasodilation)	Edinger–Westphal nucleus, medial division	III	Ciliary ganglion, pterygopalatine ganglion[b]
Lacrimal gland	Superior salivary nucleus[c]	VII[d]	Pterygopalatine ganglion
Sublingual and submandibular glands	Inferior salivary nucleus[c]	VII[e]	Submandibular ganglion
Lingual (von Ebner) glands	Inferior salivary nucleus[c]	IX	Intralingual ganglia in posterior tongue
Parotid gland	Inferior salivary nucleus[c]	IX[f]	Otic ganglion
Mucosal glands	Probably salivary nuclei[c]	VII, IX	All cranial ganglia (except ciliary ganglion)
Cranial blood vessels (vasodilation)	Probably with salivary and lacrimal preganglionic cells	VII, IX	All cranial ganglia (except ciliary ganglion)
Airways and lungs	Nucleus ambiguus	X	Ganglia in airways
Heart (pacemaker, atria)	Nucleus ambiguus (external formation)	X	Cardiac ganglia
Stomach and other abdominal organs[g]	Dorsal motor nucleus of the vagus	X	Enteric neurons

Notes:
[a] Location of cell bodies of preganglionic neurons mostly based on tracing studies in animals.
[b] Synonym sphenopalatine ganglion.
[c] Salivatory nuclei are not very well defined (see Table 1.1). The *superior salivary nucleus* is located in the lateral reticular formation medial to the oral subnucleus of the spinal trigeminal nucleus at the level (dorsolateral) of the rostral part of the facial nucleus. The *inferior salivary nucleus* extends rostrocaudally adjacent to the medial border of the rostral part of the nucleus tractus solitarii (NTS) and somewhat rostrally (Contreras *et al.* 1980; Matsuo and Kang 1998; Paxions and Watson 1998; Kim *et al.* 2004).
[d] Via greater petrosal and zygomaticotemporal nerves.
[e] Via chorda tympany and lingual nerve.
[f] Via tympanic and lesser petrosal nerves.
[g] Various functions related to contraction, relaxation, exocrine secretion, endocrine secretion (see Table 1.2; Chapter 5 and Subchapter 10.7).

III, oculomotorius nerve; VII, facial nerve; IX, hypoglossal nerve; X, vagus nerve. Modified from Blessing (1997). For eye see Reiner *et al.* (1983) and Gamlin (2000).

other types of preganglionic neurons projecting to visceral organs in the thoracic cavity (Fox and Powley 1985, 1992; Powley et al. 1992).[3]

- The preganglionic neurons in the DMNX projecting through the subdiaphragmatic vagus nerves to the gastrointestinal tract are topographically organized. The rat subdiaphragmatic vagus nerves consist of five branches, which innervate different parts of the gastrointestinal tract with some overlap (left and right gastric branches, left hepatic branch, left accessory celiac branch and right celiac branch [Prechtl and Powley 1985, 1990a]). The preganglionic neurons in the DMNX are organized in rostrocaudal cell columns. The neurons projecting in the gastric branches are located medially and the neurons projecting in the celiac branches laterally. Neurons projecting in the hepatic branch are located in a cell column between the medial (gastric) and the lateral (cecal) cell column of the left DMNX. The dendrites of these preganglionic neurons are almost confined to the cell columns projecting rostrocaudally but also extend dorsally to the subnuclei of the NTS (Fox and Powley 1992; Powley et al. 1992) (Figure 8.10).[4] Most preganglionic parasympathetic neurons innervating the lower esophageal sphincter are probably also located in a medial rostrocaudal cell column of the DMNX (Rossiter et al. 1990).
- Preganglionic neurons regulating glands of the head (lacrimal, salivary and mucosal) are located dorsal to the rostral portion of the facial nucleus in areas called superior and inferior salivary nuclei.
- In primates and birds, preganglionic parasympathetic neurons regulating the inner muscles of the eye (constrictor pupillae and ciliary muscle [accommodation]) are located in the Edinger–Westphal nucleus of the mesencephalon and project through the oculomotor nerve to the ciliary ganglion. Most neurons in the lateral Edinger–Westphal nucleus are associated with accommodation; neurons in the caudal part of this lateral nucleus are associated with pupilloconstriction. In rat, rabbit and cat most oculomotor parasympathetic preganglionic neurons do not lie in the Edinger–Westphal nucleus but ventral to the oculomotor nucleus in the ventral tegmental area and in the central grey matter (see Loewy [1990b]).
- Preganglionic parasympathetic vasodilator neurons innervating cranial blood vessels seem to be located in the same nuclei that contain the parasympathetic secretomotor neurons. This population of vasomotor neurons has not been investigated at all. In birds, preganglionic neurons generating active vasodilation of the chorioid blood vessels of the eye are situated in the medial Edinger–Westphal nucleus (see Reiner et al. [1983]; Gamlin [2000]).

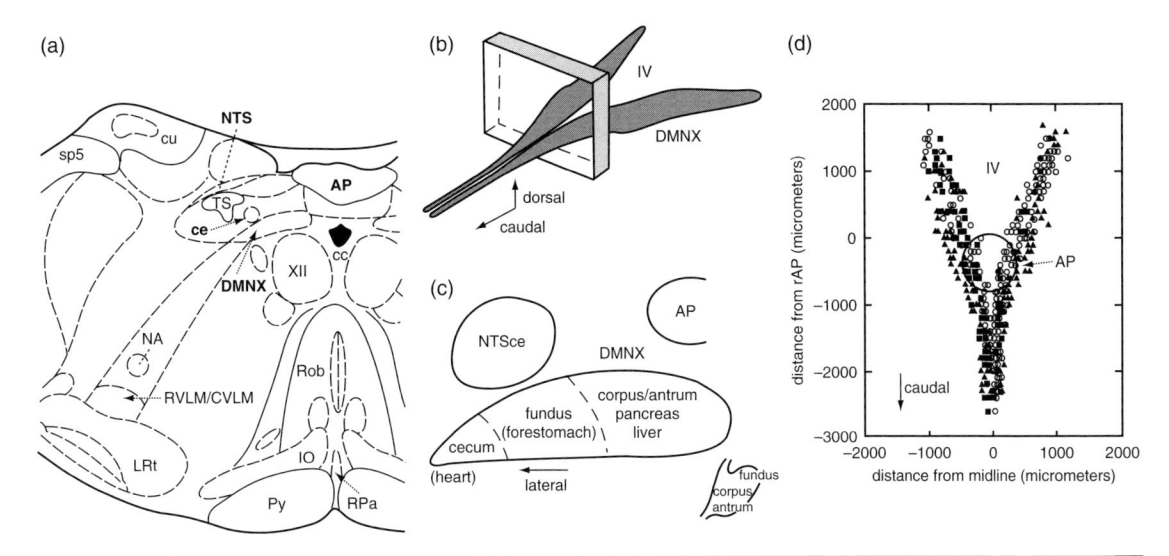

Figure 8.10 Topographic organization of the dorsal motor nucleus of the vagus (DMNX) in the rat. (a) Transverse section through the medulla oblongata at the level of the area postrema (AP) showing the location of the DMNX in relation to other nuclei, in particular the nucleus tractus solitarii (NTS), and the AP. (b) The DMNX viewed from a three-dimensional dorsocaudal perspective. The level of the transverse section in (a) and (c) is indicated. (c) Mediolateral viscerotopic organization of the DMNX at the rostral level of the AP. (d) The distribution of preganglionic neurons projecting through the gastric (circles), celiac (triangles) and hepatic (squares) branches of the subdiaphragmatic vagus nerves as viewed in the horizontal plane. The neurons were labeled with Fast Blue. Fast Blue was injected intraperitoneally in rats four to six days before perfusing the animals. In these experiments, various combinations of abdominal vagal branches (two gastric, two celiac, one hepatic) were transected leaving one or two vagal branches intact. Fast Blue is taken up by the terminals of the intact vagal preganglionic axons and transported to the cell bodies. The DMNX was cut in horizontal 100 μm thick sections. In these sections preganglionic cell bodies are visualized under ultraviolet epifluorescence and intracellularly filled via a glass micropipette with the dye Lucifer yellow. In this way cell bodies, dendrites and axons of the neurons can be visualized and reconstructed. CC, central canal; ce, central nucleus of the NTS; cu, cuneate nucleus; IO, inferior olive; LRt, lateral reticular nucleus; NA, nucleus ambiguus; Py, pyramid; Rob, raphe obscurus; RPa, raphe pallidus; RVLM/CVLM, rostral ventrolateral medulla/caudal ventrolateral medulla; sp5, spinal trigeminal tract; TS, tractus solitarius; IV, fourth ventricle; XII, hypoglossus; (a) modified from Paxinos and Watson (1998); (b, d) modified from Fox and Powley (1992); (c), modified from Hopkins et al. (1996) with permission.

8.3 | Nucleus tractus solitarii

The nucleus tractus solitarii (NTS) is located in the dorsomedial part of the medulla oblongata and extends as columns of neurons that are arranged around the tractus solitarius from rostral to caudal (Figures 8.11). Left and right NTS unite distal to the obex of the medulla oblongata (commissural nucleus); here the NTS is closely associated with the area postrema (AP). The NTS (with the AP) is the important neuronal input structure for information from gustatory receptors, from vagal afferents innervating the respiratory system (lung, airways), from vagal afferents innervating the cardiovascular system (arterial baro- and chemoreceptors; vagal afferents innervating the heart, the portal vein and caval vein) and from

Figure 8.11 Subnuclei of the nucleus tractus solitarii (NTS) showing projection of visceral afferents from gastrointestinal tract (GIT), respiratory organs (RESP) and cardiovascular organs (CARDVASC) to the subnuclei. Two transverse sections through the dorsomedial medulla oblongata (a) rostral to the area postrema (AP) and (b) at the level of the AP. The three groups of visceral afferents have distinct topographical projections to the subnuclei of the NTS. For details see text. cc, central canal; com, commissural nucleus; ce, central nucleus; dm, dorsomedial nucleus; DMNX, dorsal motor nucleus of the vagus; g, gelatinous nucleus; i, intermedial nucleus; is, interstitial nucleus; vl, ventrolateral nucleus; m, medial nucleus; pc, parvicellular nucleus; ts, tractus solitarius. IV, fourth ventricle; XII, hypoglossus nucleus. After Altschuler et al. (1989) and Herbert et al. (1990).

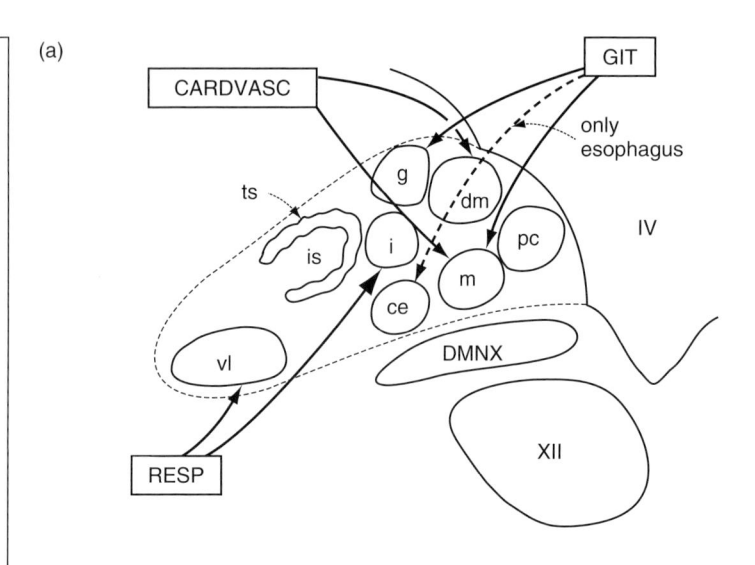

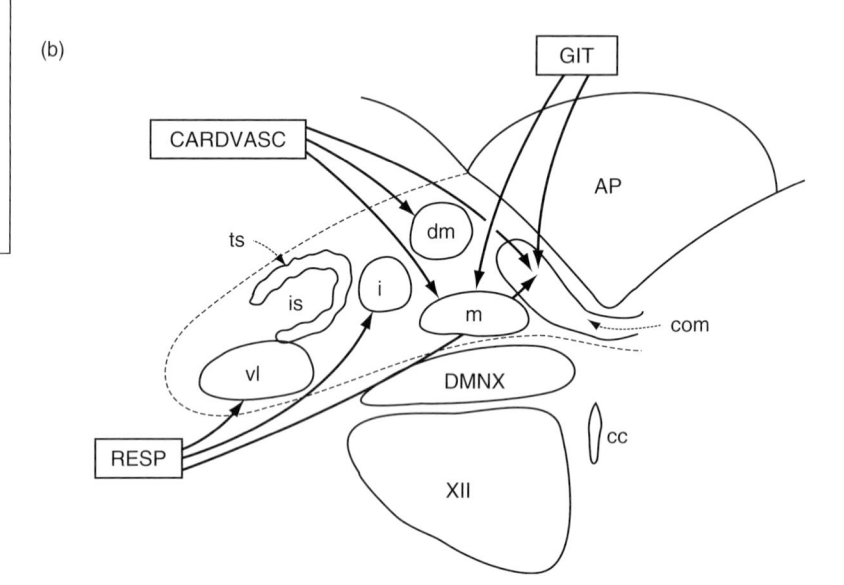

vagal afferents innervating the gastrointestinal tract. The AP is a neurohemal (circumventricular) organ of the NTS, which conveys blood-borne information (and possibly information from the cerebrospinal fluid) to the central nervous system. Thus, the afferent information from the thoracic and abdominal internal organs and from the blood, which is important for homeostatic regulation of respiration, cardiovascular system and gastrointestinal functions, is channeled through the NTS to different nuclei in the brain stem and forebrain (Figure 8.13).

The projection of gustatory and vagal visceral afferent neurons to the NTS exhibits some viscerotopic organization with respect to the rostrocaudal and mediolateral axes of the NTS and with respect to several subnuclei of the NTS, which have been defined anatomically by cytological criteria (Figure 8.11). There is controversy about the cytoarchitectonic criteria that divide the NTS into subnuclei, about the number of subnuclei and about their parcellation (see Blessing [1997]). Independent of this, it is intuitively clear that:

1. subdifferentiation of the NTS must exist,
2. this anatomical subdifferentiation must be in some way related to function and
3. different types of afferents from the three visceral organ systems must exhibit distinct projections to the subnuclei of the NTS.

Figure 8.11 follows the division of the NTS nuclei made by Loewy and Burton (1978), Altschuler *et al.* (1989) and Herbert *et al.* (1990). It shows, schematically, transverse sections at the level of the AP and at a level rostral to the AP. The gustatory afferents project to the rostral part of the NTS and the other groups of afferents to the caudal half (Figure 8.12). There is some overlap in projection between cardiovascular, respiratory and gastrointestinal afferents:

- Afferents from the subdiaphragmatic *gastrointestinal tract* project preferentially to the gelatinous, medial and commissural nuclei, but not to the central, interstitial, intermedial and ventrolateral nuclei. Afferents from the esophagus project to the central nucleus of the NTS. Figure 8.12 demonstrates the projection of afferents from the soft palate, pharynx, esophagus, stomach and cecum to the NTS. It shows that the afferent neurons from the different sections of the alimentary canal are organized topographically with limited overlap. This organization has three distinct characteristics: (1) Afferents from different organs project to different subnuclei of the NTS (Figure 8.12a–c). (2) The afferent projections have a mediolateral and caudo-rostral segregation (Figure 8.12d). (3) The organization of afferent projections is related to the mediolateral organization in the underlying DMNX of the preganglionic neurons that project to the gastrointestinal tract (Figure 8.12e). This topographical organization of projecting afferents from and efferent neurons to the alimentary tract is the anatomical basis of the neural control of the gastrointestinal tract (stomach, duodenum, small intestine; see Subchapter 10.7). The NTS subnuclei related to the gastrointestinal tract and the ventrally located DMNX are also called the dorsal vagal complex (DVC; see Subchapter 10.7).
- Afferent neurons innervating the *respiratory system* project to the commissural, intermedial and ventrolateral nuclei (Kalia and Richter 1985a, b, 1988a, b).
- *Cardiovascular afferents* (arterial baro- and chemoreceptor afferents) project to the dorsomedial, medial and commissural nuclei and to the AP (see Ciriello *et al.* 1994).

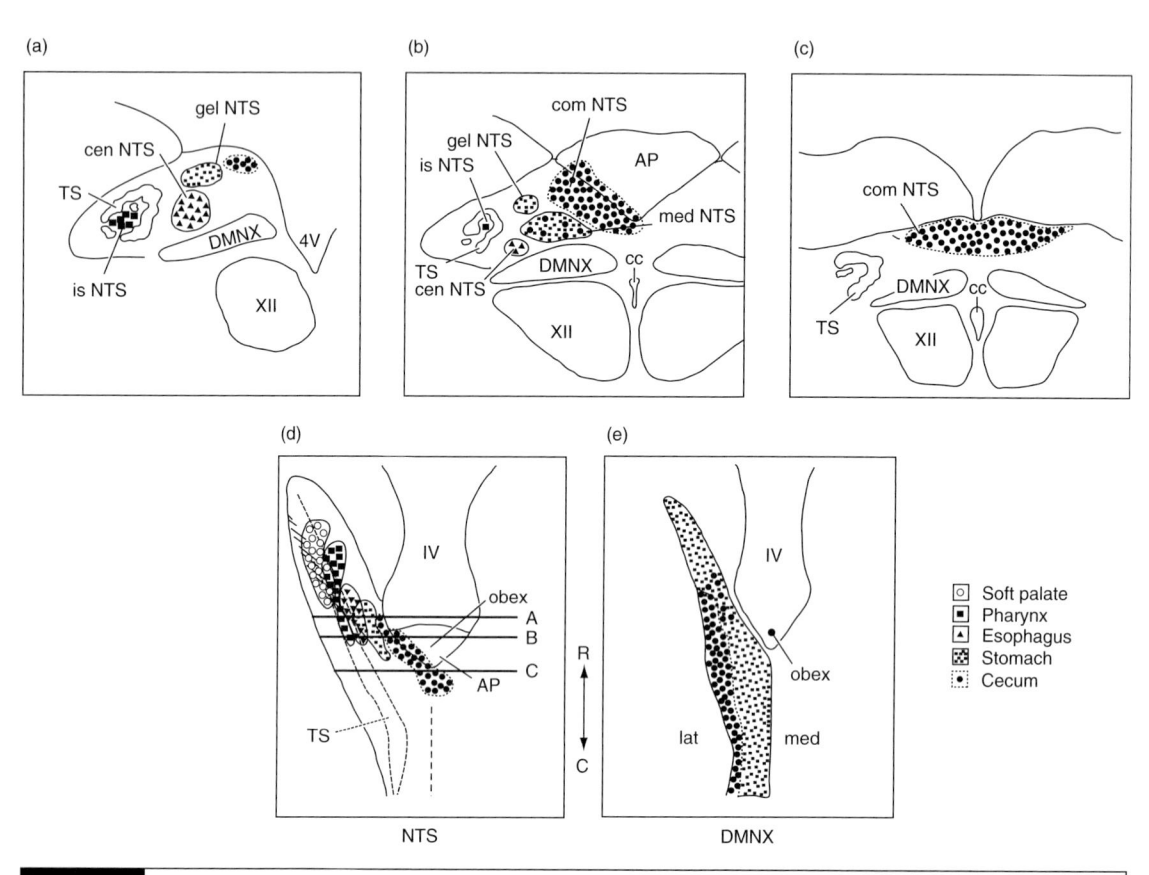

Figure 8.12 Topographic projection of afferents from different sections of the alimentary canal to the nucleus tractus solitarii (NTS). (a–c) Schematic transverse sections through the dorsomedial medulla oblongata containing the NTS and dorsal motor nucleus of the vagus (DMNX) just rostral to the area postrema (AP) (a), through the AP (b) and caudal to the AP (c). See location of the transverse section in (d). gel, gelatinosus nucleus; med, medial nucleus; 4V, fourth ventricle; For other abbreviations of subnuclei see legend of Figure 8.11. (d) Horizontal (longitudinal) section through the NTS showing rostrocaudal and mediolateral topography of the projections of the afferents. (e) Horizontal (longitudinal) section through the DMNX showing mediolateral topography of the location of preganglionic neurons projecting to the gastrointestinal tract (see Figure 8.10). Modified from Altschuler et al. (1991) with permission.

The topography of projection of visceral afferents was obtained on the basis of tracing experiments in which distinct populations of afferent neurons were labeled. Information is not available on the projections of functionally distinct groups of afferents. For example, arterial baroreceptor afferents and arterial chemoreceptor afferents have widely overlapping projection fields in the NTS (Ciriello et al. 1994). No convincing topographical functional differentiation of the second-order (and higher-order) neurons in the NTS with respect to their afferent synaptic inputs from respiratory, cardiovascular and gastrointestinal organs has been found so far, using neurophysiological and other methods. Thus, functionally identified NTS neurons appear to lack viscerotopic organization (Paton 1999; Paton and Kasparov 2000; Paton et al. 2005 for discussion and references see Blessing [1997]). This is puzzling in view of the many distinct reflex pathways associated with the gastrointestinal tract, the respiratory system and the cardiovascular system.

Neurons in the NTS either project to various regions of brain stem, hypothalamus and forebrain or are interneurons that are confined in their projection to the NTS. Both types of neuron can be excitatory or inhibitory. Furthermore, anatomical tracing studies (including those with neurotropic viruses) show (1) that neurons of the NTS are synaptically influenced from many sites in spinal cord, lower and upper brain stem, hypothalamus and telencephalon, and (2) that projection neurons in the NTS project to the same brain regions from which the NTS receives its synaptic inputs (Figure 8.13).[5]

The general functions of the NTS are as follows (Figure 8.14):

- The activity in vagal and gustatory afferents is synaptically transmitted to the projection neurons and relayed to various brain centers (function 1 in Figure 8.14).
- The synaptic transmission is locally modulated by inhibitory and excitatory interneurons in the NTS which also receive synaptic input from primary afferents. This modulation can occur pre- or postsynaptically (function 2).
- The projection neurons and interneurons in the NTS are under inhibitory and excitatory control from various brain centers (function 3).
- The neuronal circuits in the NTS are modulated by blood-borne substances via the AP, such as angiotensin, corticotropin-releasing hormone (CRH), thyrotropin-releasing hormone (TRH), gastrointestinal hormones (see Subchapter 10.7) or signals from the immune system (function 4).
- The synaptic transmission in the NTS is modulated by local signals related to the blood vessels (e.g., nitric oxide [NO], function 5; Paton and Kasparov 2000; Paton et al. 2005).

This shows that the neurons in the NTS do not only transmit activity in a relay-like fashion to other nuclei, but that NTS circuits also have integrative functions and are under powerful control of the brain. The degree of relay and integrative functions probably varies between different functional systems. Details about the integrative functions of the NTS and their underlying mechanisms with respect to specific afferent inputs (e.g., from arterial baro- or chemoreceptors or from specific gastrointestinal afferents) and with respect to specific homeostatic regulations are virtually unknown (Paton 1999; Paton and Kasparov 2000; Paton et al. 2005; see Subchapter 10.7). The bewildering functional anatomy of the NTS, with its intrinsic integrative processes, its multiple peripheral and central synaptic inputs and its multiple efferent projections, clearly shows that the mechanisms operating in the NTS during regulation of autonomic functions can only be worked out using a

ANATOMY OF CENTRAL AUTONOMIC SYSTEMS

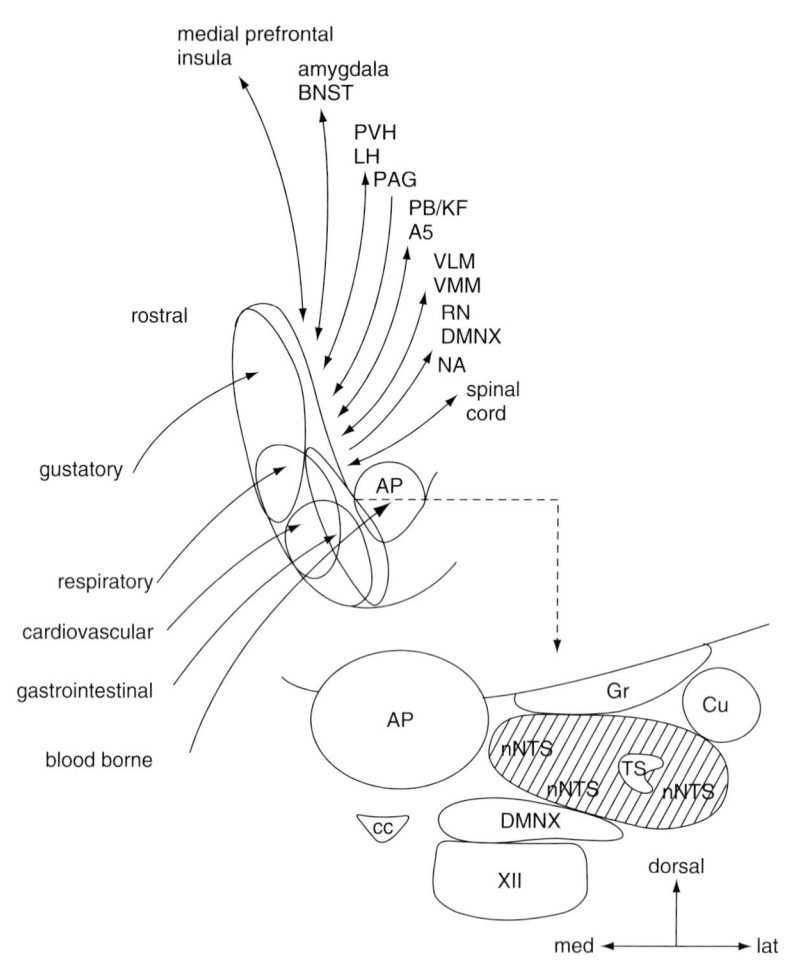

Figure 8.13 Afferent and efferent connections of the nucleus tractus solitarii (NTS) and area postrema (AP). The peripheral synaptic inputs from gustatory afferents, afferents of the respiratory system (including the trachea), cardiovascular afferents (including afferents from arterial chemoreceptors and arterial baroreceptors) and gastrointestinal afferents (including afferents from the oropharyngeal structures) are topographically organized. However, a topographic organization of synaptic inputs to the second-order neurons of the NTS from functionally different types of afferent neurons of a particular type of visceral organ seems to be absent. "Afferent" blood-borne inputs are relayed by neurons in the AP. Neurons in the NTS and AP project to various nuclei in the neuraxis and forebrain; neurons in most of these nuclei project back to the NTS/AP. *Lower right*: transverse section through the dorsomedial part of the medulla oblongata at the level of AP. The subnuclei of the NTS (nNTS), which are arranged around the tractus solitarius (TS), are not shown. A5, pontine A5 nucleus (most of the neurons are noradrenergic); BNST, bed nucleus of the stria terminalis; Cu, Gr, cuneate and gracile nucleus; cc, central canal; DMNX, dorsal motor nucleus of the vagus; LH, lateral hypothalamus; NA, nucleus ambiguus; PAG, periaqueductal grey; PB/KF, parabrachial and Kölliker-Fuse nuclei; PVH, paraventricular nuclei of the hypothalamus; RN, raphe nuclei (raphe magnus, obscurus and pallidus); VLM, ventrolateral medulla; VMM, ventromedial medulla; XII, nucleus of the hypoglossal nerve. Modified from Loewy (1990a), Saper (1995) and Blessing (1997).

multidisciplinary approach, which includes in vivo experimentation on animal models with full experimental control over the autonomic systems under study. One promising preparation to be used in these experiments is the working-heart-brain-stem preparation of the mouse and rat in which the lower brain stem is artificially

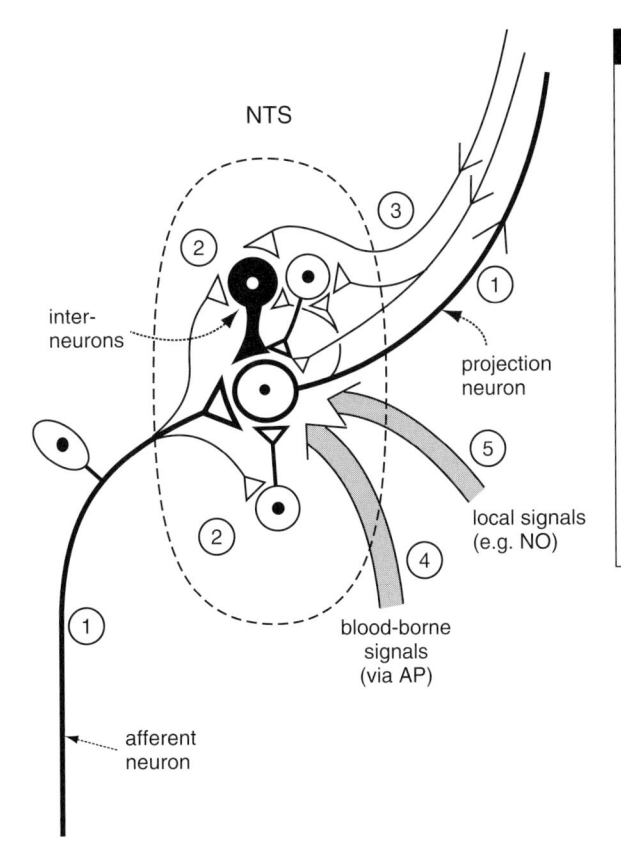

Figure 8.14 Functions of the nucleus tractus solitarii (NTS). The NTS has relay function transmitting the activity in gustatory and vagal afferents to other brain centers (*function 1*). The impulse transmission is modulated by local circuits involving excitatory and inhibitory interneurons (*function 2*), from brain centers in spinal cord, brain stem, hypothalamus and telencephalon that project to the NTS (*function 3*), by blood-borne substances via the area postrema (AP; *function 4*), and by local signals (e.g., related to blood vessels, nitric oxide [NO]; *function 5*). Black, inhibitory interneuron.

perfused but attached to the beating heart and the ventilated lung by vagal afferents (Paton 1996a, b).

8.4 Sympathetic and parasympathetic premotor neurons in brain stem and hypothalamus

Neurons in brain stem, hypothalamus and cerebral hemispheres that project to the territory of preganglionic neurons in spinal cord and brain stem have been called *autonomic premotor neurons*. The autonomic premotor neurons have been localized by retrograde labeling of their cell bodies with tracers. These neurons are sometimes called presympathetic or preparasympathetic neurons although these autonomic premotor neurons may form synapses with interneurons that are situated in close proximity to the preganglionic neurons. Thus, the term premotor neuron does not always mean that these autonomic neurons form direct synaptic contacts with the preganglionic neurons (this also applies to premotor neurons, projecting e.g. through the pyramidal tract, that form synapses with somatomotor neurons in the ventral horn and associated interneurons). Furthermore, these premotor neurons may form synapses with propriospinal neurons (e.g., in the cervical spinal cord) that project to the intermediate zone in the thoracolumbar spinal cord (see below; Jansen and Loewy 1997).

The cell bodies of the autonomic premotor neurons can be located by tracers applied to the terminal fields of the axons of these neurons (e.g., in the spinal intermediate zone, the external portion of the nucleus ambiguus, the DMNX). Alternatively, a neurotropic virus, which is transported retrogradely and transsynaptically, can be applied to the termination fields of postganglionic axons (e.g., in the kidney, rat tail artery and/or skin, heart etc.) or of preganglionic axons (e.g., in the adrenal medulla, stellate ganglion, the pterygopalatine ganglion etc.; see Tables 8.2 and 8.3). These two techniques may be applied together in the same experiment or combined with histochemical techniques that visualize the putative transmitters in the neurons (Figure 8.1). Furthermore, two different tracers or two different neurotropic viruses may be applied in the same experiment. Taking the limitations of these labeling techniques into account (Jansen et al. 1993, 1995a) it has been demonstrated that several distinct groups of neurons in the brain stem and hypothalamus project to the preganglionic neurons or the interneurons that are closely associated with them.

With a few exceptions (e.g., Siberian hamster, brown adipose tissue), most transneuronal labeling studies have been performed in the rat. Most functional studies about the central organization of the autonomic nervous system have been performed in this species in the recent past and will probably be performed in future. However, there is a need for more information about mammalian species other than rat.

8.4.1 Sympathetic premotor neurons

Sympathetic premotor neurons that project to the sympathetic preganglionic neurons and/or their interneurons (see Subchapter 8.1) are mainly found in the following nuclei of brain stem and hypothalamus (Figures 8.15, 8.16; marked by # in Table 8.2[6]):

- *Neurons in the ventrolateral medulla (VLM)*: these neurons are located in the rostral ventrolateral medulla (RVLM) and in the lateral paragigantocellular nucleus (LPGi). About half to two-thirds of the neurons in the RVLM projecting to the preganglionic neurons are adrenergic (i.e., contain phenyl-N-methyltransferase, an enzyme that converts noradrenaline into adrenaline) and belong to the C1 cell group (for review see Guyenet et al. [2001]; Subchapter 10.2, Figure 10.3). Some neurons contain peptides (e.g., enkephalin, substance P, TRH or VIP). Neurons in the LPGi are serotonergic or non-serotonergic, some are also adrenergic and belong to the C1 cell group.
- *Neurons in the ventromedial medulla (VMM)*: sympathetic premotor neurons are located in the gigantocellular reticular nucleus,

alpha and ventral (GiA and GiV) and in the parapyramidal region or nucleus (PPy) (see note 6). Many of these projecting autonomic premotor neurons are serotonergic (i.e., contain 5-hydroxytryptamine [5-HT] as putative transmitter). Some neurons contain the neuropeptides enkephalin, vasoactive intestinal peptide and/or substance P.
- *Neurons in the caudal raphe nuclei*: these nuclei consist of the raphe magnus, the raphe pallidus and the raphe obscurus. Many of these neurons are serotonergic. Some neurons contain neuropeptides such as enkephalin, substance P or other peptides.
- *A5 cell group in the caudal ventrolateral pons*: 90% of these neurons are catecholaminergic and use noradrenaline as putative transmitter.
- *Neurons in the paraventricular nucleus of the hypothalamus (PVH)*: these sympathetic premotor neurons are located in the parvocellular nuclei of the PVH and may contain oxytocin, CRH or vasopressin as putative transmitter.
- *Neurons in the lateral hypothalamus (LH)*: these neurons may contain the neuropeptide orexin as putative transmitter (Geerling *et al.* 2003).
- Propriospinal neurons located in the grey matter of the spinal segments C1 to C6 and in the lateral funiculus and lateral spinal nucleus of the segments C1 to C4 (see Subchapter 8.2.1). It is a matter of semantics to call these neurons propriospinal interneurons or sympathetic premotor neurons.

These results clearly illustrate that there are several candidates for sympathetic premotor neurons in the brain stem and hypothalamus as postulated in neurophysiological studies of sympathetic preganglionic and postganglionic neurons (see Chapters 4 and 9). The types of premotor system are characterized by their supraspinal origin, their histochemistry (types of neuropeptide colocalized in the neurons) and whether the premotor neurons synapse directly with the preganglionic neurons, with the segmental autonomic interneurons or with the propriospinal autonomic neurons (Figure 8.15).

Many sympathetic premotor neurons use glutamate (or perhaps aspartate) as excitatory transmitter. The role of adrenaline, noradrenaline or 5-HT as transmitter during ongoing regulation of the activity in the sympathetic preganglionic neurons is unclear. Action of these monoamines can be inhibitory. Inhibitory pathways from the brain stem probably use GABA too. Furthermore, it is unclear whether, and in which functional context, the neuropeptides colocalized in the sympathetic premotor neurons are used as neurotransmitters (see Subchapter 9.1).

Premotor neurons in the RVLM have been most thoroughly studied. These neurons are largely involved in cardiovascular regulation (see Subchapters 10.2 to 10.4). Sympathetic premotor neurons in the raphe nuclei may be involved in regulation of cutaneous blood flow for thermoregulation and also of brown adipose tissue (see

Figure 8.15 Premotor neurons in brain stem and hypothalamus, spinal segmental interneurons and propriospinal neurons projecting to sympathetic preganglionic neurons. Sympathetic premotor neurons project to preganglionic neurons and to spinal autonomic interneurons (not shown). They are located in the rostral ventrolateral medulla (RVLM), in the caudal raphe nuclei of the medulla oblongata (raphe magnus, pallidus and obscurus [Rob]), in the A5 area of the caudal ventrolateral pons, in the lateral hypothalamus (LH) and in the paraventricular nucleus of the hypothalamus (PVH). Autonomic interneurons are located in laminae 1, 2, 5, 6, 9 and 10 of the spinal cord. Propriospinal neurons are located in laminae 1, 5, 7 and 10 of the cervical spinal segments C1 to C6, in the lateral funiculus (LF) and in the lateral spinal nucleus (LSN) of the cervical segments C1 to C4. IO, inferior olive; PY, pyramidal tract; sp5, spinal trigeminal tract; 3V, third ventricle; 4V, fourth ventricle. For further abbreviations see Table 8.2. Modified from Jansen et al. (1995b) and Jansen and Loewy (1997) with permission.

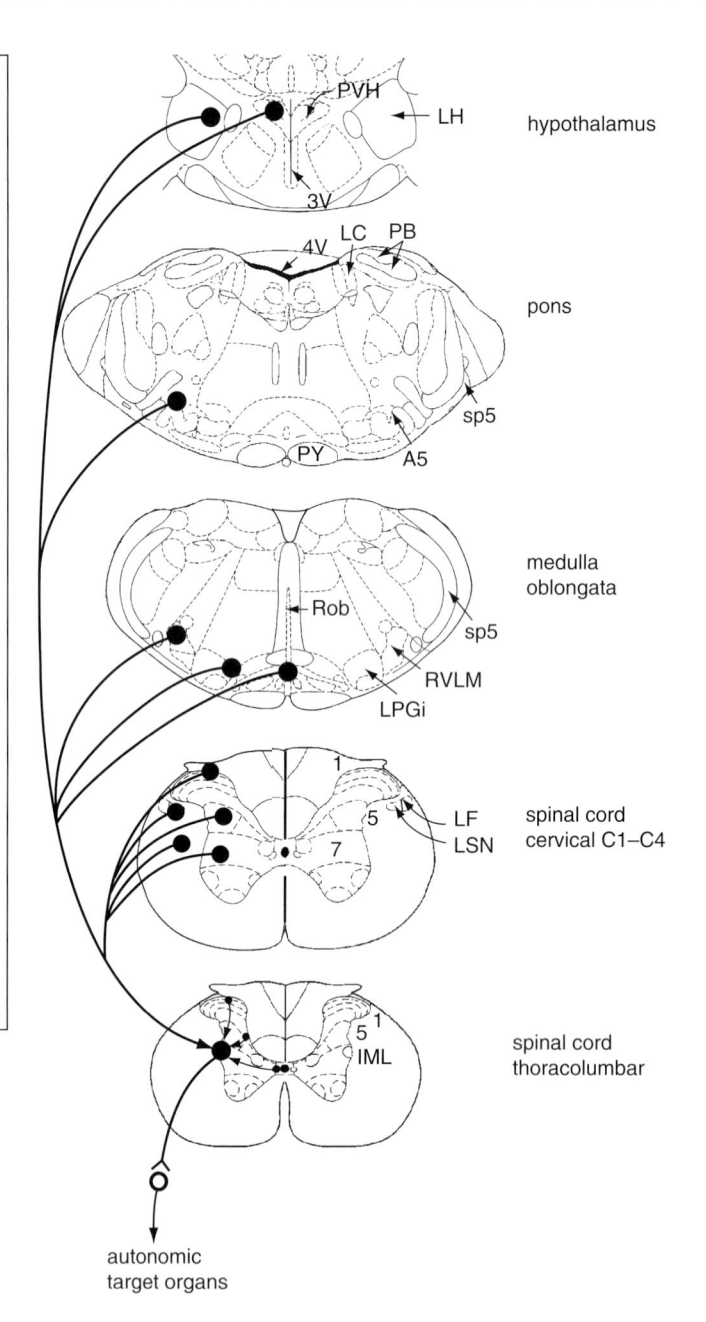

Subchapter 10.5). The functions of the other sympathetic premotor neurons have been very little studied and how the different sympathetic premotor neurons relate to the different sympathetic pathways is practically unknown.

8.4.2 Parasympathetic premotor neurons

The parasympathetic premotor neurons have the following distribution (Figure 8.17, Table 8.3):

8.4 SYMPATHETIC AND PARASYMPATHETIC PREMOTOR NEURONS | 321

Lateral Hypothalamic Area:
Orexin

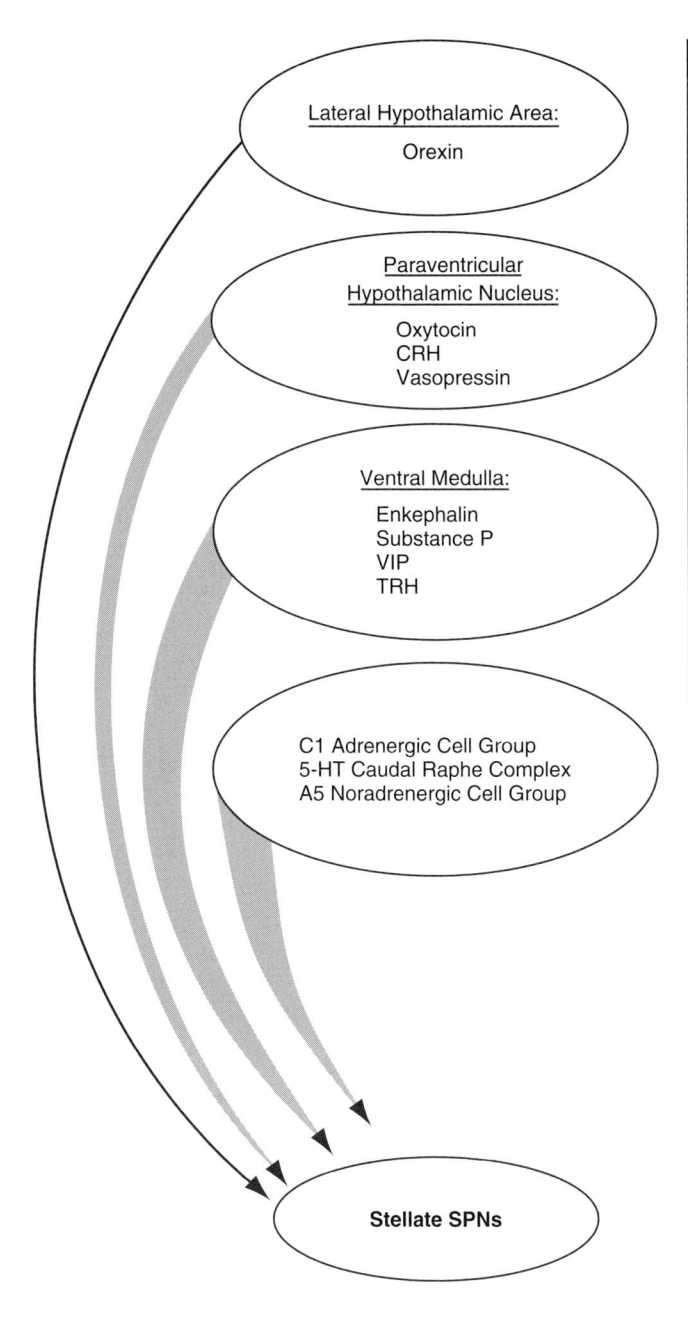

Paraventricular Hypothalamic Nucleus:
Oxytocin
CRH
Vasopressin

Ventral Medulla:
Enkephalin
Substance P
VIP
TRH

C1 Adrenergic Cell Group
5-HT Caudal Raphe Complex
A5 Noradrenergic Cell Group

Stellate SPNs

Figure 8.16 Putative transmitters in sympathetic premotor neurons. In many (all?) sympathetic premotor neurons, glutamate is the fast excitatory transmitter. Data obtained in sympathetic premotor neurons labeled by a neurotropic virus or by a tracer applied to the intermediate zone in the spinal cord. Monoamine synthesizing enzymes (e.g., tyrosine hydroxylase, phenyl-N-methyl transferase) and neuropeptides have been localized immunohisto-chemically. CRH, corticotropin-releasing hormone; 5-HT, 5-hydroxytryptamine; TRH, thyrotropin-releasing hormone; SPNs, sympathetic preganglionic neurons; VIP, vasoactive intestinal peptide. Modified from Jansen et al. (1995b), Westerhaus and Loewy, (1999) and Stornetta et al. (2001) with permission.

- Parasympathetic premotor neurons that project to the parasympathetic preganglionic neurons in the *external formation of the nucleus ambiguus*, *DMNX* or *salivary nuclei* have a similar pattern of distribution as sympathetic premotor neurons. The neurons are located in the ventromedial medulla, in the caudal raphe nuclei, some in the ventrolateral medulla (including adrenergic [C1] and noradrenergic neurons [A1]), in the ventrolateral pontine A5 area, in the paraventricular nucleus of the hypothalamus, in the lateral

Table 8.2 Sympathetic nervous system

Target	Spinal cord		VMM			VLM		Raphe N.			Pons			PAG	Hypothalamus					Telencephalon				References
	Preggl. IML IC, CA	INN RN	GiA	GiV	PPN	LPGi	RVLM	M	O	P	A5	PB KF	BN		PVH	LH	DMH	VMH	Other	BNST	MPO	CeAM	Cortex	
Adrenal Medulla	X T4–T13	I, II, V, VII C1–C4 LF, LSN	X			X	X	X	X	X	X	X			Xpc				perifornical		X			Strack et al. 1989a, b; Jansen et al. 1995a; Westerhaus & Loewy 1999, 2001; Jansen and Loewy 1997
																						X		Smith et al. 1998
CVC (tail)	X T11–L2	I, V, VII, IX, X C1–C4	X	X		X	X C1	X	X	X	LC, SC	X	X	Xdm, dl	Xpc				perifornical, ZI	X	X	X		Cano et al. 2000, 2001
Spleen	X T3–T12	X	ventromedial medulla			X	X	X	X	X	LC, SC		X		Xdpc	X	X							Schramm et al. 1993; Sly et al. 1999; Cano et al. 2000; Huang & Weiss 1999
Kidney	X T3–T12		ventromedial reticular formation			X	X C1	X	X	X	NAd		X	X	Xpc Ox	X			retrochiasmatic	X	X			
Heart	X T1–T7		ventromedial medulla				X	raphe nuclei						X					retrochias perifom				insula, infralimbic	Ter Horst et al. 1996
Stellate ganglion	X	I, II, V, VII, X C1–C4, LF, LSN		X	X	X	X	X	X	X	X LC, SC		X	Xvl, EW	Xpc	X			perifornical		X	septum	insula, frontal, infralimbic	Jansen et al. 1995a, b; Westerhaus & Loewy 1999, 2001; Jansen and Loewy 1997
Pineal gland	X T1–T3	X	X			X	X	X	X	X	X LC	X	X	X	X				SCN		X			Larsen et al. 1998

Brown adipose tissue	X	X	X	X	X X X	X	X	Xv, l	X	X	septum	Bamshad et al. 1999
White adipose tissue	X	X	X C1	X X X X	X	X	Xd, v	X X X	X	SCN	X. septum	Bamshad et al. 1998
	#		#	#			#	#			

Notes:

Preganglionic neurons, interneurons, propriospinal neurons, premotor neurons and neurons antecedent to the premotor neurons of the sympathetic system labeled by a neurotropic virus injected into the target tissue or into a sympathetic ganglion.

Sites of labeled neurons are indicated by **X** or are specified. Location of premotor neurons is indicated by # on the lower margin. Abbreviations: *Spinal cord:* IML intermediolateral nucleus, IC intercalate nucleus, CA central autonomic nucleus (dorsal commissural nucleus); INNRN, interneurons: I, II, V, VII, IX, X spinal laminae of the grey matter; LSN, lateral spinal nucleus; LF, lateral funiculus; Pregl., preganglionic. *Ventromedial medulla (VMM):* GiA, GiV, gigantocellular reticular nucleus, alpha and ventral: PPN, parapyramidal nucleus. *Ventrolateral medulla (VLM):* C1, adrenergic neurons; LPGi, lateral paragigantocellular nucleus; RVLM, rostral ventrolateral medulla. *Raphe nuclei:* M, magnus; O, obscurus; P, pallidus. *Pons:* A5, A5 area (nucleus); NAd, noradrenergic neurons; PB, parabrachial nuclei; KF, Kölliker–Fuse nucleus; BN, Barrington nucleus (pontine micturition center); LC, locus ceruleus; SC, locus subceruleus. *Periaqueductal grey matter (PAG):* EW, nucleus Edinger–Westphal; dm, dorsomedial; dl, dorsolateral; vl, ventrolateral; l, lateral; d, dorsal; v, ventral. *Hypothalamus:* LH, lateral hypothalamus; PVH, paraventricular hypothalamus (pc, parvocellular); DMH, VMH, dorso- and ventromedial hypothalamus; Ox, oxytocin; retrochias, retrochiasmatic SCN, subchiasmatic nucleus; ZI, zona incerta. *Telencephalon:* BNST, bed nucleus of the stria terminalis; MPO, medial preoptic nucleus (area); CeAM, central nucleus of the amygdala; infralimb, infralimbic cortex; ins., insula; sept., septum.

Figure 8.17 Premotor neurons for parasympathetic preganglionic neurons projecting to airways and pancreas. The premotor neurons synapse with parasympathetic preganglionic neurons and/or interneurons associated with these preganglionic neurons. The premotor neurons have been labeled by a neurotropic virus (Bartha strain of the pseudorabies virus), which was injected into the wall of the airways or the pancreas. LRN, lateral reticular nucleus; PY, pyramidal tract; sp5, spinal trigeminal tract; 3V, third ventricle; 4V, fourth ventricle. For other abbreviations see legend of Table 8.2. Data from Haxhiu et al. (1993, 1998), Loewy and Haxhiu (1993), Loewy et al. (1994) and Haxhiu and Loewy (1996).

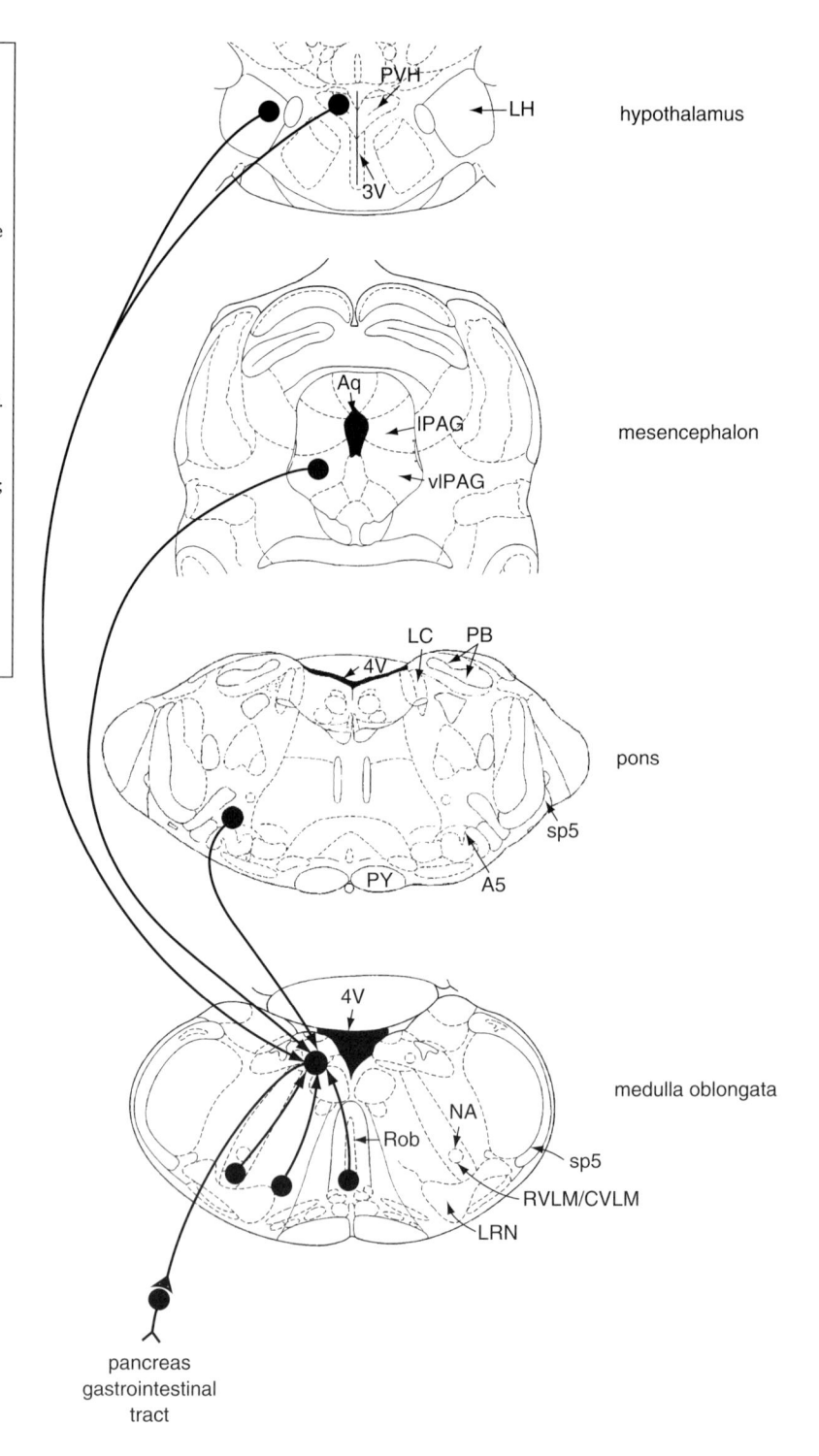

hypothalamus and in the preoptic area as well as in the central nucleus of the amygdala.
- Parasympathetic premotor neurons projecting to the *sacral parasympathetic neuronal circuits* (preganglionic neurons and interneurons) are located in the ventromedial medulla (parapyramidal nucleus), the ventrolateral medulla (RVLM, LPGi), the raphe nuclei, the pontine A5 area, Barrington's nucleus, the paraventricular nucleus of the hypothalamus, the lateral hypothalamus and the preoptic area (Table 8.3). In most virus studies of the central autonomic innervation of the pelvic organs lumbar sympathetic systems were included since the nerves containing sympathetic pre- or postganglionic axons that supply pelvic organs were not cut.

The pattern of labeled sacral premotor neurons is, first, similar for all pelvic organs and, second, similar to the pattern of premotor neurons labeled transsynaptically via cranial preganglionic parasympathetic neurons (e.g., from heart, pancreas, airways, glands; Table 8.3). Thus, no pattern of labeled preganglionic neurons characteristic for individual pelvic organs or organs innervated by the vagus nerve is seen. Furthermore, with the exception of the labeling of neurons in the nucleus gigantocellularis medialis from the cranial organs, no pattern has emerged that is distinct for these organ groups.

8.4.3 Autonomic neurons antecedent to sympathetic or parasympathetic premotor neurons

In addition to the autonomic premotor neurons there are other groups of neurons in the brain stem, hypothalamus and telencephalon that have been labeled using a retrogradely transsynaptically transported virus. Prominent groups of labeled neurons were present in the parabrachial and Kölliker-Fuse nuclei, in the periaqueductal grey matter (PAG; most of them located ventromedial, ventrolateral or dorsolateral), in the preoptic region of the hypothalamus (Westerhaus and Loewy 1999) and in the telencephalon including cortical structures (e.g., insula, infralimbic cortex, medial prefrontal cortices [Westerhaus and Loewy 2001]). These neurons have been labeled via autonomic premotor neurons. For example, labeled neurons in the PAG are infected via premotor neurons in the ventromedial medulla, ventrolateral medulla, raphe nuclei, the A5 area and the paraventricular nucleus of the hypothalamus (Farkas *et al.* 1998) (Table 8.2 and 8.3). Labeled neurons in the septum, amygdaloid complex, insula, infralimbic cortex, anterior cingulate cortex, medial prefrontal cortex etc. may be infected via premotor neurons in the hypothalamus (Westerhaus and Loewy 1999, 2001).

8.4.4 Patterns of central autonomic neurons labeled from distinct targets: strengths and limitations

Unfortunately, the results obtained with the transsynaptic tracing technique give only rather limited information about the *central*

Table 8.3 Parasympathetic nervous system

Target	Spinal cord		Dors. med. obl./NA				VMM				VLM		Raphe N.				Pons		PAG	Hypothalamus					Telencephalon			References
	IML	INNRN	NTS	AP	NA	DMNX	GiA	GV	VMM	PPN	LPGi	RVLM	M	O	P		A5	PB KF	BN		PVH	LH	DMH VMH	Other	BNST	MPO	CeAM	
Cranial systems																												
Heart	x	SPN, DCN	x	x	x	x	x	x			x	x	x	x	x		x LC, SC	x	x	Xvl, m	x	x	x	retrochiasmatic	x	x	lateral septum	Ter Horst et al. 1996
Pancreas	x	SPN	x	x			5-HT	5-HT			x	5-HT	5-HT	x	x		x SC	x	x	Xvl	x DA	x	x	perifornical	x		x	Loewy & Haxhiu 1993; Loewy et al. 1994
Airways	x	SPN, DCN DH(LaI), LaX	x	x		x	x	x		x	x	x	x	x	x		x LC, SC	x	x	Xvl	x	x	x	DH				Haxhiu et al. 1993, 1998; Haxhiu & Loewy 1996; Hadziefendic & Haxhiu 1999
Glands (SM, PTPG)	x	SPN, L1/L2	x			sup. salivary n.	x	x		x	x	x	x	x	x		x LC, SC #	x	x	Xvl ventr. tegm. #	Xpc #	x #	x #	periform periventr	x	x	x	Spencer et al. 1990; Blessing et al. 1991; Jansen et al. 1992
Sacral systems (*includes sympathetic systems)																												
Urinary bladder	x	DCN(LaI), LaX						x			x		raphe nuclei				x SC	x	x									Nadelhaft et al. 1992; Nadelhaft & Vera 1995
Urinary bladder,* EUS*	x	SPN, DCN DH(LaI), LaX								x		x		x	x		x SC	x	x	Xvm	Xpc		x	periform		x		Marson 1997
Urethra*	x	DCN, DH L6/S1, L1/L2					gigantocell. ret. n.						x				x A7, LC, SC	x	x	Xvl								Vizzard et al. 1995
Penis,* Clitoris,* Prostate*	x	DCN, LaX dors IML L5-S1, T12-L4								x	x	x	x	x	x		x LC, SC	x	x	Xvl	Xpc	x				x		Marson et al. 1993; Marson 1995; Orr & Marson 1998
Ischio/bulb. Cavernos.*	x	LS, THL								x	C1	x	raphe nuclei				x LC, SC	Xve, vm	x	x	Xpc		x			x		Marson & McKenna 1996
Uterus*	x	DCN, LaX, DH, intermed. grey	x	x	x	x				x		x	x	x	x		x LC, SC	x	x	x	x		x		medial/frontal lobe	x		Papka et al. 1998
Colon*	x	SPN, IML	x	x	x	x				#	#	#	#	#	#		x LC, SC #	x #	x #	x #	X so #	x #						Vizzard et al. 2000

Notes:
Preganglionic neurons, premotor neurons and neurons antecedent to the premotor neurons of the parasympathetic system labeled by a neurotropic virus injected into the target tissue.
Location of premotor neurons are indicated by # on the lower margins. AP, area postrema; DA, dopamine; Dors. med. obl., dorsal medulla oblongata; DCN, dorsal commissural nucleus; DH, dorsal hypothalamus; DMNX, dorsal motor nucleus of the vagus; EUS, external urethral sphincter; 5-HT, 5-hydroxytryptamine; Ischio/bulb. cavernos., musculus ischiocavernous/bulbocavernous; LaI, LaX, lamina I, lamina X of dorsal horn; LS, lumbar/sacral; NA, nucleus ambiguus; NTS, nucleus tractus solitarii; PTPG, pterygopalatine ganglion; SM, submandibular gland; SO, nucleus supraopticus; SPN, sacral preganglionic neuron; sup.sal.n., superior salivatory nucleus; THL, thoracic/lumbar; ventr. tegm., ventral tegmentum of the mesencephalon. For further abbreviations see legend of Table 8.2.

organization of functionally distinct autonomic systems. The differences between functionally distinct autonomic systems must exist, otherwise it would be impossible to explain (1) the characteristic reflex discharge patterns in the neurons of the different types of autonomic pathways and (2) the precise neuronal regulation of autonomic target organs. It is perhaps not unexpected that labeled neurons do not exhibit obvious differences in experiments in which the virus has been injected in the stellate ganglion, celiac ganglion or other sympathetic ganglia. In the rat, the stellate ganglion contains cardiomotor, cutaneous vasoconstrictor, muscle vasoconstrictor, sudomotor, pilomotor, lipomotor (innervating brown adipose tissue) neurons and possibly other functional types of neurons (Baron *et al.* 1995). The celiac ganglion contains various types of postganglionic neurons innervating the gastrointestinal tract (secretomotor, motility-regulating, vasoconstrictor neurons; see Subchapters 4.3 and 5.7), spleen (vasoconstrictor neurons, neurons innervating the immune tissue), kidney and other targets.

However, when considering those experiments in which rather distinct targets have been labeled by the virus, the situation is different. The apparent lack of central differentiation probably results because the differentiation occurs in the central microcircuits, i.e. the excitatory and inhibitory synaptic connections between neurons, the transmitters used etc. The transported virus does not discriminate between excitatory and inhibitory synapses nor does it identify their relative effectiveness. The lack of central differentiation is unlikely to result from technical deficits of the transneuronal tracing technique (e.g., in non-specific retrograde transfer of the virus between neurons via the extracellular space or via glia; see Card *et al.* [1990], Jansen *et al.* [1993], Enquist and Card [2003]). It may partly be related to the projection of vasoconstrictor neurons to most autonomically innervated target tissues. But even this argument is not entirely convincing to explain the relative uniformity of the central labeling with this technique when one compares, e.g., the neuronal networks associated with the adrenal medulla, pineal gland, heart, spleen and urinary bladder, all five having entirely different functions of their non-vasoconstrictor innervation. Furthermore, the innervation of the blood vessels in heart, spleen or urinary bladder has also different functions. The labeled central neuronal networks associated with these five target tissues are rather similar and resemble the neuronal network labeled from the cutaneous blood vessels of the rat tail, which are involved in thermoregulation (Smith *et al.* 1998; see Table 8.2).

In summary, the transsynaptic labeling studies have laid down the groundwork about the anatomy of central autonomic circuits and will be one basis for further functional and anatomical studies of these circuits. Unfortunately, few if any of these labeling studies conducted so far provide much of a basis for understanding the differential regulation of final autonomic motor pathways.

Conclusions

Neurophysiological investigations of the peripheral sympathetic and parasympathetic neurons strongly support the notion that autonomic systems have a distinct organization in the central nervous system, including spinal cord and brain stem. Anatomic investigations, using various anterograde, orthograde and transsynaptic tracing techniques, have provided an important basis for this postulated functional organization of central autonomic systems. The structural organization is and will be the basis for future functional studies of the central autonomic systems.

1. Sympathetic preganglionic neurons are located in the intermediate zone of the thoracic and upper lumbar spinal cord. These neurons exhibit some broad mediolateral topographic organization with respect to the efferent innervation of the visceral and somatic body domains. Otherwise functionally different groups of preganglionic neurons are intermingled. Most preganglionic neurons supplying somatic tissues are located laterally in the funicular and principal part of the intermediolateral nucleus. Preganglionic neurons supplying viscera have a more medial location.
2. Parasympathetic preganglionic neurons in the sacral spinal cord are also located in the intermediate zone with some organotopic organization. Neurons innervating the urogenital tract are located laterally and those innervating the hindgut medially.
3. Parasympathetic preganglionic neurons in the brain stem are located in the external formation of the nucleus ambiguus (largely cardiomotor neurons and bronchomotor neurons), in the dorsal vagus motor nucleus (DMNX; largely gastrointestinal tract), in the salivary nuclei and in the nucleus Edinger–Westphal (eye; birds and primates) or ventral to this nucleus (rat, rabbit, cat). Each of these nuclei projects to a distinct organ system. The preganglionic neurons in the DMNX exhibit a rostrocaudal columnar viscerotopic organization, those innervating the stomach being situated medially and those innervating the cecum laterally.
4. The nucleus tractus solitarii (NTS) receives peripheral synaptic inputs from gustatory afferents (rostral half) and from afferents of the gastrointestinal tract, the respiratory system and the cardiovascular system (caudal half of the NTS). Projection of visceral afferents from the latter three systems to the subnuclei of the NTS exhibits some topographic organization. This topography is particularly obvious for the different sections of the alimentary canal, showing a mediolateral and caudorostral organization.
5. Neurons in the NTS consist of projecting neurons and interneurons. The NTS neurons project to many nuclei in spinal cord, brain stem, hypothalamus and forebrain and receive synaptic input from most of these brain sites. The NTS has relay and integrative

functions. The mechanisms underlying the integrative functions are almost unknown.
6. Several groups of autonomic interneurons lie in the thoracolumbar (laminae I, II, V, VI, IX and X) and sacral spinal cord (close to the preganglionic neurons, laminae I and X, dorsal commissural nucleus) whereas propriospinal interneurons have been identified in the cervical spinal cord (lateral funiculus, lateral spinal nucleus, dorsal horn).
7. Sympathetic premotor neurons that project through the dorsolateral funiculus to the preganglionic neurons and/or the corresponding autonomic interneurons are situated in ventrolateral medulla (mainly in the rostral ventrolateral medulla), in the ventromedial medulla, in the caudal raphe nuclei, in the A5 area of the pons, in the paraventricular nucleus of the hypothalamus and in the lateral hypothalamus.
8. Parasympathetic premotor neurons that project to the preganglionic neurons innervating pancreas, trachea or salivary glands are also located in the ventral medulla in the same nuclei as sympathetic premotor neurons but additionally in the periaqueductal grey (mostly ventrolateral). The same applies to parasympathetic premotor neurons associated with the urinary bladder and probably also those associated with the other pelvic target organs. Some of these lie in Barrington's nucleus in the pons.
9. Neurons antecedent to autonomic premotor neurons are also present in various nuclei of pons (parabrachial nuclei, Kölliker–Fuse nucleus, Barrington's nucleus), periaqueductal grey, various hypothalamic nuclei and various nuclei of the telencephalon (preoptic nuclei, amygdaloid complex, insula, infralimbic cortex, medial preoptic cortex).
10. Attempts to relate the anatomical location of autonomic premotor neurons and those antecedent to them have so far failed to yield patterns specific for particular target organs.

Notes
1. A few lumbar sympathetic preganglionic neurons project to sacral paravertebral ganglia and form synapses with postganglionic neurons that project to pelvic organs (see Jänig and McLachlan [1987]). These neurons probably supply the vasculature of the pelvic organs.
2. The grey matter of the spinal cord is divided into laminae by cytoarchitectonic criteria according to Rexed (see Molander and Grant [1995]).
3. These numerical data for the rat are based on estimations described by Powley et al. (1992). In other publications of this author group estimations of afferent and efferent (preganglionic) neurons projecting in the abdominal vagal branches of the rat, based on counting of cell bodies in the DMNX labeled by the fluorescent tracer True Blue and on counting of nerve fibers under the light and the electron microscope, resulted in the following numbers: the vagal branches contain about 22 000 nerve

fibers (11 000 in each subdiaphragmatic vagal trunk); 16 000 nerve fibers are afferent and 6000 efferent. About 85% of the preganglionic neurons project through the gastric branches, 11% through the celiac branches and 4% through the hepatic branch of the subdiaphragmatic vagus nerves (Fox and Powley 1985; Prechtl and Powley 1990a; Berthoud and Neuhuber 2000) (see note 5 for the innervation territories of these branches).
4. The rostrocaudal columnar organization of the preganglionic neurons in the DMNX is based on labeling of the branches of the subdiaphragmatic vagus nerves in the rat. There does not exist a one-to-one correspondence between these vagal branches and the different parts of the gastrointestinal tract. The hepatic branch contains also preganglionic axons that innervate the antrum and duodenum. The gastric branches contain preganglionic axons innervating the proximal duodenum. The celiac branches contain preganglionic axons innervating (in addition to the small intestine, cecum and proximal colon) the distal part of the duodenum (Berthoud et al. 1991). Finally, preganglionic neurons innervating the pancreas project through the gastric branches and the hepatic branch and probably not through the celiac branches (Berthoud and Powley 1990; Berthoud et al. 1990). These three points explain why preganglionic neurons to liver and pancreas are represented more medially in the DMNX (Figure 8.10c).
5. In primates the rostral (gustatory) NTS projects viscerotopically also to the basal part of the ventromedial nucleus of the thalamus (VMb), which in turn projects to the dorsal insula (limbic sensory cortex representing interoception; see Chapter 2.6).
6. There is some confusion about the nomenclature and the limitations of these nuclei (see Blessing 1997; Mason 2001). Here I will use the nomenclature of Paxinos and Watson (1998). Under ventrolateral medulla (VLM), the rostral ventrolateral medulla (RVLM) and the lateral paragigantocellular nucleus (LPGi) are subsumed. The ventromedial medulla (VMM) includes the medial paragigantocellular reticular nucleus, alpha and ventral (GiA, GiV) and the parapyramidal region or nucleus (PPy). (See Chapter 7.2 and Figure 10.2).

Chapter 9

Spinal autonomic systems

9.1 The spinal autonomic reflex pathway as a building
   block of central integration   page 332
9.2 Spinal reflexes organized in sympathetic systems   336
9.3 Sacral parasympathetic systems   349
9.4 The spinal cord as integrative autonomic organ   362

The activity in spinal preganglionic neurons is the result of the summation of potential changes in the neuronal membrane arising from integrative processes in the spinal cord, brain stem, hypothalamus and forebrain.[1] As in the somatomotor system (Lloyd 1960), the spinal cord itself is a major highway of interaction between the brain and the autonomic target organs. In this chapter, I will concentrate on the thoracolumbar spinal autonomic (sympathetic) systems and summarize what is known and discussed about the sacral spinal autonomic (parasympathetic) systems. I will give arguments supporting the idea that the spinal cord contains intrinsic autonomic systems and that the spinal autonomic systems are coordinated via function-specific autonomic interneurons and integrated in the regulation of autonomic effector organs. This integration involves synaptic events derived from segmental spinal, propriospinal and supraspinal sources. Thus, I will put forward the idea that the spinal autonomic systems are integrated in the regulation of activity in preganglionic neurons by supraspinal centers. This idea has been borrowed from our understanding of the physiology of regulation of the activity in somatic motoneurons in the generation of movements. Alternative ideas discussed here are that spinal autonomic circuits modulate descending signals in the regulation of the activity in the preganglionic neurons or that spinal autonomic circuits are not important during normal regulation of autonomic target organs via sympathetic and sacral parasympathetic systems.

Sacral parasympathetic systems are essential in the regulation of lower urinary tract, hindgut and reproductive organs. The organization of these sacral parasympathetic systems, their functioning in the regulation of the three pelvic organ systems and the underlying peripheral and central mechanisms have been described in detail elsewhere (Loewy and Spyer 1990a; Jänig 1996b, c; de Groat 2002; McKenna and Marson 1997; McKenna 1999, 2000, 2001, 2002).

In Chapter 4, I have emphasized that many components of the reflex patterns in sympathetic neurons are likely to depend on spinal circuits (see Figures 4.3, 4.5 and 4.16). The principles of organization of the basic circuits of the parasympathetic systems in the brain stem, related to gastrointestinal function, cardiac function, function of airways, secretions of exocrine glands and to regulation of pupil diameter, are presumably organized in the same way (see Chapter 10.7).

9.1 | The spinal autonomic reflex pathway as a building block of central integration

9.1.1 Structure and inputs of the spinal autonomic reflex pathway

Sympathetic preganglionic neurons receive synaptic inputs from spinal and propriospinal interneurons and sympathetic premotor neurons in the brain stem and hypothalamus (Figure 8.15). Primary afferent neurons from skin, deep somatic tissues and viscera form spinal reflex circuits with the preganglionic neurons. These reflex circuits are disynaptic or polysynaptic although occasional monosynaptic connections of primary afferent neurons with preganglionic neurons have been reported. The autonomic circuits in brain stem and hypothalamus that project to the spinal cord connect to the preganglionic neurons mono-, di- or polysynaptically. This arrangement of preganglionic neurons, interneurons, primary afferent neurons and autonomic premotor neurons in supraspinal centers is the basic building block of spinal autonomic systems (Figure 9.1) and can be recognized for most if not all final spinal autonomic pathways. It probably applies to final cranial autonomic (parasympathetic) pathways as well.

Most autonomic premotor neurons in the supraspinal centers that connect synaptically to the spinal autonomic circuits project mainly through the dorsolateral funiculi of the spinal cord. Their projection to individual preganglionic neurons is partly bilateral. The location of the cell bodies of these neurons has been identified by experiments in which the descending neurons have been labeled retrogradely using various types of markers including neurotropic viruses (see Chapter 8.4; Tables 8.2 and 8.3).

The functions of most of these individual autonomic premotor pathways projecting to spinal sympathetic or parasympathetic circuits are unknown. Sympathetic premotor pathways related to the regulation of resistance vessels or heart (see Subchapters 10.2 and 10.3, Figures 10.4 and 10.7) and related to thermoregulation of cutaneous blood flow (see Subchapter 10.5 and Figure 10.14) have been characterized. The number of putative sympathetic

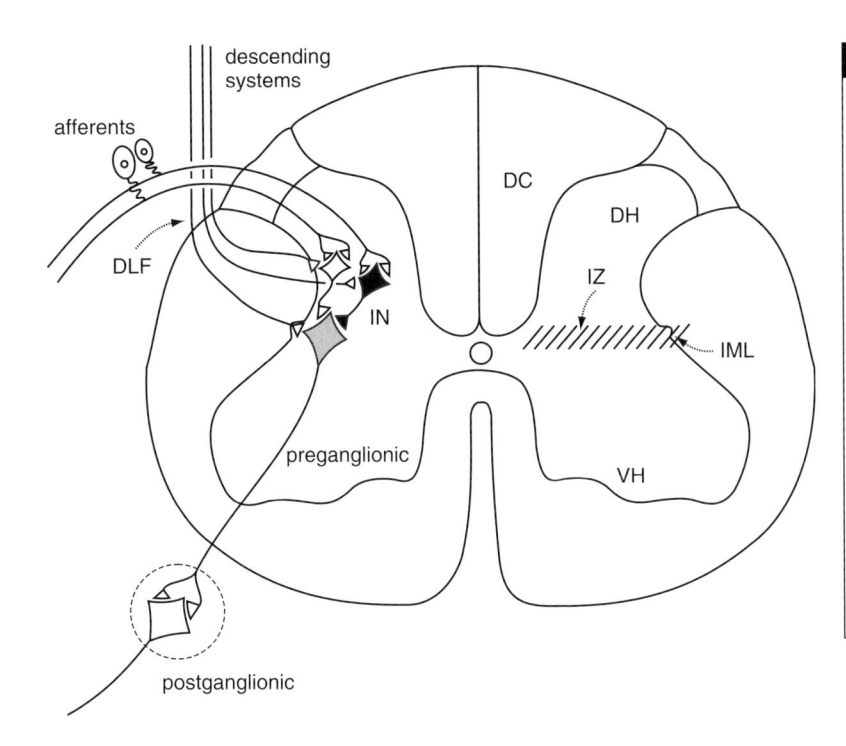

Figure 9.1 The spinal autonomic reflex pathway as the building block between supraspinal centers and final autonomic pathways. There is usually at least one (excitatory or inhibitory) interneuron (IN) between (visceral or somatic) primary afferent neurons and preganglionic neurons. Supraspinal centers in brain stem and hypothalamus project mainly through the dorsolateral funiculus (DLF) of the spinal cord and connect synaptically to the autonomic interneurons and the preganglionic neurons. DC, dorsal column; DH, dorsal horn; IML, intermediolateral nucleus; IZ, intermediate zone; VH, ventral horn.

premotor neurons in brain stem and hypothalamus identified so far morphologically and histochemically seems to be low relative to the number of postulated types of descending sympathetic premotor neurons, which appear necessary for the function of the different sympathetic systems (see Figure 4.23 for the lumbar sympathetic systems). This has also been shown for the spinal parasympathetic systems that are involved in the regulation of the storage and evacuation of the pelvic organs.

9.1.2 Synaptic transmission and transmitters

Synaptic transmission from primary afferent neurons and autonomic premotor neurons to the spinal autonomic circuits and within the spinal autonomic circuits is chemical[2]:

- Transmission from primary afferent neurons with myelinated and unmyelinated fibers to spinal interneurons is excitatory, the excitatory transmitters being amino acids (e.g., glutamate and possibly aspartate) and possibly neuropeptides (such as substance P and calcitonin gene-related peptide [CGRP]) although the functions of these peptides in synaptic transmission in the spinal cord are not well understood (Levine et al. 1993).
- Transmission from descending systems to preganglionic neurons and autonomic interneurons is excitatory or inhibitory (Deuchars et al. 1997, 2001b; Llewellyn-Smith and Weaver 2001; Whyment et al. 2004). The main *excitatory transmitters* of the autonomic premotor systems, projecting bilaterally through the dorsolateral funiculi, in fast synaptic transmission in many, if not most, of these systems are

glutamate and other excitatory amino acids. The *inhibitory transmitters* in fast synaptic transmission from sympathetic premotor neurons to preganglionic neurons are γ-aminobutyric acid (GABA) and possibly glycine. Both GABA and glycine may be co-expressed in the same neurons. The functional meaning of this co-expression is unknown (Llewellyn-Smith *et al.* 1995, 1997; Llewellyn-Smith and Weaver 2001).

- 5-Hydroxytryptamine, adrenaline, noradrenaline and dopamine and the various neuropeptides localized or colocalized in autonomic premotor neurons (see Figure 8.16) may also be transmitters. Focal electrical stimulation in spinal cord slices evokes, in addition to fast synaptic potentials elicited by release of amino acids, slow excitatory (sEPSP) and slow inhibitory postsynaptic potentials (sIPSP) in the sympathetic preganglionic neurons. The sEPSPs may be mediated by noradrenaline or adrenaline acting via α_1-adrenoceptors and the sIPSPs via α_2-adrenoceptors (Yoshimura *et al.* 1987a, b). Dopamine has inhibitory effects on preganglionic sympathetic neurons in vitro (Gladwell and Coote 1999). However, overall it is unclear how these monoamines and neuropeptides released by sympathetic premotor neurons are used as transmitters during autonomic regulation.
- Nitric oxide (NO) released by sympathetic neurons during excitation may act on nearby interneurons causing the release of the inhibitory transmitter glycine (Yang *et al.* 2004).
- Synaptic transmission from the interneurons to the preganglionic neurons is excitatory or inhibitory, probably involving excitatory amino acids (such as glutamate) and inhibitory amino acids (such as GABA and glycine).
- Synaptic transmission from descending excitatory (glutamatergic) and inhibitory (GABAergic) pathways to preganglionic neurons or associated interneurons is modulated presynaptically by adenosine. Excitatory transmission is inhibited via A_1 adenosine receptors and inhibitory transmission enhanced via A_{2A} adenosine receptors located in the presynaptic terminals. Adenosine may be released within presynaptic terminals or as the result of extracellular breakdown of adenosine triphosphate (ATP) during local ischemia and hypoxia. This mechanism of inhibiting excitatory synaptic transmission and enhancing inhibitory synaptic transmission may protect sympathetic neurons and associated interneurons against overexcitation (Deuchars *et al.* 2001a; Brooke *et al.* 2004).

9.1.3 Spinal interneurons associated with autonomic circuits

Up to now no autonomic interneuron has been fully and unequivocally identified by its function; the best examples of reasonably well-identified autonomic interneurons exist in the sacral spinal cord (see below). However, indirect evidence for the existence of functionally different types of autonomic interneurons is overwhelming, based on several physiological and morphological investigations (see Cabot et al. [1994], Cabot [1996], Weaver and Polosa [1997], Shefchyk [2001]):

- Putative interneurons were labeled retrogradely transsynaptically by compounds applied to the cut preganglionic axons or their terminals in the respective ganglia. The compounds are actively taken up by the preganglionic fibers and conveyed by retrograde axoplasmic transport mechanisms to the preganglionic cell bodies and dendrites. Here they are released by exocytosis and actively taken up by the presynaptic terminals innervating the preganglionic neurons. Such substances are, e.g., wheat germ agglutinin, the B-unit of the cholera toxin or neurotropic viruses (see Chapter 8.1). These types of studies demonstrate that autonomic interneurons are located in the grey matter of the thoracolumbar spinal cord in laminae I, II, V, VII (dorsal to the intermediolateral nucleus) and X (intermediomedial nucleus, central autonomic nucleus in rats) (see Table 8.2 and Subchapter 8.2.1).
- Propriospinal neurons labeled by a neurotropic virus from the stellate ganglion, the adrenal medulla or the tail of rats are located in various laminae of the grey matter, in the lateral spinal nucleus and in the lateral funiculus of the cervical spinal cord (Strack et al. 1989a, b; Jansen et al. 1995a, b; Jansen and Loewy 1997; Smith et al. 1998) (see Table 8.2). These propriospinal neurons may form synaptic contacts with preganglionic neurons via local interneurons, but less so directly (Tang et al. 2003, 2004).
- In the sacral spinal cord, similar morphological investigations have been conducted using pseudorabies virus injected into the urinary bladder and urethra of rats. Interneurons were found dorsal and medial to the preganglionic neurons in the lateral band, dorsal to the central canal (in the dorsal commissural nucleus), in lamina I and lamina X of the dorsal horn (de Groat et al. 1996) (see Table 8.3, literature here).
- Neurophysiological recordings demonstrate some neurons in the spinal cord close to sympathetic preganglionic neurons, which could not be activated antidromically by stimulation of preganglionic axons in the ventral root and which exhibited discharge patterns that were correlated with those in the preganglionic neurons (Barman and Gebber 1984).
- Some neurons in the intermediomedial nucleus in the cat (central autonomic nucleus in the rat) exhibit discharge patterns similar to preganglionic neurons in the intermediolateral nucleus and inhibit the preganglionic neurons (McCall et al. 1977). In vitro experiments show that these interneurons in the central autonomic area are GABAergic and form inhibitory synapses with preganglionic neurons (Deuchars et al. 2005). These inhibitory interneurons probably mediate inhibition of preganglionic neurons generated from supraspinal centers (and even the prefrontal cortex [Bacon and Smith 1993]).

- In both acutely and chronically spinal rats,[3] discharges in some neurons in the dorsal horn and in the intermediate zone in the thoracic segments T10 and adjacent segments precede the efferent discharge in the renal nerve by a latency of about 60 ms. These putative interneurons exhibit reflex patterns similar to renal sympathetic neurons, but are distinct in the electrophysiological properties compared to preganglionic neurons (Chau and Schramm 1997; Chau et al. 2000; Krassioukov et al. 2002).
- Two weeks after transection of the spinal cord about 30% to 50% of the synaptic inputs to sympathetic preganglionic neurons remain, about 40% of them being glutamatergic (excitatory) and 60% GABAergic (inhibitory) (Llewellyn-Smith et al. 1997; Llewellyn-Smith and Weaver 2001). In control rats, the proportions are closer to 50:50.
- Reflexes generated in preganglionic neurons to electrical stimulation of *myelinated* afferents in white rami in normal (and spinal) cats have relatively long and variable latencies, arguing that they are not elicited monosynaptically but di- or polysynaptically (see shortest latency responses in Figure 9.2b).
- Several functionally distinct groups of autonomic interneurons have to be postulated on the basis of neurophysiological investigations of reflexes elicited in functionally identified sympathetic neurons upon physiological stimulation of afferents in spinal cats (see Figures 4.3, 4.5 and 4.16 and see below).
- Electrical stimulation of the dorsolateral funiculus, through which the axons of most autonomic premotor neurons project, or electrical stimulation of the rostral ventrolateral medulla (RVLM), elicits monosynaptic and polysynaptic EPSPs in preganglionic neurons (in vivo experiments and in vitro experiments) (Dembowsky et al. 1985; Deuchars et al. 1997).
- Several distinct types of autonomic interneuron that are associated with preganglionic neurons innervating pelvic organs have been postulated or shown to exist on the basis of neurophysiological experimental investigations in vivo or in vitro (see Shefchyk [2001] and Figures 9.8 to 9.10).

In spite of this limited knowledge, we can hypothesize that the spinal cord circuits are involved in distinct discharge patterns in sympathetic and spinal parasympathetic neurons. We can now develop a concept about the role of spinal autonomic circuits in the regulation of activity of the final autonomic pathways by the brain.

9.2 Spinal reflexes organized in sympathetic systems

In Chapter 4, I have discussed the reflex pattern in functionally different types of sympathetic pre- and postganglionic neurons and indicated that many characteristics of these reflex patterns are dependent on spinal reflex pathways. Here I will extend this idea and show, for the lumbar sympathetic outflow, *first*, that some functionally defined sympathetic pathways are associated with several distinct spinal reflex pathways and, *second*, that there are indications for coordination of spinal autonomic reflex circuits in the spinal cord.

9.2.1 Segmental and suprasegmental reflexes elicited in preganglionic sympathetic neurons by electrical stimulation of afferents

Sympathetic preganglionic neurons exhibit short- and long-latency reflexes to electrical stimulation of spinal afferents (e.g., in the dorsal roots or white rami). These reflexes are mediated by spinal and supraspinal pathways. The spinal reflexes are most powerful when the afferents of the same or a neighboring segment in which the preganglionic neurons are located are stimulated. Although there seems to exist some general organizing principle of reflexes in the sympathetic system at the spinal segmental and supraspinal levels it is unclear from these investigations, *first*, which functional types of afferent are involved in these reflexes and, *second*, whether different types of sympathetic neuron exhibit functionally distinct types of segmental (spinal) and suprasegmental reflexes (Sato and Schmidt 1971, 1973; Sato 1972; Sato *et al.* 1997). Coote (1988) has extensively discussed the integration of spinal and supraspinal reflexes elicited in sympathetic preganglionic neurons by electrical stimulation of visceral and somatic nerves.

Spinal and supraspinal reflexes elicited by electrical stimulation of spinal afferents in functionally identified sympathetic preganglionic neurons have been examined in only one study (Bahr *et al.* 1986c):

- *Preganglionic visceral vasoconstrictor neurons* that project in the lumbar splanchnic nerves exhibit very powerful reflexes to electrical stimulation of lumbar somatic and visceral afferents (Figure 9.2). The excitatory reflexes consist of several components when classified by their latencies: short-latency reflexes are probably mediated by spinal reflex circuits and long-latency reflexes by supraspinal reflex circuits. Interestingly, the reflexes with the shortest latency have the lowest threshold and are probably elicited by stimulation of fast-conducting somatic and visceral afferents. Similar powerful short- and long-latency reflexes to electrical stimulation of afferents have been reported in the cardiac and renal nerves (Coote and Downman 1966; Coote and Sato 1978; Coote 1984; review see Sato and Schmidt [1973]).
- *Preganglionic motility-regulating neurons* that project in the lumbar splanchnic nerves also exhibit reflexes to electrical stimulation of lumbar somatic and spinal afferents (Figure 9.2c). However, these reflexes are present in only about 60% of the functionally identified motility-regulating neurons; they are much weaker than the reflexes in visceral vasoconstrictor neurons and most of them have

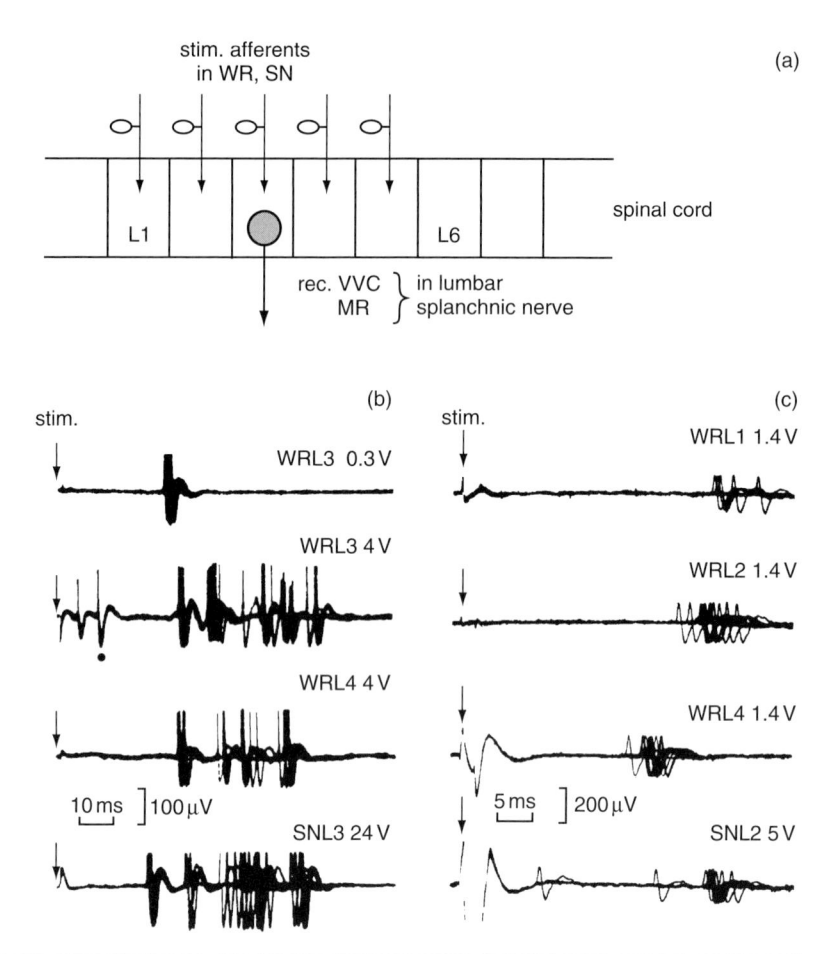

Figure 9.2 Spinal and supraspinal reflex activation of sympathetic neurons to electrical stimulation of spinal afferents. Activity was recorded from a preganglionic visceral vasoconstrictor axon (VVC in [b]) and a preganglionic axon of a motility-regulating neuron (MR in [c]), which were both isolated for extracellular unit recording from one of the lumbar splanchnic nerves. These functional classifications arise from the characteristic discharge patterns (see Subchapter 4.3). Spinal afferents in spinal nerves (SN, nerve containing somatic afferents) or in white rami (WR, containing visceral afferents) were stimulated electrically with single pulses (pulse duration 0.5 ms for SN and 0.2 ms for WR). Different segmental nerves were stimulated in each row (e.g., WRL3, etc.). (a) Experimental set-up. rec., record; stim., stimulus. (b) VVC neuron. The neuron projected through WRL_3 (axon directly excited by stimulation of WRL3, see dot). Each trace 10 times superimposed. Note the powerful short-latency (probably segmental) and long-latency (probably suprasegmental) reflex responses. (c) MR neuron. Each trace 11 times superimposed. Note the weak reflex activity and the short latency of the reflexes in the MR neuron. (b) modified from Bahr et al. (1981); (c) modified from Bahr et al. (1986c).

a short latency (i.e., are probably mediated by a spinal reflex pathway) (Bahr et al. 1986c).

Intracellular recordings from preganglionic sympathetic neurons in the spinal segments T3 of the cat (Dembowsky et al. 1985) have shown that electrical stimulation of somatic or visceral spinal afferents elicits early and late EPSPs in $\geq 80\%$ of the neurons when the spinal cord is intact, early EPSPs in almost all neurons in spinal animals and late EPSPs in about 50% of the neurons in spinal animals. The results obtained so far on functionally identified preganglionic neurons reported by Bahr *et al.* (1986c) agree with these intracellular measurements. These experiments show: (1) that spinal reflexes exist in most sympathetic preganglionic neurons and (2) that spinal and supraspinal reflexes are different in functionally distinct types of preganglionic neuron. Unfortunately, these results, which were obtained on functionally identified preganglionic sympathetic neurons, give only limited answers to the question of which spinal reflexes are associated with which type of sympathetic system because the function of the spinal afferents stimulated electrically are unknown.

9.2.2 Recovery of spinal reflexes after spinal cord transection

When the spinal cord is acutely isolated from the brain stem, almost all reflexes in sympathetic neurons, including those elicited by strong noxious stimuli, disappear for days to weeks in cats and in humans, probably as a consequence of the interruption of the descending control systems. This state is called "spinal shock"[4] (Guttmann 1976; Jichia and Frank 2000; Mathias and Frankel 2002). The recovery of functionally distinct reflexes elicited by physiological stimulation of primary afferent neurons[5] and of ongoing activity in those sympathetic systems that are under dominant supraspinal control takes weeks to months, depending on the functional type of sympathetic system, the type of (excitatory or inhibitory) reflex and the animal species studied.[6] The following data were obtained in cats (Horeyseck and Jänig 1974c; Jänig and Spilok 1978; Jänig and Kümmel 1981):

- Rate of spontaneous activity, proportion of neurons with spontaneous activity and excitatory reflexes to noxious stimuli of skin recover faster in postganglionic muscle vasoconstrictor neurons than in postganglionic cutaneous vasoconstrictor neurons. For the spontaneous activity the recovery is in the range of <40 (muscle) or <60 days (skin). For the excitatory reflexes the recovery is in the range of 30 to 50 days (muscle) and 60 to 90 days (skin).
- Inhibitory reflexes elicited in postganglionic neurons by cutaneous noxious stimuli (noxious stimuli; stimulation of hair follicles) take longer to recover than excitatory ones.
- Inhibitory reflexes elicited in cutaneous vasoconstrictor neurons by visceral stimuli (distension or contraction of urinary bladder or colon) disappear, thus never recover, after transection of the spinal cord. In chronic spinal cats these reflexes are excitatory (Figure 9.7; Jänig 1985a, 1996a).[7]
- Reflexes in sudomotor neurons (to stimulation of Pacinian corpuscles [vibration reflex] or of nociceptors) take >50 days to recover.
- Spinal excitatory reflexes and ongoing activity in some sympathetic systems that project to visceral organs and have no vasoconstrictor function (e.g., "motility-regulating" neurons in the cat

[Bartel *et al.* 1986]; see Chapter 4.3) change very little after acute transection of the spinal cord performed rostral to the preganglionic cell bodies of these systems. Spinal inhibitory reflexes in these neurons may be enhanced or depressed acutely after spinalization.

In conclusion, "spinal shock" (1) is present in vasoconstrictor systems and in the sudomotor system, (2) is particularly strong for inhibitory reflexes (e.g., in the cutaneous vasoconstrictor neurons) and generally in the cutaneous vasoconstrictor system and (3) is absent or weak for excitatory reflexes in sympathetic motility-regulating neurons.

9.2.3 Autonomic dysreflexia chronically after spinal cord transection

In the chronic stage after spinal cord transection, spinal reflexes to noxious and innocuous somatic and visceral stimuli may be very strong (Horeyseck and Jänig, 1974c; Jänig and Spilok 1978; see Lee *et al.* [1995]; Karlsson [1999]; Mathias and Frankel [2002]). For example, in animals and humans with the spinal cord interrupted at the upper thoracic or lower cervical level, leading to tetraplegia or high paraplegia in patients, there is uniform reflex activation of the vasculature in the viscera, deep somatic tissues (e.g, skeletal muscle) and skin, with an ensuing dramatic increase of arterial blood pressure, reflex activation of sweat glands (although this may be rather irregular [Guttmann 1976]), reflex activation of arrector pili muscles and other reflexes to stimulation of visceral afferents during bladder or hindgut contractions or distension or to stimulation of deep somatic afferents, e.g., during muscle spasms (see Jänig [1985a, 1996a]; Mathias and Frankel [2002]) or noxious stimuli to the skin. These spinal cardiovascular and other autonomic reflexes, together with reflex dysfunctions of urinary bladder, reproductive organs and gastrointestinal tract, are part of a condition called *autonomic dysreflexia*.

The detailed mechanisms underlying the autonomic dysreflexia are only partially known (for review and literature see Weaver *et al.* 2002):

- Transection of the spinal cord leads to loss of input to the preganglionic neurons and autonomic interneurons from descending excitatory and inhibitory supraspinal and propriospinal control systems. Initially, in the first days after spinal cord transection, the dendrites of preganglionic neurons shrink. Later these dendrites regrow again.
- The density of synapses on the somata and dendrites of the preganglionic neurons decreases by about 70% and 50%, respectively in the chronic state, arguing that interneurons do not or minimally sprout to the denervated sites of the preganglionic neurons (Figure 9.3) by two weeks after the injury. However, sprouting at longer times after injury, using electron microscopy, has not been studied.

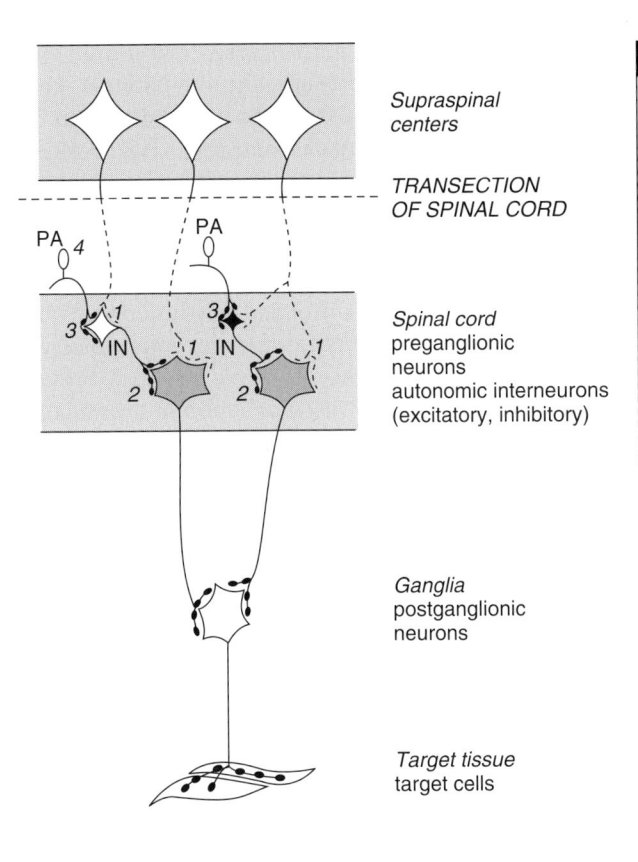

Figure 9.3 Consequences of spinal cord transection for spinal autonomic systems. *1*, denervation of preganglionic neurons and autonomic interneurons; *2*, changes of geometry of preganglionic neurons; sprouting of interneurons into the denervated territories of the preganglionic neurons seems unclear; *3*, sprouting of afferent neurons; *4*, neurochemical changes of afferent neurons. IN, interneurons; PA, primary afferent neurons.

- Most of the synapses on the preganglionic neurons deprived of their supraspinal synaptic inputs are GABAergic (about 60%), i.e. inhibitory, and about 40% are glutamatergic (Llewellyn-Smith *et al.* 1995; Llewellyn-Smith and Weaver 2001).
- Peptidergic small-diameter afferents sprout in the spinal cord distal to the lesion site, probably forming new synapses with autonomic interneurons and other neurons (Krenz *et al.* 1999; Weaver *et al.* 2001). However, only fibers containing CGRP but not those containing substance P sprout (Marsh and Weaver, 2004). This is surprising since CGRP is colocalized in most afferent neurons containing substance P (Lawson 2005). Since more primary afferent neurons contain CGRP than substance P (Lawson 1996; 2005) this would mean that only a subclass of primary afferent neurons containing CGRP sprout.
- The changes are probably produced by subsets of astrocytes in the injured spinal cord *and/or by entry of immune cells into the injured cord*. Whether macrophages (i.e. activated microglia) are involved is debated. These cells are sources of nerve growth factor and other molecules, which in turn trigger sprouting of peptidergic primary afferents (Krenz *et al.* 1999; Krenz and Weaver 2000; Marsh *et al.* 2002; Brown *et al.* 2004). However, other processes must be involved since substance P afferents do not sprout.

The changes occurring in the spinal cord cannot entirely explain the autonomic dysreflexia observed in the patients. Responses of the rat tail arteries to nerve stimulation is enhanced in chronically spinal animals. This increase of vascular responsiveness is probably related to decrease of activity in vasoconstrictor neuron following spinal cord transection since decentralization of the postganglionic neurons innervating the tail artery (by cutting the preganglionic axons) is also followed by an increased responsiveness of the tail artery to nerve stimulation. Both pre- and postjunctional changes seem to be involved in this increased vascular responsiveness, such as increased release of noradrenaline on nerve impulse activity and increased responsiveness of the vascular smooth musculature (sensitization of the contractile mechanism to calcium entering the cell from outside) but probably not an increase in number and affinity of adrenoceptors (Yeoh et al. 2004a, b). These data imply that the increased reactivity of blood vessels to nerve impulses chronically after spinal cord transection could contribute to autonomic dysreflexia of the vasculature.

Here I want to emphasize, irrespective of the plastic changes occurring in the spinal cord following spinal cord injury, that the isolated spinal cord is capable of regulating residual autonomic functions and that it is the *preservation* of these discrete spinal autonomic functions after interruption of the spinal cord that is remarkable rather than the uniformity of the discharge of neurons of functionally different sympathetic pathways. Thus, I want to put the main emphasis on the *differentiation* of the spinal autonomic circuits and that some reflexes elicited by physiological stimulation of somatic or visceral afferents are qualitatively similar in normal and spinal cats. This view does not deny the importance and prominence of hyperreflexia of spinal autonomic systems after spinal cord transection, in particular, to visceral stimuli.

9.2.4 Spinal reflexes upon physiological stimulation of afferents chronically after spinal cord transection

The vasoconstrictor pathways to skeletal muscle, viscera or skin, the sudomotor pathway and the sympathetic non-vasoconstrictor pathways to pelvic viscera and hindgut are connected to spinal reflex circuits that are linked to specific groups of afferents. These spinal autonomic reflex circuits are summarized in Figure 9.4 and in Table 9.1. They have been inferred on the basis of experimental neurophysiological investigations of the reflexes in these functionally identified sympathetic neurons in chronic spinal cats (for the vasoconstrictor pathways and the sudomotor pathway) and in acutely spinal cats (for the motility-regulating pathways). It is clear from these studies that some components of the reflex patterns that were observed in these neurons in cats with intact spinal cord are organized at spinal levels (see Chapter 4).

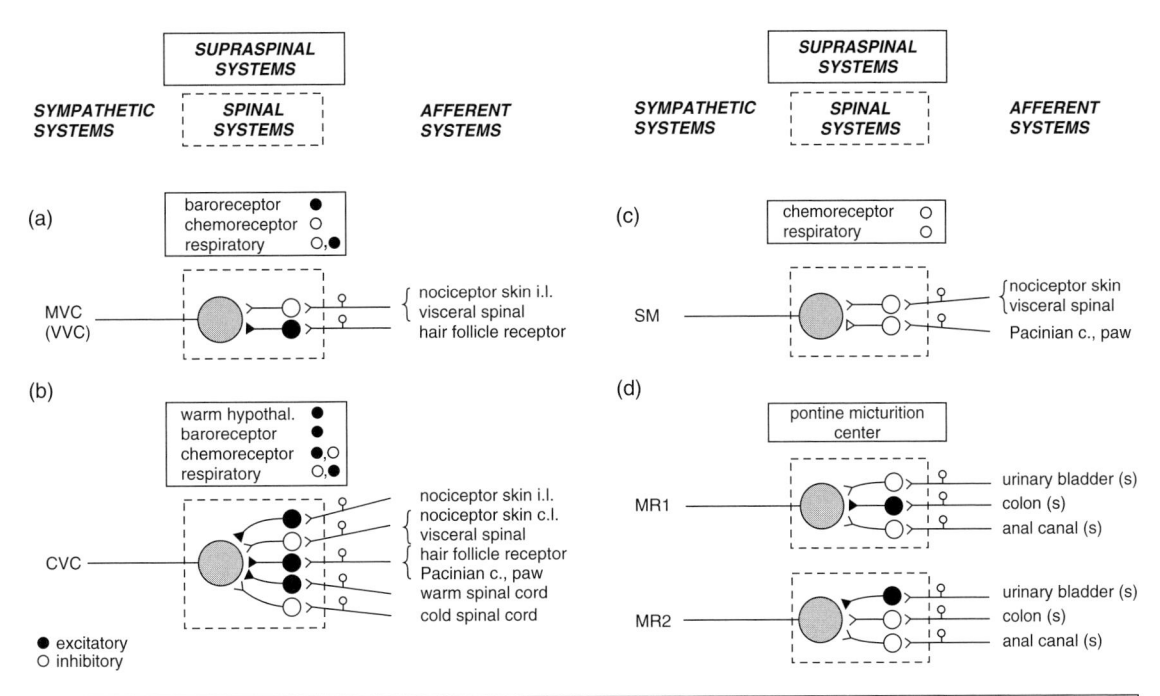

Figure 9.4 Spinal reflex pathways in lumbar sympathetic systems. These pathways were postulated on the basis of neurophysiological investigations of functionally identified pre- and postganglionic neurons projecting to skin, skeletal muscle and pelvic viscera (see Subchapters 4.1 to 4.3). They are defined by the spinal (right) and supraspinal afferent input systems and by the function (target cells, reflex patterns) of the output systems (left). The *spinal systems* are shown in the dashed-line boxes: reflex pathways are defined by the functional type of spinal afferent input. The *supraspinal systems* connecting to the spinal systems (and related to baroreceptor, chemoreceptors, respiratory generator, warm receptors or pontine micturition center) are shown in the solid-line boxes. Closed circles, inhibitory system/interneurons; open circles, excitatory system/interneurons. After chronic transection of the spinal cord, some neurons of each type of these sympathetic systems have ongoing activity. Visceral spinal, sacral afferents from pelvic organs; c.l./i.l., afferent input contralateral/ipsilateral to sympathetic outflow. (a) MVC, muscle vasoconstrictor neurons; VVC, visceral vasoconstrictor neurons. (b) CVC, cutaneous vasoconstrictor neurons. (c) SM, sudomotor neurons. (d) MR, motility-regulating neurons. Warm hypothal., warming rostral hypothalamus; Pacinian c., paw, Pacinian corpuscle in paw; s, sacral afferent input. Modified from Jänig (1996a) with permission.

Neurons of the five types of sympathetic final pathways which are demonstrated in Figure 9.4 and Table 9.1 have spontaneous activity in the spinal state (Horeyseck and Jänig 1974c; Jänig and Spilok 1978; Bartel et al. 1986). In sympathetic pathways to the vasculature and to sweat glands, spontaneous activity is not present hours to days after interruption of the spinal cord, but may recover over weeks and months (see above). In human para- or tetraplegic patients, spontaneous activity in sympathetic postganglionic neurons innervating skin or muscle nerves of the hindlimb, recorded microneurographically, is low or absent, even after many months (Stjernberg and Wallin 1983; Wallin and Stjernberg 1984; Stjernberg et al. 1986).[8] In many motility-regulating neurons innervating pelvic viscera and colon, spontaneous activity remains acutely after spinalization (Bartel et al. 1986).

Table 9.1 | *Reflex pathways in the spinal cord associated with lumbar sympathetic systems*

Sympathetic system	Afferent neuron	Reaction[a]	References
MVC	Nox. cutaneous	Excitation	3,8
	Hair follicle	Inhibition	3
	Viscera, sacral	Excitation	5,8
CVC	Nox. cutaneous	Inhibition i.l.[b]	2–8
		Excitation c.l.	2,4
	Low threshold mech. (Pacinian c., hair follicle)	Inhibition[c]	3,8
	Viscera, sacral	Excitation[d]	5,8
	Warm, spinal cord	Inhibition	6
	(cold, spinal cord	Excitation)	6
SM	Nox. cutaneous	Excitation[b]	7,8
	Pacinian c., paw	Excitation	7,8
	Viscera, sacral	Excitation	5,8
MR1	Urinary bladder, sacral	Excitation	1
	Colon, sacral	Inhibition[e]	1
	Anus, sacral	Excitation[f]	1
MR2[g]	Urinary bladder	Inhibition	1
	Colon, sacral	Excitation	1
	Anus, sacral	Excitation[f]	1

Notes:
All data obtained on chronic spinal (MVC, CVC, SM) or acute spinal (MR) cats. Spinal cord transected between segmental level T7 and T13 30 to 135 days before the experiments; c.l., contralateral; i.l., ipsilateral
Abbreviations: MVC, muscle vasoconstrictor neurons; CVC, cutaneous vasoconstrictor neurons; SM, sudomotor neurons; MR, motility-regulating neurons; Pacinian c., Pacinian corpuscle
[a] All reflexes that are present in the sympathetic neurons after transection of the spinal cord are also present, with two exceptions (see[c,d]), in animals with intact spinal cord under standardized experimental conditions
[b] Inhibition and excitation outlast stimulus
[c] In cats with intact spinal cord, stimulation of hair follicle receptors elicits excitation mostly followed by depression of activity in many CVC neurons (Horeyseck and Jänig, 1974a; Grosse and Jänig, 1976)
[d] In cats with intact spinal cord, inhibition in most CVC neurons (see Häbler *et al.* 1992)
[e] Probably more pronounced in spinal preparation (acute) than in the intact preparation
[f] Afterdischarge sometimes present, but shorter in duration than in intact preparation (Bahr *et al.* 1986a); sometimes also inhibition present in acute spinal preparation to mechanical stimulation of the anal canal
[g] This pattern seems to be rare
References: 1. Bartel *et al.* (1986); 2. Grosse and Jänig (1976); 3. Horeyseck and Jänig (1974c); 4. Jänig (1975); 5. Jänig (1985a); 6. Jänig and Kümmel (1981); 7. Jänig and Spilok (1978); 8. Kümmel (1983)

The spontaneous activity in sympathetic neurons after spinalization is not correlated in any way with respiration or with cardiovascular parameters (e.g., with the excitation of arterial baroreceptors; see Chapter 10). The source of the spontaneous activity in these sympathetic neurons after spinal cord transection is in the spinal networks associated with the spinal pathways, possibly in the activity in primary afferents (Taylor and Weaver 1993), and possibly in the preganglionic neurons themselves.

Of particular interest is the comparison of the reflex circuits of functionally related sympathetic pathways innervating skin, pelvic viscera or skeletal muscle (such as cutaneous vasoconstrictor and sudomotor; muscle vasoconstrictor and cutaneous vasoconstrictor neurons; different types of motility-regulating neurons). In the chronic spinal cat, these pairs of final pathways seem to be reciprocally organized with respect to functionally distinct afferent inputs and therefore with respect to the populations of interneurons:

- Activity in postganglionic cutaneous vasoconstrictor neurons is inhibited and activity in sudomotor neurons increased during stimulation of nociceptors of the ipsilateral hindpaw in cats (Figure 9.5a_1, a_2). Interestingly, these reciprocal reflexes are followed by a long-lasting decrease and increase of spontaneous activity in the neurons respectively. These simultaneously occurring *long-lasting* depression of spontaneous activity and afterdischarges are typical of nociceptive reflexes in these neurons in chronic spinal cats but are almost absent in animals with intact spinal cord. Stimulation of contralateral nociceptors excites the neurons of both pathways.
- Stimulation of Pacinian corpuscles in the paw also leads to reflex excitation of sudomotor neurons and reflex inhibition of some cutaneous vasoconstrictor neurons, but there are no long-lasting changes of the spontaneous activity (Figure 9.5b_1, b_2).
- Reflexes in motility-regulating neurons type 1 and type 2 are reciprocally organized with respect to the afferent inputs from the urinary bladder and from the colon. Figure 9.6 demonstrates the reciprocal reflexes in motility-regulating type 1 neurons to distension of the urinary bladder (activation) and to distension of the distal colon (inhibition). These reflexes and those elicited by mechanical stimulation of the anal canal and of the perianal skin are mediated by several distinct types of spinal sacrolumbar reflex circuits.
- Stimulation of nociceptive and non-nociceptive afferents from skin elicits reflexes that are reciprocally organized between cutaneous and muscle vasoconstrictor neurons of the *same* distal hindlimb.
- Stimulation of warm receptors in skin of chronic paraplegic humans by high ambient temperatures (Randall et al. 1966) activates sudomotor neurons innervating the trunk skin and either inhibits cutaneous vasoconstrictor neurons or activates cutaneous vasodilator neurons (increase of cutaneous blood flow). By the same token, warming the spinal cord in chronic spinal cats decreases activity in cutaneous vasoconstrictor neurons innervating the hindpaws (with increase in cutaneous blood flow) and may possibly activate cutaneous vasodilator neurons (Gregor et al. 1976; Jänig and Kümmel 1981).

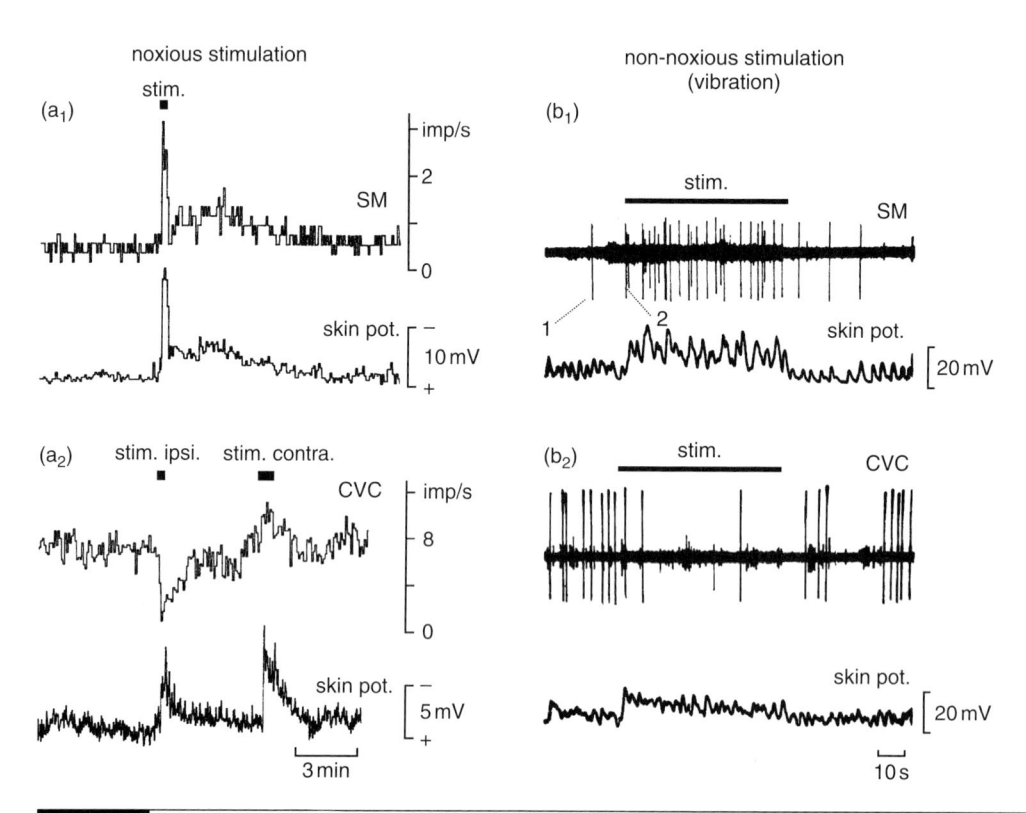

Figure 9.5 Reflexes in postganglionic cutaneous vasoconstrictor (CVC) and sudomotor (SM) neurons in chronic spinal cats. Activity was recorded from fine nerve fiber bundles isolated from the medial plantar nerve of the cat hindpaw. Skin potential was recorded simultaneously from the hairless skin of the central pad of the hindpaw indicating sweat gland activation (experimental set-up see Figure 4.10a). (a) Nociceptive reflexes to mechanical noxious stimulation of one of the toes of the ipsilateral (ipsi.) hindpaw (a_1) or of the ipsilateral or contralateral (contra.) hindpaw (a_2). Single sudomotor axon in a_1. CVC-multiunit activity in a_2. Note (1) SM and CVC neurons had ongoing activity, (2) long-lasting afterdischarge in SM neurons following noxious stimulation, (3) long-lasting inhibition of CVC activity following ipsilateral noxious stimulation and (4) absence of inhibition in CVC neurons following contralateral noxious stimulation. (a_1) 91 days and (a_2) 108 days after spinalization at the lower thoracic level. (b) Reflexes to stimulation of Pacinian corpuscles in the hindpaw by vibration. (b_1), recording from bundle with two SM axons (axons 1 and 2); (b_2), recording from bundle with one CVC axon (small signals probably also from CVC axons). Note (1) absence of afterdischarge in SM neurons and (2) inhibition of CVC neuron. (a) from Jänig and Spilok (1978); (b) from Kümmel (1983) with permission.

- However, visceral stimuli (e.g., contraction or distension of pelvic organs; mechanical stimulation of the anal skin) elicit uniform excitatory reflexes in all vasoconstrictor neurons and in sudomotor neurons accompanied by large increase of arterial blood pressure (in animals with high spinal transection) (Figure 9.7).[9]

The spinal autonomic reflex circuits, connected to the final sympathetic pathways to skin, skeletal muscle and pelvic viscera, have been postulated on the basis of neurophysiological investigations of sympathetic pre- and postganglionic neurons in spinal cats. There is strong indirect evidence (from recording of autonomic effector

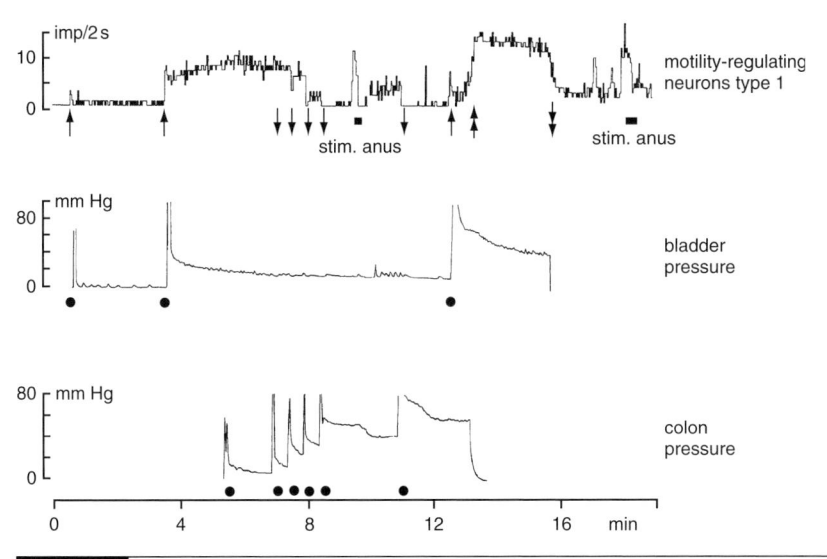

Figure 9.6 Interaction of reflex activation and inhibition of motility-regulating (MR) neurons by stimulation of sacral afferents from urinary bladder and colon in an acutely spinal cat. The spinal cord was sectioned 1 hour before the measurements at the thoracic level T8. Activity was recorded in a bundle with two preganglionic MR axons type 1 (see Subchapter 4.3) isolated from one of the lumbar splanchnic nerves. *Upper histogram*: activity in the two neurons (ordinate scale in impulses/2 s). ↑, excitation produced by filling of the urinary bladder; ↓, depression of activity produced by filling of the colon; ↟, excitation following release of pressure in the colon; ↡, decrease of activity following release of intravesical pressure; stim. anus, gentle mechanical shearing stimuli applied to the anal canal. *Middle record*: intravesical pressure measured through a urethral catheter in the urinary bladder. Filling of the bladder in steps with 10 ml each through the same catheter (dots). *Lower record*: pressure in a large flexible balloon in the distal colon connected to a catheter inserted through the anal canal. Filling of the colon balloon in steps six times with 10 ml each (dots). From Bartel et al. (1986) with permission.

responses and from clinical observations) that there exist other distinct spinal autonomic reflex circuits, and therefore also other groups of spinal autonomic interneurons, which are associated with the sympathetic systems to viscera and somatic tissues. It is postulated that these spinal reflex circuits also exhibit functional specificity with respect to the primary afferent neurons and the hypothetical interneurons.

- The sympathetic innervation of the kidney influences four functions: kidney blood flow and glomerular filtration (blood vessels), renin release (juxtaglomerular cells) and sodium reabsorbtion by the tubules, although there is only a sparse innervation of the latter (Luff et al. 1991, 1992). Various groups of mechano- and/or chemosensitive visceral afferents innervate the kidney. Based on physiological investigations, several types of *spinal reno-renal reflexes* have been postulated, which are involved in neural control of these renal functions (Kopp and DiBona 2000; DiBona and Kopp 1997).
- Stimulation of spinal visceral afferents from the heart (e.g., during exercise or during coronary hypoxia) leads to reflex activation of

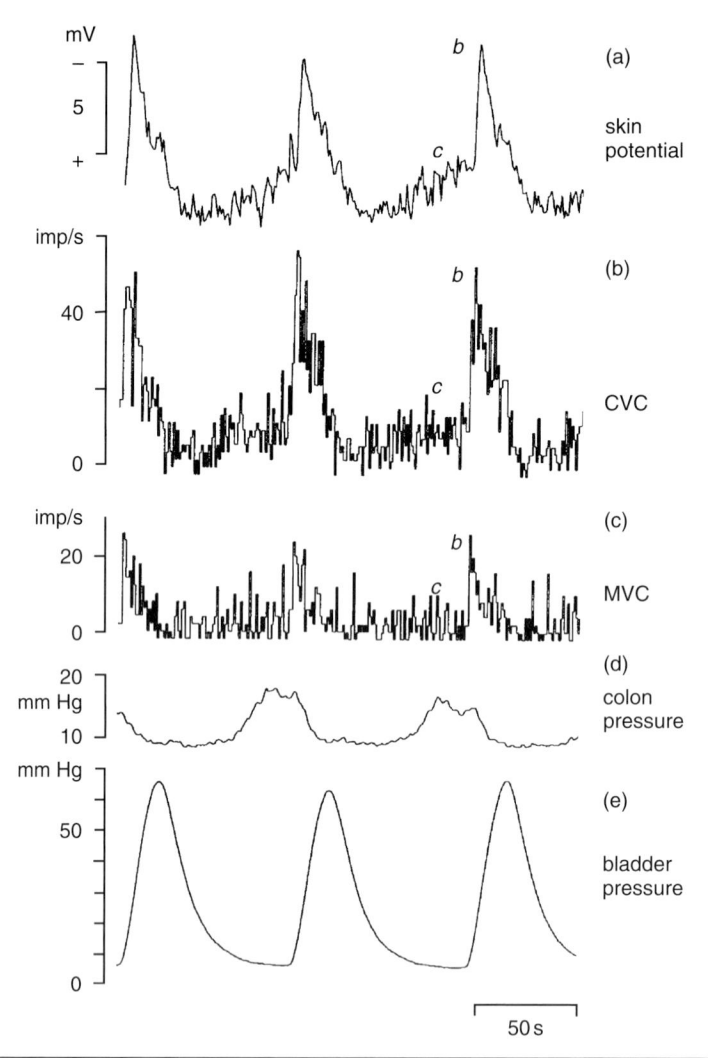

Figure 9.7 Reactions of cutaneous vasoconstrictor (CVC), sudomotor (SM, skin potential) and muscle vasoconstrictor neurons (MVC) during isovolumetric contractions of urinary bladder and colon in a chronic spinal cat. Experiment was done 135 days after interruption of the spinal cord at T10. Postganglionic activities were recorded from multiunit bundles isolated from the superficial peroneal nerve (to hairy skin) and from a muscle branch of the deep peroneal nerve. For SM activity, the skin potential was recorded from the surface of the central pad of the ipsilateral hindpaw (see Jänig and Kümmel [1981] and Figure 4.10). The urinary bladder was filled with 20 ml saline and the intravesical pressure measured through a urethral catheter. The distal colon was filled with about 40 ml saline in a flexible balloon and the intracolonic pressure recorded through an anal catheter connected with the balloon. Note that CVC, SM and MVC neurons are synchronously excited during isovolumetric contraction of the urinary bladder (indicated by b) and colon (indicated by c); that the reflex activation to urinary bladder contraction starts with the beginning of the contraction; that urinary bladder and colon contract reciprocally (due to reciprocal inhibition in the spinal cord, see Figure 9.10); and that the reflex excitation of muscle vasoconstrictor neurons is not very strong in comparison to the reflex excitation of cutaneous vasoconstrictor and SM neurons. Jänig and Kümmel, unpublished.

sympathetic cardiomotor neurons via *spinal cardio-cardiac reflexes* (Malliani 1982).
- Stimulation of spinal visceral afferents from the gastrointestinal tract (by distension or inflammatory processes) may activate, via *spinal intestino-intestinal reflex circuits*, sympathetic motility-regulating

neurons and induce immobilization of the gastrointestinal tract by inhibition of enteric circuits of the myenteric plexus. The spinal intestino-intestinal reflex circuits are connected to extraspinal intestino-intestinal reflexes relayed through prevertebral sympathetic ganglia (see Subchapters 2.3, 4.3, 5.7 and 6.5 and Figures 2.3, 5.15, 6.10 and 6.11). In isolated preparations consisting of prevertebral ganglia attached to parts of the gastrointestinal tract inhibitory intestino-intestinal reflexes can be rather powerful (see Figure 6.10). It is debated how much the sympathetic preganglionic neurons contribute to these reflexes.

- Irritation of viscera (e.g., by inflammatory processes) may generate changes of blood flow, sweating and consistency of the skin (trophic changes) in the corresponding dermatomes. These changes are supposed to be generated, at least in part, by sympathetic pathways to skin and mediated by spinal visceral afferents and spinal autonomic circuits (*viscero-cutaneous reflex circuits*; Jänig 1993). The zones of skin in which these changes occur are identical to those to which visceral pain is referred to (see Jänig and Morrison [1986]; Ness and Gebhart [1990]; Jänig and Häbler [2002]). Patients with affected viscera and visceral pain (e.g., with angina pectoris, appendicitis, inflammation of pelvic viscera) may experience spontaneous pain and mechanical hyperalgesia in these cutaneous zones (Vecchiet *et al.* 1993; Jänig and Häbler 1995, 2002; Giamberardino 1999).

In *conclusion*, distinct groups of spinal autonomic interneurons have been postulated on the basis of reflexes measured in the pre- and postganglionic neurons of functionally distinct sympathetic pathways, on the basis of reflex reactions of sympathetic effector organs or on the basis of clinical observations. The question arises as to whether or not these spinal circuits are integrated with the supraspinal ones in the regulation of the activity in the neurons of the final sympathetic pathways. This question will be discussed in the last part of this chapter.

9.3 | Sacral parasympathetic systems

The pelvic organs receive an abundant nerve supply that arises at two levels of the spinal cord, lumbar and sacral. At both levels, the innervation has afferent (sensory) as well as efferent components; the latter are classified as sympathetic and parasympathetic, respectively. The three organ systems in this region have storage and evacuation as major functions:

1. The lower urinary tract (urinary bladder, urethra) retains and evacuates urine (continence and micturition, respectively).
2. The distal part of the large bowel (distal colon, rectum and anus) stores and evacuates feces (continence and defecation) and also resorbs water and sodium ions.

3. The internal reproductive organs and erectile tissues of the external reproductive organs have basically similar functions related to the retention and transportation of semen and ova, together with, in the female, implantation of fertilized ova, storage of the developing fetus and subsequent birth of the new individual.

These specific functions are supplemented by two more general functions of the innervation of these three organ systems, namely, regulation of resistance and flow in the vasculature and sensory feedback. Both of these aspects of the innervation serve to adjust the function of these organ systems to the overall behavior of the organism.

The precision of the control of the three organs systems and their coordination in their adaptation to the actual behavior of the organism depend completely on the afferent, spinal autonomic (sympathetic and parasympathetic) and somatomotor innervation and on the central circuits in the spinal cord, brain stem, hypothalamus and telencephalon. The supraspinal mechanisms of the regulation of these organ systems are insufficiently understood, in particular those of the hindgut and the reproductive organs. However, we have at least some knowledge and ideas about the spinal reflex circuits associated with these organ systems. I will describe these spinal reflex circuits in order to give an additional argument that these circuits, and the interneurons integrated in these circuits, are essential in the regulation of the three pelvic organ systems. Furthermore, the spinal circuits will be shown to be important in the coordination of spinal autonomic systems and somatomotor systems. Details and principles about the neural regulation of these pelvic systems are described in the literature (Jänig and McLachlan 1987; de Groat and Booth 1993; de Groat et al. 1993, 1996, 1998, 2001; Blok and Holstege 1996; Jänig 1996b, c; McKenna and Marson 1997; de Groat 2002; McKenna 1999, 2000, 2001, 2002).

9.3.1 Urinary bladder and colon

Detailed neurophysiological investigations of sacral preganglionic neurons innervating the urinary tract and of other sacral neurons resulted in the postulation that at least five types of sacral interneurons are involved in the regulation of activity of these preganglionic neurons (Figure 9.8). Two types of interneuron are inhibitory; they are activated by collaterals of sacral preganglionic axons or by perineal somatic afferents, respectively. Three types of interneuron are excitatory and activated by sacral bladder afferents or by perineal somatic afferents or by neurons projecting from the pontine micturition center. It is not too far-fetched to hypothesize that these groups of interneuron are synaptically connected with each other and that they receive convergent synaptic inputs from different sources. They form networks of interneurons in the sacral spinal cord that contribute to the excitation of the preganglionic neurons regulating the lower urinary tract (Figure 9.8b).

9.3 SACRAL PARASYMPATHETIC SYSTEMS | 351

Furthermore, we have to postulate at least one type of excitatory and inhibitory interneuron, each connected synaptically to the motoneurons innervating the external urethral sphincter (Figure 9.9b, c). The excitatory interneuron receives convergent synaptic input from primary afferents projecting through the pelvic and pudendal nerves and from supraspinal systems, e.g. from an area in the pons called the lateral pontine micturition center. The inhibitory interneuron located in the dorsal commissural nucleus receives synaptic input

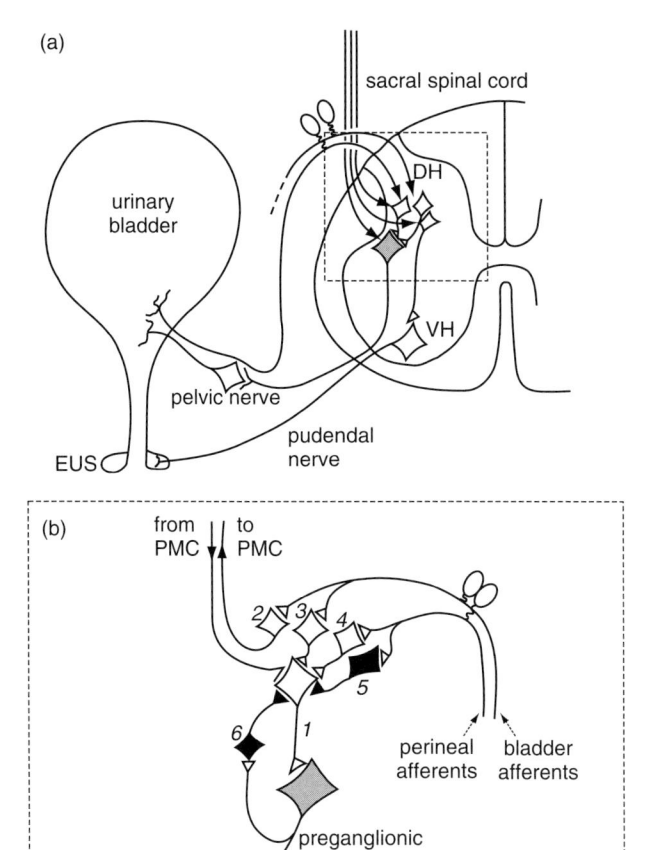

Figure 9.8 Putative role of spinal interneurons in the control of activity in the sacral parasympathetic pathway to urinary bladder. (a) General scheme. DH, dorsal horn; VH, ventral horn; EUS, external urethral sphincter. (b) Reflex pathways connected to preganglionic neurons to the urinary bladder involving six types of interneuron. The interneuron 1 mediating the command signals from the pontine micturition center (PMC), which are initiated from the urinary bladder via ascending neuron 2, receives additional excitatory synaptic input from sacral interneuron 3 activated by vesical afferents and inhibitory synaptic input from interneuron 6 activated by collaterals of vesical preganglionic neurons. Furthermore, stimulation of perineal (pudendal) afferents may activate or inhibit the vesical preganglionic neurons via two populations of interneurons (4, 5). It is likely that these populations of interneurons are synaptically interconnected and that interneurons 3 to 6 also receive synaptic input from supraspinal centers. Inhibitory neurons filled with black. Data from de Groat et al. (1982) for interneurons 1; McMahon and Morrison (1982a, b) and Coonan et al. (1999) for interneurons 2; McMahon and Morrison (1982b) for interneurons 3; Araki and de Groat (1996, 1997) for interneuron 4; Araki (1994) for interneuron 5; de Groat and Ryall (1968) and de Groat (1976) for interneuron 6. Modified after Shefchyk (2001).

from the medial pontine micturition center and possibly from sacral pelvic afferents (Figure 9.9a).

Micturition reflexes and spinal circuits
Let us undertake a brief excursion to the neural regulation of the lower urinary tract (urinary bladder [detrusor muscle], bladder neck, urethra and external urethral sphincter) as described by de Groat and coworkers (de Groat 2002) and others, in order to exemplify the importance of sacral interneurons. The urinary bladder stores and periodically evacuates urine, which is continuously produced by the kidney. Long collecting phases alternate with brief emptying phases. The ability of the bladder to collect and store urine is called *continence* and the act of emptying, *micturition*. Both processes are dependent on myogenic, peripheral neural and central neural mechanisms. The essential neural control involves afferent and efferent components and a cascade of control centers in spinal cord, brain stem, hypothalamus and forebrain. Sacral visceral afferent neurons innervating the bladder are activated during contraction and distension of the urinary bladder and encode in their activity the intravesical pressure (Figure 2.7). The detrusor muscle is innervated by a sacral parasympathetic pathway that contracts the muscle when activated using acetylcholine as the transmitter and ATP in some species. Smooth muscles of bladder neck and urethra are innervated by sacral parasympathetic neurons that relax both, using NO as transmitter. The muscle of the external urethral sphincter is part of the pelvic floor musculature and thus a striated muscle. It is innervated by sacral motoneurons situated in the lateral ventral horn in Onuf's nucleus in humans and other mammalian species that project through the pudendal nerve (Figure 9.9a).

Bladder muscle, bladder neck and urethra are additionally innervated by noradrenergic (sympathetic) neurons that have their cell bodies in the pelvic or inferior mesenteric ganglia and that are innervated by preganglionic neurons in the upper lumbar segments (in the cat segments L3 to L5 [Baron *et al.* 1985d]; in the rat L1 to L3 [Baron and Jänig 1991]). This innervation is important for continence (inhibition of bladder muscle and contraction of bladder neck and urethra) and for contraction of the proximal urethra during emission and ejaculation of semen (Figure 9.12), but not for micturition.

In the adult under physiological conditions, the urinary bladder slowly fills and accommodates to the increasing intravesical volume. This accommodation is related to the large extensibility of the smooth musculature and is actively generated by the innervation of the urinary bladder (relaxation of the detrusor muscle and activation of the bladder base and urethral smooth musculature by the sympathetic innervation; constricted external urethral sphincter). Depending on the degree of filling of the urinary bladder and the central (cortical) command signals (acting at the spinal and supraspinal neural circuits involved in regulation of micturition and continence; see Figure 9.9), micturition is initiated. The detrusor muscle contracts and bladder

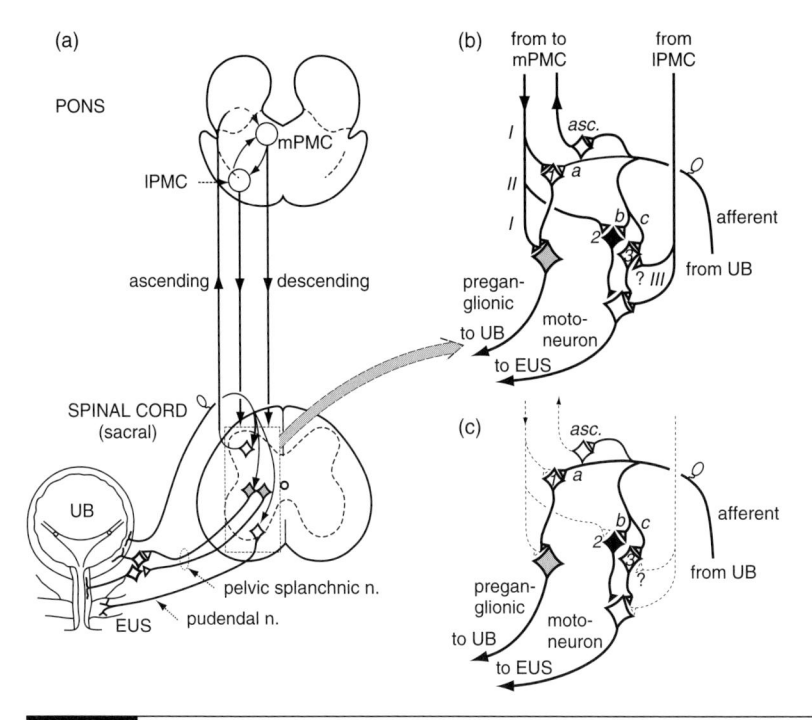

Figure 9.9 The micturition reflexes. (a) Sacral visceral afferents from the urinary bladder (UB) project to interneurons involved in micturition and ascending tract neurons (asc.) that project (probably via the periaqueductal grey) to the pontine micturition center (PMC) consisting of the lateral PMC (lPMC) and the medial PMC (mPMC, Barrington's nucleus). Neurons in the PMC project to the sacral spinal cord (descending). Sacral preganglionic neurons (grey) project to bladder body, inducing contraction of the urinary bladder. Other sacral preganglionic neurons project to urethra and bladder neck, inducing relaxation. Motoneurons project to the external urethral sphincter (EUS). (b) Micturition reflex pathways in intact spinal cord indicated in bold lines: neurons in the mPMC project to sacral preganglionic parasympathetic neurons (grey) and interneurons (excitatory in white, inhibitory in black); they activate preganglionic neurons (directly and via interneuron *1*; pathway *I*) and inhibit motoneurons (via interneuron *2*; pathway *II*). Neurons in the lPMC activate motoneurons during continence (directly and possibly also via interneuron *3*; pathway *III*). (c) Micturition reflex pathways after chronic transection of the spinal cord indicated in bold lines; dotted lines indicate degenerated pathways. Afferents from the urinary bladder form spinal reflex circuits via interneurons *1*, *2* and *3* (reflex pathways *a*, *b* and *c*).

neck, urethra and external urethral sphincter relax, resulting in the evacuation of urine. These coordinated actions are initiated by activation of sacral afferents from the bladder dome and generated by (1) reflex activation of parasympathetic neurons to the detrusor muscle, (2) reflex activation of inhibitory parasympathetic neurons to the urethra and bladder neck (leading to active relaxation of the outlet of the urinary bladder) and (3) inhibition of pudendal motoneurons. They are probably reflexly enhanced by stimulation of urethral afferents activated by urine flow during micturition.

Figure 9.9 outlines schematically the central reflex circuits involved in micturition:

- Activity in sacral vesical afferents activates via an ascending pathway (*asc* in Figure 9.9b) neurons in the medial pontine micturition center (mPMC; Barrington's nucleus). The output neurons of

Barrington's nucleus project to the sacral spinal cord and activate the preganglionic neurons to the urinary bladder directly and via interneurons (pathway *I* to interneuron *1* and preganglionic neurons in Figure 9.9b) and inhibit the motoneurons projecting to the external urethral sphincter via inhibitory interneurons (pathway *II* to interneuron *2*). The lateral pontine micturition center (lPMC) is inhibited by the mPMC.

- Chronically after interruption of the spinal cord in humans and animals, filling of the urinary bladder also leads to reflex activation of the preganglionic neurons to the urinary bladder and to micturition contractions (via pathway *a* and interneurons *1* in Figure 9.9c). However, now the pudendal motoneurons to the external urethral sphincter are simultaneously activated via the spinal reflex pathway *c* and interneuron *3* resulting in simultaneous contraction of urinary bladder and sphincter. Bladder neck and urethra also do not relax or relax only weakly. Thus, the coordinated action between bladder body and sphincter is lost. Pathway *c* connecting via the excitatory interneuron *3* to the motoneurons seems to dominate over the inhibitory pathway *b* via the inhibitory interneuron *2*. This functional state of the lower urinary tract is called *sphincter-detrusor dyssynergia*.

- It is a matter of debate whether the spinal circuits *a*, *b* and *c* represented by the interneurons *1*, *2* and *3* are important during normal micturition when the spinal cord is intact. De Groat and his coworkers (de Groat *et al.* 1993; de Groat 2002) believe that the spinobulbospinal reflex pathway via the mPMC mediates micturition in the adult and that the spinal pathways are unimportant during normal micturition. Others propagate the concept that these spinal reflex pathways are integrated in the normal micturition reflexes (Morrison 1997; Shefchyk 2001, 2002). Thus, it is discussed whether the signals from the PMC are gated by the spinal interneurons activated by vesical afferents (see Subchapter 9.4 for the concept).[10]

- Motoneurons to the external urethral sphincter are activated by neurons in the lPMC projecting to the sacral spinal cord during continence. This activation occurs by direct synaptic activation of the motoneurons or by synaptic activation of excitatory interneurons antecedent to the motoneurons (interneuron *3* in Figure 9.9b). The neurons in the lPMC are inhibited during micturition, probably from the mPMC. In animals (e.g., rats), the external urethral sphincter contracts and relaxes rhythmically during micturition resulting in rhythmic release of urine. These rhythmic contractions may be generated by alternating activation and inhibition of the motoneurons via the inhibitory interneuron *2* and the excitatory interneuron *3*.

The types and number of sacral interneurons innervating preganglionic neurons supplying the hindgut are unknown. However, they probably are in the same order as those of the neural circuits regulating the urinary bladder. More importantly, the supraspinal mechanisms involved in the regulation of the hindgut are unknown.

Interaction between regulation of urinary bladder and hindgut: role of spinal circuits

After intraluminal filling, both urinary bladder and distal colon exhibit alternating contractions in spinal cord intact animals as well as in chronic spinal animals (see Figure 9.7d, e). This indicates that the spinal neural circuits associated with both organs systems inhibit each other reciprocally. Activation of sacral afferents from the urinary bladder inhibits hindgut activity and activation of afferents from the hindgut inhibits urinary bladder activity. This reciprocal inhibition of the two organ systems may be generated by the following mechanisms and involves several types of spinal interneurons:

1. As described before, many sympathetic motility-regulating neurons projecting in the lumbar splanchnic and hypogastric nerves (preganglionic and postganglionic neurons) exhibit reciprocal reflex patterns upon stimulation of sacral afferents from the urinary bladder and colon (excitation from the urinary bladder or excitation from the colon, possibly combined with inhibition from the alternative organ tract; see Figures 4.14 and 9.6). Some of these motility-regulating neurons exhibit fast excitation or inhibition upon contraction of the pelvic evacuative organs (Bahr et al. 1986a; Bartel et al. 1986, Jänig et al. 1991). Neurons excited during bladder contraction or distension and inhibited during colon contraction or distension may innervate the colon. Neurons inhibited from the urinary bladder and/or excited from the distal colon may innervate the lower urinary tract.
2. Interneurons in the sacral spinal cord excited by stimulation of bladder afferents may inhibit the preganglionic neurons to the colon. By the same token interneurons in the sacral spinal cord excited by stimulation of the colonic afferents may inhibit preganglionic neurons to the lower urinary tract.

The diagrams in Figure 9.10 demonstrate the putative spinal neural mechanisms underlying the reciprocal inhibition between both organ systems (in [a] when the sacral afferent neurons from the urinary bladder are activated, in [b] when the sacral afferent neurons from the hindgut are activated). The spinal interneurons *postulated* to be involved in this reciprocal inhibition are two types of propriospinal (sacro-lumbar) excitatory neurons and several types of inhibitory interneurons.

Supraspinal control systems associated with the lower urinary tract (indicated in Figure 9.10 by the dotted lines) may act at the different pools of interneurons and at the preganglionic neurons. These supraspinal autonomic premotor neurons are located in the medulla oblongata (lateral paragigantocellular nucleus, ventral

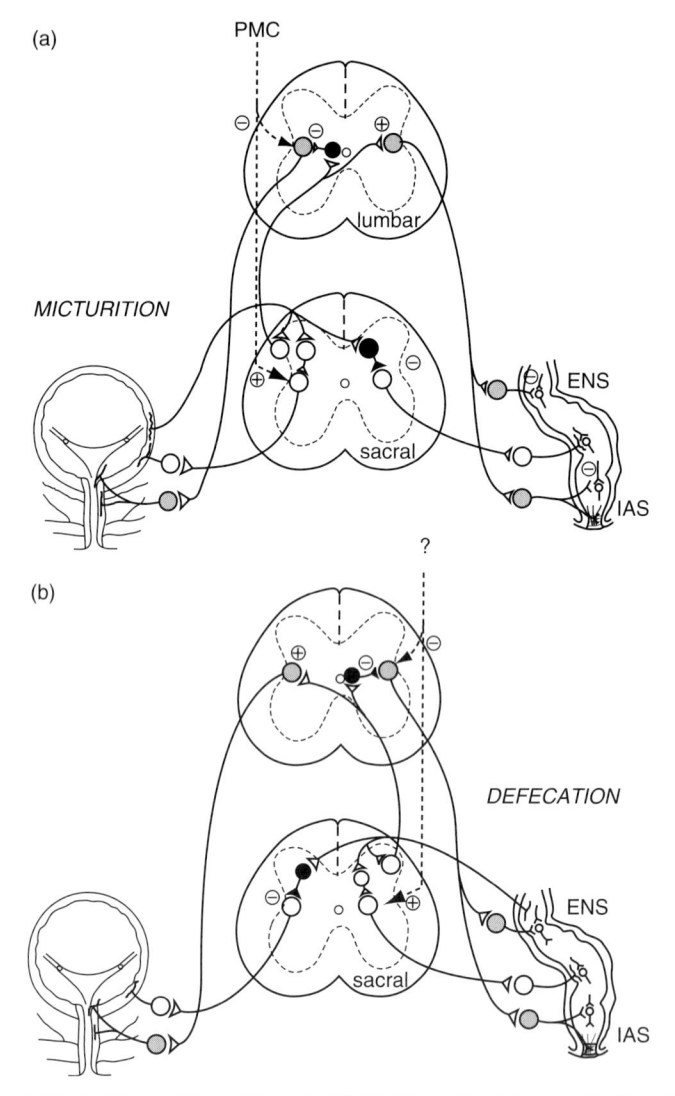

Figure 9.10 Spinal reflex pathways involved in reciprocal inhibition of urinary bladder and distal colon and in continence produced by both organ systems. Hypothetical spinal reflex pathways involved in reciprocal contraction and inhibition of urinary bladder and distal colon (Figure 9.7d, e). Demonstrated are neural circuits mediating inhibition of colon during micturition contractions (a) and inhibition of bladder during defecation contractions (b). (a) Activation of sacral bladder afferents leads to inhibition of sacral preganglionic neurons to colon, excitation of lumbar preganglionic neurons to colon (contraction of the internal anal sphincter [IAS] – *this is shown with a minus* – and inhibition of enteric nervous system [ENS]) and inhibition of lumbar preganglionic neurons to urinary bladder. (b) Activation of sacral colon afferents leads to inhibition of sacral preganglionic neurons to urinary bladder, excitation of lumbar preganglionic neurons to urinary bladder and inhibition of lumbar preganglionic neurons to colon. Supraspinal systems related to micturition (PMC, pontine micturition center) or defecation (?) (both indicated by dotted lines) control these reflex circuits by acting at the preganglionic neurons or at the interneurons (excitation of sacral neurons; inhibition of lumbar neurons). Note that postganglionic parasympathetic neurons in the pelvic ganglion interact with the enteric nervous system to produce contraction. Inhibitory interneurons in black; sympathetic neurons, shaded; +, excitation; −, inhibition.

part of the gigantocellular nucleus, raphe nuclei), in the pons (Barrington's nucleus, A5, subceruleus nucleus) and in the hypothalamus (paraventricular nucleus, lateral hypothalamus) (see Table 8.3). However, all or most of these areas also project to preganglionic neurons that innervate other functional types of autonomic target organs, such as colon, reproductive organs or other organs (see Tables 8.2 and 8.3). Neurons in Barrington's nucleus (mPMC) and in a pontine area lateral to Barrington's nucleus (lPMC) appear to be most important in the regulation of micturition and continence of the urinary bladder.

9.3.2 Reproductive organs and spinal circuits

Precise neural regulation of reproductive organs and its coordination with the regulation of motor behavior is of utmost importance for the propagation of species. It has to be adapted in the context of the actual behavior of the individual. Most importantly reproductive function in the male is entirely dependent on neural control. However, knowledge about neural mechanisms underlying the regulation of reproductive organs in the spinal cord and in supraspinal centers is still in its infancy. This applies particularly to the female, although the principles of neural regulation of reproductive organs appear to be rather similar in males and females (McKenna 2002).

In addition to the described groups of spinal interneurons associated with the regulation of the urinary tract and hindgut, we have to postulate several other types of interneuron in the sacral spinal cord and sacro-lumbar propriospinal neurons which are associated with sacral and lumbar preganglionic neurons innervating the erectile tissue and internal reproductive organs (vas deferens, prostate, seminal vesicle, proximal urethra, uterus, proximal vagina) (de Groat and Booth 1993; Jänig 1996c; McKenna and Marson 1997; de Groat 2002; McKenna 1999, 2000, 2001, 2002).

Spinal autonomic pathways involved in regulation of reproductive organs are usually divided in sacral (parasympathetic) pathways mediating erection and thoracolumbar (sympathetic) pathways mediating contraction of internal reproductive organs and secretion (emission). However, this is an oversimplification. Thoracolumbar sympathetic neurons are also involved in erection and both parasympathetic and sympathetic pathways are involved in secretion (Figure 9.12). Thus, functions assumed to be primarily associated with sacral (parasympathetic) systems are well duplicated by thoracolumbar (sympathetic) pathways (Müller 1906; Semans and Langworthy 1938; Root and Bard 1947; Bors and Comarr 1960; see Jänig and McLachlan [1987]; McKenna and Marson [1997]). This shows that the division of the spinal autonomic systems into sympathetic and parasympathetic with respect to sexual functions is questionable. Theoretically it is possible that the separation of neuron groups in lumbar autonomic and sacral autonomic neurons by the expansion of the lumbar enlargement during development is incomplete. This could be examined using genetic techniques.

The central mechanisms leading to the sequence of erection, emission and ejaculation are made up of several neural components. The central mechanisms that mediate erection are descriptively divided into spinal reflex mechanisms and supraspinal "psychogenic" mechanisms. This distinction is somewhat artificial, although clinically useful, since both mechanisms always work together under biological conditions (Coolen et al. 2004). Here I will focus mainly on spinal mechanisms underlying this sequence of reproductive reflexes since they illustrate the coordination (1) between different spinal autonomic pathways and (2) between spinal autonomic systems and somatomotor systems. This coordination requires several types of interneurons.

The supraspinal control involves the paragigantocellular nucleus, the periaqueductal grey, the paraventricular nucleus of the hypothalamus, the medial amygdala and the medial preoptic area (see Table 8.3). With the exception of the latter, these supraspinal centers receive information from pelvic somatic and visceral structures by afferent neurons projecting through the pelvic splanchnic and pudendal nerves via second-order ascending tract neurons in the sacral dorsal horn. The result of the integrative activity in the supraspinal centers is channeled through the paragigantocellular nucleus, the periaqueductal grey and the paraventricular nucleus of the hypothalamus to the spinal reflex circuits, the first having inhibitory effects and the latter two excitatory effects (McKenna 2001, 2002).

Erection
The sacral preganglionic neurons involved in the generation of erection are situated in the intermediate zone of the spinal cord, probably lateral to the preganglionic neurons that are associated with the hindgut and dorsomedial to preganglionic neurons associated with the lower urinary tract (Figure 9.11, left). These preganglionic neurons have dendrites projecting medially to the interneurons in the dorsal commissural nucleus (DCN), laterally receiving synaptic input from descending systems in the brain stem and hypothalamus, and dorsally along the lateral dorsal horn. Sacral afferent neurons innervating the penis, which are important to elicit reflex erections, project to the superficial dorsal horn, medially through the dorsal horn to the dorsal commissural nucleus and also laterally to the preganglionic neurons (Figure 9.11, dotted on the left).

Mechanical stimulation of afferent receptors in the penis or in the tissue surrounding the penis elicits erection of the penis via a reflex pathway in the spinal cord (Figure 9.11, right). The reflex pathway in the sacral spinal cord is at least disynaptic and involves one interneuron. It functions in males with transected spinal cords as long as the interruption is rostral to the sacral segments and as long as spinal shock is over.[11]

Mechanical stimulation of penile afferents also activates supraspinal centers that control the spinal reflex mechanisms, resulting in

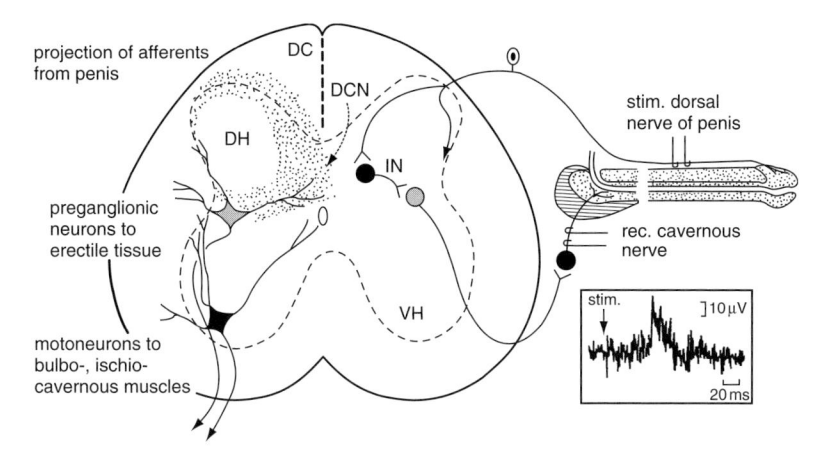

Figure 9.11 The spinal sacral reflex pathway mediating penile erection (right) and its morphological components (left). *Left side*: anatomy of connections in sacral spinal cord. Dots, sites of projection of penile afferents in the pudendal nerve determined by horseradish peroxidase tracing in the cat. They project to the superficial dorsal horn (DH), medially to the dorsal commissural nucleus (DCN, which contains the interneurons) and laterally towards the preganglionic neurons. Preganglionic neurons (shaded) located in the lateral part of the intermediate zone have dendrites within regions of afferent projection (medially and dorsally) and laterally into the region of termination of descending systems. Motoneurons (black) innervating the bulbo- and ischiocavernous muscles are situated laterally in the ventral horn (VH; Onuf's nucleus in humans) and send dendrites laterally, medially and dorsally. IN, interneuron. *Right side*: diagram of the spinal components of the erection reflex in the rat. Electrical stimulation of penile afferents with single impulses (arrow in inset) activates postganglionic neurons projecting in the cavernous nerve and innervating erectile tissues (reflex response in inset). From de Groat and Booth (1993) with permission.

sexual sensations when centers in the forebrain are activated. The supraspinal centers can also be triggered by imaginary, visual, auditory and olfactory stimuli and induce erection, apparently independently of stimulation of afferents from the penis. This "psychogenic" erection is produced not only via the sacral efferent innervation of the erectile tissue (pathway 1 in Figure 9.12) but also via the thoracolumbar sympathetic innervation (pathway 2 in Figure 9.12). Some patients with destroyed sacral spinal cord (e.g., following an accident) or in whom the spinal cord is interrupted between upper lumbar and sacral segments can still induce erection psychogenically (Kuhn 1950). This erection can only be produced by activation of thoracolumbar sympathetic neurons (pathway 2 in Figure 9.12). It is unclear to what degree excitation of neurons of this sympathetic pathway contributes to erection in normal conditions. However, these observations on human patients are fully consistent with experiments performed on animals (Müller 1902, 1906; Semans and Langworthy 1938; Root and Bard 1947; Dail 1993).

It is unclear whether sympathetic preganglionic neurons and parasympathetic preganglionic neurons generating erection converge on the same postganglionic neurons in the pelvic ganglion that project through the cavernous nerve to the erectile tissues (pathway 2 and 1 in Figure 9.12). Sympathetic neurons of pathway 2 must be separate from sympathetic vasoconstrictor neurons, which can obviously override the dilation of the erectile tissue in various mental

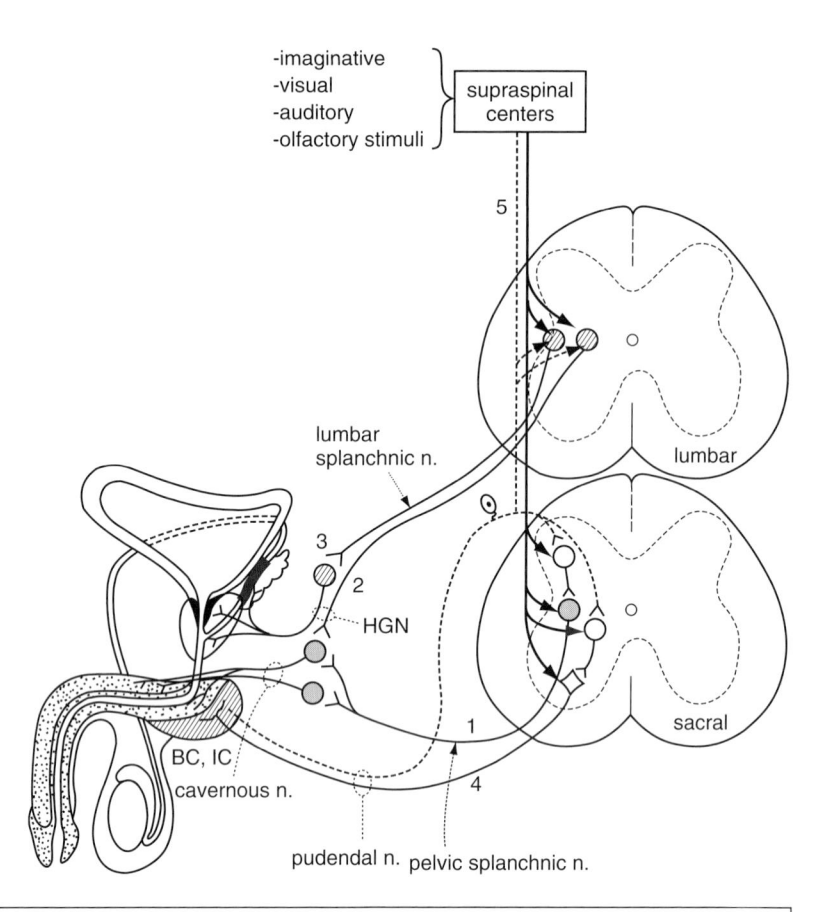

Figure 9.12 Peripheral neural pathways controlling erection, emission and ejaculation. 1, Efferent parasympathetic pathway to erectile tissue (grey neuron). 2, Efferent sympathetic pathway to erectile tissue (grey neuron). Preganglionic sympathetic axons in lumbar splanchnic and hypogastric (HGN) nerves and preganglionic parasympathetic axons in pelvic splanchnic nerve. These may both synapse with the same postganglionic neurons in the pelvic plexus/ganglion that project through the cavernous nerve to the erectile tissue. 3, Efferent sympathetic pathway to internal reproductive organs (hatched neuron, vas deferens, seminal vesicle, prostate, proximal urethra [internal vesical sphincter]). 4, Motor axons to bulbo- and ischiocavernous muscles (BC, IC). 5, Ascending spinal systems to and descending spinal system from supraspinal control centers. Sacro-lumbar interneurons and ascending tract neurons mediating the afferent activity to supraspinal centers are not shown. Vasoconstrictor neurons to blood vessels of the reproductive organs that pass through lumbar splanchnic and hypogastric nerves as well as through the sympathetic chain and the pelvic splanchnic and pudendal nerves are not shown.

and bodily conditions (e.g., during mental stress or exercise). Whether sympathetic neurons projecting though the cavernous nerve are different from sympathetic neurons innervating the internal reproductive organs is unknown (pathway 3 in Figure 9.12).

Emission and ejaculation

Emission and ejaculation are the high point of the male sexual act. Emission is generated by thoracolumbar sympathetic pathways innervating smooth muscle and secretory epithelia of the internal reproductive organs (epididymis, vas deferens, seminal vesicles, ampulla, prostate, proximal urethra). The sympathetic preganglionic neurons project though the lumbar splanchnic nerves to postganglionic neurons in the caudal part of the inferior mesenteric ganglion (superior hypogastric plexus) and in the pelvic plexus (ganglion; see pathway 3 in Figure 9.12). Ejaculation is generated by pudendal motoneurons innervating bulbo- and ischiocavernous muscles, the external urethral sphincter and other muscles of the pelvic floor.

As sacral afferents from the reproductive organs become excited during copulation, thoracolumbar sympathetic pathways to the internal reproductive organs are activated in bursts (Figure 9.13b). Contraction of the internal reproductive organs propels the semen into the posterior urethra. Contraction of the proximal urethra prevents reflux of the secretion into the urinary bladder. The activation of sympathetic neurons to the internal reproductive organs requires powerful enhancement from supraspinal brain centers. After a delay, ejaculation starts. This is also triggered by activation of pudendal afferents and possibly pelvic splanchnic afferents from the urethra and consists of rhythmic contractions of the bulbo- and ischiocavernous muscles (Figure 9.13), which enclose the proximal erectile tissue, and of the muscles of the pelvic floor. These rhythmic contractions expel the secretion from the posterior urethra through the anterior urethra.

During the ejaculation phase, the excitation of the parasympathetic and sympathetic innervation of the reproductive organs becomes maximal due to continuous afferent feedback and strong supraspinal excitation of the spinal reflex arcs. The synchronous maximal excitation of autonomic neurons and pudendal motoneurons is rhythmic and generated by spinal interneuronal circuits. It is represented by the model of the *urethro-genital reflex* in the rat (Figure 9.13; McKenna *et al.* 1991; McKenna and Marsden 1997) showing that stimulation of urethral afferents leads to synchronous rhythmic activation of pudendal motoneurons to striated perineal muscles, of parasympathetic neurons innervating the erectile tissue through the cavernous nerve and of sympathetic neurons innervating internal reproductive organs (hypogastric nerve in Figure 9.13b) and possibly the erectile tissue. Discharges in autonomic neurons always precede the discharges in pudendal motoneurons innervating the bulbo- or ischiocavernous muscles (see insets in Figure 9.13). The rhythmic contractions are produced by a spinal pattern generator involving spinal interneurons. The segmental interneurons and propriospinal sacro-lumbar neurons of the spinal pattern generator that coordinates excitation of autonomic and somatomotor neurons are unknown.

The spinal reflex mechanisms in females leading to engorgement of the erectile tissues, rhythmic contractions of the striated perineal muscles and rhythmic contractions of vagina and uterus are probably similar to those in males. This includes strong rhythmic activation of postganglionic neurons projecting in the cavernous nerve and of preganglionic neurons projecting in the pelvic or in the hypogastric nerve (McKenna *et al.* 1991; McKenna 2002).

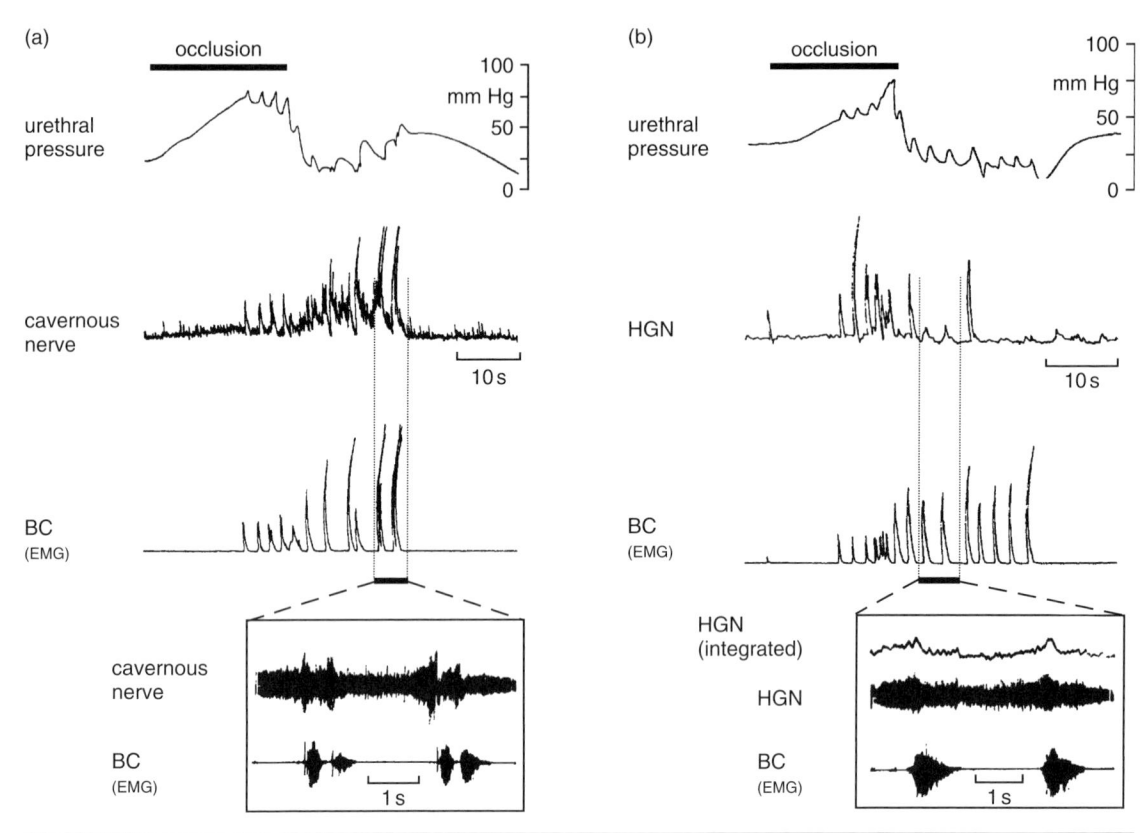

Figure 9.13 The urethro-genital (UG) reflex (coitus reflex) in the male rat elicited by stimulation of urethral afferents. Urethral afferents were stimulated by distension of the urethra (increase of intraurethral pressure) caused by infusion of saline into the urethra closed at the urethral meatus (bar above upper records) in urethane-anesthetized rats. (a) Simultaneous recording of intraurethral pressure, activity in the cavernous nerve and of electromyogram (EMG) recorded from the bulbocavernous (BC) muscle. Inset: original recordings from cavernous nerve and of EMG (during time period indicated by bar). (b) Simultaneous recording of intraurethral pressure, activity in the hypogastric nerve (HGN) and EMG recorded from the BC muscle. Inset: original recordings from HGN (plus integrated HGN activity) and of EMG during time period indicated by bar. Mechanical stimulation of the urethra leads to repetitive synchronous bursting discharges in pudendal motoneurons to the BC muscle (and other pelvic floor muscles [not shown]), in postganglionic neurons projecting into the cavernous nerve, which are activated by upper lumbar (HGN) and sacral preganglionic neurons. The synchronous bursting discharges in autonomic and somatomotor neurons outlast the urethral stimulation. Both signals were rectified and integrated with a time constant of 50 ms. Modified from McKenna et al. (1991) with permission.

9.4 The spinal cord as integrative autonomic organ

9.4.1 A concept developed in research on the somatomotor system

In mammals, the spinal cord is able to generate several motor acts such as locomotion, standing, flexor reflexes, paw shake, scratching reflexes etc. These motor patterns were already known to Sherrington (1906) and his forerunners. They can be elicited by physiological stimulation of distinct types of primary afferent neurons innervating skin, skeletal muscle or other deep somatic structures. They are mediated by distinct reflex circuits in the spinal cord, which include interneurons between the primary afferent neurons and the motoneurons. In the second half of the nineteenth century, Michael Foster had already stated, in his *Textbook of Physiology* (1879; cited in Liddell [1960]),

> ... the [spinal] cord contains a number of more or less complicated mechanisms capable of producing, as reflex results, co-ordinated movements(?) altogether similar to those which are called forth by the will. Now it must be an economy to the body, that the will should make use of these mechanisms already present, by acting directly on their centers, rather than it should have recourse to a special apparatus of its own of a similar kind.

The general idea behind this statement was that the brain is using these spinal neuronal machineries as building blocks to generate the various movements, i.e. that the spinal reflex circuits are integral components of the cortically and subcortically generated movement patterns. Sherrington (1906) elaborated extensively on this idea. It was entirely clear to him that a motor reflex as such is a tool to study the central organization of neural regulation of movements and therefore so-to-speak an experimental artifact or fiction. However, it was also clear to him that these spinal reflex circuits are not a developmental leftover that exists in parallel to supraspinal circuits that project via descending pathways directly to the motoneurons and generate the coordinated movements. On the contrary, these spinal circuits appeared to be integrated into the supraspinally elicited movement programs. They appear to be essential to understand the regulation of activity in the motoneurons to skeletal muscles by the brain. Thus, the supraspinally generated motor command signals target the spinal circuits at the level of the interneurons post- and presynaptically and incorporate so-to-speak the "spinal motor programs" in the execution of the centrally generated motor commands.

The concept about control of movement by the brain, including voluntary control, as formulated by Sherrington and his pupils, was the basis for systematic experimental investigations of spinal reflex circuits and their connections with the supraspinal circuits that are involved in motor control. These experimental investigations included neurophysiological, anatomical, behavioral, pharmacological and other techniques. They were strictly anchored in the different forms of physiological movement. The most systematic and seminal experimental approaches were conducted, since the end of the 1950s, by Anders Lundberg and his group, notably Elzbieta Jankowska and Hans Hultborn. The following conceptual discussion is fully based on the thoughts and ideas developed by this group (Lundberg 1971, 1975, 1979; Baldissera *et al.* 1981; Jankowska and Lundberg 1981; Jankowska 1992, 2001; McCrea 1994, 2001; Hultborn 2001).

This systematic experimental work characterized more than ten functionally distinct groups of spinal interneurons in the motor system. Most of these interneurons were originally named according to their main primary afferent synaptic input (e.g., Ia interneuron, Ib interneuron, group II interneuron) or according to rather restricted reflex functions (e.g., Renshaw interneuron, lumbar and cervical propriospinal interneurons, flexor reflex interneuron). However, it soon became clear that these interneurons are neither relay neurons mediating spinal reflexes nor relay neurons for supraspinally generated signals in pathways to the motoneurons. Thus, the original functional labels for these interneurons, i.e. according to their synaptic afferent input or according to their projection, turned out to be too narrow and sometimes misleading. The most remarkable characteristics of these interneurons are (Figure 9.14b, c):

- They receive convergent synaptic inputs from various primary afferent neurons innervating skin, skeletal muscle and joints.
- They receive convergent synaptic inputs from various descending pathways originating in brain stem and cortex.
- Reciprocal inhibitory synaptic connections exist between functionally related interneurons (e.g., Ia interneurons, Renshaw cells).
- Interneurons exhibit divergent projections to various interneurons and motoneurons.
- The same type of interneuron is involved in more than one motor function, i.e. has a multifunctional character.
- Reflexes generated by a distinct group of primary afferent neuron can dynamically change and may be reversed according to the motor task (e.g., during locomotion; reflex reconfiguration).

These findings led to the idea that the spinal interneuronal networks of the motor system (and their synaptic connections) represent dynamic movements that include several muscles and joints. Thus, the population of interneurons and their synaptic connections with primary afferent neurons, motoneurons and interneurons represent *spinal motor programs*. These spinal motor programs are not static in the sense that *one* population of interneuron is involved in *only one* motor program (e.g., spinal locomotion), but rather is dynamic. Thus, individual groups of interneurons may be part of several spinal motor programs depending on the synaptic afferent inputs and on the behavioral state of the organism. The different groups of interneuron are under multiple excitatory and inhibitory control from descending pathways originating in brain stem and cortex (Figure 9.14b, c). The synaptic connections of supraspinal systems with spinal somatic interneurons outnumber the direct connections of these pathways to the motoneurons by an order of magnitude. Although this situation is almost incalculable as far as the role of the individual types of interneurons and of descending pathways in the generation of distinct movements are concerned, it is intuitively clear that (1) this complex system of interneurons could provide economic control of movements by the brain and (2) the spinal networks of interneurons are integrated in the initiation and maintenance of these movements by the brain.

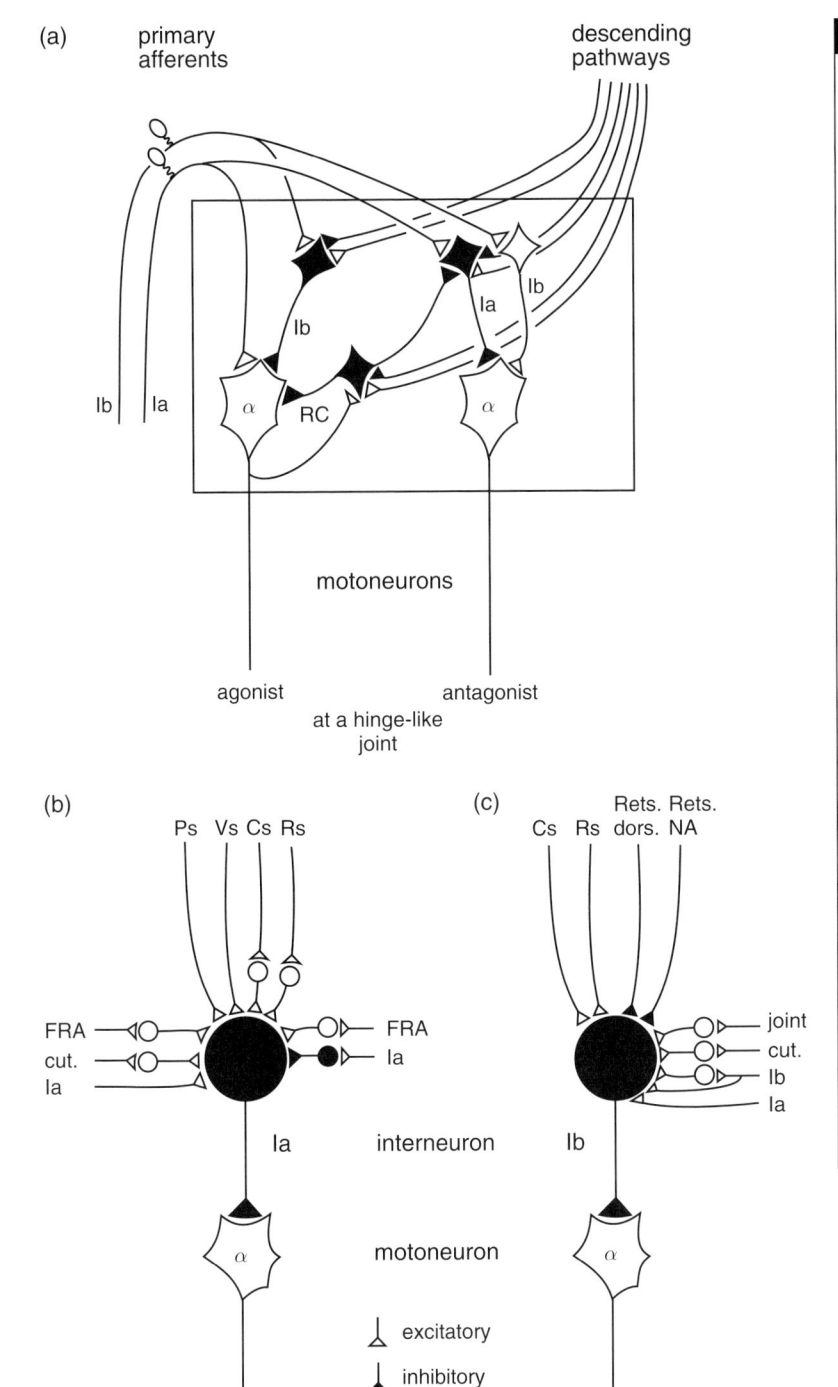

9.4 THE SPINAL CORD AS INTEGRATIVE AUTONOMIC ORGAN | 365

Figure 9.14 Spinal motor programs involving Ia interneurons, Ib interneurons and Renshaw cells (RC) to regulate the activity in α-motoneurons acting antagonistically at a hinge-like joint. (a) Synaptic connections between (1) Ia and Ib afferents, (2) four groups of interneurons (Ia inhibitory, Ib inhibitory, Ib excitatory, RC [inhibitory]), (3) α-motoneurons and (4) descending excitatory and inhibitory systems converging on the inhibitory interneurons. This type of arrangement represents the spinal motor programs used by the supraspinal centers in the regulation of activity in motoneurons during movements. Inhibitory neurons and synapses in black. (b) Convergent synaptic input to Ia interneurons from ipsi- (left) and contralateral (right) spinal afferents (Ia from muscle spindles, cutaneous [cut.], flexor reflex afferents [FRA]) and from supraspinal centers (Cs, corticospinal; Rs, rubrospinal; Vs, vestibulospinal) or propriospinal neurons (Ps). (c) Convergent synaptic input to Ib interneurons from spinal afferents (Ia, Ib from Golgi-tendon organs, joint, cutaneous) and from supraspinal centers (Cs, Rs, reticulospinal dorsalis and noradrenergic [Rets. dors., Rets. NA]). Modified from Hultborn (2001).

Figure 9.14b, c demonstrates in two examples the convergent synaptic inputs of Ia and Ib interneurons from primary afferents, descending systems and interneurons. Figure 9.14a demonstrates the synaptic connections of the three most extensively studied inhibitory interneurons (the Renshaw cell [RC], the Ia interneuron and the Ib interneuron) with motoneurons that innervate antagonistic skeletal muscles acting at a hinge-like joint (e.g., the knee joint), their synaptic inputs from Ia and Ib afferents and from motoneuron collaterals, and their excitatory and inhibitory synaptic inputs from supraspinal centers. It is immediately clear, even from this grossly simplified diagram, that the functional plasticity of this neuronal network regulating the activity in motoneurons is high. Supraspinal systems mostly do not influence the activity in single pools of motoneurons, and therefore not single skeletal muscles, but do influence the coordinated activity in groups of motoneurons by acting at the pools of interneurons associated with these motoneurons.

9.4.2 Integration of autonomic circuits and supraspinal centers

Functional characteristics of the discharge pattern in neurons of the sympathetic and sacral parasympathetic systems are dependent on spinal reflex circuits and supraspinal mechanisms (see Chapter 4). For example:

- Discharge pattern and ongoing activity ("tone") in vasoconstrictor neurons regulating resistance vessels (e.g., muscle vasoconstrictor and visceral vasoconstrictor neurons, renal vasoconstrictor neurons) largely depend on activity generated in the medulla oblongata, on reflexes associated with cardiovascular afferents (notably baroreceptor and chemoreceptor afferents; Guyenet 1990; Koshiya et al. 1993) and on the coupling between respiratory neurons and sympathetic premotor neurons in the medulla oblongata (McAllen 1987; Richter and Spyer 1990; Guyenet and Koshiya, 1992; Häbler et al. 1994b; Häbler and Jänig 1995; see Subchapters 10.3, 10.4 and 10.6).
- The discharge pattern and ongoing activity in most cutaneous vasoconstrictor neurons is dependent on the hypothalamus (thermoregulation), on spinal circuits (nociceptive and non-nociceptive somato-sympathetic and viscero-sympathetic reflexes, thermoregulatory reflexes) and on the medulla oblongata (cardiovascular reflexes and coupling to the central respiratory generator mediated by the rostral ventrolateral medulla; see Subchapters 10.3, 10.4 and 10.6).
- The discharge pattern and ongoing activity in motility-regulating neurons is dependent on sacro-lumbar reflex pathways and relatively independent of brain stem and hypothalamus.
- Discharge patterns in sympathetic neurons innervating the kidney that are involved in the regulation of blood volume and extracellular sodium concentration are dependent on: (1) the hypothalamus (paraventricular nucleus); (2) cardiovascular reflexes mediated by the medulla oblongata; (3) coupling to the central respiratory generator mediated by the rostral ventrolateral medulla; (see Subchapters 10.3, 10.4 and 10.6); and (4) specific spinal reflex circuits receiving distinct afferent inputs from the kidney (Badoer 2001; Yang and Coote 2003; Yang et al. 2004).

These findings do not rule out, *first*, that spinal circuits are important for the regulation of activity in those spinal autonomic systems that are normally under predominant supraspinal control (e.g., vasoconstrictor pathways) and, *second*, that supraspinal controls are important for the regulation of activity in those spinal autonomic systems that normally depend predominantly on spinal circuits (e.g., motility-regulating pathways). Furthermore, these findings are completely compatible with experimental results showing (1) that up to 50% (or more) of the synapses on sympathetic preganglionic neurons are formed by descending systems (Llewellyn-Smith and Weaver 2001) and (2) that some EPSPs and IPSPs elicited monosynaptically in sympathetic preganglionic neurons from descending axons are relatively large (McLachlan and Hirst 1980; Deuchars et al. 1997), this being probably more a property of the preganglionic neuron input impedance than of the synapses.

Thus, the results reported so far suggest that the autonomic circuits in the spinal cord are important for integrating information from the periphery and from supraspinal brain structures for the spinal autonomic outflows. They clearly demonstrate that the spinal cord contains neural circuits, which consist of preganglionic neurons, putative interneurons (which includes segmental and propriospinal interneurons) and the synaptic connections of these neurons with the afferent inflow from the periphery and with autonomic premotor neurons projecting to the spinal cord. Different types of spinal final autonomic pathways may receive connections from the same spinal circuit and an individual final autonomic pathway may be associated with several spinal circuits (Figure 9.4, Table 9.1). Furthermore, functionally distinct pools of interneurons may be shared by two or more final autonomic pathways.

Here I *hypothesize* that integration of spinal and supraspinal circuits, which leads to the characteristic discharge patterns in sympathetic neurons and parasympathetic sacral neurons, occurs in a similar way as in the somatomotor system. The experiments described in this chapter are fully consistent with this hypothesis but they do not prove it:

- Spinal autonomic circuits are integrative mechanisms for the control of autonomic target organs. These may be called "*spinal autonomic subroutines*" or "*spinal autonomic motor programs.*" They are preprogrammed and require for appropriate functioning distinct afferent inputs from the periphery and synaptic input from supraspinal centers.
- Higher (supraspinal) centers "use" these spinal integrative mechanisms in the control of the peripheral sympathetic pathways. Signals in the supraspinal autonomic premotor neurons are finally shaped by the spinal circuits before they are channeled into the final peripheral sympathetic and parasympathetic pathways. This shaping by the spinal autonomic motor programs varies between sympathetic systems, as there are also many monosynaptic

connections between supraspinal autonomic premotor neurons and preganglionic sympathetic neurons (Strack *et al.* 1989b; Morrison *et al.* 1991; Zagon and Smith 1993; McAllen *et al.* 1994). For example, the fast baroreceptor reflexes may be mediated monosynaptically by sympathetic premotor neurons in the lower brain stem to sympathetic preganglionic vasoconstrictor and cardiomotor neurons. However, it is not excluded (1) that baroreceptor reflexes are additionally mediated by inhibitory spinal interneurons and by excitation of spinal interneurons and (2) that baroreceptor reflexes are mediated by direct inhibition via inhibitory sympathetic premotor neurons projecting through the dorsolateral funiculus (McLachlan and Hirst 1980; Deuchars *et al.* 1997; Stornetta *et al.* 2004) (see Subchapter 10.3).

- Thermoreceptor reflexes, somatosympathetic reflexes, micturition reflexes, defecation reflexes, erection reflexes etc. may be preferentially mediated by spinal autonomic interneurons that receive convergent synaptic input from supraspinal neurons and from primary afferent neurons (Figures 9.8, 9.9, 9.10). Although not readily visible in the neural regulation of several autonomic target organs (e.g., cardiovascular regulation, thermoregulatory change of blood flow through skin, regulation of evacuative organs, regulation of gastrointestinal functions), the spinal components are probably very important in these control mechanisms.
- Sacrolumbar reflexes related to the sympathetic outflow that is involved in the regulation of pelvic organs (continence of urinary bladder and hindgut; emission of semen by the internal reproductive organs) are mediated by several types of propriospinal interneurons (Bartel *et al.* 1986). The supraspinal control of these spinal circuits is unknown.
- Spinal circuits may set the gain for supraspinal reflexes or supraspinal systems may regulate the sensitivity of spinal reflex circuits. For example,
  - activation (by supraspinal systems) of sympathetic cardiomotor neurons during exercise may be enhanced and maintained by the spinal afferent feedback from the heart and from the exercising skeletal muscles;
  - inhibition of cutaneous vasoconstrictor neurons during warming of the hypothalamus is enhanced non-linearly by activation of spinal thermosensitive circuits (Grewe *et al.* 1995), as is also predicted from thermoregulatory changes of skin blood flow during hypothalamic and spinal cord warming (Simon 1974; Simon *et al.* 1986);
  - nociceptive and non-nociceptive spinal reflexes in sudomotor neurons are enhanced by supraspinal systems; etc.

Preganglionic neurons, autonomic interneurons, spinal afferent neurons and sympathetic premotor neurons in brain stem and hypothalamus may be synaptically interconnected in a manner similar to motoneurons, somatic afferents, spinal interneurons and

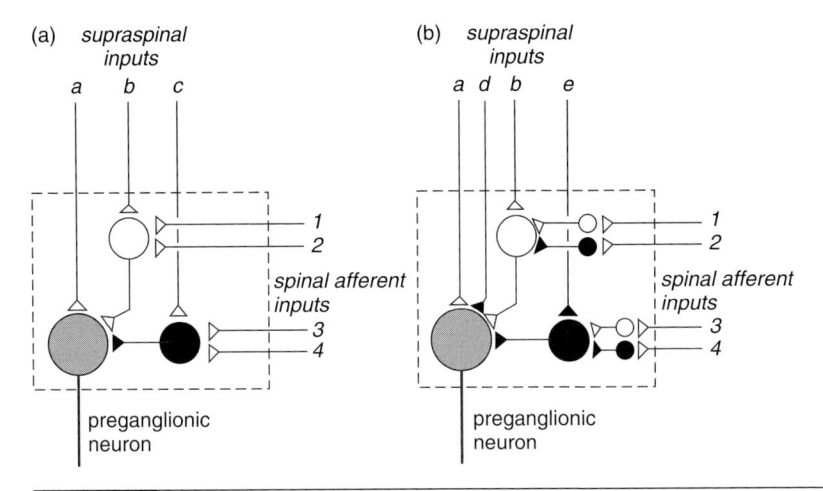

Figure 9.15 Hypothetical arrangement of autonomic interneurons and their synaptic connections with spinal afferent neurons and descending systems in the regulation of activity in preganglionic neurons. Spinal afferent inputs 1 to 4 synapse with autonomic interneurons; descending systems a, b, c, d and e synapse with interneurons or preganglionic neurons. These descending systems and interneurons are either excitatory (white) or inhibitory (black). By analogy to the somatomotor system, this type of arrangement could represent the spinal autonomic motor programs used by the supraspinal autonomic centers to regulate the activity in preganglionic neurons during various autonomic activities. Synaptic connections between interneurons are not shown. Modified from Jänig (1996a) with permission.

supraspinal systems in the somatomotor system (see Baldissera *et al.* [1981]; Jankowska and Lundberg [1981]; Schomburg [1990]), as demonstrated by the two hypothetical examples in Figure 9.15. Synaptic input from autonomic premotor neurons in brain stem and hypothalamus can act directly on the preganglionic neurons or on autonomic interneurons. This synaptic input can be excitatory or inhibitory. Spinal afferent input from body tissues acts on spinal (excitatory and inhibitory) interneurons, which may also receive excitatory and/or inhibitory synaptic input from supraspinal centers. Thus, spinal autonomic reflex circuits may gate synaptic input from supraspinal centers and synaptic input from supraspinal centers may gate spinal reflex circuits. This shows that the neuronal regulation of target organs by spinal autonomic systems would have considerable flexibility by using this principle of organization that has been worked out in the somatomotor system.

We have no direct data about synaptic connections of autonomic interneurons with each other. However, we can use the analogy of the reciprocal synaptic connections of Renshaw interneurons, Ia interneurons and motoneurons (Figure 9.14), or interneurons associated with the flexor-reflex pathways constituting "spinal motor units," which are important in the regulation of the activity in alpha- and gamma-motoneurons (Baldissera *et al.* 1981; Schomburg 1990). We conceptualize that similar circuits exist for the spinal sympathetic and parasympathetic systems, e.g. for cutaneous vasoconstrictor and sudomotor system in the coordinated regulation of sweating and blood flow through hairless skin, for the motility-regulating systems in the coordinated regulation of evacuative organs (lower urinary tract and hindgut) and for parasympathetic neurons to urinary bladder and distal colon in the coordinated regulation of the two organ tracts. Indeed the experimental work that has so far been done on autonomic interneurons in the sacral spinal cord clearly supports this contention (Figure 9.6; Shefchyk 2001). It has not been shown so far, directly or indirectly, that different pools of autonomic interneurons are synaptically connected with each other to form networks of interneurons that are responsible for generation of the typical discharge patterns in peripheral autonomic pathways.

9.4.3 Coordination of activity in somatomotor systems and spinal autonomic systems

In the introduction of this book, I have described that the regulation of the autonomic nervous system is closely integrated with the regulation of the activity in the somatic motoneurons. The two systems (together with the neuroendocrine motor system in the hypothalamus) are hierarchically organized in spinal cord, brain stem and hypothalamus, have common synaptic inputs from afferent systems, the cerebral cortex and the behavioral state system (see Figure 2), and constitute the basis for behavior (Swanson 2000, 2003; Watts and Swanson 2002). This will be further elaborated in Chapter 11. The discussion of the spinal autonomic systems and the putative interneurons associated with these systems have shown that spinal circuits, and therefore specific interneurons, may be important in the coordination of activity in somatic motor systems and activity in spinal autonomic systems. This is reflected in coupling of somatomotor and sympathetic outflows from the spinal cord (Chizh *et al.* 1998). Putative interneurons involved in this coordination may be activated by supraspinal command signals. This coordination between somatomotor and spinal autonomic functions could be important in several functional contexts. Examples are:

- Regulation of pelvic organs (coordination between external sphincters and smooth muscle activity of urinary tract [Figures 9.8, 9.9, 9.10], hindgut and reproductive organs [Figure 9.13]).
- Regulation of sweat glands and blood flow through skin during somatosensory discrimination and manipulation (see vibration reflex in Subchapter 4.2).
- Regulation of cardiovascular system (heart rate, blood flow through skeletal muscle) during cortically induced and maintained muscle contractions (exercise; see Figures 11.3, 11.4) (Victor *et al.* 1995).
- Initiation of shivering, cutaneous vasoconstriction and inhibition of the sudomotor pathway during body cooling.

Future experiments, using in vitro or in vivo animal models, will have to dissect out which groups of interneurons are involved in this coordination of somatomotor and autonomic motor activity and what the underlying mechanisms are.

Conclusions

1. The spinal cord is an autonomic integrative organ in its own right, which determines several components of the discharge pattern in the spinal autonomic pathways in animals with intact spinal cord. Preganglionic neurons, autonomic interneurons and primary afferent neurons form spinal autonomic reflex circuits, which are integrated in the regulation of preganglionic activity by supraspinal centers. Neurons projecting from brain stem and hypothalamus to the spinal cord synapse with autonomic interneurons and with the preganglionic neurons.
2. Analysis of the discharge patterns in the autonomic neurons under standardized experimental conditions leads to the following conclusions:
   a. The spinal cord contains distinct autonomic reflex pathways, which are integrated with supraspinal reflex pathways during normal regulation of the autonomic target organs.
   b. The discharge pattern in the different types of sympathetic neuron consists of components that are associated with integration in the spinal cord and integration in lower brain stem, upper brain stem and hypothalamus.
   c. Reflex integration in the spinal cord is related to distinct afferent inputs from skin, viscera, spinal cord (thermoreceptors) and other structures.
   d. Signals in supraspinal systems are integrated with spinal circuits at the preganglionic neuron, leading to reciprocal facilitation.
   e. Spinal circuits may be important in the coordination of somatomotor functions and spinal autonomic functions.
3. By analogy with motor systems, it is hypothesized that spinal autonomic interneurons and preganglionic neurons constitute spinal autonomic motor programs that are integrated in the regulation of autonomic target organs.
4. Spinal circuits, spinal afferent inflows and descending influences from brain stem and hypothalamus often work together in determining the activity of the preganglionic sympathetic neurons. The systems may either be under predominant control of the lower brain stem (e.g., the muscle and visceral vasoconstrictor pathways innervating resistance vessels), of the hypothalamus (e.g., the cutaneous vasoconstrictor pathways) or of the circuits in the spinal cord (e.g., the sympathetic motility-regulating and secretomotor pathways innervating the gastrointestinal tract or pelvic organs). However, in *all* spinal autonomic systems, the

spinal component may be essential for this integration because it may set the excitability of the preganglionic neurons and/or shape the supraspinal signals according to "spinal autonomic programs".

5. This organization is particularly exemplified in the regulation of evacuation and continence of urinary bladder and distal colon and in the regulation of the sequence of erection, emission and ejaculation of male reproductive organs. The functioning of these pelvic organs depends on multiple sacral and sacro-lumbar reflex circuits and several supraspinal integrative centers that interact with these spinal circuits.

6. In conclusion, experimental approaches that aim to clarify the central neural circuitry of the spinal autonomic systems should start at the level of the spinal cord and use functionally identified peripheral neurons of autonomic final pathways as the reference point. This is important as functionally different types of preganglionic neuron are likely to be controlled in different ways.

Notes

1. Bypassing integrative processes, changes in excitability (conductance of membrane) of preganglionic neurons can also occur due to local low oxygen tension and spillover of neuropeptides released in the dorsal horn or may be intrinsic to the preganglionic neurons (as activity dependent processes).
2. Sympathetic preganglionic neurons may be electrically coupled as has been shown electrophysiologically in vitro on thoracolumbar spinal cord slices of rats aged 8 to 14 days (Logan et al. 1996; Nolan et al. 1999). It is unclear whether this coupling occurs between functionally related preganglionic neurons and whether it contributes to synchronized firing of sympathetic neurons in vivo (see Subchapter 10.2.5).
3. Using the term spinal rat (or any other animal) I mean an animal in which the spinal cord has been completely transected (in our experiments at the thoracic segmental level T_8/T_9).
4. Spinal shock describes the temporary loss or depression of reflex activity mediated by the spinal cord below the lesion of the spinal cord. The term "spinal shock" was introduced by Hall (1841) in order to differentiate hypotension following blood loss from hypotension following contusion of the spinal cord with subsequent interruption of communication between supraspinal centers and spinal cord.
5. With "functionally distinct reflexes" I mean reflexes that are well defined according to the function (receptive properties) of the primary afferent neurons and the function (target cells) of the sympathetic system. These reflexes are not necessarily functionally meaningful.
6. Electrical stimulation of afferent axons in nerves elicits short- and long-latency reflexes in preganglionic neurons, depending on the afferent axons stimulated (A- or C-fibers) and on the central pathways (spinal or supraspinal). Spinal reflexes are preserved acutely after transection of the spinal cord rostral to the preganglionic neurons recorded from (e.g., those projecting into the cardiac or renal nerves) and do not appear to exhibit "spinal shock" (Coote and Downman 1966; Coote and Sato 1978; Coote 1984). It is important to emphasize that this experimental situation is entirely different from that when afferent fibers are activated

physiologically (e.g., by noxious or non-noxious stimulation of the skin, by contraction or distension of a visceral organ): electrical stimulation leads to synchronous non-physiological activation of afferent neurons; physiological stimulation activates afferent neurons asynchronously.

7. This experimental finding does not argue that the spinal inhibitory pathway activated by visceral afferents does not exist. It argues that this pathway cannot be activated after spinalization. We have studied, in the cat, the viscero-sympathetic reflexes in cutaneous and muscle vasoconstrictor neurons to distension or contraction of urinary bladder or distal colon 60 to 130 days after spinalization at the thoracic segmental level T8. During this time period we did not observe inhibitory reflexes in cutaneous vasoconstrictor neurons to these visceral stimuli. We have not studied the time course of recovery of the excitatory viscero-sympathetic reflexes (Jänig and Kümmel, unpublished).

8. This discrepancy between experimental studies on animals (mainly cats) and studies on paraplegic patients is puzzling. However, when the spinal autonomic circuits are no longer under supraspinal control after interruption of the spinal cord, postganglionic sympathetic neurons exhibit asynchronous activity. Because of the low signal-to-noise ratio in the microneurographic recordings and since single unit recordings from postganglionic cutaneous vasoconstrictor axons and postganglionic sudomotor axons are very difficult to obtain with this recording technique it is probably difficult, if not impossible, to detect this asynchronous spontaneous activity in the unmyelinated postganglionic axons. Recently, Macefield and Wallin (1999a, b) succeeded in recording routinely from single muscle and cutaneous vasoconstrictor axons in humans, using high impedance tungsten metal electrodes (see Subchapter 3.5).

9. Note in Figure 9.7 that the contractions of urinary bladder and colon are alternating. This is probably generated by activation of reciprocal inhibitory spinal reflex pathways connected to preganglionic neurons innervating urinary bladder and colon (see Subchapter 9.3.1 and Figure 9.10). Stimulation of bladder afferents leads to strong synchronous reflex activation of the sympathetic vasoconstrictor systems (CVC, MVC) and the sudomotor system (skin pot.) (marked b in Figure 9.7). Stimulation of colon afferents leads to a smaller synchronous reflex activation of the three sympathetic pathways (marked c in Figure 9.7).

10. The *arguments* supporting the concept of de Groat and coworkers are: (1) Lesion of the PMC abolishes micturition. (2) Electrical or chemical stimulation of the PMC triggers micturition. (3) In cats, electrical stimulation of Aδ-afferents in the pelvic splanchnic nerve elicits long-latency reflexes in the preganglionic neurons innervating the urinary bladder mediated by the PMC. Short-latency reflexes are absent. Thus, Aδ-fibers form the afferent limb of the pontine micturition reflexes. (4) In chronic spinal cats, the afferent limb of the micturition reflexes elicited by electrical stimulation of the pelvic splanchnic nerve are C-fibers but not Aδ-fibers.

These results have considerable *implications*: (1) Sacral afferent C-fibers innervating the urinary bladder are not involved in micturition reflexes in the adult cat. This conclusion is consistent with investigations of sacral afferents innervating the urinary bladder, showing that most Aδ-fibers have low thresholds to distension of the urinary bladder and that only a few afferent C-fibers can be activated by distension of the urinary bladder at high intravesical pressures of >30 mmHg (see Figure 2.7b; Häbler et al.

1990a). (2) In chronic spinal cats, afferent C-fibers innervating the urinary bladder and their synaptic connections in the spinal cord must undergo plastic changes, serving now as afferent limb in the micturition reflexes. Capsaicin, a neurotoxin that disrupts the function of nociceptive C-fibers that have the vanilloid-receptor VLR_1 (now $TRPV_1$-receptor), prevents the micturition reflex contractions when given systemically in chronic spinal cats, but not in normal cats (see de Groat et al. [1993]; de Groat [2002]).

Activation of the spinal parasympathetic pathway, which relaxes the urethra by releasing NO during distension of the urinary bladder (see Figure 9.9), appears to be dependent on spinal reflex pathways in rats that are modulated by supraspinal centers.

11. (1) The erectile tissue of the penis (corpus cavernosum and corpus spongiosum) and the erectile tissue in females consist of trabecular tissue and sinusoids, which contain smooth muscle cells. It is vascularly supplied by branches of the internal pudendal artery via helicine arteries. The postganglionic neurons that innervate these vascular sections of the erectile tissue are cholinergic or noradrenergic. The cholinergic neurons are located in the pelvic ganglion and project through the cavernous nerve (Figure 9.12) to the erectile tissue. The noradrenergic neurons are located in the sacral sympathetic chain ganglia (and possibly the inferior mesenteric ganglion). (2) Activation of the cholinergic postganglionic neurons dilates the helicine arteries and relaxes sinusoids and trabecles of the erectile tissue. Pressure in and volume of the erectile tissue increase leading to elongation of the penis. The venous outflow of the erectile tissue of the penis is reduced by passive compression of the emissary veins. The pressure in the erectile tissue reaches values that are about 10 to 20 mmHg below arterial blood pressure. It is further increased above arterial blood pressure by contraction of the bulbo- and ischiocavernous muscles (generated by activation of pudendal motoneurons); this enhances rigidity of the penis in humans and generates rigidity in animals. Activation of noradrenergic neurons supplying the erectile tissue leads to subsidence of erections (detumescence). (3) Acetylcholine and the vasoactive intestinal peptide are colocalized in the postganglionic neurons generating the dilation of the erectile tissue. Additionally, the neurons synthesize and release the radical NO during activation. These three substances relax the branches of the internal pudendal artery, helicine arteries and sinusoids and work cooperatively together in the initiation, potentiation and maintenance of the active dilation, NO released by the postganglionic neurons being the principal mediator of penile erection in humans and animals. The quantitative importance of the three compounds in the generation of erection varies between species. Finally, NO released by the endothelium of the blood vessels in the erectile tissue may be important to mediate erection too (Burnett et al. 1992; Andersson 2001).

Chapter 10

Regulation of organ systems by the lower brain stem

10.1 General functions of the lower brain stem	page 377
10.2 Sympathetic premotor neurons in the ventrolateral medulla oblongata	378
10.3 Baroreceptor reflexes and blood pressure control	398
10.4 Arterial chemoreceptor reflexes in sympathetic cardiovascular neurons	410
10.5 Sympathetic premotor neurons in the caudal raphe nuclei	414
10.6 Coupling between regulation of autonomic pathways and regulation of respiration	420
10.7 Vagal efferent pathways and regulation of gastrointestinal functions	440

Sympathetic and parasympathetic premotor neurons are present in the lower brain stem (medulla oblongata and pons) in addition to those located more rostrally (see Chapter 8.4). Their cell bodies are located in the ventrolateral medulla (mainly rostral ventrolateral medulla), the ventromedial medulla, the caudal raphe nuclei and the A5 area of the ventrolateral pons (see Figures 8.15 and 8.17 and Tables 8.2 and 8.3). Premotor neurons form synapses with preganglionic neurons (and/or local interneurons associated with the preganglionic neurons) (Chapters 8, 9). How do autonomic premotor neurons in the lower brain stem function so as to contribute to the characteristic discharge patterns of the neurons of the peripheral autonomic pathways? Are they specialized for the behavioral repertoire of the organism? Are there discharge patterns similar to those seen in the neurons of the final autonomic pathways (see Chapter 4)?

On the basis of the knowledge we have about the central autonomic systems, I can partially answer these questions for some systems in the lower brain stem. These answers, incomplete as they may be, clearly indicate that we will be able to unravel the maze of the neural organization of autonomic systems in the lower brain stem in the near future. One of the first decisive and important steps to reach this aim, particularly for the cardiovascular system, was and still is to record the effects of microinjection of pharmacological agents into defined regions of the lower brain stem on effector responses (e.g., arterial blood pressure, heart rate, blood flow through an organ). However, the final ways to achieve this aim are:

1. experimental studies on single central neurons that combine neurophysiological, histological, neurochemical and neuropharmacological methods;
2. correlation of these data with the firing patterns of neurons in functional autonomic pathways in the periphery (see Chapter 4);
3. the alignment of these studies with autonomic physiological behavior (cardiovascular responses [e.g., heart rate, arterial blood pressure, blood flow through organs], thermoregulatory responses, gastrointestinal responses, responses of pelvic organs, etc.).

To formulate it somewhat apodictically (Guyenet 2000) only "a cellular and neurochemical approach combined with a precise knowledge of the input–output functions of individual cell groups will advance our understanding" of autonomic integration (see also Blessing [1997]). This includes various experimental techniques some of them having been mentioned in Chapters 3 and 8 and, to emphasize and repeat it, the anchor of all investigations is the physiology of the regulation (cardiovascular system; respiration; gastrointestinal tract). Using this approach, this aim has been achieved to some degree for some central autonomic neurons involved in the regulation of vascular resistance and of cardiac output.

Here I will concentrate mainly on three topics:

1. Sympathetic premotor neurons in the rostral ventrolateral medulla, which have been most thoroughly studied in the context of cardiovascular regulation and its coordination with the regulation of respiration.
2. The role of neurons in the caudal raphe nuclei in the regulation of activity in cutaneous vasoconstrictor neurons during thermoregulation and of activity in sympathetic neurons innervating the brown adipose tissue in the rat (lipomotor neurons).
3. The function of the dorsal vagus complex (nucleus tractus solitarii [NTS] and dorsal motor nucleus of the vagus) in the regulation of gastrointestinal functions.

Neurons in the ventromedial medulla and raphe nuclei are also important in the endogenous control of nociceptive impulse transmission in the spinal dorsal horn and caudal trigeminal nucleus and therefore important in protection of body tissues during noxious events. This function is closely associated with the regulation of peripheral autonomic pathways (in particular sympathetic ones) and is reviewed in the context of "Autonomic nervous system and body protection" elsewhere (Mason 2001; Fields *et al.* 2005; Jänig 2005b; Jänig and Levine 2005).

10.1 | General functions of the lower brain stem

The lower brain stem contains the neuronal mechanisms for the homeostatic control of the cardiovascular system, the respiratory system and the gastrointestinal tract (Figure 10.1). This global regulation of body systems is closely coordinated in different time domains and according to the external and internal demands on the organism. The regulation of respiration, cardiac output and peripheral resistance is adjusted in the time domain of seconds, so as to precisely adjust the gas transport into the body and to the peripheral tissues in the body. Ingestion of food and fluid and respiration are precisely regulated in the time domain of milliseconds in order to prevent nutrients and fluid getting into the lungs. Intake, digestion and absorption of nutrients are closely coordinated with blood flow through the gastrointestinal tract in the time domain of hours to guarantee their transport to liver and body tissues. These differentiated control systems, their coordination and their adaptation to the external and internal challenges of the body require:

1. Specialized primary afferent neurons, which continuously inform the lower brain stem about the internal state of these systems (see Chapter 2).
2. Specialized final autonomic pathways to the three systems (and somatic motor pathways to the respiratory muscles); these may share the same premotor systems.
3. Distinct synaptic connections between the afferent and the efferent pathways to these three systems via distinct groups of interneurons.
4. Complex and differentiated neuronal control of these reflex pathways by centers in the upper brain stem, hypothalamus and telencephalon.

It is clear that the different groups of interneurons in the lower brain stem must be synaptically interconnected in a functionally specific way, otherwise the closely coordinated control of the three systems would not be possible. We have only limited knowledge about the different types of interneurons, their transmitters, their synaptic connections and their connections with supramedullary brain centers. The specific components of these neuronal control systems become visible in the distinct reflexes elicited in autonomic neurons by physiological stimulation of different groups of afferent neurons (e.g., from the cardiovascular, gastrointestinal or respiratory system). Specific reflexes obtained under standardized experimental conditions change quantitatively and qualitatively according to the state of the control systems and therefore to the command signals from suprapontine centers. Details about the specific reflexes, their integration in homeostatic regulation and their anatomical basis have been described elsewhere (Spyer 1981, 1994; Guyenet 1990, 2000; Loewy

Figure 10.1 Three global homeostatic control systems and their coordination in the medulla oblongata. Regulation of the cardiovascular system, the respiratory system and the gastrointestinal tract (GIT). Each system receives specific afferent inputs from the periphery and is under the control of upper brain stem, hypothalamus and telencephalon. The three systems are closely integrated.

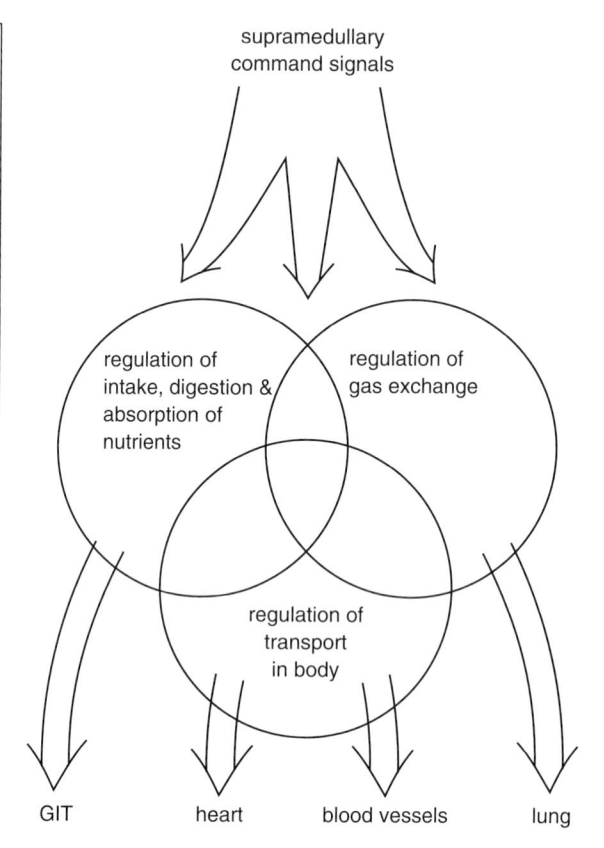

and Spyer 1990a; Guyenet and Koshiya 1992; Ritter *et al.* 1992; Dampney 1994; Sun 1995; Guyenet *et al.* 1996; Richter 1996; Blessing 1997 [see extensive literature here]; Jordan 1997a; Kirchheim *et al.* 1998; Rekling and Feldman 1998; St.-John 1998; Chapleau and Abboud 2001 [see extensive discussion of baroreceptor reflexes here]; Richter and Spyer 2001; Pilowsky and Goodchild 2002; Dampney and Horiuchi 2003; Feldman *et al.* 2003). Some general principles of this organization will be described here for the arterial baroreceptor reflexes, arterial chemoreceptor reflexes, reflexes in cutaneous vasoconstrictor neurons and lipomotor neurons innervating brown adipose tissue, respiratory rhythmicities in peripheral autonomic activity and gastrointestinal reflexes. This description is restricted to the short time domain of the functioning of these systems yet not the long ones related to the adaptation of these systems during environmental load (e.g., thermal load, stress, exercise, fluid loss etc.).

10.2 | Sympathetic premotor neurons in the ventrolateral medulla oblongata

Neurophysiological recording from sympathetic nerves in anesthetized animals and from muscle nerves in humans using microneurography, under resting conditions shows that many sympathetic neurons have spontaneous activity and that the discharge pattern of this activity appears to be rather similar in different nerves. The activity shows rhythmic changes that are correlated with the arterial pulse pressure wave ("cardiac rhythmicity" of the activity; see Figures 4.1, 4.6 and 4.7 in Chapter 4) and with the central respiration generator. But rhythmic changes of sympathetic activity also occur independent of cardiovascular and respiratory parameters and seem to be generated by neural oscillators in the brain stem (Gebber 1990; Barman and Gebber 2000; see Subchapters 10.2.5, 10.6.2). Activity with cardiovascular, respiratory and similar rhythmicities occurs in sympathetic neurons innervating resistance vessels (such as skeletal muscle, kidney and general visceral vasoconstrictor neurons) and sympathetic cardiomotor neurons. Vasoconstrictor neurons project in practically all peripheral nerves and dominate the activity in them. For this reason the mass discharge patterns are rather similar in different peripheral nerves.

Spontaneous activity and its rhythmic changes in sympathetic cardiovascular neurons, being involved in blood pressure control, but not necessarily in other types of sympathetic neuron are dependent on the lower brain stem. Both disappear *acutely* after transection of the spinal cord at the cervical level, paralleled by a decrease in arterial blood pressure to values of 50 to 60 mmHg (for discussion of spontaneous activity in sympathetic neurons after spinalization see Subchapter 9.2). They are little affected after transection of the brain stem at a suprapontine level. However, this is a very controversial issue and results depend very much on the preparation, whether the animal is anesthetized or awake and on the animal species. Many parasympathetic cardiomotor neurons are also spontaneously active and exhibit the same rhythmicities in their activity. What is the origin of this resting activity in sympathetic and parasympathetic cardiovascular neurons? Which mechanisms underlie these typical and universally observed phasic changes of the spontaneous activity?

10.2.1 The discovery of the cardiovascular center in the medulla oblongata

With the introduction of experimental physiology and medicine, physiologists became interested in the central origin of the spontaneous activity in sympathetic vasoconstrictor neurons and in sympathetic and parasympathetic cardiomotor neurons, which are responsible for the resting level of arterial blood pressure. Based on experiments in which he performed ablation experiments in the medulla oblongata and transected the spinal cord in rabbits Claude Bernard (1858) came to the conclusion that the activity in sympathetic neurons has a supraspinal origin. In Carl Ludwig's laboratory in Leipzig, Owsjannikow (1874) concluded that the cardiovascular center in rabbits is localized in the medulla oblongata about 4 to 8 mm rostral to the most distal part of the fourth ventricle. He performed ablation experiments in curarized and non-anesthetized rabbits in which he measured the arterial resting blood pressure and blood pressure changes upon electrical and mechanical stimulation of the sciatic nerve. Dittmar (1873) extended these experiments in Ludwig's laboratory on curarized rabbits. He showed that removal of dorsal or medial parts of the medulla oblongata (raphe nuclei and adjacent parts of the ventromedial and dorsomedial medulla) or interruption of the medulla oblongata rostral to the facial nuclei did not change the resting blood pressure level or the increase of blood pressure generated by sciatic nerve stimulation. Bilateral destruction of the ventrolateral part of the medulla oblongata at the level of and distal to the facial nucleus reduced the blood pressure and prevented the reflex increase of the blood pressure. About 70 years later Alexander (1946) performed systematic experiments in cats, lesioning and electrically stimulating different sites of the medulla oblongata. He defined the excitatory and inhibitory cardiovascular centers in the medulla oblongata. His description dominated the textbooks for many decades.

The systematic investigation of the organization of the cardiovascular center in the medulla oblongata, using various methods, started in the second half of the 1960s:

- Schläfke and Loeschcke (1967) showed that cooling the ventrolateral medulla in the anesthetized cat dramatically reduced arterial blood pressure and respiration.
- A few year later, it was shown in the cat that application of the inhibitory amino acids γ-aminobutyric acid (GABA) or glycine to the ventrolateral surface of the medulla oblongata or bilateral lesion of this area also generated a dramatic fall of arterial blood pressure (Guertzenstein and Silver 1974; Feldberg and Guertzenstein 1976). These experiments clearly demonstrated that neurons responsible for or mediating the spontaneous activity in sympathetic vasoconstrictor neurons are present in the ventrolateral medulla.
- Amendt *et al.* (1979) showed in the cat that the rostral ventrolateral medulla contains neurons that project to the preganglionic neurons in the nucleus intermediolateralis of the spinal cord. They retrogradely labeled the cell bodies of these neurons by the tracer horseradish peroxidase, which was transported in the axons to the cell bodies of the sympathetic premotor neurons after its application to the nucleus intermediolateralis.
- At the beginning of the 1980s Roger Dampney and his group demonstrated in the rabbit that excitation of neurons in the rostral ventrolateral medulla (RVLM) by ionophoretic application of the excitatory amino acid glutamate increased arterial blood pressure, and that the sites of stimulation in the RVLM corresponded very closely with those of spinally projecting neurons. This group was the first showing that bilateral lesion of the RVLM abolished the baroreceptor reflex mediated by vasoconstrictor neurons to resistance vessels. Finally, based on immunohistochemical investigations in

rabbits and rats, they proposed that bulbospinal neurons probably involved in the generation of activity in the vasoconstrictor neurons (so-called "sympathetic vasomotor tone") and in mediating the baroreceptors reflex contain adrenaline (i.e., are C1 neurons) (Dampney et al. 1982; Goodchild et al. 1984; for early reviews see Dampney [1981], Dampney et al. [1985]).

- These experimental investigations were extended by Donald Reis and his group systematically studying responses of arterial blood pressure and heart rate to electrical and chemical stimulation of, or topic injection of the inhibitory amino acid GABA (or its specific antagonist bicuculline) into or injection of the sodium channel blocker tetrodotoxin into the RVLM. They compared the sites from which the responses could be elicited with the location of adrenergic (C1) neurons. As Dampney and coworkers did they came to the conclusion that neurons in the RVLM (probably adrenaline-synthesizing neurons) are responsible for the activity in sympathetic vasomotor neurons to maintain arterial blood pressure, are under inhibitory control and may mediate the baroreceptor reflex (Ross et al. 1984a, b; Ruggiero et al. 1989).

- The first neurophysiological recordings were made in the 1980s from bulbospinal neurons in the RVLM, which had the same discharge properties as muscle vasoconstrictor neurons and which were therefore the candidates to generate the activity in these cardiovascular sympathetic neurons (Brown and Guyenet 1984, 1985; Sun and Guyenet 1985; McAllen 1987; Sun and Spyer 1991).

10.2.2 The ventrolateral medulla oblongata

Sympathetic preganglionic neurons and their related interneurons obtain their supraspinal synaptic inputs from various nuclei in the brain stem and hypothalamus (Figure 8.15; Table 8.2). The sympathetic premotor neurons in the RVLM have been more thoroughly studied than the sympathetic and parasympathetic premotor neurons in the other nuclei. Therefore, I will concentrate preferentially on bulbospinal neurons in the RVLM (and on parasympathetic preganglionic cardiomotor neurons). The RVLM belongs to a complex network of neurons in the ventrolateral part of the medulla oblongata (VLM), which is important in the homeostatic regulation of the cardiovascular system and respiration. The VLM contains neuronal circuits, which are part of a network generating the respiratory rhythm in respiratory and cardiovascular neurons and the spontaneous activity in sympathetic cardiovascular neurons, possibly parasympathetic cardiomotor and bronchomotor neurons. These neuronal circuits mediate various cardiovascular and respiratory reflexes, and integrate both homeostatic control systems. Therefore the neurons that are involved in blood pressure regulation and in regulation of respiration are anatomically closely associated in the VLM (for review see Dampney [1994]; Pilowsky and Goodchild [2002]). Functionally, these neurons are sympathetic premotor neurons, parasympathetic preganglionic neurons, respiratory premotor neurons and various groups of excitatory and inhibitory interneurons related to these output pathways.

The information about the neurons in the VLM as components of neuronal circuits being involved in cardiovascular regulation (or regulation of respiration) was obtained by using multiple methodical approaches:

- Effect of chemical activation (e.g., by glutamate or another excitatory amino acid) or inhibition (e.g., by GABA agonists such as muscimol) of spatially circumscribed neuron populations on arterial blood pressure (or respiration).
- Effect of local blockade of excitatory or inhibitory synaptic transmission on arterial blood pressure (or respiration).
- Recording the expression of *c-fos* mRNA or protein in neuron populations that are activated by physiological stimulation of afferents (e.g., arterial baroreceptors or chemoreceptors).[1]
- The effect of selective destruction of a neurochemically specific population of neurons by a toxin that is conjugated to an enzyme or agonist that is reacting with its specific receptor in the membrane of the neurons. Toxin–enzyme complex or toxin–agonist–receptor complex is taken up by the neurons and leads to their destruction.[2]
- Recording of activity in single functionally identified neurons in the VLM with respect to cardiovascular parameters and activity in peripheral cardiovascular neurons.
- Determination of the morphology (location of cell body, orientation and projection of dendrites, projection and collateralization of axon), immunohistochemistry (putative fast transmitters, such as glutamate or GABA; monoamines, such as adrenaline, noradrenaline or 5-hydroxytryptamine (5-HT); colocalized peptides) and pharmacology of the neurophysiologically and functionally identified neurons.[3]

Figure 10.2 illustrates, for the rat and in simplified form, the functional anatomical organization of the VLM and the nucleus ambiguus (NA) between the caudal pole of the facial nucleus and the caudal pole of the lateral reticular nucleus as described by investigators who have studied its anatomy, the neural control of respiration and the neural control of arterial blood pressure or of perfusion of organs with blood. This anatomical organization applies with some modifications to other mammalian species too (e.g., cat, rabbit or dog and probably also humans). The division of the VLM into subnuclei was and is still somewhat controversial since the anatomical borders, as reported in Paxinos and Watson (1998), which are based on cytological and other anatomical criteria, do not necessarily correspond to the functional subdivision of the VLM (see Bieger and Hopkins [1987], Guyenet [1990] and Blessing [1997] for critical discussion). The nomenclature used in Figure 10.2 (as well as in this Chapter; see also Chapter 8) is based on Paxinos and Watson (1998). Here the VLM is defined for practical reasons as the region ventrolateral and ventral to the nucleus ambiguus (NA, compact and subcontact parts) but not including it, as shown in Paxinos and Watson (1998).[4] Figure 10.2 illustrates the anatomical structures that are relevant for the topics discussed in this Chapter. These structures have been projected on a parasagittal section of the medulla oblongata (a, d), on three coronal (transverse) sections (b_1: caudal to the facial nucleus; b_2: just rostral to the obex; B_3: about 1 mm caudal to the obex) and on a dorsal view of the VLM (c)[5]:

- The sympathetic (bulbospinal) premotor neurons, which generate spontaneous activity and mediate many reflexes in neurons of sympathetic pathways innervating blood vessels in skeletal muscle, kidney and other viscera and in sympathetic cardiomotor neurons are located in the RVLM.[6] It is situated caudal to the facial nucleus, rostral to the lateral reticular nucleus and ventral to the rostral (compact and subcompact) part of the NA. It extends close to the ventral medullary surface (Figure 10.2a). The RVLM overlaps dorsally with the rostral part of the external formation of the NA and of the compact and subcompact formation of the NA (Figure 10.2b_1). It lies rostral to most parasympathetic cardiomotor neurons, which are situated in the more distal part of the external formation of the NA, and ventral to the distal part of the NA, which contains the motoneurons innervating the larynx (Figure 10.2a, b_2).
- The RVLM extends caudally into the caudal VLM (CVLM). The CVLM is divided into a rostral part (the rCVLM) and a caudal part (cCVLM). The rCVLM is also sometimes called intermediate VLM (IVLM, which was originally defined in the rabbit; see Dampney [1994]). The CVLM is located between the NA and lateral reticular nucleus (LRN in Figure 10.2a). The CVLM contains various groups of inhibitory and excitatory interneurons, which are important for the regulation of activity in sympathetic premotor neurons of the RVLM and which project directly or indirectly to the RVLM. Chemical stimulation of the neurons in the CVLM is followed by a decrease of arterial blood pressure. Therefore, this part of the VLM is also called the ventrolateral depressor area (Figure 10.3).
- The most caudal part of the VLM is also called the caudal pressure area (CPA) since stimulation of the neurons in this area (e.g., by the excitatory amino acid glutamate) increases the blood pressure and activity in "sympathetic" nerves (renal and splanchnic nerves) whereas inhibition of the neurons in the CPA decreases the arterial blood pressure and abolishes activity in these nerves. The CPA is situated between the caudal pole of the LRN and the medullary dorsal horn close to the transition between medulla oblongata and spinal cord (Gordon and McCann 1988; Possas *et al.* 1994; Campos

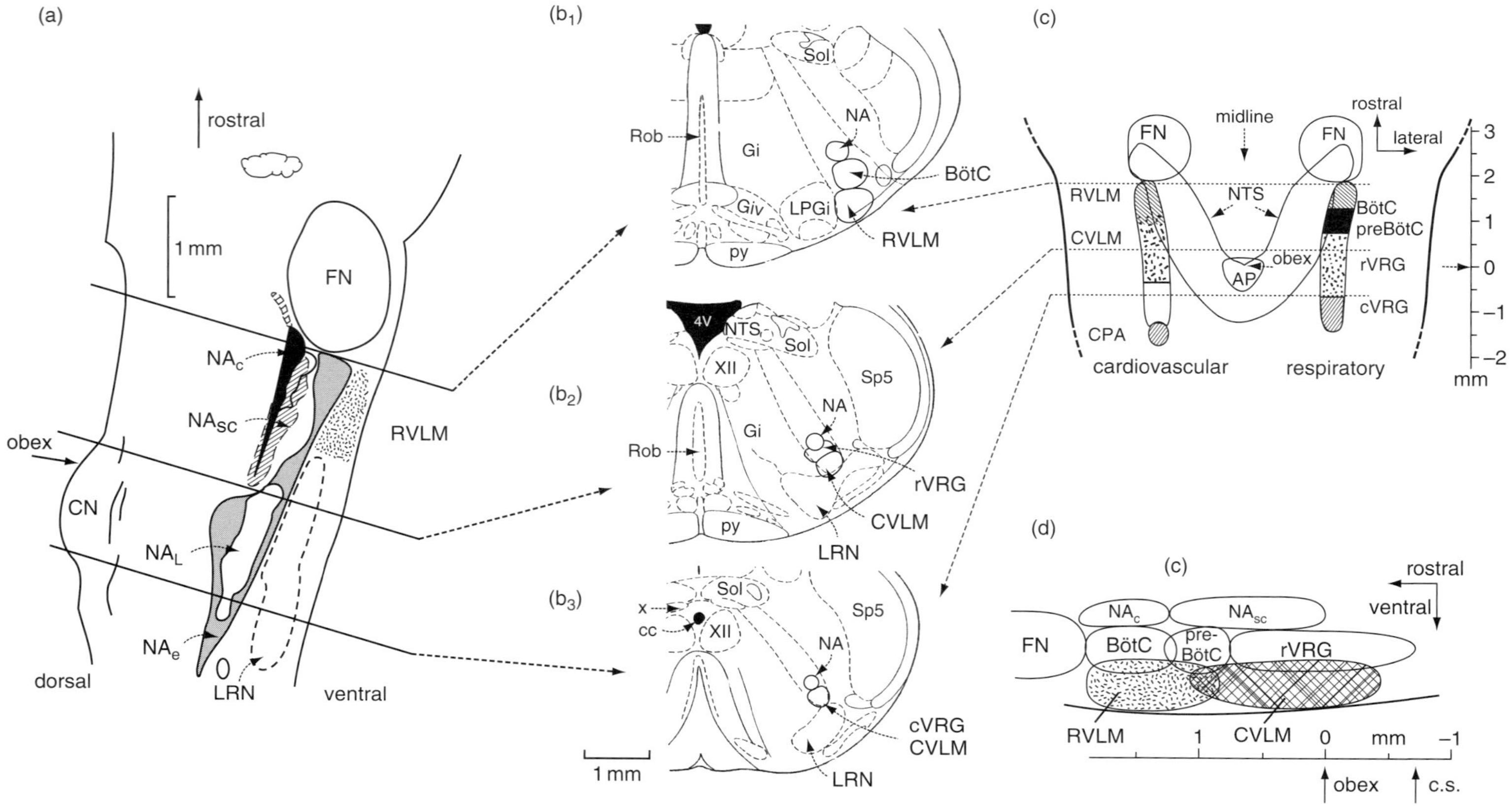
(a)
rostral
1 mm
FN
RVLM
NA_c
NA_{SC}
obex
CN
NA_L
NA_e
LRN
dorsal
ventral
(b_1)
Rob
Gi
Sol
NA
BötC
Giv
LPGi
RVLM
py
(b_2)
4V
NTS
Sol
XII
Sp5
NA
Gi
Rob
rVRG
CVLM
py
LRN
(b_3)
Sol
Sp5
x
cc
XII
NA
cVRG
CVLM
LRN
1 mm
(c)
midline
rostral
lateral
FN
FN
RVLM
CVLM
NTS
BötC
preBötC
obex
rVRG
AP
cVRG
CPA
cardiovascular
respiratory
3
2
1
0
−1
−2
mm
(d)
(c)
rostral
ventral
NA_c
NA_{sc}
FN
BötC
pre-BötC
rVRG
RVLM
1
CVLM
0
mm
−1
obex
c.s.

Figure 10.2 Anatomical organization of the ventrolateral medulla (VLM) in the rat caudal to the facial nucleus (FN). Location of subdivisions of the nucleus ambiguus (NA), of the rostral ventrolateral medulla (RVLM), caudal ventrolateral medulla (CVLM) and cell groups involved in regulation of respiration. (a) Major subdivisions of the NA and location of the RVLM in a schematic drawing of a parasagittal section of the rat medulla oblongata. (b) Coronal (transverse) sections through the medulla oblongata just caudal to the facial nucleus (b_1; i.e. at the level 12.8 mm caudal to bregma) according to the atlas of Paxinos and Watson [1998]), 300 μm rostral to the obex (b_2; 13.24 mm caudal to bregma) and about 0.8 mm caudal to the obex (b_3; 14.3 mm caudal to bregma). (c) Dorsal view of the medulla oblongata showing the location of neurons being involved in cardiovascular regulation (left) and in regulation of respiration (right) and the outline of the nucleus tractus solitarii (NTS) and area postrema (AP). (d) Parasagittal diagram of the VLM showing in schematic outline the topographical relation between NA (NA_C, compact NA; NA_{SC}, subcompact NA), groups of respiratory neurons (Bötzinger complex [BötC], preBötzinger complex [preBötC], rostral ventral respiratory group [rVRG]) and RVCL/CVLM (same scale as in [c]). (1) The motoneurons of esophagus, pharynx and larynx are situated in the compact (NA_C), semicompact (NA_{SC}) and loose (NA_L) formation of the NA. (2) The external formation of the NA (NA_e) is anatomically not very well defined; it contains the parasympathetic cardiomotor neurons (in particular ventrolateral at the level of the obex and extending rostrally) and the parasympathetic bronchomotor and secretomotor neurons innervating the trachea and bronchi (in particular at the level of the obex, but also throughout NA_C, NA_{SC} and NA_L). (3) Respiratory neurons are located between the NA_C/NA_{SC} cell column and the RVLM/CVLM cell column: (i) the caudal ventral respiratory group (cVRG) in the caudal third and further caudal containing expiratory neurons; (ii) the rVRG in the intermediate part rostral and caudal to the obex containing inspiratory neurons; (iii) the preBötC about midway between FN and lateral reticular nucleus (LRN) containing respiratory interneurons; (iv) the BötC caudal to the facial nucleus containing expiratory neurons. (4) The RVLM is located caudal to the FN, rostral to the LRN and ventral to the NA_e. It overlaps with the BötC and preBötC (see [c, d]). (5) The CVLM is situated at the level of the preBötC and further caudal. (6) The caudal pressure area (CPA) is located at the most caudal part of the VLM. CC, central canal; CN, cuneate nucleus; Gi, gigantocellular nucleus; Giv, ventral part of the gigantocellular nucleus; LPGi, lateral paragigantocellular nucleus; py, pyramidal tract; Rob, raphe obscurus; Sol, solitary tract; Sp5, spinal trigeminal nucleus; 4 V, fourth ventricle; X, dorsal motor nucleus of the vagus; XII, hypoglossal nucleus; (a) modified from Bieger and Hopkins (1987) with permission; (b_1–b_3) modified from Paxinos and Watson (1998), (c) reconstructed from Paxinos and Watson (1998), after Dobbins and Feldman (1994) and Sun and Panneton (2002); (d) modified from Rekling and Feldman (1998).

and McAllen 1999; Natarajan and Morrison 2000; Sun and Panneton 2002, 2004).
- Respiratory neurons are also located in a cell column in the VLM that extends from the caudal border of the facial nucleus to about 1.5 mm caudal to the obex (Figure 10.2b, c). These neurons and their excitatory and inhibitory synaptic connections are part of a respiratory network, including the dorsal respiratory group in the medulla oblongata and centers in the caudal and rostral pons (ponto-medullary respiratory network; St-John 1998; Rybak et al. 2004; St-John and Paton 2004), that is responsible for the generation of the different types of respiratory pattern as present in the activity of motoneurons projecting in the phrenic nerve, glossopharyngeal nerve, superior laryngeal nerve, hypoglossal nerve or the pharyngeal branch of the vagus nerve (Monnier et al. 2003). This ventrolateral cell column is anatomically interposed between the NA (compact and subcompact part and external formation) and the RVLM/CVLM cell column. However, it overlaps with both rostrocaudal cell columns and there is considerable intermingling between the neurons in the three columns (Pilowsky et al. 1990). It is divided into the following groups of neurons (Dobbins and Feldman 1994; Bianchi et al. 1995; Rekling and Feldman 1998; Monnier et al. 2003):
  a. Most neurons in the Bötzinger complex are expiratory neurons. Many of these neurons are glycinergic propriobulbar neurons inhibiting respiratory premotor neurons (bulbospinal inspiratory neurons in the rostral ventral respiratory group [rVRG] and/or bulbospinal expiratory neurons in the caudal ventral respiratory group [cVRG]) and respiratory motoneurons. Some neurons in the Bötzinger complex are bulbospinal neurons inhibiting inspiratory phrenic motoneurons (Schreihofer et al. 1999; Ezure et al. 2003).
  b. The preBötzinger complex contains various types of respiratory interneurons that may develop pacemaker activity under certain conditions and propriobulbar neurons (and a few respiratory premotor neurons). Theses neurons form circuits that are responsible for the generation of respiratory gasping in animals with intact medulla oblongata isolated from the pons or an equivalent rhythm in in vitro slice or block preparations of the medulla oblongata. However, they are not responsible for generation of the normal breathing rhythm (eupnea) and it is debated whether pacemaker discharges of these neurons occur at all during eupnea (St-John 1998; St-John and Paton 2004; see Subchapter 10.6.1).
  c. The rVRG contains mainly bulbospinal inspiratory premotor neurons that project to and excite phrenic motoneurons.
  d. The cVRG, which extends from the obex to the cervical segment C1, contains mainly bulbospinal expiratory premotor neurons

that project to and excite motoneurons (or corresponding interneurons) in the thoracic and upper lumbar spinal cord.

It is important to emphasize, explicitly by alluding to Figure 10.2, that *the anatomical substrates of the neural homeostatic regulation of respiration and of the cardiovascular system are present in the same column of neurons of the VLM*. The neurons in the VRG are part of the pontomedullary respiratory network that adapts respiration to various needs of the organism (for details see Feldman [1986]; Richter [1996]; Rekling and Feldman [1998]; St.-John 1998; Feldman and McCrimmon [2003]). The "cardiovascular" neurons in the VLM appear to be organized in a *continuous column* of neurons extending from the RVLM down to the CPA. Functionally different types of cardiovascular neurons *overlap* in their location. However, as shown in Figure 10.2, the rostrocaudal boundaries of cardiovascular and respiration-related groups of neurons do not exactly coincide (e.g., RVLM and Bötzinger complex; CVLM and rVRG). Furthermore, it must be kept in mind that the dendrites of cardiovascular and respiratory neurons intertwine. Finally, as already mentioned, the strict anatomical borders of nuclei as described by Paxinos and Watson (1998) do not need to be identical with the exact location of the functionally determined subdivisions of neuron assemblies[7] (see Figure 10.2c, d; for details see Pilowsky and Goodchild [2002]; Schreihofer and Guyenet [2002]; Wang *et al.* [2002]; Weston *et al.* [2003]):

1. The RVLM region overlaps with the more dorsally located Bötzinger complex and also with part of the preBötzinger complex.
2. In the rostral part of the CVLM (sometimes called the lVLM) are located sympathoinhibitory interneurons, most of which are excited by baroreceptor afferents (see Subchapter 10.3). This region contains most neurons of the preBötzinger complex and neurons of the rVRG, both groups of respiratory neurons being more dorsally located than the cardiovascular neurons.
3. In the caudal part of the CVLM are located mainly sympathoinhibitory interneurons and probably sympathoexcitatory interneurons which are independent of the baroreceptor afferents. This region contains respiratory neurons of the rVRG.
4. In the most caudal part of the VLM, the CPA, are located sympathoexcitatory interneurons. This region overlaps with the caudal part of the cVRG. Cardiovascular neurons in the CPA have not been studied in detail (Sun and Panneton 2002).

In the rat, the VLM contains catecholaminergic neurons that either synthesize adrenaline (C1 neurons) or noradrenaline (A1 neurons). The C1 neurons contain the enzyme phenylethanolamine-N-methyl-transferase (PNMT), in addition to the other enzymes to synthesize noradrenaline (tyrosine hydroxylase [TH] and

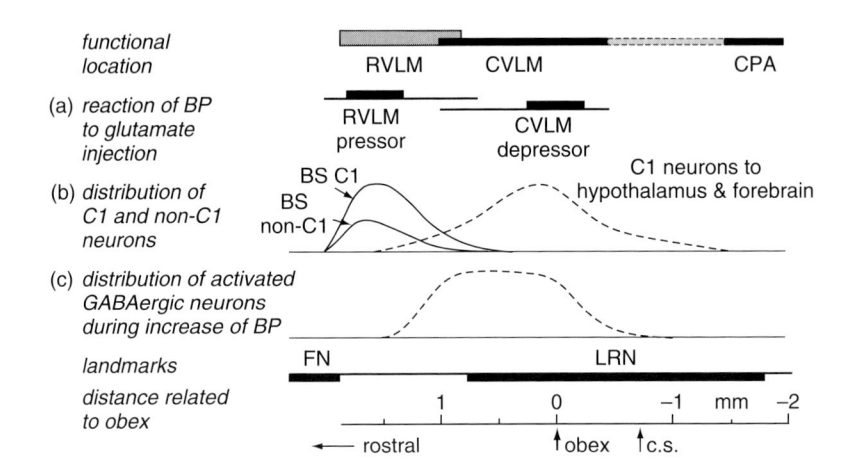

Figure 10.3 Schematic rostrocaudal distribution of the location of cardiovascular neurons in the ventrolateral medulla (VLM) with respect to the obex in the rat. The ventral landmarks are the facial nucleus (FN) and the lateral reticular nucleus (LRN). The dorsal landmarks are the obex and the calamus scriptorius (c.s.) (see note 5). Location of the rostral ventrolateral medulla (RVLM), caudal ventrolateral medulla (CVLM) and caudal pressure area (CPA) as in Figure 10.2. (a) Effect of microinjection of glutamate on the arterial blood pressure (BP). (b) Distribution of adrenergic (C1) and non-adrenergic (non-C1) bulbospinal (BS) neurons. Dotted lines, distribution of C1 neurons projecting to hypothalamus or telencephalic nuclei. (c) Distribution of GABAergic neurons activated during increase of BP. These neurons express c-Fos protein during activation (see note 1). After Schreihofer and Guyenet (2002) and Sun and Panneton (2002) for the CPA. This figure and Figure 10.2c, d are to the same distance scale. For references see text.

dopamine-β-hydroxylase [DβH]), which is necessary to convert noradrenaline into adrenaline by methylation. The A1 neurons lack PNMT, but contain TH and DβH. The non-aminergic neurons in the VLM lack the enzymes PNMT, TH and DβH necessary to synthesize adrenaline and noradrenaline. Immunoreactivity against TH and PNMT is used to identify the neurons as C1, A1 or non-catecholaminergic neurons.

All catecholaminergic neurons in the RVLM and in the rostral part of the CVLM are C1 neurons. Most catecholaminergic neurons in the caudal part of the CVLM are C1 neurons, a few are A1 neurons. Caudal to the obex, an increasing portion of the catecholaminergic neurons in the VLM do not express PNMT but TH (Figure 10.3) and are therefore considered to be A1 neurons. In other species, such as guinea pig, sheep, rabbits or cat, PNMT-immunoreactive neurons (i.e., C1 neurons) seem to be absent in the VLM (Halliday and McLachlan 1991). In the rat, all spinally projecting catecholaminergic (TH-positive) neurons in the VLM are adrenergic (PNMT-positive) and project to the intermediate zone of the spinal cord where the sympathetic preganglionic neurons and autonomic interneurons are located. About two-thirds of the excitatory sympathetic premotor neurons in the RVLM of the rat are C1 neurons. The C1 and A1 neurons in the CVLM project to the hypothalamus and forebrain structures (Schreihofer and Guyenet 1997; Verberne et al. 1999; Guyenet et al. 2001) (for numbers of bulbospinal C1 and non-C1 neurons in the RVLM see note 10).

10.2.3 The RVLM as sympathetic cardiovascular premotor nucleus

What is the evidence that the RVLM contains sympathetic premotor neurons (in addition to other types of projecting neurons and interneurons) that are important in the regulation of arterial blood pressure (see Dampney [1994]; Schreihofer and Guyenet [2002])?

- Bilateral lesions of the RVLM in the anesthetized animal are followed by an acute decrease of arterial blood pressure, sometimes to the same level as after transection of the spinal cord at the cervical spinal level (Dampney and Moon 1980; see Dampney [1994]; Blessing [1997]).[8]
- Bilateral inhibition of RVLM neurons by microinjection of an inhibitory amino acid or an analog of these amino acids (e.g., muscimol [a GABA agonist]) also leads to a fall in arterial blood pressure that is comparable to that after destruction of the RVLM (Ross et al. 1984b; Sun and Reis 1996; Morrison 1999; Schreihofer et al. 2000).
- Stimulation of the cell bodies in the RVLM with glutamate generates an increase in arterial blood pressure and excites muscle, visceral and renal vasoconstrictor neurons (Ross et al. 1984b; Guyenet and Brown 1986; Schreihofer et al. 2000).
- The RVLM contains neurons that project to the intermediate zone (in particular the nucleus intermediolateralis) of the thoracolumbar spinal cord. Some of these sympathetic premotor neurons form monosynaptic connections with the preganglionic neurons (Milner et al. 1988; Zagon and Smith 1993; McAllen et al. 1994; Pyner and Coote 1998). Other connections are di- or polysynaptic (via spinal autonomic interneurons).
- The discharge characteristics of the bulbospinal neurons in the RVLM are similar to those recorded in muscle and visceral vasoconstrictor neurons (see Figure 4.1). The activity of most sympathetic premotor neurons in the RVLM exhibits cardiac rhythmicity and some respiratory rhythmicity (see Subchapter 10.6); they are inhibited by stimulation of arterial baroreceptors and excited by stimulation of arterial chemoreceptors (see Chapter 10.5), nociceptors, trigeminal and other receptors (Figure 10.4; see Guyenet [1990]; Dampney [1994]; Sun [1995]). A subgroup seems to behave like cutaneous vasoconstrictor neurons (McAllen 1992; McAllen and May 1994a).

In the rat, the catecholaminergic (C1) and non-catecholaminergic sympathetic premotor neurons in the RVLM probably use glutamate as primary transmitter (Stornetta et al. 2002). Various neuropeptides (e.g., enkephalin, substance P, galanin, calbindin) may be colocalized in C1 neurons showing that these sympathetic premotor neurons are histochemically heterogeneous. For example some catecholaminergic (C1) and most non-catecholaminergic barosensitive bulbospinal premotor neurons express enkephalin (Stornetta et al. 2001). Neuropeptide Y (NPY) is colocalized in many C1 neurons projecting

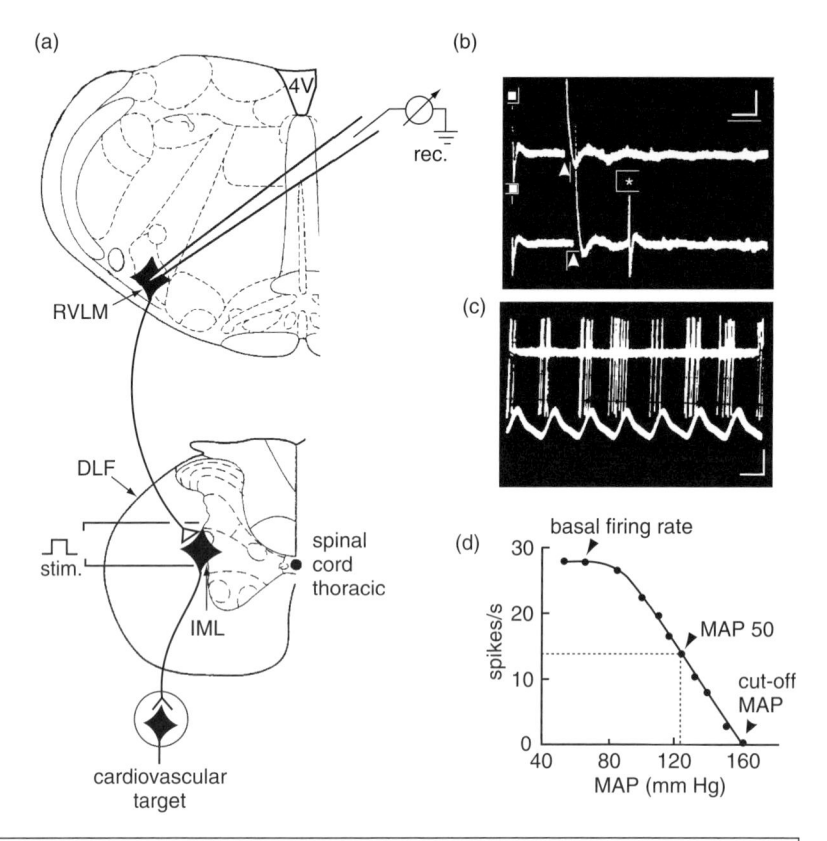

Figure 10.4 Identification and baroreceptor sensitivity of a sympathetic (bulbospinal) premotor neuron in the rostral ventrolateral medulla (RVLM) in the rat. (a) Recording from a neuron in the RVLM with a microelectrode. Identification of the neuron by electrical stimulation of its axon projecting to the upper thoracic spinal cord. DLF, dorsolateral funiculus; IML, intermediolateral nucleus. (b) Identification of the RVLM neuron as a bulbospinal neuron. The oscilloscope recording is triggered by spontaneous action potentials in the RVLM neuron (square). Electrical stimulation of the axon in the IML (arrow head in lower trace) with single pulses activates the neuron (marked by star) when the time interval between the orthodromic spontaneous action potential and the electrical stimulus is long. The evoked antidromically conducting action potential collides with the orthodromically conducted spontaneous action potential when the time interval between the spontaneous action potential and electrical stimulus is short (upper trace). Each recording five times superimposed. (c) Pulse-synchronous discharge (cardiac rhythmicity) of the RVLM neuron with the arterial pulse pressure wave (lower trace). Increased blood pressure leads to phasic inhibition of activity (six consecutive recordings superimposed). (d) Relation between discharge rate of an RVLM neuron (ordinate) and mean arterial blood pressure (MAP, abscissa). Time scale in (b) 5 ms and in (c) 120 ms. Vertical scale in (b) and (c) 0.5 mV. In (c) the phasic inhibition of the activity in the bulbospinal RVLM neuron starts about 50 ms after the onset of the systole. This delay of inhibition is due to the conduction time in axons of the baroreceptors, nucleus tractus solitarii neurons and neurons in the caudal ventrolateral medulla (Jeske et al. 1993; see Figure 10.7). Modified from Sun and Guyenet (1985) with permission.

to the hypothalamus. The function of adrenaline and of the colocalized neuropeptides in the sympathetic premotor neurons is unknown (see Dampney [1994]; Sun [1995]; Pilowsky and Goodchild [2002]) and there is some debate about whether adrenaline is released in the spinal cord (see Guyenet et al. [2001]).[9]

Glutamate and GABA (in interneurons) seem to be the fast primary transmitters to the sympathetic preganglionic neurons (Llewellyn-Smith et al. 1992, 1995, 1998). Whether descending inhibitory pathways

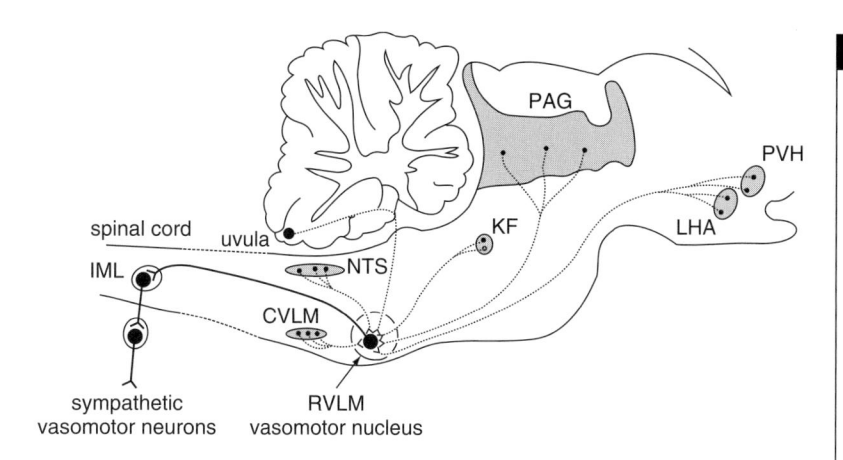

Figure 10.5 Major extrinsic synaptic inputs to sympathetic (bulbospinal) premotor neurons of the rostral ventrolateral medulla (RVLM). CVLM, caudal ventrolateral medulla; IML, intermediolateral nucleus; KF, Kölliker–Fuse nucleus; LHA, lateral hypothalamic area; NTS, nucleus tractus solitarii; PAG, periaqueductal grey of the mesencephalon; PVH, paraventricular nucleus of the hypothalamus. The sympathetic premotor neurons also receive various synaptic inputs from interneurons intrinsic to the RVLM. Modified from Dampney (1994).

from the RVLM that use GABA and/or glycine exist is very much debated but cannot be excluded (see Miura et al. [1994]; Deuchars et al. [1997]; Stornetta et al. [2004]) (see Subchapter 10.3.4 and note 9).

Sympathetic premotor neurons in the RVLM obtain multiple synaptic inputs from various nuclei in the brain stem and hypothalamus (CVLM, NTS, parabrachial nuclei and Kölliker–Fuse nucleus in the dorsolateral pons, periaqueductal grey [PAG], lateral hypothalamic area [LH] and paraventricular nucleus of the hypothalamus [PVH], uvula [lobule IXb of posterior vermis of the cerebellum] [Silva-Carvalho et al. 1991]) and probably from interneurons within the RVLM. Interneurons are more frequent in the RVLM than sympathetic (bulbospinal) premotor neurons (Schreihofer et al. 1999), but seem to be rare in the subretrofacial nucleus of the cat (see note 6; Polson et al. 1992) (Figure 10.5). Neurons in these nuclei mediate various reflexes elicited by stimulation of cardiovascular afferents (see Subchapters 10.3 and 10.5), pulmonary afferents, trigeminal afferents, nociceptive afferents and other afferents. Furthermore, they involve the RVLM in various complex homeostatic control systems and behaviors that are controlled from the upper brain stem, hypothalamus and cerebral hemispheres (e.g., defense behaviors, thermoregulation, exercise, etc.; for review see Dampney [1994]).

The synaptic inputs to RVLM neurons are excitatory or inhibitory and use glutamate, GABA, glycine, angiotensin II and possibly other yet unknown substances as primary transmitters. Many neuropeptides are colocalized in neurons that form synapses with the bulbospinal RVLM neurons. The nature of the function of these peptides is only now beginning to be characterized.

Sympathetic premotor neurons in the RVLM are topographically organized with respect to the functional type of vascular target innervated by the sympathetic neurons but *not* with respect to the body region. This has been demonstrated in the *cat* for the subretrofacial nucleus in the RVLM (Lovick 1987; McAllen et al. 1995, 1997). Subpopulations of neurons in this nucleus were microstimulated by ionophoretic application of the excitatory amino acid glutamate and activity was recorded from postganglionic neurons innervating

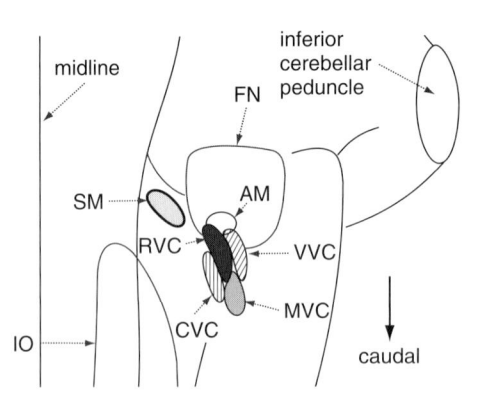

Figure 10.6 Viscerotopic organization of sympathetic premotor neurons in the subretrofacial cell column of the rostral ventrolateral medulla (RVLM) of the cat. Ventral view of the cat left medulla showing the surface structures, inferior olive (IO) and facial nucleus (FN). The premotor neurons are organized according to the functional type of the neurons and not according to body region. The approximate locations of the premotor neurons overlap. AM, sympathetic premotor neurons associated with the adrenal medulla (release of adrenaline); CVC, cutaneous vasoconstrictor; MVC, muscle vasoconstrictor; RVC, renal vasoconstrictor; VVC, visceral vasoconstrictor (mesenteric vascular bed). Also indicated is the location of sympathetic sudomotor premotoneurons (SM). Derived from Lovick (1987), Dampney and McAllen (1988) and McAllen and May (1994a, b). Modified from McAllen et al. (1995) with permission.

different vascular beds or effector responses were recorded. Sympathetic renal vasoconstrictor premotor neurons (RVC in Figure 10.6) are located most rostrally and sympathetic muscle premotor vasoconstrictor neurons (MVC) most caudally; sympathetic visceral premotor vasoconstrictor neurons (which regulate the mesenteric vascular bed; VVC) are located lateral to the renal neurons and sympathetic cutaneous vasoconstrictor premotor neurons (CVC) medial to the muscle vasoconstrictor neurons (Figure 10.6). Sympathetic cardiomotor premotor neurons are situated in a territory rostromedial to and overlap with the sympathetic premotor neurons controlling blood flow through skeletal muscle (Lovick 1987; Campos and McAllen 1997; not shown in Figure 10.6). Furthermore, sympathetic premotor neurons controlling the adrenal medulla are located in the most rostral part of the RVLM (Lovick 1987; see Pyner and Coote [1998]). Thus, subpopulations of sympathetic vasoconstrictor premotor neurons projecting to the respective functional types of preganglionic neurons in the spinal cord exhibit distinct though overlapping locations in the RVLM. Finally, sympathetic premotor neurons controlling sweat glands (SM) appear to be situated rostromedially to the cardiovascular premotor neurons (McAllen et al. 1995).

A similar topographic organization is also present in the RVLM of rabbits (Ootsuka and Terui 1997). Furthermore, using morphological tracing methods, it has been shown that this type of organization may also apply to the rat (Pyner and Coote 1998) although systematic functional studies in this species have not been done.[10]

The functional topographic organization of the sympathetic premotor neurons in the RVLM does not collide with the idea that (1) different groups of sympathetic premotor neurons converge on the same type of preganglionic neuron and/or associated interneurons and (2) the same type of sympathetic premotor neurons diverge on functionally different types of preganglionic neurons.

10.2.4 Origin of spontaneous activity in bulbospinal neurons of the RVLM

An extensively discussed, controversial and unsolved issue is the origin of spontaneous (tonic) activity in the sympathetic premotor neurons of the RVLM. This spontaneous activity determines the ongoing activity in sympathetic cardiovascular neurons, which innervate resistance vessels (MVC, VVC, RVC neurons) and the heart, and therefore arterial blood pressure. Based on experimental results, alternative theories are discussed (Dampney *et al.* 2000):

1. In vitro, the spontaneous activity of sympathetic premotor neurons in the RVLM is the result of intrinsic pacemaker properties of the neurons. Experiments on transverse sections of the lower brain stem at the level of the RVLM have shown that some RVLM neurons exhibit activity and generate action potentials after complete blockade of all synaptic inputs to the cells. This firing fulfilled the criteria of intrinsic pacemaker properties[11] (Sun *et al.* 1988a, b; see Guyenet [1990]; Sun [1995]). Using extracellular recordings from sympathetic premotor neurons in the RVLM in a working-heart-brain-stem preparation (WHBP; Paton 1996a, b) of rats three to four weeks of age Allen could show that these neurons can generate activity (regular firing, bursting activity) without any synaptic input (Andrew Allen, personal communication). However, other intracellular measurements in presympathetic RVLM neurons, performed in in vivo and in vitro experiments, could not confirm that these neurons have intrinsic pacemaker properties (Lipski *et al.* 1996, 1998). The ionic channels maintaining tonic activity in the neurons of the RVLM may involve low-voltage-activated calcium channels (T-type channels). Blockade of these channels (bilaterally in the RVLM) reduces spontaneous activity in the splanchnic nerve and arterial blood pressure, but does not change reflexes mediated by the RVLM (Miyawaki *et al.* 2003).

   The RVLM neurons may only show pacemaker-like activity under extreme pathophysiological conditions, e.g. when all synaptic inputs to these neurons are blocked or fail. In that case the capacity of the RVLM neurons to generate pacemaker activity may be a last "line of defense" to generate tonic activity in cardiovascular neurons to maintain arterial blood pressure (Lipski *et al.* 2002).

2. The spontaneous activity of the sympathetic premotor neurons in the RVLM is the result of the emergent properties of the neuronal networks and therefore depends on the inhibitory and excitatory synaptic connections between the neurons of the network, the

underlying synaptic currents and the intrinsic chemosensitivity of the neurons (related to arterial P_{CO_2}, P_{O_2} and pH). Thus, the generation of spontaneous activity in the sympathetic premotor neurons of the RVLM (and therefore the so-called vasomotor and cardiomotor tone) would be generated by a neuronal network that is coupled synaptically with the RVLM neurons (Lipski *et al.* 2002). Evidence for the existence of such a neural network associated with the RVLM is missing. The CPA in the caudal part of the CVLM (Figure 10.2) may be important for the generation of activity of sympathetic premotor neurons in the RVLM. Inhibition of the activity in neurons of the CPA decreases arterial blood pressure and activity in the renal nerve to the same degree as transection of the cervical spinal cord (Possas *et al.* 1994; Horiuchi and Dampney 2002) and reduces the activity of the sympathetic premotor neurons in the RVLM. Whether this is generated by reduction of excitatory synaptic activity to or by synaptic disinhibition of the neurons in the RVLM is controversial (Natarajan and Morrison 2000; Horiuchi and Dampney 2002).[12] However, what is the origin of the activity of the neurons in the CPA?

3. The RVLM neurons receive multiple excitatory glutamatergic synaptic inputs. Additionally, many neurons that synapse with RVLM neurons contain neuropeptides. Furthermore, substances, which are probably of non-neural cellular origin, such as nitric oxide (NO) and angiotensin II are released in the RVLM (angiotensin II also from nerve terminals). It is hypothesized that neuropeptides and local substances of unclear cellular origin modulate the excitability of the RVLM bulbospinal neurons and their synaptic inputs in a paracrine way and contribute to or are mainly responsible for the level of spontaneous activity ("tone") in sympathetic cardiovascular neurons (see Guyenet and Stornetta [1997]).

Several points in regard to the spontaneous (tonic) activity in autonomic neurons and its underlying mechanism and origin need to be emphasized:

- Practically all experimental approaches addressing the origin of spontaneous activity in autonomic neurons have concentrated so far on activity in sympathetic neurons supplying resistance vessels (in skeletal muscle, viscera including kidney) or heart.
- Many vasoconstrictor neurons innervating blood vessels in skin or skeletal muscle and sudomotor neurons develop spontaneous activity weeks and months after transection of the spinal cord at the cervical level (see Chapter 9.2).
- Intracellular measurements in sympathetic preganglionic neurons of cats with spinal cords acutely transected at the cervical level show subthreshold spontaneous synaptically evoked activity (mainly excitatory postsynaptic potentials) (Dembowsky *et al.* 1985). The ongoing synaptic activity may be related to electrical

coupling between sympathetic preganglionic neurons by gap junctions. In slices, these neurons may exhibit spontaneous membrane oscillations that are transmitted to neighboring preganglionic neurons by electrical coupling. This may be a mechanism of synchronization of activity in sympathetic preganglionic neurons (Logan *et al.* 1996; Nolan *et al.* 1999).
- Spontaneous activity in cutaneous vasoconstrictor neurons probably depends, under normal conditions, on the activity in both premotor neurons of the RVLM and in premotor neurons of the caudal raphe nuclei (McAllen, personal communication; see Chapter 10.5 and Figure 10.14).
- Sympathetic premotor neurons have a target-specific viscerotopic organization in the RVLM (Figure 10.6). This organization may imply that activity in functionally different sympathetic neurons, which are involved in cardiovascular regulation or regulation of the adrenal medulla, can be selectively altered from the functionally different groups of neurons in the RVLM. However, this needs to be shown.
- Spontaneous activity in sympathetic non-vasoconstrictor neurons (Table 4.3) probably has an entirely different central origin compared to the origin of spontaneous activity in vasoconstrictor neurons or sympathetic cardiomotor neurons. Rather spontaneous activity in sympathetic motility-regulating neurons may originate in the spinal cord (Bartel *et al.* 1986).
- The origin of spontaneous activity in parasympathetic neurons including parasympathetic cardiomotor neurons (Table 4.6) is entirely unknown.

In *conclusion*, it is believed by some groups that the activity in vasoconstrictor neurons innervating resistance vessels and in sympathetic cardiomotor neurons originates in neurons of the RVLM (bulbospinal neurons and/or local interneurons synaptically connected with the RVLM neurons), yet the underlying mechanism of its generation is still unsolved. It may turn out that ongoing activity even in these cardiovascular neurons originates in multiple nuclei of brain stem (e.g., raphe nuclei, catecholaminergic nuclei [A2, A5]), hypothalamus (e.g., paraventricular nucleus) or even the spinal cord (see Subchapter 9.2). But what it shows is that even for the spontaneous activity in sympathetic cardiovascular neurons involved in blood pressure control the underlying mechanism is practically unknown (see, e.g., Horiuchi *et al.* [2004b]; Sved [2004]). Thus, a key problem of neural regulation of arterial blood pressure in health and disease (hypertension!) in which the physiologists have been interested since Claude Bernard and Carl Ludwig in the middle of the nineteenth century remains unsolved up to the present time.

The results show furthermore that the discussion about the origin of spontaneous activity in sympathetic neurons should not be restricted to the RVLM. Spontaneous activity in other functional types of sympathetic and parasympathetic neuron most likely has different origins, depending on the functional type of neuron. Therefore, I find it inappropriate to use the terms sympathetic or parasympathetic "tone" in a generalized way. Spontaneous activity in autonomic systems should always be specified with respect to the functional system involved and this most likely also applies to the systems represented in the RVLM (Figure 10.6).

10.2.5 Bursting activity in sympathetic nerves

Activity in "sympathetic" nerves occurs in bursts, which are generated by synchronous firing of many pre- and postganglionic axons. These bursts may occur in rhythms related to the cardiac and respiratory rhythms (see Chapters 4.1, 10.6), or in rhythms of 10 Hz, 2 to 6 Hz and 0.2 to 0.4 Hz (low-frequency rhythms below the respiratory rhythm). The latter three rhythms are independent of the cardiac and respiratory rhythmicities (see multiunit activity in the splanchnic nerve [SPL] of the rat in Figure 10.15b_1–b_3 and c_1; see note 22), they are also present in experimental conditions in which the arterial baroreceptors have been denervated and the central respiratory generator has been silenced. The bursts are not immediately seen in single sympathetic neurons because the rate of spontaneous activity in these neurons is too low (see Table 6.2).

The synchronous discharges of sympathetic cardiovascular neurons are believed to be generated by neuronal circuits in the brain stem that oscillate; multiple oscillators would be responsible for the rhythms with distinct frequencies (*network oscillator hypothesis*). It is hypothesized (1) that the 2 to 6 Hz oscillator is normally entrained by the phasic (pulsatile) activation of arterial baroreceptors and (2) that the oscillator working at a frequency similar to the breathing frequency of central breathing is entrained by the central respiratory network or that both oscillators are entrained by a third oscillator (for reviews see Gebber [1990]; Malpas [1998]; Barman and Gebber [2000]).[13] The origin of the multiple rhythms not strictly related to respiration or pulsatile activation of arterial baroreceptors and their underlying mechanisms are entirely unknown. The generation of the rhythmic changes of bursting activity in sympathetic nerves should not be confused with the origin of the spontaneous (tonic) activity in sympathetic neurons.

The discussion of the origin of the rhythmic bursting discharge in populations of sympathetic neurons is extremely controversial. For me it resembles the controversy about the rhythmogenesis of the drive in neurons of the respiratory network in mature mammals. Respiratory rhythmicity may depend on pacemaker activity in early developmental (immature) states of the respiratory network. In fact pacemaker activity (in the preBötzinger complex) has only been shown to exist in slices of the medulla oblongata of immature rats. This activity appears to be essential for gasping in the neonatal and adult rat when the pontine part of the respiratory network is damaged or during severe hypoxia. So far no evidence for pacemakers has been found in vivo for eupneic respiration as well as for the generation of bursting sympathetic activity. Thus, in adult and neonatal rats, pacemakers are not important or not necessary for normal respiratory rhythm generation (eupnea) (see Richter [1996]; Rekling and Feldman [1998]; Smith et al. [2000]; Rybak et al. [2002, 2003, 2004]).

Using autocorrelation, coherence and spectral analysis, it has been shown that the discharges of single sympathetic postganglionic neurons innervating the rat tail exhibit a dominant rhythm independent of the respiratory rhythm. The frequency of this dominant rhythm is similar to that of the respiratory rhythm and varies between individual postganglionic neurons (Johnson and Gilbey 1996, Chang et al. 1999). Häbler et al. (1999a, 2000) could not confirm the existence of this dominant respiration-independent rhythm in cutaneous vasoconstrictor neurons in the same preparation under the same experimental conditions using autocorrelation and spectral analysis. Coupling of the activity in the neurons to respiration did not change or increased during whole-body warming.

The multiple dominant rhythms in the activity of the sympathetic neurons can be synchronized (entrained) by the respiratory generator (e.g., during increased central respiratory drive) and transiently synchronized by activation of primary afferent neurons (e.g., from the lung, from cutaneous nociceptors [Chang et al. 2000, Staras et al. 2001]).

It is hypothesized that the dominant rhythms in the activity of the postganglionic neurons are generated by multiple oscillators with different intrinsic frequencies, which are probably located in the lower brain stem and possibly also in the spinal cord (Chizh et al. 1998). These oscillators are normally weakly coupled or uncoupled. The system of multiple oscillators associated with a functional homogeneous sympathetic output system (here the cutaneous vasoconstrictor pathway involved in thermoregulation) and the respiratory oscillator show dynamic coupling depending on the functional state of the organism (e.g., increased afferent input from body tissues, changed signals from the forebrain, increased respiratory drive). It is hypothesized that this dynamic coupling contributes to or is mainly responsible for the bursting activity in groups of sympathetic nerves (here postganglionic neurons projecting in the dorsal or ventral collector nerves of the rat tail). Furthermore, it may be responsible for smooth regulation of the target (here blood flow through the skin) during various functional conditions of the organism (Chang et al. 1999, 2000).

The network oscillator hypothesis is an interesting idea. It is a subject of controversial discussions to explain the rhythmic changes of activity in sympathetic nerves at frequencies that are different from those generated by pulsatile stimulation of arterial baroreceptors and by the central respiratory generator. The controversies about the existence and nature of the rhythms of the activity in sympathetic neurons, which are not related to rhythmic activation of arterial baroreceptors or to central respiration, will continue. It is necessary to work out the central anatomical, physiological and molecular basis of these rhythms.

10.3 | Baroreceptor reflexes and blood pressure control

10.3.1 Detection of baroreceptor reflexes

Arterial baroreceptor reflexes have fascinated cardiovascular physiologists since the second half of the nineteenth century. Cyon and Ludwig (1866) were the first to report that stimulation of the central end of the cut aortic nerve in the rabbit causes decrease in arterial blood pressure and heart rate by action of the afferent impulses on the vasomotor center in the lower brain stem. They coined the term "nervus depressor" for the aortic nerve. They assumed that the afferent nerve fibers in the aortic (depressor) nerve were linked to cardiac receptors and measure intracardiac pressure. They concluded from their experiments that the decrease of arterial blood pressure following stimulation of the aortic nerve is generated by decrease of the activity (tonus) in the vascular nerves. They missed detecting that the bradycardia is also generated by a reflex. These experimental results were confirmed by several experimenters in the next decades. In the 1920s Heymans and Ladon (see Heymans and Neil [1958]) performed their famous cross-perfusion experiments on dogs, finally proving that the bradycardia is a reflex elicited via the vasomotor centers by activation of afferent fibers innervating the aortic arch.

The situation was more difficult for the reflexes elicited by stimulation of the baroreceptors in the carotid sinus nerve. Up to the first half of the 1920s, rise of blood pressure and heart rate after occlusion of the common carotid arteries were thought to be generated by anemia of those parts of the brain that are supplied by the carotid arteries (for detailed description see Heymans and Neil [1958]). This situation changed dramatically when Hering performed his experiments on dogs in which he showed that electrical stimulation of the carotid sinus nerve (which later became known as Hering's nerve) or mechanical stimulation of the carotid sinus decreases heart rate and arterial blood pressure. He formulated the concept of the baroreceptor reflex, i.e. that decrease in heart rate and blood pressure are generated by stimulation of carotid sinus afferents and mediated by reflexes via the vasomotor center (Hering 1927). It then took about another 30 years to work out details about the baroreceptor reflexes, in particular the functional characteristics of the arterial baroreceptors and of their reflex effects (see Heymans and Neil [1958] for a detailed and careful description of this development). In the 1980s various laboratories began to explore the organization of the baroreceptor reflexes (see Guyenet [1990]; Dampney [1994]; Pilowsky and Goodchild [2002]).

10.3.2 Function of arterial baroreceptor reflexes

Arterial baroreceptor afferents with *myelinated nerve fibers* encode in their activity the instantaneous level of arterial blood pressure and the change of arterial blood pressure. They rapidly adapt in less than an hour to a change of the mean level of the arterial blood pressure, i.e. they undergo acute resetting. Thus, their main function is to encode in their activity the pressure range between diastolic and systolic blood pressures irrespective of whether the mean pressure is low or high (Coleridge et al. 1981; Dorward et al. 1982; for review see Koushanpour [1991]). Arterial baroreceptors with *unmyelinated nerve fibers*, which have been less well studied, behave differently: they do not adapt to long-term changes of arterial blood pressure, i.e. do not exhibit resetting. Thus, these baroreceptor afferents may be responsible for signaling systemic arterial blood pressure to the NTS in their rate of activity and therefore also for long-term changes in the level of the mean arterial blood pressure (Seagard et al. 1992). Vagal mechanosensitive cardiovascular afferent neurons innervating the heart (notably the right atrium), which are involved in volume regulation, may be important for regulation of the level of the arterial blood pressure (Kirchheim et al. 1998).

Since the 1970s it was believed that the function of the baroreceptor reflexes is to minimize the fluctuations of arterial blood pressure during various actions of the body (e.g., exercise, change of position of the body in gravity of earth, mental activity, etc.), but not to regulate the level of blood pressure (Cowley et al. 1973; Persson et al. 1988; for review see Cowley [1992]). This belief was mainly based on continuous measurements of arterial blood pressure over three to five days in dogs with chronic denervation of the arterial (sinoaortic) baroreceptors. These dogs exhibit tremendous fluctuation of mean arterial blood pressure during their normal behavior but the average level of arterial blood pressure over 24 hours did not change or increased very little (Cowley et al. 1973). Later these experiments were reproduced using the same animal model (dogs) and also in rats (Norman et al. 1981; Persson et al. 1988). Additional denervation of the cardiopulmonary receptors leads to a long-term increase of arterial blood pressure (Persson et al. 1988). However, denervation of the arterial (sinoaortic) baroreceptors in baboons leads chronically to a small increase of arterial blood pressure (Shade et al. 1990).

The dogma that arterial baroreceptors are not important in the regulation of long-term mean arterial blood pressure was seriously challenged if not refuted by Thrasher (Thrasher 2002, 2004, 2005). He used dogs in which the aortic baroreceptors and the carotid sinus baroreceptors on one side were chronically denervated but leaving the carotid sinus baroreceptors on the other side intact. These dogs exhibit normal levels of mean arterial blood pressure. However, unloading of the remaining intact arterial baroreceptors by occlusion of the common carotid artery proximal to the innervated carotid sinus over seven days generated an immediate increase of arterial blood pressure and heart rate, both remaining high during the occlusion of the common carotid artery of seven days. Both returned to normal control levels after removal of the occlusion. This experiment clearly shows that arterial baroreceptors are *also* important to signal mean long-term arterial blood pressure to the lower brain stem (see Thrasher [2004] for critical interpretation of his key experiments in relation to the key experiments conducted by Cowley et al. [1973]).[14]

Several arterial baroreceptor reflex pathways are related to the parasympathetic and sympathetic cardiomotor neurons and to the different types of vasoconstrictor neuron innervating resistance vessels (mainly in skeletal muscle, kidney, other viscera). Phasic stimulation of the arterial baroreceptors in the carotid sinus and aortic arch with each heart beat leads to phasic activation of the parasympathetic cardiomotor neurons (McAllen and Spyer 1978b) and to phasic inhibition of the sympathetic cardiomotor and vasoconstrictor neurons innervating resistance vessels. This is reflected in phasic changes of the activity in these peripheral cardiovascular neurons linked to the pulse pressure wave. This has been recorded in anesthetized and awake animals and humans (see cardiac rhythmicity of the activity in muscle vasoconstrictor neurons in Figures 4.1, 4.6 and 4.7), but not in other autonomic neurons that do not innervate resistance vessels (e.g., most cutaneous vasoconstrictor neurons innervating blood vessels in distal skin of the extremities; sympathetic neurons innervating gland cells or non-vascular smooth musculature; see Chapter 4). It is also reflected in phasic changes of the activity in sympathetic premotor neurons of the RVLM (Figure 10.4).[15]

Some neuronal components of the baroreceptor pathways have been worked out. Figure 10.7 demonstrates a model of these baroreceptor reflex pathways that is commonly accepted. The diagram shows the essential circuitry for the baroreceptor reflex pathways, although there may exist alternative pathways as discussed below. Baroreceptor afferent neurons project to the caudal NTS (Figure 8.11). However, as mentioned already in Chapter 8, there exists no *obvious* viscerotopic organization of the second-order neurons in the caudal NTS with respect to the synaptic inputs from arterial baroreceptors, arterial chemoreceptors and gastrointestinal receptors (Paton 1999; Paton and Kasparov 2000). However, neurophysiological and morphological investigations of neurons in the NTS, using whole cell patch clamp recording in a working-heart-brain-stem preparation of the rat, clearly show that subgroups of NTS neurons are functionally almost entirely specific with respect to their afferent synaptic input from arterial baroreceptors, arterial chemoreceptors or from receptors of the gastrointestinal tract. Only a few NTS neurons receive convergent excitatory synaptic inputs from arterial baro- and chemoreceptors (Deuchars et al. 2000; Paton et al. 2000, 2001). Figure 10.8 demonstrates one of the first intracellular recordings from an NTS neuron that was activated by stimulation of arterial baroreceptors in the carotid sinus (Figure 10.8c, increase of carotid sinus pressure [CSP]) and identified morphologically.

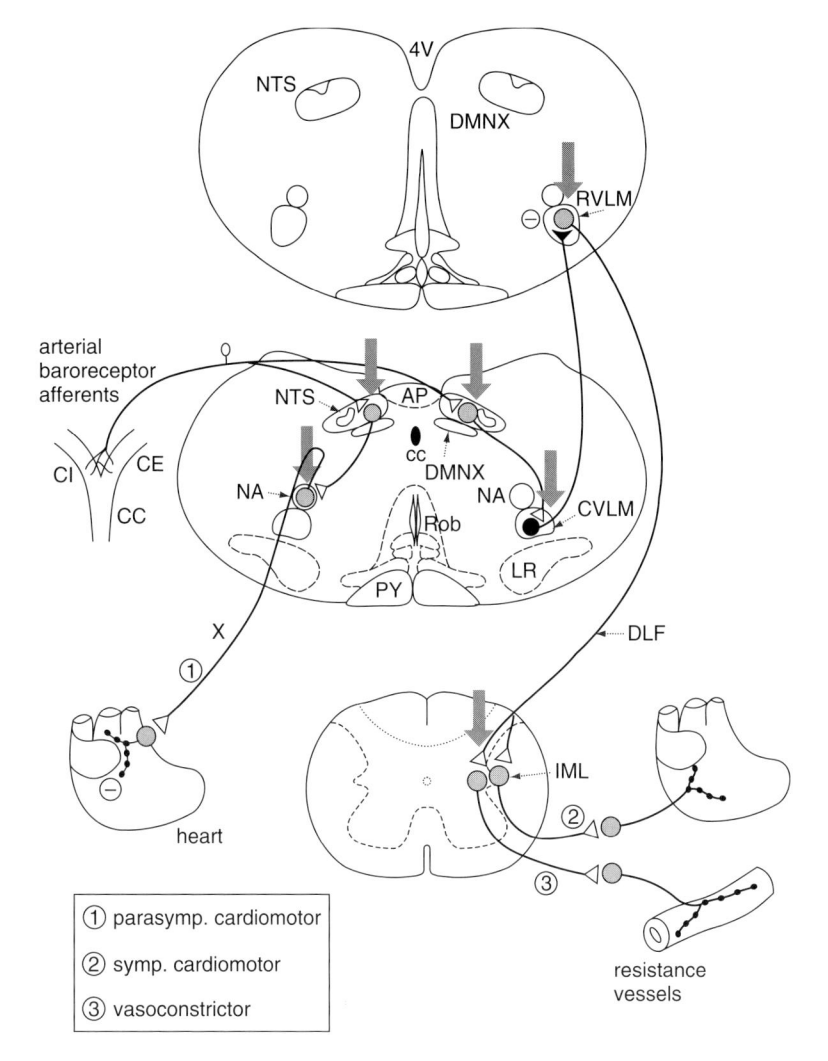

Fig. 10.7 Arterial baroreceptor reflexes. The output pathways are the parasympathetic cardiomotor (PCM) neurons (to pacemaker cells and atria and atrioventricular nodes of the heart), sympathetic cardiomotor (SCM) neurons (to pacemaker cells, atria and ventricles) and vasoconstrictor neurons (VC) innervating resistance vessels. Arterial baroreceptor afferents in the carotid bifurcation and aortic arch project bilaterally to the nucleus tractus solitarii (NTS). Their activity encode the mean arterial blood pressure and the rate of change of blood pressure. *Left*: parasympathetic baroreceptor reflex pathway to the heart. Excitatory second-order neurons of the NTS project to the preganglionic PCM neurons in the external formation of the nucleus ambiguus (NA). Activation of the baroreceptors leads to activation of the PCM neurons and subsequently to a decrease of heart rate (by inhibition of the pacemaker cells). *Right*: sympathetic baroreceptor pathways to the heart and the resistance blood vessels in skeletal muscle, kidney and other viscera (muscle, visceral and renal vasoconstrictor [MVC, VVC, RVC] neurons). These baroreceptor pathways consist of a chain of at least three neurons between the baroreceptor afferents and the preganglionic neurons in the intermediolateral nucleus (IML). Sympathetic premotor neurons are located in the rostral ventrolateral medulla (RVLM), project through the dorsolateral funiculus (DLF) of the spinal cord to the preganglionic SCM and VC neurons and excite these neurons. These sympathetic premotor pathways, projecting through the DLF, are probably distinct for different final cardiovascular pathways (see Figure 10.6). Excitatory second-order neurons of the NTS project to inhibitory interneurons in the caudal ventrolateral medulla (CVLM). These inhibitory interneurons project to the sympathetic premotor neurons in the RVLM and inhibit them. Thus activation of arterial baroreceptors leads to decrease of activity in the SCM and VC neurons by inhibition in the neurons in the RVLM. The transmitter mediating this inhibition is GABA (γ-aminobutyric acid). The transmitter at all other central synapses of the baroreceptor pathways is glutamate. Ongoing activity in the SCM and VC neurons has its origin in the RVLM and elsewhere (see text). Signal transmission through the baroreceptor pathways can be modulated (inhibited, enhanced) from other centers in brain stem, hypothalamus and telencephalon at all synapses (see shaded arrows). AP, area postrema; CC, common carotid artery; cc, central canal; CE, external carotid artery; CI, internal carotid artery; DMNX, dorsal motor nucleus of the vagus nerve; LR, lateral reticular nucleus; PY, pyramidal tract; Rob, raphe obscurus; 4V, fourth ventricle. Modified from Guyenet (1990) with permission.

- Some neurons in the CVLM that project to the RVLM receive monosynaptic glutamatergic contacts from neurons in the NTS (Aicher et al. 1995; Weston et al. 2003).
- The CVLM contains GABAergic neurons that project to the RVLM. These neurons are activated by stimulation of arterial baroreceptors and exhibit a high degree of pulse-modulated activity ("cardiac rhythmicity") (Figure 10.10) as do bulbospinal neurons in the RVLM (see Figure 10.4) and pre- and postganglionic vasoconstrictor neurons involved in blood pressure regulation (see Figures 4.1, 4.6, 4.7) (Jeske et al. 1993; Schreihofer and Guyenet 2002, 2003).
- Sustained increase of arterial blood pressure in awake or anesthetized rats and rabbits (e.g., produced by an α_1-adrenoceptor agonist) induces expression of c-fos mRNA and protein in neurons of the CVLM (see note 1). Many of these neurons contain GABA and project to the RVLM (Li and Dampney 1994; Polson et al. 1995; Minson et al. 1997; Chan and Sawchenko 1998; Schreihofer and Guyenet 2002, 2003).
- Stimulation of arterial baroreceptors by sustained increase of arterial blood pressure activates second-order neurons in the NTS dorsomedial and medial to the tractus solitarii and in the commissural nucleus (Figure 8.11). This activation is most likely monosynaptic. Some of these NTS neurons project to the CVLM; they are glutamatergic (Czachurski et al. 1988; Deuchars et al. 2000; Weston et al. 2003). The baroreceptor-sensitive NTS neurons that project to the CVLM may be those with pulse-modulated activity. But this population of neurons represents only a minority of the baroreceptor-sensitive neurons in the NTS (Deuchars et al. 2000).[16]

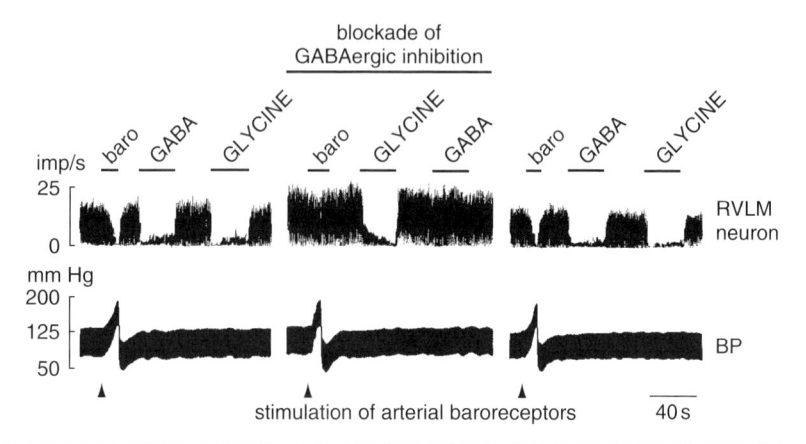

Figure 10.9 Baroreceptor reflex and GABA-mediated inhibition of a sympathetic premotor neuron in the RVLM. Recording from a single vasomotor neuron in the RVLM in the anesthetized rat (see Figure 10.4) with a multibarrel glass microelectrode allowing ionophoretic ejection of GABA, glycine (both inhibitory transmitters) or bicuculline (GABA receptor antagonist) into the RVLM during recording within 40 μm of the cell body. The neuron is spontaneously active. Raising arterial blood pressure (BP, produced by constriction of the descending aorta) inhibits the activity of the neuron. Application of GABA or glycine inhibits the discharge too. During application of bicuculline, both the inhibition during BP increase and that during GABA application are blocked but not the inhibition during application of glycine. Modified from Sun and Guyenet (1985) with permission.

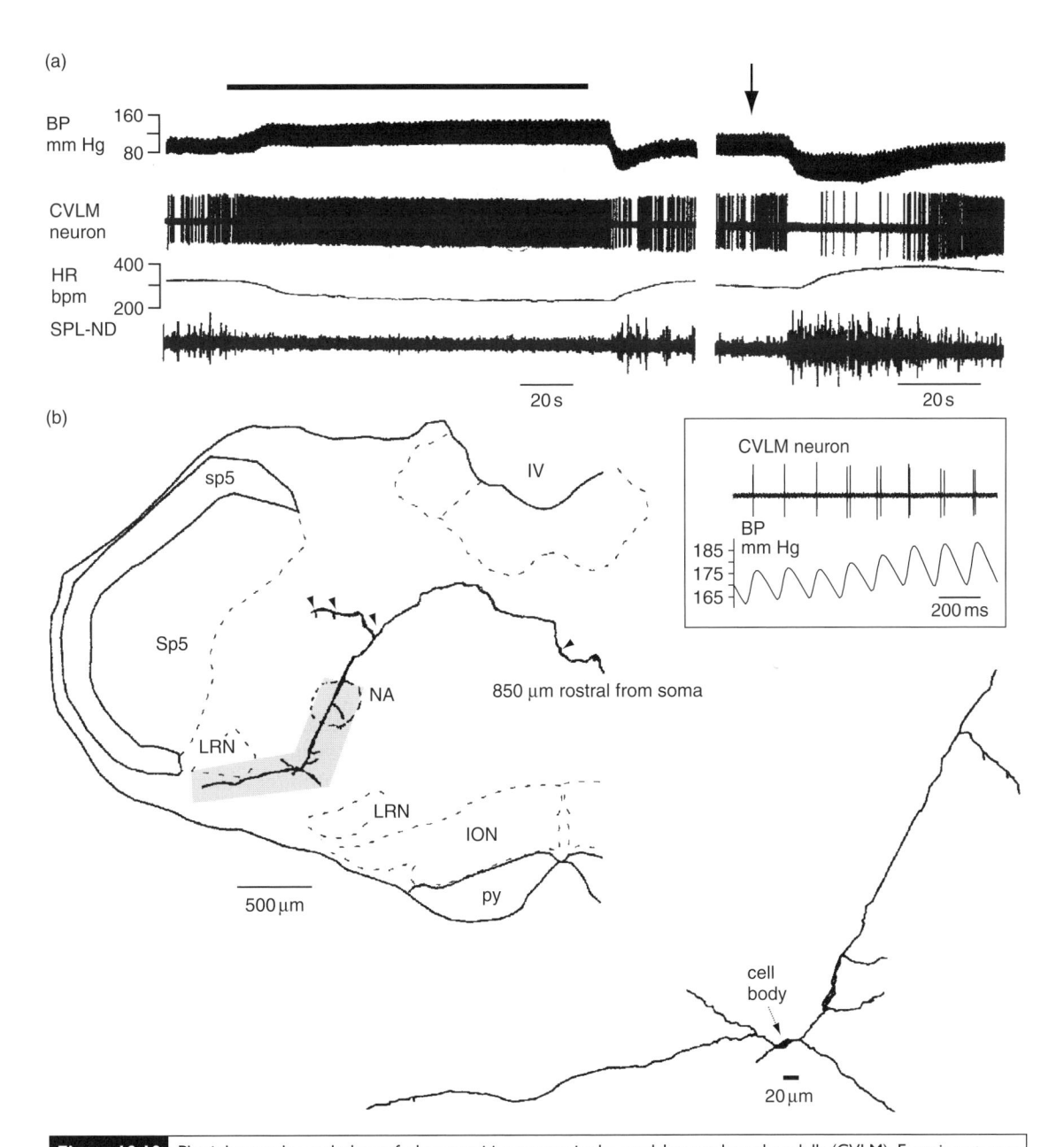

Figure 10.10 Physiology and morphology of a barosensitive neuron in the caudal ventrolateral medulla (CVLM). Experiment on a chloralose-anesthetized, artificially ventilated and immobilized rat. (a) Activation of arterial baroreceptors (bar; during increase of arterial blood pressure [BP] produced by constriction of the subdiaphragmatic aorta) activated the neuron. The activity in the major splanchnic nerve (splanchnic nerve discharge [SPL-ND], which occurs in visceral vasoconstrictor neurons) was inhibited and the heart rate (HR) decreased. Decrease of BP by intravenous injection of nitroprusside (5 μg/kg), which dilates blood vessels (arrow), leads to unloading of arterial baroreceptors, decrease of activity in the CVLM neuron, increase of HR and increase of SPL-ND. *Inset:* pulse modulation of activity of a CVLM neuron during constriction of the descending aorta (see high BP; CVLM neuron different from that in [a]). (b) The CVLM neuron in (a) was filled through the microelectrode with biotinamide (using a juxtacellular labeling method; see note 3). The biotinamide distributed intracellularly through the axon and dendrites of the neuron. The medulla oblongata was removed after perfusing the rat with the fixative formaldehyde and cut in 30 μm thick coronal serial sections. The labeled neuron was reconstructed from the serial sections. The axon and its collaterals (see arrow heads indicating branching points) projected dorsomedially and rostrally (into the direction of the RVLM). The lower right is a magnified view of this reconstructed neuron (note the small cell body; axon reconstructed maximally 860 μm rostrally). The cell body was located 1.3 mm caudal to the facial nucleus. Most neurons with these physiological and morphological properties are GABAergic (i.e. inhibitory). Coronal section through the rostral part of the CVLM (intermediate VLM). ION, inferior olive nucleus; LRN, lateral reticular nucleus; NA, nucleus ambiguus; py, pyramid; Sp5, spinal trigeminal nucleus; sp5, spinal trigeminal tract; IV, fourth ventricle. Modified from Schreihofer and Guyenet (2003) with permission. For technical details see here.

The model of the baroreceptor pathways linked to the sympathetic and parasympathetic cardiovascular final pathways, as outlined in Figure 10.7, was synthesized on the basis of many experiments in several laboratories. It is a minimal model and there remain several open questions to be solved as far as synaptic transmission in the different "relay nuclei" (NTS, CVLM, RVLM, intermediolateral nucleus [IML]; see Figure 10.7), the specificity of these transmissions and alternative reflex pathways are concerned.

Experimental studies using *c-fos* gene expression show that there is evidence for additional baroreceptor pathways in the medulla oblongata: (1) Barosensitive NTS neurons may project directly to the RVLM and inhibit, via local inhibitory (GABAergic) interneurons, sympathetic premotor neurons. (2) Inhibitory (GABAergic) interneurons may project to the CVLM and inhibit excitatory interneurons that project to the RVLM and activate sympathetic premotor neurons (Dampney and Horiuchi 2003). It has to be shown how important these alternative pathways are in relation to the pathway as outlined in Figure 10.7.[17]

Inhibition in the spinal cord

Experiments on rats and cats show that activity in sympathetic preganglionic neurons can be inhibited during stimulation of arterial baroreceptors by activation of spinal inhibitory interneurons. This possibility is not supported by most investigators (see Pilowsky and Goodchild [2002]) and has apparently been refuted in experiments conducted by Goodchild *et al.* (2000). However, I believe that it cannot be ignored that inhibition of activity in sympathetic neurons generated by stimulation of arterial baroreceptors may also occur at the level of the spinal cord, *first*, in view of several pieces of (mostly older) experimental data and, *second*, in view that such an important control system may be redundant and multiple (Coote 1988):

- In the cat activity generated in silent sympathetic preganglionic neurons by ionophoretic application of glutamate can be inhibited during stimulation of arterial baroreceptors (Coote *et al.* 1981).
- Activity elicited synaptically in sympathetic preganglionic neurons by electrical stimulation of descending fibers in the dorsolateral funiculus of the spinal cord was inhibited during stimulation of arterial baroreceptors (with intact or with blocked RVLM). The inhibition is mediated by GABA and possibly glycine (Lewis and Coote 1995, 1996).
- Fast hyperpolarizing postsynaptic potentials reminiscent of inhibitory postsynaptic potentials (IPSPs) were elicited in sympathetic preganglionic neurons during stimulation of arterial baroreceptors (McLachlan and Hirst 1980; Coote, personal communication). These inhibitory potentials may also be elicited by activation of GABAergic neurons in the medulla oblongata that project to the preganglionic neurons (e.g., in the RVLM [Miura *et al.* 1994; Deuchars *et al.* 1997]). In rabbits some 10% of the sympathetic

premotor neurons in the RVLM are excited by stimulation of arterial baroreceptors (Li *et al.* 1991).[18] Recently it has been shown that the lower brain stem contains GABAergic neurons that project to the intermediolateral nucleus of the thoracic spinal cord and form synapses with preganglionic neurons. Most of these GABAergic sympathetic premotor neurons also contain the inhibitory transmitter glycine and some 5-HT. The cell bodies of these neurons are located medial to the RVLM (in the gigantocellular reticular nucleus, ventral and alpha and the medial reticular formation) (Stornetta *et al.* 2004).

The baroreceptor-induced reflex inhibition of sympathetic preganglionic neurons operating at the spinal level may duplicate the baroreceptor-induced reflex inhibition of sympathetic premotor neurons in the RVLM; the way both baroreceptor pathways are integrated during regulation of activity in preganglionic cardiovascular neurons is unknown.

10.3.5 Modulation of baroreceptor pathways

The baroreceptor pathways worked out so far (Figure 10.7) are clearly defined by and can only be understood in the limitations of the experimental approaches used. Under physiological ("closed-loop") conditions[19] these baroreceptor reflex pathways are component parts of many complex regulations. They are therefore subject to many modifications during these complex regulations. This is reflected in the observation that the relays of the baroreceptor pathways are under powerful modulatory control exerted by other centers in the spinal cord, lower and upper brain stem, hypothalamus and telencephalon. This is graphically expressed in Figure 10.11 for the baroreceptor pathways to sympathetic preganglionic neurons. These influences lead functionally to enhancement or depression of the baroreceptor reflexes, potentially at each synaptic relay in the NTS, CVLM, RVLM, and IML (Figure 10.11):

- In Subchapter 8.3 I have described that the NTS is anatomically reciprocally connected with various nuclei in spinal cord, brain stem, hypothalamus and telencephalon (Figure 8.13). Electrical stimulation of the hypothalamic defense area depresses arterial baroreceptor-induced reflexes by inhibition in the NTS and probably elsewhere (Spyer 1981, 1994; Paton and Kasparov 2000). Activation of neurons (by disinhibition; blockade of GABAergic inhibition by bicuculline) in the defense area (perifornical area) and in the dorsomedial hypothalamus increases the threshold of the baroreceptor control of heart rate and renal nerve activity and shifts the set point of the baroreceptor reflexes to higher pressure levels without changing their gain (Dampney, unpublished results). It is hypothesized that this shift occurs by acting in the NTS. Noxious stimulation of hindlimb, forelimb or cornea attenuates the baroreceptor reflexes. This inhibition involves GABAergic interneurons in the NTS that are activated via NK_1 receptors (Boscan and Paton 2002; Boscan et al. 2002). Only the baroreceptor pathway to parasympathetic cardiomotor neurons but not to sympathetic cardiomotor neurons is involved (Pickering et al. 2003), arguing that the NTS neurons are distinct for both baroreceptor pathways.

- The inhibitory GABAergic neurons in the CVLM receive tonic excitatory (glutamatergic) synaptic input leading to tonic inhibition of the bulbospinal neurons in the RVLM independent of the phasic inhibition exerted by activation of arterial baroreceptors. The increase of arterial blood pressure following blockade of the inhibitory (GABAergic) interneurons in the CVLM is significantly higher than the blood pressure increase following denervation of the arterial baroreceptors (which removes the activation of GABAergic interneurons generated by activity in the arterial baroreceptors [Natarajan and Morrison 2000]). Furthermore, the inhibitory barosensitive neurons in the CVLM integrate activity from somatic and visceral nociceptive afferent neurons and possibly other afferent neurons. This integration is responsible for some aspects of somato- and viscerosympathetic reflexes in muscle and visceral vasoconstrictor neurons (Schreihofer and Guyenet 2003). It is likely that only a subpopulation of GABAergic interneurons in the CVLM that project to the RVLM and inhibit the sympathetic premotor neurons are activated by stimulation of arterial baroreceptors. Other subpopulations of inhibitory CVLM neurons that also project to the RVLM are not activated by stimulation of arterial baroreceptors (Cravo et al. 1991). These subpopulations of GABAergic neurons in the CVLM that project to the RVLM will be characterized functionally and histochemically in future investigations (Wang et al. 2003).

- Sympathetic premotor neurons in the RVLM are also tonically excited by interneurons in the CVLM. The tonic activity in these excitatory interneurons is synaptically maintained by activity of neurons in the so-called caudal pressure area (CPA) in the VLM, which is located in the most caudal part of the VLM (Figure 10.2; Natarajan and Morrison 2000; see also Horiuchi and Dampney [2002]; Dampney et al. [2003]).

- Neurons of the respiratory network (central respiratory generator [CRG] in Figure 10.11) influence the neurons of the baroreceptor reflex pathway synaptically (i.e., the parasympathetic preganglionic cardiomotor neurons, the sympathetic premotor neurons in the RVLM). The mechanisms of this integration between respiratory and cardiovascular neurons are largely unknown. However, one type of coupling may occur between glutamatergic neurons expressing NK_1-receptors in the preBötzinger complex and GABAergic neurons in the CVLM (Wang et al. 2003; see Chapter 10.6).

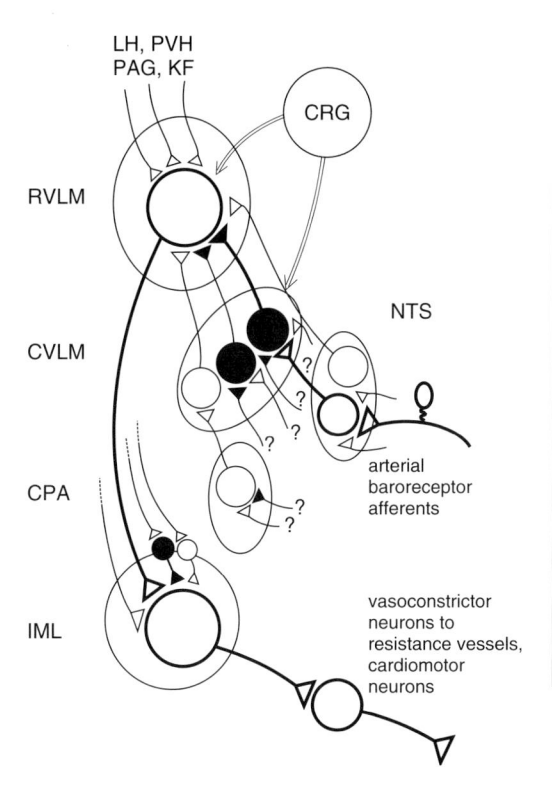

Figure 10.11 Neural influences on the nuclei of the baroreceptor reflex pathway to sympathetic cardiovascular neurons. The neurons of the baroreceptor reflex pathway are outlined in bold. Excitatory neurons/synapses, open symbols. Inhibitory neurons/synapses, closed symbols. CVLM, caudal ventrolateral medulla; CPA, caudal pressure area; CRG, central respiratory generator; IML, intermediolateral cell column; LH, lateral hypothalamus; KF, Kölliker–Fuse nucleus; NTS, nucleus tractus solitarii; PAG, periaqueductal grey; PVH, paraventricular nucleus of the hypothalamus; RVLM, rostral ventrolateral medulla; Modified from Pilowsky and Goodchild (2002).

In *conclusion*, the example of the arterial baroreceptor reflexes illustrates paradigmatically the complexity in the organization of these cardiovascular reflex pathways in the medulla oblongata. It illustrates furthermore that it is difficult to interpret the role of the baroreceptor reflexes during normal closed-loop cardiovascular regulations. Although many laboratories have concentrated and/or are concentrating on the mechanisms underlying the baroreceptor reflexes, several controversial issues remain to be explored. These address, e.g., the function of unmyelinated baroreceptor afferents, the convergence of baroreceptor afferents on single NTS neurons, the functional types of barosensitive NTS neurons (related to parasympathetic and sympathetic cardiovascular neurons, to regulation of vasopressin release), the modulation of barosensitive NTS neurons, the projection of barosensitive NTS neurons, the neurochemistry of barosensitive NTS neurons, the differentiation of barosensitive CVLM neurons from other inhibitory CVLM neurons, alternative baroreceptor pathways etc.

Similar complexities, as they exist for the baroreceptor pathways, are present for the autonomic chemoreceptor pathways (see Chapter 10.4) and other reflex pathways related to various types of cardiovascular, pulmonary, gastrointestinal, vestibular and somatic afferents. The organization of most of them is still unknown or incompletely known. However, on the positive side, I want to emphasize that these complexities are the basis for the high adaptability of the homeostatic cardiovascular regulation and its distortion during disease (see Randall [1984]; Eckberg and Sleight [1992]; Rowell [1993]; Kirchheim *et al.* [1998]; Mathias and Bannister [2002]).

10.4 | Arterial chemoreceptor reflexes in sympathetic cardiovascular neurons

Reductions of oxygen tension in the blood are detected by peripheral chemoreceptors in the carotid and aortic bodies. These chemoreceptors are innervated by afferent fibers in the aortic and carotid sinus nerves that project to the commissural nucleus of the NTS. Most of these fibers are unmyelinated. The primary, neurally mediated responses to decrease in arterial oxygen tension (PaO_2) are increased ventilation, arousal, aversive responses and autonomic adjustments that compensate for direct (non-neural) vasodilating effects of hypoxia on the blood vessels and redistribute blood to essential organs including brain, heart and kidneys (Marshall 1994). Large reductions of PaO_2 are also detected directly by the brain (Reis *et al.* 1994, 1997) and cause sympathetic excitation. Sympathetic neurons are either excited by brief stimulation of arterial chemoreceptors (muscle, visceral, renal vasoconstrictor, sudomotor, inspiratory neurons) or inhibited (many cutaneous vasoconstrictor neurons) or not affected (e.g., motility-regulating neurons) in the cat (see Chapter 4; Figures 4.1, 4.4 and 4.11). Most sympathetic premotor neurons in the RVLM are excited following stimulation of arterial chemoreceptors; some are inhibited (McAllen 1992). What are the reflex pathways for the primary excitatory responses elicited in sympathetic vasoconstrictor and cardiomotor neurons and in sympathetic premotor neurons in the RVLM?

A subgroup of neurons in the NTS can be activated by stimulation of arterial chemoreceptors. These NTS neurons do not receive convergent synaptic input from gastrointestinal tract afferents, very few can be activated or inhibited by stimulation of arterial baroreceptors and some are activated or inhibited by pharyngeoesophageal stimulation. Morphology and neurophysiological characteristics are similar to other functional types of NTS neurons (Paton *et al.* 2001). The projection of chemoreceptive NTS neurons is unknown.

The following description is based on experiments performed on rats by Guyenet's group (Koshiya *et al.* 1993; Guyenet *et al.* 1996; Guyenet 2000):

- Stimulation of arterial chemoreceptors by brief anoxia (ventilation of the rat with 100% N_2 for 4 to 12 seconds)[20] activates several types of peripheral sympathetic neurons, e.g. that projecting in the major splanchnic nerve (all activated neurons probably being visceral vasoconstrictor neurons), and many sympathetic premotor neurons in the RVLM in parallel to activation of the respiratory network reflected in an increase of rate and amplitude of phrenic nerve activity (Figure 10.12a_1, b_1). This reflex activation of

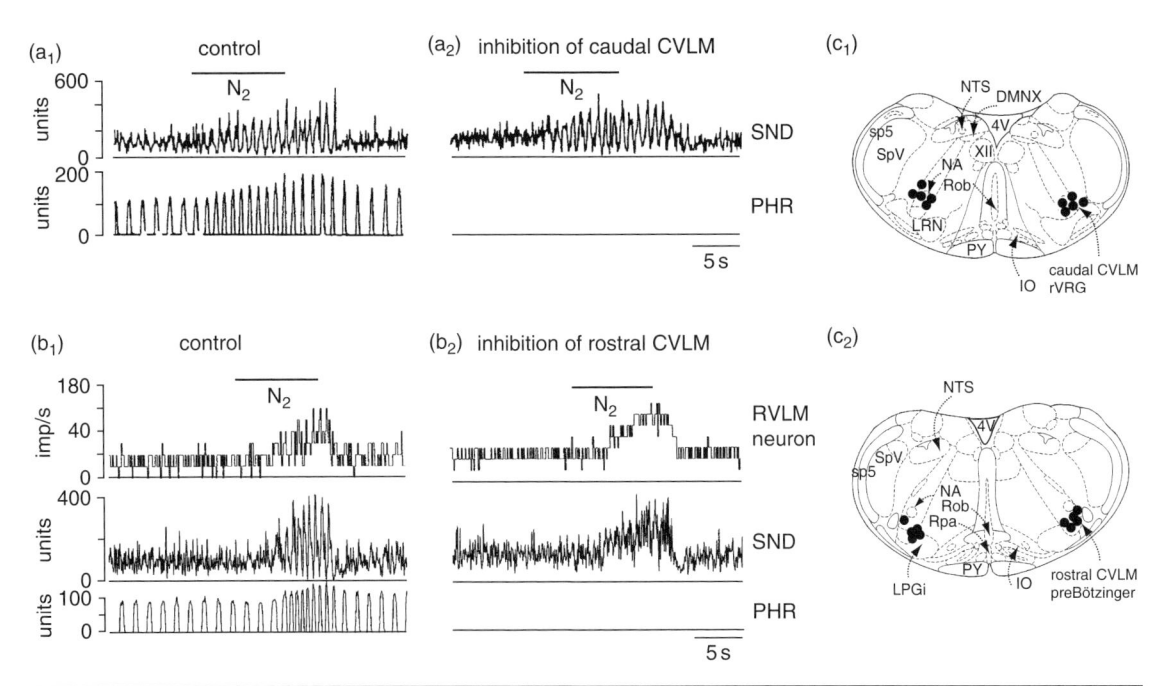

Figure 10.12 Role of the rostral ventrolateral medulla (RVLM) and caudal ventrolateral medulla (CVLM) in chemoreceptor-mediated activation of vasoconstrictor neurons projecting to the viscera. Experiments on urethane-anesthetized, vagotomized and artificially ventilated rats. Arterial carotid chemoreceptors were stimulated by ventilating the rats with 100% N_2 for 4 to 12 s. Integrated activity was recorded from the phrenic nerve (indicating activation of inspiratory neurons; phrenic nerve discharge, PHR) and from the major splanchnic nerve (mainly activity in visceral vasoconstrictor neurons; splanchnic nerve discharge, SND) in (a) and also from a sympathetic premotor neuron in the RVLM in (b). (a_1, b_1) Stimulation of the arterial chemoreceptors activates phrenic motoneurons, RVLM neurons and visceral vasoconstrictor neurons. During the activation of RVLM and sympathetic neurons, respiratory rhythmicity is pronounced. This activation is eliminated after denervation of the arterial chemoreceptors (not shown). (a_2) Inhibition of neurons in the CVLM (location see Figures 10.2c, d, 10.3, 10.11) by bilateral injection of the GABA agonist muscimol into the CVLM eliminates the phrenic nerve activity (by inhibition of the phrenic premotor neurons in the rostral ventral respiratory group [rVRG]) but not the rhythmic reflex activation of sympathetic neurons. The arterial baroreceptor reflexes are also eliminated and the resting sympathetic activity increases (due to inhibition of the GABAergic interneurons). (b_2) After bilateral inhibition of the neurons in the rostral part of the CVLM (which contains the preBötzinger complex; see Figure 10.2c, d), by injection of muscimol, the reflex activation of the visceral vasoconstrictor neurons and of the RVLM neurons is unaffected but the rhythmic activation and the PHR are eliminated (probably because the neurons of the respiratory rhythmic generator are inhibited). (c_1, c_2) Typical injection sites of muscimol into the caudal CVLM for the effect shown in (a_2) ([c_1]) and into the rostral CVLM/preBötzinger region (also called intermediate CVLM by Dampney [1994]) for the effect shown in (b_2) ([c_2]). DMNX, dorsal motor nucleus of the vagus nerve; IO, inferior olive; LPGi, lateral paragigantocellular nucleus; LRN, lateral reticular nucleus; NA, nucleus ambiguus; NTS, nucleus tractus solitarii; Rob, Rpa, raphe obscurus, pallidus; Sp5, spinal trigeminal nucleus; sp5, spinal trigeminal tract; 4V, fourth ventricle; XII, hypoglossus nucleus. PY, pyramid; (a_1, a_2) and (c_1) modified from Koshiya et al. (1993). (b_1), (b_2) and (c_2) modified from Koshiya and Guyenet (1996) with permission.

sympathetic neurons exhibits a rhythmicity correlated with the phrenic nerve discharges.
- Inhibition of the neurons in the CVLM, located caudal to the preBötzinger complex (see Figure 10.2b–d), by topical injection of the GABA agonist muscimol, eliminates the phrenic nerve discharges (probably by inhibiting the respiratory premotor neurons in the rVRG) but does not block the reflex activation of sympathetic premotor neurons in the RVLM (not shown) and therefore

not the reflex activation of sympathetic neurons (Figure 10.12a$_2$). The respiratory rhythmicity of the activity in the sympathetic neurons is still preserved during the hypoxia-induced activation of the sympathetic neurons, probably because the component neurons of the ponto-medullary respiratory network that are essential to generate the respiratory rhythm are situated more rostrally than the CVLM and are not inhibited (Figure 10.12a$_2$).

- After inhibition of the neurons in the rostral CVLM by injection of muscimol, hypoxia-induced reflex activation of the sympathetic neurons, as well as of the sympathetic premotor neurons in the RVLM, is still present. However, the respiratory rhythmicity in this reflex response and the phrenic nerve discharges are eliminated (probably since the neurons of the respiratory rhythm generator are inhibited) (Figure 10.12b$_2$).

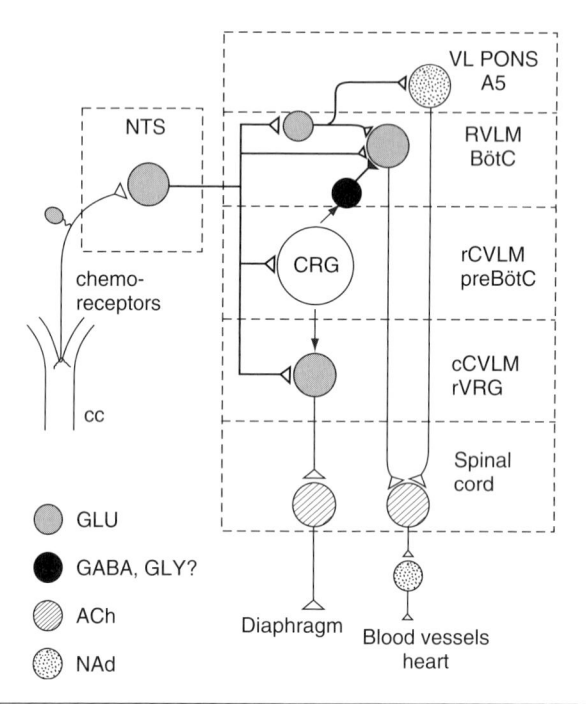

Figure 10.13 Hypothetical chemoreceptor reflex circuits linked to sympathetic cardiovascular neurons innervating resistance vessels and heart. These pathways are based on experiments shown in Figure 10.12. The locations of the respiratory neurons and the presympathetic premotor neurons in the ventrolateral medulla (VLM) are described in Figure 10.2. Chemoreceptor afferent neurons from the carotid and aortic bodies activate second-order neurons in the commissural nucleus of the nucleus tractus solitarii (NTS). These second-order neurons project to phrenic premotor neurons in the rostral ventral respiratory group (rVRG) (in the caudal ventrolateral medulla [CVLM]), to neurons in the preBötzinger complex (preBötC in the rostral CVLM [rCVLM]), to sympathetic premotor neurons in the rostral ventrolateral medulla (RVLM) (monosynaptically and/or disynaptically; C1 [adrenergic] and non-C1 neurons) and (possibly disynaptically) to noradrenergic neurons in the A5 area of the ventrolateral (VL) pons. Rhythmic respiratory discharges in the sympathetic premotor neurons and peripheral sympathetic cardiovascular neurons are probably mediated via the preBötC, possibly by inhibitory interneurons. The transmitters used by the neurons are glutamate (Glu), GABA (γ-aminobutyric acid; or glycine), acetylcholine (ACh) or noradrenaline (NAd). BötC, Bötzinger complex: cc, common carotid artery; CRG, central respiratory generator. Modified from Guyenet (2000) with permission.

These and other experiments (in particular those of Dampney's group using Fos expression of neurons in the lower brain stem during stimulation of arterial chemoreceptors [Hirooka et al. 1997; Dampney and Horiuchi 2003]) show that the primary reflex activation of sympathetic neurons generated by stimulation of arterial chemoreceptors is not only mediated via the activation of respiratory neurons (e.g., in the preBötzinger complex) but also by reflex pathways to the sympathetic premotor neurons in the RVLM that are *independent* of the respiratory neurons. Further experiments performed by Guyenet's group have shown that sympathetic premotor neurons in the A5 area of the ventrolateral pons (see Figure 8.15) may also be involved in integration of the primary chemoreceptor reflex activation of the sympathetic neurons. These A5 neurons are noradrenergic.

Figure 10.13 depicts the putative primary chemoreceptor reflex pathways to the (phrenic) inspiratory motoneurons innervating the diaphragm and to the sympathetic pathways innervating resistance vessels and heart. The main groups of respiratory and cardiovascular neurons in the different sections of the VLM (see Figure 10.2b–d) and in the ventrolateral pons are shown. The reflex pathways between the chemoreceptor afferents, projecting to the NTS, and the sympathetic premotor neurons are unknown but involve at least one interneuron. These interneurons are probably glutamatergic. The synaptic connection between the neurons of the respiratory rhythm neurons in the preBötzinger complex and the sympathetic premotor neurons in the RVLM, which is responsible for the increased respiratory modulation of the chemoreceptor reflex response, is possibly mediated by an inhibitory (GABAergic) interneuron.

As is the case with the baroreceptor reflex pathways (Figure 10.7), all synapses of the chemoreceptor pathways are under modulatory control of other nuclei in the lower brain stem, upper brain stem, hypothalamus and telencephalon. Furthermore, the primary chemoreceptor reflexes linked to respiratory and sympathetic premotor neurons are modified by secondary cardiopulmonary reflexes connected to arterial baroreceptors, volume receptors and various types of pulmonary afferents as well as the chemosensitivity of the neurons in the medulla oblongata (to changes of $PaCO_2$ and pH; to large changes of PaO_2) (Marshall 1994).

As described in Chapter 4, most cutaneous vasoconstrictor neurons innervating hairy skin of the distal extremities (Blumberg et al. 1980) and all cutaneous vasoconstrictor neurons innervating hairless skin of the cat hindpaw (Jänig and Kümmel 1977) are inhibited following stimulation of arterial chemoreceptors (generated by ventilation of the cat with a hypoxic gas mixture of 8% O_2 in N_2 or by brief stimulation of arterial chemoreceptors with a bolus of CO_2-saturated saline; Figures 4.4a_2, b_2, 4.11c). This inhibition is accompanied by an increase in cutaneous blood flow (Jänig and Koltzenburg 1991b). The physiological significance of these powerful inhibitory chemoreceptor reflexes in the cutaneous vasoconstrictor neurons is unknown. However, they seem to be important in diving animals (see

Subchapter 11.3). The inhibitory chemoreceptor reflexes in the cutaneous vasoconstrictor neurons are correlated (1) with inhibitory reflexes to noxious stimulation of skin areas innervated by the cutaneous vasoconstrictor neurons and (2) with weak or absent inhibitory baroreceptor reflexes (see Chapter 4 and Figures 4.4 and 4.11). The central pathway(s) mediating these inhibitory chemoreceptor reflexes in the cutaneous vasoconstrictor neurons are unknown. However, they are most likely different from those mediating the excitatory chemoreceptor reflexes in cardiovascular sympathetic neurons that are involved in blood pressure regulation (see Figure 10.13).

Cutaneous vasoconstrictor neurons in the cat, which are normally inhibited following stimulation of arterial chemoreceptors, do exhibit excitatory chemoreceptor reflexes under changed experimental conditions. This is illustrated by the following groups of experiments performed on cats:

1. Inhibition of activity in cutaneous vasoconstrictor neurons during stimulation of arterial chemoreceptors (by ventilating the animals with a gas mixture of 8% O_2 in N_2) is reversed to excitation after midcollicular (mesencephalic) decerebration, which removes reflex pathways mediated by the upper mesencephalon and hypothalamus (Gregor and Jänig 1977).
2. Weeks to months after experimental lesion of cutaneous nerves in the cat (cutting and ligating the superficial peroneal nerve; cross-anatomizing the central stump of the superficial peroneal nerve to the distal stump of a muscle nerve or the tibial nerve) many cutaneous vasoconstrictor neurons behave like muscle vasoconstrictor neurons: they are excited by stimulation of arterial chemoreceptors and exhibit strong inhibitory baroreceptor reflexes. These reflex changes occur in both cutaneous vasoconstrictor neurons projecting in the lesioned nerve as well as in cutaneous vasoconstrictor neurons projecting in intact skin nerves of the same extremity (Blumberg and Jänig 1983b, 1985; Jänig and Koltzenburg 1991b).

Both experiments demonstrate quite clearly that the central chemoreceptor circuits linked to the peripheral cutaneous vasoconstrictor pathways are complex and may change plastically depending on the functional conditions.[21]

10.5 | Sympathetic premotor neurons in the caudal raphe nuclei

Cooling the skin and/or core of the body activates cutaneous vasoconstrictor neurons, resulting in decrease of cutaneous blood flow and of heat transfer, and (in the rat) activation of lipomotor neurons (which innervate brown adipose tissue [BAT]), resulting in non-shivering thermogenesis. Both autonomic responses are part of the body's thermoregulatory cold-defense response, which is integrated in the preoptic area of the hypothalamus (Kazuyuki et al. 1998; Nagashima et al. 2000). The sympathetic premotor neurons mediating the responses of cutaneous vasoconstrictor neurons and lipomotor neurons to body cooling are most likely located in the caudal raphe nuclei (raphe magnus, raphe pallidus) of the medulla oblongata (Fig. 10.2).

10.5.1 Premotor neurons innervating preganglionic cutaneous vasoconstrictor neurons

Several experimental investigations demonstrate indirectly that the raphe nuclei of the medulla oblongata contain sympathetic premotor neurons for the cutaneous vasoconstrictor pathway:

- Many neurons in the raphe nuclei of the medulla oblongata project to the spinal cord, some of them project to the preganglionic neurons in the intermediolateral cell column (Morrison and Gebber 1985); they may form monosynaptic connections with the preganglionic neurons (Bacon et al. 1990).
- Transneuronal labeling of postganglionic cutaneous vasoconstrictor neurons innervating the rat tail (which is a thermoregulatory organ) with a neurotropic virus shows that many neurons are labeled in the caudal raphe nuclei (in addition to other labeled sympathetic premotor neurons, e.g. in the RVLM; Smith et al. 1998; see Table 8.2).
- Electrical and chemical stimulation of neurons in the caudal raphe nuclei generate vasoconstriction in skin independent of the RVLM (Blessing et al. 1999; Blessing and Nalivaiko 2000) and activate cutaneous vasoconstrictor neurons, but not those vasoconstrictor neurons involved in blood pressure control (such as renal vasoconstrictor neurons; Rathner and McAllen 1999).
- Many neurons in the ventral raphe nuclei of the medulla oblongata express the transcription factor c-Fos after decrease of body temperature by whole-body cooling (Morrison et al. 1999). Expression of c-Fos indicates an increase in electrical activity in these cells (see note 1).

Some neurons in the nuclei raphe magnus and pallidus that project to the intermediolateral nucleus of the spinal cord (IML; Figure 10.14a, b_1–b_2) are activated during cooling of the skin. This activation is enhanced by decreasing body core temperature (Figure 10.14b_3). Some bulbospinal neurons projecting to the IML are also inhibited by body cooling (Figure 10.14c_1), but many are not affected (Figure 10.14c_2). The bulbospinal raphe neurons that are excited during cooling exhibit a discharge pattern to other afferent inputs that is similar to that seen in cutaneous vasoconstrictor neurons innervating the rat tail or rat hindpaw (Häbler et al. 1994b, 2000; Owens et al. 2002); i.e., no or weak inhibition by stimulation of arterial baroreceptors; mostly inhibition (some excitation) by stimulation of nociceptors; respiratory rhythmicity.

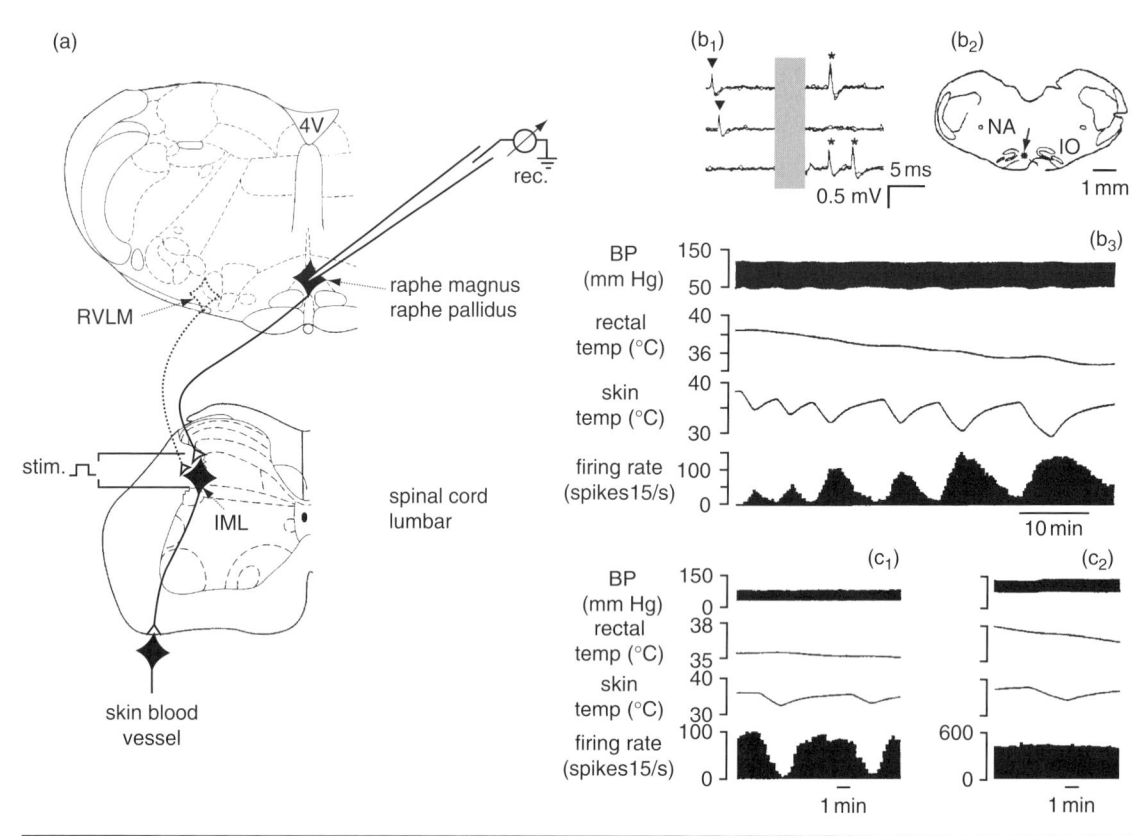

Figure 10.14 Responses of sympathetic premotor neurons in the caudal raphe nuclei of the medulla oblongata to cooling of the body surface in the rat. (a) Experimental setup for recording and identification of sympathetic premotor neurons in the caudal raphe nuclei with a microelectrode. The neuron was identified by electrical stimulation of its axon projecting to the upper lumbar spinal cord. IML, intermediolateral nucleus; 4V, fourth ventricle. Putative preganglionic cutaneous vasoconstrictor neurons also receive supraspinal synaptic input from premotor neurons in the rostral ventrolateral medulla (RVLM) (see Figure 10.6).
(b_1) Identification of a sympathetic premotor neuron in the raphe nuclei as a bulbospinal neuron: electrical stimulation of the axon of the neuron in the IML with single pulses (upper trace; stim. in [a]) or with a pair of pulses (lower trace) activates the neuron at constant latency (antidromically evoked spikes marked by stars). When the electrical stimulus occurs at ≤ 8 ms after a spontaneous orthodromically conducted action potential, the antidromic spike evoked by electrical stimulation was canceled by the orthodromic spike (second trace). This collision does not occur at a time interval of ≥ 9 ms between spontaneous spike and electrical stimulus. Each recording three times superimposed. (b_2) Marked recording site for this neuron (indicated by arrow) illustrated on a transverse section of the medulla oblongata just caudal to the facial nucleus (see Figure 10.2). NA, nucleus ambiguus. (b_3) Activation of the neuron in (b_1/b_2) by lowering skin temperature on the trunk and rectal temperature. Upper trace, arterial blood pressure (BP). (c_1, c_2) Two additional raphe neurons, which are either inhibited or unaffected by skin cooling (same type of experiment as in [b_1–b_3]). Modified from Rathner et al. (2001) with permission.

These spinally projecting neurons in the nucleus raphe magnus and pallidus constitute a distinct class of cutaneous vasoconstrictor premotor neuron, being mainly involved in the regulation of cutaneous blood flow during thermoregulation. They are independent of the sympathetic premotor neurons in the RVLM. However, sympathetic preganglionic cutaneous vasoconstrictor neurons also receive synaptic input (either directly or via interneurons) from sympathetic premotor neurons in the RVLM (Figure 10.14a; see Figure 10.6). Thus, the functional characteristics of pre- and postganglionic cutaneous vasoconstrictor neurons depend on the discharge characteristics of both types of identified sympathetic premotor neurons in the medulla oblongata (and of other types of sympathetic premotor neurons that have not yet been identified [see Table 8.2]) and spinal interneurons [Figure 8.9]). This is consistent with what is actually measured in cutaneous vasoconstrictor neurons, in particular those innervating hairy skin, i.e. most of them exhibit no or weak cardiac rhythmicity in their activity and inhibitory reflexes to stimulation of arterial chemoreceptors and some 20% exhibit strong cardiac rhythmicity in their activity and excitatory chemoreceptor reflexes (Blumberg et al. 1980).

Activity in these sympathetic premotor neurons during thermoregulatory load (cooling or warming) is partly regulated via inhibitory GABAergic neurons. Blockade of this GABAergic transmission (by ionophoretic injection of the $GABA_A$-receptor antagonist bicuculline) during warming of the preoptic area of the hypothalamus eliminates the inhibition of these neurons and therefore the concomitantly occurring increase of cutaneous blood flow (Tanaka et al. 2002). The reflex pathways mediating the excitatory response to cooling (and the reflexes to noxious stimulation) are so far unknown. However, they may pass through the hypothalamus as well as through a spinobulbar pathway.

10.5.2 Premotor neurons innervating preganglionic lipomotor neurons

Lipomotor neurons innervating BAT in the rat are also under the control of sympathetic premotor neurons in the caudal raphe nuclei (mainly raphe pallidus) of the medulla oblongata. Excitation of postganglionic lipomotor neurons directly activates the lipocytes of the BAT, probably synaptically via β_3-adrenoceptors. The activation of lipocytes leads to activation of an uncoupling protein in these cells and subsequently to production of heat. This mechanism of nonshivering thermogenesis is important in small mammals (like rat, mouse, guinea pig) and in hibernating animals, but less so in large-sized animals and in adult humans, except in the newborn. Brown adipose tissue is present between the scapulae and in smaller amounts in the thoracic and abdominal cavities (e.g., around the heart, adrenal glands and kidneys; Smith and Horwitz 1969). Postganglionic axons of lipomotor neurons are present, in addition to some vasoconstrictor axons, in high concentration in nerves innervating the interscapular BAT and in low numbers in the splanchnic nerves. Therefore, sympathetic multiunit activity in the nerves to the BAT consists mainly of activity in lipomotor axons, and activity in thoracic or lumbar splanchnic nerves is dominated by activity in visceral vasoconstrictor axons that are involved in regulation of resistance vessels and therefore in blood pressure regulation.

Figure 10.15[22] demonstrates recordings from these nerves in the anesthetized rat (in addition to recording of arterial blood pressure and of phrenic nerve activity) during various interventions in the

Figure 10.15 Activation of sympathetic lipomotor neurons innervating brown adipose tissue (BAT) by hypothermia and disinhibition of the nucleus raphe pallidus. (a) Arrangement of recording electrodes and electrodes for ionophoresis to inhibit activity of neurons in the rostral ventrolateral medulla (RVLM) or to block GABAergic transmission in the nucleus raphe pallidus (Rpal). Simultaneous recording of phrenic nerve activity (PHR; lower trace, original signal; upper trace, integrated signal), arterial blood pressure (BP), activity in the major splanchnic nerve (SPL), and activity in a nerve to the interscapular BAT in two (b, c) anesthetized, paralyzed and artificially ventilated rats. The measurements were done sequentially. DLF, dorsolateral funiculus; IML, intermediolateral nucleus. (b_1, c_1) Control. Body core temperature 37 °C. (b_2) Hypothermia (rectal temperature 34.4 °C). (b_3) Disinhibition of neurons in the nucleus raphe pallidus by microinjection of bicuculline in normothermia (antagonist to GABA). (c_2) Inhibition of neurons in the RVLM by microinjection of muscimol (GABA agonist). Note elimination of PHR nerve activity (probably due to inhibition of neurons of the nearby preBötzinger complex [see Figure 10.2]), of SPL nerve activity and decrease of BP. (c_3) Subsequent disinhibition of neurons in the Rpal by microinjection of bicuculline. Note activation of BAT and small activation of SPL. Modified from Morrison (1999) with permission.

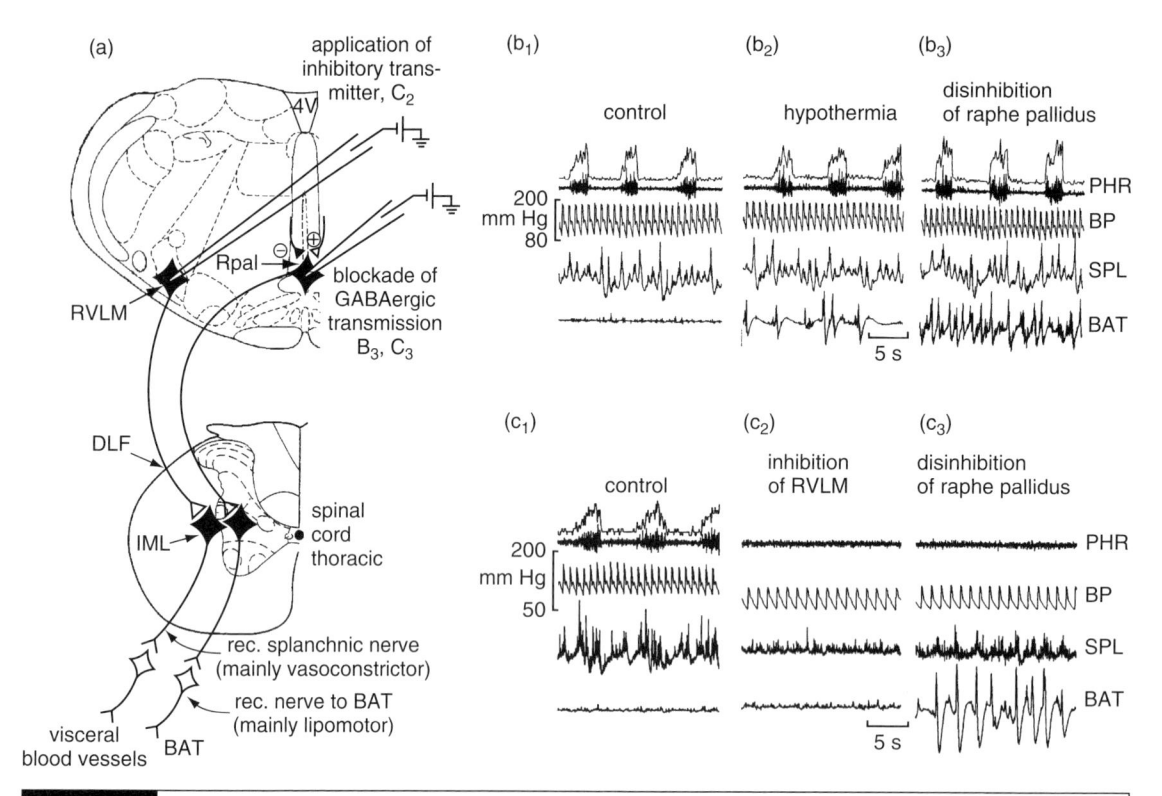

nucleus raphe pallidus and during hypothermia) (Morrison 1999, 2001b; Morrison et al. 1999):

- Activity in lipomotor neurons is low or absent whereas vasoconstrictor activity is high under normal thermoneutral regulatory conditions (body core temperature of 37 °C; Figure 10.15b_1, c_1).
- Hypothermia (exposure to cold with decrease in body core temperature) activates lipomotor neurons without changing visceral vasoconstrictor activity (Figure 10.15b_2).
- Disinhibition of the neurons in the nucleus raphe pallidus by ionophoretic application of the $GABA_A$-receptor antagonist

bicuculline activates lipomotor neurons without changing vasoconstrictor activity and arterial blood pressure (Figure 10.15b$_3$).

- Inhibition of neurons in the RVLM by the GABA agonist muscimol eliminates phrenic nerve activity (by inhibiting the neurons of the respiratory network [probably in the nearby preBötzinger complex], see Figure 10.2) and vasoconstrictor activity (by silencing the sympathetic premotor neurons in the RVLM that are involved in blood pressure regulation) with a subsequent fall in arterial blood pressure. Note there was no effect on the lipomotor neurons (Figure 10.15c$_2$).
- Disinhibition of the neurons in the nucleus raphe pallidus by ionophoretic application of the GABA$_A$-receptor antagonist bicuculline, after inhibition of the RVLM neurons, activates the lipomotor neurons but not the vasoconstrictor neurons (Figure 10.15c$_3$; note some small activation in the splanchnic nerve in Figure 10.15c$_3$, which occurs synchronously with the activity in the lipomotor neurons innervating the interscapular BAT; these neurons are presumably lipomotor neurons innervating BAT in the viscera).

These key experiments show that a second class of sympathetic premotor neuron in the nucleus raphe pallidus innervates

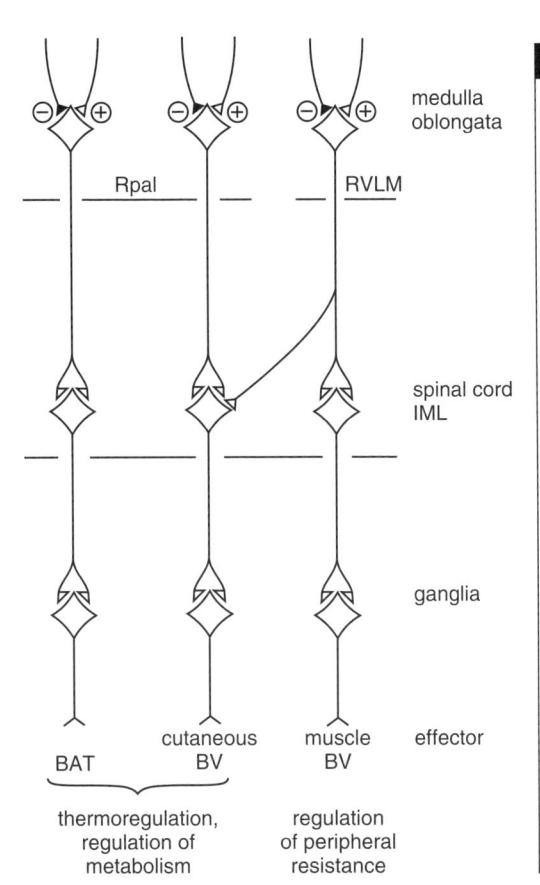

Figure 10.16 Three major bulbospinal pathways to sympathetic preganglionic neurons. Lipomotor neurons supplying brown adipose tissue (BAT) are activated by sympathetic premotor neurons in the raphe pallidus (Rpal). Vasoconstrictor neurons (visceral and muscle vasoconstrictor, VVC, MVC) and cardiomotor neurons are activated by sympathetic premotor neurons in the rostral ventrolateral medulla (RVLM). Cutaneous vasoconstrictor neurons involved in thermoregulation are strongly activated by premotor neurons in the Rpal and weakly by premotor neurons in the RVLM. There are possibly other bulbospinal pathways in the raphe pallidus nucleus to sympathetic preganglionic cardiomotor neurons and to preganglionic neurons innervating the kidney which have not been characterized yet. BV, blood vessel; IML, intermediolateral nucleus. Modified from Morrison (1999).

preganglionic lipomotor neurons. These sympathetic premotor neurons are most likely to be different from those in the same raphe nucleus that innervate preganglionic cutaneous vasoconstrictor neurons (see Figure 10.15a).

Figure 10.16 shows the three major types of sympathetic premotor neuron that have so far been functionally identified in the medulla oblongata of the rat. In the cat, premotor neurons in the RVLM are subdifferentiated with respect to different vascular beds and the heart (Figure 10.6). In future other sympathetic premotor neurons may be functionally identified in the medulla oblongata (and in the pons and hypothalamus) as postulated from tracing experiments performed with neurotropic viruses (Figure 8.15 and Table 8.2) and from neurophysiological investigations of the neurons in the peripheral sympathetic pathways (Chapter 4; see Figure 4.23). For example, the nucleus raphe pallidus contains sympathetic premotor neurons that are activated during stress (via the dorsomedial hypothalamus) and generate tachycardia. These sympathetic premotor neurons are normally silent and converge with sympathetic premotor neurons in the RVLM on sympathetic preganglionic cardiomotor neurons or their associated interneurons in the thoracic spinal cord (Zaretsky *et al.* 2003a, b). The nucleus raphe pallidus may also contain sympathetic premotor neurons that are also under hypothalamic control and converge on preganglionic neurons to the kidney (Horiuchi *et al.* 2004b).

10.6 | Coupling between regulation of autonomic pathways and regulation of respiration

10.6.1 The respiratory network

Respiration is the coordinated alternation of the three respiratory phases inspiration, postinspiration (stage 1 expiration) and expiration (stage 2 expiration) (Richter 1982, 1996; Richter and Ballantyne 1983; Schwarzacher *et al.* 1991), which are reflected in the discharge pattern of phrenic motoneurons (see PHR in Figures 10.18 and 10.20). Normal respiration (eupnea) is characterized by an augmenting pattern of phrenic activity during inspiration and fully dependent in its pattern and rhythm on the ponto-medullary respiratory network of neurons (St.-John 1998; St.-John and Paton 2003, 2004). Such coordination of the respiratory phases is important for breathing (and its adaptation during muscular activity), swallowing, vocalization, thermoregulation and other motor actions. Activity in the three respiratory phases is controlled by specific classes of neuron of the ponto-medullary respiratory network. The output neurons of this network of respiratory neurons project to inspiratory and expiratory motoneurons (or to interneurons closely associated with these motoneurons) in the spinal cord (inspiratory or expiratory bulbospinal premotor neurons) or to motoneurons in the medulla oblongata that are involved in the regulation of muscles of the upper respiratory tract (pharynx, larynx). Interneurons, proprio-brain-stem neurons and output neurons of the central respiratory network are located in the VRG (Figure 10.2), in the dorsal respiratory group (not shown in Figure 10.2), in the so-called pneumotaxic center in the rostral pons close to the medial parabrachial nucleus and the Kölliker–Fuse nucleus and in scattered clusters of neurons in the pontine reticular formation between the pneumotaxic center and the rostral portion of the ventral respiratory group in the medulla oblongata (St. John 1998). Most inspiratory bulbospinal premotor neurons are located in the rVRG, the expiratory bulbospinal premotor neurons are located in the so-called Bötzinger complex and the interneurons (consisting of at least four types that are not respiratory premotor neurons) are located in the preBötzinger complex (and also in other parts of the VRG).

The functioning of the central respiratory network, the mechanism(s) of generating the respiratory rhythm and current theories on the respiratory rhythm generation are discussed extensively in the literature and are still very controversial (Feldman 1986; von Euler 1986; Bianchi *et al.* 1995; Richter 1996; Rekling and Feldman 1998; St.-John 1998; Smith *et al.* 2000; Richter and Spyer 2001; Feldman *et al.* 2003; St.-John and Paton 2004). Here it suffices to mention and emphasize the following:

1. Generation of rhythms and pattern of respiration (eupnea) and their adaptation to various behavioral and environmental conditions requires the ponto-medullary respiratory network. Its functioning is fully dependent on the pontine pneumotaxic center. Mechanisms responsible for rhythm generation in eupnea are still unknown. However, they may fully depend on the inherent properties of the (mostly) inhibitory synaptic connections between neurons of the ponto-medullary respiratory network; they require a tonic input to the respiratory network neurons (mainly generated by the chemosensitivity [to arterial P_{CO2}] of the respiratory neurons and by neurons of the reticular formation), and they may be entirely independent of pacemaker activity of groups of neurons of the ponto-medullary respiratory network. The generation of respiratory patterns (shape and amplitude of the respiratory drive) is dependent on multiple hypothetic pattern generators; details of the underlying mechanisms are unknown.
2. Generation of the eupneic respiratory rhythm by pacemaker neurons in the preBötzinger complex is unlikely. The pacemaker idea has been propagated based on experiments conducted on slice preparations that include the preBötzinger complex (Smith *et al.* 1991). However, it has been convincingly shown that pacemaker activity in neurons of the preBötzinger complex or neighboring neurons of the ventral respiratory group of neurons

appears when the medulla oblongata is isolated from the pons and/or is under severe hypoxia. Under these conditions the activity in the phrenic nerve is steeply rising, decrementing and shows a low frequency of phrenic bursts. This pacemaker activity and the subsequent explosive activation of inspiratory neurons is the underlying mechanism of gasping. Gasping may be a respiratory mechanism of last defense during hypoxia.

3. Neurons of the ponto-medullary respiratory network are classified by their activities in reference to the activity of phrenic nerve, i.e. in reference to the three phases of respiration. There exist at least eight types of respiration-modulated neurons.[23] These neurons have distinct (mostly inhibitory) synaptic connections, leading to membrane potential changes generated by distinct ionic currents for each type of neuron. This organization is the base for the coordinated activity in motoneurons supplying inspiratory or expiratory muscles or oropharyngeal muscles in various functional conditions. We do not know yet how these assemblies of functionally distinct types of neurons (including those that have the potential to generate pacemaker activity) of the ponto-medullary respiratory network function together to generate the different types of pattern and rhythm of respiration. However, taking existing data obtained on neurons of the respiratory network in vivo and in the working-heart-brainstem preparation, during normal respiration (eupnea) and during respiration when the rostral or entire pontine part of the network has been removed or deactivated, a computational model of the ponto-medullary respiratory network has been simulated (Rybak *et al.* 2004). This model reproduces several findings that have been observed experimentally, e.g. all changes of eupnea during increased or decreased respiratory drive, during an increase or decrease of activity in vagal lung afferents or switching of eupnea to apneusis or gasping (generated by pacemaker neurons in the preBötzinger complex) when the pontine parts of the respiratory network are inactivated. Interaction between modeling and experimental studies will finally explain the long-lasting enigma on how automatic ventilation is generated by the ponto-medullary respiratory network and how this network is controlled voluntarily (St.-John 1998; Rybak *et al.* 2004).

4. Only little is known about the synaptic connections between respiratory circuits and cardiovascular circuits and therefore about the neuronal mechanisms that underlie the precise coordination of regulation of respiration and regulation of cardiovascular parameters. However, it is fair to hypothesize that this important neuronal integration does not occur by a common respiratory-cardiovascular neuronal network. It is achieved by integration of distinct reflex pathways that are defined by their afferent input, output neurons, synaptic connections, neurochemistry etc.

10.6.2 Respiratory generator and activity in autonomic systems

The neurons in the medulla oblongata involved in regulation of blood pressure and respiration are closely organized in columns (nuclei) of the ventrolateral medulla as demonstrated schematically in Figure 10.2. This tight anatomical and functional organization is reflected in the integration of both control systems leading to a temporal adaptation under all behavioral conditions. The general framework of the integration between the two control systems is outlined in a simplified scheme in Figure 10.17. The ponto-medullary respiratory network and the cardiovascular neural network in the lower brain overlap, i.e. they are connected via common interneurons (it is at present impossible to make any statement about the coupling of the ponto-medullary respiratory network to neural circuits related to regulation of gastrointestinal functions, regulation of body temperature or other autonomic non-cardiovascular function). Both neural networks receive multiple afferent inputs from the periphery (vagal afferents from cardiovascular organs, lung, gastrointestinal tract via the NTS; spinal afferents from the same organ systems and somatic tissues via the spinal cord). The output neurons of the ponto-medullary respiratory network project to respiratory premotor or motor neurons and to autonomic premotor neurons or preganglionic neurons. The output neurons of the cardiovascular network project to parasympathetic and sympathetic premotor (or preganglionic) neurons, however not to respiratory motor neurons. Thus, integration between neural regulation of respiration and of the cardiovascular system occurs principally on more than one level. I presume that this applies to other autonomic systems too (not included in Figure 10.17).

The integration of the two control systems is reflected in respiratory oscillations of autonomic effector organ responses observed in animals and humans under standardized experimental conditions. They can be shown to exist for heart rate (respiratory sinus arrhythmia), volume of the nasal mucosa, vascular resistance in skeletal muscle, pupil diameter and tracheal air flow resistance. These neurally mediated oscillations have a central component (coupling between respiratory neurons and autonomic neurons in the medulla oblongata) and a peripheral (reflex) component. Both the parasympathetic and the sympathetic final pathways are potentially involved in the generation of the respiratory oscillations in effector organ regulations (for references and discussion see Häbler et al. [1994a]; for airway resistance see Paton and Dutschmann [2002]; for humans see Eckberg [2003]).

Recordings from sympathetic nerves in anesthetized animals (e.g., cardiac, splanchnic, renal nerves in dog, cat, rabbit, piglet) in which the vagal afferents from lungs and heart and arterial baroreceptors have been interrupted show that sympathetic neurons are

Figure 10.17 Cardiorespiratory integration in the medulla oblongata: a general conceptual framework. The ponto-medullary respiratory network (St.-John 1998; St.-John and Paton 2004) and the cardiovascular neural network overlap, i.e. both neural networks have common interneurons. The output neurons of the ponto-medullary respiratory network activate respiratory premotor neurons and are synaptically connected with parasympathetic and sympathetic premotor and preganglionic neurons, the latter probably being weaker than the first one. The output neurons of the cardiovascular neural network connect synaptically to parasympathetic and sympathetic premotor and preganglionic neurons, but not to respiratory premotor neurons.

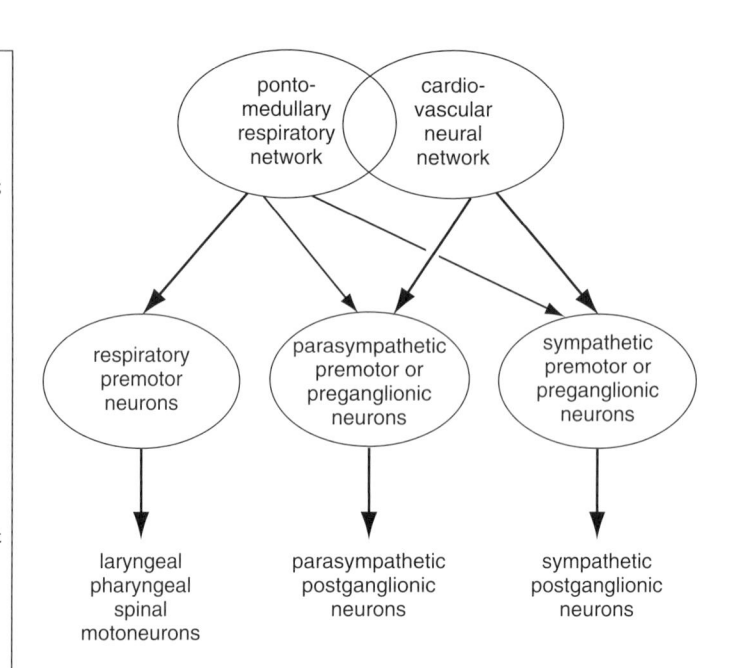

activated during inspiration (for review and literature see Häbler et al. [1994a]). Detailed analysis of this respiratory rhythmicity, in the anesthetized and artificially ventilated cat, in which vagal afferents and afferents from arterial baroreceptors have been interrupted, show that the mass activity in "sympathetic" nerves, which originates predominantly in vasoconstrictor neurons regulating resistance vessels, exhibits a distinctive pattern, consisting of activation in inspiration, depression in postinspiration and sometimes a depression in early inspiration (Bainton et al. 1985; Richter and Spyer 1990) (Figure 10.18; details see below and in Figures 10.20 and 10.21).[24] The respiratory pattern in the sympathetic activity is due to coupling between the central respiratory network and the sympathetic system, probably in the ventrolateral medulla oblongata (Figure 10.17; Richter and Spyer 1990; Richter et al. 1991; Guyenet and Koshiya 1992; see Häbler et al. [1994a]). Thus, sympathetic discharge monitors aspects of the respiratory network during early inspiration, inspiration, postinspiration and expiration phase II (Richter and Ballantyne, 1983).

The pattern of respiratory rhythmicity in the activity of the autonomic neurons is modified by feedback in cardiovascular and pulmonary afferents so that respiratory changes in the activity of these afferent neurons (e.g., arterial baroreceptor afferents) can lead in itself to respiratory changes of activity in the autonomic neurons and mask the respiratory changes generated by the central respiratory rhythm generator (Figures 10.19a, b, 10.20b, 10.23). The profile of sympathetic activity during the respiratory cycle shows that this pattern is not due to a general irradiation of inspiratory activity to

10.6 COUPLING BETWEEN AUTONOMIC PATHWAYS AND RESPIRATORY NETWORK | 425

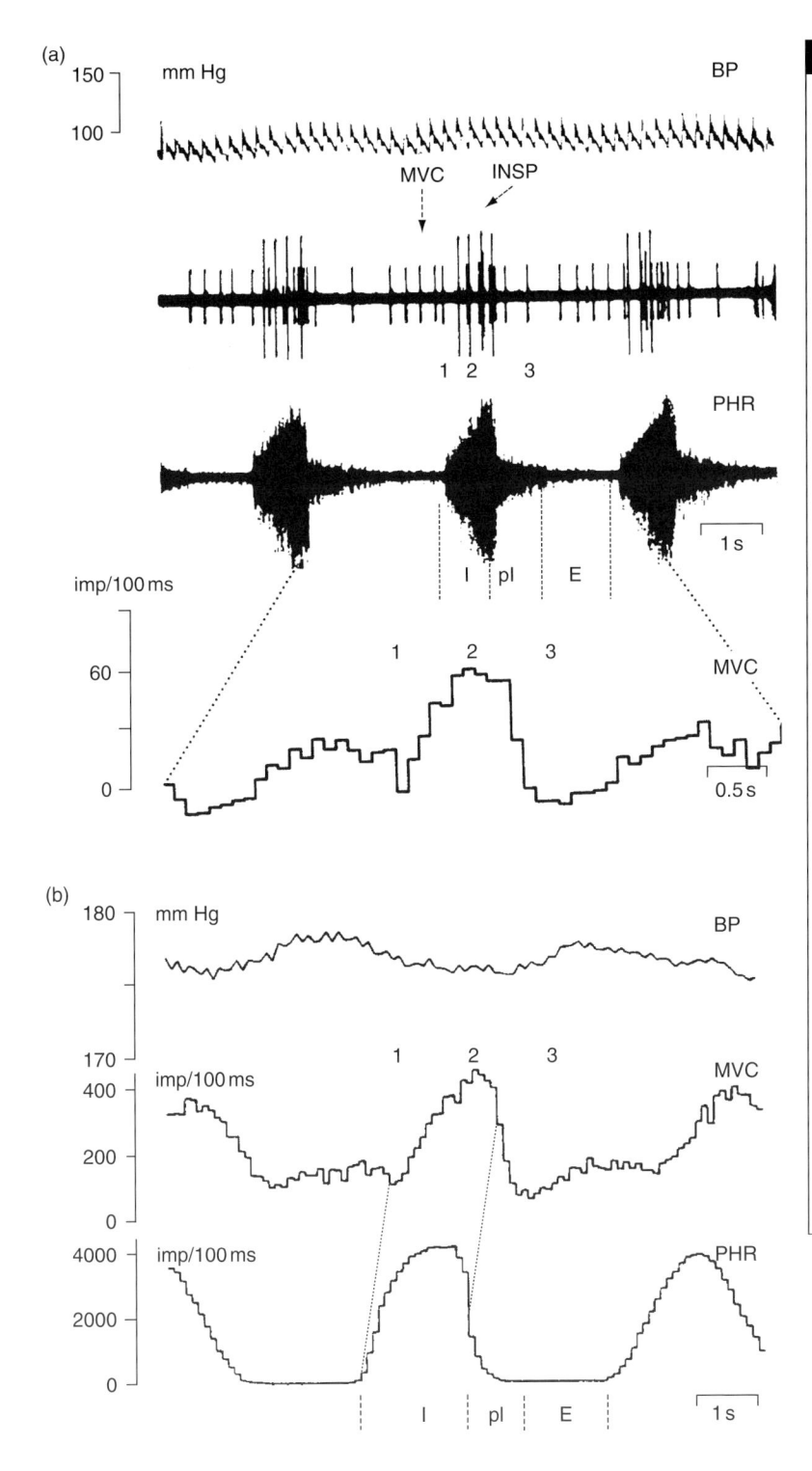

Figure 10.18 Respiratory profile in the activity of pre- (a) and postganglionic (b) muscle vasoconstrictor (MVC) neurons in the cat. (a) Recording from a bundle, isolated from the cervical sympathetic trunk, with a single preganglionic MVC axon and a single axon of a preganglionic inspiratory neuron (INSP) in relation to phrenic nerve activity (PHR). Both vagus and aortic nerves were sectioned; in this way activity in cardiovascular and pulmonary afferents was eliminated. Upper trace, blood pressure (BP). Lower trace, histogram of the activity in the MVC neuron in 100 respiratory double cycles superimposed (bin width 100 ms) with respect to the end of inspiration on an expanded time scale (see PHR). (b) Recording from a multiunit bundle isolated from a muscle branch of the deep peroneal nerve. Vagus, aortic and carotid sinus nerves were sectioned. Activity in 100 respiratory double cycles superimposed (bin width 100 ms) with respect to the end of inspiration. Profile of activity in postganglionic neurons shifted to the left by 400 ms (peripheral conduction time) with respect to PHR (dotted lines). Upper trace, blood pressure (BP). 1, early inspiration; 2, inspiration (I); 3, postinspiration (pI); E, expiration. The cats were anesthetized, ventilated and immobilized. Boczek-Funcke et al. (unpublished observation).

sympathetic premotor neurons, but probably the result of a specific synaptic coupling between neurons of the respiratory network and central sympathetic neurons as will be argued in the next subsection.

Respiratory changes of activity are also present in sympathetic postganglionic neurons innervating skeletal muscle or skin of humans. Figure 10.19 demonstrates one of the first microneurographic recordings from bundles of postganglionic muscle and cutaneous vasoconstrictor axons in muscle and skin nerves of an awake human subject, performed in the laboratory of Karl-Erik Hagbarth in Uppsala (Sweden). In the muscle vasoconstrictor neurons, the respiratory groupings of discharges exhibit cardiac rhythmicity (Fig. 10.19b). Respiratory as well as pulsatile groupings of the multiunit activity are supposed to be generated by unloading of arterial baroreceptors (decrease of baroreceptor activity) produced by the rhythmic respiratory and pulsatile decrease of arterial blood pressure. Activity in cutaneous vasoconstrictor neurons is also loosely correlated with respiration (Figure 10.19c) and during deep breathing cutaneous vasoconstrictor neurons are strongly activated in the inspiratory phase (Figure 10.19d). This leads to rhythmic vasoconstriction of cutaneous blood vessels in the hands and feet.

10.6.3 Early studies of coupling between regulation of respiration and cardiovascular system

The first evidence for the existence of a neural connection between the regulation of respiration and the regulation of autonomic systems was found by Traube (1865) and Hering (1869) in their experiments on the cardiovascular system. After stopping artificial ventilation in vagotomized and partially curarized dogs or cats Traube was the first to observe the occurrence of large blood pressure waves. It was Hering who noticed that each blood pressure wave (later called Traube–Hering waves) was correlated with a rudimentary ineffective movement of the respiratory muscles. The waves occurred not only during asphyxia but also during artificial ventilation with a hypercapnic gas mixture. Using high-frequency artificial ventilation, centrally generated Traube–Hering waves of the blood pressure could be easily distinguished from blood pressure waves produced passively by each artificial inflation. Hering pointed out that a "certain degree of venosity" of the arterial blood (decrease of PaO_2 and increase of $PaCO_2$ with activation of the central respiratory generator [CRG]) was needed to evoke respiratory blood pressure waves. This, he wrote, could be accomplished by restricting the ventilation of the lungs. The heart rate mostly did not change appreciably during the blood pressure waves in vagotomized animals. Nevertheless, by isolation of the peripheral circulation from the heart, Hering excluded the heart as the cause of the observed pressure waves and showed that the vasculature alone was capable of generating respiratory blood pressure waves. From these observations he concluded that "also the vascular system performs respiratory movements." The observed blood pressure waves were thought

10.6 COUPLING BETWEEN AUTONOMIC PATHWAYS AND RESPIRATORY NETWORK | 427

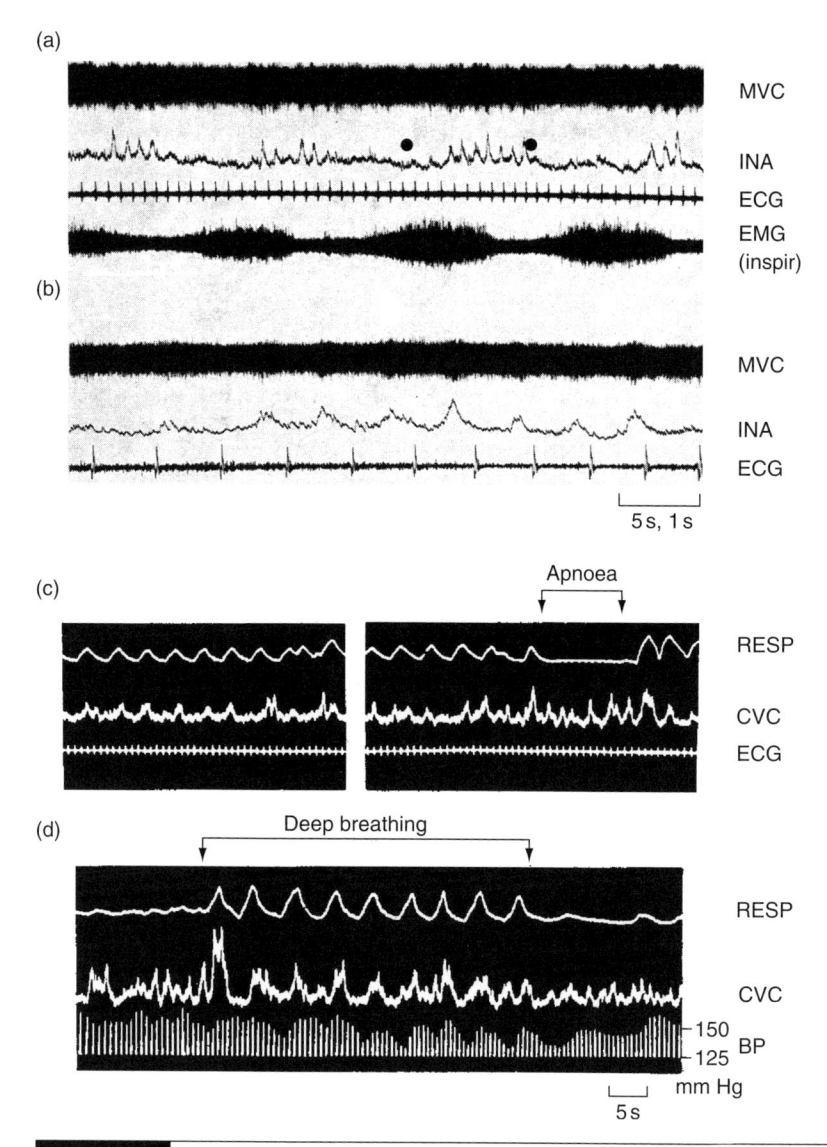

Figure 10.19 Multiunit activity in sympathetic muscle (MVC) and cutaneous vasoconstrictor (CVC) neurons recorded microneurographically in muscle or skin nerves of healthy human beings. (a, b) Respiratory and cardiac grouping of discharges in MVC neurons (deep branch of the peroneal nerve). (a) Simultaneous recording of neural activity (MVC), integrated MVC activity (INA), electrocardiogram (ECG) and electromyogram from intercostal inspiratory muscle (EMG, inspir). (b) Part of record in (a) (between the dots) illustrating on expanded time scale the temporal relation between ECG and grouped neural discharges in the MVC neurons. Note that some MVC activity occurs during inspiration. Time scale of 5 s for (a) and of 1 s for (b). (c, d) Relation between integrated nerve activity (INA) in bundles of CVC fibers and respiration (RESP) (recorded by a strain gauge attached to a rubber band placed around the thorax). Simultaneous recording of respiratory movements (inspiration up), INA and ECG (c) or arterial blood pressure (d; left radial artery). (c) Normal respiration (left, median nerve) or voluntary apnea (right, superficial peroneal nerve). (d) Voluntary deep inspiration. Note correlation between inspiration and INA in (c) left and (d). (a, b) Modified from Hagbarth and Vallbo (1968). (c, d) Modified from Hagbarth et al. (1972) with permission.

to be caused by the rhythmic activity of the respiratory center in the medulla oblongata. In similar experiments on animals with intact vagus nerves, Traube (1865) observed an increase of heart rate during inspiration. He correctly attributed these heart rate changes to a decrease of activity in the vagal supply to the heart. With these experiments, Traube demonstrated that the sympathetic innervation of resistance vessels and the parasympathetic innervation of the heart showed a reciprocal coupling to respiration.

Respiratory sinus arrhythmia was first studied systematically by Anrep *et al.* (1936a, b) on animals. They proved its neural origin and showed that it is largely mediated by the vagus nerves. Anrep and coworkers postulated two contributing mechanisms: a central one being dependent on the strength of respiratory drive and a peripheral (reflex) one.

Twenty to 30 years later, the mechanisms underlying Hering–Traube blood pressure waves were studied by Koepchen and coworkers in a series of experiments on anesthetized dogs (and also on some rabbits) (Koepchen and Thurau 1959; Koepchen *et al.* 1968; Seller *et al.* 1968). They perfused a vascularly isolated gracilis muscle with intact innervation at constant flow with O_2-saturated blood at body temperature and recorded the perfusion pressure, which is a measure for the vascular resistance. Fluctuations of perfusion pressure synchronous to phrenic nerve activity (PHR) or to integer multiples of PHR were observed, mostly with the increase in resistance during inspiration. The amplitude of these fluctuations paralleled the depth of breathing, being largest when inspiration was prevented by tracheal occlusion. These fluctuations of vascular resistance were still present when ventilation was stopped by muscular paralysis with curare and when the arterial baroreceptors were denervated and the vagus nerves were sectioned in order to eliminate coupling by peripheral reflexes. This observation indicates that a central mechanism was operating.

10.6.4 Studies of functionally identified sympathetic neurons in animals

Studies of sympathetic activity in whole nerves in animals and microneurographic recordings from bundles of postganglionic axons in human nerves may give the impression that the sympathetic neurons exhibit in their activity a more or less uniform pattern of respiratory modulation characterized by a peak in inspiration. However, this is not the case when one analyzes the respiratory rhythmicity of activity in *functionally identified sympathetic pre- and postganglionic neurons* in more detail in the anesthetized and artificially ventilated cat under various experimental conditions (increased/decreased respiratory drive, with/without rhythmic feedback in vagal [pulmonary] and/or arterial baroreceptor afferents). This analysis shows that the pattern of respiratory rhythmicity depends on the function of the sympathetic neuron and indicates that the coupling between the respiratory rhythm generator and the presympathetic neurons in the medulla oblongata is probably not uniform (Boczek-Funcke et al. 1992a, b; Häbler et al. 1994a; Häbler and Jänig 1995; Jänig and Häbler 2003). Figures 10.20 and 10.21 demonstrate the main results of these studies:

- *Muscle, visceral*, a few *cutaneous vasoconstrictor neurons* and *sympathetic cardiomotor neurons* exhibit respiratory patterns in their activity which is typical for the uniform pattern seen in the "sympathetic" nerves (Figures 10.18, 10.20a, b, 10.21a). They consist of two components, a central and a peripheral (reflex) one: (1) The *central component* is activated during inspiration (*ii* in Figure 10.20b) with depression of activity during postinspiration and sometimes also in early inspiration (see *1* in Figure 10.18). The amplitudes of activation and depression of activity increase with increasing respiratory drive. This pattern is clearly seen in vagotomized and baroreceptor-denervated animals (Figure 10.18). (2) The *peripheral component* consists of a peak of activity during the declining phase of the respiratory blood pressure waves, associated with unloading of arterial baroreceptors (*i* in Figures 10.20b and 10.21a). In artificially ventilated animals, with intact vagus and baroreceptor nerves (and normal blood gases and pH), the activity is in many cases only related to the declining phase of the respiratory blood pressure waves, whereas the activation related to central inspiration is small or undetectable. Two factors seem to contribute to this limited evidence of coupling to central inspiration in normocapnia. *First*, the inspiratory drive to the central sympathetic neurons appears to be rather weak in normocapnia. *Second*, in ventilated cats with intact vagus nerves the rising phase of the respiratory blood pressure wave coincides with central inspiration. Thus, the activation of sympathetic premotor neurons in the RVLM during inspiration may be prevented by reflex inhibition via the baroreceptor reflex.
- There is no evidence from these studies that additional peripheral reflexes from other cardiovascular or pulmonary afferents directly contribute to respiration-related phasic modulation in muscle and visceral vasoconstrictor activity other than by synchronizing central respiration. In *apneic hypocapnia* (decreased $PaCO_2$) generated by an increase of tidal volume, the respiration-related rhythm in the activity of muscle vasoconstrictor neurons is due to the arterial baroreceptor reflex and not, or at most only to a small extent, to activation of cardiopulmonary vagal afferents. The neural rhythm in vasoconstrictor neurons that is related to the respiratory blood pressure waves in cats, in which only the arterial baroreceptors from the carotid sinuses are intact, is almost totally abolished by bilateral carotid occlusion when vagus nerves remained intact.[25]
- The respiratory modulation of activity in most *cutaneous vasoconstrictor neurons* is different from that in muscle and visceral vasoconstrictor neurons in the cat. No peripheral (reflex) component

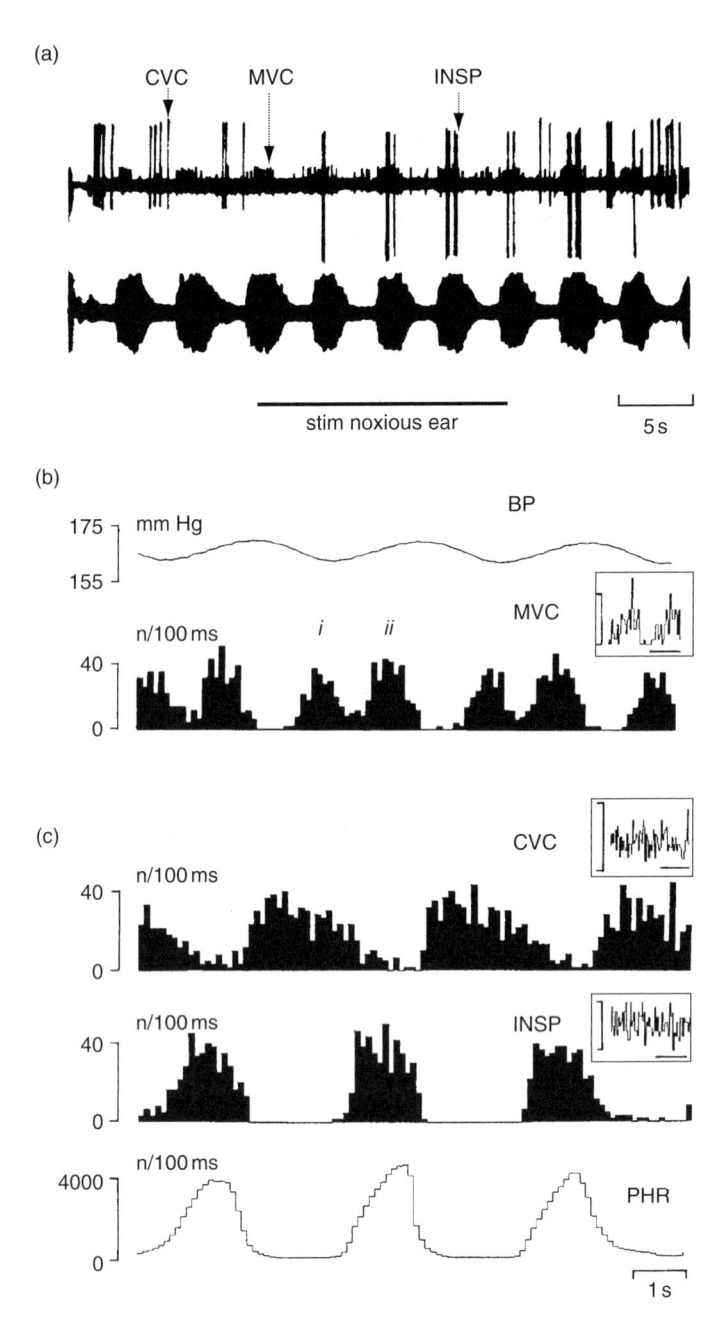

Figure 10.20 Respiratory modulation of activity in different types of functionally identified sympathetic preganglionic neurons projecting to head or neck in the cat. (a) Responses of preganglionic neurons (upper trace) to noxious stimulation of ear skin and correlation of activity with phrenic nerve activity (lower trace). Cutaneous vasoconstrictor (CVC) neuron (inhibited; activity in expiration). "Inspiratory" (INSP) neuron (activated: activity in inspiration). Muscle vasoconstrictor (MVC) neuron (activated; activity in inspiration and expiration). (b, c) Typical patterns of respiratory modulation of the activity in single preganglionic MVC, CVC and INSP neurons. Vagus and baroreceptor nerves intact. One hundred respiratory double cycles superimposed (bin width 100 ms) with respect to the end of inspiration (see phrenic nerve activity [PHR] in [c]). *Insets*: post-R-wave histograms of the activity ("cardiac rhythmicity"; 500 cardiac double cycles superimposed, bin width 8 ms, times scale 200 ms, ordinate scales 20 imp/8 ms). Upper trace in (b), mean arterial blood pressure. (b) The activity in the MVC neuron shows two peaks, the first peak (*i*) being due to unloading of arterial baroreceptors during the falling phase of the second-order blood pressure waves (BP) and the second one (*ii*) being due to a

of respiratory modulation has been found that is related to rhythmic activation of arterial baroreceptor afferents or cardiopulmonary afferents. Thus, the observed patterns are due to central mechanisms. In normocapnia, most cutaneous vasoconstrictor neurons (about 80% of the postganglionic cutaneous vasoconstrictor neurons, about 40% of the preganglionic cutaneous vasoconstrictor neurons) show no respiratory modulation. The remainder of neurons exhibits some modulation in their activity, showing a weak maximum either in expiration (CVC in Figure 10.20a, c) or in inspiration (Figure 10.21). During increased respiratory drive (produced by increasing the P_{CO2} in the inspired air), some cutaneous vasoconstrictor neurons that show no respiratory modulation in their activity in normocapnia exhibit a peak of activity that occurs either in inspiration or in expiration phase II or in postinspiration, whereas others maintain their unmodulated pattern. The cutaneous vasoconstrictor neurons exhibiting a peak of activity in postinspiration appear to be actively inhibited during inspiration (Figure 10.20c), which is particularly seen when the inspiratory drive is high (e.g., during hypercapnia). This respiratory activity pattern in cutaneous vasoconstrictor neurons resembles the pattern observed in parasympathetic cardiomotor neurons, which are inhibited during inspiration and excited during postinspiration (Gilbey *et al*. 1984). Cutaneous vasoconstrictor neurons exhibiting an inspiratory peak of activity preferentially innervate hairless skin. The activation during inspiration in these neurons is not followed by a depression of activity in postinspiration.

- Discharges in *inspiratory neurons*, which are present only in the thoracic sympathetic outflow to the head and upper neck, are confined to central inspiration and absent in postinspiration and expiration (INSP in Figures 10.18, 10.20a, c, 10.21). These neurons are silenced during hyperventilation (hypocapnia) and enhanced in their discharge during hypercapnia.
- Activity in *sudomotor neurons* has its maximum in postinspiration, which is enhanced during increased respiratory drive and similar to that observed in some cutaneous vasoconstrictor neurons (see Figure 10.21).
- Most *motility-regulating neurons* show no respiratory modulation in their activity. A few, however, discharge preferentially in postinspiration, similar to sudomotor and some cutaneous vasoconstrictor neurons.

Figure 10.20 (cont.)
central coupling to inspiratory neurons. Peak *i* matches the high degree of pulse rhythmicity in the activity of the MVC neuron (inset) and is increased with an increase of amplitude of the second-order blood pressure waves. Peak *ii* is increased during hypercapnia and decreased during hyperventilation. Recorded during light hypercapnia. (c) The CVC neuron shows a pronounced expiratory peak of activity with its maximum in postinspiration. The INSP neuron discharges exclusively during inspiration. Both neurons showed no pulse rhythmicity in their activity (see insets). The CVC and INSP neuron recorded simultaneously in normocapnia. Modified from Häbler *et al.* (1994b) with permission.

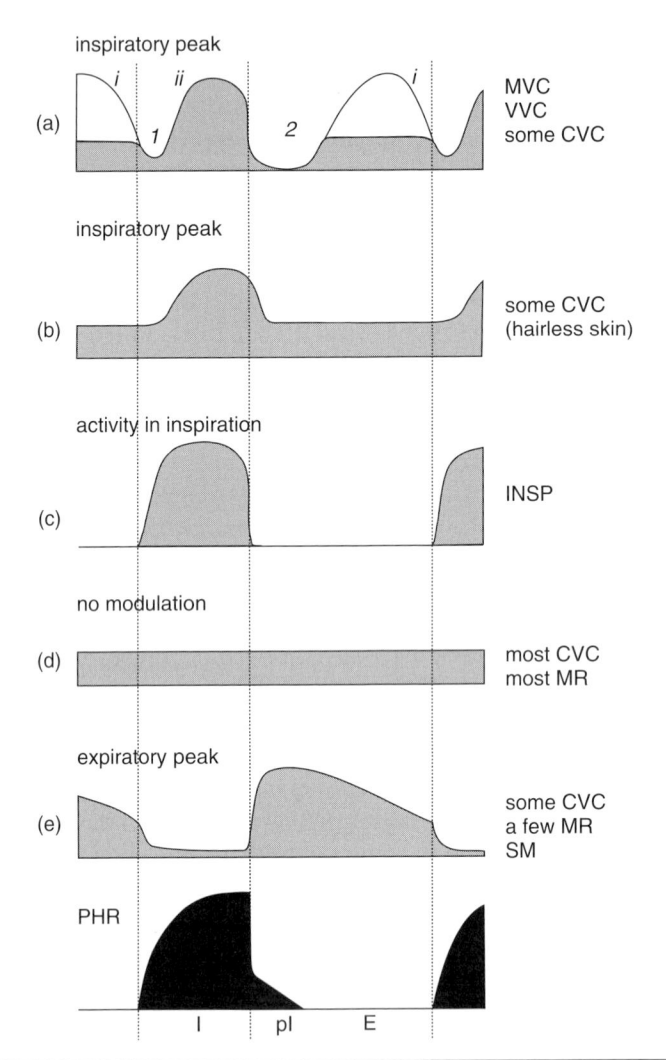

Figure 10.21 Synopsis of the various patterns of respiratory modulation exhibited by sympathetic neurons in the cat under conditions of normal and increased respiratory drive. (a) Pattern present in almost all neurons in the cat that exhibit a high degree of cardiac rhythmicity in their activity (muscle vasoconstrictor [MVC], visceral vasoconstrictor [VVC], some cutaneous vasoconstrictor [CVC] neurons). The baroreceptor mediated peak of activity (*i*) occurs when ventilation pump and central respiration are entrained to each other. Inspiratory drive-dependent peak of activity (*ii*) is preceded by a small and variable depression of activity (*1*) in early inspiration and followed by a pronounced depression in postinspiration (*2*). These components have not been demonstrated in humans so far. (b) An inspiratory peak of activity is exhibited by a few CVC neurons (mainly supplying hairless skin). (c) Neurons discharging only in inspiration (INSP) are present in the thoracic preganglionic outflow projecting into the cervical sympathetic trunk. Most of these neurons are not under baroreceptor control (see Fig. 10.20a, c). (d) Respiratory modulation is absent in many CVC neurons and most motility-regulating (MR) neurons under normal respiration and neutral ambient temperature. (e) Expiratory pattern of activity with a depression during inspiration and activation during postinspiration is shown by some CVC and a few motility-regulating (MR) neurons, and in sudomotor (SM) neurons. This pattern is also seen in parasympathetic cardiomotor neurons in the cat. PHR, activity in phrenic nerve: I, inspiration; pI, postinspiration; E, phase II expiration. Modified from Häbler *et al.* (1994b) with permission. For references see there and text.

In conclusion, in the cat, the respiratory profiles of activity in sympathetic neurons are not uniform but vary according to the functional type of neuron (Figure 10.21), indicating that the central coupling between the neurons of the respiratory network and neurons associated with the sympathetic nervous system varies with the function of the sympathetic subsystem. This coupling probably depends on specific synaptic connections between respiratory neurons and pre-sympathetic neurons in the medulla oblongata (e.g., in the VLM, caudal raphe nuclei, lateral paragigantocellular nuclei [McAllen 1987; Dampney and McAllen 1988; Haselton and Guyenet 1989; Guyenet 1990; Guyenet and Koshiya 1992]).

There are some species differences between the rat and larger mammals, such as cats, dogs, rabbits and piglets (for discussion and literature see Häbler *et al.* 1994a). The pattern of respiratory modulation that is most frequently found in the rat is depression of activity during inspiration and activation during postinspiration (Häbler *et al.* 1994b, 1996, 2000), whereas in larger animals the majority of sympathetic neurons shows the reverse. Similar patterns have been observed in bulbospinal neurons of the RVLM of the rat (Haselton and Guyenet 1989). Contrary to what has been observed in cats, the patterns of respiratory modulation in rats are similar in postganglionic muscle and cutaneous vasoconstrictor neurons of the hindlimb and tail and in thoracic preganglionic neurons projecting to head and neck (Häbler *et al.* 1996, 1999a, 2000). Therefore, respiratory modulation cannot be used as a functional marker for sympathetic neurons in the rat projecting to skin or skeletal muscle (Häbler *et al.* 1993c, 1994a). However, for rats it is unknown whether non-vasoconstrictor neurons (e.g., pilomotor, sudomotor, motility-regulating neurons) exhibit distinct respiratory profiles in their activity.

10.6.5 Studies of functionally identified sympathetic neurons in humans

Muscle vasoconstrictor neurons
Bursts of sympathetic activity in human muscle vasoconstrictor neurons are generated by pulsatile unloading of arterial baroreceptors and occur in groups related to the respiration (Figure 10.19a, b). They usually occur only during the falling phase of the respiratory blood pressure fluctuations (Figure 10.22a; Hagbarth and Vallbo 1968; Eckberg *et al.* 1985, 1988). Therefore it has been concluded that this rhythm is secondary to the blood pressure changes induced by respiration (Hagbarth and Vallbo 1968; Eckberg *et al.* 1985; Wallin and Fagius 1988). This rhythm appears to be analogous to the peripheral component of respiratory modulation in the activity of muscle vasoconstrictor neurons in cats when respiratory drive to the sympathetic premotor neurons is relatively low (Figure 10.22b). Thus, the data obtained in conscious human subjects match those obtained in anesthetized cats.

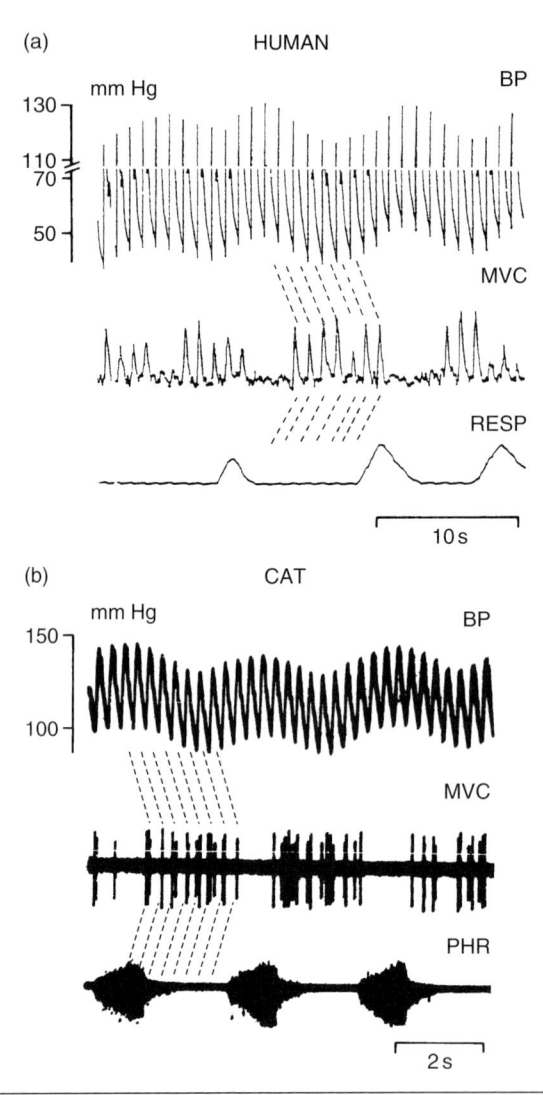

Figure 10.22 Comparison of the firing patterns of postganglionic muscle vasoconstrictor (MVC) activity in the human (a) and in the cat (b). Traces from above show pulsatile arterial blood pressure (BP), neural activity (original record of MVC activity in [b] and integrated MVC activity in [a], respectively) and central respiration (phrenic nerve activity [PHR] and pneumograph signal [RESP; inspiration upward], respectively). (a) Obtained from a conscious human being under resting conditions. (b) Obtained from an anesthetized and artificially ventilated cat in normocapnia with intact vagus and buffer nerves and synchronized cycles of central respiration and artificial ventilation. The dotted lines indicate the close temporal relationship of the MVC action potentials and integrated MVC bursts, respectively, to each pulse pressure wave. Note that activity in the postganglionic axons is delayed by about 400 ms (cat) and 1.5 s (human) due to central and peripheral conduction time. Muscle vasoconstrictor activity occurs mainly during the declining phase of the ventilatory oscillations of blood pressure, which coincide here with expiration. There is some, but only little, discharge during inspiration, which is probably due to the coincidence of low inspiratory drive and the rising phase of the ventilatory blood pressure oscillations in the record on the human. (a) modified from Eckberg et al. (1985) with permission. (b) from Häbler et al. (1994b) with permission.

A central respiratory component in MVC activity (i.e. activation during inspiration, equivalent to the activation of MVC neurons in cats) is absent or could not be detected in human subjects (Eckberg et al. 1985),[26] probably because (1) respiratory drive to MVC neurons under resting conditions is relatively weak (e.g., from the sympathetic premotor neurons in the RVLM; McAllen [1987]) and (2) the rising phase of the arterial blood pressure waves occurring during inspiration (activating arterial baroreceptors followed by inhibition of activity in muscle vasoconstrictor neurons) probably counteracts the inspiratory activation of these neurons.[27]

Cutaneous vasoconstrictor neurons and sudomotor neurons

In humans sympathetic activity in nerves supplying skin, which is a mixture of vasoconstrictor and sudomotor activity depending on the central thermoregulatory state, occurs in irregular bursts (Hagbarth et al. 1972; Wallin and Fagius 1988), which are independent of the pulsatile blood pressure oscillations. Thus, a baroreceptor-mediated reflex component of respiratory modulation is mostly not detectable in skin sympathetic activity (Hagbarth et al. 1972).

During normal quiet breathing, bursts in cutaneous sympathetic activity are loosely and variably coupled to respiration (mostly to the inspiratory phase [Figure 10.19c]; Hagbarth et al. 1972; Hallin and Torebjörk 1974). With increased vasoconstrictor activity at low ambient temperature respiratory modulation of cutaneous sympathetic activity becomes more pronounced (Bini et al. 1980a, b). Activity of single postganglionic cutaneous vasoconstrictor or sudomotor axons innervating hairy skin of the foot in humans (superficial peroneal nerve) was investigated for respiratory rhythmicity during body cooling (skin temperature $\leq 22\,°C$; only cutaneous vasoconstrictor neurons active) or during body heating (skin temperature 31 to 33 $°C$; sudomotor neurons active [subjects were sweating] and activity in cutaneous vasoconstrictor neurons low or absent). Under these conditions about 50% of the sudomotor neurons and two-thirds of the cutaneous vasoconstrictor neurons exhibited respiratory rhythmicity in their activity. The probability of discharge in both types of sympathetic neurons was higher in inspiration than in expiration (Macefield and Wallin 1999b).

From plethysmographic measurements, it is known that deep inspiration is followed by vasoconstriction in the skin of the fingers (Bolton et al. 1936; Gilliat 1948). This is probably due to bursts of activity in skin sympathetic neurons occurring in response to deep breaths (Figure 10.19d). Activation of cutaneous vasoconstrictor neurons during deep inspiration does not correlate with the systolic blood pressure changes (Figure 10.19d). This could theoretically be due to central coupling of skin sympathetic premotor neurons to inspiratory neurons. However, since the inspiration-induced vasoconstriction was still found in patients with spinal cords transected at a high thoracic level, it was suggested that this reflex is spinal

(Gilliat *et al.* 1948), the involved afferents being unknown. However, Hagbarth *et al.* (1972) also found some neural "reflex" components with long latencies following a deep breath and concluded that both spinal and supraspinal mechanisms may be involved, arguing that a central component of respiratory modulation contributes to the response of the skin sympathetic activity.

In conclusion, in humans, muscle vasoconstrictor, cutaneous vasoconstrictor and sudomotor neurons exhibit some respiratory modulation of activity. To analyze the precise temporal relationship of activity in these sympathetic neurons in relation to the different phases of central respiration in humans more closely, EMG-recordings from respiratory muscles under various experimental conditions (e.g., hypercapnia, hyperthermia, hypothermia etc.) are warranted in these studies (see Figure 10.19a).

10.6.6 Models of integration between autonomic and respiratory generators

The classical view on integration between autonomic and respiratory generators held by Traube (1865) and Hering (1869) was that vasomotor and respiratory centers are situated separately in the lower brain stem. With respect to cardiorespiratory coupling both authors gave the primacy to the respiratory system and thought that the rhythmic activity "irradiates" from the respiratory to the cardiovascular center. This concept was based on experiments performed on dogs and cats in which the coupling between the two systems was studied under extreme conditions, such as asphyxia and high CO_2 load. It appears that the term "irradiation" is now outdated, yet the concept behind it is still implicit in the term "respiratory modulation" of sympathetic activity (see Koepchen [1983]; Koepchen *et al.* [1987]). From the observations of respiratory blood pressure waves, a non-specific "common central rhythmicity," which is not primarily related to respiration but generates rhythmic activity in various peripheral output systems, e.g. to respiratory muscles and cardiovascular target organs, was postulated (Koepchen 1962). Later several oscillators for the respiratory, the sympathetic and other output systems were proposed (Koepchen *et al.* 1981; Koepchen 1983). These common oscillators were thought to be entrained. The neural matrix comprising the oscillators was thought to process information from cardiovascular afferents, pulmonary afferents and other types of afferent neurons and from the central chemosensitive sites (e.g., on the dendrites of the respiratory neurons). The output of the respiratory network was believed to "irradiate" backward into the reticular formation, which then modulates the activity of the sympathetic output. Barman and Gebber (1976) suggested that the respiratory rhythm and the respiratory modulation of sympathetic activity are generated by two independent oscillators which are normally entrained to each other.

The idea of a common central oscillator or independent but entrained central oscillators was challenged experimentally on single sympathetic neurons as described above and by Bachoo and

Polosa (1985, 1986, 1987) showing quite clearly that respiratory modulation of activity in these neurons is locked to particular respiratory phases (for detailed discussion see Häbler et al. [1994a]).

The most elaborate concept of the generation of respiratory rhythmicity in autonomic neurons has been developed recently by Richter, Spyer and coworkers (Richter and Spyer 1990, Richter et al. 1991). Based on their working model, it is proposed that coupling between the respiratory network and the neurons related to the cardiovascular autonomic systems occurs in the lower brain stem via ramp-inspiratory, early-inspiratory and postinspiratory neurons of the respiratory network. According to this model, sympathetic premotor neurons (e.g. in the RVLM) would be influenced in the following way:

1. activation by ramp-inspiratory neurons (which activate inspiratory [bulbospinal] premotor neurons);
2. inhibition by postinspiratory neurons (which inhibit ramp-inspiratory and early-inspiratory neurons); and
3. inhibition by early-inspiratory neurons (which inhibit postinspiratory and ramp-inspiratory neurons).

Thus, this model can fully account for the respiratory activity profile in muscle, visceral, renal vasoconstrictor and sympathetic cardiomotor neurons which dominates the respiratory rhythmicity of the activity in all "sympathetic" nerves in cat (see Figures 10.18, 10.21). It is supported by the finding that many sympathetic premotor neurons in the RVLM change their activity during the respiratory cycle, being activated during inspiration, although details about the depression of activity in early inspiration and postinspiration have not been worked out. As emphasized by Richter and Spyer (1990) this model also accounts for the inhibition of parasympathetic cardiomotor neurons in inspiration and activation or disinhibition in postinspiration. In order to account for the various patterns of respiratory modulation seen in the activity of functionally different types of sympathetic neurons, the nature of the coupling between neurons of the central respiration generator and the presympathetic neurons must vary according with the function of the sympathetic neurons.

Any attempt to explain the neural mechanisms of respiratory patterns in the activity of autonomic neurons rests on the comparison between rhythmic discharges in phrenic nerve and autonomic neurons as well as in sympathetic premotor neurons in the RVLM. Thus, the explanations are based on analogy and at best on indirect arguments derived from investigations of neurons of the respiratory network with respect to the phrenic nerve. Direct evidence from systematic intracellular measurements in sympathetic premotor neurons in the medulla oblongata does not exist. As proposed in Figures 10.17 and 10.23 the coupling between respiratory system and the sympathetic cardiovascular pathways may occur at the level of the sympathetic premotor neurons in the RVLM. However, respiratory modulation at the level of the preganglionic neurons in the spinal cord might also be possible.

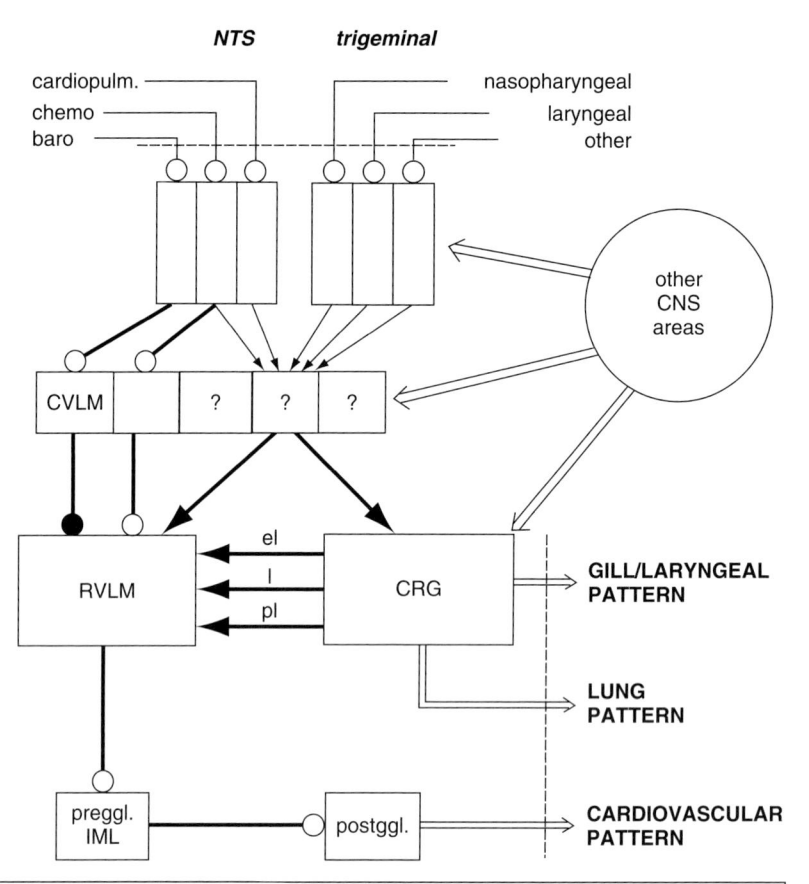

Figure 10.23 Integration of the central respiratory generator (CRG) and the rostral ventrolateral medulla (RVLM) and the different inputs from cardiovascular and pulmonary afferents as well as other types of afferents. The CRG determines the respiratory lung pattern as well as the laryngeal rhythm (which is phylogenetically equivalent to the gill rhythm in fish [see Richter et al. 1991]). The respiratory rhythm in the autonomic cardiovascular neurons is determined by the integration between the respiratory network and the cardiovascular neurons in the RVLM (constituting the so-called "common cardiorespiratory network" [Richter and Spyer 1990; Richter et al. 1991]). This system is modulated by various reflex pathways linked to cardiovascular, pulmonary, trigeminal, laryngeal, pharyngeal and other afferents. The reflexes may influence the sympathetic premotor neurons in the RVLM either directly or via the CRG. Other areas of brain stem, hypothalamus and telencephalon modulate RVLM, CRG and reflex pathways. CNS, central nervous system; CVLM, caudal ventrolateral medulla; IML, intermediolateral nucleus; NTS, nucleus tractus solitarii; preggl., preganglionic; postggl., postganglionic; el, early inspiration; I, inspiration; pl, postinspiration.

Several groups of cardiovascular, pulmonary, laryngeal, pharyngeal and other afferents can influence sympathetic premotor neurons, parasympathetic cardiomotor neurons and respiratory premotor neurons. Detailed analysis of the pathways underlying these reflexes has only been performed for a few systems (see baroreceptor reflexes and chemoreceptor reflexes in Subchapters 10.3 and 10.4). As demonstrated for the chemoreceptor reflex in sympathetic cardiovascular neurons, reflex activation or inhibition are mediated by neural pathways that are independent of the central respiratory rhythm generator as well as via the respirator rhythm generator (Figure 10.23; Guyenet and Koshiya 1992; Häbler et al. 1997a; Guyenet 2000; for discussion see Häbler et al. [1994a]).

The general idea of a "common cardiorespiratory network," as propagated by Richter, Spyer and coworkers, is that the neurons and their synaptic connections in the VLM (which includes the RVLM, lVLM/CVLM and VRG) represent a complex sensorimotor program that guarantees at any moment and under any physiological condition a precisely coordinated regulation of the respiratory pump, the arterial blood pressure and the perfusion of the lung. This includes the coordinated neural regulation of right and left heart as well as of peripheral resistance. This idea requires distinct inhibitory and excitatory synaptic connections between different groups of neurons in the "common cardiorespiratory network." It requires furthermore that the functionally distinct types of afferent neuron involved in this coordinated regulation (see Figure 10.23) form distinct reflex circuits with the output neurons of this network. These reflex circuits may include neurons of the central respiratory rhythm generator or may be independent of it. For example, the baroreceptor pathways can normally function independently of the central respiratory rhythm generator, but clearly interact with it as well. This is therefore still a controversial issue (Seller *et al.* 1968; Boczek-Funcke *et al.* 1991).

Overall, the distinct component parts of this neuronal network that integrates regulation of respiration and regulation of the cardiovascular system are unknown. In the rat, a main coupling may occur from the preBötzinger complex via glutamatergic interneurons (that express the NK_1-receptor) to inhibitory GABAergic and/or glycinergic interneurons in the CVLM that inhibit bulbospinal neurons in the RVLM (Wang *et al.* 2003). This coupling would imply that activity in vasoconstrictor neurons that are under the control of the RVLM is inhibited during inspiration. It has been shown in the rat that most barosensitive postganglionic neurons are depressed in their activity during inspiration and that this depression is enhanced during increased respiratory drive (Häbler *et al.* 1993c, 1994a; Bartsch *et al.* 1999).

Command signals generated in supramedullary centers (e.g., the periaqueductal grey, the hypothalamus, the telencephalon) normally do not activate cardiovascular sympathetic premotor, cardiovascular parasympathetic premotor neurons or respiratory premotor neurons (nor the final autonomic and somatic motor neurons) but instead engage with the "common cardiorespiratory network" (i.e., the cardiorespiratory motor program), e.g., during exercise or defense behaviors.

The idea of a "common cardiorespiratory network" should not be generalized to all sympathetic or parasympathetic systems. Most types of sympathetic and parasympathetic neurons do not have cardiovascular functions. Future experimental work is likely to show that other sensorimotor programs exist in the lower brain stem that are involved in thermoregulation, regulation of energy balance (see Subchapter 10.5) or regulation of gastrointestinal functions (see Subchapter 10.7). These probably include premotor pathways to motor neurons involved in body movements, sympathetic premotor neurons to cutaneous vasoconstrictor, lipomotor, sudomotor and motility-regulating pathways and parasympathetic pathways innervating the gastrointestinal tract. By the same token it will be shown that the central autonomic systems that are not involved in cardiovascular regulation are also coupled to the regulation of respiration. However, the underlying mechanisms of this coupling will turn out to be different from that between the neural network related to respiration and the neural network related to the cardiovascular system.

10.7 | Vagal efferent pathways and regulation of gastrointestinal functions

10.7.1 Vagal afferents and preganglionic neurons

Vagal afferent neurons of the gastrointestinal tract, which have their cell bodies in the inferior nodose ganglion of the vagus nerve project topographically to the NTS, showing a distinct mediolateral as well as rostrocaudal arrangement of the projections to its subnuclei (see Figures 8.10 and 8.11). These visceral afferents monitor several parameters (see Subchapters 2.3 and 2.4):

1. Mechanical events related to contraction or distension of the gastrointestinal tract or to shearing stimuli exerted at the mucosal surface.
2. Chemical events related to decrease in pH, changes of glucose or lipid concentration or other chemical events.
3. Toxic events related to the ingested food that may be harmful for the gastrointestinal tract and for the organism; putative afferents monitoring these toxic events may be particularly associated with the most powerful defense line in the body, the gut-associated lymphoid tissue (GALT; Shanahan 1994).

The various functional types of vagal visceral afferent neurons innervating the gastrointestinal tract and their transduction mechanisms are only partially known and incompletely understood. The excitation of some types of vagal afferents innervating the small intestine (probably those activated by chemical stimuli such as lipids) is mediated by cholecystokinin (CCK) or 5-HT (and possibly other hormones; Richards *et al.* 1996; Hillsley *et al.* 1998; Hillsley and Grundy 1998; Kreis *et al.* 2002). Many afferent neurons terminate within ganglia of the myenteric plexus (Berthoud and Powley 1992; Phillips *et al.* 1997; Berthoud and Neuhuber 2000). These afferents are possibly mechanosensitive (Zagorodnyuk and Brookes 2000; Zagorodnyuk *et al.* 2001) (for description of vagal visceral afferent neurons see Subchapter 2.3).

Further afferent signals from the gastrointestinal tract to the lower brain stem (acting via the area postrema [AP]; see Subchapters 8.3 and 10.7.4) or to neurons in the arcuate nucleus in the hypothalamus are hormones. These hormones are involved in regulation of satiety (the peptide pancreatic polypeptide [PP] released from the pancreas to the AP; glucagon-like peptide [GLP-1] from the mucosa of the small intestine to the AP; the peptide YY (PYY) released from the mucosa of the duodenum to the arcuate nucleus [Schwartz and Morton 2002]), initiation of food intake (the peptide ghrelin released from the mucosa of the stomach to the arcuate nucleus [Rosicka et al. 2002]) or nausea (CCK to the AP [Moran and Schwartz 1994; Baldwin et al. 1998]).

Parasympathetic preganglionic neurons projecting to the gastrointestinal tract are situated in the dorsal motor nucleus of the vagus (DMNX). They are topographically arranged in rostrocaudally oriented cell columns according to the organ section they innervate (e.g., stomach and cecum; see Figures 8.10). These preganglionic neurons have various functions related to:

1. regulation of motility of the gastrointestinal tract (several types of excitatory and inhibitory [non-cholinergic non-adrenergic] motoneurons);
2. regulation of exocrine secretion (e.g., secretomotor neurons innervating gastric parietal cells secreting proton ions, exocrine pancreas, submucosal glands or mucosa cells);
3. regulation of hormone release (e.g., gastrin, secretin, insulin, glucagon); or possibly
4. regulation of the inner defense line of the body that is connected with the GALT (but this is rather hypothetical).

Thus, there are several functionally distinct types of vagal and spinal visceral afferent neurons and several functionally distinct types of preganglionic neuron innervating the gastrointestinal tract (see Figures 5.14 and 5.15). These are the peripheral substrates for the specific regulation of different functions of the gastrointestinal tract and their coordination by the brain (Travagli et al. 2006).

10.7.2 Distinct intestino-intestinal reflex circuits in the dorsal vagal complex

An intimate anatomical relationship exists between the nuclei in the NTS and the DMNX implying that afferents from the gastrointestinal tract form disynaptic contacts with the preganglionic neurons projecting to the gastrointestinal tract. Therefore the NTS (including AP) and DMNX are also called collectively the dorsal vagal complex (DVC) (see Figure 10.25). Some afferents form monosynaptic contacts with the dendrites of the preganglionic neurons in the DMNX that project dorsally into the NTS (Rinaman et al. 1989). However, these monosynaptic contacts seem to be rare compared to the disynaptic contacts via interneurons in the NTS.

Systematic neurophysiological analysis of preganglionic neurons in the DMNX that project in one of the gastric branches of the abdominal vagus nerves have shown that several functional types of preganglionic neuron can be identified by their reflex responses to physiological stimuli applied to stomach, duodenum or small intestine (e.g., distension of organ sections [esophagus, stomach, duodenum], intraluminal chemical stimulation).

Figure 10.24 demonstrates two representative experiments. The preganglionic neurons exhibit various combinations of responses to the physiological stimuli. These patterns are correlated with other properties of the neurons, such as the rate of spontaneous activity, the morphology of dendrites and peptide content. At least four functional types of preganglionic neurons have been identified in the DMNX (see also Chapter 4.8), indicating the functional differentiation of the parasympathetic pathways that are represented in the DMNX and project to the gastrointestinal tract (Zhang et al. 1992, 1995a, 1998; Fogel et al. 1996; Browning et al. 1999).

Synaptic transmission from vagal afferents to the NTS neurons is excitatory and the transmitter is glutamate. In the rostrocaudal area of the NTS where vagal afferents project to some 40% to 50% of the NTS neurons are synaptically activated by electrical stimulation of the subdiaphragmatic vagal nerve. Inhibitory responses are practically absent. These neurons do not receive convergent synaptic input from arterial baro- or chemoreceptor afferents. Surprisingly, some 20% of the NTS neurons receiving synaptic afferent input from the gastrointestinal tract are activated by myelinated fibers (Paton et al. 2000) although only $\leq 1\%$ of the afferent fibers in the subdiaphragmatic vagus nerve are myelinated in the rat (Gabella 1976).

The predominant synaptic transmission between the NTS neurons and the preganglionic neurons in the DMNX projecting to the gastrointestinal tract is inhibitory, the transmitter being mainly GABA. Other synaptic connections between NTS neurons and DMNX neurons are excitatory, the transmitter probably being glutamate. Gastric relaxation induced by esophageal distension is mediated by a reflex pathway involving the central nucleus of the NTS (see Figure 8.11) and the DMNX. The NTS neurons involved in this reflex are noradrenergic and this receptive relaxation reflex is significantly reduced after blockade of α-adrenoceptors (α_1, α_2) in the DMNX. Seventy-five percent of neurons in the DMNX projecting to the stomach are either excited or inhibited by noradrenaline. Thus, noradrenaline is believed to be the transmitter mediating this reflex in the DMNX (Fukuda et al. 1987; Rogers et al. 2003, Martinez-Pena y Valencuela et al. 2004).

For most of these parasympathetic pathways, it is at present uncertain or a matter of speculation which target cells or functionally distinct enteric circuits they innervate (see also Chapter 5). However, it is reasonable to assume that circuits of the enteric nervous system, which are related to non-vascular smooth muscles, exocrine glands, endocrine glands or even the vasculature, receive differential signals from parasympathetic preganglionic neurons in the DMNX. Thus, it is also not far-fetched to assume that the number of functional types of preganglionic parasympathetic neurons in the DMNX is higher than four (see Table 1.2). Based on cyto- and chemoarchitectonic criteria, the DMNX in the human contains nine types of neurons (Huang et al. 1993).

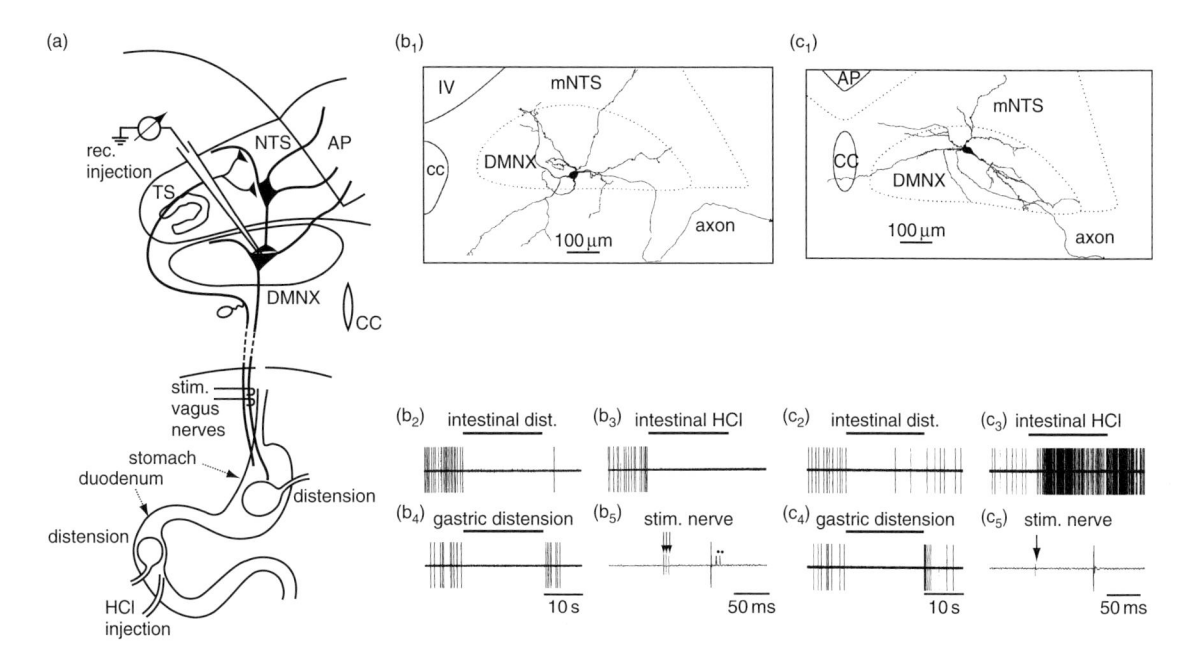

Figure 10.24 Responses of parasympathetic preganglionic neurons in the dorsal motor nucleus of the vagus (DMNX) of the rat to mechanical and chemical stimulation of the gastrointestinal tract and morphological identification of the neurons. (a) Experimental setup. The neurons in the DMNX were recorded extracellularly with glass microelectrodes of 50 to 70 MΩ filled with neurobiotin (2%). After functional characterization of the neuron the micropipette was advanced until the neuron was impaled (drop in membrane potential of 20 to 40 mV) and neurobiotin was injected intracellularly. The neurons were identified by electrical stimulation of their axons in the gastric branches of the abdominal vagus nerves (see [b_5] and [c_5]) with pulses of 0.5 ms duration. For this purpose both gastric branches were put on a stimulation electrode. Vagal afferents were activated by distension of the stomach or duodenum by balloons (intraluminal pressure 12 to 13 mmHg) or by intraluminal injection of HCl into the duodenum. After physiological identification of the neurons by their reflex responses neurobiotin was injected through the recording electrode into the neurons. The rats were perfused, the medulla oblongata was removed and its dorsomedial part containing nucleus tractus solitarii (NTS) and DMNX was cut into serial sections and the neuron was visualized. (b) Spontaneously active preganglionic neuron, which was inhibited by the mechanical and chemical stimuli (b_2 to b_4). (b_1) Demonstrates the morphology of the neuron. (b_5) Demonstrates the antidromic responses of the neuron to a train of three electrical stimuli at 150 Hz to its axon. The responses to the second and third stimulus were only electrotonically conducted (dots) since the action potentials of the axons did not actively invade the soma of the neuron. (c) Spontaneously active preganglionic neuron, which was inhibited by mechanical stimulation and excited by chemical stimulation (c_2 to c_4). (c_1) Demonstrates the morphology of the neuron. (c_5) Demonstrates the antidromic response of the neuron to electrical stimulation of its axon. Note that the dendrites of the neurons project into the NTS and close to the area postrema (AP). CC, central canal; mNTS, medial nucleus of the NTS; TS, tractus solitarii; IV, fourth ventricle. Modified from Zhang et al. (1998) with permission.

10.7.3 Integration in the dorsal vagal complex: a concept

In order to account for the many types of functional reflexes in the preganglionic neurons in the DMNX (as defined by the excitatory or inhibitory effect on the different target cells [different types of smooth muscle cells, exocrine glands, endocrine glands etc.]) by the different types of vagal afferent neuron from the gastrointestinal tract, Powley and coworkers have created an interesting hypothesis on the spatial organization of the DVC (NTS and DMNX) (Powley et al. 1992). This hypothesis is based on the mediolateral representation of the gastrointestinal tract in the DMNX (Figure 8.10 and 8.12) and on the rostrocaudal organization of the projections of vagal afferents to the NTS (Figure 8.12). Powley and coworkers propose that these two layers of the DVC form a sensorimotor lattice as illustrated in Figure 10.25. The two components of the DVC are fused. The proposed sensorimotor lattice would allow for many specific reflexes elicited in the parasympathetic preganglionic neurons by stimulation of distinct types of afferents from the gastrointestinal tract and for various combinations (patterns) of reflexes (Powley *et al.* 1992). The functionally distinct reflex circuits formed in the DVC between the afferents from and the preganglionic neurons to the gastrointestinal tract are the basic building blocks for the brain to control gastrointestinal functions. The idea of the sensorimotor lattice of the DVC as proposed by Powley is a heuristic model and an approximation and may not apply to all situations. For example, esophageal distension elicits proximal gastric relaxation, the basis being the "receptive relaxation reflex" arc consisting of vagal afferents from the esophagus, second-order neurons in the pars centralis of the NTS (NTS_{cen}) that receive synaptic input from these afferents and vagal preganglionic neurons projecting to the stomach (probably two pathways). It has been shown by Rogers *et al.* (1999) that the neurons in the NTS_{cen} project extensively throughout the rostrocaudal DMNX.

Anatomical studies show that several nuclei in the brain stem and forebrain have reciprocal connections with the circuits of the DVC. This situation is conceptually quite similar to the role of spinal autonomic circuits in the control of spinal autonomic final pathways by supraspinal centers (see Chapter 9). Thus, these reflex pathways are under modulatory control of neurons in supramedullary brain centers (so-called "executive" neurons; e.g., in the paraventricular nucleus of the hypothalamus [PVH], the central nucleus of the amygdala [CNA], the bed nucleus of the stria terminalis [BNST], etc.; see Figure 10.26) which also receive detailed afferent information from the gastrointestinal tract (via the NTS) and from other body domains. Executive neurons and basic autonomic circuits in the DVC associated with the gastrointestinal tract represent the *internal state of the organism* as far as the gastrointestinal tract is concerned. This internal state is adapted to the behavior of the organism by cortex and limbic system structures, which monitor and represent the *external state of the organism*. However, the internal state also modulates the central representation of the external state, leading to changes of sensory perception, body feelings and experience of emotions. This concept shows that there is a close integration between the homeostatic regulation of gastrointestinal tract functions and the higher nervous system functions related to body perception, emotions and adaptation of behavior. The basis of this integration is the highly specific autonomic reflex pathways in the spinal cord, brain stem and forebrain. This general concept of the control of gastrointestinal function by the brain has been propagated and worked out by Rogers and coworkers (Rogers and Hermann 1992, 1995; Travagli and Rogers 2001). It is the biological basis for the changes of gastrointestinal function during various behaviors including stress.

10.7 VAGAL EFFERENT PATHWAYS | 445

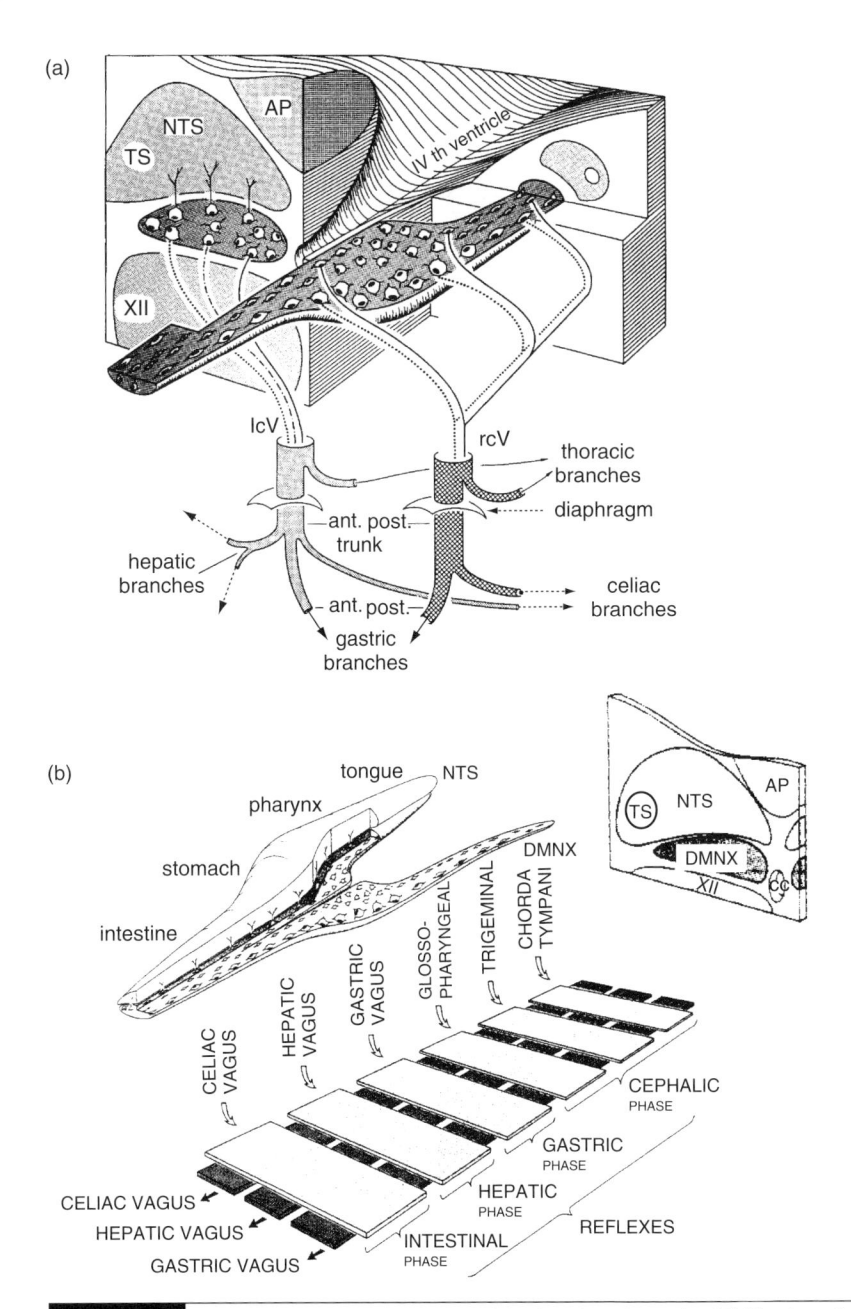

Figure 10.25 Sensorimotor spatial organization of the dorsal vagal complex (DVC): a hypothesis. (a) Schematic view of the DVC: nucleus tractus solitarii (NTS), area postrema (AP), dorsal motor nucleus of the vagus (DMNX) of the medulla oblongata seen from a caudal and lateral perspective. The DVC is shown in a transverse (frontal) section on the left and the DMNX in a longitudinal section on the right. The medial cell columns in the DMNX project in the anterior (ant.) and posterior (post.) gastric branches of the abdominal vagus nerves; the lateral cell columns project in the celiac branches; the intermediate cell column on the left projects in the hepatic branch. The nucleus ambiguus, which is positioned ventrolateral and parallel to the nucleus hypoglossus (XII), is not shown. (b) The sensorimotor lattice of the DVC. Upper left, layers of NTS and DMNX. Upper right, transverse section through the DVC. Lower: mediolaterally organized cell groups in the NTS receiving distinct afferent inputs from different parts of the nasopharyngeal or gastrointestinal tract; rostrocaudally organized columns of preganglionic neurons in the DMNX that project through the different branches of the subdiaphragmatic vagus nerve. This sensorimotor lattice may be the basis for various cephalic, gastric, hepatic or intestinal reflexes. cc, central canal; lcV, left cervical vagus; rcV, right cervical vagus; TS, tractus solitarius. Modified from Powley et al. (1992) with permission.

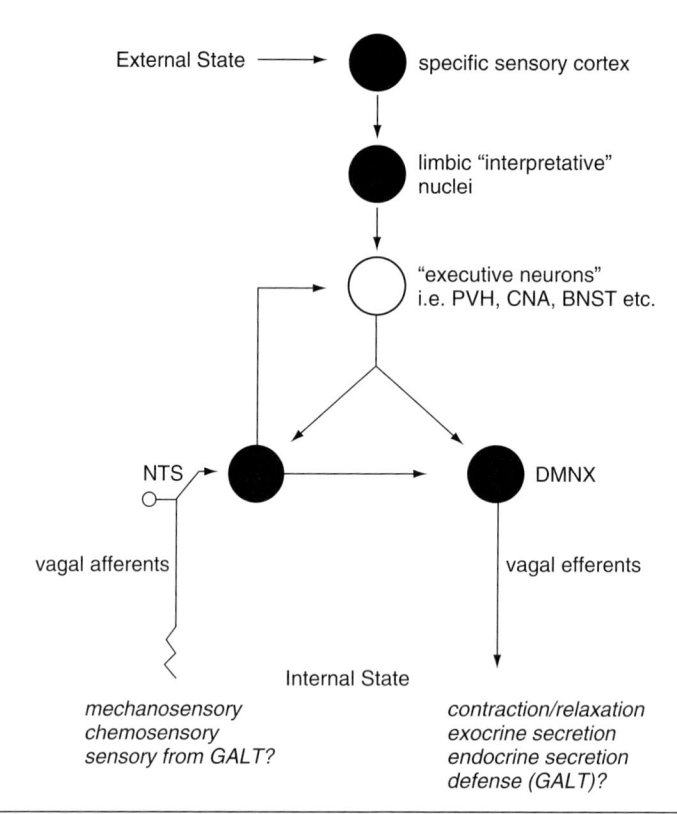

Figure 10.26 Proposed relationship between the gastrointestinal vago-vagal reflex pathways, forebrain autonomic "executive" neuronal circuits, limbic "interpretive" neuronal circuits and exterosensory systems. Several functionally specific vago-vagal reflex pathways are the basic neuronal building blocks. They are mediated through the medulla oblongata. Vagal afferents measure mechano-, chemo- and other sensory events (e.g., sensing events in the gut-associated lymphoid tissue, GALT) and project to the nucleus tractus solitarii (NTS); vagal preganglionic neurons are located in the dorsal motor nucleus of the vagus (DMNX) and are involved in regulation of motility, exocrine secretion, endocrine secretion and other events (e.g., those associated with the GALT). Forebrain "executive" centers (e.g., paraventricular nucleus of the hypothalamus, PVH; central nucleus of the amygdala, CNA; bed nucleus of the stria terminalis, BNST) evaluate the state of the internal milieu (by way of inputs from vagal and other visceral afferents) as well as the current or anticipated behavioral state (via input from limbic nuclei that evaluate the significance of exteroceptive signals). These executive centers adapt the internal state (e.g., gastrointestinal functions) to the behavioral state of the organism. Modified from Rogers and Hermann (1992) with permission.

10.7.4 The area postrema (AP) and the integration of endocrine signals

The reflex neural circuits in the DVC are also modulated by blood-borne signals, which reach the neurons of the DVC by way of the AP and neighboring parts of the DVC. These signals are, for example, hormones from the gastrointestinal tract, such as insulin, GLP-1, PP or PYY. In the AP and neighboring medial parts of the DVC the blood–brain barrier is absent. Thus, these parts of the DVC do not have a vascular diffusion barrier (Figure 10.27) allowing the hormonal signals (peptides) to have free access to the neural circuits of the DVC. The dendrites of both NTS neurons and DMNX neurons project into the area that lacks a blood–brain barrier. The circuits in the DVC are furthermore influenced by neuronal circuits that use neuroendocrine hormones as transmitters or

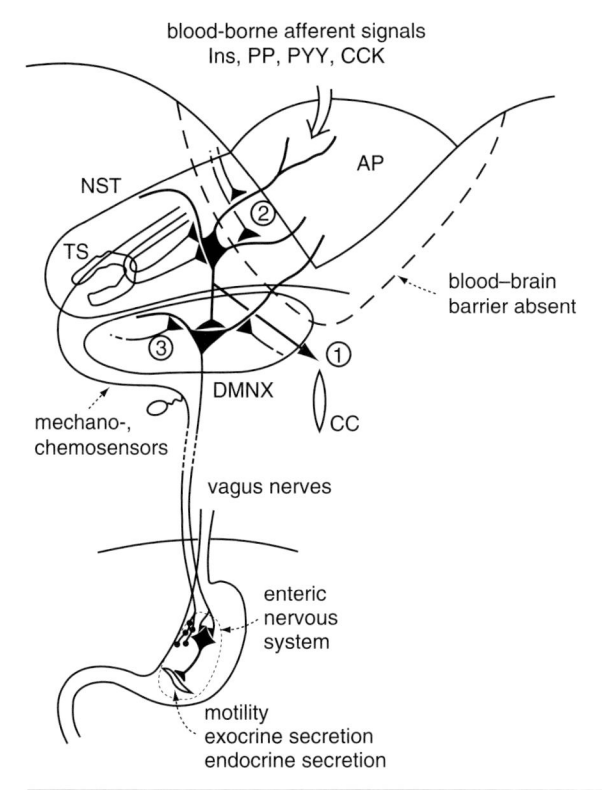

Figure 10.27 Basic vago-vagal reflex circuit involved in the regulation of gastrointestinal functions by the brain are modulated by hormones via the area postrema. Signals in the blood (e.g., hormones) have access to the vago-vagal circuits via the area postrema (AP) and neighboring areas in which the blood–brain barrier is absent. The dorsal vagal complex (DVC) consists of the nucleus tractus solitarii (NTS), the AP and the dorsal motor nucleus of the vagus (DMNX). Neurons in the NTS form inhibitory or excitatory synapses with preganglionic neurons in the DMNX and are synaptically activated from the periphery by mechano- and/or chemosensitive vagal afferent neurons. The preganglionic neurons project through the vagus nerves to the enteric nervous system and have various functions. The NTS neurons project additionally to various regions of the brain stem, hypothalamus and forebrain (*1*; see Figure 8.13). Both neurons in the NTS and neurons in the DMNX are under multiple synaptic control from brain stem, hypothalamus and forebrain (*2* and *3*). CC, central canal; CCK, cholecystokinin; Ins, insulin; PP, pancreatic peptide; PYY, peptide YY; TS, tractus solitarius. Designed after Rogers *et al.* (1995) and Travagli and Rogers (2001).

neuromodulators (such as thyrotropin-releasing hormone [TRH] or corticotropin-releasing hormone [CRH]). The following examples illustrate this integration of neuroendocrine and autonomic functions at the level of the medulla oblongata in the regulation of gastrointestinal functions:

- *Thyrotropin-releasing hormone* induces gastric acid secretion, enhances feeding behavior, activates metabolism, enhances the neural signals to produce cutaneous vasoconstriction and induces shivering by asynchronous firing of motoneurons. This response pattern is typically induced during cold stress and is an expression of the coordination of *thermoregulation* and *metabolic control*. Neurons in the raphe obscurus and in the parapyramidal nucleus of the ventromedial medulla (in which TRH and 5-HT are colocalized) project to the DVC and form synapses with neurons in the NTS and in the

DMNX (Rogers et al. 1980; Yang et al. 2000b). These neurons are activated during cold stress. Thus TRH, both circulatory and neuronally released, appears to be the organizing principle in the metabolic and autonomic responses to cold stress (see Rogers et al. [1995]; Travagli and Rogers [2001]).

- The *PYY* released from endocrine cells in the ileum in response to fatty acids in the lumen produces, via the AP and the DVC, decreased acid secretion and decreased gastric motility. Circulating PYY inhibits neurons in the DMNX that are linked to cholinergic motoneurons of the enteric nervous system innervating smooth musculature or parietal cells. The humoral effect of PYY antagonizes the activation of the preganglionic neurons by TRH/5-HT neurons in the raphe obscurus (see above) (Yang et al. 2000a; see references here).

- *Corticotropin-releasing hormone* injected into the cysterna magna (fourth ventricle) or in the DMNX elicits the same responses of the gastrointestinal tract as *stress* (decreased motility and acid secretion, decreased transit time of the small bowel, increased transit time of the large bowel). Both CRH- and stress-induced effects in the stomach and small bowel are mediated by CRH_2 receptors in the DMNX neurons and by activation of inhibitory (non-cholinergic non-adrenergic) pathways in the enteric nervous system. The neurons of the DMNX are activated during stress by neurons of the paraventricular nucleus of the hypothalamus and by neurons in Barrington's nucleus (see Chapter 8), both releasing CRH (Taché et al. 2001; Lewis et al. 2002).

- *Pancreatic polypeptide* (PP) is released from endocrine cells of the pancreas during fasting in anticipation of feeding (e.g., in the cephalic phase of the activation of stomach and duodenum) and postprandially (after a meal). Its release is primarily under reflex neural control by the DMNX. About 30% of the pancreas-projecting neurons in the DMNX respond to PP, half of them being excited and half of them inhibited (Browning et al. 2005). Pancreatic peptide in physiological concentrations increases gastric motility, gastric acid secretion and gastric transit. These effects are produced by activation of parasympathetic secretomotor and motility-regulating neurons in the DMNX via the AP (see Rogers et al. [1995]; Browning et al. [2005]).

- *Glucagon-like peptide 1* is released by the L cells of the distal gut into the circulation. It has multiple functions (e.g., glucose-dependent stimulation of insulin secretion, inhibition of gastric emptying and acid secretion, reduction of caloric intake, enhancement of satiety). Some of these functions are probably mediated by the AP. Glucagon-like peptide 1 always has excitatory effects on neurons in the DMNX and enhances the currents in the neurons elicited by electrical stimulation of the NTS (Browning et al. 2005).

- Infectious disease states involving activation of the immune system and production of cytokines may lead to autonomic disorders including gastric stasis, nausea, emesis and anorexia. These changes associated with the gastrointestinal tract can be mimicked by the *cytokine tumor necrosis factor α (TNFα)*, which is released from macrophages, T lymphocytes and glia cells during the disease states. Tumor necrosis α inhibits gastric function by acting on neurons in the NTS and DMNX possibly via the AP (Hermann and Rogers 1995; Emch et al. 2000, 2002; Hermann et al. 2002; see Travagli and Rogers [2001]). Alternatively, TNFα may excite vagal afferents, which then excite synaptically the neurons in NTS and DMNX (see Subchapter 2.3, Jänig 2005b).

These examples clearly demonstrate the close integration of autonomic (parasympathetic) systems and neuroendocrine systems in the regulation of gastrointestinal functions.

Conclusions

Homeostatic regulation of arterial blood pressure, respiration and gastrointestinal functions are represented in the lower brain stem. These control systems require temporally precise coordination and adaptation to somatic body functions and are therefore closely integrated. This integration is reflected in the anatomy and physiology of the neural substrates of these homeostatic regulations in the lower brain stem. Included in this integration are the final autonomic pathways (in addition to the enteric nervous system), described in Chapters 4 to 8, and the spinal autonomic circuits (Chapter 9). The circuits in the medulla oblongata are under the control of the upper brain stem, hypothalamus and telencephalon. They are essential building blocks of the complex autonomic control systems integrated with the behavioral repertoires represented in the supramedullary brain centers (see Chapter 11).

1. Neurons involved in regulation of arterial blood pressure (regulation of cardiac output and peripheral resistance) and respiration are situated in rostrocaudally organized columns of neurons in the ventrolateral medulla oblongata (VLM), which extend from the facial nucleus to caudal to the obex. The VLM includes:
   - The rostral ventrolateral medulla (RVLM) containing sympathetic premotor neurons and associated interneurons.
   - The caudal ventrolateral medulla (CVLM) containing excitatory and inhibitory interneurons that mediate various types of cardiovascular reflexes.
   - The caudal pressor area (CPA).
   - The ventral respiratory groups of neurons (VRG) consisting of the caudal VRG (mostly interneurons), the rostral VRG (largely inspiratory premotor neurons), the preBötzinger complex and the Bötzinger complex (largely expiratory premotor neurons).

   The VLM is closely associated with parasympathetic preganglionic cardiomotor and bronchomotor neurons in the external

formation of the nucleus ambiguus. The respiratory neurons in the VLM form together with the respiratory neurons in the dorsal respiratory group and, in the pons, the ponto-medullary respiratory network.
2. The RVLM is a sympathetic cardiovascular premotor nucleus mediating homeostatic reflexes to cardiovascular sympathetic preganglionic neurons (muscle, visceral, renal vasoconstrictor neurons, some cutaneous vasoconstrictor neurons, cardiomotor neurons). Spontaneous activity in the neurons of these cardiovascular pathways possibly originates, at least in part, in the RVLM and/or in neuronal networks associated with the RVLM.
3. Discrete arterial baroreceptor reflexes involve parasympathetic cardiomotor neurons and sympathetic cardiovascular neurons:
    - The pathway to the parasympathetic cardiomotor neurons is disynaptic between the baroreceptor input to the nucleus tractus solitarii (NTS) and the preganglionic neurons. The transmitter at both synapses is glutamate.
    - The pathway to the sympathetic preganglionic neurons consists of four synapses between the baroreceptor input to the NTS and the preganglionic neurons in the spinal cord: interneurons in the NTS project to inhibitory interneurons in the rostral part of the CVLM, these project to the sympathetic premotor neurons in the RVLM. The transmitter of the inhibitory interneuron is GABA, the transmitter at the other synapses is glutamate.
    - Parallel baroreceptor pathways to sympathetic preganglionic neurons may include inhibitory interneurons in the spinal cord, the transmitter being GABA (and possibly glycine), and possibly other pathways (e.g., related to a long descending inhibitory pathway). The circuits of these putative baroreceptor pathways in the medulla oblongata are unknown.
    - All components of the baroreceptor reflexes are under modulatory influence from other nuclei in the lower and upper brain stem, hypothalamus and telencephalon.
4. Excitatory arterial chemoreceptor reflexes to sympathetic cardiovascular preganglionic neurons are mediated both via the respiratory network and independently of the respiratory network.
5. The caudal raphe nuclei of the medulla oblongata (raphe obscurus and pallidus) contain sympathetic premotor neurons to preganglionic cutaneous vasoconstrictor neurons or to preganglionic lipomotor neurons (supplying brown adipose tissue). These premotor neurons are involved in thermoregulation and regulation of energy balance.
6. The activity of many sympathetic neurons exhibits respiratory rhythmicity. This activity profile is related to the phases of respiration and varies with the function of the sympathetic neurons. The following activity patterns are present in the cat:
    - Sympathetic cardiovascular neurons are excited during inspiration and inhibited during postinspiration and sometimes early inspiration.

- Cutaneous vasoconstrictor neurons exhibit several patterns, the most common being inhibition during inspiration and excitation during postinspiration (like parasympathetic cardiomotor neurons); some cutaneous vasoconstrictor neurons are only excited during inspiration and some show no respiratory rhythmicity.
- Sudomotor neurons are excited during postinspiration. Most motility-regulating neurons do not exhibit respiratory rhythmicity.
- Inspiratory neurons projecting to the head (possibly to the nasal mucosa) are only excited during inspiration.

7. Respiratory modulation of activity in sympathetic neurons (muscle vasoconstrictor, cutaneous vasoconstrictor, sudomotor neurons) is also present in humans, showing that there are differences between functionally different types of neurons. However, this needs to be studied systematically under controlled experimental conditions.

8. The respiratory pattern in cardiovascular autonomic neurons is probably mediated by the "common cardiorespiratory network," which is represented in the VLM. This network represents a sensorimotor program, consisting of various functionally distinct reflex pathways, that closely coordinates the regulation of respiration and arterial blood pressure under all physiological conditions. Various types of cardiovascular and pulmonary vagal afferent neurons influence the cardiovascular autonomic neurons both independent of the respiratory network and via this network. The respiratory modulation in the other types of sympathetic neurons is generated by other mechanisms, probably also in the medulla oblongata.

9. Neural control of the functions of the gastrointestinal tract is exerted by multiple reflex circuits consisting of vagal afferents from the gastrointestinal tract that project to the NTS, interneurons in the NTS that project to the dorsal motor nucleus of the vagus (DMNX) and parasympathetic preganglionic neurons in the DMNX that project to the gastrointestinal tract. These reflex circuits are specified by distinct functional types of vagal afferent neurons and by several functional types of preganglionic neuron. The reflex arcs are under modulatory control of supramedullary centers in the upper brain stem and hypothalamus, and involved in behavior by activity from the forebrain.

10. Parasympathetic pathways to the gastrointestinal tract in the dorsal vagal complex (DVC; NTS plus DMNX) are under the control of hormones from the gastrointestinal tract (e.g., peptide YY, pancreatic polypeptide, cholecystokinin [CCK]) and other hormones (corticotrophin releasing hormone [CRH], thyrotropin releasing hormone [TRH]) via the area postrema. The hormonal modulation of the circuits in the DVC occurs in relation to satiety, anticipation of feeding, cold and other stressors, and defense against the entry of toxic substances via the gastrointestinal tract.

Notes

1. Genes, such as the *c-fos* gene, that are activated rapidly within minutes, transiently, and without requiring new protein synthesis are often described as cellular immediate early genes (IEGs). Under unstimulated (resting) conditions *c-fos* mRNA and protein are nearly undetectable. Neural stimuli that activate the cAMP, Ca^{2+}, protein kinase C or other intracellular pathways produce a rapid induction of *c-fos* gene expression. Detection of c-Fos and other IEG-encoded proteins and *in situ* hybridization for the mRNA of these proteins using immunohistochemistry have become standard tools to map neurons and neural circuits that are activated (synaptically, by drugs or otherwise) (see article by Schulman and Roberts in Squire et al. [2003]; Dampney and Horiuchi [2003]).
2. Bulbospinal catecholaminergic neurons (including the C1 neurons in the RVLM) can be selectively destroyed by using the ribosomal inactivating protein (toxin) saporin conjugated to an anti-dopamine-β-hydroxylase (DβH) antibody. This immunotoxin injected in the upper thoracic segments in rats (close to the intermediolateral cell column) binds to DβH, which is expressed in the plasma membrane of the terminals of adrenergic and noradrenergic neurons. The toxin–DβH complex is internalized and retrogradely transported to the cell bodies of these catecholaminergic neurons in the lower brain stem. The neurons are selectively destroyed by blockade of protein synthesis without affecting other bulbospinal neurons (Guyenet et al. 2001). By the same token can neurons with NK_1-receptors be selectively ablated by a conjugate of the toxin saporin with a selective NK_1-receptor agonist (e.g. substance P). The saporin–agonist–NK1-receptor complex is internalized by the neurons and leads to their destruction (Mantyh et al. 1997).
3. Morphological identification: activity is recorded intracellularly from the neurons, to functionally characterize them, and then the neurons are filled with a marker (e.g., horseradish peroxidase, Lucifer yellow). In a recently developed technique identified neurons can be labeled juxtacellularly by a dye (e.g. biotinamide) released from a microelectrode positioned extracellularly close to the cell body or a large dendrite of the neuron (Pinault 1996; Pilowsky and Makeham 2001). The labeled neurons are reconstructed from serial sections of the perfused and fixated preparation after the physiological experiment.
4. The demarcation of the VLM and its subsections may differ between different investigators. Most workers place the dorsal border of the VLM as the dorsal border of the nucleus ambiguus (NA), others use the NA pars compacta (thus, including the parasympathetic cardiomotor and bronchomotor neurons). For practical reasons the medial border is taken as a line drawn from the pars compacta of the NA to the point where the pyramidal tract touches the medullary surface laterally; the lateral border is taken as a line from the pars compacta of the NA to the ventral edge of the trigeminal nucleus; the ventral border is the ventral surface of the medulla oblongata. This area includes neurons of the ventral respiratory group (Pilowsky, personal communication).
5. Several landmarks are used to locate the groups of neurons in the medulla oblongata: (1) In the stereotaxic atlas of Paxinos and Watson (1998) the rostrocaudal position is located with respect to the bregma. This is the point of the skull corresponding to the junction of the coronal and sagittal sutures. (2) A ventral landmark for the VLM and the column of respiratory neurons in the ventrolateral column is the caudal border of the facial

nucleus. (3) Dorsal landmarks commonly used are the calamus scriptorius or the obex. The calamus scriptorius is the caudal part of the floor of the fourth ventricle towards its apex. This part presents the appearance of a pen nib; therefore it is called calamus scriptorius. The obex is the point on the midline of the dorsal surface of the medulla oblongata that marks the caudal angle of the fourth ventricle. Anatomically it corresponds to a small transverse medullary fold and is therefore rostral to the calamus scriptorius. Most workers define the obex operationally as the rostral border of the area postrema in the midline (Figures 10.2, 10.3).

6. In the cat, the sympathetic premotor neurons involved in regulation of arterial blood pressure are located in a compact nucleus of the RVLM, sometimes called the subretrofacial nucleus (Polson *et al.* 1992), that is not present in the rat. For numbers of bulbospinal neurons in the RVLM in rat and cat that are related to sympathetic preganglionic neurons see note 10.

7. The delineation of the cell columns in the transverse sections in Figure 10.2b were only done for didactic reasons and according to Paxinos and Watson [1998].

8. There exist conflicting data concerning the effect of bilateral lesion of the RVLM on arterial blood pressure in rats: (1) In the awake rat after bilateral electrolytic lesion of the RVLM mean arterial blood pressure and heart rate were not decreased five days after lesioning. Blockade of the renin–angiotensin system by inhibition of the angiotensin-converting enzyme did not change mean arterial blood pressure and heart rate. Blockade of impulse transmission in autonomic ganglia reduced both (Cochrane and Nathan 1989). (2) After bilateral lesion of the RVLM in the anesthetized rat, the blood pressure may not entirely fall acutely to the level that is observed after complete transection of the cervical spinal cord or after complete blockade of transmission in autonomic ganglia. This has several reasons: (a) the extent of the lesion or the inhibition of the neurons in the RVLM is not complete; (b) other sympathetic premotor neurons may contribute to the activity in vasoconstrictor neurons; (c) the renin–angiotensin system and the vasopressin system may be indirectly or reflexly activated (Cochran and Nathan 1993).

    It is important to emphasize that decrease of activity in vasoconstrictor neurons innervating resistance vessels or in sympathetic cardiomotor neurons upon bilateral lesion of the RVLM does not prove that the ongoing activity originates *only* in the RVLM. In fact, multiple nuclei in the brain stem and hypothalamus and even the spinal cord may contribute to the generation of activity in these cardiovascular neurons (see Subchapter 10.2.4).

9. After the toxin saporin conjugated to an anti-DβH antibody is injected in the upper thoracic segments in rats (close to the intermediolateral cell column; see note 2) about 75% to 85% of the C1 neurons in the RVLM (together with other bulbospinal adrenergic neurons and with >95% of bulbospinal noradrenergic neurons [for example in the area A5]) are lost (Schreihofer and Guyenet 2000). Rats with depleted bulbospinal C1 cells in the RVLM have normal arterial blood pressure, probably normal level of sympathetic activity in the major splanchnic nerve and almost normal baroreceptor reflexes. However, they exhibit reduced excitatory reflexes in sympathetic neurons to stimulation of arterial chemoreceptors and to stimulation of the RVLM (Schreihofer *et al.* 2000; Madden and Sved 2003). Thus, the integrity of the C1 neurons is not essential for the generation of activity in sympathetic vasoconstrictor neurons that are involved in

blood pressure regulation and maintenance of arterial blood pressure under resting conditions. These functions are primarily maintained by non-catecholaminergic bulbospinal neurons in the RVLM (Guyenet et al. 2001). However, the C1neurons may well be essential for the regulation of arterial blood pressure under extreme conditions when the system is under long-term load.

10. In the rat, the RVLM contains about 600 to 800 bulbospinal sympathetic premotor neurons on both sides. From these bulbospinal neurons about 400 to 550 neurons are adrenergic (C1) and the remaining neurons non-adrenergic (assuming that two-thirds of these bulbospinal neurons are adrenergic and one-third non-adrenergic [Sved et al. 1994; Schreihofer and Guyenet 1997]). These estimated numbers are based on counts of neurons retrogradely labeled from the spinal cord (upper thoracic segments) in serial transverse sections through the VLM (830 neurons: Schreihofer and Guyenet 2000; Guyenet et al. 2001; 800 neurons: Phillips et al. 2001; 760 neurons: Sved et al. 1994; 560 neurons: Tucker et al. 1987). Assuming that these sympathetic premotor neurons are target-specific (Figure 10.6) it follows that in the range of 100 to 150 sympathetic premotor neurons determine the activity in each sympathetic pathway to the different vascular beds (muscle, mesenteric, renal, skin), to the heart and to the adrenal medulla. Although an even distribution of sympathetic premotor neurons is unlikely it shows that a large number of sympathetic preganglionic neurons (in the rat about 12 000 neurons on both sides, of which 20% or less may have a cardiovascular function or be related to the adrenal medullae [Strack et al. 1988; see Table 4.5 for the estimation of the proportion of preganglionic muscle or visceral vasoconstrictor projecting in different "sympathetic nerves"]) is (directly or indirectly) innervated by a small number of RVLM neurons.

    The subretrofacial nucleus in the RVLM of cat contains about 2000 sympathetic premotor neurons on both sides (Polson et al. 1992; McAllen, personal communication).

11. Intrinsic pacemaker properties of neurons are characterized by the following criteria (Guyenet 1990): (1) The spikes result from gradual interspike membrane depolarizations as opposed to regularly occurring excitatory postsynaptic potentials (EPSPs). (2) After hyperpolarizing the neurons by negative current injection pacemaker activity stops but EPSPs are still present. (3) The pacemaker activity can be reset by a brief hyperpolarizing current or by an action potential generated by a brief depolarizing pulse or by a synaptically induced EPSP. (4) The pacemaker activity does not depend on synaptic release of a neuromodulator.

12. The network idea to explain the generation of spontaneous activity in cardiovascular neurons appears to be least likely. It may turn out to be difficult to prove experimentally that the spontaneous activity in sympathetic cardiovascular neurons is generated by a network of neurons connected by excitatory and inhibitory synapses, independent of synaptic and other inputs from outside of the network. For example, the extracellular fluid matrix of the neurons provides continuous chemical inputs to the neurons.

13. Using spectral analysis of integrated activity in the renal nerve of conscious rats only rhythms related to fluctuation of heart rate, frequency of respiration and frequency of heart beat and their harmonics were found. After denervation of the arterial baroreceptors all frequency components (except an occasional component related

to central breathing) in the activity of the phrenic nerve were abolished and the power spectrum was almost flat (Kunitake and Kannan 2000). The discrepancy in results compared with other investigations (see Gebber [1990]; Malpas [1998]; Barman and Gebber [2000]) is considered to be related to the method of signal processing for spectral analysis.

14. In the experiments conducted by Thrasher and Shifflet (2001) one set of arterial baroreceptors was left intact. Chronic occlusion of the common carotid artery proximal to the intact baroreceptor afferents abolished the rhythmic diastolic–systolic changes of the arterial blood pressure in the innervated carotid sinus, but did not lead to a significant decrease of level of arterial blood pressure in the carotid sinus distal to the occlusion. Thus, in this experiment the rhythmic activation of the arterial baroreceptors was eliminated. In the experiments conducted by Cowley et al. (1973) all arterial baroreceptors were denervated. Chronic denervation may lead to plastic changes in the central baroreceptor pathways (e.g., in the NTS), which may explain why the mean arterial blood pressure in this preparation did not change. This clearly shows the model of Cowley et al. (1973) is *not* a model of chronic baroreceptor unloading (Thrasher 2004). It is worth mentioning that the interpretation given by Cowley and coworkers to their experiments were generally accepted by the scientific community. This is documented in the textbooks of physiology for students all over the world and needs to be revised. This situation demonstrates how careful we have to be in the interpretation of experimental results as far as the normal functioning (here arterial baroreceptor reflexes) is concerned.

15. In the experiment documented in Figure 10.4c the phasic inhibition of the activity in the bulbospinal RVLM neuron starts about 50 ms after the onset of the systole. This delay of inhibition is due to the conduction time in axons of the baroreceptors, NTS neurons and neurons in the CVLM (Jeske et al. 1993; see Figure 10.7).

16. Most NTS neurons that are activated by stimulation of arterial baroreceptors have rather weak pulse rhythmicity in their activity or some even none (Spyer 1981; Mifflin et al. 1988; Rogers et al. 1993). I find this rather puzzling and an unsolved problem for the following reasons: (1) Strong pulse rhythmicity of the activity in arterial baroreceptors with myelinated axons. (2) Strong pulse rhythmicity of the activity in all neurons of the baroreceptor reflex pathways downstream to the NTS (CVLM neurons: Figure 10.10; RVLM: Figure 10.4; pre- and postganglionic muscle and visceral vasoconstrictor neurons: Figures 4.1; 4.6, 4.7).
(3) In anesthetized cats pulsatile stimulation of only relatively few intact baroreceptor afferents innervating the common carotid artery (e.g., after transection of both carotid sinus nerves and both vagoaortic nerves) seem to be enough to generate cardiac rhythmicity of the activity in muscle vasoconstrictor neurons (Jänig, unpublished). The traditional explanation is that there is considerable convergence of baroreceptor afferents on NTS neurons and of NTS neurons (or interneurons synaptically connected to NTS neurons) on neurons in the CVLM. I remain skeptical whether the model outlined in Figure 10.7 can sufficiently explain the experimental observations made on sympathetic cardiovascular neurons.

17. Based on experiments performed on *cats*, Barman and Gebber propagate the idea that neurons in the so-called lateral tegmental field (LTF) of the

medulla oblongata (which may correspond to the dorsal part of the medullary reticular nucleus in the rat, located medial to the spinal trigeminal nucleus; see Figure 10.2b) mediate baroreceptor reflexes to neurons in the RVLM. These LTF neurons excite neurons in the RVLM and are inhibited upon stimulation of arterial baroreceptors. Thus, according to these authors, presympathetic neurons in the RVLM are disfacilitated during stimulation of arterial baroreceptors by removal of excitatory drive from LTF neurons. Details about the putative baroreceptor reflex pathway via the LTF neurons are unknown. The depression of activity in the LTF neurons generated by stimulation of arterial baroreceptors does not occur by direct inhibition of the LTF since selective blockade of the N-methyl-D-aspartate (NMDA)-receptors in the LTF prevents the inhibition of LTF neurons (and of the activity in sympathetic cardiovascular neurons) generated by stimulation of arterial baroreceptors (for review see Barman et al. [2001]). The view of Barman and Gebber is not accepted or ignored by most investigators. However, also this putative alternative baroreceptor pathway to sympathetic premotor neurons should not be dismissed.

18. This idea appears to be at variance with the findings of Verberne et al. (1999): (1) Spinally projecting neurons in the RVLM in rats, that are activated by stimulation of arterial baroreceptors and would qualify as putative sympathoinhibitory GABAergic neurons were not found in the RVLM. (2) Neurons in the VLM that were activated by stimulation of arterial baroreceptors were located in the CVLM and did not project to the spinal cord. (3) Baroreceptor-activated neurons were intermixed with C1 neurons in the caudal VLM that project to the hypothalamus (or possibly to telencephalic structures) and are inhibited by baroreceptor stimulation.

    Furthermore, Pilowsky et al. (1994) did not find IPSPs in lumbar preganglionic neurons in cats with intact arterial baroreceptors that are correlated with blood pressure changes (increase of arterial blood pressure and activation of arterial baroreceptors). However, it must be kept in mind that preganglionic vasoconstrictor neurons to skeletal muscle and viscera in the lumbar spinal cord, which are involved in blood pressure control, (1) comprise only about 10% to 15% of all preganglionic neurons (Table 4.5) and (2) have rather slowly conducting axons (mean 3.4 ± 1.9 m/s [mean $\pm$ SD] for preganglionic muscle vasoconstrictor axons and 2.8 ± 2.5 m/s for preganglionic visceral vasoconstrictor axons [Jänig and Szulczyk 1980; Jänig 1985a; Bahr et al. 1986c]). Thus, the cell bodies of these neurons are probably also very small. For these reasons it is difficult to detect these neurons in the spinal cord and to record intracellularly.

19. See note 1 in Chapter 3.
20. This is a short-lasting strong experimental anoxic adequate stimulus that does not occur under physiological conditions but stimulates specifically arterial chemoreceptors.
21. This switch from inhibitory chemoreceptor reflexes to excitatory ones, in the anesthetized cat, in cutaneous vasoconstrictor neurons supplying the cat hindpaw, following decerebration of the animal or peripheral nerve lesion, is accompanied by a switch from inhibitory to excitatory reflexes in cutaneous vasoconstrictor neurons to noxious stimulation of the hindpaw (heating of the contralateral hindpaw with water $>45\,°C$ or noxious mechanical stimulation of the ipsilateral hindpaw [Jänig 1975; Blumberg and Jänig 1985]). These experimental results are reminiscent of

observations in patients with complex regional pain syndrome (CRPS). Many patients having CRPS I (previously called reflex sympathetic dystrophy; Stanton-Hicks et al. 1995; Jänig and Stanton-Hicks 1996; Harden et al. 2001) exhibit chronic changes in thermoregulatory reflexes and respiration-related reflexes in cutaneous vasoconstrictor neurons innervating the distal extremities. The changed reflexes in cutaneous vasoconstrictor neurons are most likely due to changes of central cutaneous vasoconstrictor circuits and related to the injury that triggers the development of CRPS type I. They may reverse after successful treatment of this pain disease (Wasner et al. 1999, 2001; Jänig and Baron 2002, 2003). Patients with CRPS II, which sometimes develops after a trauma with nerve lesion (previously called causalgia), may also exhibit dramatic changes at their distal extremities that are related to the sympathetic nervous system (cutaneous vasoconstrictor neurons, sudomotor neurons). These sympathetically maintained changes developing after nerve lesions have already been described by Mitchell (1872), who coined the term causalgia. The experimental approach to study the plastic changes of reflexes in cutaneous vasoconstrictor neurons following nerve lesions was designed on the basis of these clinical observations (Jänig 1985b, 1990b; Jänig and McLachlan 1994).

22. Note that, in the experiments documented in Figure 10.15, the multiunit activity in the splanchnic nerve (SPL, most of it occurring in visceral vasoconstrictor axons) exhibits typical bursting activity (Figure 10.15b_1–b_3, c_1). The multiunit activity in the nerve to the interscapular brown adipose tissue (BAT; most of it occurring in lipomotor axons) also exhibits bursting activity during hypothermia (Figure 10.15b_2) and during disinhibition of the nucleus raphe pallidus (Figure 10.15b_3, c_3). The frequency of this bursting activity is different from that in the splanchnic nerve (Figure 10.15b_2, b_3) (Morrison 1999) (see Chapter 10.2.5 for discussion of the bursting activity).

23. The nomenclature of these respiratory neurons varies somewhat between groups. The main types are: ramp-inspiratory, late inspiratory, pre-inspiratory, early inspiratory, postinspiratory, augmenting expiratory, inspiratory-expiratory phase-spanning, expiratory-inspiratory phase-spanning neurons (Richter 1996; St.-John 1998; Richter and Spyer 2001; Rybak et al. 2004).

24. In the rat the pattern of respiratory modulation of activity in sympathetic neurons is different. It mostly consists of depression of activity in inspiration and sometimes of enhanced activity in postinspiration depending on the respiratory drive (Häbler et al. 1994a, b, 1996, 1999a, 2000; Bartsch et al. 1996).

25. Daly and coworkers showed that activation of slowly adapting pulmonary afferents by lung inflation leads to a decrease of vascular resistance in the vascularly isolated dog hindlimb (Daly et al. 1967; Daly and Robinson 1968; Daly 1991). Thus, lung inflation is followed by a decrease of activity in muscle vasoconstrictor neurons and subsequently a decrease in vascular resistance. The most plausible explanation of this observation is that activation of lung afferents leads to a decrease of activity in inspiratory neurons (Hering–Breuer reflex) and this in turn leads to a decrease of activity in the muscle vasoconstrictor neurons (Gerber and Polosa 1978). This interpretation is contradicted by Daly et al. (1987). However, there probably is another simple reason explaining the apparent difference between the results of Boczek-Funcke et al. (1992b, c)

and those of Daly and coworkers: in the experiments conducted by Boczek-Funcke *et al.* lung stretch afferents were stimulated phasically at low intrapulmonary pressure whereas in the experiments of Daly *et al.* lung afferents were stimulated tonically at high intrapulmonary pressure. Thus, the results obtained in both types of experiment do not contradict each other.

26. The early recording of Hagbarth and coworkers in humans (Hagbarth and Vallbo 1968; see Figure 10.19a) clearly demonstrates that activity of MVC neurons occurs during inspiration (compare activity in MVC axons and EMG of inspiratory muscle in Figure 10.19a and take into account that postganglionic activity is delayed by about 2 s with respect to the EMG activity due to conduction time of the action potentials in the axons of the postganglionic muscle vasoconstrictor neurons innervating the peroneal muscles).

27. This interpretation of the changes of activity in muscle vasoconstrictor neurons with respiration in the human has been challenged by Seals *et al.* (1990, 1993) and Macefield and Wallin (1995). Their studies show that maximal activity in muscle vasoconstrictor neurons occurs in late expiration and early inspiration and minimal activity in late inspiration and early expiration. This pattern is seen during normal breathing, hyperpnea, passive positive pressure ventilation as well as in lung denervated human subjects and does not depend on unloading of baroreceptors. These data appear to be at variance with *all* published animal data showing that the central component of respiratory modulation in muscle vasoconstrictor neurons has an inspiratory peak in cats and a postinspiratory peak in rats (Häbler *et al.* 1994a).

Chapter 11

Integration of autonomic regulation in upper brain stem and limbic-hypothalamic centers: a summary

11.1 Functions of the autonomic nervous system: Cannon and Hess	page 460
11.2 General aspects of integrated autonomic responses	469
11.3 Autonomic responses activated quickly during distinct behavioral patterns	474
11.4 Emotions and autonomic reactions	491
11.5 Integrative responses and the hypothalamus	498
11.6 Synopsis: the wisdom of the body revisited	507
11.7 Future research questions	510

The final chapter will describe how integrative neural control of most body functions is vital to keep the body able to survive and act in its environment. The autonomic nervous system is involved in virtually all of these functions (see Tables 11.1 and 11.2). I want to make clear:

- that the power and range of this integrative control of body function in mammals are dependent on the mesencephalon, hypothalamus and cerebral hemispheres;
- that the mastermind of the integration of autonomic, somatomotor and endocrine systems is located in the telencephalon;
- that the functionally differentiated autonomic pathways are the slaves of this mastermind, and
- that the "wisdom of the body" is to be found within these regions of the brain.

I will strictly adhere to the autonomic systems, described in the preceding chapters, in this description and will emphasize some critical points. I do not intend to describe the mechanisms underlying these integrative control systems in detail (as this would require another book). The chapter will not cover (1) how stress involves the autonomic nervous system in body protection (see Goldstein [1995, 2000]; Chrousos [1998]; McEwen [2001a]) and (2) neural mechanisms underlying emotional and motivational processes (Le Doux 1996; Panksepp 1998; Davidson et al. 2003; Morris and Dolan 2004).

The chapter will start with some critical reflections on concepts about the functioning of the autonomic nervous system that were propagated by Walter Bradford Cannon and Walter Rudolf Hess[1] and are still influential in physiology and medicine. The chapter will finish with a synopsis in which I want to show that we are, again, at a new beginning in the exploration of the autonomic nervous system, with new experimental tools that have been developed in the last 10 to 20 years, and that the main focus will be on the central nervous system.

11.1 Functions of the autonomic nervous system: Cannon and Hess

In the introduction of this book, I alluded to the very wide range of vital functions that are under the control of the autonomic nervous system. Here I will critically discuss the general concepts of the function of the autonomic nervous system as they have evolved from the experimental work of Walter Bradford Cannon and Walter Rudolf Hess and others in the first half of the last century and that still have considerable impact on our conceptual approach to this system.

The reason why I focus on Cannon and Hess is clear: both analyzed the functions of the autonomic nervous system from an integrative point of view, i.e. implicit in their descriptions and reasoning was the assumption that circuits in the central nervous system are responsible for the integrative action of the autonomic nervous system. The development of our knowledge on anatomy, physiology and pharmacology of the peripheral autonomic nervous system has been extensively described in the literature (Gaskell 1916; Sheehan 1936, 1941; Kuntz 1954; Pick 1970). Aspects related to the development of knowledge on special topics and ideas have been discussed in the earlier chapters of this book.

11.1.1 Cannon and the concept of the sympathico-adrenal system

In the second half of the nineteenth century, *Claude Bernard* formulated the idea that:

> It is the fixity of the 'milieu intérieur' which is the condition of free and independent life and that all the vital mechanisms, however varied they may be, have only one object, that of preserving constant the conditions of life in the internal environment
>
> (quoted from Cannon, 1939, p. 38).

The American physiologist *Walter Bradford Cannon* was strongly influenced by the ideas of Claude Bernard. He called the milieu intérieur the *fluid matrix* of the body. Cannon described the coordinated physiological processes that maintain the steady state of the organism with the term *homeostasis* (see Subchapter 11.2). He was convinced that the automatic corrections of the physiological parameters of the body are the primary function of the autonomic nervous system, in particular the sympathetic nervous system.

Cannon's ideas about how homeostasis is achieved and the role the autonomic nervous system plays were first presented in a review "Organization for physiological homeostasis" (Cannon 1929a) and culminated in his famous and influential book *The Wisdom of the Body* (Cannon 1939). The title for this book was taken from the late E. H. Starling who gave a Harvey Lecture with the same title before the Royal College of Physicians in London in 1923. Starling declared that, by understanding the wisdom of the body, we shall attain the *mastery of disease and pain which enables us to relieve the burden of mankind* (quoted from the preface to Cannon's book *The Wisdom of the Body*, 1939), and that tenet became also the belief of Cannon.

Cannon was originally influenced by ideas about the role of the sympathetic nervous system in strong emotions, pain and stress. As an undergraduate he was in William James' philosophy course at Harvard. James contended that the emotional state of the organism is associated with afferent feedback from the body, notably from vascular and visceral structures (James 1994). He believed that the brain triggers bodily changes by activity in the autonomic nervous system (particular the sympathetic nervous system) and that the activity initiated in the afferent neurons from the regulated organs leads the emotions (see Subchapter 11.4).

Cannon was intrigued by this theory, which states that the brain generates activity of the internal organs (including blood vessels) via the autonomic nervous system and of skeletal muscle and that the various emotional states are brought about by afferent signals from these organs. The consequence of this idea is that different emotions are generated by different patterns of activity in afferent neurons from the internal organs. Interestingly, this would strictly require that the autonomic efferent pathways are functionally specific. If this were not the case, it would not be possible to generate the distinct basic emotions by functionally distinct patterns of afferent discharge arising from internal organs.

However, Cannon critically argued:

> If various strong emotions can thus be expressed in the diffuse activities of a single division of the autonomic nervous system ... it would appear that the bodily conditions which have been assumed, by some psychologists, to distinguish emotions from one another must be sought for elsewhere than in the viscera. We do not "feel sorry because we cry", as James contended, but we cry because, when we are sorry or overjoyed or violently angry or full of tender affection, – when any of these diverse emotional states is present – there are nervous discharges by sympathetic channels to various viscera, including the *lachrymal* glands [tear formation is generated by activation of parasympathetic neurons!]. And in terror and rage and intense elation, for example, the responses in the viscera seem too uniform [and therefore the discharges of

sympathetic neurons to various target organs] to offer a satisfactory means of distinguishing emotional states which in man, at least, are subjectively very different. For this reason I am inclined to urge that the visceral changes merely contribute to an emotional complex more or less indefinite, but still pertinent, feelings of disturbance, in organs which we are not usually conscious of

(Cannon 1914a, p. 280).

Instead Cannon proposed (Cannon 1914a) that the different emotional states are represented in the brain rather than being peripheral in origin, and are expressed by changes of activity in sympathetic and parasympathetic neurons.

Later in the 1920s, this reasoning led to the famous experiments conducted by Philip Bard on diencephalic cats in Cannon's laboratory, in which he had removed the cortex and most of the structures of the limbic system. These cats exhibited a behavior during stimulation of the skin that was phenomenologically very much reminiscent of a cat in rage (see Figure 4.12) as was so beautifully described by Charles Darwin (1872) in the last century in his book *The Expression of the Emotions in Man and Animals*. Bard and Cannon called the behavior of their diencephalic cats "sham rage behavior" (Bard 1928). These cats exhibited reactions with typical somatomotor components (tail arched backward, everted claws, hissing) and responses in target organs that are under control of the sympathetic nervous system (piloerection, dilatation of pupil, sweating of paw pads). Had they recorded other autonomic parameters, they would have observed increase in arterial blood pressure, heart rate and blood flow through skeletal muscle and decreased blood flow through viscera and skin, decreased motility of the gastrointestinal tract and increased secretion of adrenaline and noradrenaline by the adrenal medulla. These are all effects generated by activation of sympathetic pathways (see Table 1.2), but they do not involve the entire sympathetic outflow. These changes are also associated with activation of the adrenal cortex via the anterior pituitary gland and an increase of corticosterone in the blood. On the basis of their experiments, Bard and Cannon created the *thalamic theory of emotions* (Cannon 1927, 1929b; Bard 1932).

Cannon obtained the first experimental experience of the powerful influence the sympathetic nervous system can have on body functions when he studied movements of the stomach and intestines in conscious cats using X-rays. He was surprised to see that the movements of the gastrointestinal tract ceased under strong emotional stimuli, since the cats were frightened after having been fixated and forced to swallow barium solution. However, when the cats were pacified or asleep the movements of the intestine recommenced. He attributed the cessation of the movements of the gastrointestinal tract to the activation of the sympathetic nervous system (Cannon and Murphy 1906; Cannon 1911). Then, over the years, in experiments on cats, dogs and rabbits, Cannon studied the role of the sympathetic nervous system in maintaining homeostasis during various disturbances, such as hemorrhage, hypoglycemia, hypoxia, low and high body temperature, muscle exercise, emotional disturbances. On the basis of these studies, he formulated the concept of a fundamental role of the sympathetic nervous system in maintaining homeostasis: The sympathetic nervous system acts promptly and directly to prevent changes of the internal environment. It exhibits a widespread discharge through the sympathetic channels and different sympathetic outflows act simultaneously in one direction. It is organized for diffuse effects (Cannon 1939).

This generalization was extensively discussed by Cannon in his book *The Wisdom of the Body* (1939). Cannon obviously did not believe that individual sympathetic preganglionic neurons make only functional synaptic contacts with postganglionic neurons subserving the same function but, rather, that they diverge widely and form contacts with postganglionic neurons with many different functions.[2] Generalized activation of the sympathetic nervous system included activation of the adrenal medulla causing the secretion of adrenaline and noradrenaline into the blood. It was assumed that the circulating adrenaline and noradrenaline reinforce the nervous effects on the target organs and mobilize glucose and free fatty acids from their stores, decrease the time for blood clotting, enhance gas exchange in the lung (by relaxation of the smooth muscles of the airways and subsequent reduction of airway resistance) and decrease fatigue of skeletal muscle.[3] These broad functional effects were conceptualized under the term *sympathico-adrenal system* (see Subchapter 4.5).

In contrast, the parasympathetic functions were thought to be more specific. This system serves to conserve body energies and the stability and constancy of the internal milieu of the body. It influences special viscera differentially and its discharges are directed to specific organs only. Different types of parasympathetic neurons are therefore not bound to act simultaneously but separately depending on the organ. Individual parasympathetic preganglionic neurons influence one target organ only. The effects of the sympathetic nervous system and of the parasympathetic nervous system are generally opposite in the same organ (Cannon 1939). *This view of Cannon's, his contemporaries and followers is grossly at variance with our modern view as outlined particularly in Chapter 4: Activation of lacrimal glands and pelvic organs is not related to conservation of body energies. Sympathetic neurons are functionally as specialized as parasympathetic ones. Functionally different types of sympathetic neurons do not act simultaneously but separately, dependent on the organ. Individual sympathetic neurons influence one group of target cells only. The effect of sympathetic neurons and parasympathetic neurons are not opposite on most target cells (see Table 1.2).*

Cannon's view of the autonomic nervous system was that of a system designed to preserve life during grave physical crises requiring extreme effort. The sympathetic division of the autonomic nervous system was considered to mobilize bodily forces during struggle, the cranial (parasympathetic) division to preserve body energies and the sacral (parasympathetic) division to function in emptying of the hollow organs and reproduction of the species (Cannon 1928, 1929a). Animals from which he had removed the entire sympathetic paravertebral chains survived, suggesting that the sympathetic nervous system might not be important at all. However, these animals lived in the protected environment of the laboratory. They would not have been able to adapt to environmental extremes, or even to maintain physiological stability in terms of body temperature, adequate arterial blood pressure for cerebral perfusion, constant fluid volume, and so forth, under more normal conditions although this has never been experimentally tested as far as I know (Cannon et al. 1929). In this context I remind the reader that the regulations exerted by the autonomic nervous system and the endocrine system were "designed" by nature (through natural selection during development) to cope with the extremes (e.g., living in the tropics, in Siberia, in the desert; running marathon etc.).

Cannon was aware that the autonomic nervous system is active during small disturbances. Cannon's idea of synchronized sympathetic activity in the *fright, fight and flight responses* (Cannon 1929a, 1939) is what we would call today the *defense reaction* (see Subchapter 11.3); however, this idea was readily picked up by the scientific and clinical community, and even by lay people. The coordinated response was taken to indicate that activity of all parts of the sympathetic system was linked so as to occur in an *all-or-none* fashion without differentiation between the different effector organs. This activation of the sympathetic system was thought to be generally protective and the level of arousal to be expressed in the level of sympathetic discharge or *sympathetic tone*.

Cannon himself was surprised that the same unified action of the sympathetic nervous system could be useful in circumstances as diverse as hypoglycemia, hypotension, hypothermia and so forth. He was aware that the unified system apparently produced responses that, although physiologically meaningful in certain states of the body, were useless in others (e.g., sweating in hypoglycemia, rise of blood sugar in asphyxia; pupillodilation in fear and love; Cannon 1939). But, he contented himself by assuming that the appearance of inappropriate features in the total complex of sympathico-adrenal function makes sense in the context of its *emergency function* (*Notfallfunktion*, Cannon 1928) if one considers

> first, that it is on the whole, a unitary system; second, that it is capable of producing effects in many different organs; and third, that among these effects are different combinations which are of the utmost utility in correspondingly different conditions of need
>
> (Cannon, 1939, p. 298).

Great emphasis was placed on the adrenal medulla in Cannon's research (Cannon 1914b). In his animals, under experimental conditions in which the body's ability to recover from major disturbances was tested, Cannon measured many effects that were responses to *catecholamines (adrenaline and noradrenaline)* released from the adrenal medulla. It is almost universally stated in textbooks that adrenaline and noradrenaline released from the adrenal medulla act on the same effector organs as the sympathetic postganglionic neurons and thereby enhance and support the effects of sympathetic neurons on target tissues. This statement is misleading as far as the function of the sympathetic nervous system is concerned, at least under normal conditions in higher vertebrates, not only because it assumes widespread and uniform actions of the sympathetic nervous system on target tissues but also because catecholamines from the adrenal medulla often do *not* have the same effect on the target tissues as sympathetic nerve activity. About 92% to 98% of circulating noradrenaline originates from sympathetic nerve terminals (Esler *et al.* 1990), and all circulating adrenaline is released by the adrenal medulla. Adrenaline released from the adrenal medulla under physiological conditions is primarily a *metabolic hormone* and chiefly serves to catalyze the mobilization of glucose and lactic acid from glycogen and of free fatty acids from adipose tissue (Table 1.2, Figure 4.18); in physiological concentrations, it does not support the effect of sympathetic postganglionic neurons on target tissues (see Subchapter 4.5; Celander 1954; Silverberg *et al.* 1978; Cryer 1980; Shah *et al.* 1984). In fact the physiological concentration of circulating adrenaline and noradrenaline is too low to affect the low affinity receptors on normally innervated tissues (see Subchapter 4.5).[4]

Cannon's was a puzzling way of arguing, given that the precise and distinct control of, for example, body temperature, cerebral perfusion and so forth, by the autonomic nervous system was already known at the time of Cannon. Therefore, and as already mentioned in the introduction, such control systems could not work if Cannon's concept about the sympathico-adrenal system were true! Cannon's argument was even more surprising given the enormous amount of detailed experimental work described by Langley between 1890 and 1920, which supported the principle that each organ and tissue is innervated by distinct sympathetic and parasympathetic pathways (Langley 1903a, 1921). Moreover, Langley's conclusions were strengthened by further experiments in which he studied the regeneration of lesioned preganglionic axons (e.g., in the cervical sympathetic trunk to the superior cervical ganglion) and found orderly restitution of function (i.e. topographically correct and functionally appropriate synaptic connections; Langley 1897).

11.1.2 Hess and the dichotomous organization of the autonomic nervous system

In the 1920s, the Swiss physiologist Walter Rudolf Hess, influenced by Karplus and Kreidle (see Akert [1981]), started his famous experiments in which he observed the behavior elicited in conscious cats by local electrical stimulation of the hypothalamus. He implanted electrodes stereotactically and correlated the type of behavior (defensive behavior [Abwehrreaktion; see Subchapter 11.3], submissive, nutritive, sexual and evacuative behaviors and particular autonomic reactions, such as piloerection, sweating, pupillodilation, micturition, defecation) that he could evoke from the anatomical sites he stimulated in freely moving cats. After the experiments, he left the electrodes in position and perfused the brains of the animals in order to locate the tips of the electrodes in the diencephalon, in particular in the hypothalamus. Over several years, Hess probed the whole diencephalon and constructed maps for the different autonomic reactions, which he documented as integral components of the different elicited behaviors. Hess interpreted his results to mean that the hypothalamus integrates the activity of the autonomic nervous system so as to adapt body organs to somatomotor behavior (for English translation of the key publications of Hess, see Akert [1981]).

Similar experiments were later conducted with chickens by von Holst and St. Paul (1960, 1962). These experimenters showed that electrical stimulation of the different areas of the hypothalamus via implanted electrodes leads to species-specific (instinctive) behavior of the birds, including vocalization, which could not be distinguished from their natural behavior. These authors concluded that the different components of behavior, including the adaptive changes in the body that are dependent on the autonomic nervous system and the neuroendocrine system, are represented in the diencephalon.

On the basis of his experiments, Hess propagated generalizations similar to those of Cannon about mechanisms of the autonomic nervous system. He believed that the cranial division of the parasympathetic nervous system promotes the conservation of energy and aids in the recovery of the body after stress (i.e. has *trophotropic* functions), while the sympathetic nervous system has *ergotropic* functions, mobilizing bodily energy and adapting the body to challenges from outside (Hess 1948). These observations are based on the evidence that stimulation of the parasympathetic system leads to an activation of gastrointestinal tract and pelvic organs, whereas stimulation of the sympathetic branch of the autonomic nervous system mediates all those reactions that are also seen during defensive behavior elicited from the caudal hypothalamus and the central grey matter of the midbrain (Hess 1954; in Akert [1981]). This generalizing belief is very similar to that of Cannon.

Hess later transferred the idea of a dichotomy of the functional organization of the autonomic nervous system to the hypothalamus (Hess 1954). He hypothesized that the rostral parts of the hypothalamus integrate somatic, autonomic and endocrine reactions and controls to promote recovery and conservation of energy, digestion, excretion and evacuation of waste as well as reproductive functions. He thought that these functions were associated with the excitation of the parasympathetic nervous system, and the entire process was subsumed under the umbrella term *trophotropic reaction*. He further hypothesized that activation of the caudal parts of the hypothalamus causes general excitation of the sympathetic nervous system, mobilization of body energy and enhancement of performance capacity (*ergotropic reaction*). This concept requires that the hypothalamus consists of two functionally and anatomically different systems. Thus, the unifying concept of the antagonistic function of the sympathetic and parasympathetic nervous system was extended to the hypothalamus. Today it is known that the concept itself is far too general to explain the complexities of the central control of autonomic functions.

11.1.3 The consequences of the generalizing concepts of Cannon and Hess

Both Cannon and Hess had enormous impact on the scientific community and in clinical medicine. Their influence was to some extent positive because it focused clinical practice, clinical research and research in systems physiology on the effects the autonomic nervous system has in regulating body functions.

The ideas of Cannon and Hess were soon accepted by physiology and pharmacology. Generalizations about the actions of the sympathetic nervous system also led to a change in the connotation of the terms *sympathetic* and *parasympathetic* from that originally defined by Langley. Langley's anatomical definitions were clear (see Subchapters 1.1 and 2.1). However, now the terms implied particular types of function, that is, "sympathetic" function and "parasympathetic" function, in line with the generalists' ideas. This is best demonstrated by the commonly used term *sympathico-adrenal system* (Cannon 1939). However, modern research shows that the generalizations made by Cannon and Hess have no justification (other than ontogenetic) and to lump all the sympathetic systems functionally together is obsolete.

The concept that the sympathetic nervous system operates in a more or less unitary way was strengthened by the increase of knowledge about the receptors (notably the adrenoceptors and muscarinic acetylcholine receptors). Drugs that interfere with or mimic actions of autonomic transmitter substances were developed for potential therapeutic use. This trend continues in modern molecular pharmacology:

1. A plethora of receptors in the membranes of the autonomic neurons and their target cells have been detected, cloned and their molecular structure analyzed. For example, at present nine types of adrenoceptors and five types of muscarinic receptors have been identified on the basis of their molecular structure, their coupling to intracellular signaling pathways and their genes (Alexander *et al.* 2004). The function of these different receptors in the neural regulation of autonomic target tissues is only known for some tissues and specifically for the subtypes of adrenoceptors α_1, α_2, β_1, β_2 (see Table 1.2). Unfortunately current evidence suggests that most of these receptors are not accessed by neuronally released transmitter as this often produces its effects by interacting with localized "junctional" receptors in restricted regions of the

postsynaptic cell (see Chapter 7). The identification of molecularly distinct receptors will permit the development of specific antagonists for the identification of the postjunctional receptors relevant for neural control of effector organs.

2. Many neuropeptides and other substances have been detected in autonomic neurons, which are colocalized with the classical transmitters in the vesicles or the cytoplasm of the presynaptic terminals and which may have neurotransmitter functions (see Subchapter 1.4). Molecular techniques are being used to demonstrate the distribution of mRNA for a range of identified receptors/binding sites, as well as for the pathways for synthesis and the peptides themselves (Alexander *et al.* 2004). Again when specific antagonists have been developed, it may be possible to determine whether or not these peptides play a physiological role.

3. The almost exponential expansion of molecular pharmacology has not necessarily led to a better understanding of how autonomic systems work in regulating the target organs under physiological conditions. Thus, conclusions cannot be drawn from this type of research as far as the biological meaning and significance of autonomic regulations are concerned.

I do not want to dispute the scientific achievements of both Cannon and Hess. Both were, over several decades, dedicated experimental scientists and observers; both were dedicated and gifted teachers (Brooks *et al.* 1975; Akert 1981; Benison *et al.* 1987). Hess was awarded the Nobel Prize for his scientific work. Cannon was somewhat unfortunate not to receive it, but this did not affect his fame. However, one wonders why both propagated their unifying and simplifying concepts on how the autonomic nervous system functions, independent of the experimental tools that were rather limited. Indeed one wonders why the unifying concepts were (and still are) attractive for the medical and the public (lay) community. Reading their work carefully, in particular the book *The Wisdom of the Body* (Cannon 1939) and both of Hess' books *The Organization of the Autonomic Nervous System* (Hess 1948) and *The Diencephalon* (Hess 1954), one is left with the impression that both were aware that the autonomic nervous system is differentiated in its organization, in particular the sympathetic nervous system. Otherwise many observations in their experiments could not have been explained. And, as mentioned above, Cannon questioned for this reason the James–Lange theory of emotions and asked himself how a system functionally as uniform in its reactions as the sympathetic nervous system can be involved in so many differentiated functions. I find the conceptual situation even more puzzling for two reasons. *First*, both of them knew Langley's work, although he is barely mentioned in Cannon's book (Cannon 1939) and not discussed in Hess' book on the autonomic nervous system despite his very important achievements. Perhaps they did not recognize or ignored the important messages of Langley, namely that there are distinct spinal autonomic (in particular sympathetic) pathways to the effector systems. *Second*, both had extensive knowledge about the normal regulation of target systems by the sympathetic nervous system (they were teaching organ regulations over tens of years to generations of medical students!). This regulation of organs requires precise and specific autonomic nervous systems otherwise one could not understand how the regulations are brought about.

11.2 | General aspects of integrated autonomic responses

Life of complex organisms in a continuously changing environment is only possible if the internal milieu of the body (1) remains constant in relatively narrowly confined limits, (2) can be quickly adapted within these limits to transient brief environmental perturbations, such as the requirement for physical exertion, temperature changes, loss of blood during an accident etc., and (3) can be semi-permanently adapted to slow changes in the environment, e.g. during the changing seasons, during long-term thermal loads (i.e. in the arctic or tropics), during pregnancy, at high altitude etc. The regulation of the internal (extracellular and intracellular) parameters for constancy in vertebrates, in particular mammals, is called *homeostasis* (Cannon 1929a, 1939). It is dependent on functioning autonomic and neuroendocrine systems. The homeostatic control systems are represented in the brain stem and hypothalamus.

11.2.1 Homeostasis and allostasis

The concept of homeostasis was formulated by Cannon (1929a, 1939) based on Claude Bernard's idea of the constancy of the internal milieu (Bernard 1957, 1974). This idea developed at a time that was scientifically dominated by the idea of the thermodynamic equilibrium. The formulation of the concept of homeostasis became a milestone in our understanding of how the body is controlled by the autonomic nervous system and the endocrine systems. This concept was then applied to the regulation of the intracellular milieu of single cells and also to the regulation of higher brain functions.

Homeostasis is defined by the stability of physiological systems that maintain life; it applies strictly to a limited number of systems such as regulation of pH, concentration of different ions in the extracellular fluid, osmolality of extracellular fluid, glucose levels and arterial oxygen tension that are truly essential for life and are therefore maintained within a narrow range for the current life history stage (McEwen and Wingfield 2003). This restricted definition does not allow to understand how autonomic and endocrine controls enable the organism to adapt to severe and unpredictable environmental changes. And in fact physiological systems operate within a *dynamic range* of steady states, which change according to the demands on the organism. Thus, the concept of homeostatic control does not make clear in which way this adaptation during fast and slow changes in the environment occurs. The temporary deviations from the "set point" of the homeostatic control systems during adaptation to internal or external loads impinging on the organism are dependent on the forebrain. The adaptation of the homeostatic control systems to these external and internal perturbations is called *allostasis*.

The concept of allostasis, which was first formulated by Sterling and Eyer (1988) and has been elaborated by McEwen (1998, 2000, 2001b; McEwen and Wingfield 2003), encompasses the dynamic temporary adjustments of the physiological control of body parameters. Parameters that are regulated within quite narrow limits under practically all physiological conditions, as mentioned above, are examples of homeostatic regulation in the original sense and are very similar between individuals. Parameters such as arterial blood pressure, heart rate, body core temperature, concentration of circulating hormones (e.g., glucocorticoids, adrenaline, sexual hormones), sleep–wake cycle, energy metabolism and regulation of other parameters vary more widely within and between individuals. Here allostatic adaptations minimize the duration and the magnitude of the changes of the parameters from the steady state. Thus, allostasis means achieving physiological stability through change of state. The process of altered and sustained activity during allostasis leads to change of "set points" and boundaries of homeostatic control and is called allostatic state. This process is fully dependent on the autonomic and endocrine regulation of body functions by the brain, in particular the forebrain (for discussion and literature see Berntson and Cacioppo [2000]; McEwen and Wingfield [2003]; Schulkin [2003a, b]).

Adaptation of homeostatic regulation during changing environments is a temporary process. If the allostatic adaptations are not switched off once they are no longer needed (i.e., if they fail to habituate), if they occur too frequently, or if they fail to occur at all, systemic diseases such as cardiovascular diseases, type II diabetes, obesity and metabolic syndromes etc. may develop (Henry and Grim 1990; Björntop 1997; Folkow *et al.* 1997; Henry 1997; Henry and Stephens 1977; McEwen 2001a).

11.2.2 Behavioral patterns and autonomic responses

Homeostatic control of the cardiovascular system, of the respiratory pump and of the intake of nutrients as well as their moment-to-moment coordination is represented in the lower brain stem. The integrative neural networks receive inputs from many functionally different types of afferents (cardiovascular, respiratory and gastrointestinal), which trigger distinct reflexes when stimulated. The individual responses are predominantly based on anatomically and physiologically defined reflex pathways, some of which have been described in Chapter 10. It is the *coordination between these functionally distinct reflex pathways that demonstrates how these integrative networks work*. This is illustrated by the coordination of the neural regulation of the cardiovascular and respiratory systems (Subchapter 10.6), by the precise interactive control of breathing and the entrance to the gastrointestinal tract and in several functions discussed in this chapter in which the upper brain stem and hypothalamus are involved (Tables 11.1 and 11.2). These coordinated response patterns can be triggered or inhibited by physiological stimulation of afferents (trigeminal afferents from the nasal and oropharyngeal cavities, arterial chemoreceptors, arterial baroreceptors, unmyelinated vagal afferents from the atria and ventricles of the heart etc.).

The integrative networks in the medulla oblongata and pons are under the powerful control of the mesencephalon, hypothalamus and cerebral hemispheres. They are involved in all elementary behavioral repertoires that are represented in the mesencephalon and hypothalamus (see Table 11.2). All activities of the body that are initiated from the telencephalon, including those controlled by the neuronal programs in the hypothalamus and mesencephalon, co-opt the lower brain stem centers.

Mesencephalon, hypothalamus and limbic system contain the neuronal programs that generate response patterns to adjust the behavior of the organism in a continuously changing environment. These programs utilize repertoires of *somatomotor*, *autonomic* and *hormonal adjustments*, which are organized so that the motor behaviors are accompanied by changes in the viscera, metabolism etc., so as to optimize net performance. The brain is continuously informed at all levels about the environment via the sensory systems and other forms of feedback from the body (e.g., hormonal or physical signals) and selects the most appropriate reaction patterns to cope with the current environmental challenge.

Figure 11.1 depicts the situation schematically. The neuronal programs that regulate and integrate autonomic, endocrine and

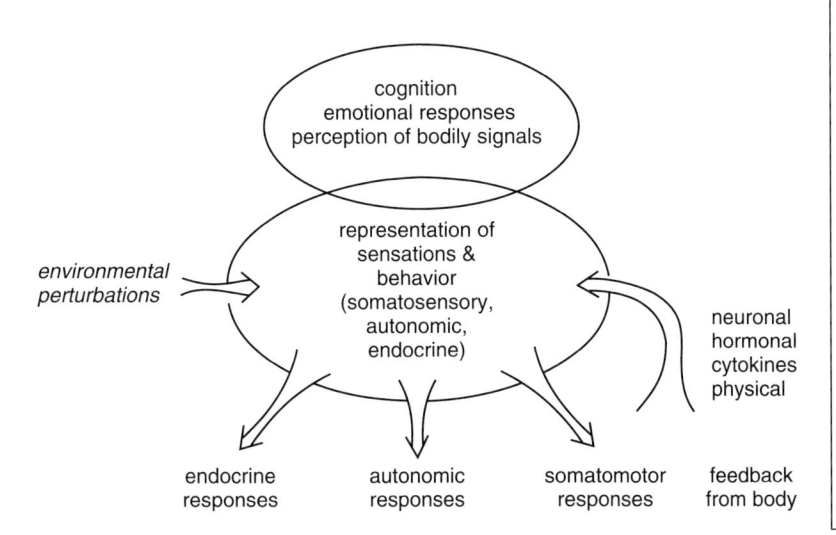

Figure 11.1 Representation of body functions in forebrain and mesencephalon. Activation of these representations generates coordinated motor, autonomic and neuroendocrine responses, body sensations and experience of emotions. The central representations receive continuous feedback information from the body (neuronal, hormonal [including cytokines], physicochemical [e.g., glucose level in the blood, temperature of the blood]) and from the environment. The central circuits representing body sensations and emotions act back on the neural circuits representing motor functions of the body (somatomotor, autonomic, neuroendocrine).

motor functions are at the core of the system. Their activation leads to specific responses (see lower part of Figure 11.1) and to the unconscious and conscious perceptions of bodily changes correlated with the autonomic and endocrine functions, such as hunger, thirst, respiratory effort, satiety, muscle effort etc. (see Subchapter 2.6). These interoceptive sensations from the body are associated with the perception of the environmental situation and emotions. Both are represented in the telencephalon (limbic system and neocortex [mainly insula and orbitofrontal cortex]). Here I want to emphasize, by alluding to Figure 11.1:

- Autonomic and endocrine control systems are integrated and represented in the hypothalamus and cerebral hemispheres.
- Sensations and emotions are related to the functional state of body tissues. They are generated in close association with the homeostatic autonomic and endocrine control systems.
- The telencephalic representations act back on the central representation of the autonomic and endocrine control systems.

The latter feedback has seldom been investigated but is a matter of controversial discussion. It is important for understanding the fast adaptation of homeostatic regulations during external or internal perturbations ("stress"). The temporary adjustments that occur during challenge from the environment (or occurring as a consequence of cortical processes related to past experience and anticipation of future events) is called allostasis (see above). Understanding how the telencephalon adapts may turn out to be the key to clarifying the mechanisms underlying psychosomatic and other behaviorally determined diseases.

Here I will divide the behavioral patterns for practical reasons into three groups: (1) Patterns that are quickly activated in minutes, seconds or less and involve somatomotor and autonomic systems (Table 11.1). (2) Patterns of autonomic responses occurring during experimentally generated emotions. (3) Patterns of somatomotor, autonomic and neuroendocrine responses that involve the hypothalamus and are mostly activated more slowly (Table 11.2).

The main point I want to make is that the optimal activation of autonomic and endocrine responses in the behavioral patterns requires the telencephalon.

I will not discuss the following three groups of experimental approaches and refer the reader to the literature:

1. The Russian school beginning with Ivan Pavlov. In their experimentation on awake (non-anesthetized) animals, mostly dogs, Pavlov and his followers, notably Bykov, have clearly shown that autonomic systems are under the control of the telencephalon. Pavlov studied the neural regulation of digestive glands (salivary, gastric, pancreatic secretion). Using the paradigm of the conditioned reflex he has clearly shown that the parasympathetic pathways supplying the gastrointestinal tract are under cortical

Table 11.1 Patterns of activity in neurons of sympathetic and parasympathetic pathways during different behavioral reactions organized in the upper brain stem and hypothalamus

Organ	Pathway	Diving response	Exercise	Defense Reaction					Vigilance	Freezing reaction
				Confrontation	Flight	Quiescence	Tonic immobility		
Heart	SCM	↓↓	↑↑ (tonic)	↑↑	↑↑	∅/↓?	↓↓	?	?
	PCM	↑↑↑	↓↓ (initial)	↓↓	↓↓	↑↑	↑↑↑	↑↑	↑
Skeletal Muscle	MVC	↑↑↑	↑	↑	↑	↑	↓↓	↑↑	↑
	MVD**	∅	∅	↑↑?	↑↑	∅	↑?/∅	∅	∅
Skin	CVC (av a.)	↓↓	↑↑	↑↑*	↑↑	↑?	↓↓	?	?
	CVC (nutrit.)	↑↑↑	↑↑	↑↑	↑↑	↑?	↓↓	?	?
Kidney	VVC	↑↑↑	↑↑	↑↑	↑↑	↑?	↓↓	↑↑	↑?
GIT	VVC	↑↑↑	↑↑	↑↑	↑↑	↑?	↓↓	↑↑	?
Art. BP		∅	↑	↑	↓	↓↓	↓↓↓	↑↑	∅
Card. Output		↓↓↓	↑↑	↑↑	↑↑	↓↓	↓↓	↓	↓
Respiration		↓↓↓	↑↑	↑/∅	↑↑	↓	↓↓↓	↓	↓

Notes:

These patterns of activity can be quickly recruited by the forebrain or reflexly by activating afferents. Not included are non-cardiovascular changes occurring during these behavioral reactions/responses. Modified from Folkow (2000).

CVC, cutaneous vasoconstrictor neurons (to arteriovenous anastomoses [av a.]; nutritional [nutrit.] vessels); MVC, muscle vasoconstrictor neurons; MVD, muscle vasodilator neurons; GIT, gastrointestinal tract; PCM/SCM, parasympathetic/sympathetic cardiomotor neurons; VVC, visceral vasoconstrictor neurons; ↑, ↑↑, ↑↑↑/↓, ↓↓, ↓↓↓, increase/decrease of activity (or of cardiovascular and respiratory parameters); ∅, relatively unchanged; ?, effect unknown; Art. BP, mean arterial blood pressure. Modified from Folkow (2000), Bandler et al. (2000a) and Keay and Bandler (2004).

*Extracranial vasodilation in facial skin;

**Probably only present in some species (see Subchapter 4.2).

control (see Figure 10.26). Later it was shown by the Russian School (Bykov and his pupils), using the same experimental paradigm, that urinary secretion of the kidney and the cardiac output is also under cortical control involving sympathetic systems (Bykov 1959; Ádám 1967, 1998; see Dworkin [1993, 2000] for extensive discussion and literature).

2. The psychophysiologists have rendered arguments that the cortex is involved in the control of the cardiovascular system (in particular the heart) and the gastrointestinal tract via autonomic pathways. This is, e.g., expressed in the formation of the cardiac-somatic hypothesis by Obrist (Obrist 1981). This work fully supports the points I am going to make in Subchapter 11.3 showing that regulation of autonomic target organs during body challenge in vivo is influenced by the cortex (Berntson and Cacioppo 2000, see here for older literature).

3. Brain processes underlying behavior, cognition and emotions are interdependent with bodily states that are generated by autonomic and endocrine systems. This interdependence is studied in humans by applying modern neuroimaging techniques to the brain. The overall aims of these studies are to explore the central representations of autonomic and endocrine regulation in the forebrain, to explore how cognitive and emotional processes are dependent on and interact with the autonomically determined body states, and to show that the brain processes underlying subjective experience and motivational behaviors require the autonomic and endocrine control of the body (and the afferent feedback from the periphery) (Damasio 1994, 1999; Critchley and Dolan 2004).

11.3 Autonomic responses activated quickly during distinct behavioral patterns

Table 11.1 summarizes response patterns of autonomic and respiratory systems for seven behavioral reactions, including exercise, that are organized in the brain stem and hypothalamus and recruited quickly. The main emphasis is on cardiovascular parameters and the peripheral autonomic pathways involved are listed on the left. These response patterns also affect other organ systems, such as the gastrointestinal tract, the pelvic organs, the motor components of the eye etc. that are not listed in Table 11.1 (Folkow 1987, 2000). Integrated in the generation of these patterns are the various spinal and bulbar reflexes related to the cardiovascular system, the respiratory system, the gastrointestinal tract and the evacuative organs as they have been described in the preceding chapters. The telencephalon can initiate these autonomic patterns within seconds and maintain them for some time or can shift quickly between response patterns (e.g., from an active defense reaction to a passive defense reaction or to tonic immobility). Thus, signals generated in the telencephalon have direct access to the neural circuits of the hypothalamus and brain stem responsible for the generation of the autonomic responses characteristic for these behavioral reactions.

11.3.1 Autonomic changes during diving

In terms of magnitude and temporal precision, the most spectacular pattern of autonomic response is the *diving response*, which is used both for protection and for food seeking in diving species (Elsner *et al.* 1966; Blix and Folkow 1983; Butler and Jones 1997). Figure 11.2a illustrates the changes of blood flow and heart rate (heart beat is indicated by the pulsatile blood flow changes and the dots during diving) through the abdominal aorta and renal artery in a harbor seal during a forced dive in which the animal remained quietly restrained under water for 8 minutes (Elsner *et al.* 1966). This seal had experienced many dives of this type and was therefore habituated. The start of the dive is immediately followed by a decrease of heart rate from about 100 beats per minute to about 7 per minute. Blood flow through the abdominal aorta and renal artery almost ceased, indicating that the peripheral vascular beds (viscera, including kidney, skeletal muscle) had virtually closed down. On surfacing, heart beat and blood flow through both arteries increased within a few seconds to their baseline levels as before diving.

Figure 11.2b shows the changes of heart rate of a harp seal during an entirely unrestrained dive (Casson and Ronald 1975). This animal had been trained to dive on command. The bradycardia (heart rate decreased from 125 to 30 beats per minute) occurred a few seconds *before* the actual dive and the heart rate increased to its pre-dive level about 5 to 15 seconds *before* the seal surfaced. These measurements clearly show that the dramatic changes of cardiovascular parameters during initiation, maintenance and termination of diving depend on neural signals from the telencephalon. These signals activate, probably via the lateral hypothalamus, the neural circuits in the medulla oblongata that are involved in the regulation of the heart, blood vessels and lung. The central command signals probably do not target the final autonomic and respiratory pathways directly (e.g., parasympathetic cardiomotor neurons in the nucleus ambiguus, sympathetic premotor neurons in the rostral ventolateral medulla, bulbospinal inspiratory neurons), but rather they utilize the circuits of the neural autonomic and respiratory motor programs that are involved in the regulation of cardiovascular (and other) target organs and respiration.

The cortical signals inducing the diving response are facilitated by stimulation of trigeminal mechano- and other receptors (from the face, nasopharyngeal mucosa), by excitation of arterial chemoreceptors (generated by decreased arterial oxygen tension) and by respiratory arrest in expiration. Excitation of trigeminal receptors by mechanical, cold and other stimuli triggers reflex bradycardia too but this bradycardia is much weaker than that induced by the cortical signal in the

Figure 11.2 Cardiovascular adjustments during diving responses initiated by central signals from the telencephalon. (a) Blood flow in abdominal aorta and renal artery of a harbor seal before, during and after a forced dive of 8 min duration during which the animal remained quietly restrained under water. Note that the blood flows and heart rate decrease within 1 s after the beginning of the dive, remain low throughout the dive and increase within 2 s after termination of the dive (surface). Heart beats are indicated by the pulsatile changes of blood flow and by dots during dive. (b) Change of heart rate in a harp seal trained and conditioned to dive. Note the anticipatory decrease in heart rate (bradycardia) before diving and the anticipatory increase in heart rate before surfacing. (a) Modified from Elsner et al. (1966) and (b) modified from Casson and Ronald (1975) with permission.

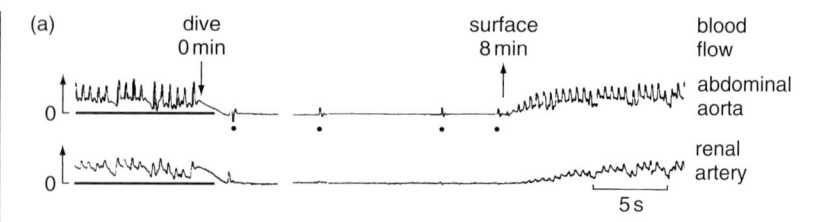

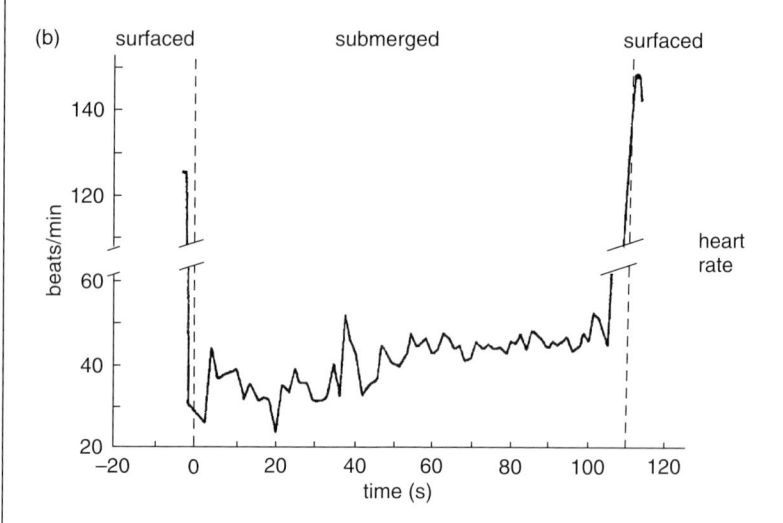

awake animal. The autonomic pattern of the diving response can be elicited in all mammalian species and birds (even those that never dive normally) although the autonomic responses are usually much weaker than those occurring in diving animals (seals, whales, penguins). The neural mechanisms underlying the fast signaling from the telencephalon to the circuits *in the brain stem* are unknown.

The observations on diving animals show that the cardiovascular system is transformed in a few seconds into a "heart-brain circulation system". Major vascular beds (of skeletal muscle and viscera) are virtually shut off by strong activation of vasoconstrictor neurons innervating resistance vessels. Frequency of the heart is profoundly reduced by activation of parasympathetic cardiomotor neurons and withdrawal of sympathetic activity. In the skin, the apical non-nutritional arteriovenous anastomoses, which are under powerful nervous control, are *left open* during submersion by inhibition of activity in cutaneous vasoconstrictor neurons innervating these anastomoses, whereas the cutaneous vasoconstrictor neurons supplying nutritional blood vessels are activated.[5] This helps O_2-containing blood in large venous "depots" to be diverted to the heart pump, without peripheral O_2 consumption, for subsequent nutritional delivery to brain and myocardium in a situation where cardiac output is reduced 20-fold. Furthermore, opening of arteriovenous anastomoses counteracts increase of resistance in skeletal muscle and viscera and takes away load from the heart.

11.3.2 Autonomic changes during centrally generated muscle effort

Any dynamic or tonic action involving skeletal muscle that is initiated by higher centers in the telencephalon is accompanied by fast and precise adjustments of the cardiovascular and respiratory systems to supply the exercising skeletal muscle with O_2 and nutrients. These adjustments are reflected in increases of heart rate, arterial blood pressure and ventilation, in addition to the local metabolic changes in the skeletal muscle, and to the secondary changes involved in regulation of body temperature, extracellular fluid volume and acid–base balance. Here I will describe for the first seconds to minutes after start of isometric exercise in humans that the cortical signal activating the motoneurons and their associated neural circuits (see Subchapter 9.4) also access the neural circuits involved in regulation of the cardiovascular and respiratory systems.[6]

Figure 11.3 shows an experiment on an experimentally immobilized and ventilated healthy human being in whom arterial blood pressure and heart rate were continuously measured. Before the experiment the subject had been systematically trained to undergo graded isometric dorsiflexions of both ankles. Furthermore, he had been trained to be mentally relaxed during immobilization; therefore, he did not exhibit any sign of anxiety or stress during immobilization, as reported after recovery from paralysis. On verbal instruction by the experimenter, the subject attempted to grade his efforts to 25%, 50% and 100% of the maximum effort. Zero percent effort was a sham maneuver in which the subject was encouraged by the experimenter to initiate contractions but he did not attempt to generate them. The graded attempted effort to contract the muscles, but without actual contraction due to the immobilization, was correlated with a graded increase in arterial blood pressure and heart rate. These cardiovascular responses were not correlated with the muscle mass that the subject was trying to contract but with the intensity of the central signal. Thus, the attempt to contract a large or a small muscle mass with the same effort leads to the same cardiovascular changes. The cardiovascular changes started within a few seconds (see inset of Figure 11.3). This experiment clearly shows that the central signal initiating muscle contraction activates neural circuits regulating cardiovascular organs and activates these circuits within a second or less.

The central signal generated in the cortex is also important to *maintain* the cardiovascular and respiratory changes. This is shown in Figure 11.4. Tension of the biceps (b) or triceps surae (a) of the forearm was maintained at about 20% of maximum tension over three minutes. During this period, diastolic and systolic blood pressure, heart rate and ventilation increased (solid line curves). Stimulation of the Ia-fibers from the muscle spindles of the biceps muscle using a vibrator applied to its tendon (a stimulus that does not generate a sensation from the muscle) monosynaptically activates

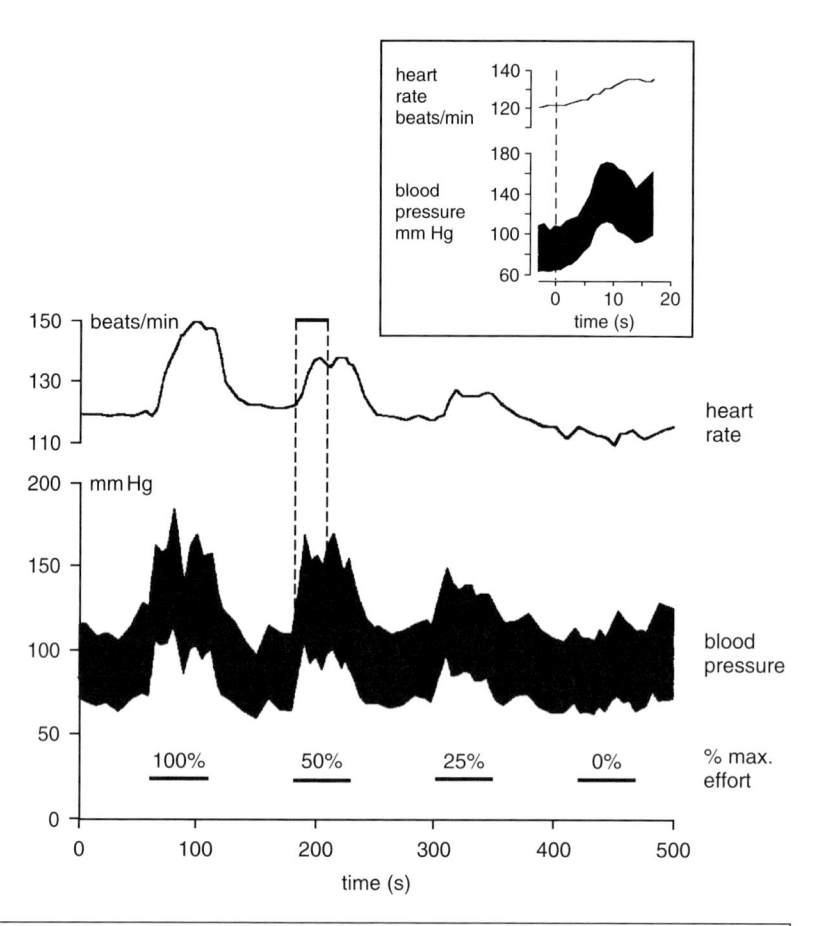

Figure 11.3 Responses of heart rate and arterial blood pressure in a paralyzed human subject to a sequence of graded attempted contractions of the dorsiflexors of the ankle. The subject was paralyzed and ventilated. Paralysis was generated by intravenous infusion of atracurium leading to neuromuscular blockade. The dose of atracurium was five times greater than required for surgical paralysis. Measurements were started about 50 minutes after the onset of paralysis. Attempted contractions were graded for 100%, 50% and 25% of the maximal central effort for the duration indicated by the bars. The effect of a sham maneuver was included (0% effort, verbal encouragements but no attempted contraction). *Inset:* response of heart rate and blood pressure to the attempted contraction of 50% of maximal effort on an expanded time scale. The dashed line indicates the request to start the attempted effort. Note the rapid increase of heart rate and blood pressure after start of the effort. Arterial blood pressure and heart rate were monitored continuously with a servo-controlled device to maintain a constant volume in a finger using a cuff around the finger. The high heart rate is explained by partial blockade of the cholinergic transmission from the parasympathetic cardiomotor neurons to the pacemaker cells. Modified from Gandevia et al. (1993) with permission.

the motoneurons innervating the biceps muscle and disynaptically inhibits motoneurons of the triceps surae (the antagonist of the biceps muscle). The central signal to contract the triceps muscle is inhibited by the afferent Ia-input from the biceps muscle (Figure 11.4a) and the central signal to contract the biceps muscle is enhanced by the afferent Ia-input from the biceps muscle (Figure 11.4b). As a consequence of the activation of the Ia-fibers, either the strength of the central signal to the triceps increases or to the biceps decreases in order to maintain the muscle tension constant at 20% of maximum tension. These quantitative changes of the cortical signal are

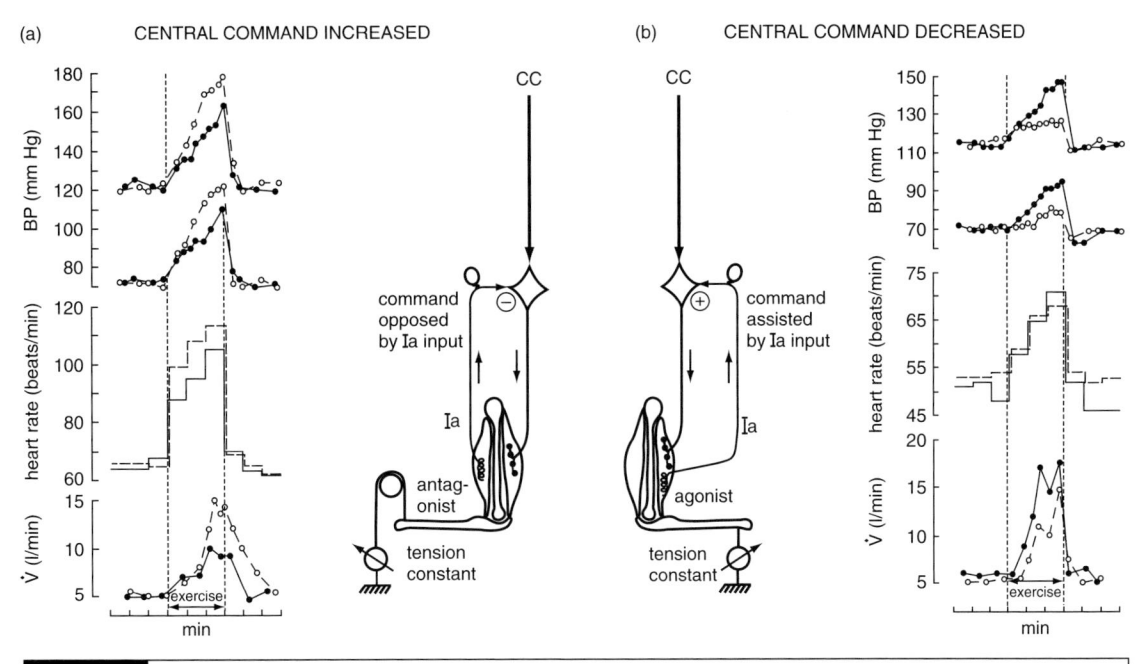

Figure 11.4 Changes of arterial blood pressure, heart rate and ventilation during changes of central command to generate isometric contractions at constant muscle tension of the biceps or triceps surae of the forearm. Tension of the biceps (b) or triceps surae (a) generated by central command is adjusted to 20% of the maximal tension during 3 minutes. Increase in muscle tension is accompanied by increase in systolic and diastolic blood pressure (BP), heart rate and ventilation (closed circles and solid line curves). The Ia-fibers of the muscle spindles in the biceps were activated by a vibrator attached to its tendon that oscillated at 100 Hz and at an amplitude of a few μm (this stimulus generates no sensation). Activation of Ia-fibers activates motoneurons to the biceps muscle monosynaptically and leads to a decrease of the central command to maintain a biceps contraction at 20% of its maximum (b). It inhibits motoneurons to the triceps surae (antagonist) and leads to an increase of the central command to maintain a triceps contraction at 20% of its maximum (a). Changes of arterial blood pressure, heart rate and ventilation during exercise were smaller when the central command decreased (b) and larger when the central command increased (a; open circles and dashed-line curves). This shows that the central command signal generated by the telencephalon has access to the cardiovascular centers in the medulla oblongata regulating heart and peripheral resistance and to the respiratory neural network. Note the persistent and slow increase of heart rate and blood pressure after start of the effort. CC, central command. Modified from Goodwin et al. (1972) with permission.

associated with changes of blood pressure, heart rate and ventilation. An increased central activation is followed by a larger increase of blood pressure, heart rate and ventilation (Figure 11.4a) and a decreased central activation is followed by a smaller increase of these cardiovascular parameters than in the controls (Figure 11.4b, compare dotted line curves with solid line curves).

We can conclude unambiguously from these experiments that central (cortical) signals initiating and maintaining isometric muscle contraction access in parallel the neural circuits involved in regulation of the cardiovascular and respiratory systems. The cortical signals are important to initiate these changes quickly and to maintain them. Fast changes of heart rate are probably generated by reduced activity in parasympathetic cardiomotor neurons (i.e., these neurons are inhibited by the cortex); slower increase of heart rate and blood pressure (during long-lasting isometric exercise) involve particularly changes in the activation of sympathetic cardiovascular neurons. The central signals are integrated, during the static phase, with the afferent signals from the exercising skeletal muscle (small-diameter myelinated [Aδ] and unmyelinated [C] afferent nerve fibers sensing muscle contraction and metabolic changes; Mitchell 1985; Rowell 1993), from the lung and from cardiovascular targets. The neural pathways by which the central (telencephalic) signals influence the neural regulation centers of the cardiovascular and respiratory systems are unknown.

11.3.3 Defense reactions integrated in the mesencephalon

Responses of the autonomic nervous system to events that endanger the integrity of the body (e.g., threatening or stressful stimuli, acute blood loss, invasion of the body by microorganisms and their toxins) are well orchestrated and their functional specificity resides in the individual responses associated with functionally discrete autonomic pathways. They enable the organism to avoid real or perceived danger and are presumably protective and adaptive under normal biological conditions. They are associated with the activation of the hypothalamo–pituitary–adrenal system, the sympatho-adrenal system, and the somatomotor system. The stereotyped patterns of these protective reactions are organized in the upper brain stem and hypothalamus and are under control of the telencephalon (limbic system, neocortex). These patterns can be recruited quickly by afferent signals from the body, by visual, acoustic or olfactory stimuli, or by both sensory signals and signals from the telencephalon working together, or by signals from the telencephalon alone (e.g., during imagined or subjectively attributed dangerous environments).

The general pattern of reaction when an individual is in pain and under stress can best be exemplified by the different types of *defense behavior*. These are integrated responses consisting of autonomic, endocrine and motor components and sensory adjustments. The periaqueductal grey (PAG) in the midbrain is the *final common pathway for the defense behavior* as far as the somatomotor and autonomic responses are concerned. It contains the executive neural systems for brain structures located rostrally such as the hypothalamus, amygdala, other structures of the limbic system and the prefrontal cortex (Panksepp 1998; for critical discussion and older literature see Bandler [1988]).

Types of defense reaction elicited from the PAG

Microinjection of the excitatory amino acid glutamate with a micropipette in the PAG of rats excites cell bodies of neurons in a small volume but not axons passing through. This way of stimulating neurons in the PAG elicits typical defense reactions that are characterized by their somatomotor and autonomic responses and sensory changes. The organization of three types of defense reactions has been described by Bandler and coworkers for the rat. They called these reactions (1) confrontational defense, (2) flight or (3) quiescence. The organization of these defense reactions has been worked out in the rat. There are good reasons to assume that the same principle of organization applies to all mammals (Bandler *et al.* 1991; Bandler and Shipley 1994; Bandler and Keay 1996; Bandler *et al.* 2000a, b). The characteristics of the three defense patterns are (Figure 11.5, Table 11.1):

1. *Confrontational defense* is characterized by hypertension, tachycardia, decreased blood flow through limb muscles and viscera and increased blood flow in the face. It is activated by stimulation of the rostral part of the dorsolateral and lateral PAG. This defense behavior is accompanied by an endogenous non-opioid analgesia.
2. *Flight* is characterized by hypertension, tachycardia, increased blood flow in the limb muscles and decreased blood flow in the face. It is initiated in the caudal part of the dorsolateral and lateral PAG. This defensive behavior is also accompanied by an endogenous non-opioid analgesia.
3. *Quiescence* (hyporeactivity) is characterized by hypotension, bradycardia and endogenous opioid analgesia. It is activated by stimulation of the ventrolateral PAG.[7]

It is likely that, in addition to the cardiovascular changes, other autonomically mediated changes (e.g., related to the gastrointestinal tract or the pelvic organs) occur during the three types of defense reaction elicited from the PAG columns. The behavioral patterns elicited in this way occur independently of the forebrain (hypothalamus and telencephalon), thus the neural circuits responsible for the generation of these patterns are present in the rostrocaudally organized cell columns of the PAG.

It is important to emphasize that one component of the patterned defense reaction is the modification of sensory systems, leading, e.g., to analgesia during defense behavior in vivo. But also functional changes of non-nociceptive somatic sensory systems, the visual system and the acoustic system do occur (see Bandler [1988]). The neural circuits in the PAG are integral components of the typical defensive behaviors generated by the telencephalon via the hypothalamus. It is hypothesized that the cortically generated behavior patterns are the basis for *active behavioral coping strategies* and *passive behavioral coping strategies* in mammals (Keay and Bandler 2001, 2004).

The systemic cardiovascular changes (and probably other autonomic changes in, e.g., skin blood flow, piloerection, sweating, gastrointestinal motility, adrenal medullary secretion, pupil size etc.) are generated by activation or inhibition of specific sympathetic and parasympathetic pathways. Some of these changes are listed in Table 11.1 for these types of defense behavior.

Systematic anatomical tracing studies of the afferent and efferent connections of the cell columns in the PAG show that these cell columns project to multiple sites in the hypothalamus, thalamus, midbrain, pons, medulla oblongata and spinal cord and that they receive descending synaptic inputs from the cortex, amygdala and

*active coping strategies
evoked from the lPAG and the dlPAG*

CONFRONTATIONAL DEFENSE/THREAT
hypertension and tachycardia
extracranial vasodilation
hindlimb and renal vasoconstriction
non-opioid-mediated analgesia

FLIGHT
hypertension and tachycardia
hindlimb vasodilation extracranial
and renal vasoconstriction
non-opioid-mediated analgesia

*passive coping
strategies
evoked from the vlPAG*

QUIESCENCE/HYPOREACTIVITY
hypotension
bradycardia
opioid-mediated analgesia

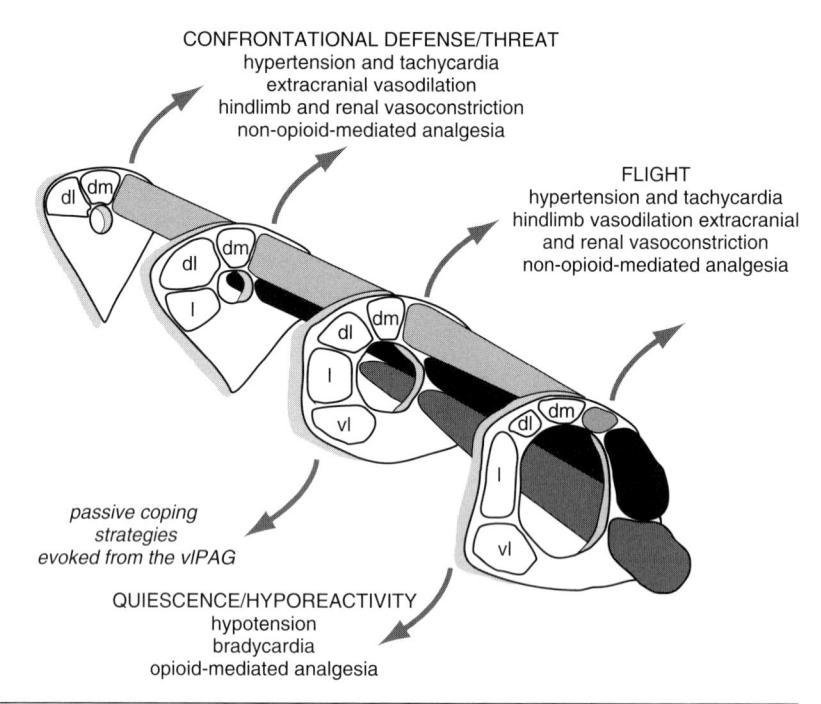

Figure 11.5 Representation of defense behaviors in the dorsolateral, lateral and ventrolateral periaqueductal grey. Schematic illustration of the dorsolateral (dl), lateral (l) and ventrolateral (vl) columns within the rostral, intermediate and caudal periaqueductal grey (PAG). The dorsomedial (dm) neural PAG column is indicated too. Stimulation of neuron populations in the dlPAG, lPAG and vlPAG by microinjections of the excitatory amino acid glutamate, which excites only cell bodies of neurons but not axons, evokes distinct defense behaviors: *confrontational defense* is elicited from the rostral portion of the dlPAG and lPAG; *flight* is elicited from the caudal part of the dlPAG and lPAG; *quiescence* (cessation of spontaneous motor activity) is elicited from the vlPAG in the caudal portion of the PAG. These defense behaviors include typical autonomic cardiovascular reactions (changes of blood pressure, heart rate, blood flows) and sensory changes (non-opioid- or opioid-mediated analgesia). The representations of confrontational defense and flight are the basis for active coping strategies produced by the cortex. The representation of quiescence is the basis for passive coping strategies produced by the cortex. Modified from Bandler and Shipley (1994) and Bandler et al. (2000a).

hypothalamus and ascending synaptic inputs from spinal cord, medulla oblongata and pons. With the exception of the cortex and amygdala, these connections are reciprocally organized (Bandler et al. 2000a; Carrive 1993; Carrive and Morgan 2004).

Efferent projections of the PAG cell columns to the medulla oblongata

Neurons in the lateral and ventrolateral PAG columns project directly to various autonomic centers in the medulla oblongata (rostral and caudal ventrolateral medulla [RVLM, CVLM], rostral and caudal ventromedial medulla [VMM] including raphe nuclei and paramedian reticular formation). Neurons in the dorsolateral PAG column project indirectly, via the cuneiform nucleus, to the same regions of the medulla oblongata (Figure 11.6) (van Blockstaele *et al.* 1991; Henderson *et al.* 1998). These areas contain (see Chapter 10):

- the sympathetic premotor neurons and other neurons involved in regulation of the cardiovascular system, skin blood flow and other autonomic parameters;
- the parasympathetic preganglionic neurons involved in regulation of the heart and gastrointestinal tract and
- the neurons of the respiratory network (particularly in the ventral respiratory group).
- The VMM contains the neurons controlling transmission of nociceptive impulses in the dorsal horn and caudal trigeminal nucleus (Mason 2001; Fields *et al.* 2005).

The dorsolateral, lateral and ventrolateral PAG columns project to the *same* regions in the medulla oblongata. Thus, these anatomical tracing studies do not reveal how stimulation of the neurons in the PAG cell columns leads to the distinct autonomic reaction patterns that are characteristic for the three types of defense behavior as elicited by PAG stimulation (Table 11.1).

Afferent projections to the PAG cell columns from the body (upper part of Figure 11.6)

Particularly second-order neurons in the superficial layer (lamina I) but possibly also in deep layers (lamina V and deeper) of the spinal and medullary dorsal horn project somatotopically to the lateral column of the PAG (see also Figures 2.11 and 2.12). These projections originate preferentially from the body surface. About 50% of the projecting neurons are located in the segments C1 to C3. In the sacral spinal cord, the projecting neurons are located within the sacral parasympathetic nucleus (intermediolateral cell column) (Keay *et al.* 1997).

The ventrolateral column of the PAG receives direct projections from second-order neurons in the spinal cord (superficial and deep dorsal horn), but not from those in the medullary dorsal horn. About 50% of these projecting neurons are located in deep laminae of the spinal segments C1 to C3. The projections to the ventrolateral PAG are *not* somatotopically organized; they originate preferentially in the deep somatic and visceral body domains. The ventrolateral PAG receives additionally direct projections from the nucleus tractus solitarii (NTS). Other projections to the lateral and ventrolateral PAG originate in the dorsomedial, ventromedial and ventrolateral medulla. These anatomical connections are consistent with functional studies showing that deep injury (skeletal muscle, viscera), ischemia (during hemorrhage) or injurious stimuli exciting vagal cardiopulmonary afferents activate preferentially the neurons in the ventrolateral PAG. Injurious stimuli exciting only deep spinal afferents preferentially activate neurons in the caudal ventrolateral PAG and those exciting vagal afferent neurons activate neurons in the rostral ventrolateral PAG (Keay *et al.* 1994, 1997, 2002; Clement *et al.* 1996, 2000).

Figure 11.6 Afferent inputs from the body to the cell columns in the periaqueductal grey (PAG) and efferent outputs to autonomic nuclei in the medulla oblongata from these cell columns. Second-order neurons in the spinal and medullary (trigeminal) dorsal horn (mainly lamina I but also deep laminae) project topographically to the lPAG. Second-order neurons in the spinal dorsal horn (DH) and second-order neurons in the caudal nucleus tractus solitarii (NTS) project to the vlPAG; this projection exhibits no topography. The dlPAG does not receive direct afferent input from the body. The lPAG and vlPAG project to the rostral ventrolateral medulla (RVLM), the caudal ventrolateral medulla (CVLM) and the ventromedial medulla (VMM; including raphe nuclei and paramedian reticular formation). The dlPAG projects to these nuclei via the cuneiform nucleus (cnf). AP, area postrema; CU, cuneate nucleus; cc, central canal; dl, dorsolateral; dm, dorsomedial; DMNX, dorsal motor nucleus of the vagus; GR, gracile nucleus; l, lateral; py, pyramidal tract; Sol, soleus tract; Sp5C, spinal trigeminal nucleus, caudal part; 4V, fourth ventricle; vl, ventrolateral. Modified from Bandler et al. (2000a) and Keay and Bandler (2004) with permission.

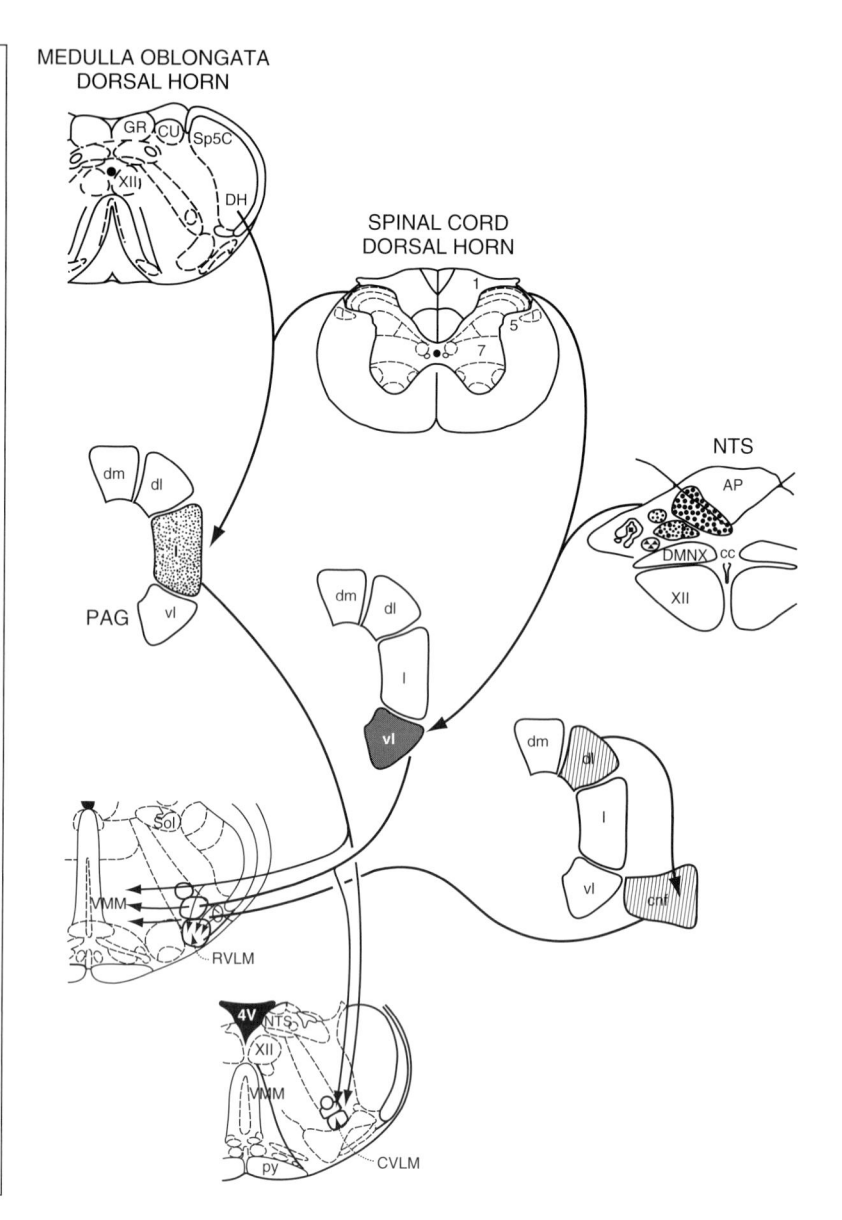

The dorsolateral PAG does not receive direct synaptic connections from the spinal cord, the NTS or other sites of the medulla oblongata.

Projections from cortex and hypothalamus to the PAG cell columns

Cortical structures and subcortical telencephalic structures (e.g., the central and basal nucleus of the amygdala and the medial preoptic area) have powerful projections to the PAG and hypothalamus. These afferent projections from the forebrain also have a topic (columnar) organization, those from the neocortex being spatially more discrete

11.3 AUTONOMIC RESPONSES ACTIVATED QUICKLY

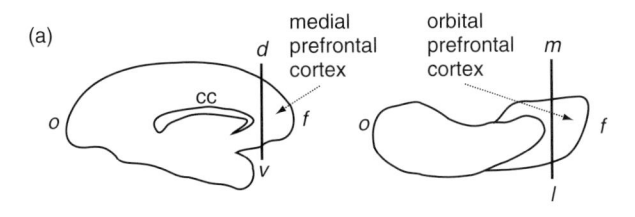

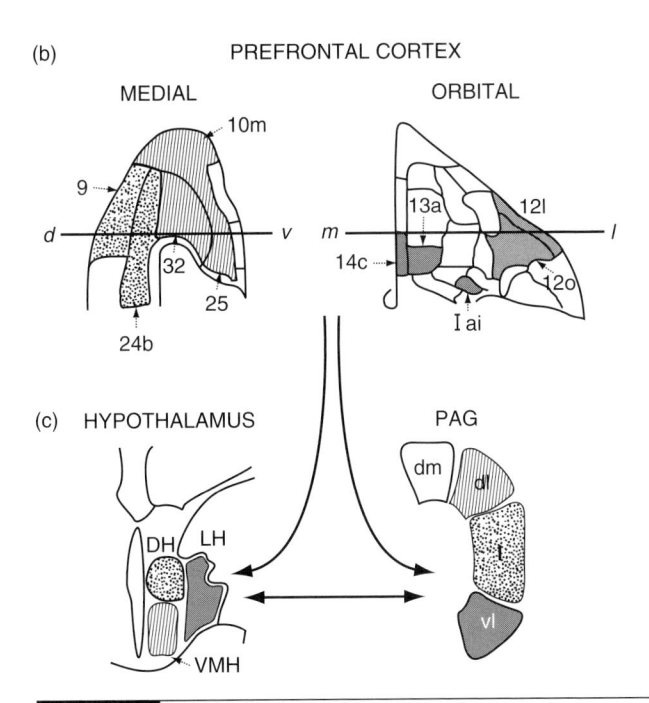

Figure 11.7 Connections of the orbital and medial prefrontal cortex (PFC) with the cell columns in the periaqueductal grey and with the hypothalamus in the macaque monkey. (a) Location of the medial and orbital PFC in the monkey. Right, horizontal section; left, sagittal section. The dorsoventral (d, v) and mediolateral (m, l) sections correspond to those in (b). (b) Three "visceromotor" networks are distinguished in the orbital and medial PFC and indicated by different shadings in the upper part. The numbers refer to areas of Brodmann according to cytoarchitectonic criteria. The orbitoinsular PFC (areas 12l, 12o, Iai, 13a, 14c) projects preferentially to the ventrolateral periaqueductal grey (vlPAG) and the lateral hypothalamus (LH). The medial wall PFC areas (areas 10 m, 32 and 25) project preferentially to the dorsolateral column of the PAG (dlPAG) and the ventromedial hypothalamus (VMH). The dorsomedial convexity of the PFC (areas 9 and 24b) projects preferentially to the lateral column of the PAG (lPAG) and the dorsal hypothalamus (DH). Corresponding hypothalamic regions and PAG columns communicate with each other reciprocally. a, anterior; cc, corpus callosum; d, dorsal; f, frontal; l, lateral; m, medial; o, occipital; v, ventral. Modified from Öngür et al. (1998) and Bandler et al. (2000a) with permission.

than those from subcortical structures. Furthermore, many projections from subcortical structures are denser than those from the cortex (An et al. 1998; Bandler et al. 2000a).

Figure 11.7 demonstrates in the monkey the anatomical connections between the medial or orbital prefrontal cortex, the dorsolateral, lateral and ventrolateral PAG and the lateral, dorsal and ventromedial hypothalamus. There is specificity of the anatomical projections from the medial and orbital prefrontal cortex to the three cell columns of the PAG and the three hypothalamic regions.

Corresponding regions in the hypothalamus and PAG communicate with each other reciprocally. This basic principle of anatomical connections between prefrontal cortex, hypothalamus and PAG has been described for the monkey and the rat (Floyd et al. 2000, 2001; Öngür and Price 2000). These (and other) anatomical studies demonstrate the powerful control of PAG and hypothalamus by the telencephalon. However, they do not reveal the underlying mechanisms of this control.[8]

The attractive element of the idea of Bandler and his collaborators (notably Carrive, Keay and Shipley) and others is that the rostrocaudally organized cell columns in the PAG contain the neural networks that enable the forebrain structures to coordinate, on a moment-to-moment basis, the somatic, autonomic and antinociceptive mechanisms and other sensory mechanisms during stress and pain. These fast neuronal adjustments are critical for the survival of the organism. Primitive strategies to cope with threatening situations seem to be represented in the longitudinal columns of the PAG. Noxious events related to the environment and occurring at the body surface are associated with active coping strategies (e.g., confrontational defense and flight) and are represented in the dorsolateral and lateral cell columns. Noxious events in the deep body domains are associated with passive coping strategies (quiescence) and are represented in the ventrolateral cell column. The dorsolateral PAG cell column that does not receive direct afferent input from second-order neurons (dorsal horn, NTS) is probably involved in active coping strategies initiated from the cortex. These fast neuronal protective mechanisms are coordinated with hypothalamic mechanisms that control homeostatic body functions including the hypothalamo–pituitary axis and the immune system (Table 11.2). The latter may be mediated by a special sympathetic pathway (see Subchapter 4.6).

The fast neuronally directed protective adjustments of body functions require precisely functioning sympathetic and parasympathetic pathways as described in Chapter 4.

11.3.4 Autonomic responses during conditioned emotional responses

An important experiment, combining measurements of cardiovascular parameters with those of somatomotor parameters during conditioned emotional responses, was conducted by Orville Smith and coworkers on male baboons. In a first step, they asked whether the cardiovascular changes occurring during an experimentally produced emotional state of the animals (using a conditioned emotional response paradigm) are integrated in the hypothalamus. They tested this by making controlled local lesions in the hypothalamus in animals that had developed emotional responses to the conditioning stimulus. They determined whether the autonomic changes could be attenuated without affecting either the "emotionality" of the animals (as indicated by their motor behavior during the

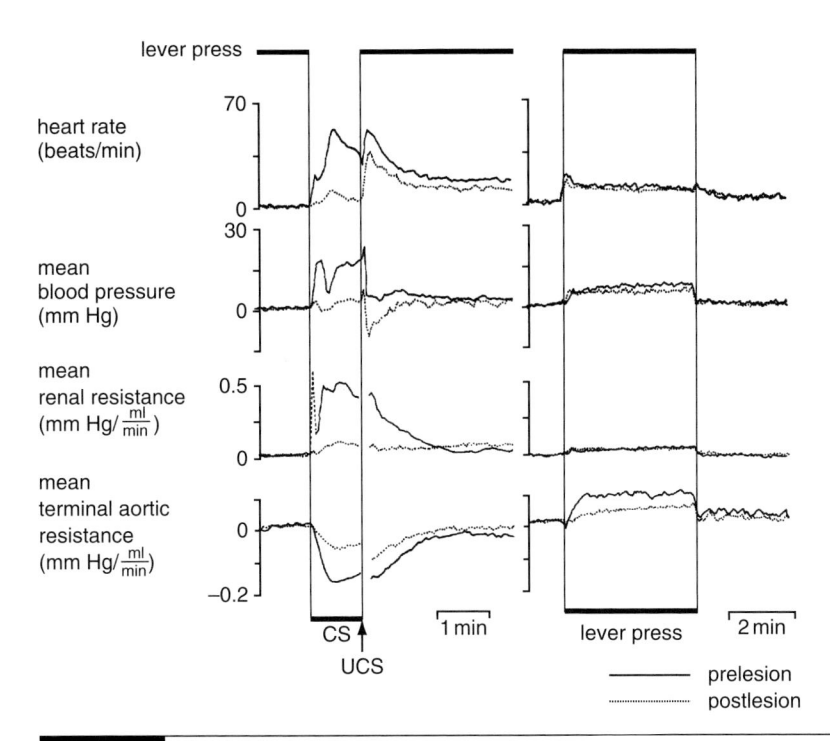

Figure 11.8 Cardiovascular reactions during a conditioned emotional response (left) and during lever pressing (right) before (solid-line curves) and after hypothalamic lesions (dotted-line curves) in monkeys. The conditioned emotional response was elicited by a conditioned stimulus (CS, auditory signal of 2900 Hz, 80 dB interrupted 2.5 times per second) and terminated by an unconditioned stimulus (UCS, electrical shock applied to the abdominal skin; 7 to 12 mA, 1 s duration). Upper bar: lever pressing; cessation of lever pressing during the CS indicates that the monkeys felt emotionally. This response is the same before and after bilateral lesion in the perifornical region in the lateral hypothalamus. Ordinate scales of cardiovascular responses indicate the changes compared to the baselines in the minute preceding the conditioned emotional response (left) or preceding the lever pressing (right). The data were averaged from five prelesion and five postlesion trials in each of six monkeys (baboons). Modified from Smith et al. (1980) with permission.

conditioned emotional response) or other cardiovascular response patterns during, e.g., exercise or eating that are also organized in the hypothalamus. As a second step, they tried to work out the afferent and efferent connections of the hypothalamic area that integrates the cardiovascular changes during the conditioned emotional response (Smith et al. 1979, 1980, 2000; Smith and DeVito 1984). Here I will only discuss the first aspect.[9]

During the conditioning trial, characteristic cardiovascular changes occurred in the monkeys (solid-line curves on the left side in Figure 11.8). Heart rate, arterial blood pressure and renal vascular resistance increased whereas resistance in the terminal aorta (reflecting hindlimb resistance) decreased. The increases in heart rate and blood pressure were generated by decreased activity in parasympathetic cardiomotor neurons (particularly for the fast component) and increased activity in sympathetic cardiomotor and vasoconstrictor neurons. The slow increases in renal vascular resistance

Figure 11.9 Cardiovascular response during exercise or during eating before (solid-line curves) and after bilateral lesion of the perifornical region in the hypothalamus (dotted-line curves) of monkeys. For exercise the monkeys turned a wheel with their legs. For details see legend of Figure 11.8. Modified from Smith et al. (1980) with permission.

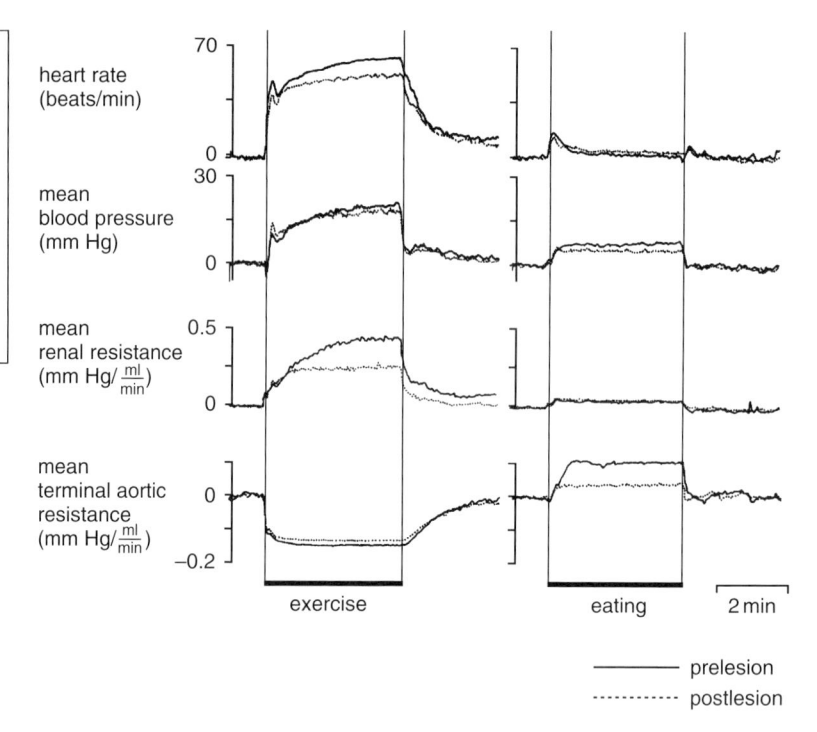

and decreased hindlimb resistance were assumed to be generated by adrenaline released by the adrenal medulla (via β_2-adrenoceptors in the vascular bed of skeletal muscle and α_1-adrenoceptors in the vascular bed of the kidney). This may or may not be the case, particularly in the kidney (see Subchapter 4.4).[10] During the conditioned stimuli, the monkeys stopped pressing the lever (see upper bar in Figure 11.8, left); this is the emotional motor response of the animals (they had been conditioned and expected the unconditioned electrical stimulus). The pattern of cardiovascular response was different from that during lever pressing alone (solid-line curves on right side of Figure 11.8), eating or exercise (solid-line curves in Figure 11.9).

Electrical stimulation of the perifornical region in the lateral hypothalamus produced the same pattern of cardiovascular response as that occurring during the conditioned emotional response. After bilateral lesions of this region, the cardiovascular changes during the conditioned stimulation were largely abolished (see dotted-line curves in Figure 11.8 left) whereas those occurring during lever pressing (dotted-line curves in Figure 11.8 right), exercise or eating did not change significantly (dotted-line curves in Figure 11.9). The monkeys with the hypothalamic lesions continued to stop lever pressing during the conditioned stimuli, indicating that they had not "forgotten" the significance of this stimulus. Thus, they still showed emotional behavior during the conditioning stimuli but on this occasion without the characteristic cardiovascular changes.

We can draw the following general conclusions from this experiment:

- The hypothalamus contains an area (perifornical area in the lateral hypothalamus) that controls the cardiovascular response pattern associated with emotional behavior.
- Lesioning of this area abolishes these cardiovascular responses but does not affect the cardiovascular responses during other behaviors, such as exercise or eating.
- After destruction of the perifornical region in the lateral hypothalamus the monkeys still responded in an emotional way during the presentation of the conditioned stimulus; thus, the two components of the conditioned emotional response have been *separated* by the hypothalamic lesion.
- The cardiovascular response pattern elicited during the conditioned emotional response was mediated by sympathetic and parasympathetic pathways to the cardiovascular system.
- Other changes mediated by non-vascular autonomic pathways (e.g., those to the gastrointestinal tract, to the pelvic organs, to skin) may also be generated during the conditioned emotional response.

Do the cardiovascular changes during the conditioned emotional response have underlying mechanisms comparable to those elicited by central signals during exercise? Both groups of cardiovascular changes are elicited from the telencephalon, those generated by the conditioned emotional response precede the motor responses (they occur in anticipation of the somatomotor changes in posture and locomotion) and those generated during isometric exercise start almost immediately with the real or attempted motor activity (see Figures 11.3 and 11.4). This has been experimentally studied in baboons, which were fully instrumented to measure cardiovascular parameters to be transmitted by telemetry and which lived entirely unrestrained in their natural habitat in groups of four or five (Smith *et al.* 2000):

- The monkeys exhibited cardiovascular responses during postural or locomotor changes *without emotional content* (e.g., changes from quiet sitting to standing, walking or running) that started *within a second* of the motor activity but never longer before. The explanation must be that somatomotor and cardiovascular responses are both initiated by the same central signal (see Figure 11.3).
- Changes of motor activity during a behavior *with emotional content* (e.g., changes from quiet standing to "aggressive walking" of an Alpha [dominant] male towards a subordinate Beta male in the group) were *preceded* by cardiovascular changes by up to 10 seconds. These anticipatory changes in the unrestrained baboons are equivalent to those during the conditioned emotional response (Figure 11.8 left).
- After bilateral lesions of the perifornical region in the lateral hypothalamus with ibotenic acid (which destroys neuronal cell bodies but not axons), these Alpha males exhibited the same

aggressive behavior, but without any anticipatory cardiovascular changes.
- The cardiovascular responses occurring with postural or locomotor changes without emotional context were unchanged in the baboons with bilateral perifornical hypothalamic lesions.

These results support that the central efferent signals and their pathways initiating the cardiovascular changes during the conditioned emotional response, or during postural locomotor changes in an emotional context, differ from those initiating cardiovascular changes during motor behavior without emotional context.

11.3.5 Tonic immobility, freezing and vigilance

The autonomic (and somatomotor) responses during *tonic immobility* in animals are extreme (see Table 11.1). The animal drops as if dead. Its state is characterized by complete immobility (the musculature is flaccid), apnea (cessation of respiration), decreased heart rate and blood pressure and decreased consciousness (relative lack of responsiveness to external stimuli). Tonic immobility response is used by animals in a confrontation with a predator to increase the chances of survival (e.g., by sheep against a poaching dog). This response is therefore sometimes also called "playing-dead response" or "play possum." This reaction is quickly activated in dangerous situations. Electrical stimulation of the anterior cingulate gyrus in the cat produces the same pattern of responses: decreased heart rate, blood pressure, respiration and skeletal muscle tone and increased blood flow in skeletal muscle. It is assumed that activity in other types of vasoconstrictor neurons (e.g., those to skin or viscera) also decrease (Löfving 1961). The response pattern elicited from the anterior cingulate gyrus is probably relayed through the central nucleus of the amygdala (Applegate *et al.* 1983; Cox *et al.* 1987; Leite-Panissi *et al.* 2003), the lateral hypothalamus and the depressor area of the medulla oblongata. In fact Hess had already described how electrical stimulation of some sites in the lateral anterior hypothalamus leads to adynamia, reduced reactivity, decreased arterial blood pressure and bradycardia in awake cats (Hess 1944).

In humans, the tonic immobility probably corresponds to *the neurally mediated syncope* that occurs without any obvious peripheral pathophysiology. This syncope is characterized by quick loss of consciousness and of motor tone and decrease of peripheral resistance, heart rate and arterial blood pressure. Decreased vasoconstrictor activity and increased cardiovagal activity seem to be the primary autonomic events (Grubb and Karas 2002; Hainsworth 2002). The neurally mediated syncope can be generated by emotional stimuli; therefore, it is also called *emotional fainting/syncope*.

It is generally assumed that loss of consciousness during the neurally mediated syncope is the consequence of decreased arterial blood pressure (due to centrally generated bradycardia and decreased peripheral resistance). However, it could well be that the three events, namely decreased activity in vasoconstrictor neurons, increased activity in parasympathetic cardiomotor neurons and loss of consciousness, are generated by the *same* central mechanism. This would not be at variance with the long-held experience that syncope can be triggered by many peripheral stimuli exciting afferents (e.g., from heart, lung, gastrointestinal tract, urogenital tract, carotid sinus) (Freeman and Rutkove 2000). Whether the tonic immobility and the neurally mediated syncope in humans correspond to each other and are generated by similar central mechanisms is currently controversial (Freeman and Rutkove 2000; Mosqueda-Garcia *et al.* 2000; van Dijk 2003).

The *freezing reaction* can also be characterized by complete immobility, yet high preparedness for sudden motor action and intense alertness. In this respect, the freezing reaction is fundamentally different from tonic immobility ("playing-dead") reaction. The autonomic responses are decreased heart rate (probably generated by activation of parasympathetic cardiomotor neurons), increased muscle vasoconstrictor activity and decreased respiration. Thus, the autonomic reactions of the freezing reaction resemble those of the diving response, although they are much stronger in the latter.

The *vigilance reaction* in animals is characterized by immobility, a large decrease in heart rate (with decreased cardiac output; probably generated by activation of parasympathetic cardiomotor neurons), increased arterial blood pressure (due to increased peripheral resistance generated by activation of muscle and visceral vasoconstrictor neurons), decreased respiration and behavioral signs of increased vigilance. This response can be elicited, in rabbits, by electrical stimulation of the dorsolateral hypothalamus ("hypothalamic vigilance area") or the ventrolateral PAG ("periaqueductal vigilance area") (McCabe *et al.* 1994; Duan *et al.* 1996, 1997). As far as the autonomic pattern is concerned, the vigilance reaction is unlikely to be identical with quiescence, the freezing reaction or tonic immobility (Table 11.1).

11.4 Emotions and autonomic reactions

11.4.1 Basic emotions and autonomic response patterns

Emotional feelings and the corresponding emotional expressions generated by the somatomotor system are highly integrated components of behavior in humans and animals, which are important (externally and internally) for reproduction, selection, genome protection and regulation of social behavior (Darwin 1872). An influential theory of emotions that was first propagated by James (James 1994; see Meyers [1986]) states that activation of emotions is closely associated with the afferent feedback from the periphery of the body. This theory became known as the James–Lange theory of emotions (see Subchapter 11.1).[11] According to this theory, the brain triggers bodily changes by the activity in the autonomic and somatomotor systems and the perception of these changes generated by the afferent feedback from the viscera and somatic tissues leads to the emotional responses. This theory is still mentioned in textbooks of psychology. It is important to emphasize that this peripheral theory of generation of emotions was not based on data obtained in experimentation but on deduction and introspection.

In its original version, the James–Lange theory is no longer tenable, in particular, since it cannot be refuted experimentally. It appears to be impossible to design an experiment in which the perception of emotions can be investigated without afferent feedback from the body (viscera and deep somatic structures). However, it is quite widely accepted that the activity in afferents from the somatic and visceral body domains shapes and amplifies the emotions. The afferent activity might be generated by activation of the efferent autonomic systems. From this point of view, the James–Lange theory of emotions is of course not at variance with the idea that different basic emotions (or groups of related affected states) can be characterized by specific autonomic motor patterns (see below) in addition to the somatomotor patterns.

There is some, although not generally accepted, consensus that there exist six basic emotions that are the product of evolution: anger, fear, disgust, sadness, surprise, happiness (Figure 11.10). This idea was formulated in Charles Darwin's famous book *The Expression of Emotions in Man and Animals* (1872). The term "basic emotion" should not be taken too literally. Each emotion is, according to Ekman and Panksepp (see Ekman and Davidson [1994]), not a single discrete separable affective state but a group of related affective states. These states are universal and the result of evolution. They unfold and develop in specific environments. They are represented in central circuits and are not the result of associative learning (for extensive discussion see Ekman and Davidson [1994]). The relatively invariant expression of these basic emotions by the motor system (above all by the facial muscles in humans and non-human primates) is represented in the central programs of the limbic system and neocortex (notably the amygdaloid complex and the orbitofrontal cortex; see Aggleton [2000]). These central programs are also responsible for the internal experience of the emotions (see Ekman and Davidson [1994]).

Ekman and his coworkers studied American actors, American college students, Americans in old age and indigenous people from West Sumatra, whose cultural background is entirely different from the Western one. The participants followed muscle-by-muscle instructions and coaching to produce facial configurations that resemble the facial expression of the different types of *basic emotions* (Figure 11.10). The *patterns of autonomic responses* (changes in heart rate, skin temperature [dependent on cutaneous vasoconstrictor

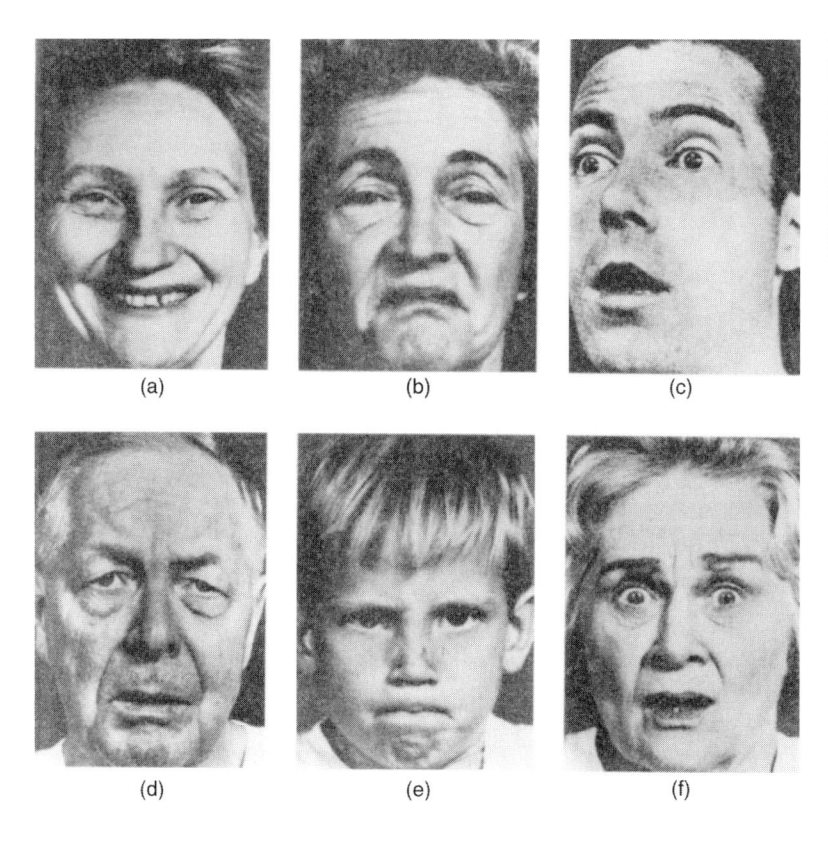

Figure 11.10 Facial expression of the six basic emotions. (a) Happiness; (b) disgust; (c) surprise; (d) sadness; (e) anger; (f) fear. From Darwin (1998, 3rd edition edited by Paul Ekman) with permission.

activity] and skin conductance [dependent on sudomotor activity]) were measured and the subjective experience reported by the experimental subjects during these activities was recorded (Ekman *et al.* 1983; Levenson *et al.* 1990, 1991, 1992; Levenson 1993). In the study of elderly Americans, autonomic activity was also measured while the subjects attempted to relive emotional experiences occurring during produced facial configurations. The patterns of autonomic reactions were found mainly to be specific for each basic emotion. This specificity was independent of cultural background, age and profession, and the three parameters (expression of emotions, relived subjective emotions and autonomic patterns) correlated significantly with each other. The three outcomes, i.e., the subjective emotional feelings, the somatomotor expression of the emotions and the autonomic responses of the emotions, were therefore concluded to be parallel (not sequential) "read-outs" of the same brain regions (Figure 11.11).

The authors came to the conclusion that the autonomic patterns that are specific for the different groups of affective states are functionally distinct adaptive responses which have developed during evolution. The authors express their view by stating that

> there is an innate affect program for each emotion that once activated directs for each emotion changes in the organism's biological state by

Figure 11.11 Changes of autonomic parameters during the six basic emotions (see Figure 11.10). The facial motor expression of the basic emotions were generated experimentally under visual control. The experimental persons did not know the type of emotion expressed. Increase of heart rate (dependent on changes in activity of parasympathetic cardiomotor neurons), decrease of skin temperature of the fingertips (dependent on skin blood flow and therefore on activity in cutaneous vasoconstrictor neurons) and increase of skin conductance (dependent on activity of sweat glands and therefore on activity in sudomotor neurons) were measured simultaneously. The relived subjective emotions experienced by the experimental subjects were reported afterwards. The patterns of the autonomic reactions and the type of relived emotions are highly correlated with each other. Data from 12 experimental subjects. Mean + SEM. Modified from Levenson et al. (1992) with permission.

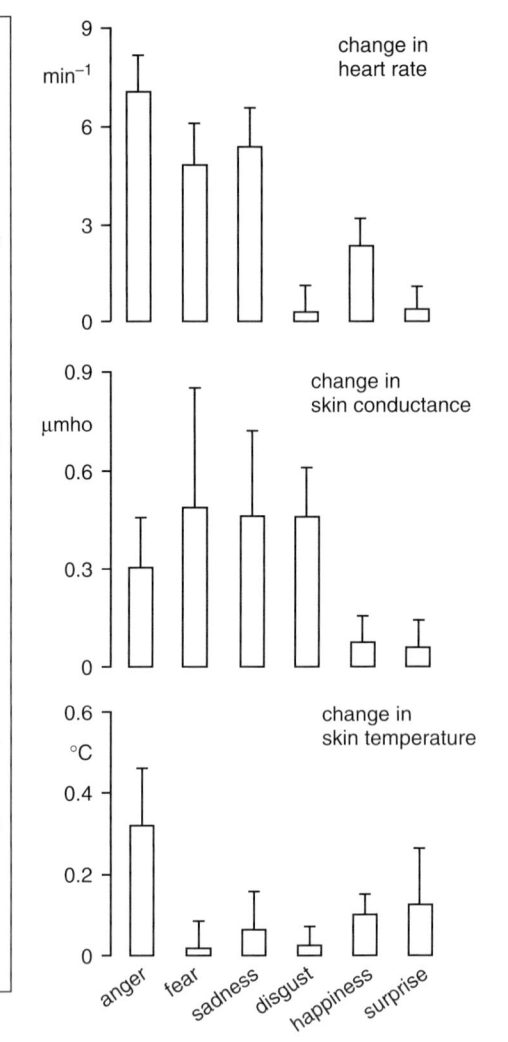

providing instructions to multiple response systems including facial muscles, skeletal muscles and the autonomic nervous system
(Levenson et al. 1990, 1991; Ekman 1992).

Finally, they come to the important conclusion that a general arousal model of the sympathetic nervous system, as originally propagated by Cannon (1928, 1939), cannot account for the differentiated autonomic responses seen during the expression of the basic emotions.

11.4.2 Central representation of emotional states and patterns of autonomic reactions

Ekman and coworkers (Figure 11.11) have shown that there is a correlation between the type of basic emotion and the autonomic responses. These data are fully compatible with the findings discussed extensively in this book showing that the autonomic, and in particular the sympathetic pathways in the periphery and centrally are functionally specific. The findings are also fully compatible with the generally accepted notion that the basic emotions (consisting of the characteristic motor expressions and the corresponding subjective emotional responses) are distinct affective states of the brain, each having also a distinct pattern of autonomic (and possibly endocrine) responses. Future experimental investigations on humans, using modern imaging methods and more refined methods to measure autonomic responses, will have to show that emotional feeling states, autonomic response patterns and patterns of activation of forebrain areas generated during the six basic emotions correlate with each other (see Anders et al. [2004]).

Using functional imaging, it can be shown that the perception of the different basic emotions from visual stimuli (faces) or auditory stimuli is correlated with characteristic patterns of activation of the forebrain (Damasio et al. 2003). Forebrain structures representing expression of emotions are the orbitofrontal cortex, the anterior and posterior cingulate cortex, the insular cortex, the anterior parietal cortex, parts of the anterior basal ganglia and the amygdala (see Phan et al. [2002]; Damasio et al. [2003]; Morris and Dolan [2004]).

As an example, functional imaging studies show that subregions of the cingulate cortex exhibit distinct responses during the basic emotions happiness, sadness or fear. These regions are distinct in their cytoarchitectonic structure, in their projections to subcortical nuclei involved in regulation of autonomic motor activity or regulation of skeletomotor activity and in their synaptic inputs (e.g., from these subcortical centers or from the midline intralaminar thalamic nuclei).

- Activity during *sadness* is greatest in the subgenual anterior cingulate cortex. This area regulates autonomic motor activity and projects to amygdala, parabrachial nuclei and the periaqueductal grey.
- Activity during *happiness* is greatest in the perigenual anterior cingulate cortex (which is located more rostrally and dorsally to the subgenual anterior cingulate cortex). This part of the cingulate cortex projects to the basal and accessory basal nuclei of the amygdala.
- During *fear* activity in the anterior midcingulate cortex is highest. This cortex is the rostral cingulate motor area and receives dense synaptic input from the amygdala.

These examples demonstrate that basic emotions may be related to specific forebrain circuits (for discussion and references see Vogt [2005]).

The amygdala orchestrates mainly the expression of negative emotions, such as anger and fear, for the somatomotor, autonomic and endocrine responses. This has been shown in the rat by LeDoux and coworkers (LaBar and LeDoux 2001). Using fear conditioning in rats as an experimental model system, this group has worked out the neural circuits in the forebrain that coordinate the behavioral reactions, including those occurring in the body involving the autonomic nervous system and neuroendocrine systems, to threatening stimuli. Figure 11.12 outlines the structures in the forebrain, hypothalamus and brain stem that are activated during fear conditioning. Any environmental or intracerebral stimulus that activates these neural centers leads to somatomotor, autonomic and endocrine responses that are characteristic for the different types of defense behavior, as has been described in Subchapter 11.3, and also for the basic emotions fear and anger. The amygdala is a key region in the limbic system that is important to recognize dangerous environmental situations and initiate the appropriate behavioral changes. The following components of the fearful behavior in which the amygdala is involved can be distinguished (LaBar and LeDoux 2001):

- Direct subcortical connections from the thalamus (lateral and medial geniculate for the visual and auditory system respectively) mediate a pre-attentive generation of fear to archetypical biologically aversive somatosensory, visual or acoustic stimuli. This activation is fast and occurs without awareness. There is no precise representation of the stimulus or environmental situation (e.g., a spider or a crocodile) (connection *1* in Figure 11.12) but only prepared response components (contours, colors etc.) are analyzed.
- Processed and discriminated stimuli reach the amygdala from unimodal association cortices (visual, acoustic etc). Via these synaptic connections, neutral conditioned stimuli (e.g., touch, visual, acoustic stimuli) can be paired with biologically important stimuli. The synaptic activation of neurons in the lateral amygdala by the conditioned stimuli is enhanced. Now the conditioned stimuli can activate the neurons and trigger the fear reaction (connection *2* in Figure 11.12).
- The conditioned fear reaction (and therefore the activation of the neurons in the lateral amygdala by the conditioned stimuli) is coactivated in the context of past experience and expectations (i.e. with regard to temporal and spatial contexts in which the stimulus occurs) with the hippocampus and medial prefrontal cortex (and probably other prefrontal areas). This leads to enhancement or extinction of the synaptic activation in the lateral and basal nuclei of the amygdala and of the fear reactions (connections *3* and *4* in Figure 11.12).
- The central nucleus of the amygdala is the output nucleus to neural circuits in the hypothalamus and brain stem that are involved in somatomotor, autonomic and neuroendocrine responses resulting in the typical defensive and arousal responses. Furthermore, the nucleus basalis (Meynert) is activated by the central nucleus of the amygdala, which activates the neocortex resulting in cortical arousal and attention.
- Monkeys with bilateral lesions of the amygdala (or of the prefrontal cortex) are no longer able to learn recognizing harmful

11.4 EMOTIONS AND AUTONOMIC REACTIONS | 497

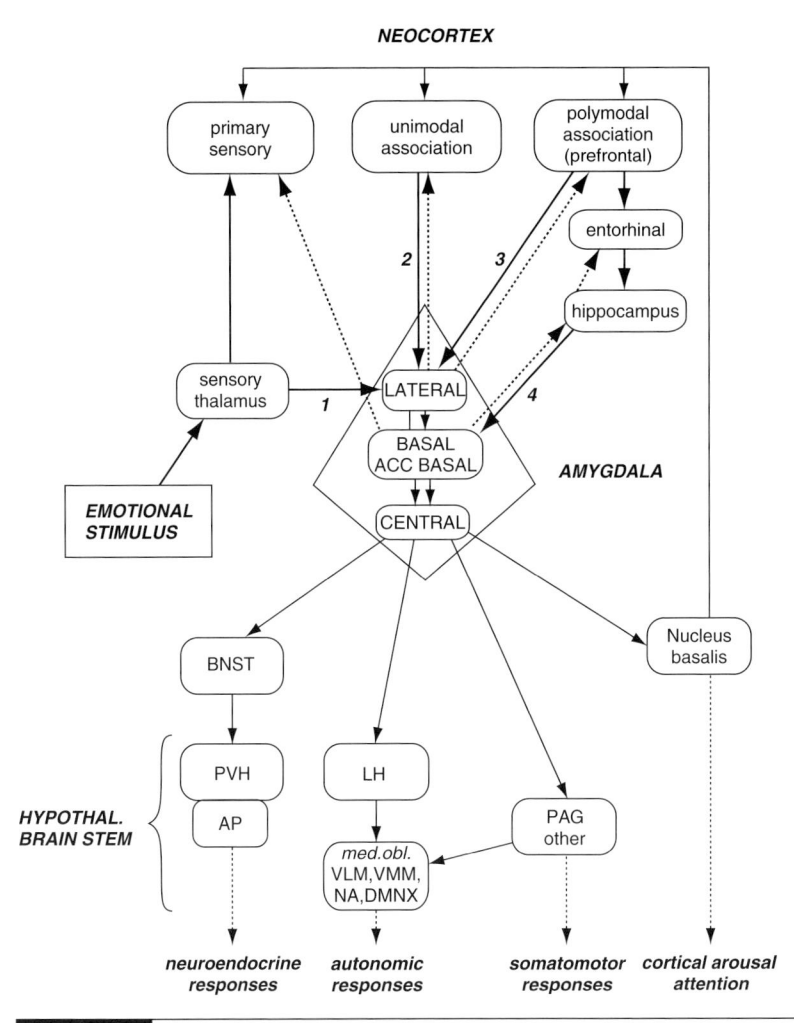

Figure 11.12 Amygdala, fear conditioning and autonomic systems. Emotional (somatosensory, visual, acoustic or olfactory) stimuli in the environment signaling danger elicit fear behavior consisting of the defensive motor behavior, central emotional feeling state, autonomic response pattern and endocrine responses. The lateral nucleus of the amygdala receives afferent information about the environmental stimuli via synaptic input directly from the thalamus (pathway 1) and from unimodal and polymodal association cortices (pathways 2 and 3) as well as from the hippocampus (pathway 4 to the basolateral amygdala). These synaptic inputs are processed in parallel during fear conditioning. Simple cues that do not require discrimination are relayed via pathway 1. The emotional event is discriminated from other events and evaluated in the context of past and expectations and relayed by pathways 2 to 4 to the amygdala. The amygdala virtually projects to every cortex region and to the hippocampus. The somatomotor, endocrine and autonomic responses elicited by fear conditioning are mediated by the central (output) nucleus of the amygdala. Cortical arousal and attention are generated from the central nucleus via the nucleus basalis (Meynert). ACC BASAL, accessory basal nucleus; AP, anterior pituitary gland; BNST, bed nucleus of the stria terminalis; DMNX, dorsal motor nucleus of the vagus; LH, lateral hypothalamus; med. obl., medulla oblongata; NA, nucleus ambiguus; PVH, paraventricular nucleus of the hypothalamus; PAG, periaqueductal grey; VLM, ventrolateral medulla. Modified from LaBar and LeDoux (2001) with permission.

situations in their environment. They are not able to acquire the "meaning" of exteroceptive (visual, auditory, somatosensory or olfactory) stimuli in social context of the behavior of the other members of the group and cannot associate these stimuli with their own affective states. Humans with bilateral lesions of the amygdala are not able to recognize faces that signal anger or fear yet they have no or little difficulty to recognize the emotional content of faces signaling happiness or surprise. These patients exhibit changed patterns of cortical activation when shown faces signaling anger or fear (Damasio et al. 2003). This shows that the amygdala is not only important for somatomotor, autonomic and neuroendocrine expression and for recognition of the emotional states of anger and fear, but also for the generation of the corresponding emotional feelings.

How do these experimental findings connect to the results of Ekman and coworkers as described above in terms of the regulation of autonomic pathways? Is the central nucleus of the amygdala, which orchestrates the autonomic and neuroendocrine responses during fear conditioned behavior in rats, also responsible for generating the autonomic response patterns during anger and fear in humans? The finding that humans with bilateral lesions of the amygdala recognize the emotional content of faces expressing happiness or surprise but no longer those showing anger or fear argues that the amygdala is not involved in the former recognition process. It would be interesting to know whether the autonomic response patterns during the basic emotions of happiness, surprise, disgust or sadness are normal in these patients with lesioned amygdalas whereas those occurring during anger and fear are absent. These studies have yet to be done.

11.5 | Integrative responses and the hypothalamus

The volume of the hypothalamus is about 1% of the brain (Swanson 1995). An individual in whom the periventricular and medial parts of the hypothalamus are destroyed (the latter extending about 800 μm lateral to the midline in rats), but with intact spinal cord, brain stem and cerebral hemispheres, is not able to survive because the basic control of the fluid matrix, temperature, food intake and metabolism, protection and reproduction fail (Table 11.2). The hypothalamus is directly or indirectly connected to every part of the brain (including the brain stem and spinal cord). Anatomically at least 2000 projections of the hypothalamic nuclei have been recognized due to the enormous development of neuroanatomical methodology. Experimental anatomical work, using modern tracing techniques (conducted particularly by Larry Swanson and coworkers), tells us that the hypothalamus consists of about 50 to 100 differentiable groups of neurons (nuclei), each being characterized by the projection of their axons, their synaptic inputs and their histochemistry (Risold et al. 1997; Swanson 2000; Petrovich et al. 2001; Watts and Swanson 2002; Thompson and Swanson 2003). Thus, we have detailed knowledge about the microstructure of the hypothalamus (at least of the periventricular and medial parts) and about the multiple physiological functions of the hypothalamus; however, it is difficult to connect the microstructure and the physiology.[12]

11.5.1 Anatomy and functions of the hypothalamus

Figure 11.13 illustrates the location of the hypothalamus as the ventral part of the diencephalon (interbrain; the dorsal part being the thalamus) that is situated between the brain stem and the cerebral hemispheres. The preoptic nuclei are sometimes considered to belong to the forebrain ganglia. Based on cytoarchitectonic and topographical criteria, the hypothalamus is conventionally divided into three longitudinal zones (periventricular zone, medial zone and lateral zone). Rostrocaudally the hypothalamus is divided, for practical reasons, into the preoptic, anterior, tuberal and mammillary part (Saper 2004). Based on the patterns of connections between neuron assemblies and on the functional systems involved (Swanson 2000; Thompson and Swanson 2003), the hypothalamus is divided into the

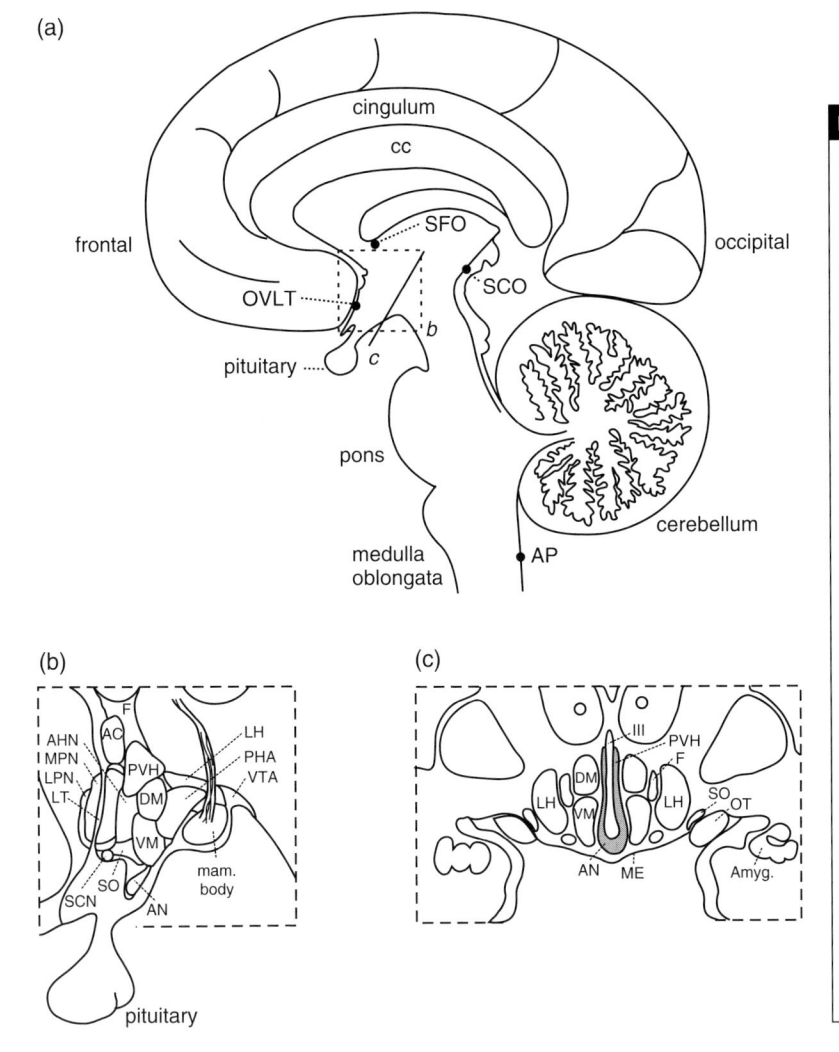

Figure 11.13 Location and components of the hypothalamus. (a) Sagittal section through the brain. The circumventricular organs are indicated (AP, area postrema; cc, corpus callosum; OVLT, organum vasculosum laminae terminalis; SFO, subfornical organ; SCO, subcommissural organ). (b, c) Magnified parts as indicated in (a) showing the nuclei of the hypothalamus. AC, anterior commissure; AHN, anterior hypothalamic nucleus; Amyg., amygdala; AN, arcuate nucleus; DM, dorsomedial nucleus; F, fornix; LH, lateral hypothalamic area; LPN, lateral preoptic nucleus; LT, lamina terminalis; mam. body, mammillary body; ME, median eminence; MPN, medial preoptic nucleus; OT, optic tract; PHA, posterior hypothalamic area; PVH, paraventricular nucleus; SCN, suprachiasmatic nucleus; SO, supraoptic nucleus; VM, ventromedial nucleus; VTA, ventral tegmental area; III, third ventricle. Modified from Iversen et al. (2000) with permission.

neuroendocrine motor zone, the periventricular region generating visceromotor patterns, the suprachiasmatic nucleus generating behavioral states, the behavior control column in the medial hypothalamus, and the lateral hypothalamus. Thus, the longitudinal zones are roughly identified with integrative functions, the classical medial zone being further divided into three parts:

- The *periventricular zone* contains mainly the endocrine motor neurons projecting to the posterior pituitary gland or to the median eminence called the *neuroendocrine motor zone* (Figure 11.14a). The magnocellular neurons secreting either vasopressin or oxytocin are located in the magnocellular division of the paraventricular nucleus, in the supraoptic nucleus and in some accessory supraoptic cell groups in between. They project to the posterior pituitary gland. In the median eminence, the axons of the secretomotor neurons release excitatory or inhibitory releasing hormones (RH, IRH) to the anterior pituitary gland. The parvicellular neurons projecting to the median eminence and releasing corticotropin-RH, growth hormone-RH, dopamine, somatostatin or thyrotropin-RH are located in a continuous zone around the third ventricle, including the anterior periventricular, the parvicellular (neuroendocrine), paraventricular and the arcuate nuclei. The parvicellular neurons secreting gonadotropin-RH are located more rostrally due to their unusual embryologic origin (Markakis and Swanson 1997; Thompson and Swanson 2003).

- The *medial zone* contains several anatomically defined nuclei of neurons (Figure 11.13b, c) that constitute, according to Swanson and coworkers, the *behavior control column* (Figure 11.14d). These nuclei project to the somatomotor centers and autonomic centers in brain stem and spinal cord, to the lateral hypothalamus and to the thalamus. This column includes the medial preoptic nucleus (MPO, lateral part), the anterior hypothalamic nucleus (AHN), large parts of the paraventricular nucleus (dorsal, lateral and ventral parvicellular parts of the PVH), the ventromedial nucleus (VMH), the supraoptic nucleus (SO), the posterior hypothalamic area and some other small premammillary nuclei. The neuron assemblies in the behavior control column determine, *first*, the somatomotor programs in the brain stem and spinal cord that underlie elementary behaviors related to defense, reproduction, ingestive behavior (nutritive, fluid) and thermoregulatory behavior, *second*, the autonomic motor programs adapting the body organs during these behaviors (cardiovascular and respiratory systems, gastrointestinal tract, pelvic organs, immune system) and, *third*, the neurendocrine output systems (Figure 11.14d). Single nuclei or subnuclei cannot be identified to have any of these integrative functions. However, groups of nuclei constituting networks can be roughly identified with particular functions:
    - *Defensive (agonistic) behavior* is organized in the AHN, the dorsomedial part of the ventromedial nucleus (VMHdm), the

ventrolateral part of the ventromedial nucleus (VMHvl) and the dorsal part of the premammillary nuclei (PMd).
- *Reproductive behavior* is organized in the MPO, the VMHvl and the ventral part of the premammillary nuclei (PMv). Both the defensive and the reproductive behavior are the basis for the social behavior.
- *Ingestive behavior* is organized by those parts of the paraventricular nucleus that do not belong to the periventricular neuroendocrine motor zone (dorsal part; lateral part).
- *Thermoregulatory behavior* is organized in the preoptic region, the anterior hypothalamus and posterior parts of the hypothalamus. Precise data about anatomical nuclei involved in this behavior do not exist (Kanosue et al. 1998; Nagashima et al. 2000).

• In the *periventricular region*, the rostral part of the medial zone of the hypothalamus contains several small nuclei that constitute, together with the dorsomedial nucleus (DMH), the *hypothalamic visceral pattern generator (HVPG)*. The rostral nuclei involved are the median preoptic nucleus (MePO), the parastrial nucleus (PS), the anterior dorsal preoptic nucleus (ADP), the anterior ventral preoptic nucleus (AVP), and the anterior ventral periventricular nucleus (AVPV). The medial part of the medial preoptic nucleus (MPNm) and the presuprachiasmatic nucleus (PSCN) probably also belong to this neural network (Figure 11.14c):
- Each component of the HVPG innervates differentially the pools of neuroendocrine motor neurons, pre-autonomic neurons located in the dorsal, ventral and lateral parts of the PVH, possibly other pre-autonomic neurons in the hypothalamus and possibly, e.g. via the PVH and the lateral hypothalamus or directly, autonomic circuits in the periaqueductal grey, pons and the medulla oblongata.
- The synaptic inputs to the nuclei of the HVPG consist of four components: (1) Sensory inputs from the NTS via the ventrolateral medulla and parabrachial nuclei, from the spinal and medullary dorsal horn, most of them also via the parabrachial nuclei, from the subfornical organ (osmolality, angiotensin II). (2) Synaptic input from the suprachiasmatic nucleus (SCN), which determines the behavioral state during circadian rhythm of the organism. (3) Synaptic input from the behavior control column, each nucleus having a distinct pattern of projection. (4) Synaptic input from the cerebral hemispheres related to cognitive and affective (emotional) processes (infralimbic area [area 25 in Figure 11.7] of the medial prefrontal cortex; ventral subiculum, medial and posterior amygdala nucleus, ventral lateral septal nucleus, bed nucleus of the stria terminalis).
- The location of the HVPG between the various input systems, related to reflexes, behavioral state, behavioral context and cerebral (cognitive and emotional) activity, and the output system to

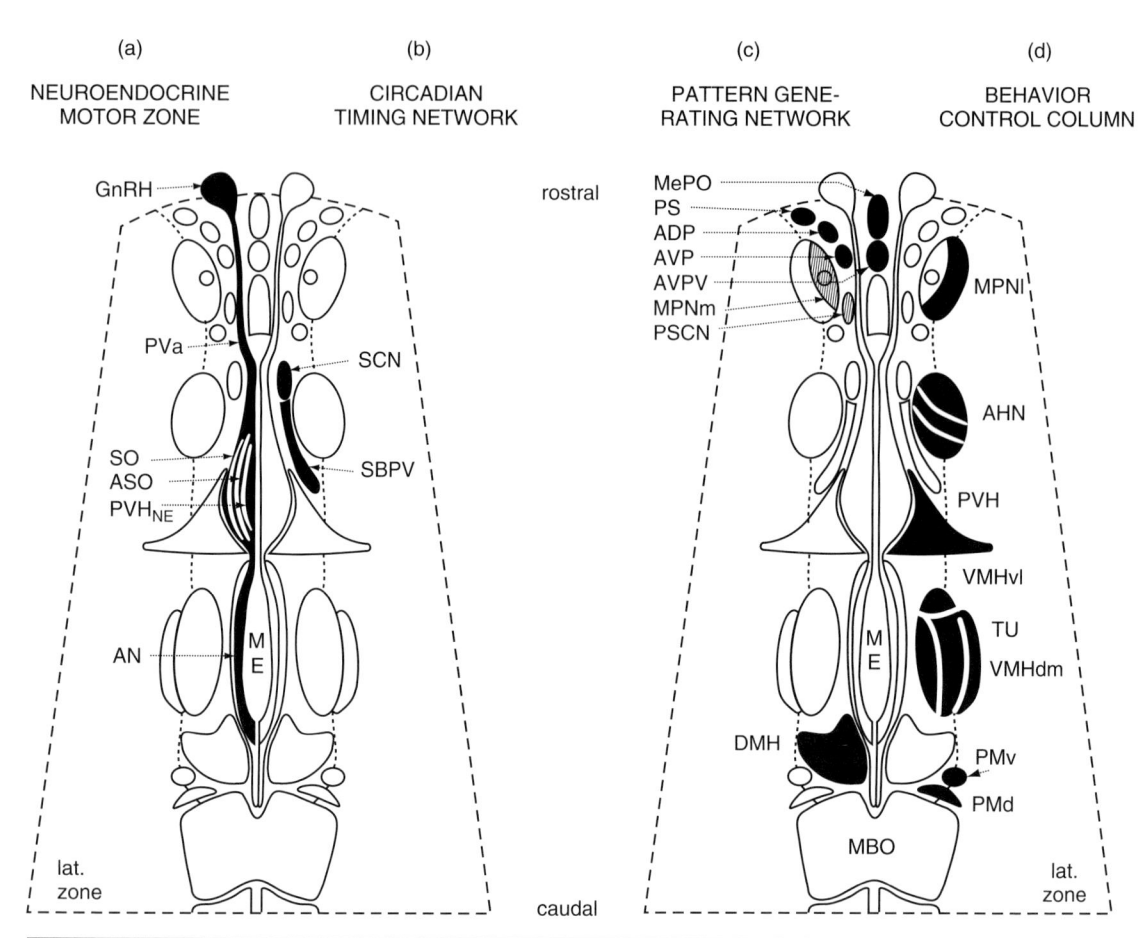

Figure 11.14 Organization of the hypothalamus into functionally different types of networks or zones. This division is based on systematic morphological studies using tracing techniques conducted on rats. The results of these studies were interpreted in the context of hypothalamic functions. The nuclei belonging to the respective network or zone are indicated in black. (a) The neuroendocrine motor zone. This zone is centered in the periventricular zone of the hypothalamus and contains the endocrine motor neurons related to the posterior pituitary or the anterior pituitary gland. (b) The circadian timing network consisting of the suprachiasmatic nucleus (SCN) and the subparaventricular zone (SBPV), which organizes the temporal structure of hypothalamic functions. (c) The hypothalamic visceral motor pattern generator network coordinating neuroendocrine and autonomic response patterns. (d) The behavior control column representing regulation of defensive behavior, ingestive behavior (nutrition, fluid balance), reproductive behavior and thermoregulatory behavior. *Abbreviations*: (a) AN, arcuate nucleus; ASO, accessory supraoptic cell groups; GnRH, nucleus synthesizing gonadotropin-releasing hormone; PVH_{NE}, neuroendocrine part of the paraventricular nucleus; PVa, periventricular nucleus anterior. (c) ADP, anterodorsal preoptic nucleus; AVP, anteroventral preoptic nucleus; AVPV, anteroventral periventricular nucleus; DMH, dorsomedial nucleus; MePO, median preoptic nucleus; MPNm, medial part of medial preoptic nucleus; PS, parastrial nucleus; PSCN, presuprachiasmatic preoptic nucleus. (d) AHN, anterior hypothalamic nucleus; MPNl, lateral part of the medial preoptic nucleus; PMd, dorsal premammillary nucleus; PMv, ventral premammillary nucleus; PVH, paraventricular nucleus; TU, tuberal nucleus; VMHdm, dorsomedial part of the ventromedial nucleus; VMHvl, ventrolateral part of the ventromedial nucleus. MBO, mamillary body; ME, median eminence. Modified from Thompson and Swanson (2003) with permission.

the neuroendocrine and autonomic circuits allows a selection of variable patterns of autonomic and endocrine responses.
- The *suprachiasmatic nucleus (SCN)* in the medial zone is the dominant *mammalian circadian pacemaker*. Its endogenous rhythm is entrained by the retina via the retino-hypothalamic tract to the light–dark cycle of the environment and determines the temporal organization of virtually all body functions. Most of its projections are confined to intrahypothalamic nuclei, some of them via the subparaventricular zone (SPVZ); some projections are to the basal forebrain and the thalamus (Moore 2003). The SCN nuclear complex is the *circadian timing network* of the body and involved in the organization of the *behavioral states of the organism* (sleep and wakefulness, arousal) (Figure 11.14b).
- The lateral zone of the hypothalamus does not contain well-defined nuclei of neurons, but appears to be a *liaison structure* or crossroad connecting forebrain structures, medial and periventricular zone of the hypothalamus, brain stem and spinal cord. The lateral hypothalamus has been little explored.

The hypothalamic nuclei receive multiple neural afferent inputs via various pathways (involving the spinal and medullary dorsal horn, the NTS, the ascending A1 area in the medulla oblongata, the parabrachial nuclei) and non-neural inputs from the body via the circumventricular organs (Figure 11.13a; plasma osmolality, angiotensin II, cytokines), via neurons that are thermosensitive or sensitive to hormones of peripheral endocrine organs (adrenal cortex, thyroid gland, sexual glands, gastrointestinal endocrine glands [insulin, glucagon, ghrelin, pancreatic peptide YY, glucagon-like peptide 1, cholecystokinin], adipose tissue [leptin]). Furthermore they receive multiple synaptic inputs from various structures of the forebrain related to cognitive and emotional activity (direct projections from the nuclei in the amygdala, septal nuclei, bed nuclei of the stria terminalis, subiculum, prefrontal cortex [medial and orbital prefrontal cortex, infralimbic cortex]).

11.5.2 Functional model of the hypothalamus

The anatomical characterization of the hypothalamus in longitudinal (rostrocaudal) zones arranged mediolaterally, by the different types of nuclei (in the periventricular and medial hypothalamus) and by their afferent and efferent neural connections does not reveal the mechanisms by which such a small part of the brain is able to regulate the body functions. Swanson and coworkers have proposed a general model to explain the mechanisms underlying the basic integrative functions of the hypothalamus (Figure 11.15). This explanatory model helps us to understand how somatomotor, autonomic and neuroendocrine responses are integrated to control body temperature, the fluid matrix, nutritional state, reproduction and defense and to adapt these complex regulations to the needs of the organism. These needs change during challenges from the environment that are related

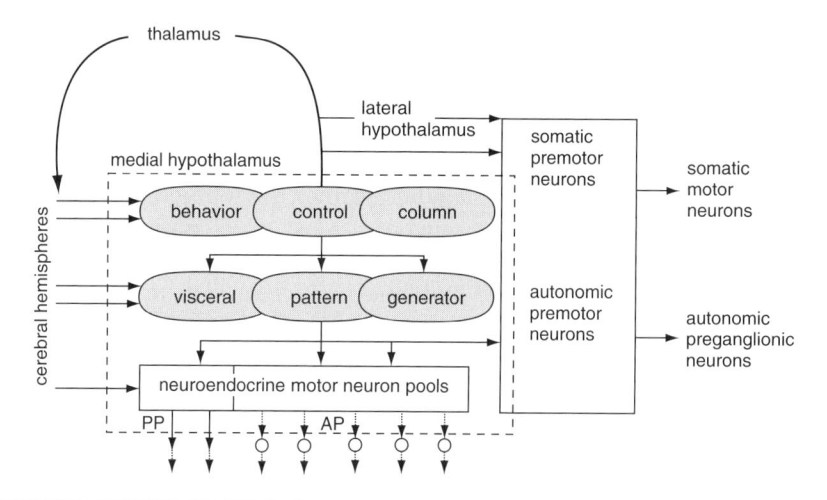

Figure 11.15 A model of the functional organization of the hypothalamus. 1. Neurons of the behavior control column project (see Figure 11.14d), *first*, to regions in brain stem, and even spinal cord, that are associated with somatic and autonomic motor systems (descending projections), *second*, to thalamocortical loops (ascending projections), *third*, to the adjacent lateral hypothalamic area (both being involved with modulation of the behavior state) and *fourth*, to nuclei of the visceromotor pattern generator (HVPG) network. 2. Neurons of each nucleus of the HVPG project to characteristic combinations of neuroendocrine motor neuron pools and pools of preautonomic neurons in the paraventricular nucleus. Neurons in the behavior control column, in the HVPG and in the neuroendocrine motor zone are under multiple influence of the cerebral hemispheres. AP, anterior pituitary gland; PP, posterior pituitary gland. Modified from Swanson (2000) and Thompson and Swanson (2003) with permission.

to the search for nutritive substances and water, to environmental temperature changes, to an aggressor, to toxic substances or to reproductive cues (mating and raising the next generation).

The model as outlined in Figure 11.15 includes the four functional components of the hypothalamus as discussed before: (1) the neuroendocrine motor zone; (2) the periventricular zone representing the HVPG; (3) the behavior control column (medial nuclei) representing the basic functional states (regulation of defense, nutrition, body fluid, body temperature, reproduction and temporal organization) and (4) the lateral hypothalamus as an intersection of communication between cerebral hemispheres and circuits regulating body functions (somatomotor, neuroendocrine, autonomic):

- The behavior control column has descending projections to somatomotor and autonomic circuits, ascending projections to thalamocortical circuits and projections to the lateral hypothalamus.
- The behavior control column projects topographically and in a functionally specific way (i.e., related to the basic hypothalamic functions represented in the nuclei of this control column [see Table 11.2]) to the HVPG. This guarantees a coordinated activation of neuroendocrine and autonomic circuits (in the PVH and probably in other nuclei).
- The behavioral state (circadian, diurnal, ultradian, mensual, seasonal rhythms) is modulated in the nuclei of the behavior control

Table 11.2 Integrative functions of the hypothalamus

Function	Behavior	Nuclei in hypothalamus	Afferent systems hormones, cytokines	Autonomic systems	Endocrine systems, hormones
Thermoregulation	Thermoregulatory behavior	Preopt. region, AH, posterior H, OVLT (pyrogenic zone, fever)	Peripheral thermoreceptors, central thermosens. (preopt.), cytokines (fever)	SyNS (skin [CVC, SM]). (BAT [rodents])	TRH/Thyr (anterior pituitary)
Reproduction, sexual behavior	Sexual behavior and sexual orientation	MPNl (male; human dimorphic), VMl (female), TU, PMv	Afferents from sexual organs, afferents from sensory systems	SyNS (thor.-lumb.), PaNS (sacral) (sexual organs)	GnRH, FSH/LH (anterior pituitary)
Volume-, osmoregulation (fluid homeostasis)	Drinking behavior, thirst	PVHmc, SOmc, AVPV, MePO, OVLT, SFO	Osmorecept. in OVLT & liver, volume rec. ri. atrium (vagal), angiotensin II via SFO	NTS, SyNS (kidney)	Vasopressin (posterior pituitary)
Regul. of nutrition, regul. of metabolism	Nutritive behavior, hunger/satiety	AN, PVHd, v, l, VMH (insulin secretion)	Vagal afferents & hormones from GIT (CCK, ghrelin, glucagon, GLP-I, insulin), leptin from BAT, glucose	Enteric nervous system, PaNS (DMNX), SyNS (BAT [rodents], WAT)	Orexin
Temporal organization of body functions	Sleep-waking behavior, circadian/endogenous rhythm of body functions	SCN, SBPV	Afferents from retina (retino-hypothalamic tract)	SyNS, PaNS, SyNS to pineal gland	Melatonin (pineal gland)
Body protection (acute, e.g. during pain and stress)	Defense behavior (fight, flight, quiescence)	AHN, VMHdm, PMd, PAGdl, l, vl	Nociceptive afferents	SyNS, PaNS (cardiovascular system [CMN, MVC, VVC etc.])	CRH/ACTH (anterior pituitary gland), adrenaline (SA system)

Table 11.2 (cont.)

Function	Behavior	Nuclei in hypothalamus	Afferent systems hormones, cytokines	Autonomic systems	Endocrine systems, hormones
Immune defense	Defense of toxic substances and situations (sickness behavior)	PVH, ?	Cytokines	SyNS (to immune tissue)	CRH/ACTH (anterior pituitary gland), adrenaline (SA system)

Notes:

Listed are the behavior, the nuclei in the hypothalamus, the afferent systems, the autonomic (parasympathetic and sympathetic) pathways (shaded) and the endocrine systems (including the hormones) that are involved in these hypothalamic functions. The function "body protections" includes fast defense behaviors (see Table 11.1) as well as immune defense.

Abbreviations: AH, anterior hypothalamus; AHN, anterior hypothalamic nucleus; AN, arcuate nucleus; AVPV, anteroventral periventricular nucleus; BAT, brown adipose tissue; CCK, cholecystokinin; CRH/ACTH, cortocotropin-RH/adrenocorticotropic hormone; CVC, cutaneous vasoconstrictor neurons; CMN, cardiomotor neurons (symp., parasymp.); d, dl, dm, dorsal, dorsolateral, dorsomedial; DMNX, dorsal motor nucleus of the vagus; FSH/LH, follicle-stimulating hormone/luteinizing hormone; GIT, gastrointestinal tract; GLP-1, glucagon-like peptide 1; GnRH, gonadotropin RH; H, hypothalamus; l, lateral; mc, magnocellular; MePO, median preoptic nucleus; MPNl, lateral part of the medial preoptic nucleus; MVC, muscle vasoconstrictor neurons; NTS, nucleus tractus solitarii; OVLT, organum vasculosum laminae terminalis (osmosensors); PAG, periaqueductal grey; PaNS, parasympathetic nervous system; PM, premammillary nucleus; PVH, paraventricular nucleus of the hypothalamus; PYY, pancreatic peptide YY; RH, releasing hormone; SA system, sympatho-adrenal system (adrenal medulla); SBPV, subparaventricular zone; SCN, suprachiasmatic nucleus; SFO, subfornical organ (angiotensin sensibility); SM, sudomotor neurons; SO$_{mc}$, magnocellular part of supraoptic nucleus; SyNS, sympathetic nervous system:thor.-lumb., thoracolumbar; TRH/Thyr, thyrotropin-RH/thyroxine; TU, tuberal n; v, vl, ventral, ventrolateral; WAT, white adipose tissue; VMH, ventromedial nucleus; VVC, visceral vasoconstrictor neurons.

References: Gore and Roberts (2003); McKinley et al. (2004); Moore (2003); Stricker and Verbalis (2003); Woods and Stricker (2003)

column and of the HVPG by the SCN without changing the functions of these nuclei.
- The HVPG network consists of six discrete nuclei (or nodes) that are reciprocally connected with each other. This visceral pattern generating network determines the patterns of activation of neuroendocrine channels (pools of neuroendocrine motor neurons) and of final autonomic pathways (Table 11.2). The mechanisms underlying the generation of the function-specific pattern by this network are unknown.
- The hypothalamic complex consisting of behavior control column, HVPG, periventricular neuroendocrine motor zone and preautonomic circuits (in the lateral hypothalamus and PVH) is under multiple control by the cerebral hemispheres. This powerful hemispheric input coordinates the hypothalamic functions with the cognitive and emotional state of the organism.

This general model, based on anatomical tracing studies and functional considerations, is highly speculative. However, it helps to understand the complex control of integrative functions of the hypothalamus. It will serve as a platform for future functional studies, using in vivo and in vitro animal models.

11.6 | Synopsis: the wisdom of the body revisited

11.6.1 Autonomic nervous system and behavior

As already mentioned in the introduction of this book the behavior of vertebrates (including humans) consists of somatomotor, autonomic and neuroendocrine responses generated by the target cells of the motor neurons of these three systems. The pattern of activity in these output channels is dependent on: (1) The behavior control programs in spinal cord, brain stem and hypothalamus that orchestrate the somatomotor, autonomic and endocrine output channels. (2) The afferent feedback from the body (hard-wired, hormonal as well as humoral) to all levels of central integration. (3) The input from the cerebral hemispheres related to cognitive and emotional processes. (4) Intrinsic processes in the central nervous system that determine the behavioral state of the organism (sleep stages, arousal) and the activity in output channels that are not dependent on afferent activity or activity in the cerebral hemispheres (e.g., the resting activity of neurons of peripheral autonomic pathways occurring either randomly or in relation to intrinsic oscillators, such as the respiratory pattern generator in the medulla oblongata, the circadian pattern generator in the SCN, etc.). This basic conceptual plan underlying the behavior of the individual is graphically depicted in Figure 2 in the introduction of this book. This concept has been developed by Swanson (Swanson 2000, 2003) and serves as a framework for the understanding of integrative action of the autonomic and neuroendocrine systems by the brain. As has been detailed in Chapters 3 and 4 and repeatedly emphasized and exemplified in Chapter 5 (enteric nervous system), Chapter 9 (spinal cord) and Chapter 10 (medulla oblongata), reflexes, generated in functionally distinct autonomic neurons, are components of the readouts of the central "sensorimotor" programs.

11.6.2 Integrative autonomic responses and forebrain

In this book I have described the organization of the peripheral and central nervous systems, focusing on the functional specificity of the final autonomic pathways with respect to the target organs, defined on the basis of neurophysiological investigations. I have tried to identify the principles of organization. Wherever we looked in the periphery, in the spinal cord, in the lower brain stem or, more generally, in the upper brain stem and hypothalamus, we concluded that the regulation of the target organs by the autonomic nervous system can only be understood if the generation of the electrical signals in the peripheral autonomic pathways by the central nervous system is function specific. In view of the many function-specific peripheral autonomic pathways and their characteristic signal transmission described in Chapters 3 to 8, the question arises, *how is the central nervous system able to distinguish between these many autonomic channels and to activate them differentially?*

The discharge patterns generated in the neurons of the peripheral autonomic pathways clearly demonstrate that the central nervous system is capable of differential activation. What are the principles of organization in spinal cord, brain and hypothalamus that enable the autonomic systems to function precisely during the behavioral repertoire that is available to the forebrain? How does the forebrain master these many autonomic machines? Looking at the microstructure of the central systems related to autonomic regulations we are astounded by the differentiation, yet we are barely able to bring these complex anatomical central structures together with the functions they are involved in.

There are four sets of independent data from different disciplines that teach us that the secret of the precise autonomic control systems in relation to the motor behavior of the organism has to be found in the brain:

1. Precise regulation of the inner milieu and organ systems. This regulation is traditionally described with the concept of homeostasis.
2. The specificity of the neurophysiological response patterns of the peripheral autonomic neurons and of the central neurons associated with the peripheral autonomic pathways (e.g., sympathetic or parasympathetic premotor neurons in the brain stem; autonomic interneurons in the spinal cord).
3. The detailed microstructure (location, projection of axons, synaptic inputs, histochemistry) of neuron assemblies in the spinal cord, brain stem and hypothalamus that are involved in regulating the activity in the peripheral autonomic pathways.

4. The temporary adaptation of the regulation of the inner milieu during extreme challenges (e.g., during extreme climatic and other environmental changes during bacterial or viral infection), a process called allostasis (see Subchapter 11.2). Adaptation of the autonomic control systems during external load occurs in anticipation of somatomotor actions. This anticipatory adaptation of autonomic control is dependent on the cerebral hemispheres and reflected in the dense reciprocal communication between autonomic centers in brain stem and hypothalamus on one side and in the cortical and subcortical structures of the cerebral hemispheres on the other side.

The principles of autonomic integrative action already apply to the sympathetic and parasympathetic systems in the spinal cord (Chapter 9). It contains many autonomic sensorimotor programs that underlie the coordinated activation of spinal autonomic pathways to physiological stimulation of spinal afferents. These sensorimotor programs are represented in pools of interneurons that are synaptically connected to preganglionic neurons and with each other. Spinal afferent systems form distinct synapses with these pools of interneurons. Preautonomic neurons in brain stem and hypothalamus access these spinal autonomic programs by forming synapses with the interneurons and/or preganglionic neurons (see Subchapter 9.4). A further characteristic already seen at the level of the spinal cord is that some of these systems generate intrinsic activity independent of the afferent and supraspinal inputs. Finally, the spinal cord is able to coordinate activity in the autonomic systems and the somatomotor systems (see sacral spinal systems; Subchapter 9.3). Thus, three of the four components of neural regulation of behavior by the brain as defined by Swanson are already present in the spinal cord: autonomic sensorimotor programs coordinated with somatic sensorimotor programs, multiple sensory feedbacks and intrinsic generators of activity.

The same applies to the lower brain stem (the reader is reminded of the discussion on the origin of spontaneous activity and its pattern in peripheral autonomic neurons and the coordination between autonomic and somatic sensorimotor programs during the regulation of respiration and gastrointestinal function in Chapter 10), yet this coordination is incomplete and does not instigate behavior. Full coordination of autonomic, somatomotor and sensory changes can be generated from the PAG and from Barrington's nucleus in the defense behavior and in the evacuation of pelvic organs (urinary bladder and hindgut), respectively. Furthermore, the basic neural machinery coordinating autonomic and somatomotor responses is probably present in the mesencephalon and pons for the diving response, the tonic immobility, the freezing reaction, the vigilance reaction and other types of response pattern. But these fast initiated coordinated response patterns too either cannot be elicited on their own or only in a rudimentary way by physiological afferent stimuli when the hypothalamus and forebrain have been removed or inactivated. In order to be fully effective these mechanisms that lead to coordinated motor and autonomic responses need differential activation from the hypothalamus. Chronic hypothalamic cats exhibit apparently normal homeostatic behaviors consisting of coordinated neuroendocrine, autonomic and somatomotor responses (see Table 11.2).[13]

Both Cannon and Hess and others like Bard, Brooks, Fulton (Fulton 1949)[14] and Ranson (Ranson and Clark 1959; Ranson and Magoun 1939) were fascinated by the integrative behavior of the autonomic nervous system throughout their scientific life. The brain's role in integrative action of the autonomic nervous system remained a mystery. However, they knew intuitively, and believed, that the secret of autonomic regulation is buried in the integration between signals generated in the cerebral hemispheres that are related to the environment (which includes memory and anticipation in the context of cognition and emotions), the behavioral state of the organism, the afferent feedback from the body and the efferent autonomic and neuroendocrine pathways to the body. From this viewpoint, it is clear why the afferent feedback systems from deep somatic and visceral tissues cannot be considered to be part of the autonomic systems, although they are indispensable for the function of the autonomic systems (see Subchapter 2.1). Furthermore, it is now clear that a dichotomous concept of the autonomic nervous system into sympathetic and parasympathetic, being antagonistically organized, could never explain how the brain directs the control of the body's organs.

11.7 | Future research questions

In the field of the autonomic nervous system the exciting research developments appear to be in the subcellular world and in the world of the brain organizing the multicellular machines and their neural programs to a unity in somatomotor, autonomic and neuroendocrine behavior and in perception of the body. Today we would rephrase the title of Cannon's book *The Wisdom of the Body* to "The Wisdom of the Brain to Direct the Body."

The next most interesting and promising scientific activities in the field of the autonomic system research are[15]:

1. Working out the cellular and subcellular mechanisms by which each type of autonomic postganglionic neuron influences its target cells and tissues in the periphery. What are the functions of junctional and extrajunctional receptors and of the intracellular postreceptor pathways connected to them in the regulation of autonomic target tissue? Working out the cellular mechanisms by which functionally distinct autonomic pathways, which are directed by the brain, influence the target tissues. These are two distinct sets of problems. Much of the understanding will probably come from neural network analysis of the effects of firing patterns once one knows how the target tissues function and how the various postganglionic neurons influence these (Subchapters 7.2, 7.3).

2. Autonomic pre- and postganglionic neurons contain many neuropeptides colocalized with classical transmitters (acetylcholine, noradrenaline, adenosine triphosphate). What are the functions of these neuropeptides in the neural regulation of autonomic target organs? (Subchapters 1.4, 6.6).
3. The role of interstitial cells of Cajal (ICC), which generate pacemaker activity, to mediate activity in autonomic postganglionic fibers to autonomic target cells (smooth muscle cells, secretory epithelial cells) in non-gastrointestinal cavity organs (Subchapter 5.4).
4. Connected to these fundamental questions is research in the tradition of the research conducted in the last 20 years (see Chapters 4 to 7) that will give us leads to understand why autonomic control runs so smoothly under healthy conditions, or, to put it the other way round, why the failure of autonomic control in disease has catastrophic consequences. To mention a few examples, we do not understand:
   - The mechanisms by which the sympathetic pathways to the kidney regulate resistance, renin release and sodium reabsorption differentially (Subchapter 4.4).
   - The mechanisms by which cutaneous vasodilation in the proximal extremities or trunk during body warming or in skin of the face during emotional stimuli is generated by the sympathetic innervation. After all this vasodilation is unlikely to be generated by withdrawal of activity in vasoconstrictor neurons (Subchapter 4.2).
   - The mechanisms by which the different functions of the gastrointestinal tract related to motility, secretion and defense are regulated via parasympathetic and sympathetic pathways (Subchapters 5.7, 10.7).
   - The mechanisms by which activity in parasympathetic sacral pathways and sympathetic lumbar pathways is integrated in the control of each of the pelvic organs (Subchapters 4.3, 9.3).
   - The organization of parasympathetic pathways supplying tracheal smooth musculature (excitatory and inhibitory bronchomotor pathways) or tracheal glandular tissue (tracheal secretomotor pathway) (Subchapter 4.8).
   - The function of the parasympathetic cardiomotor neurons with unmyelinated fibers located in the most lateral part of the dorsal vagal motor nucleus (Subchapter 4.8).
   - The function of the many silent sympathetic preganglionic neurons converging, together with spontaneously active preganglionic neurons, on postganglionic muscle or cutaneous vasoconstrictor neurons (Subchapter 4.7).
   - The function of non-nicotinic (muscarinic or peptidergic) transmission onto postganglionic vasoconstrictor neurons innervating skin or skeletal muscle or of non-cholinergic transmission onto postganglionic vasodilator and secretomotor neurons innervating pelvic organs (Subchapters 6.6).

- Mechanisms underlying the role of adrenaline released by the adrenal medulla in nociceptor sensitization, control of inflammation and memory consolidation (Subchapter 4.5).

5. A special and up to now entirely unsolved problem is the interaction between the autonomic nervous system and the immune system (see Subchapters 4.6 and 5.6). Unraveling this communication is important so that we learn how the brain controls and orchestrates protective body reactions and is involved in the regulation of inflammation and hyperalgesia (see Subchapters 4.5, 4.6 and 5.6). It is unlikely that this problem can be approached by measuring plasma levels of adrenaline and noradrenaline. It appears more likely that a special sympathetic pathway exists by which the brain sends its signals to the immune and related cells. Connected to this problem is the question as to the nature of the afferent messages from the immune tissue to the brain. Vagal afferents may be important in this function, in addition to cytokines (Subchapter 2.3).

6. An important problem not addressed in this book is the development, maintenance and plasticity of the functional specificity of the peripheral autonomic pathways. The origin of the autonomic neurons from the neural crest, which has been extensively studied using various methods, is established (Le Douarin and Kalcheim 1999). Details of the migration routes and differentiation of precursor cells and the signaling molecules involved have to be worked out (Anderson 1993; Ernsberger *et al.* 2005). Considerable progress has been made in the search for the molecular signals that are important in the differentiation of autonomic neurons by the characterization of BMP (bone morphogenetic protein) growth factors and Phox2 (paired-like homeobox-containing) transcription factors (Ernsberger 2000; Goridis and Rohrer 2002). However, questions as to the development of the functional diversity of autonomic neurons are entirely unclear (Ernsberger 2001). Which conditions enable the different functional classes of autonomic neurons to "know" what they are and to what target cells they should be connected to? What are the long-term trophic interactions between target cells and postganglionic neurons to maintain their functional specificity?

7. As it is obvious from the descriptions in Chapters 8 to 11, we have just begun to unravel the central circuits in the spinal cord, brain stem and hypothalamus that control the peripheral autonomic pathways. Functional anatomical studies, using tracing, histochemical, genetic and other methods, have resulted in an enormous wealth of data. However, these data do not tell us how the circuits work in the control of the autonomic pathways. Any attempt to conduct this research conceptually on the basis of a unifying concept of the functioning of sympathetic and parasympathetic nervous system is going to fail. A few examples related to Chapters 9 and 10 illustrate this, and each of them can be detailed for the various autonomic systems:

- I have hypothesized in Chapter 9 that spinal integration is important in order to understand central regulation of autonomic systems. This has to be shown for the various autonomic systems, i.e. it has to be shown how signals in supraspinal autonomic premotor neurons and autonomic spinal circuits interact in this regulation. Future research will show that this spinal integration varies between different autonomic systems. An important aspect in the experimental work, and in the interpretation of the experimental data, is the increased encephalization of vertebrates and therefore the increasing descending control of spinal autonomic circuits by supraspinal centers in primates compared to rats. For example, the relative volume of the spinal cord relative to the major divisions of the central nervous system is 35% in the rat and 2% in the human (Swanson 1995).
- The NTS is the afferent input gate for all information from the viscera via the vagus nerves. It consists of a collection of integrative nuclei. The way this integration and its modulation by other nuclei works is barely understood (Subchapter 8.3).
- Integration between regulation of respiration and regulation of autonomic systems is still an enigma (Subchapter 10.6).
- The origin of resting activity in sympathetic neurons involved in cardiovascular regulation has now been discussed for almost 150 years and is still unclear, not to mention the origin of resting activity in other functional types of sympathetic and in parasympathetic systems (Subchapter 10.2).
- The baroreceptor reflex pathways are the best-explored central autonomic pathways. However, the way these reflex pathways work in phasic and tonic control of arterial blood pressure is not clear. Furthermore, it is not clear whether there exist alternative pathways. Reflex pathways related to other cardiovascular afferents, including chemoreceptor afferents are little explored (Subchapters 10.3, 10.4).
- Functionally distinct classes of sympathetic premotor neurons, projecting to sympathetic preganglionic neurons (or interneurons) regulating resistance vessels (blood pressure control), cutaneous blood vessels (temperature control) or brown adipose tissue in the rat, have been shown to exist in the rostral ventrolateral medulla and in the caudal raphe nuclei. What are the functional characteristics of other types of autonomic premotor neurons projecting to autonomic spinal circuits? (Subchapters 10.2, 10.4).
- Research on central regulation of gastrointestinal functions via the dorsal vagal motor complex is still in its infancy. Unraveling the functioning of this structure is necessary to understand how forebrain and hypothalamus control gastrointestinal functions (Subchapter 10.7).

8. As described in this last chapter, the mesencephalon and hypothalamus are almost entirely unexplored neural territories as far as the cellular mechanisms underlying their functions are

concerned. However, detailed behavioral studies, studies of systems' physiology and functional anatomical studies have paved the way for studies on the single neuron level, combined with studies addressing whole systems, that will clarify the neural mechanisms underlying integrative functions represented in the mesencephalon and hypothalamus as listed in Tables 11.1 and 11.2. These functions are related to:
   - fast behavioral reactions such as the diving response, exercise, defense reactions, tonic immobility, vigilance or freezing reaction and
   - complex hypothalamic functions, such as thermoregulation, volume- and osmoregulation, regulation of reproduction, regulation of nutrition and metabolism, regulation of body protection including immune defense, regulation of temporal organization of all body functions.
9. This book does not describe how external or internal stress involves the autonomic nervous system in the protection of the body (Goldstein 1995, 2000). Research in this field will show that regulation of autonomic systems related to the cardiovascular system (heart, blood vessels), body temperature, metabolism, gastrointestinal tract and regulation of the nociceptive system are highly coordinated to protect the body (Mason 2001).
10. To work out which parts of the cerebral hemispheres represent the autonomic control systems, the neural connections between these central representations and the circuits in spinal cord, brain stem and hypothalamus, and how this communication functions during distinct autonomic regulations is a big challenge. Everyday and clinical experience suggests that maintaining health and recovery from disease is very much dependent on mental activity, i.e., on activity in the forebrain related to cognition and emotions. However, we have no neurobiological concept as to how this is brought about. It is not far-fetched to predict that systematic research on the forebrain structures in relation to the peripheral autonomic pathways and the corresponding circuits in spinal cord, brain stem and hypothalamus will solve the problem. This will result in a reformulation of the concept of allostasis (Subchapter 11.2) and in the formulation of the neurobiological basis of psychosomatic relations. It will provide answers about how the brain influences the body in health and disease. The autonomic and endocrine systems will be in the center of this concept.

Conclusions

In Chapters 8 to 10 I have described that spinal cord and medulla oblongata contain neural circuits linked to the final autonomic pathways. The circuits are integral components of regulation of the cardiovascular system, the gastrointestinal tract, the respiratory tract, the pelvic organs, the eye and some other targets. In this chapter the role of mesencephalon and the hypothalamic-limbic system in regulation of body functions by the autonomic nervous system has been summarized.

1. The hypothalamus and cerebral hemispheres contain the representations of elementary motivational behaviors. These representations receive multiple afferent feedback and are responsible for the integration of autonomic, neuroendocrine and somatomotor regulations. In parallel they generate affective body states and emotional feelings closely associated with the autonomic control systems that are represented in the telencephalon. The telencephalic structures representing emotions are connected with the central representations involved in autonomic and neuroendocrine control.
2. The dorsolateral, lateral and ventrolateral cell columns in the periaqueductal grey of the mesencephalon contain the neural circuits representing the autonomic and somatomotor components of defense behaviors, confrontation, flight and quiescence. These circuits are reciprocally connected with the autonomic centers in the medulla oblongata (rostral and caudal ventrolateral medulla, ventromedial medulla, raphe nuclei), with the hypothalamus (lateral, dorsal, medial) and with the medial and orbital prefrontal cortex. They are quickly activated by the cortex during dangerous situations in the environment or in the body and represent the basic neural machinery for active and passive coping.
3. Coordinated autonomic responses are quickly generated by signals from the telencephalon during diving, freezing, tonic immobility, exercise and other behaviors. These autonomic responses occur in anticipation of the somatomotor responses demonstrating that the cortical signals have direct access to the autonomic centers, independent of the afferent feedback, but in conjunction with it. The underlying mechanisms of these cortical effects are unknown but probably involve the lateral hypothalamus and midbrain.
4. Conditioned emotional responses are accompanied by distinct cardiovascular changes mediated by autonomic pathways. Lesions in the perifornical region of the lateral hypothalamus in monkeys abolish cardiovascular changes without affecting emotional responses.
5. The basic emotions in humans are accompanied by autonomically mediated response patterns characteristic for each emotion. Emotional states, somatomotor patterns (facial expression) and autonomic patterns are connected. The adaptive autonomic patterns during anger and fear are generated and mediated by the amygdala. The central structures involved in the generation of the autonomic patterns of the other basic emotions (happiness, surprise, sadness, disgust) are unknown.

6. The hypothalamus contains the neural structures that integrate and coordinate autonomic, neuroendocrine and somatomotor responses to basic (elementary) behaviors such as defensive, reproductive, ingestive and thermoregulatory behavior. The organization of these behaviors depends on the neuroendocrine motor zone, the periventricular zone constituting the hypothalamic visceral pattern generator, the behavior control column in the medial hypothalamus (representing hypothalamic functional states), the suprachiasmatic nucleus representing the behavior states and the lateral hypothalamus as the intersection of communication to and from the brain stem, spinal cord and cerebral hemispheres. The hypothalamic centers that are responsible for the homeostatic control mechanisms are under multiple control of the cerebral hemispheres. This is the basis for the adaptation of these control systems to external and internal challenges.
7. The homeostatic regulation of body function by the autonomic nervous system and its adaptation to short- and long-term changes in the environment or in the body (allostasis) require functionally differentiated peripheral autonomic pathways and specific central pathways. The underlying central mechanisms of these homeostatic regulations are still a puzzle.
8. Future research focussing on the brain to unravel these mechanisms will make clear how the "wisdom of the brain" directs the regulation of body fluids and organs. This research will focus on:
   - The subcellular and cellular mechanisms by which peripheral autonomic pathways regulate target cells.
   - Regulation of the immune system by the sympathetic nervous system, this being the basis to understand the mechanisms by way of which the brain is involved in the control of body protections.
   - The neural circuits in spinal cord, brain stem and hypothalamus that are at the base of autonomic regulations.
   - The representations of autonomic regulation in the forebrain and its connection with the circuits in spinal cord, brain stem and hypothalamus. This will result in the formulation of the neurobiological basis of psychosomatic relations.

Notes

1. Subchapter 11.1 is my tribute to Walter Bradford Cannon and Rudolf Hess (see my foreword of this book).
2. We still do not know whether or not this is true for weak synaptic inputs by preganglionic axons to sympathetic postganglionic neurons. As described in Subchapter 6.2, these synaptic inputs may be incorrect connections that are functionally not important under in vivo conditions.
3. As described in Subchapter 4.5, these functions are poorly or not at all mediated by nerves, but are the functions of the adrenal medulla. They do not reinforce the effects of sympathetic nerves at all, they may complement them.

4. Part of Cannon's problem may have been the condition of the animals on which they worked. The anesthetics of the time often depleted the adrenal medulla of catecholamines and removed reflex control of autonomic outflow. This meant that the only responses that could be detected were the enormous ones produced by extreme stimuli (like asphyxiation).
5. In this context I remind the reader that many cutaneous vasoconstrictor neurons are inhibited by stimulation of arterial chemoreceptors or trigeminal receptors in the anesthetized cat (see Figures 4.4 and 4.11 and Subchapter 4.1) and a few are excited by these stimuli, these possibly innervating nutritional blood vessels. Muscle and visceral vasoconstrictor neurons are excited (Figures 4.1, 4.2).
6. This was called cardio-somatic coupling by Obrist (Obrist 1981, Berntson and Cacioppo 2000).
7. The stereotype of confrontational defense is the defense behavior of the cat described by Darwin in his book *The Expression of the Emotions in Man and Animals* (1872) (see Figure 4.12), the sham-rage behavior occurring spontaneously or evoked by cutaneous stimuli in hypothalamic cats (Bard 1928), and the defense behavior (Abwehrreaktion) elicited by electrical stimulation of the caudal perifornical area of the hypothalamus in cats (Hess and Brügger 1943; Hess 1948; Akert 1981). Hess has described sites in the hypothalamus from which he could evoke flight ("Fluchttrieb") or a sequence of confrontational defense followed by flight by electrical stimulation (Hess 1948). Later Hunsperger (1956), in his careful study of the "affective" defense reaction in awake freely moving cats (which he considered to be an intermediary reaction between attack and flight), showed that electrical stimulation of the midbrain central grey at sites that correspond approximately to the dorsolateral and lateral cell columns elicits attack or flight (depending on the stimulus strength). He concluded that the periaqueductal grey represents a relay for the affective reaction evoked from the perifornical region of the hypothalamus. He extended this study using systematic lesions in the periaqueductal grey, hypothalamus or amygdala in cats and came to the conclusion that these three subcortical structures are important in the shaping of the defense behavior elicited from the cortex (Fernandez de Molina and Hunsperger 1962).
8. The morphological studies on monkeys demonstrate that the viscerosensory and visceromotor prefrontal cortices connected to the autonomic regulations of body organs are much more differentiated than believed (for discussion see Öngür et al. [1998] and Öngür and Price [2000]).
9. In the *conditioned emotional response* paradigm a neutral stimulus (here a tone as conditioned stimulus) is paired with an aversive stimulus (here electrical shocks applied to the abdominal skin as unconditioned stimulus). The animals learn that the conditioned stimulus is followed by the aversive stimulus and as a consequence the originally neutral conditioned stimulus generates a state of fear in the animal. This emotional state of fear is indicated by the suppression of an appetitive behavior performed by the animal (here suppression of lever pressing to obtain apple juice as reward).

The monkeys were first trained with operant techniques to perform mild dynamic exercise (turning a wheel with their legs) and to press a small lever; in doing so they are rewarded by juice. Then the conditioned emotional response was elicited in the monkeys by the classical conditioning paradigm. The procedure began by presenting the signal for lever pressing. After stable lever pressing was reached, an auditory signal was applied for 1 minute as a conditioned stimulus. This stimulus was terminated by an unconditioned stimulus consisting of electrical shocks of 7 to 15 mA, applied for 1 to 2 seconds to the abdominal skin. The command signal for lever pressing remained constant throughout the conditioned stimulus and afterwards. Only *one* classical conditioning trial was applied during a one-day training period at intervals of two to four days. After the training procedures, the animals were instrumented in order to measure heart rate, arterial blood pressure, renal blood flow (resistance) and terminal aortic blood flow (resistance), the latter mainly as a measure of blood flow through the skeletal musculature of the hindlimbs. Furthermore, electrodes were stereotactically placed bilaterally in the perifornical regions of the lateral hypothalamus for electrical stimulation or lesioning (for details see Smith et al. [1979]).

10. The argument that the decrease in renal blood flow is generated by adrenaline released by the adrenal medulla is based on two measurements: a chronically denervated kidney showed the same slow response (1) during the conditioned emotional response and (2) when 5 µg adrenaline was injected intravenously (Smith et al. 1979). The problem is that the vascular bed of a denervated kidney develops denervation hyperreactivity and that 5 µg adrenaline is a very high concentration compared to the physiologically occurring plasma concentrations of adrenaline (see Subchapter 4.5).

11. Based on the examination of peripheral physiological changes that occur during emotions the Dane Carl S. Lange came to the conclusion that vascular changes (dilation of blood vessels and increase of blood volume in the organs) are the primary effects that trigger emotions (Lange 1887).

12. The reasoning in the rest of this subchapter is largely based on the experimental work of and the concepts developed by Larry Swanson and his group (Swanson 1987, 2000, 2003; Watts and Swanson 2002; Card et al. 2003; Thomson and Swanson 2003).

13. Philip Bard, together with his student Rioch, has conducted systematic studies on four chronically diencephalic cats (cerebral cortex and most subcortical structures of the cerebral hemispheres were removed). These animals could be kept alive for 28, 13, 1.5 and 1 months. The authors showed that these animals exhibit, in addition to the so-called sham-rage behavior, various types of behavior that are typical for the integrative hypothalamic functions (Table 11.2; thermoregulation, urination, defecation, reproduction, sleep) (Bard and Rioch 1937). Later Bard and his student Macht studied also chronic pontile (bulbospinal), low mesencephalic or high mesencephalic cats (1 to 5 months after transection of the brain stem). These animals required utmost care because of the absence of the hypothalamically organized control systems (Bard and Macht 1957).

14. Ten percent of Fulton's textbook, *Physiology of the Nervous System*, was devoted to the autonomic nervous system and the hypothalamus.

15. This is my personal and somewhat subjective view.

References

Ádám, G. (1967). *Interoception and Behavior: An Experimental Study*. Tanslated from the Hungarian by R. de Chantel, revised by H. Slucki. Budapest: Akadémiai Kiadó.

Ádám, G. (1998). *Visceral Perception: Understanding Internal Cognition*. New York: Plenum Press.

Adams, D. J. and Harper, A. A. (1995). Electrophysiological properties of autonomic ganglia. In *The Autonomic Nervous System*, Vol. 6, *Autonomic Ganglia*, ed. E. M. McLachlan. Chur, Switzerland: Harwood Academic Publisher, pp. 153-212.

Ader, A. and Cohen, N. (1993). Psychoneuroendocrinology: conditioning and stress. *Annu. Rev. Physiol.*, **44**, 53-85.

Aggleton, J. P. (ed.) (2000). *The Amygdala: A Functional Analysis*. Oxford: Oxford University Press.

Aicher, S. A., Kurucz, O. S., Reis, D. J. and Milner, T. A. (1995). Nucleus tractus solitarius efferent terminals synapse on neurons in the caudal ventrolateral medulla that project to the rostral ventrolateral medulla. *Brain Res.*, **693**, 51-63.

Akasu, T. and Koketsu, K. (1986). Muscarinic transmission. In *Autonomic and Enteric Ganglia*, ed. K. Karczmar, K. Koketsu and S. Nishi. New York, London: Plenum Press, pp. 161-180.

Akasu, T. and Nishimura, T. (1995). Synaptic transmission and function of parasympathetic ganglia. *Prog. Neurobiol.*, **45**, 459-522.

Akert, K. (1981). *Biological Order in Brain Organization. Selected Works of W. R. Hess*. Berlin, Heidelberg, New York: Springer-Verlag.

Akoev, G. N. (1981). Catecholamines, acetylcholine and excitability of mechanoreceptors. *Prog. Neurobiol.*, **15**, 269-294.

Al Chaer, E. D., Lawand, N. B., Westlund, K. N. and Willis, W. D. (1996a). Visceral nociceptive input into the ventral posterolateral nucleus of the thalamus: a new function for the dorsal column pathway. *J. Neurophysiol.*, **76**, 2661-2674.

Al Chaer, E. D., Lawand, N. B., Westlund, K. N. and Willis, W. D. (1996b). Pelvic visceral input into the nucleus gracilis is largely mediated by the postsynaptic dorsal column pathway. *J. Neurophysiol.*, **76**, 2675-2690.

Al Chaer, E. D., Westlund, K. N. and Willis, W. D. (1997). Nucleus gracilis: an integrator for visceral and somatic information. *J. Neurophysiol.*, **78**, 521-527.

Al Chaer, E. D., Feng, Y. and Willis, W. D. (1999). Comparative study of viscerosomatic input onto postsynaptic dorsal column and spinothalamic tract neurons in the primate. *J. Neurophysiol.*, **82**, 1876-1882.

Alexander, R. S. (1946). Tonic and reflex functions of medullary sympathetic cardiovascular centers. *J. Neurophysiol.*, **9**, 205-217.

Alexander, S. P. H., Mathie, A. and Peters, J. A. (2004). Guide to receptors and channels, 1st edition. *Br. J. Pharmacol.*, **141 Suppl 1**, S1-S126.

Altschuler, S. M., Bao, X. M., Bieger, D., Hopkins, D. A. and Miselis, R. R. (1989). Viscerotopic representation of the upper alimentary tract in the rat: sensory ganglia and nuclei of the solitary and spinal trigeminal tracts. *J. Comp. Neurol.*, **283**, 248-268.

Altschuler, S. M., Ferenci, D. A., Lynn, R. B. and Miselis, R. R. (1991). Representation of the cecum in the lateral dorsal motor nucleus of the vagus nerve and commissural subnucleus of the nucleus tractus solitarii in rat. *J. Comp. Neurol.*, **304**, 261–274.

Altschuler, S. M., Rinaman, L. and Miselis, R. R. (1992). Viscerotopic representation of the alimentary tract in the dorsal and ventral vagal complexes in the rat. In *Neuroanatomy and Physiology of Abdominal Vagal Afferents*, ed. S. Ritter, R. C. Ritter and C. D. Barnes. Boca Raton: CRC Press, pp. 22–53.

Amann, R., Dray, A. and Hankins, M. W. (1988). Stimulation of afferent fibres of the guinea-pig ureter evokes potentials in inferior mesenteric ganglion neurones. *J. Physiol.*, **402**, 543–553.

Amendt, K., Czachurski, J., Dembowsky, K. and Seller, H. (1979). Bulbospinal projections to the intermediolateral cell column: a neuroanatomical study. *J. Auton. Nerv. Syst.*, **1**, 103–107.

An, X., Bandler, R., Öngür, D. and Price, J. L. (1998). Prefrontal cortical projections to longitudinal columns in the midbrain periaqueductal gray in macaque monkeys. *J. Comp. Neurol.*, **401**, 455–479.

Anders, S., Lotze, M., Erb, M., Grodd, W. and Birbaumer, N. (2004). Brain activity underlying emotional valence and arousal: a response-related fMRI study. *Hum. Brain Mapp.*, **23**, 200–209.

Anderson, C. R., McAllen, R. M. and Edwards, S. L. (1995). Nitric oxide synthase and chemical coding in cat sympathetic postganglionic neurons. *Neuroscience*, **68**, 255–264.

Anderson, D. J. (1993). Molecular control of cell fate in the neural crest: the sympathoadrenal lineage. *Annu. Rev. Neurosci.*, **16**, 129–158.

Anderson, E. A., Wallin, B. G. and Mark, A. L. (1987). Dissociation of sympathetic nerve activity in arm and leg muscle during mental stress. *Hypertension*, **9, III**, 114–119.

Anderson, R. L., Gibbins, I. L. and Morris, J. L. (1996). Non-noradrenergic sympathetic neurons project to extramuscular feed arteries and proximal intramuscular arteries of skeletal muscles in guinea-pig hindlimbs. *J. Auton. Nerv. Syst.*, **61**, 51–60.

Andersson, K. E. (2001). Pharmacology of penile erection. *Pharmacol. Rev.*, **53**, 417–450.

Andreev, N. Y., Dimitrieva, N., Koltzenburg, M. and McMahon, S. B. (1995). Peripheral administration of nerve growth factor in the adult rat produces a thermal hyperalgesia that requires the presence of sympathetic postganglionic neurones. *Pain*, **63**, 109–115.

Andrew, D. and Craig, A. D. (2001a). Spinothalamic lamina I neurones selectively responsive to cutaneous warming in cats. *J. Physiol.*, **537**, 489–495.

Andrew, D. and Craig, A. D. (2001b). Spinothalamic lamina I neurons selectively sensitive to histamine: a central neural pathway for itch. *Nat. Neurosci.*, **4**, 72–77.

Andrew, D. and Craig, A. D. (2002a). Quantitative responses of spinothalamic lamina I neurones to graded mechanical stimulation in the cat. *J. Physiol.*, **545**, 913–931.

Andrew, D. and Craig, A. D. (2002b). Responses of spinothalamic lamina I neurons to maintained noxious mechanical stimulation in the cat. *J. Neurophysiol.*, **87**, 1889–1901.

Andrews, P. L. R. (1986). Vagal afferent innervation of the gastrointestinal tract. *Prog. Brain Res.*, **67**, 65–86.

Anrep, G. V., Pascual, W. and Rössler, R. (1936a). Respiratory variations of the heart rate. I. The reflex mechanism of the sinus arrhythmia. *Proc. Royal Soc. Lond.*, **119 (Series B)**, 191–217.

Anrep, G. V., Pascual, W. and Rössler, R. (1936b). Respiratory variations of the heart rate. II. The central mechanism of the sinus arrhythmia and the interrelationships between central and reflex mechanism. *Proc. Royal Soc. Lond.*, **119 (Series B)**, 218–230.

Apkarian, A. V. and Shi, T. (1994). Squirrel monkey lateral thalamus. I. Somatic nociresponsive neurons and their relation to spinothalamic terminals. *J. Neurosci.*, **14**, 6779–6795.

Apodaca, G. (2004). The uroepithelium: not just a passive barrier. *Traffic*, **5**, 117–128.

Apodaca, G., Kiss, S., Ruiz, W. et al. (2003). Disruption of bladder epithelium barrier function after spinal cord injury. *Am. J. Physiol. Renal Physiol.*, **284**, F966–F976.

Appenzeller, O. and Oribe, E. (1997). *The Autonomic Nervous System. An Introduction to Basic and Clinical Concepts*, 5th edn. Amsterdam: Elsevier.

Appenzeller, O. (ed.) (1999). *Handbook of Clinical Neurology*, Vol. 74, *The Autonomic Nervous System, part I: Normal Functions*. Amsterdam: Elsevier.

Appenzeller, O. (ed.) (2000). *Handbook of Clinical Neurology*, Vol. 75, *The Autonomic Nervous System, part II: Dysfunctions*. Amsterdam: Elsevier.

Applegate, C. D., Kapp, B. S., Underwood, M. D. and McNall, C. L. (1983). Autonomic and somatomotor effects of amygdala central N. stimulation in awake rabbits. *Physiol. Behav.*, **31**, 353–360.

Araki, I. (1994). Inhibitory postsynaptic currents and the effects of GABA on visually identified sacral parasympathetic preganglionic neurons in neonatal rats. *J. Neurophysiol.*, **72**, 2903–2910.

Araki, I. and de Groat, W. C. (1996). Unitary excitatory synaptic currents in preganglionic neurons mediated by two distinct groups of interneurons in neonatal rat sacral parasympathetic nucleus. *J. Neurophysiol.*, **76**, 215–226.

Araki, I. and de Groat, W. C. (1997). Developmental synaptic depression underlying reorganization of visceral reflex pathways in the spinal cord. *J. Neurosci.*, **17**, 8402–8407.

Bachoo, M. and Polosa, C. (1985). Properties of a sympatho-inhibitory and vasodilator reflex evoked by superior laryngeal nerve afferents in the cat. *J. Physiol.*, **364**, 183–198.

Bachoo, M. and Polosa, C. (1986). The pattern of sympathetic neurone activity during expiration in the cat. *J. Physiol.*, **378**, 375–390.

Bachoo, M. and Polosa, C. (1987). Properties of the inspiration-related activity of sympathetic preganglionic neurones of the cervical trunk in the cat. *J. Physiol.*, **385**, 545–564.

Bacon, S. J., Zagon, A. and Smith, A. D. (1990). Electron microscopic evidence of a monosynaptic pathway between cells in the caudal raphe nuclei and sympathetic preganglionic neurons in the rat spinal cord. *Exp. Brain Res.*, **79**, 589–602.

Bacon, S. J. and Smith, A. D. (1993). A monosynaptic pathway from an identified vasomotor centre in the medial prefrontal cortex to an autonomic area in the thoracic spinal cord. *Neuroscience*, **54**, 719–728.

Badoer, E. (2001). Hypothalamic paraventricular nucleus and cardiovascular regulation. *Clin. Exp. Pharmacol. Physiol.*, **28**, 95–99.

Bahns, E., Ernsberger, U., Jänig, W. and Nelke, A. (1986a). Functional characteristics of lumbar visceral afferent fibres from the urinary bladder and the urethra in the cat. *Pflügers Arch.*, **407**, 510–518.

Bahns, E., Ernsberger, U., Jänig, W. and Nelke, A. (1986b). Discharge properties of mechanosensitive afferents supplying the retroperitoneal space. *Pflügers Arch.*, **407**, 519–525.

Bahns, E., Halsband, U. and Jänig, W. (1987). Responses of sacral visceral afferents from the lower urinary tract, colon and anus to mechanical stimulation. *Pflügers Arch.*, **410**, 296–303.

Bahr, R., Blumberg, H. and Jänig, W. (1981). Do dichotomizing afferent fibers exist which supply visceral organs as well as somatic structures? A contribution to the problem of referred pain. *Neurosci. Lett.*, **24**, 25–28.

Bahr, R., Bartel, B., Blumberg, H. and Jänig, W. (1986a). Functional characterization of preganglionic neurons projecting in the lumbar splanchnic nerves: neurons regulating motility. *J. Auton. Nerv. Syst.*, **15**, 109–130.

Bahr, R., Bartel, B., Blumberg, H. and Jänig, W. (1986b). Functional characterization of preganglionic neurons projecting in the lumbar splanchnic nerves: vasoconstrictor neurons. *J. Auton. Nerv. Syst.*, **15**, 131–140.

Bahr, R., Bartel, B., Blumberg, H. and Jänig, W. (1986c). Secondary functional properties of lumbar visceral preganglionic neurons. *J. Auton. Nerv. Syst.*, **15**, 141–152.

Bainton, C. R., Richter, D. W., Seller, H., Ballantyne, D. and Klein, J. P. (1985). Respiratory modulation of sympathetic activity. *J. Auton. Nerv. Syst.*, **12**, 77–90.

Baker, D. G., Coleridge, H. M., Coleridge, J. C. and Nerdrum, T. (1980). Search for a cardiac nociceptor: stimulation by bradykinin of sympathetic afferent nerve endings in the heart of the cat. *J. Physiol.*, **306**, 519–536.

Baldissera, F., Hultborn, H. and Illert, M. (1981). Integration in spinal neuronal systems. In *Handbook of Physiology, Section 1, The Nervous System*, Vol. II, *Motor Control, part I*. ed. V. B. Brooks. Bethesda: American Physiological Society, pp. 509–595.

Baldwin, B. A., Parrott, R. F. and Ebenezer, I. S. (1998). Food for thought: a critique on the hypothesis that endogenous cholecystokinin acts as a physiological satiety factor. *Prog. Neurobiol.*, **55**, 477–507.

Bamshad, M., Aoki, V. T., Adkison, M. G., Warren, W. S. and Bartness, T. J. (1998). Central nervous system origins of the sympathetic nervous system outflow to white adipose tissue. *Am. J. Physiol.*, **275**, R291–R299.

Bamshad, M., Song, C. K. and Bartness, T. J. (1999). CNS origins of the sympathetic nervous system outflow to brown adipose tissue. *Am. J. Physiol.*, **276**, R1569–R1578.

Bandler, R. (1988). Brain mechanisms of aggression as revealed by electrical and chemical stimulation: suggestion of a central role for the midbrain periaqueductal grey region. *Prog. Psychobiol. Physiol. Psychol.*, **13**, 67–153.

Bandler, R. and Shipley, M. T. (1994). Columnar organization in the midbrain periaqueductal gray: modules for emotional expression? *Trends Neurosci.*, **17**, 379–389.

Bandler, R. and Keay, K. A. (1996). Columnar organization in the midbrain periaqueductal gray and the integration of emotional expression. *Prog. Brain Res.*, **107**, 285–300.

Bandler, R., Carrive, P. and Zhang, S. P. (1991). Integration of somatic and autonomic reactions within the midbrain periaqueductal grey: viscerotopic, somatotopic and functional organization. *Prog. Brain Res.*, **87**, 269–305.

Bandler, R., Keay, K. A., Floyd, N. and Price, J. (2000a). Central circuits mediating patterned autonomic activity during active vs. passive emotional coping. *Brain Res. Bull.*, **53**, 95–104.

Bandler, R., Price, J. L. and Keay, K. A. (2000b). Brain mediation of active and passive emotional coping. *Prog. Brain Res.*, **122**, 333–349.

Bao, J. X., Gonon, F. and Stjärne, L. (1993). Frequency- and train length-dependent variation in the roles of postjunctional alpha 1- and alpha 2-adrenoceptors for the field stimulation-induced neurogenic contraction of rat tail artery. *Naunyn-Schmiedeberg's Arch. Pharmacol.*, **347**, 601–616.

Barcroft, H., Brod, Z., Hejl, Z., Hirsjärvi, E. A. and Kitchen, A. H. (1960). The mechanism of the vasodilatation in the forearm muscle during stress (mental arithmetic). *Clin. Sci.*, **19**, 577–586.

Bard, P. (1928). A diencephalic mechanism for the expression of rage with special reference to the sympathetic nervous system. *Am. J. Physiol.*, **84**, 490–515.

Bard, P. (1932). An emotional expression after decortication with some remarks on certain theoretical views. Part I. *Psychol. Rev.*, **41**, 309–329.

Bard, P. (1960). Anatomical organization of the central nervous system in relation to control of the heart and blood vessels. *Physiol. Rev.*, **40 (Suppl 4)**, 3–26.

Bard, P. and Macht, M. B. (1957). The behavior of chronically decerebrate cats. In *Ciba Foundation Symposium on the Neurological Basis of Behavior*, ed. G. E. W. Wolstenholme and M. O'Connor. Boston: Little, Brown and Co., pp. 55–71.

Bard, P. and Rioch, D. McK. (1937). A study of four cats deprived of neocortex and additional portions of the forebrain. *Bull. Johns Hopkins Hosp.*, **60**, 73–147.

Barker, D. and Saito, M. (1981). Autonomic innervation of receptors and muscle fibres in cat skeletal muscle. *Proc. R. Soc. London Biol.*, **212**, 317–332.

Barman, S. M. and Gebber, G. L. (1976). Basis for synchronization of sympathetic and phrenic nerve discharges. *Am. J. Physiol.*, **231**, 1601–1607.

Barman, S. M. and Gebber, G. L. (1984). Spinal interneurons with sympathetic nerve-related activity. *Am. J. Physiol.*, **247**, R761–R767.

Barman, S. M. and Gebber, G. L. (2000). "Rapid" rhythmic discharges of sympathetic nerves: sources, mechanisms of generation, and physiological relevance. *J. Biol. Rhythms*, **15**, 365–379.

Barman, S. M., Orer, H. S. and Gebber, G. L. (2001). The role of the medullary lateral tegmental field in the generation and baroreceptor reflex control of sympathetic nerve discharge in the cat. *Ann. N. Y. Acad. Sci.*, **940**, 270–285.

Barnes, P. J. (ed.) (1997). *The Autonomic Nervous System*, Vol. 7, *Autonomic Control of the Respiratory System*. Amsterdam: Harwood Academic Publishers.

Baron, R. and Jänig, W. (1988). Neurons projecting rostrally in the hypogastric nerve of the cat. *J. Auton. Nerv. Syst.*, **24**, 81–86.

Baron, R. and Jänig, W. (1991). Afferent and sympathetic neurons projecting into lumbar visceral nerves of the male rat. *J. Comp. Neurol.*, **314**, 429–436.

Baron, R., Jänig, W. and McLachlan, E. M. (1985a). On the anatomical organization of the lumbosacral sympathetic chain and the lumbar splanchnic nerves of the cat – Langley revisited. *J. Auton. Nerv. Syst.*, **12**, 289–300.

Baron, R., Jänig, W. and McLachlan, E. M. (1985b). The afferent and sympathetic components of the lumbar spinal outflow to the colon and pelvic organs in the cat: I. The hypogastric nerve. *J. Comp. Neurol.*, **238**, 135–146.

Baron, R., Jänig, W. and McLachlan, E. M. (1985c). The afferent and sympathetic components of the lumbar spinal outflow to the colon and pelvic organs in the cat. II. The lumbar splanchnic nerves. *J. Comp. Neurol.*, **238**, 147–157.

Baron, R., Jänig, W. and McLachlan, E. M. (1985d). The afferent and sympathetic components of the lumbar spinal outflow to the colon and pelvic organs in the cat. III. The colonic nerves, incorporating an analysis of all components of the lumbar prevertebral outflow. *J. Comp. Neurol.*, **238**, 158–168.

Baron, R., Jänig, W. and Kollmann, W. (1988). Sympathetic and afferent somata projecting in hindlimb nerves and the anatomical organization of the lumbar sympathetic nervous system of the rat. *J. Comp. Neurol.*, **275**, 460–468.

Baron, R., Jänig, W. and With, H. (1995). Sympathetic and afferent neurones projecting into forelimb and trunk nerves and the anatomical organization of the thoracic sympathetic outflow of the rat. *J. Auton. Nerv. Syst.*, **53**, 205–214.

Barraco, I. R. A. (ed.) (1994). *Nucleus of the Solitary Tract*. Boca Raton: CRC Press.

Bartel, B., Blumberg, H. and Jänig, W. (1986). Discharge patterns of motility-regulating neurons projecting in the lumbar splanchnic nerves to visceral stimuli in spinal cats. *J. Auton. Nerv. Syst.*, **15**, 153–163.

Bartness, T. J. and Bamshad, M. (1998). Innervation of mammalian white adipose tissue: implications for the regulation of total body fat. *Am. J. Physiol.*, **275**, R1399–R1411.

Bartsch, T., Häbler, H. J. and Jänig, W. (1996). Functional properties of post-ganglionic sympathetic neurones supplying the submandibular gland in the anaesthetized rat. *Neurosci. Lett.*, **214**, 143–146.

Bartsch, T., Häbler, H. J. and Jänig, W. (1999). Hypoventilation recruits preganglionic sympathetic fibers with inspiration-related activity in the superior cervical trunk of the rat. *J. Auton. Nerv. Syst.*, **77**, 31–38.

Bartsch, T., Häbler, H. J. and Jänig, W. (2000). Reflex patterns of preganglionic sympathetic fibers projecting to the superior cervical ganglion in the rat. *Auton. Neurosci.*, **83**, 66–74.

Baumann, T. K., Simone, D. A., Shain, C. N. and LaMotte, R. H. (1991). Neurogenic hyperalgesia: the search for the primary cutaneous afferent fibers that contribute to capsaicin-induced pain and hyperalgesia. *J. Neurophysiol.*, **66**, 212–227.

Bayliss, W. M. (1901). On the origin from the spinal cord of the vaso-dilator fibres of the hind-limb, and on the nature of these fibres. *J. Physiol.*, **26**, 173–209.

Bayliss, W. M. and Starling, E. H. (1899). The movements and innervation of the small intestine. *J. Physiol.*, **24**, 99–143.

Bayliss, W. M. and Starling, E. H. (1900). The movements and innervation of the large intestine. *J. Physiol.*, **26**, 107–118.

Beckstead, R. M., Morse, J. R. and Norgren, R. (1980). The nucleus of the solitary tract in the monkey: projections to the thalamus and brain stem nuclei. *J. Comp. Neurol.*, **190**, 259–282.

Bell, C., Jänig, W., Kümmel, H. and Xu, H. (1985). Differentiation of vasodilator and sudomotor responses in the cat paw pad to preganglionic sympathetic stimulation. *J. Physiol.*, **364**, 93–104.

Belmonte, C. and Cervero, F. (eds.) (1996). *Neurobiology of Nociception*. Oxford, New York, Tokyo: Oxford University Press.

Benison, S., Barger, A. C. and Wolfe, E. L. (1987). *Walter B. Cannon. The Life and Times of a Young Scientist*. Cambridge Mass. London England: The Belknarp Press of Harvard University Press.

Bennett, T. and Gardiner, S. M. (eds.) (1996). *The Autonomic Nervous System, Vol. 8, Nervous Control of Blood Vessels*. Amsterdam: Harwood Academic Publishers.

Berkley, K. J., Robbins, A. and Sato, Y. (1988). Afferent fibers supplying the uterus in the rat. *J. Neurophysiol.*, **59**, 142–163.

Berkley, K. J., Hotta, H., Robbins, A. and Sato, Y. (1990). Functional properties of afferent fibers supplying reproductive and other pelvic organs in pelvic nerve of female rat. *J. Neurophysiol.*, **63**, 256-272.

Berkley, K. J., Robbins, A. and Sato, Y. (1993). Functional differences between afferent fibers in the hypogastric and pelvic nerves innervating female reproductive organs in the rat. *J. Neurophysiol.*, **69**, 533-544.

Bernard, C. (1858). *Leçons Sur la Physiologie et la Pathologie du Système Nerveau [On the Physiology and Pathophysiology of the Nervous System]*. Paris: Bailliere.

Bernard, C. (1957) [1865]. *Introduction à l'Étude de la Médicine Experimentale. [An Introduction to the Study of Experimental Medicine]*. New York: Dover Publications. Paris: Baillière.

Bernard, C. (1974) [1878]. *Lecons sur les Phénomènes de la Vie Communes aux Animaux et aux Végétaux [Lectures on the Phenomena of Life Common to Animals and Plants]*. Translated by H. E. Hoff, R. Guillemin and L. Guillemin. Paris (Springfield Illinois): B. Ballière et Fils (Thomas).

Bernard, J. F. and Bandler, R. (1998). Parallel circuits for emotional coping behaviour: new pieces in the puzzle. *J. Comp. Neurol.*, **401**, 429-436.

Bernard, J. F. and Besson, J. M. (1990). The spino(trigemino)pontoamygdaloid pathway: electrophysiological evidence for an involvement in pain processes. *J. Neurophysiol.*, **63**, 473-490.

Bernard, J. F., Huang, G. F. and Besson, J. M. (1994). The parabrachial area: electrophysiological evidence for an involvement in visceral nociceptive processes. *J. Neurophysiol.*, **71**, 1646-1660.

Berne, C. and Fagius, J. (1986). Skin sympathetic activity during insulin-induced hypoglycemia. *Diabetologia*, **29**, 855-860.

Berntson, G. G. and Cacioppo, J. T. (2000). From homeostasis to allodynamic regulation. In *Handbook of Psychophysiology*, 2nd edn., eds. J. T. Cacioppo, L. G. Tassinary and G. G. Berntson. Cambridge: Cambridge University Press, pp. 459-481.

Berthoud, H. R. and Neuhuber, W. L. (2000). Functional and chemical anatomy of the afferent vagal system. *Auton. Neurosci.*, **85**, 1-17.

Berthoud, H. R. and Powley, T. L. (1990). Identification of vagal preganglionics that mediate cephalic phase insulin response. *Am. J. Physiol.*, **258**, R523-R530.

Berthoud, H. R. and Powley, T. L. (1992). Vagal afferent innervation of the rat fundic stomach: morphological characterization of the gastric tension receptor. *J. Comp. Neurol.*, **319**, 261-276.

Berthoud, H. R., Fox, E. A. and Powley, T. L. (1990). Localization of vagal preganglionics that stimulate insulin and glucagon secretion. *Am. J. Physiol.*, **258**, R160-R168.

Berthoud, H. R., Carlson, N. R. and Powley, T. L. (1991). Topography of efferent vagal innervation of the rat gastrointestinal tract. *Am. J. Physiol.*, **260**, R200-R207.

Berthoud, H. R., Patterson, L. M., Willing, A. E., Mueller, K. and Neuhuber, W. L. (1997). Capsaicin-resistant vagal afferent fibers in the rat gastrointestinal tract: anatomical identification and functional integrity. *Brain Res.*, **746**, 195-206.

Besedovsky, H. O. and del Rey, A. (1992). Immune-neuroendocrine circuits: integrative role of cytokines. *Front. Neuroendocrinol.*, **13**, 61-94.

Besedovsky, H. O. and del Rey, A. (1995). Immune-neuroendocrine interactions: facts and hypotheses. *Endocr. Rev.*, **17**, 64-102.

Bester, H., Chapman, V., Besson, J. M. and Bernard, J. F. (2000). Physiological properties of the lamina I spinoparabrachial neurons in the rat. *J. Neurophysiol.*, **83**, 2239–2259.

Beyak, M. J. and Grundy, D. (2005). Vagal afferents innervating the gastrointestinal tract. In *Advance in Vagal Afferent Neurobiology*, ed. B. Undem and D. Weinreich. Boca Raton: CRC Press, pp. 315–350.

Bianchi, A. L., Denavit-Saubie, M. and Champagnat, J. (1995). Central control of breathing in mammals: neuronal circuitry, membrane properties, and neurotransmitters. *Physiol. Rev.*, **75**, 1–45.

Bieger, D. and Hopkins, D. A. (1987). Viscerotopic representation of the upper alimentary tract in the medulla oblongata in the rat: the nucleus ambiguus. *J. Comp. Neurol.*, **262**, 546–562.

Bielefeld, T. K. and Gebhart, G. F. (2005). Visceral pain: basic mechanisms. In *Wall and Melzack's Textbook of Pain*, 5th edn., ed. S. B. McMahon and M. Koltzenburg. Amsterdam, Edinburgh: Elsevier Churchill Livingstone, pp. 721–736.

Bini, G., Hagbarth, K. E., Hynninen, P. and Wallin, B. G. (1980a). Thermoregulatory and rhythm-generating mechanisms governing the sudomotor and vasoconstrictor outflow in human cutaneous nerves. *J. Physiol.*, **306**, 547–552.

Bini, G., Hagbarth, K. E., Hynninen, P. and Wallin, B. G. (1980b). Regional similarities and differences in thermoregulatory vaso- and sudomotor tone. *J. Physiol.*, **306**, 553–565.

Bini, G., Hagbarth, K. E. and Wallin, B. G. (1981). Cardiac rhythmicity of skin sympathetic activity recorded from peripheral nerves in man. *J. Auton. Nerv. Syst.*, **4**, 17–24.

Birder, L. A. (2005). More than just a barrier: urothelium as a drug target for urinary bladder pain. *Am. J. Physiol. Renal. Physiol.*, **289**, F489–F495.

Björntop, P. (1997). Behavior and metabolic disease. *Int. J. Behav. Med.*, **3**, 285–302.

Blackman, J. G. (1974). Function of autonomic ganglia. In *The Peripheral Nervous System*, ed. J. I. Hubbard. New York: Plenum Press, pp. 257–276.

Blackman, J. G., Ginsborg, B. L. and Ray, C. (1963). Synaptic transmission in the sympathetic ganglion of the frog. *J. Physiol.*, **167**, 355–373.

Blair, D. A., Glover, W. E., Greenfield, A. D. M. and Roddie, I. C. (1959). Excitation of cholinergic vasodilator nerves to human skeletal muscles during emotional stress. *J. Physiol.*, **148**, 633–646.

Blessing, W. W. (1997). *The Lower Brain Stem and Bodily Homeostasis*. New York, Oxford: Oxford University Press.

Blessing, W. W. and Nalivaiko, E. (2000). Regional blood flow and nociceptive stimuli in rabbits: patterning by medullary raphe, not ventrolateral medulla. *J. Physiol.*, **524**, 279–292.

Blessing, W. W., Li, Y. W. and Wesselingh, S. L. (1991). Transneuronal transport of herpes simplex virus from the cervical vagus to brain neurons with axonal inputs to central vagal sensory nuclei in the rat. *Neuroscience*, **42**, 261–274.

Blessing, W. W., Yu, Y. H. and Nalivaiko, E. (1999). Raphe pallidus and parapyramidal neurons regulate ear pinna vascular conductance in the rabbit. *Neurosci. Lett.*, **270**, 33–36.

Blix, A. S. and Folkow, B. (1983). Cardiovascular adjustments to diving in mammals and birds. In *Handbook of Physiology, Section 2: The Cardiovascular System* Vol. III: *Peripheral Circulation*, ed. J. T. Shepherd and F. M. Abboud. Bethesda: American Physiological Society, pp. 917–945.

Blok, B. F. and Holstege, G. (1996). The neuronal control of micturition and its relation to the emotional motor system. *Prog. Brain Res.*, **107**, 113–126.

Blumberg, H. and Jänig, W. (1982). Changes in unmyelinated fibers including sympathetic postganglionic fibers of a skin nerve after peripheral neuroma formation. *J. Auton. Nerv. Syst.*, **6**, 173–183.

Blumberg, H. and Jänig, W. (1983a). Enhancement of resting activity in postganglionic vasoconstrictor neurones following short-lasting repetitive activation of preganglionic axons. *Pflügers Arch.*, **396**, 89–94.

Blumberg, H. and Jänig, W. (1983b). Changes of reflexes in vasoconstrictor neurons supplying the cat hindlimb following chronic nerve lesions: a model for studying mechanisms of reflex sympathetic dystrophy? *J. Auton. Nerv. Syst.*, **7**, 399–411.

Blumberg, H. and Jänig, W. (1985). Reflex patterns in postganglionic vasoconstrictor neurons following chronic nerve lesions. *J. Auton. Nerv. Syst.*, **14**, 157–180.

Blumberg, H. and Wallin, B. G. (1987). Direct evidence of neurally mediated vasodilatation in hairy skin of the human foot. *J. Physiol.*, **382**, 105–121.

Blumberg, H., Jänig, W., Rieckmann, C. and Szulczyk, P. (1980). Baroreceptor and chemoreceptor reflexes in postganglionic neurones supplying skeletal muscle and hairy skin. *J. Auton. Nerv. Syst.*, **2**, 223–240.

Blumberg, H., Haupt, P., Jänig, W. and Kohler, W. (1983). Encoding of visceral noxious stimuli in the discharge patterns of visceral afferent fibres from the colon. *Pflügers Arch.*, **398**, 33–40.

Boczek-Funcke, A., Häbler, H. J., Jänig, W. and Michaelis, M. (1991). Rapid phasic baroreceptor inhibition of the activity in sympathetic preganglionic neurones does not change throughout the respiratory cycle. *J. Auton. Nerv. Syst.*, **34**, 185–194.

Boczek-Funcke, A., Dembowsky, K., Häbler, H. J. et al. (1992a). Classification of preganglionic neurones projecting into the cat cervical sympathetic trunk. *J. Physiol.*, **453**, 319–339.

Boczek-Funcke, A., Dembowsky, K., Häbler, H. J., Jänig, W. and Michaelis, M. (1992b). Respiratory-related activity patterns in preganglionic neurones projecting into the cat cervical sympathetic trunk. *J. Physiol.*, **457**, 277–296.

Boczek-Funcke, A., Häbler, H. J., Jänig, W. and Michaelis, M. (1992c). Respiratory modulation of the activity in sympathetic neurones supplying muscle, skin and pelvic organs in the cat. *J. Physiol.*, **449**, 333–361.

Boczek-Funcke, A., Dembowsky, K., Häbler, H. J., Jänig, W. and Michaelis, M. (1993). Spontaneous activity, conduction velocity and segmental origin of different classes of thoracic preganglionic neurons projecting into the cat cervical sympathetic trunk. *J. Auton. Nerv. Syst.*, **43**, 189–200.

Bogduk, N. (1983). The innervation of the lumbar spine. *Spine*, **8**, 286–293.

Bogduk, N., Windsor, M. and Inglis, A. (1988). The innervation of the cervical intervertebral discs. *Spine*, **13**, 2–8.

Bolme, P. and Fuxe, K. (1970). Adrenergic and cholinergic nerve terminals in skeletal muscle vessels. *Acta Physiol. Scand.*, **78**, 52–59.

Bolme, B., Novotny, J., Uvnäs, B. and Wright, P. G. (1970). Species distribution of sympathetic cholinergic vasodilator nerves in skeletal muscle. *Acta Physiol. Scand.*, **78**, 60–64.

Bolter, C. P., Wallace, D. J. and Hirst, G. D. (2001). Failure of Ba^{2+} and Cs^+ to block the effects of vagal nerve stimulation in sinoatrial node cells of the guinea-pig heart. *Auton. Neurosci.*, **94**, 93–101.

Bolton, B., Carmichael, E. A. and Stürup, G. (1936). Vaso-constriction following deep inspiration. *J. Physiol.*, **86**, 83–94.

Bors, E. H. and Comarr, A. E. (1960). Neurological disturbances of sexual function with special reference to 529 patients with spinal cord injury. *Urol. Survey*, **10**, 191–222.

Bos, J. D. (ed.) (1989). *Skin Immune System*. Boca Raton: CRC Press.

Bos, J. D. and Kapsenberg, M. L. (1986). The skin immune system. Its cellular constituents and their interactions. *Immunol. Today*, **7**, 235–240.

Boscan, P. and Paton, J. F. (2002). Integration of cornea and cardiorespiratory afferents in the nucleus of the solitary tract of the rat. *Am. J. Physiol. Heart Circ. Physiol.*, **282**, H1278–H1287.

Boscan, P., Kasparov, S. and Paton, J. F. (2002). Somatic nociception activates NK1 receptors in the nucleus tractus solitarii to attenuate the baroreceptor cardiac reflex. *Eur. J. Neurosci.*, **16**, 907–920.

Bosnjak, Z. J. and Kampine, J. P. (1982). Intracellular recordings from the stellate ganglion of the cat. *J. Physiol.*, **324**, 273–283.

Bosnjak, Z. J. and Kampine, J. P. (1984). Peripheral neural input to neurons of the middle cervical ganglion in the cat. *Am. J. Physiol.*, **246**, R354–R358.

Bosnjak, Z. J. and Kampine, J. P. (1985). Electrophysiological and morphological characterization of neurons in stellate ganglion of cats. *Am. J. Physiol.*, **248**, R288–R292.

Boucsein, W. (1992). *Electrodermal Activity*. New York: Plenum Press.

Boyd, H. D., McLachlan, E. M., Keast, J. R. and Inokuchi, H. (1996). Three electrophysiological classes of guinea pig sympathetic postganglionic neurone have distinct morphologies. *J. Comp. Neurol.*, **369**, 372–387.

Bramich, N. J., Edwards, F. R. and Hirst, G. D. (1990). Sympathetic nerve stimulation and applied transmitters on the sinus venosus of the toad. *J. Physiol.*, **429**, 349–375.

Bramich, N. J., Brock, J. A., Edwards, F. R. and Hirst, G. D. (1993). Responses to sympathetic nerve stimulation of the sinus venosus of the toad. *J. Physiol.*, **461**, 403–430.

Bramich, N. J., Brock, J. A., Edwards, F. R. and Hirst, G. D. (1994). Ionophoretically applied acetylcholine and vagal stimulation in the arrested sinus venosus of the toad, *Bufo marinus*. *J. Physiol.*, **478**, 289–300.

Bramich, N. J., Cousins, H. M., Edwards, F. R. and Hirst, G. D. (2001). Parallel metabotropic pathways in the heart of the toad, *Bufo marinus*. *Am. J. Physiol. Heart Circ. Physiol.*, **281**, H1771–H1777.

Brock, J. A. and Cunnane, T. C. (1988). Electrical activity at the sympathetic neuroeffector junction in the guinea-pig vas deferens. *J. Physiol.*, **399**, 607–632.

Brock, J. A. and Cunnane, T. C. (1992). Impulse conduction in sympathetic nerve terminals in the guinea-pig vas deferens and the role of the pelvic ganglia. *Neuroscience*, **47**, 185–196.

Brock, J. A. and Cunnane, T. C. (1993). Neurotransmitter release mechanisms at the sympathetic neuroeffector junction. *Exp. Physiol.*, **78**, 591–614.

Brock, J. A. and van Helden, D. F. (1995). Enhanced excitatory junction potentials in mesenteric arteries from spontaneously hypertensive rats. *Pflügers Arch.*, **430**, 901–908.

Brock, J. A., McLachlan, E. M. and Rayner, S. E. (1997). Contribution of alpha-adrenoceptors to depolarization and contraction evoked by continuous asynchronous sympathetic nerve activity in rat tail artery. *Br. J. Pharmacol.*, **120**, 1513–1521.

Brodal, P. (1998). *The Central Nervous System. Structure and Function.* New York, Oxford: Oxford University Press.

Brooke, R. E., Pyner, S., McLeish, P. *et al.* (2002). Spinal cord interneurones labelled transneuronally from the adrenal gland by a GFP-herpes virus construct contain the potassium channel subunit Kv3.1b. *Auton. Neurosci.*, **98**, 45–50.

Brooke, R. E., Deuchars, J. and Deuchars, S. A. (2004). Input-specific modulation of neurotransmitter release in the lateral horn of the spinal cord via adenosine receptors. *J. Neurosci.*, **24**, 127–137.

Brookes, S. J. (2001). Classes of enteric nerve cells in the guinea-pig small intestine. *Anat. Rec.*, **262**, 58–70.

Brookes, S. and Costa, M. (eds.) (2002). *The Autonomic Nervous System*, Vol. 14, *Innervation of the Gastrointestinal Tract.* London, New York: Francis and Taylor.

Brooks, C. M., Koizumi, K. and Pinkston, J. O. (eds.) (1975). *The Life and Contributions of Walter Bradford Cannon 1871–1945.* New York: State University of New York, Downstate Medical Center.

Brown, A. M. (1967). Cardiac sympathetic adrenergic pathways in which synaptic transmission is blocked by atropine sulfate. *J. Physiol.*, **191**, 271–288.

Brown, A. M. (1969). Sympathetic ganglionic transmission and the cardiovascular changes of the defense reaction in the cat. *Circ. Res.*, **24**, 843–849.

Brown, A., Ricci, M. J. and Weaver, L. C. (2004). NGF message and protein distribution in the injured rat spinal cord. *Exp. Neurol.*, **188**, 115–127.

Brown, D. L. and Guyenet, P. G. (1984). Cardiovascular neurons of brain stem with projections to spinal cord. *Am. J. Physiol.*, **247**, R1009–R1016.

Brown, D. L. and Guyenet, P. G. (1985). Electrophysiological study of cardiovascular neurons in the rostral ventrolateral medulla in rats. *Circ. Res.*, **56**, 359–369.

Browning, K. N., Renehan, W. E. and Travagli, R. A. (1999). Electrophysiological and morphological heterogeneity of rat dorsal vagal neurones which project to specific areas of the gastrointestinal tract. *J. Physiol.*, **517**, 521–532.

Browning, K. N., Coleman, F. H. and Travagli, R. (2005). Effects of pancreatic polypeptide on pancreas-projecting rat dorsal motor nucleus of the vagus neuron. *Am. J. Physiol. Gastrointest. Liver Physiol.*, **289**, G209–G219.

Bruce, A. N. (1910). Über die Beziehung der sensiblen Nervenendigungen zum Entzündungsvorgang [On the relation between sensory nerve endings and inflammation]. *Arch. Exptl. Pathol. Pharmakol.*, **63**, 424–433.

Bruce, A. N. (1913). Vaso-dilator axon-reflexes. *Q. J. Exp. Physiol.*, **6**, 339–354.

Burnett, A. L., Lowenstein, C. J., Bredt, D. S., Chang, T. S. and Snyder, S. H. (1992). Nitric oxide: a physiologic mediator of penile erection. *Science*, **257**, 401–403.

Burns, A. J., Lomax, A. E., Torihashi, S., Sanders, K. M. and Ward, S. M. (1996). Interstitial cells of Cajal mediate inhibitory neurotransmission in the stomach. *Proc. Natl. Acad. Sci. U.S.A.*, **93**, 12008–12013.

Burnstock, G. and Hoyle, C. H. V. (eds.) (1992). *The Autonomic Nervous System*, Vol. 1, *Autonomic Neuroeffector Mechanisms.* Chur, Switzerland: Harwood Academic Publishers.

Burnstock, G. and Sillito, A. M. (eds.) (2000). *The Autonomic Nervous System*, Vol. 13, *Nervous Control of the Eye.* Amsterdam: Harwood Academic Publishers.

Busse, R., Edwards, G., Feletou, M. *et al.* (2002). EDHF: bringing the concepts together. *Trends Pharmacol. Sci.*, **23**, 374–380.

Butler, P. J. and Jones, D. R. (1997). Physiology of diving of birds and mammals. *Physiol. Rev.*, **77**, 837–899.

Bykov, K. M. (1959) [1944]. *The Cerebral Cortex and the Internal Organs*. [Translated from Russian and edited by R. Hodes and A. Kilbey.] Moscow: Foreign Language Publishing House.

Cabot, J. B. (1990). Sympathetic preganglionic neurons: cytoarchitecture, ultrastructure, and biophysical properties. In *Central Regulation of Autonomic Functions*, eds. A. D. Loewy and K. M. Spyer. New York, Oxford: Oxford University Press, pp. 44–67.

Cabot, J. B. (1996). Some principles of the spinal organization of the sympathetic preganglionic outflow. *Prog. Brain Res.*, **107**, 29–42.

Cabot, J. B., Alessi, V., Carroll, J. and Ligorio, M. (1994). Spinal cord lamina V and lamina VII interneuronal projections to sympathetic preganglionic neurons. *J. Comp. Neurol.*, **347**, 515–530.

Cajal, S. R. (1995) [1911]. Histologie du Système Nerveux de l'Homme et des Vertèbrès. [Histology of the Nervous System of Man and Vertebrates]. *Maloine*, Vol. 2, edited and translated by L. W. Swanson and N. Swanson. Oxford: Oxford University Press.

Campbell, G. D., Edwards, F. R., Hirst, G. D. S. and O'Shea, J. E. (1989). Effects of vagal stimulation and applied acetylcholine on pacemaker potentials in the guinea-pig heart. *J. Physiol.*, **415**, 57–68.

Campos, R. R. and McAllen, R. M. (1997). Cardiac sympathetic premotor neurons. *Am. J. Physiol.*, **272**, R615–R620.

Campos, R. R. and McAllen, R. M. (1999). Tonic drive to sympathetic premotor neurons of rostral ventrolateral medulla from caudal pressor area neurons. *Am. J. Physiol.*, **276**, R1209–R1213.

Canning, B. J. and Mazzone, S. B. (2005). Reflexes initiated by activation of the vagal afferent nerves innervating the airways and lung. In *Advances in Vagal Afferent Neurobiology*, ed. B. J. Undem and D. Weinreich, Boca Raton: CRC, Taylor & Francis, pp. 403–430.

Cannon, W. B. (1911). *The Mechanical Factors of Digestion*. London: Edward Arnold.

Cannon, W. B. (1914a). The interrelations of emotions as suggested by recent physiological researches. *Am. J. Physiol.*, **25**, 252–282.

Cannon, W. B. (1914b). The emergency function of the adrenal medulla in pain and the major emotions. *Am. J. Physiol.*, **33**, 356–372.

Cannon, W. B. (1927). The James-Lange theory of emotions: a critical examination and an alternative theory. *Am. J. Psychol.*, **39**, 106–124.

Cannon, W. B. (1928). Die Notfallfunktion des sympathico-adrenalen Systems [The emergency function of the sympathico-adrenal system]. *Ergebn. Physiol.*, **27**, 380–406.

Cannon, W. B. (1929a). Organization for physiological homeostasis. *Physiol. Rev.*, **9**, 399–431.

Cannon, W. B. (1929b). *Bodily Changes in Pain, Hunger, Fear and Rage*. New York: Appleton.

Cannon, W. B. (1933). A method of stimulating autonomic nerves in the anesthetized cat with observation on the motor and sensory effects. *Am. J. Physiol.*, **105**, 366–372.

Cannon, W. B. (1939). *The Wisdom of the Body*, 2nd revised and enlarged edition. New York: Norton.

Cannon, W. B. and Murphy, F. T. (1906). The movements of the stomach and intestine in some surgical conditions. *Ann. Surg.*, **43**, 512–536.

Cannon, W. B., Newton, H. F., Bright, E. M., Menkin, V. and Moore, R. M. (1929). Some aspects of the physiology of animals surviving complete exclusion of sympathetic nerve impulses. *Am. J. Physiol.*, **89**, 84–107.

Cano, G., Card, J. P., Rinaman, L. and Sved, A. F. (2000). Connections of Barrington's nucleus to the sympathetic nervous system in rats. *J. Auton. Nerv. Syst.*, **79**, 117–128.

Cano, G., Sved, A. F., Rinaman, L., Rabin, B. S. and Card, J. P. (2001). Characterization of the central nervous system innervation of the rat spleen using viral transneuronal tracing. *J. Comp. Neurol.*, **439**, 1–18.

Card, J. P., Rinaman, L., Schwaber, J. S. *et al.* (1990). Neurotropic properties of pseudorabies virus: uptake and transneuronal passage in the rat central nervous system. *J. Neurosci.*, **10**, 1974–1994.

Card, J. P., Swanson, L. W. and Moore, R. Y. (2003). The hypothalamus: an overview of regulatory systems. In *Fundamental Neuroscience*, 2nd edn., eds. L. R. Squire, F. E. Bloom, S. K. McConnell, *et al.* San Diego: Academic Press, pp. 897–909.

Carrive, P. (1993). The periaqueductal gray and defensive behavior: functional representation and neuronal organization. *Behav. Brain Res.*, **58**, 27–47.

Carrive, P. and Morgan, M. M. (2004). Periaqueductal grey. In *The Human Nervous System*, ed. G. Paxinos and J. K. Mai. Amsterdam: Elsevier Academic Press, pp. 393–423.

Cassell, J. F. and McLachlan, E. M. (1986). The effect of a transient outward current (I_A) on synaptic potentials in sympathetic ganglion cells of the guinea-pig. *J. Physiol.*, **374**, 273–288.

Cassell, J. F., Clark, A. L. and McLachlan, E. M. (1986). Characteristics of phasic and tonic sympathetic ganglion cells of the guinea-pig. *J. Physiol.*, **372**, 457–483.

Cassell, J. F., McLachlan, E. M. and Sittiracha, T. (1988). The effect of temperature on neuromuscular transmission in the main caudal artery of the rat. *J. Physiol.*, **397**, 31–49.

Casson, D. M. and Ronald, K. (1975). The harp seal, *Pagophilus groenlandicus* (Erxleben, 1777). XIV. Cardiac arrythmias. *Comp. Biochem. Physiol. A*, **50**, 307–314.

Causing, C. G., Gloster, A., Aloyz, R. *et al.* (1997). Synaptic innervation density is regulated by neuron-derived BDNF. *Neuron*, **18**, 257–267.

Cechetto, D. F. (1995). Supraspinal mechanisms of visceral pain. In *Visceral Pain*, ed. G. F. Gebhardt. Seattle: IASP Press, pp. 261–290.

Cechetto, D. F. and Saper, C. B. (1990). Role of the cerebral cortex in autonomic function. In *Central Regulation of Autonomic Functions*, ed. A. D. Loewy and K. M. Spyer. New York, Oxford: Oxford University Press, pp. 208–223.

Celander, O. (1954). The range of control exercised by the sympatho-adrenal system. *Acta Physiol. Scand. Suppl.*, **116**, 1–132.

Cervero, F. (1982). Afferent activity evoked by natural stimulation of the biliary system in the ferret. *Pain*, **13**, 137–151.

Cervero, F. (1994). Sensory innervation of the viscera: peripheral basis of visceral pain. *Physiol. Rev.*, **74**, 95–138.

Cervero, F. (1995). Mechanisms of visceral pain: past and present. In *Visceral Pain*, ed. G. F. Gebhart. Seattle: IASP Press, pp. 25–41.

Cervero, F. (1996). Visceral nociceptors. In *Neurobiology of Nociceptors*, ed. C. Belmonte and F. Cervero. Oxford, New York, Toronto: Oxford University Press, pp. 220–240.

Cervero, F. and Connell, L. A. (1984). Distribution of somatic and visceral primary afferent fibers within the thoracic spinal cord of the cat. *J. Comp. Neurol.*, **230**, 88–98.

Cervero, F. and Jänig, W. (1992). Visceral nociceptors: a new world order? *Trends Neurosci.*, **15**, 374–378.

Cervero, F. and Morrison, J. F. B. (eds.) (1986). Visceral Sensation. *Prog. Brain Res.* Vol. 67, Amsterdam: Elsevier.

Cervero, F. and Sann, H. (1989). Mechanically evoked responses of afferent fibers innervating the guinea-pig's ureter: an in vitro study. *J. Physiol.*, **412**, 245–266.

Cervero, F. and Tattersall, J. E. (1986). Somatic and visceral sensory integration in the thoracic spinal cord. *Prog. Brain Res.*, **67**, 189–205.

Chan, R. K. and Sawchenko, P. E. (1998). Organization and transmitter specificity of medullary neurons activated by sustained hypertension: implications for understanding baroreceptor reflex circuitry. *J. Neurosci.*, **18**, 371–387.

Chandler, M. J., Zhang, J., Qin, C. and Foreman, R. D. (2002). Spinal inhibitory effects of cardiopulmonary afferent inputs in monkeys: neuronal processing in high cervical segments. *J. Neurophysiol.*, **87**, 1290–1302.

Chang, H. S., Staras, K. and Gilbey, M. P. (2000). Multiple oscillators provide metastability in rhythm generation. *J. Neurosci.*, **20**, 5135–5143.

Chang, H. S., Staras, K., Smith, J. E. and Gilbey, M. P. (1999). Sympathetic neuronal oscillators are capable of dynamic synchronization. *J. Neurosci.*, **19**, 3183–3197.

Chapleau, M. W. and Abboud, F. (eds.) (2001). *Neuro-Cardiovascular Regulation: from Molecules to Man.* New York: The New York Academy of Sciences.

Chau, D. and Schramm, L. P. (1997). Sympathetically correlated activity of dorsal horn neurons in spinally transected rats. *J. Neurophysiol.*, **77**, 2966–2974.

Chau, D., Johns, D. G. and Schramm, L. P. (2000). Ongoing and stimulus-evoked activity of sympathetically correlated neurons in the intermediate zone and dorsal horn of acutely spinalized rats. *J. Neurophysiol.*, **83**, 2699–2707.

Cheng, Z. and Powley, T. L. (2000). Nucleus ambiguus projections to cardiac ganglia of rat atria: an anterograde tracing study. *J. Comp. Neurol.*, **424**, 588–606.

Cheng, Z., Powley, T. L., Schwaber, J. S. and Doyle, F. J., III. (1999). Projections of the dorsal motor nucleus of the vagus to cardiac ganglia of rat atria: an anterograde tracing study. *J. Comp. Neurol.*, **410**, 320–341.

Chizh, B. A., Headley, P. M. and Paton, J. F. (1998). Coupling of sympathetic and somatic motor outflows from the spinal cord in a perfused preparation of adult mouse in vitro. *J. Physiol.*, **508**, 907–918.

Choate, J. K., Edwards, F. R., Hirst, G. D. and O'Shea, J. E. (1993a). Effects of sympathetic nerve stimulation on the sino-atrial node of the guinea-pig. *J. Physiol.*, **471**, 707–727.

Choate, J. K., Klemm, M. and Hirst, G. D. S. (1993b). Sympathetic and parasympathetic neuromuscular junctions in the guinea-pig sino-atrial node. *J. Auton. Nerv. Syst.*, **44**, 1–15.

Christensen, J. (1994). The motility of the colon. In *Physiology of the Gastrointestinal Tract*, ed. L. R. Johnson. New York: Raven Press, pp. 991–1024.

Christian, E. P. and Weinreich, D. (1988). Long-duration spike afterhyperpolarizations in neurons from the guinea pig superior cervical ganglion. *Neurosci. Lett.*, **84**, 191–196.

Chrousos, G. P. (1998). Stressors, stress, and neuroendocrine integration of the adaptive response. The 1997 Hans Selye Memorial Lecture. *Ann. New York Acad. Sci.*, **851**, 311-335.

Ciriello, J., Hochstenbach, S. L. and Roder, S. (1994). Central projections of baroreceptor and chemoreceptor afferent fibers in the rat. In *Nucleus of the Solitary Tract*, ed. I. R. A. Barraco. Boca Raton: CRC Press, pp. 35-50.

Clayton, E. C. and Williams, C. L. (2000). Adrenergic activation of the nucleus tractus solitarius potentiates amygdala norepinephrine release and enhances retention performance in emotionally arousing and spatial memory tasks. *Behav. Brain Res.*, **112**, 151-158.

Clement, C. I., Keay, K. A., Owler, B. K. and Bandler, R. (1996). Common patterns of increased and decreased fos expression in midbrain and pons evoked by noxious deep somatic and noxious visceral manipulations in the rat. *J. Comp. Neurol.*, **366**, 495-515.

Clement, C. I., Keay, K. A., Podzebenko, K., Gordon, B. D. and Bandler, R. (2000). Spinal sources of noxious visceral and noxious deep somatic afferent drive onto the ventrolateral periaqueductal gray of the rat. *J. Comp. Neurol.*, **425**, 323-344.

Clutter, W. E., Bier, D. M., Shah, S. D. and Cryer, P. E. (1980). Epinephrine plasma metabolic clearance rates and physiologic thresholds for metabolic and hemodynamic actions in man. *J. Clin. Invest.*, **66**, 94-101.

Cochrane, K. L. and Nathan, M. A. (1989). Normotension in conscious rats after placement of bilateral electrolytic lesions in the rostral ventrolateral medulla. *J Auton. Nerv. Syst.*, **26**, 199-211.

Cochrane, K. L. and Nathan, M. A. (1993). Cardiovascular effects of lesions of the rostral ventrolateral medulla and the nucleus reticularis parvocellularis in rats. *J. Auton. Nerv. Syst.*, **43**, 69-81.

Coderre, T. J., Basbaum, A. I. and Levine, J. D. (1989). Neural control of vascular permeability: interaction between primary afferents, mast cells, and sympathetic efferents. *J. Neurophysiol.*, **62**, 48-58.

Coleridge, H. M. and Coleridge, J. C. (1980). Cardiovascular afferents involved in regulation of peripheral vessels. *Annu. Rev. Physiol.*, **42**, 413-427.

Coleridge, H. M., Coleridge, J. C. G. (1997). Afferent nerves in the airways. In *Autonomic Control of the Respiratory System*, ed. P. J. Barnes. Amsterdam: Harwood Academic Publishers GmbH, pp. 39-58.

Coleridge, H. M., Coleridge, J. C., Kaufman, M. P. and Dangel, A. (1981). Operational sensitivity and acute resetting of aortic baroreceptors in dogs. *Circ. Res.*, **48**, 676-684.

Coleridge, J. C. and Coleridge, H. M. (1984). Afferent vagal C fibre innervation of the lungs and airways and its functional significance. *Rev. Physiol. Biochem. Pharmacol.*, **99**, 1-110.

Contreras, R. J., Gomez, M. M. and Norgren, R. (1980). Central origins of cranial nerve parasympathetic neurons in the rat. *J. Comp. Neurol.*, **190**, 373-394.

Cooke, H. J. (1994). Neuroimmune signaling in regulation of intestinal ion transport. *Am. J. Physiol.*, **266**, G167-G178.

Cooke, H. J. (1998). "Enteric tears": chloride secretion and its neural regulation. *News Physiol. Sci.*, **13**, 269-274.

Cooke, H. J. and Reddix, R. A. (1994). Neural regulation of intestinal electrolyte transport. In *Physiology of the Gastrointestinal Tract*, 3rd edn., ed. L. R. Johnson. New York: Raven Press, pp. 2083-2132.

Coolen, L. M., Allard, J., Truit, W. A. and McKenna, K. E. (2004). Central regulation of ejaculation. *Physiol. Behav.*, **83**, 203-213.

Coonan, E. M., Downie, J. W. and Du, H. J. (1999). Sacral spinal cord neurons responsive to bladder pelvic and perineal inputs in cats. *Neurosci. Lett.*, **260**, 137–140.

Coote, J. H. (1984). Spinal and supraspinal reflex pathways of cardio-cardiac sympathetic reflexes. *Neurosci. Lett.*, **46**, 243–247.

Coote, J. H. (1988). The organisation of cardiovascular neurons in the spinal cord. *Rev. Physiol. Biochem. Pharmacol.*, **110**, 147–285.

Coote, J. H. and Downman, C. B. (1966). Central pathways of some autonomic reflex discharges. *J. Physiol.*, **183**, 714–729.

Coote, J. H. and Sato, A. (1978). Supraspinal regulation of spinal reflex discharge into cardiac sympathetic nerves. *Brain Res.*, **142**, 425–437.

Coote, J. H., Macleod, V. H., Fleetwood-Walker, S. M. and Gilbey, M. P. (1981). Baroreceptor inhibition of sympathetic activity at a spinal site. *Brain Res.*, **220**, 81–93.

Costa, M., Brookes, S. J., Steele, P. A. *et al.* (1996). Neurochemical classification of myenteric neurons in the guinea-pig ileum. *Neuroscience*, **75**, 949–967.

Coupland, R. E. (1965). Electron microscopic observations on the structure of the rat adrenal medulla. I. The ultrastructure and organization of chromaffin cells in the normal adrenal medulla. *J. Anat.*, **99**, 231–254.

Cousins, H. M., Edwards, F. R., Hirst, G. D. and Wendt, I. R. (1993). Cholinergic neuromuscular transmission in the longitudinal muscle of the guinea-pig ileum. *J. Physiol.*, **471**, 61–86.

Cousins, H. M., Edwards, F. R. and Hirst, G. D. (1995). Neuronally released and applied acetylcholine on the longitudinal muscle of the guinea-pig ileum. *Neuroscience*, **65**, 193–207.

Coutinho, S. V., Su, X., Sengupta, J. N. and Gebhart, G. F. (2000). Role of sensitized pelvic nerve afferents from the inflamed rat colon in the maintenance of visceral hyperalgesia. *Prog. Brain Res.*, **129**, 375–387.

Cowley, A. W. Jr. (1992). Long-term control of arterial blood pressure. *Physiol. Rev.*, **72**, 231–300.

Cowley, A. W. Jr., Liard, J. F. and Guyton, A. C. (1973). Role of the baroreceptor reflex in daily control of arterial blood pressure and other variables in dogs. *Circ. Res.*, **32**, 564–576.

Cox, G. E., Jordan, D., Paton, J. F., Spyer, K. M. and Wood, L. M. (1987). Cardiovascular and phrenic nerve responses to stimulation of the amygdala central nucleus in the anaesthetized rabbit. *J. Physiol.*, **389**, 541–556.

Craig, A. D. (1996). An ascending general homeostatic afferent pathway originating in lamina I. *Prog. Brain Res.*, **107**, 225–243.

Craig, A. D. (2002). How do you feel? Interoception: the sense of the physiological condition of the body. *Nat. Rev. Neurosci.*, **3**, 655–666.

Craig, A. D. (2003a). Pain mechanisms: labeled lines versus convergence in central processing. *Annu. Rev. Neurosci.*, **26**, 1–30.

Craig, A. D. (2003b). A new view of pain as a homeostatic emotion. *Trends Neurosci.*, **26**, 303–307.

Craig, A. D. (2003c). Interoception: the sense of the physiological condition of the body. *Curr. Opin. Neurobiol.*, **13**, 500–505.

Craig, A. D. (2004a). Lamina I, but not lamina V, spinothalamic neurons exhibit responses that correspond with burning pain. *J. Neurophysiol.*, **92**, 2604–2609.

Craig, A. D. (2004b). Distribution of trigeminothalamic and spinothalamic lamina I terminations in the macaque monkey. *J. Comp. Neurol.*, **477**, 119–148.

Craig, A. D. and Blomqvist, A. (2002). Is there a specific lamina I spinothalamocortical pathway for pain and temperature sensations in primates? *J. Pain*, **3**, 95–101.

Craig, A. D. and Kniffki, K. D. (1985). Spinothalamic lumbosacral lamina I cells responsive to skin and muscle stimulation in the cat. *J. Physiol.*, **365**, 197–221.

Craig, A. D. and Mense, S. (1983). The distribution of afferent fibers from the gastrocnemius-soleus muscle in the dorsal horn of the cat, as revealed by the transport of horseradish peroxidase. *Neurosci. Lett.*, **41**, 233–238.

Craig, A. D., Heppelmann, B. and Schaible, H. G. (1988). The projection of the medial and posterior articular nerves of the cat's knee to the spinal cord. *J. Comp. Neurol.*, **276**, 279–288.

Craig, A. D., Bushnell, M. C., Zhang, E. T. and Blomqvist, A. (1994). A thalamic nucleus specific for pain and temperature sensation. *Nature*, **372**, 770–773.

Craig, A. D., Krout, K. and Andrew, D. (2001). Quantitative response characteristics of thermoreceptive and nociceptive lamina I spinothalamic neurons in the cat. *J. Neurophysiol.*, **86**, 1459–1480.

Cravo, S. L., Morrison, S. F. and Reis, D. J. (1991). Differentiation of two cardiovascular regions within caudal ventrolateral medulla. *Am. J. Physiol.*, **261**, R985–R994.

Crawford, J. P. and Frankel, H. L. (1971). Abdominal 'visceral' sensation in human tetraplegia. *Paraplegia*, **9**, 153–158.

Critchley, H. and Dolan, R. J. (2004). Central representation of autonomic states. In *Human Brain Function*. 2nd edn., eds. R. S. J. Frackowiak, K. J. Friston, C. D. Frith *et al*. Amsterdam: Elsevier Academic Press, pp. 397–417.

Crowcroft, P. J., Holman, M. E. and Szurszewski, J. H. (1971). Excitatory input from the distal colon to the inferior mesenteric ganglion in the guinea-pig. *J. Physiol.*, **219**, 443–461.

Cryer, P. E. (1980). Physiology and pathophysiology of the human sympathoadrenal neuroendocrine system. *New Engl. J. Med.*, **303**, 436–444.

Cunnane, T. C. and Stjärne, L. (1982). Secretion of transmitter from individual varicosities of guinea-pig and mouse vas deferens: all-or-none and extremely intermittent. *Neuroscience*, **7**, 2565–2576.

Cyon, E. and Ludwig, C. (1866). Die Reflexe eines der sensiblen Nerven des Herzens auf die motorischen der Blutgefässe [The reflex of one of the heart nerves on the motor nerves to blood vessels]. *Ber. Sächs. Ges. Wiss.*, **18**, 307–328.

Czachurski, J., Dembowsky, K., Seller, H., Nobiling, R. and Taugner, R. (1988). Morphology of electrophysiologically identified baroreceptor afferents and second order neurones in the brainstem of the cat. *Arch. Ital. Biol.*, **126**, 129–144.

Dado, R. J., Katter, J. T. and Giesler, G. J., Jr. (1994). Spinothalamic and spinohypothalamic tract neurons in the cervical enlargement of rats. I. Locations of antidromically identified axons in the thalamus and hypothalamus. *J. Neurophysiol.*, **71**, 959–980.

Dail, W. G. (1993). Autonomic innervation of male reproductive genitalia. In *The Autonomic Nervous System*, Vol. 3, *Nervous Control of the Urogenital System*, ed. C. A. Maggi. Chur, Switzerland: Harwood Academic Publishers, pp. 69–101.

Dalsgaard, C. J. and Elfvin, L. G. (1982). Structural studies on the connectivity of the inferior mesenteric ganglion of the guinea pig. *J. Auton. Nerv. Syst.*, **5**, 265–278.

Dalsgaard, C. J., Hökfelt, T., Elfvin, L. G., Skirboll, L. and Emson, P. (1982). Substance P-containing primary sensory neurons projecting to the inferior mesenteric ganglion: evidence from combined retrograde tracing and immunohistochemistry. *Neuroscience*, **7**, 647–654.

Daly, M.d. (1991). Some reflex cardioinhibitory responses in the cat and their modulation by central inspiratory neuronal activity. *J. Physiol.*, **439**, 559–577.

Daly, M.d. and Robinson, B. H. (1968). An analysis of the reflex systemic vasodilator response elicited by lung inflation in the dog. *J. Physiol.*, **195**, 387–406.

Daly, M.d., Hazzledine, J. L. and Ungar, A. (1967). The reflex effects of alterations in lung volume on systemic vascular resistance in the dog. *J. Physiol.*, **188**, 331–351.

Daly, M.d., Ward, J. and Wood, L. M. (1987). The peripheral chemoreceptors and cardiovascular-respiratory integration. In *The Neurobiology of the Cardiorespiratory System*, ed. E. W. Taylor. Manchester: Manchester University Press, pp. 342–368.

Damasio, A. R. (1994). *Descartes' Error: Emotion, Reason and the Human Brain*. New York: Avon Books.

Damasio, A. R. (1999). *The Feeling of What Happens: Body and Emotion in the Making of Consciousness*. Harcourt Brace: New York.

Damasio, A. R., Adolphs, R. and Damasio, H. (2003). The contributions of the lesion method to the functional neuroanatomy of emotion. In *Handbook of Affective Sciences*, eds. R. J. Davidson, K. R. Scherer and H. H. Goldsmith. Oxford, New York: Oxford University Press, pp. 66–92.

Dampney, R. A. (1981). Brain stem mechanisms in the control of arterial pressure. *Clin. Exp. Hypertens.*, **3**, 379–391.

Dampney, R. A. (1994). Functional organization of central pathways regulating the cardiovascular system. *Physiol. Rev.*, **74**, 323–364.

Dampney, R. A. and Horiuchi, J. (2003). Functional organisation of central cardiovascular pathways: studies using c-fos gene expression. *Prog. Neurobiol.*, **71**, 359–384.

Dampney, R. A. and McAllen, R. M. (1988). Differential control of sympathetic fibres supplying hindlimb skin and muscle by subretrofacial neurones in the cat. *J. Physiol.*, **395**, 41–56.

Dampney, R. A. and Moon, E. A. (1980). Role of ventrolateral medulla in vasomotor response to cerebral ischemia. *Am. J. Physiol.*, **239**, H349–H358.

Dampney, R. A., Goodchild, A. K., Robertson, L. G. and Montgomery, W. (1982). Role of ventrolateral medulla in vasomotor regulation: a correlative anatomical and physiological study. *Brain Res.*, **249**, 223–235.

Dampney, R. A., Goodchild, A. K. and Tan, E. (1985). Vasopressor neurons in the rostral ventrolateral medulla of the rabbit. *J. Auton. Nerv. Syst.*, **14**, 239–254.

Dampney, R. A., Tagawa, T., Horiuchi, J. et al. (2000). What drives the tonic activity of presympathetic neurons in the rostral ventrolateral medulla? *Clin. Exp. Pharmacol. Physiol.*, **27**, 1049–1053.

Dampney, R. A., Horiuchi, J., Tagawa, T. et al. (2003). Medullary and supramedullary mechanisms regulating sympathetic vasomotor tone. *Acta Physiol. Scand.*, **177**, 209–218.

Dantzer, R., Bluthe, R. M., Gheusi, G. et al. (1998). Molecular basis of sickness behavior. *Ann. New York Acad. Sci.*, **856**, 132–138.

Dantzer, R., Konsman, J. P., Bluthe, R. M. and Kelley, K. W. (2000). Neural and humoral pathways of communication from the immune system to the brain: parallel or convergent? *Auton. Neurosci.*, **85**, 60–65.

Darwin, C. (1998) [1872]. *The Expression of the Emotions in Man and Animals. With an Introduction, Afterword and Commentaries by Paul Ekman*, 3rd edn. London: Harper Collins.

Davidson, R. J., Scherer, K. R. and Goldsmith, H. H. (eds.) (2003). *Handbook of Affective Sciences*. New York, Oxford: Oxford University Press.

Davies, P. J., Ireland, D. R. and McLachlan, E. M. (1996). Sources of Ca2+ for different Ca^{2+}-activated K^+ conductances in neurones of the rat superior cervical ganglion. *J. Physiol.*, **495**, 353–366.

Davies, P. J., Ireland, D. R., Martinez-Pinna, J. and McLachlan, E. M. (1999). Electrophysiological roles of L-type channels in different classes of guinea pig sympathetic neuron. *J. Neurophysiol.*, **82**, 818–828.

Davis, K. D., Meyer, R. A. and Campbell, J. N. (1993). Chemosensitivity and sensitization of nociceptive afferents that innervate the hairy skin of monkey. *J. Neurophysiol.*, **69**, 1071–1081.

Davis, M. J. and Hill, M. A. (1999). Signaling mechanisms underlying the vascular myogenic response. *Physiol. Rev.*, **79**, 387–423.

de Groat, W. C. (1976). Mechanisms underlying recurrent inhibition in the sacral parasympathetic outflow to the urinary bladder. *J. Physiol.*, **257**, 503–513.

de Groat, W. C. (1987). Neuropeptides in pelvic afferent pathways. *Experientia*, **43**, 801–813.

de Groat, W. C. (1989). Neuropeptides in pelvic afferent pathways. In *Regulatory Peptides*, ed. J. M. Polak. Basel: Birkhauser Verlag AG, pp. 334–361.

de Groat, W. C. (2002). Neural control of the urinary bladder and sexual organs. In *Autonomic Failure*, 4th edn., eds. C. J. Mathias and R. Bannister. New York, Oxford: Oxford University Press, pp. 151–165.

de Groat, W. C. and Booth, A. M. (1993). Neural control of penile erection. In *The Autonomic Nervous System, Vol. 3, Nervous Control of the Urogenital System*, ed. C. A. Maggi. Chur, Switzerland: Harwood Academic Publishers, pp. 467–524.

de Groat, W. C. and Ryall, R. W. (1968). Recurrent inhibition in sacral parasympathetic pathways to the bladder. *J. Physiol.*, **196**, 579–591.

de Groat, W. C., Booth, A. M., Milne, R. J. and Roppolo, J. R. (1982). Parasympathetic preganglionic neurons in the sacral spinal cord. *J. Auton. Nerv. Syst.*, **5**, 23–43.

de Groat, W. C., Booth, A. M. and Yoshimura, N. (1993). Neurophysiology of micturition and its modification in animal models in human disease. In *The Autonomic Nervous System, Vol. 3, Nervous Control of the Urogenital System*. ed. C. A. Maggi. Chur, Switzerland: Harwood Academic Publishers, pp. 227–290.

de Groat, W. C., Vizzard, M. A., Araki, I. and Roppolo, J. H. (1996). Spinal interneurons and preganglionic neurons in sacral autonomic reflex pathways. *Prog. Brain Res.*, **107**, 97–111.

de Groat, W. C., Araki, I., Vizzard, M. A. *et al.* (1998). Developmental and injury induced plasticity in the micturition reflex pathway. *Behav. Brain Res.*, **92**, 127–140.

de Groat, W. C., Fraser, M. O., Yoshiyama, M. *et al.* (2001). Neural control of the urethra. *Scand. J. Urol. Nephrol.*, **Suppl (207)**, 35–43.

Delius, W., Hagbarth, K. E., Hongell, A. and Wallin, B. G. (1972). General characteristics of sympathetic activity in human muscle nerves. *Acta Physiol. Scand.*, **84**, 65–81.

Dembowsky, K., Czachurski, J. and Seller, H. (1985). Morphology of sympathetic preganglionic neurons in the thoracic spinal cord of the cat: an intracellular horseradish peroxidase study. *J. Comp. Neurol.*, **238**, 453–465.

Demir, S. S., Clark, J. W. and Giles, W. R. (1999). Parasympathetic modulation of sinoatrial node pacemaker activity in rabbit heart: a unifying model. *Am. J. Physiol.*, **276**, H2221–H2244.

Denton, K. M., Luff, S. E., Shweta, A. and Anderson, W. P. (2004). Differential neural control of glomerular ultrafiltration. *Clin. Exp. Pharmacol. Physiol.*, **31**, 380–386.

Deuchars, S. A., Spyer, K. M. and Gilbey, M. P. (1997). Stimulation within the rostral ventrolateral medulla can evoke monosynaptic GABAergic IPSPs in sympathetic preganglionic neurons in vitro. *J. Neurophysiol.*, **77**, 229–235.

Deuchars, J., Li, Y. W., Kasparov, S. and Paton, J. F. (2000). Morphological and electrophysiological properties of neurones in the dorsal vagal complex of the rat activated by arterial baroreceptors. *J. Comp. Neurol.*, **417**, 233–249.

Deuchars, S. A., Brooke, R. E. and Deuchars, J. (2001a). Adenosine A1 receptors reduce release from excitatory but not inhibitory synaptic inputs onto lateral neurons. *J. Neurosci.*, **21**, 6308–6320.

Deuchars, S. A., Brooke, R. E., Frater, B. and Deuchars, J. (2001b). Properties of interneurones in the intermediolateral cell column of the rat spinal cord: role of the potassium channel subunit Kv3.1. *Neuroscience*, **106**, 433–446.

Deuchars, S. A., Milligan, C. J., Stornetta, R. L. and Deuchars, J. (2005). GABAergic neurons in the central region of the spinal cord: a novel substrate for sympathetic inhibition. *J. Neurosci.*, **25**, 1063–1070.

DiBona, G. F. and Kopp, U. C. (1997). Neural control of renal function. *Physiol. Rev.*, **77**, 75–197.

Dickens, E. J., Hirst, G. D. and Tomita, T. (1999). Identification of rhythmically active cells in guinea-pig stomach. *J. Physiol.*, **514**, 515–531.

Dietz, N. M., Rivera, J. M., Eggener, S. E. et al. (1994). Nitric oxide contributes to the rise in forearm blood flow during mental stress in humans. *J. Physiol.*, **480**, 361–368.

Dittmar, C. (1873). Über die Lage des sogenannten Gefässcentrums in der Medulla oblongata [On the location of the so-called vascular center in the medulla oblongata]. *Sitzungsber. Akad. Wiss. Wien, Math.-Naturw., Abt. 2*, **25**, 449–469.

Dobbins, E. G. and Feldman, J. L. (1994). Brainstem network controlling descending drive to phrenic motoneurons in rat. *J. Comp. Neurol.*, **347**, 64–86.

Dockray, G. J., Green, T. and Varro, A. (1989). The afferent peptidergic innervation of the upper gastrointestinal tract. In *Nerves and the Gastrointestinal Tract*, eds. M. V. Singer and H. Goebel. Dordrecht: Kluwer Academic Publishers, pp. 105–122.

Dodt, C., Gunnarsson, T., Elam, M., Karlsson, T. and Wallin, B. G. (1995). Central blood volume influences sympathetic sudomotor nerve traffic in warm humans. *Acta Physiol. Scand.*, **155**, 41–51.

Dodt, C., Lönnroth, P., Fehm, H. L. and Elam, M. (1999). Intraneural stimulation elicits an increase in subcutaneous interstitial glycerol levels in humans. *J. Physiol.*, **521**, 545–552.

Donnerer, J., Schuligoi, R. and Stein, C. (1992). Increased content and transport of substance P and calcitonin gene-related peptide in sensory nerves innervating inflamed tissue: evidence for a regulatory function of nerve growth factor in vivo. *Neuroscience*, **49**, 693–698.

Dorward, P. K., Andresen, M. C., Burke, S. L., Oliver, J. R. and Korner, P. I. (1982). Rapid resetting of the aortic baroreceptors in the rabbit and its implications for short-term and longer term reflex control. *Circ. Res.*, **50**, 428–439.

Dorward, P. K., Burke, S. L., Jänig, W. and Cassell, J. (1987). Reflex responses to baroreceptor, chemoreceptor and nociceptor inputs in single renal sympathetic neurones in the rabbit and the effects of anaesthesia on them. *J. Auton. Nerv. Syst.*, **18**, 39–54.

Dostrovsky, J. O. and Craig, A. D. (2005). Ascending projection systems. In *Wall and Melzack's Textbook of Pain*. 5th edn., ed. S. B. McMahon and M. Koltzenburg. Edinburgh: Elsevier Churchill Livingstone, pp. 187–204.

Downing, J. E. and Miyan, J. A. (2000). Neural immunoregulation: emerging roles for nerves in immune homeostasis and disease. *Immunol. Today*, **21**, 281–289.

Drummond, P. D. (1995). Mechanisms of physiological gustatory sweating and flushing in the face. *J. Auton. Nerv. Syst.*, **52**, 117–124.

Duan, Y. F., Winters, R., McCabe, P. M. *et al.* (1996). Behavioral characteristics of defense and vigilance reactions elicited by electrical stimulation of the hypothalamus in rabbits. *Behav. Brain Res.*, **81**, 33–41.

Duan, Y. F., Winters, R., McCabe, P. M. *et al.* (1997). Functional relationship between the hypothalamic vigilance area and PAG vigilance area. *Physiol. Behav.*, **62**, 675–679.

Dun, N. J. (1983). Peptide hormones and transmission in sympathetic ganglia. In *Autonomic Ganglia*, ed. L. G. Elfvin. Chichester: J. Wiley & Sons, pp. 345–366.

Dunn, W. R., Brock, J. A. and Hardy, T. A. (1999). Electrochemical and electrophysiological characterization of neurotransmitter release from sympathetic nerves supplying rat mesenteric arteries. *Br. J. Pharmacol.*, **128**, 174–180.

Dworkin, B. R. (1993). *Learning and Physiological Regulation*. Chicago: The University of Chicago Press.

Dworkin, B. R. (2000). Interoception. In *Handbook of Psychophysiology*, 2nd edn., ed. J. T. Cacioppo, L. G. Tassinary and G. G. Berntson. Cambridge: Cambridge University Press, pp. 482–506.

Ebbeson, S. O. E. (1968a). Quantitative studies of superior cervical sympathetic ganglia in a variety of primates including man. I. The ratio of preganglionic neurons. *J. Morphol.*, **124**, 117–132.

Ebbeson, S. O. E. (1968b). Quantitative studies of superior cervical sympathetic ganglia in a variety of primates including man. II. Neuronal packing density. *J. Morphol.*, **124**, 181–186.

Eckberg, D. L. (2003). The human respiratory gate. *J. Physiol.*, **548**, 339–352.

Eckberg, D. L. and Sleight, P. (1992). *Human Baroreceptor Reflexes in Health and Disease*. Oxford: Clarendon Press.

Eckberg, D. L., Nerhed, C. and Wallin, B. G. (1985). Respiratory modulation of muscle sympathetic and vagal cardiac outflow in man. *J. Physiol.*, **365**, 181–196.

Eckberg, D. L., Rea, R. F., Andersson, O. K. *et al.* (1988). Baroreflex modulation of sympathetic activity and sympathetic neurotransmitters in humans. *Acta Physiol. Scand.*, **133**, 221–231.

Ectors, L. (1941). Contribution á l'étude des réactions pilomotrices. *Arch. Int. Physiol.*, **51**, 443–455.

Edwards, F. R., Bramich, N. J. and Hirst, G. D. (1993). Analysis of the effects of vagal stimulation on the sinus venous of the toad. *Philos. Trans. R. Soc. London B Biol. Sci.*, **341**, 149-162.

Edwards, F. R., Hirst, G. D., Klemm, M. F. and Steele, P. A. (1995). Different types of ganglion cell in the cardiac plexus of guinea-pigs. *J. Physiol.*, **486**, 453–471.

Edwards, S. L., Anderson, C. R., Southwell, B. R. and McAllen, R. M. (1996). Distinct preganglionic neurons innervate noradrenaline and adrenaline cells in the cat adrenal medulla. *Neuroscience*, **70**, 825-832.

Ekman, P. (1992). Facial expression of emotion: new findings, new questions. *Psychol. Sci.*, **3**, 34-38.

Ekman, P. and Davidson, R. J. (eds.) (1994). *The Nature of Emotions*. Oxford: Oxford University Press.

Ekman, P., Levenson, R. W. and Friesen, M. V. (1983). Autonomic nervous system activity distinguishes between emotions. *Science*, **221**, 1208-1210.

Elfvin, L. G. (ed.) (1983). *Autonomic Ganglia*. Chichester: J. Wiley & Sons.

Elfvin, L. G., Lindh, B. and Hökfelt, T. (1993). The chemical neuroanatomy of sympathetic ganglia. *Annu. Rev. Neurosci.*, **16**, 471-507.

Eliasson, S., Folkow, B., Lindgren, P. and Uvnäs, B. (1951). Activation of sympathetic vasodilator nerves to the skeletal muscles in the cat by hypothalamic stimulation. *Acta Physiol. Scand.*, **23**, 333-351.

Ellison, G. D. and Zanchetti, A. (1973). Diffuse and specific activation of sympathetic cholinergic fibers of the cat. *Am. J. Physiol.*, **225**, 142-149.

Elsner, R., Franklin, D. L., Van Citters, R. L. and Kenney, D. W. (1966). Cardiovascular defense against asphyxia. *Science*, **153**, 941-949.

Emch, G. S., Hermann, G. E. and Rogers, R. C. (2000). TNF-alpha activates solitary nucleus neurons responsive to gastric distension. *Am. J. Physiol. Gastrointest. Liver Physiol.*, **279**, G582–G586.

Emch, G. S., Hermann, G. E. and Rogers, R. C. (2002). Tumor necrosis factor-alpha inhibits physiologically identified dorsal motor nucleus neurons in vivo. *Brain Res.*, **951**, 311–315.

Enquist, L. W. and Card, J. P. (2003). Recent advances in the use of neurotropic viruses for circuit analysis. *Curr. Opin. Neurobiol.*, **13**, 603-606.

Eppel, G. A., Malpas, S. C., Denton, K. M. and Evans, R. G. (2004). Neural control of renal medullary perfusion. *Clin. Exp. Pharmacol. Physiol.*, **31**, 387-396.

Ernsberger, U. (2000). Evidence for an evolutionary conserved role of bone morphogenetic protein growth factors and phox2 transcription factors during noradrenergic differentiation of sympathetic neurons. Induction of a putative synexpression group of neurotransmitter-synthesizing enzymes. *Eur. J. Biochem.*, **267**, 6976-6981.

Ernsberger, U. (2001). The development of postganglionic sympathetic neurons: coordinating neuronal differentiation and diversification. *Auton. Neurosci.*, **94**, 1-13.

Ernsberger, U., Esposito, L., Partimo, S. *et al.* (2005). Expression of neuronal markers suggests heterogeneity of chick sympathoadrenal cells prior to invasion of the adrenal anlagen. *Cell Tissue Res.*, **319**, 1-13.

Esler, M., Jennings, G., Lambert, G. *et al.* (1990). Overflow of catecholamine neurotransmitters to the circulation: source, fate, and functions. *Physiol. Rev.*, **70**, 963-985.

Evans, R. J. and Cunnane, T. C. (1992). Relative contributions of ATP and noradrenaline to the nerve evoked contraction of the rabbit jejunal artery. Dependence on stimulation parameters. *Naunyn-Schmiedeberg's Arch. Pharmacol.*, **345**, 424–430.

Evans, R. J. and Surprenant, A. (1992). Vasoconstriction of guinea-pig submucosal arterioles following sympathetic nerve stimulation is mediated by the release of ATP. *Br. J. Pharmacol.*, **106**, 242–249.

Ezure, K., Tanaka, I. and Kondo, M. (2003). Glycine is used as a transmitter by decrementing expiratory neurons of the ventrolateral medulla in the rat. *J. Neurosci.*, **23**, 8941–8948.

Fagius, J. and Sundlöf, G. (1986). The diving response in man: effects on sympathetic activity in muscle and skin nerve fascicles. *J. Physiol.*, **377**, 429–443.

Fagius, J. and Wallin, B. G. (1993). Long-term variability and reproducibility of resting human muscle nerve sympathetic activity at rest, as reassessed after a decade. *Clin. Auton. Res.*, **3**, 201–205.

Fagius, J., Wallin, B. G., Sundlöf, G., Nerhed, C. and Englesson, S. (1985). Sympathetic outflow in man after anaesthesia of the glossopharyngeal and vagus nerves. *Brain*, **108**, 423–438.

Farkas, E., Jansen, A. S. and Loewy, A. D. (1998). Periaqueductal gray matter input to cardiac-related sympathetic premotor neurons. *Brain Res.*, **792**, 179–192.

Feigl, E. O. (1998). Neural control of coronary blood flow. *J. Vasc. Res.*, **35**, 85–92.

Feldberg, W. and Guertzenstein, P. G. (1976). Vasodepressor effects obtained by drugs acting on the ventral surface of the brain stem. *J. Physiol.*, **258**, 337–355.

Feldman, J. L. (1986). Neurophysiology of breathing in mammals. In *Handbook of Physiology, The Nervous System*, Vol. IV, ed. V. B. Mountcastle and F. E. Bloom, Bethesda: American Physiological Society, pp. 463–524.

Feldman, J. L. and McCrimmon, D. R. (2003). Neural control of breathing. In *Fundamental Neuroscience*, 2nd edn., ed. L. R. Squire, F. E. Bloom, S. K. McConnell *et al.* San Diego: Academic Press, pp. 967–990.

Feldman, J. L., Mitchell, G. S. and Nattie, E. E. (2003). Breathing: rhythmicity, plasticity, chemosensitivity. *Annu. Rev. Neurosci.*, **26**, 239–266.

Fernandez de Molina, A. and Hunsperger, R. W. (1962). Organization of the subcortical system governing defence and flight reactions in the cat. *J. Physiol.*, **160**, 200–213.

Fielden, R., Sutton, T. J. and Taylor, E. M. (1980). Effect of tilting on the pressor responses to McN-A-343, a muscarinic sympathetic ganglion stimulant. *Br. J. Pharmacol.*, **71**, 287–295.

Fields, H. L., Basbaum, A. I. and Heinricher, M. M. (2005). Central nervous system mechanisms of pain modulation. In *Wall and Melzack's Textbook of Pain*, 5th edn., eds. S. B. McMahon and M. Koltzenburg. Amsterdam, Edinburgh: Elsevier Churchill Livingstone, pp. 125–142.

Fleming, W. W. and Westfall, D. P. (1988). Adaptive supersensitivity. In *Handbook of Experimental Pharmacology*, Vol. 90/I, *Catecholamines I*, ed. U. Trendelenburg and N. Weiner. Berlin, Heidelberg, New York: Springer-Verlag, pp. 509–559.

Floyd, N. S., Price, J. L., Ferry, A. T., Keay, K. A. and Bandler, R. (2000). Orbitomedial prefrontal cortical projections to distinct longitudinal columns of the periaqueductal gray in the rat. *J. Comp. Neurol.*, **422**, 556–578.

Floyd, N. S., Price, J. L., Ferry, A. T., Keay, K. A. and Bandler, R. (2001). Orbitomedial prefrontal cortical projections to hypothalamus in the rat. *J. Comp. Neurol.*, **432**, 307–328.

Foerster, O. (1927). *Die Leitungsbahnen des Schmerzgefühls und die chirurgische Behandlung der Schmerzzustände [Pain Pathways and the Surgical Treatment of Pain]*. Berlin, Wien: Urban und Schwarzenberg.

Fogel, R., Zhang, X. and Renehan, W. E. (1996). Relationships between the morphology and function of gastric and intestinal distention-sensitive neurons in the dorsal motor nucleus of the vagus. *J. Comp. Neurol.*, **364**, 78–91.

Folkow, B. (1955). The control of blood vessels. *Physiol. Rev.*, **35**, 629–663.

Folkow, B. (1987). Psychosocial and central nervous influences in primary hypertension. *Circulation*, **76**, I10–I19.

Folkow, B. (2000). Perspectives on the integrative function of the "sympatho-adrenomedullary system". *Auton. Neurosci.*, **83**, 101–115.

Folkow, B. and von Euler, U. S. (1954). Selective activation of noradrenaline and adrenaline producing cells in the cat's adrenal gland by hypothalamic stimulation. *Circ. Res.*, **2**, 191–195.

Folkow, B. and Neil, E. (1971). *Circulation*. New York: Oxford University Press.

Folkow, B. and Nilsson, H. (1997). Transmitter release at adrenergic nerve endings: Total exocytosis or fractional release? *News Physiol. Sci.*, **12**, 32–36.

Folkow, B., Schmidt, T. and Uvnäs-Moberg, K. (eds.) (1997). Stress, health and the social environment. *Acta Physiol. Scand.*, **161**, **Suppl. 640**, 1–179.

Forehand, C. J. (1987). Ultrastructural analysis of the distribution of synaptic boutons from labeled preganglionic axons on rabbit ciliary neurons. *J. Neurosci.*, **7**, 3274–3281.

Forehand, C. J. and Purves, D. (1984). Regional innervation of rabbit ciliary ganglion cells by the terminals of preganglionic axons. *J. Neurosci.*, **4**, 1–12.

Foreman, R. D. (1989). Organization of the spinothalamic tract as a relay for cardiopulmonary sympathetic afferent fiber activity. *Prog. Sens. Physiol.*, **9**, 1–51.

Foreman, R. D. (1999). Mechanisms of cardiac pain. *Annu. Rev. Physiol.*, **61**, 143–167.

Foster, M. (1879). Textbook of Physiology. In *The Discovery of Reflexes*, ed. E. G. T. Liddell. Oxford: Clarendon Press, pp. 98–101.

Fox, E. A. and Powley, T. L. (1985). Longitudinal columnar organization within the dorsal motor nucleus represents separate branches of the abdominal vagus. *Brain Res.*, **341**, 269–282.

Fox, E. A. and Powley, T. L. (1992). Morphology of identified preganglionic neurons in the dorsal motor nucleus of the vagus. *J. Comp. Neurol.*, **322**, 79–98.

Frayn, K. N. and MacDonald, I. A. (1996). Adipose tissue circulation. In *The Autonomic Nervous System, Vol. 8, Nervous Control of Blood Vessels*, ed. T. Bennett and S. M. Gardiner. Amsterdam: Harwood Academic Publishers, pp. 505–539.

Freeman, R. and Rutkove, S. (2000). Syncope. In *Handbook of Clinical Neurology, Vol. 75, The Autonomic Nervous System, part II: Dysfunctions*, ed. O. Appenzeller. Amsterdam: Elsevier, pp. 203–228.

Freyburger, W. A., Gruhzit, C. C. and Moe, G. K. (1950a). Pressor pathways not blocked by tetraethylammonium. *Am. J. Physiol.*, **163**, 290–293.

Freyburger, W. A., Gruhzit, C. C., Rennick, B. R. and Moe, G. K. (1950b). Action of tetraethylammonium on pressor response to asphyxia. *Am. J. Physiol.*, **163**, 554–560.

Fukai, K. and Fukuda, H. (1985). Three serial neurones in the innervation of the colon by the sacral parasympathetic nerve of the dog. *J. Physiol.*, **362**, 69-78.

Fukami, H. and Bradley, R. M. (2005). Biophysical and morphological properties of parasympathetic neurons controlling the parotid and von Ebner salivary glands in rats. *J. Neurophysiol.*, **93**, 678-686.

Fukuda, A., Minami, T., Nabekura, J. and Oomura, Y. (1987). The effects of noradrenaline on neurones in the rat dorsal motor nucleus of the vagus, in vitro. *J. Physiol.*, **393**, 213-231.

Fulton, J. F. (1949). *Physiology of the Nervous System*, 3rd edn. New York: Oxford University Press.

Furness, J. B. (2005). *The Enteric Nervous System*. Oxford: Blackwell Science Ltd.

Furness, J. B. and Clerc, N. (2000). Responses of afferent neurons to the contents of the digestive tract, and their relation to endocrine and immune responses. *Prog. Brain Res.*, **122**, 159-172.

Furness, J. B. and Costa, M. (1980). Types of nerves in the enteric nervous system. *Neuroscience*, **5**, 1-20.

Furness, J. B. and Costa, M. (1987). *The Enteric Nervous System*. London: Churchill Livingstone.

Furness, J. B., Morris, J. L., Gibbins, I. L. and Costa, M. (1989). Chemical coding of neurons and plurichemical transmission. *Annu. Rev. Pharmacol. Toxicol.*, **29**, 289-306.

Furness, J. B., Bornstein, J. C., Murphy, R. and Pompolo, S. (1992). Roles of peptides in transmission in the enteric nervous system. *Trends Neurosci.*, **15**, 66-71.

Furness, J. B., Kunze, W. A., Bertrand, P. P., Clerc, N. and Bornstein, J. C. (1998). Intrinsic primary afferent neurons of the intestine. *Prog. Neurobiol.*, **54**, 1-18.

Furness, J. B., Clerc, N., Vogalis, F. and Stebbing, M. J. (2003a). The enteric nervous system and its extrinsic connections. In *Texbook of Gastroenterology*, ed. T. Yamada, D. H. Alpers, L. Laine, C. Owyang and D. W. Powell. Philadelphia: Lippincott Williams & Wilkins, pp. 12-34.

Furness, J. B., Stebbing, M. J., Kunze, W. A. A. and Clerc, N. (2003b). Sensory neurons of the gastrointestinal tract. In *Textbook of Gastroenterology*, ed. T. Yamada, D. H. Alpers, L. Laine, C. Owyang and D. W. Powell. Philadelphia: Lippincott Williams & Wilkins, pp. 34-47.

Furness, J. B., Jones, C., Nurgali, K. and Clerc, N. (2004). Intrinsic primary afferent neurons and nerve circuits within the intestine. *Prog. Neurobiol.*, **72**, 143-164.

Gabella, G. (1976). *Structure of the Autonomic Nervous System*. London: Chapman and Hall.

Gamlin, P. D. R. (2000). Functions of the Edinger-Westphal nucleus. In *The Autonomic Nervous System*, Vol. 13, *Nervous Control of the Eye*, ed. G. Burnstock and A. M. Sillito. Amsterdam: Harwood Academic Publisher, pp. 117-154.

Gamlin, P. D. R. and Clarke, R. J. (1995). Single-unit activity in the primate nucleus reticularis tegmenti pontis related to vergence and ocular accommodation. *J. Neurophysiol.*, **73**, 2115-2119.

Gamlin, P. D. R., Zhang, Y., Clendaniel, R. A. and Mays, L. E. (1994). Behavior of identified Edinger-Westphal neurons during ocular accommodation. *J. Neurophysiol.*, **72**, 2368-2382.

Gandevia, S. C., Killian, K., McKenzie, D. K. et al. (1993). Respiratory sensations, cardiovascular control, kinaesthesia and transcranial stimulation during paralysis in humans. *J. Physiol.*, **470**, 85–107.

Garrett, J. R., Ekström, J. and Anderson, L. C. (eds.) (1999). *Frontiers in Oral Biology*, Vol. 11, Basel: Karger.

Gaskell, W. H. (1916). *The Involuntary Nervous System*. London: Longmans.

Gauriau, C. and Bernard, J. F. (2002). Pain pathways and parabrachial circuits in the rat. *Exp. Physiol.*, **87**, 251–258.

Gauriau, C. and Bernard, J. F. (2004a). Posterior triangular thalamic neurons convey nociceptive messages to the secondary somatosensory and insular cortices in the rat. *J. Neurosci.*, **24**, 752–761.

Gauriau, C. and Bernard, J. F. (2004b). A comparative reappraisal of projections from the superficial laminae of the dorsal horn in the rat: the forebrain. *J. Comp. Neurol.*, **468**, 24–56.

Gebhart, G. F. (ed.) (1995). *Visceral Pain. Progress in Pain Research and Management*. Seattle: IASP Press.

Gebhart, G. F. (1996). Visceral polymodal receptors. *Prog. Brain Res.*, **113**, 101–112.

Gebhart, G. F. and Randich, A. (1992). Vagal modulation of nociception. *Am. Pain Soc. J.*, **1**, 26–32.

Gebber, G. L. (1990). Central determinants of sympathetic nerve discharge. In *Central Regulation of Autonomic Functions*, ed. A. D. Loewy and K. M. Spyer. New York, Oxford: Oxford University Press, pp. 126–144.

Geerling, J. C., Mettenleiter, T. C. and Loewy, A. D. (2003). Orexin neurons project to diverse sympathetic outflow systems. *Neuroscience*, **122**, 541–550.

Gerber, U. and Polosa, C. (1978). Effects of pulmonary stretch receptor afferent stimulation on sympathetic preganglionic neuron firing. *Can. J. Physiol. Pharmacol.*, **56**, 191–198.

Gerfen, C. R. and Sawchenko, P. E. (1984). An anterograde neuroanatomical tracing method that shows the detailed morphology of neurons, their axons and terminals: immunohistochemical localization of an axonally transported plant lectin, *Phaseolus vulgaris* leucoagglutinin (PHA-L). *Brain Res.*, **290**, 219–238.

Gershon, M. D. (1994). Functional anatomy of the enteric nervous system. In *Physiology of the Gastrointestinal Tract*, 3rd edn., ed. L. R. Johnson. New York: Raven Press, pp. 381–422.

Giamberardino, M. A. (1999). Recent and forgotten aspects of visceral pain. *Eur. J. Pain*, **3**, 77–92.

Gibbins, I. L. (1990). Target-related patterns of co-existence of neuropeptide Y, vasoactive intestinal peptide, enkephalin and substance P in cranial parasympathetic neurons innervating the facial skin and glands of guinea-pigs. *Neuroscience*, **38**, 541–560.

Gibbins, I. L. (1991). Vasomotor, pilomotor and secretomotor neurons distinguished by size and neuropeptide content in superior cervical ganglia of mice. *J. Auton. Nerv. Syst.*, **34**, 171–183.

Gibbins, I. L. (1992). Vasoconstrictor, vasodilator and pilomotor pathways in sympathetic ganglia of guinea-pigs. *Neuroscience*, **47**, 657–672.

Gibbins, I. L. (1994). Comparative anatomy and evolution of the autonomic nervous system. In *The Autonomic Nervous System,* Vol. 4, *Comparative Physiology and Evolution of the Autonomic Nervous System*, ed. S. Nilsson and S. Holmgren. Chur: Harwood Academic Publishers, pp. 1–67.

Gibbins, I. L. (1995). Chemical neuroanatomy of sympathetic ganglia. In *The Autonomic Nervous System*, Vol. 6, *Autonomic Ganglia*, ed. E. M. McLachlan. Luxembourg: Harwood Academic Publishers, pp. 73–122.

Gibbins, I. L. (1997). Autonomic pathways to cutaneous effectors. In *The Autonomic Nervous System*, Vol. 12, *Autonomic Innervation of the Skin*, ed. J. L. Morris and I. L. Gibbins. Chur, Switzerland: Harwood Academic Publishers, pp. 1–56.

Gibbins, I. L. (2004). Peripheral autonomic pathways. In *The Human Nervous System*, 2nd edn., ed. G. Paxinos and J. K. Mai. Amsterdam, San Diego, London: Elsevier Academic Press, pp. 134–189.

Gibbins, I. L. and Morris, J. L. (1987). Co-existence of neuropeptides in sympathetic, cranial autonomic and sensory neurons innervating the iris of the guinea-pig. *J. Auton. Nerv. Syst.*, **21**, 67–82.

Gibbins, I. L. and Morris, J. L. (1990). Sympathetic noradrenergic neurons containing dynorphin but not neuropeptide Y innervate small cutaneous blood vessels of guinea-pigs. *J. Auton. Nerv. Syst.*, **29**, 137–149.

Gibbins, I. L., Rodgers, H. F., Matthew, S. E. and Murphy, S. M. (1998). Synaptic organisation of lumbar sympathetic ganglia of guinea pigs: serial section ultrastructural analysis of dye-filled sympathetic final motor neurons. *J. Comp. Neurol.*, **402**, 285–302.

Gibbins, I. L., Jobling, P., Messenger, J. P., Teo, E. H. and Morris, J. L. (2000). Neuronal morphology and the synaptic organisation of sympathetic ganglia. *J. Auton. Nerv. Syst.*, **81**, 104–109.

Gibbins, I. L., Jobling, P. and Morris, J. L. (2003a). Functional organization of peripheral vasomotor pathways. *Acta Physiol. Scand.*, **177**, 237–245.

Gibbins, I. L., Jobling, P., Teo, E. H., Matthew, S. E. and Morris, J. L. (2003b). Heterogeneous expression of SNAP-25 and synaptic vesicle proteins by central and peripheral inputs to sympathetic neurons. *J. Comp. Neurol.*, **459**, 25–43.

Gibbins, I. L., Teo, E. H., Jobling, P. and Morris, J. L. (2003c). Synaptic density, convergence, and dendritic complexity of prevertebral sympathetic neurons. *J. Comp. Neurol.*, **455**, 285–298.

Gilbey, M. P., Jordan, D., Richter, D. W. and Spyer, K. M. (1984). Synaptic mechanisms involved in the inspiratory modulation of vagal cardioinhibitory neurones in the cat. *J. Physiol.*, **356**, 65–78.

Gilliat, R. W. (1948). Vaso-constriction in the finger after deep inspiration. *J. Physiol.*, **107**, 76–88.

Gilliat, R. W., Guttmann, L. and Whitteridge, D. (1948). Inspiratory vasoconstriction in patients after spinal injuries. *J. Physiol.*, **107**, 67–75.

Giuliano, F., Allard, J., Compagnie, S. *et al.* (2001). Vaginal physiological changes in a model of sexual arousal in anesthetized rats. *Am. J. Physiol. Regul. Integr. Comp. Physiol.*, **281**, R140–R149.

Gladwell, S. J. and Coote, J. H. (1999). Inhibitory and indirect excitatory effects of dopamine on sympathetic preganglionic neurones in the neonatal rat spinal cord in vitro. *Brain Res.*, **818**, 397–407.

Goehler, L. E., Gaykema, R. P., Hansen, M. K. *et al.* (2000). Vagal immune-to-brain communication: a visceral chemosensory pathway. *Auton. Neurosci.*, **85**, 49–59.

Gola, M. and Niel, J. P. (1993). Electrical and integrative properties of rabbit sympathetic neurones re-evaluated by patch clamping non-dissociated cells. *J. Physiol.*, **460**, 327–349.

Goldstein, D. S. (1995). *Stress, Catecholamines, and Cardiovascular Disease*. New York, Oxford: Oxford University Press.

Goldstein, D. S. (2000). *The Autonomic Nervous System in Health and Disease*. New York: Marcel Dekker.

Golenhofen, K., Hensel, H. and Ruef, J. (1962). Sustained dilatation in human muscle blood vessels under the influence of adrenaline. *J. Physiol.*, **160**, 189–199.

Goodchild, A. K., Moon, E. A., Dampney, R. A. and Howe, P. R. (1984). Evidence that adrenaline neurons in the rostral ventrolateral medulla have a vasopressor function. *Neurosci. Lett.*, **45**, 267–272.

Goodchild, A. K., van Deurzen, B. T. M., Sun, Q. J., Chalmers, J. and Pilowsky, P. (2000). Spinal GABA receptors do not mediate the sympathetic baroreceptor reflex in the rat. *Am. J. Physiol. Regul. Integr. Comp. Physiol.*, **279**, R320–R331.

Goodwin, G. M., McCloskey, D. I. and Mitchell, J. H. (1972). Cardiovascular and respiratory responses to changes in central command during isometric exercise at constant muscle tension. *J. Physiol.*, **226**, 173–190.

Gordon, F. J. and McCann, L. A. (1988). Pressor responses evoked by microinjections of L-glutamate into the caudal ventrolateral medulla of the rat. *Brain Res.*, **457**, 251–258.

Gore, A. C. and Roberts, J. L. (2003). Neuroendocrine systems. In *Fundamental Neuroscience*, 2nd edn., ed. L. R. Squire, F. E. Bloom, S. K. McConnell et al. San Diego: Academic Press, pp. 1031–1065.

Goridis, C. and Rohrer, H. (2002). Specification of catecholaminergic and serotonergic neurons. *Nat. Rev. Neurosci.*, **3**, 531–541.

Gould, D. J. and Hill, C. E. (1996). Alpha-adrenoceptor activation of a chloride conductance in rat iris arterioles. *Am. J. Physiol.*, **271**, H2469–H2476.

Granit, R. (1981). Comments on history of motor control. In *Handbook of Physiology, Section I, The Nervous System*, Vol. II, *Motor Control, part I*. ed. V. B. Brooks. Bethesda: American Physiological Society, pp. 1–16.

Grasby, D. J., Gibbins, I. L. and Morris, J. L. (1997). Projections of sympathetic non-noradrenergic neurons to skeletal muscle arteries in guinea-pig limbs vary with the metabolic character of muscles. *J. Vasc. Res.*, **34**, 351–364.

Gray's Anatomy (1995). 38th edn. Edinburgh: Churchill Livingstone.

Graziano, A. and Jones, E. G. (2004). Widespread thalamic terminations of fibers arising in the superficial medullary dorsal horn of monkeys and their relation to calbindin immunoreactivity. *J. Neurosci.*, **24**, 248–256.

Green, P. G., Miao, F. J. P., Jänig, W. and Levine, J. D. (1995). Negative feedback neuroendocrine control of the inflammatory response in rats. *J. Neurosci.*, **15**, 4678–4686.

Green, P. G., Jänig, W. and Levine, J. D. (1997). Negative feedback neuroendocrine control of inflammatory response in the rat is dependent on the sympathetic postganglionic neuron. *J. Neurosci.*, **17**, 3234–3238.

Greenfield, A. D. (1966). Survey of the evidence for active neurogenic vasodilation in man. *Fed. Proc.*, **25**, 1607–1610.

Greger, R. and Windhorst, U. (eds.) (1996). *Comprehensive Human Physiology. From Cellular Mechanisms to Integration*, Vol. 1 and 2. Berlin, Heidelberg, New York: Springer Verlag.

Gregor, M. and Jänig, W. (1977). Effects of systemic hypoxia and hypercapnia on cutaneous and muscle vasoconstrictor neurones to the cat's hindlimb. *Pflügers Arch.*, **368**, 71–81.

Gregor, M., Jänig, W. and Riedel, W. (1976). Response pattern of cutaneous postganglionic neurones to the hindlimb on spinal cord heating and cooling in the cat. *Pflügers Arch.*, **363**, 135–140.

Grewe, W., Jänig, W. and Kümmel, H. (1995). Effects of hypothalamic thermal stimuli on sympathetic neurones innervating skin and skeletal muscle of the cat hindlimb. *J. Physiol.*, **488**, 139–152.

Griffith, W. H., III, Gallagher, J. P. and Shinnick-Gallagher, P. (1980). An intracellular investigation of cat vesical pelvic ganglia. *J. Neurophysiol.*, **43**, 343–354.

Grigg, P., Schaible, H. G. and Schmidt, R. F. (1986). Mechanical sensitivity of group III and IV afferents from posterior articular nerve in normal and inflamed cat knee. *J. Neurophysiol.*, **55**, 635–643.

Grosse, M. and Jänig, W. (1976). Vasoconstrictor and pilomotor fibres in skin nerves to the cat's tail. *Pflügers Arch.*, **361**, 221–229.

Grubb, B. P. and Karas, B. (2002). Neurally mediated syncope. In *Autonomic Failure*, 4th edn., ed. C. J. Mathias and R. Bannister. Oxford: Oxford University Press, pp. 437–447.

Grundy, D. (1988). Speculation on the structure/function relationship for vagal and splanchnic afferent endings supplying the gastrointestinal tract. *J. Auton. Nerv. Syst.*, **22**, 175–180.

Grundy, D. and Scratcherd, T. (1989). Sensory afferents from the gastrointestinal tract. In *Handbook of Physiology, Section 6: The Gastrointestinal System*, Vol. 1: *Motility and Circulation*, ed. J. D. Wood. Bethesda: American Physiological Society, pp. 593–620.

Grundy, D., Salih, A. A. and Scratcherd, T. (1981). Modulation of vagal efferent fibre discharge by mechanoreceptors in the stomach, duodenum and colon of the ferret. *J. Physiol.*, **319**, 43–52.

Guertzenstein, P. G. and Silver, A. (1974). Fall in blood pressure produced from discrete regions of the ventral surface of the medulla by glycine and lesions. *J. Physiol.*, **242**, 489–503.

Guth, L. and Bernstein, J. J. (1961). Selectivity in the re-establishment of synapses in the superior cervical sympathetic ganglion of the cat. *Exp. Neurol.*, **4**, 59–69.

Guttmann, L. (1976). *Spinal Cord Injuries. Comprehensive Management and Research.* Oxford: Blackwell Scientific Publications.

Guyenet, P. G. (1990). Role of the ventral medulla oblongata in blood pressure regulation. In *Central Regulation of Autonomic Functions*, ed. A. D. Loewy and K. M. Spyer. New York, Oxford: Oxford University Press, pp. 145–167.

Guyenet, P. G. (2000). Neural structures that mediate sympathoexcitation during hypoxia. *Respir. Physiol.*, **121**, 147–162.

Guyenet, P. G. and Brown, D. L. (1986). Nucleus paragigantocellularis lateralis and lumbar sympathetic discharge in the rat. *Am. J. Physiol.*, **250**, R1081–R1094.

Guyenet, P. G. and Koshiya, N. (1992). Respiratory-sympathetic integration in the medulla oblongata. In *Central Neural Mechanisms in Cardiovascular Regulation*, ed. G. Kunos and J. Ciriello. Boston: Birkhäuser, pp. 226–247.

Guyenet, P. G. and Stornetta, R. L. (1997). Central nervous system regulation of the sympathetic and cardiovagal vasomotor outflows. In *Anesthesia*, ed. T. L. Yaksh, M. Maze, C. Lynch *et al.* Philadelphia, New York: Lippincott-Raven, pp. 1205–1232.

Guyenet, P. G., Koshiya, N., Huangfu, D. et al. (1996). Role of medulla oblongata in generation of sympathetic and vagal outflows. *Prog. Brain Res.*, **107**, 127-144.

Guyenet, P. G., Schreihofer, A. M. and Stornetta, R. L. (2001). Regulation of sympathetic tone and arterial pressure by the rostral ventrolateral medulla after depletion of C1 cells in rats. *Ann. New York Acad. Sci.*, **940**, 259-269.

Häbler, H. J. and Jänig, W. (1995). Coordination of sympathetic and respiratory systems: neurophysiological experiments. *Clin. Exp. Hypertens.*, **17**, 223-235.

Häbler, H. J., Jänig, W. and Koltzenburg, M. (1990a). Activation of unmyelinated afferent fibres by mechanical stimuli and inflammation of the urinary bladder in the cat. *J. Physiol.*, **425**, 545-562.

Häbler, H. J., Jänig, W., Koltzenburg, M. and McMahon, S. B. (1990b). A quantitative study of the central projection patterns of unmyelinated ventral root afferents in the cat. *J. Physiol.*, **422**, 265-287.

Häbler, H. J., Hilbers, K., Jänig, W. et al. (1992). Viscero-sympathetic reflex responses to mechanical stimulation of pelvic viscera in the cat. *J. Auton. Nerv. Syst.*, **38**, 147-158.

Häbler, H. J., Jänig, W. and Koltzenburg, M. (1993a). Myelinated primary afferents of the sacral spinal cord responding to slow filling and distension of the urinary bladder. *J. Physiol.*, **463**, 449-460.

Häbler, H. J., Jänig, W. and Koltzenburg, M. (1993b). Receptive properties of myelinated primary afferents innervating the inflamed urinary bladder of the cat. *J. Neurophysiol.*, **69**, 395-405.

Häbler, H. J., Jänig, W., Krummel, M. and Peters, O. A. (1993c). Respiratory modulation of the activity in postganglionic neurons supplying skeletal muscle and skin of the rat hindlimb. *J. Neurophysiol.*, **70**, 920-930.

Häbler, H. J., Jänig, W., Krummel, M. and Peters, O. A. (1994a). Reflex patterns in postganglionic neurons supplying skin and skeletal muscle of the rat hindlimb. *J. Neurophysiol.*, **72**, 2222-2236.

Häbler, H. J., Jänig, W. and Michaelis, M. (1994b). Respiratory modulation of activity in sympathetic neurones. *Prog. Neurobiol.*, **43**, 567-606.

Häbler, H. J., Bartsch, T. and Jänig, W. (1996). Two distinct mechanisms generate the respiratory modulation in fibre activity of the rat cervical sympathetic trunk. *J. Auton. Nerv. Syst.*, **61**, 116-122.

Häbler, H. J., Boczek-Funcke, A., Michaelis, M. and Jänig, W. (1997a). Responses of distinct types of sympathetic neuron to stimulation of the superior laryngeal nerve in the cat. *J. Auton. Nerv. Syst.*, **66**, 97-104.

Häbler, H. J., Wasner, G., Bartsch, T. and Jänig, W. (1997b). Responses of rat postganglionic sympathetic vasoconstrictor neurons following blockade of nitric oxide synthesis in vivo. *Neuroscience*, **77**, 899-909.

Häbler, H. J., Wasner, G. and Jänig, W. (1997c). Interaction of sympathetic vasoconstriction and antidromic vasodilatation in the control of skin blood flow. *Exp. Brain Res.*, **113**, 402-410.

Häbler, H. J., Bartsch, T. and Jänig, W. (1999a). Rhythmicity in single fiber postganglionic activity supplying the rat tail. *J. Neurophysiol.*, **81**, 2026-2036.

Häbler, H. J., Timmermann, L., Stegmann, J. U. and Jänig, W. (1999b). Involvement of neurokinins in antidromic vasodilatation in hairy and hairless skin of the rat hindlimb. *Neuroscience*, **89**, 1259-1268.

Häbler, H. J., Bartsch, T. and Jänig, W. (2000). Respiratory rhythmicity in the activity of postganglionic neurones supplying the rat tail during hyperthermia. *J. Auton. Nerv. Syst.*, **83**, 75–80.

Hadziefendic, S. and Haxhiu, M. A. (1999). CNS innervation of vagal preganglionic neurons controlling peripheral airways: a transneuronal labeling study using pseudorabies virus. *J. Auton. Nerv. Syst.*, **76**, 135–145.

Hagbarth, K. E. and Vallbo, A. B. (1968). Pulse and respiratory grouping of sympathetic impulses in human muscle nerves. *Acta Physiol. Scand.*, **74**, 96–108.

Hagbarth, K. E., Hallin, R. G., Hongell, A., Torebjörk, H. E. and Wallin, B. G. (1972). General characteristics of sympathetic activity in human skin nerves. *Acta Physiol. Scand.*, **84**, 164–176.

Hainsworth, R. (2002). Syncope and fainting: classification pathophysiological basis. In *Autonomic Failure*, 4th edn., eds. C. J. Mathias and R. Bannister. Oxford: Oxford University Press, pp. 428–436.

Hall, M. (1841). *On the Diseases and Derangements of the Nervous System, in their Primary Forms and their Modifications by Age, Sex Constitution, Heredity, Disposition, Excesses, General Disorder, and Organic Disease*. London: Bailliere.

Halliday, G. M. and McLachlan, E. M. (1991). A comparative analysis of neurons containing catecholamine-synthesizing enzymes and neuropeptide Y in the ventrolateral medulla of rats, guinea-pigs and cats. *Neuroscience*, **43**, 531–550.

Hallin, R. G. and Torebjörk, H. E. (1974). Single unit sympathetic activity in human skin nerves during rest and various manoeuvres. *Acta Physiol. Scand.*, **92**, 303–317.

Hamblin, P. A., McLachlan, E. M. and Lewis, R. J. (1995). Sub-nanomolar concentrations of ciguatoxin-1 excite preganglionic terminals in guinea pig sympathetic ganglia. *Naunyn-Schmiedeberg's Arch. Pharmacol.*, **352**, 236–246.

Harden, R. N., Baron, R. and Jänig, W. (eds.) (2001). *Complex Regional Pain Syndrome*. Seattle: IASP Press.

Hargreaves, K. M., Roszkowski, M. T. and Swift, J. Q. (1993). Bradykinin and inflammatory pain. *Agents Actions Suppl.*, **41**, 65–73.

Harhun, M. I., Pucovsky, V., Povstyan, O. V., Gordienko, D. V. and Bolton, T. B. (2005). Interstitial cells in the vasculature. *J. Cell Mol. Med.*, **9**, 232–243.

Haselton, J. R. and Guyenet, P. G. (1989). Central respiratory modulation of medullary sympathoexcitatory neurons in rat. *Am. J. Physiol.*, **256**, R739–R750.

Haupt, P., Jänig, W. and Kohler, W. (1983). Response pattern of visceral afferent fibres, supplying the colon, upon chemical and mechanical stimuli. *Pflügers Arch.*, **398**, 41–47.

Haxhiu, M. A. and Loewy, A. D. (1996). Central connections of the motor and sensory vagal systems innervating the trachea. *J. Auton. Nerv. Syst.*, **57**, 49–56.

Haxhiu, M. A., Jansen, A. S., Cherniack, N. S. and Loewy, A. D. (1993). CNS innervation of airway-related parasympathetic preganglionic neurons: a transneuronal labeling study using pseudorabies virus. *Brain Res.*, **618**, 115–134.

Haxhiu, M. A., Erokwu, B., Bhardwaj, V. and Dreshaj, I. A. (1998). The role of the medullary raphe nuclei in regulation of cholinergic outflow to the airways. *J. Auton. Nerv. Syst.*, **69**, 64–71.

Heel, K. A., McCauley, R. D., Papadimitriou, J. M. and Hall, J. C. (1997). Review: Peyer's patches. *J. Gastroenterol. Hepatol.*, **12**, 122–136.

Hellmann, K. (1963). The effect of temperature changes on the isolated pilomotor muscles. *J. Physiol.*, **169**, 621–629.

Henderson, C. G. and Ungar, A. (1978). Effect of cholinergic antagonists on sympathetic ganglionic transmission of vasomotor reflexes from the carotid baroreceptors and chemoreceptors of the dog. *J. Physiol.*, **277**, 379–385.

Henderson, L. A., Keay, K. A. and Bandler, R. (1998). The ventrolateral periaqueductal gray projects to caudal brainstem depressor regions: a functional-anatomical and physiological study. *Neuroscience*, **82**, 201–221.

Hendry, I. A. and Hill, C. E. (eds.) (1992). *The Autonomic Nervous System*, Vol. 2, *Development, Regeneration and Plasticity*, Chur: Harwood Academic Publishers.

Henry, J. P. (1997). *Culture and High Blood Pressure*. Hamburg, Münster: LIT Publishing Company.

Henry, J. P. and Grim, C. E. (1990). Psychosocial mechanisms of primary hypertension. *J. Hypertens.*, **8**, 783–793.

Henry, J. P. and Stephens, P. M. (1977). *Stress, Health and the Social Environment: a Sociobiologic Approach to Medicine*. Heidelberg, Berlin: Springer.

Hensel, H. (1981). *Thermoreception and Temperature Regulation*. London, New York: Academic Press.

Hensel, H. (1982). *Thermal Sensations and Thermoreceptors in Man*. Springfield Illinois: Charles C. Thomas Publ.

Herbert, H., Moga, M. M. and Saper, C. B. (1990). Connections of the parabrachial nucleus with the nucleus of the solitary tract and the medullary reticular formation in the rat. *J. Comp. Neurol.*, **293**, 540–580.

Hering, E. (1869). Über den Einfluß der Atmung auf den Kreislauf. Erste Mitteilung. Über Atembewegungen der Gefäß systeme [On the influence of the respiration on the circulation system. First report. On the respiratory movement of the vascular systems]. *Sitzungsber. Akad. Wiss. Wien, Math.-Naturw., Abt. 2*, **60**, 829–856.

Hering, H. E. (1927). *Die Karotissinusreflexe [The Carotid Sinus Reflexes]*. Dresden, Leipzig: Verlag von Theodor Steinkopf.

Hermann, G. and Rogers, R. C. (1995). Tumor necrosis factor-alpha in the dorsal vagal complex suppresses gastric motility. *NeuroImmunoModulation*, **2**, 74–81.

Hermann, G. E., Tovar, C. A. and Rogers, R. C. (2002). LPS-induced suppression of gastric motility relieved by TNFR:Fc construct in dorsal vagal complex. *Am. J. Physiol. Gastrointest. Liver Physiol.*, **283**, G634–G639.

Hertz, A. F. (1911). *The Sensibility of the Alimentary Canal*. Oxford: Oxford University Press.

Hess, W. R. (1944). Hypothalamische Adynamie [Hypothalamic adynamia]. *Helv. Physiol. Acta*, **2**, 137–147.

Hess, W. R. (1948). *Die Organisation des vegetativen Nervensystems [The Organization of the Autonomic Nervous System]*. Basel: Benno Schwabe & Co.

Hess, W. R. (1954). *Das Zwischenhirn, Syndrome, Lokalisationen, Funktionen [The Diencephalon, Syndromes, Localizations, Functions]*, 2nd edn. Basel: Benno Schwabe.

Hess, W. R. and Brügger, M. (1943). Das subcortikale Zentrum der affektiven Abwehrreaktion [The subcortical center of the affective defense reaction]. *Helv. Physiol. Acta*, **1**, 33–52.

Heuckeroth, R. O., Enomoto, H., Grider, J. R. *et al.* (1999). Gene targeting reveals a critical role for neurturin in the development and maintenance of enteric, sensory, and parasympathetic neurons. *Neuron*, **22**, 253–263.

Heymans, C. and Neil, E. (1958). *Reflexogenic Areas of the Cardiovascular System*. London: J. & A. Churchill Ltd.

Hill, C. E., Hendry, I. A. and Sheppard, A. (1987). Use of the fluorescent dye, fast blue, to label sympathetic postganglionic neurones supplying mesenteric arteries and enteric neurones of the rat. *J. Auton. Nerv. Syst.*, **18**, 73–82.

Hill, C. E., Klemm, M., Edwards, F. R. and Hirst, G. D. (1993). Sympathetic transmission to the dilator muscle of the rat iris. *J. Auton. Nerv. Syst.*, **45**, 107–123.

Hill, C. E., Eade, J. and Sandow, S. L. (1999). Mechanisms underlying spontaneous rhythmical contractions in irideal arterioles of the rat. *J. Physiol.*, **521**, 507–516.

Hill, M. A., Zou, H., Potocnik, S. J., Meininger, G. A. and Davis, M. J. (2001). Invited review: arteriolar smooth muscle mechanotransduction: Ca^{2+} signaling pathways underlying myogenic reactivity. *J. Appl. Physiol.*, **91**, 973–983.

Hillsley, K. and Grundy, D. (1998). Sensitivity to 5-hydroxytryptamine in different afferent subpopulations within mesenteric nerves supplying the rat jejunum. *J. Physiol.*, **509**, 717–727.

Hillsley, K., Kirkup, A. J. and Grundy, D. (1998). Direct and indirect actions of 5-hydroxytryptamine on the discharge of mesenteric afferent fibres innervating the rat jejunum. *J. Physiol.*, **506**, 551–561.

Himms-Hagen, J. (1991). Neural control of brown adipose tissue: thermogenesis, hypertrophy, and atrophy neuroeffector junctions. *Front. Neuroendocrinol.*, **12**, 38–93.

Hirooka, Y., Polson, J. W., Potts, P. D. and Dampney, R. A. (1997). Hypoxia-induced Fos expression in neurons projecting to the pressor region in the rostral ventrolateral medulla. *Neuroscience*, **80**, 1209–1224.

Hirshberg, R. M., Al Chaer, E. D., Lawand, N. B., Westlund, K. N. and Willis, W. D. (1996). Is there a pathway in the posterior funiculus that signals visceral pain? *Pain*, **67**, 291–305.

Hirst, G. D. (2001). An additional role for ICC in the control of gastrointestinal motility? *J. Physiol.*, **537**, 1

Hirst, G. D. and Edwards, F. R. (1989). Sympathetic neuroeffector transmission in arteries and arterioles. *Physiol. Rev.*, **69**, 546–604.

Hirst, G. D. and McLachlan, E. M. (1984). Post-natal development of ganglia in the lower lumbar sympathetic chain of the rat. *J. Physiol.*, **349**, 119–134.

Hirst, G. D. and McLachlan, E. M. (1986). Development of dendritic calcium currents in ganglion cells of the rat lower lumbar sympathetic chain. *J. Physiol.*, **377**, 349–368.

Hirst, G. D. and Neild, T. O. (1980). Some properties of spontaneous excitatory junction potentials recorded from arterioles of guinea-pigs. *J. Physiol.*, **303**, 43–60.

Hirst, G. D. and Ward, S. M. (2003). Interstitial cells: involvement in rhythmicity and neural control of gut smooth muscle. *J. Physiol.*, **550**, 337–346.

Hirst, G. D. S., Holman, M. E. and McKirdy, H. C. (1974). Two types of neurones in the myenteric plexus of duodenum in the guinea-pig. *J. Physiol.*, **236**, 303–326.

Hirst, G. D., Holman, M. E. and McKirdy, H. C. (1975). Two descending nerve pathways activated by distension of guinea-pig small intestine. *J. Physiol.*, **244**, 113–127.

Hirst, G. D. S., Edwards, F. R., Bramich, N. J. and Klemm, M. F. (1991). Neural control of cardiac pacemaker potentials. *News Physiol. Sci.*, **6**, 185–190.

Hirst, G. D. S., Bramich, N. J., Edwards, F. R. and Klemm, M. (1992). Transmission at autonomic neuroeffector junctions. *Trends Neurosci.*, **15**, 40–46.

Hirst, G. D., Choate, J. K., Cousins, H. M., Edwards, F. R. and Klemm, M. F. (1996). Transmission by post-ganglionic axons of the autonomic nervous system: the importance of the specialized neuroeffector junction. *Neuroscience*, **73**, 7–23.

Hoffmeister, B., Hussels, W. and Jänig, W. (1978). Long-lasting discharge of postganglionic neurones to skin and muscle of the cat's hindlimb after repetitive activation of preganglionic axons in the lumbar sympathetic trunk. *Pflügers Arch.*, **376**, 15–20.

Hohmann, E. L., Elde, R. P., Rysavy, J. A., Einzig, S. and Gebhard, R. L. (1986). Innervation of periosteum and bone by sympathetic vasoactive intestinal peptide-containing nerve fibers. *Science*, **232**, 868–871.

Holman, M. E., Coleman, H. A., Tonta, M. A. and Parkington, H. C. (1994). Synaptic transmission from splanchnic nerves to the adrenal medulla of guinea-pigs. *J. Physiol.*, **478**, 115–124.

Holzer, P. (1992). Peptidergic sensory neurons in the control of vascular functions: mechanisms and significance in the cutaneous and splanchnic vascular beds. *Rev. Physiol. Biochem. Pharmacol.*, **121**, 49–146.

Holzer, P. (1995). Chemosensitive afferent nerves in the regulation of gastric blood flow and protection. *Adv. Exp. Med. Biol.*, **371B**, 891–895.

Holzer, P. (1998a). Neural emergency system in the stomach. *Gastroenterology*, **114**, 823–839.

Holzer, P. (1998b). Neurogenic vasodilatation and plasma leakage in the skin. *Gen. Pharmacol.*, **30**, 5–11.

Holzer, P. (2002a). Sensory neurone responses to mucosal noxae in the upper gut: relevance to mucosal integrity and gastrointestinal pain. *Neurogastroenterol. Motil.*, **14**, 459–475.

Holzer, P. (2002b). Control of gastric functions by extrinsic sensory neurons. In *The Autonomic Nervous System*, Vol. 14, *Innervation of the Gastrointestinal Tract*, ed. S. Brookes and M. Costa, London, New York: Taylor and Francis, pp. 103–170.

Holzer, P. (2003). Afferent signalling of gastric acid challenge. *J. Physiol. Pharmacol.*, **54 Suppl 4**, 43–53.

Holzer, P. and Maggi, C. A. (1998). Dissociation of dorsal root ganglion neurons into afferent and efferent-like neurons. *Neuroscience*, **86**, 389–398.

Hopkins, D. A., Bieger, D., deVente, J. and Steinbusch, W. M. (1996). Vagal efferent projections: viscerotopy, neurochemistry and effects of vagotomy. *Prog. Brain Res.*, **107**, 79–96.

Horeyseck, G. and Jänig, W. (1974a). Reflexes in postganglionic fibres within skin and muscle nerves after mechanical non-noxious stimulation of skin. *Exp. Brain Res.*, **20**, 115–123.

Horeyseck, G. and Jänig, W. (1974b). Reflexes in postganglionic fibres within skin and muscle nerves after noxious stimulation of skin. *Exp. Brain Res.*, **20**, 125–134.

Horeyseck, G. and Jänig, W. (1974c). Reflex activity in postganglionic fibres within skin and muscle nerves elicited by somatic stimuli in chronic spinal cats. *Exp. Brain Res.*, **21**, 155–168.

Horeyseck, G., Jänig, W., Kirchner, F. and Thämer, V. (1976). Activation and inhibition of muscle and cutaneous postganglionic neurones to hindlimb during hypothalamically induced vasoconstriction and atropine-sensitive vasodilation. *Pflügers Arch.*, **361**, 231–240.

Hori, T., Katafuchi, T., Take, S., Shimizu, N. and Niijima, A. (1995). The autonomic nervous system as a communication channel between the brain and the immune system. *NeuroImmunoModulation*, **2**, 203–215.

Horiuchi, J. and Dampney, R. A. L. (2002). Evidence for tonic disinhibition of RVLM sympathoexcitatory neurons from the caudal pressor area. *Auton. Neurosci.*, **99**, 102–110.

Horiuchi, J., Killinger, S. and Dampney, R. A. (2004a). Contribution to sympathetic vasomotor tone of tonic glutamatergic inputs to neurons in the RVLM. *Am. J. Physiol. Regul. Integr. Comp. Physiol.*, **287**, R1335–R1343.

Horiuchi, J., McAllen, R. M., Allen, A. M. et al. (2004b). Descending vasomotor pathways from the dorsomedial hypothalamic nucleus: role of medullary raphe and RVLM. *Am. J. Physiol. Regul. Integr. Comp. Physiol.*, **287**, R824–R832.

Horowitz, B., Ward, S. M. and Sanders, K. M. (1999). Cellular and molecular basis for electrical rhythmicity in gastrointestinal muscles. *Annu. Rev. Physiol.*, **61**, 19–43.

Hosoya, Y., Sugiura, Y., Okado, N., Loewy, A. D. and Kohno, K. (1991). Descending input from the hypothalamic paraventricular nucleus to sympathetic preganglionic neurons in the rat. *Exp. Brain Res.*, **85**, 10–20.

Hoyle, C. H. V., Milner, P. and Burnstock, G. (2002). Neuroeffector transmission in the intestine. In *The Autonomic Nervous System*, Vol. 14, *Innervation of the Gastrointestinal Tract*, ed. S. Brookes and M. Costa. London, New York: Taylor and Francis, pp. 295–340.

Hua, L. H., Strigo, I. A., Baxter, L. C., Johnson, S. C. and Craig, A. D. (2005). Anterior-posterior somatotopy of innocuous cooling activation focus in human dorsal insular cortex. *Am. J. Physiol. Regul. Intrgr. Comp. Physiol.*, **289**, R319–R325.

Huang, J. and Weiss, M. L. (1999). Characterization of the central cell groups regulating the kidney in the rat. *Brain Res.*, **845**, 77–91.

Huang, X. F., Törk, I. and Paxinos, G. (1993). Dorsal motor nucleus of the vagus nerve: a cyto- and chemoarchitectonic study in the human. *J. Comp. Neurol.*, **330**, 158–182.

Huizinga, J. D., Thuneberg, L., Vanderwinden, J. M. and Rumessen, J. J. (1997). Interstitial cells of Cajal as targets for pharmacological intervention in gastrointestinal motor disorders. *Trends Pharmacol. Sci.*, **18**, 393–403.

Huizinga, J. D., Ambrous, K. and Der-Silaphet, T. (1998). Co-operation between neural and myogenic mechanisms in the control of distension-induced peristalsis in the mouse small intestine. *J. Physiol.*, **506**, 843–856.

Huizinga, J. D. and Faussone-Pellegrini, M. S. (2005). About the presence of interstitial cells of Cajal outside the musculature of the gastrointestinal tract. *J. Cell Mol. Med.*, **9**, 468–473.

Hultborn, H. (2001). State-dependent modulation of sensory feedback. *J. Physiol.*, **533**, 5–13.

Hummel, T., Sengupta, J. N., Meller, S. T. and Gebhart, G. F. (1997). Responses of T2-4 spinal cord neurons to irritation of the lower airways in the rat. *Am. J. Physiol.*, **273**, R1147–R1157.

Hunsperger, R. W. (1956). Affektreactionen auf elektrische Reizung im Hirnstamm der Katze [Affect reactions to electrical stimulation in the brain stem of the cat]. *Helv. Physiol. Acta*, **14**, 70–92.

Iino, S., Ward, S. M. and Sanders, K. M. (2004). Interstitial cells of Cajal are functionally innervated by excitatory motor neurones in the murine intestine. *J. Physiol.*, **556**, 521–530.

Inoue, T. (1980). Efferent discharge patterns in the ciliary nerve of rabbits and the pupillary light reflex. *Brain Res.*, **186**, 43–53.

Ireland, D. R. (1999). Preferential formation of strong synapses during re-innervation of guinea-pig sympathetic ganglia. *J. Physiol.*, **520**, 827–837.

Ireland, D. R., Davies, P. J. and McLachlan, E. M. (1999). Calcium channel subtypes differ at two types of cholinergic synapse in lumbar sympathetic neurones of guinea-pigs. *J. Physiol.*, **514**, 59–69.

Ivanov, A. and Purves, D. (1989). Ongoing electrical activity of superior cervical ganglion cells in mammals of different size. *J. Comp. Neurol.*, **284**, 398–404.

Iversen, S., Iversen, L. and Saper, C. B. (2000). The autonomic nervous system. In *The Principles of Neural Science*, 4th edn., ed. E. R. Kandel, J. H. Schwartz and T. M. Jessel. New York: McGraw-Hill, pp. 960–981.

Izumi, H. (1999). Nervous control of blood flow in the orofacial region. *Pharmacol. Ther.*, **81**, 141–161.

Izzo, P. N. and Spyer, K. M. (1997). Parasympathetic innervation of the heart. In *The Autonomic Nervous System*, Vol. 11, *Central Nervous Control of Autonomic function*, ed. D. Jordan. Amsterdam: Harwood Academic Publishers, pp. 109–127.

Izzo, P. N., Deuchars, J. and Spyer, K. M. (1993). Localization of cardiac vagal preganglionic motoneurones in the rat: immunocytochemical evidence of synaptic inputs containing 5- hydroxytryptamine. *J. Comp. Neurol.*, **327**, 572–583.

Jack, J. B. B., Noble, D. and Tsien, R. W. (1975). *Electrical Current Flow in Excitable Cells*. Oxford: Clarendon Press.

James, W. (1884). What is an emotion? *Mind*, **9**, 188–205.

James, W. (1994) [1894]. The physical bases of emotion. 1894. *Psychol. Rev.*, **101**, 205–210.

Jamieson, J., Boyd, H. D. and McLachlan, E. M. (2003). Simulations to derive membrane resistivity in three phenotypes of guinea pig sympathetic postganglionic neuron. *J. Neurophysiol.*, **89**, 2430–2440.

Jan, L. Y. and Jan, Y. N. (1982). Peptidergic transmission in sympathetic ganglia of the frog. *J. Physiol.*, **327**, 219–246.

Jan, Y. N., Jan, L. Y. and Kuffler, S. W. (1979). A peptide as a possible transmitter in sympathetic ganglia of the frog. *Proc. Natl. Acad. Sci. U.S.A.*, **76**, 1501–1505.

Jan, L. Y., Jan, Y. N. and Brownfield, M. S. (1980a). Peptidergic transmitters in synaptic boutons of sympathetic ganglia. *Nature*, **288**, 380–382.

Jan, Y. N., Jan, L. Y. and Kuffler, S. W. (1980b). Further evidence for peptidergic transmission in sympathetic ganglia. *Proc. Natl. Acad. Sci. U.S.A.*, **77**, 5008–5012.

Jancsó, N. (1960). Role of the nerve terminals in the mechanism of inflammatory reactions. *Bull. Millard Fillmore Hosp. (Buffalo, NY)*, **7**, 53–77.

Jancsó, N., Jancsó-Gábor, A. and Szolcsányi, J. (1967). Direct evidence for neurogenic inflammation and its prevention by denervation and by pretreatment with capsaicin. *Br. J. Pharmacol.*, **31**, 138–151.

Jancsó, N., Jancsó-Gábor, A. and Szolcsányi, J. (1968). The role of sensory nerve endings in neurogenic inflammation induced in human skin and in the eye and paw of the rat. *Br. J. Pharmacol.*, **33**, 32–41.

Jänig, W. (1975). Central organization of somatosympathetic reflexes in vasoconstrictor neurones. *Brain Res.*, **87**, 305–312.

Jänig, W. (1985a). Organization of the lumbar sympathetic outflow to skeletal muscle and skin of the cat hindlimb and tail. *Rev. Physiol. Biochem. Pharmacol.*, **102**, 119–213.

Jänig, W. (1985b). Causalgia and reflex sympathetic dystrophy: in which way is the sympathetic nervous system involved. *Trends Neurosci.*, **8**, 471–477.

Jänig, W. (1986). Spinal cord integration of visceral sensory systems and sympathetic nervous system reflexes. *Prog. Brain Res.*, **67**, 255–277.

Jänig, W. (1988a). Pre- and postganglionic vasoconstrictor neurons: differentiation, types, and discharge properties. *Annu. Rev. Physiol.*, **50**, 525–539.

Jänig, W. (1988b). Integration of gut function by sympathetic reflexes. *Bailliere's Clin. Gastroenterol.*, **2**, 45–62.

Jänig, W. (1990a). Functions of the sympathetic innervation of the skin. In *Central Regulation of Autonomic Function*, ed. A. D. Loewy and K. M. Spyer. New York, Oxford: Oxford University Press, pp. 334–348.

Jänig, W. (1990b). The sympathetic nervous system in pain: physiology and pathophysiology. In *Pain and the Sympathetic Nervous System*, ed. M. Stanton-Hicks. Boston, Dordrecht, London: Kluwer Academic Publishers, pp. 17–89.

Jänig, W. (1993). Spinal visceral afferents, sympathetic nervous system and referred pain. In *New Trends in Referred Pain and Hyperalgesia, Pain Research and Clinical Management*, Vol. 7, ed. L. Vecchiet, D. Albe-Fessard, U. Lindblom and M. A. Giamberardino. Amsterdam: Elsevier Science Publishers, pp. 83–92.

Jänig, W. (1995a). Ganglionic transmission in vivo. In *The Autonomic Nervous System*, Vol. 6, *Autonomic Ganglia*, ed. E. M. McLachlan. Chur: Harwood Academic Publishers, pp. 349–395.

Jänig, W. (1995b). The sympathetic nervous system in pain. *Eur. J. Anaesthesiol.*, **12 (Suppl. 10)**, 53–60.

Jänig, W. (1996a). Spinal cord reflex organization of sympathetic systems. *Prog. Brain Res.*, **107**, 43–77.

Jänig, W. (1996b). Regulation of the lower urinary tract. In *Comprehensive Human Physiology*, Vol. 2, ed. R. Greger and U. Windhorst. Berlin, Heidelberg: Springer-Verlag, pp. 1611–1624.

Jänig, W. (1996c). Behavioral and neurovegetative components of reproductive functions. In *Comprehensive Human Physiology*, Vol. 2, ed. R. Greger and U. Windhorst. Berlin, Heidelberg: Springer-Verlag, pp. 2253–2263.

Jänig, W. (1996d). Neurobiology of visceral afferent neurons: neuroanatomy, functions, organ regulations and sensations. *Biol. Psychol.*, **42**, 29–51.

Jänig, W. (2002). Pain in the sympathetic nervous system: pathophysiological mechanisms. In *Autonomic Failure*, 4th edn., ed. R. Bannister and C. J. Mathias. New York, Oxford: Oxford University Press, pp. 99–108.

Jänig, W. (2005a). Vegetatives Nervensystem. In *Physiologie des Menschen*, 29th edn., ed. R. F. Schmidt, F. Lang and G. Thews. Heidelberg Berlin: Springer Medizin Verlag, pp. 425–458.

Jänig, W. (2005b). Vagal afferents and visceral pain. In *Advances in Vagal Afferent Neurobiology*, ed. B. Undem and D. Weinreich. Boca Raton: CRC Press, pp. 461–489.

Jänig, W. and Baron, R. (2001). The role of the sympathetic nervous system in neuropathic pain: clinical observations and animal models. In *Neuropathic*

Pain: Pathophysiological and Treatment, ed. P. T. Hansson, H. L. Fields, R. G. Hill and P. Marchettini. Seattle: IASP Press, pp. 125-149.

Jänig, W. and Baron, R. (2002). Complex regional pain syndrome is a disease of the central nervous system. *Clin. Auton. Res.*, **12**, 150-164.

Jänig, W. and Baron, R. (2003). Complex regional pain syndrome: mystery explained? *Lancet Neurol.*, **2**, 687-697.

Jänig, W. and Häbler, H. J. (1995). Visceral-autonomic integration. In *Visceral Pain; Progress in Pain Research and Management*, Vol. 5, ed. G. F. Gebhart. Seattle: IASP Press, pp. 311-348.

Jänig, W. and Häbler, H. J. (1999). Organisation of the autonomic nervous system: structure and function. In *Handbook of Clinical Neurology*, Vol. 74, *The Autonomic Nervous System, part I: Normal Functions*, ed. O. Appenzeller. Amsterdam: Elsevier, pp. 1-52.

Jänig, W. and Häbler, H. J. (2000a). Specificity in the organization of the autonomic nervous system: a basis for precise neural regulation of homeostatic and protective body functions. *Prog. Brain Res.*, **122**, 351-367.

Jänig, W. and Häbler, H. J. (2000b). Sympathetic nervous system: contribution to chronic pain. *Prog. Brain Res.*, **129**, 451-468.

Jänig, W. and Häbler, H. J. (2002). Physiologie und pathophysiologie viszeraler Schmerzen [Physiology and pathophysiology of visceral pain]. *Schmerz.*, **16**, 429-446.

Jänig, W. and Häbler, H. J. (2003). Neurophysiological analysis of target-related sympathetic pathways – from animal to human: similarities and differences. *Acta Physiol. Scand.*, **177**, 255-274.

Jänig, W. and Koltzenburg, M. (1990). On the function of spinal primary afferent fibres supplying colon and urinary bladder. *J. Auton. Nerv. Syst.*, **30 Suppl**, S89-S96.

Jänig, W. and Koltzenburg, M. (1991a). What is the interaction between the sympathetic terminal and the primary afferent fiber? In *Towards a New Pharmacotherapy of Pain*, ed. A. I. Basbaum and J.-M. Besson. Chichester: Dahlem Workshop Reports John Wiley & Sons, pp. 331-352.

Jänig, W. and Koltzenburg, M. (1991b). Plasticity of sympathetic reflex organization following cross-union of inappropriate nerves in the adult cat. *J. Physiol.*, **436**, 309-323.

Jänig, W. and Koltzenburg, M. (1991c). Receptive properties of sacral primary afferent neurons supplying the colon. *J. Neurophysiol.*, **65**, 1067-1077.

Jänig, W. and Koltzenburg, M. (1993). Pain arising from the urogenital tract. In *The Autonomic Nervous System*, Vol. 3., *Nervous Control of the Urogenital Tract*, ed. C. A. Maggi. Chur, Switzerland: Harwood Academic Publisher, pp. 523-576.

Jänig, W. and Kümmel, H. (1977). Functional discrimination of postganglionic neurones to the cat's hindpaw with respect to the skin potentials recorded from the hairless skin. *Pflügers Arch.*, **371**, 217-225.

Jänig, W. and Kümmel, H. (1981). Organization of the sympathetic innervation supplying the hairless skin of the cat's paw. *J. Auton. Nerv. Syst.*, **3**, 215-230.

Jänig, W. and Levine, J. D. (2005). Autonomic-endocrine-immune responses in acute and chronic pain. In *Wall and Melzack's Textbook of Pain*, 5th edn., ed. S. B. McMahon and M. Koltzenburg. Edinburgh: Elsevier Churchill Livingstone, pp. 205-218.

Jänig, W. and Lisney, S. J. (1989). Small diameter myelinated afferents produce vasodilation but not plasma extravasation in rat skin. *J. Physiol.*, **415**, 477-486.

Jänig, W. and McLachlan, E. M. (1986a). The sympathetic and sensory components of the caudal lumbar sympathetic trunk in the cat. *J. Comp. Neurol.*, **245**, 62–73.

Jänig, W. and McLachlan, E. M. (1986b). Identification of distinct topographical distributions of lumbar sympathetic and sensory neurons projecting to end organs with different functions in the cat. *J. Comp. Neurol.*, **246**, 104–112.

Jänig, W. and McLachlan, E. M. (1987). Organization of lumbar spinal outflow to distal colon and pelvic organs. *Physiol. Rev.*, **67**, 1332–1404.

Jänig, W. and McLachlan, E. M. (1992a). Characteristics of function-specific pathways in the sympathetic nervous system. *Trends Neurosci.*, **15**, 475–481.

Jänig, W. and McLachlan, E. M. (1992b). Specialized functional pathways are the building blocks of the autonomic nervous system. *J. Auton. Nerv. Syst.*, **41**, 3–13.

Jänig, W. and McLachlan, E. M. (1994). The role of modifications in noradrenergic peripheral pathway after nerve lesions in the generation of pain. In *Pharmacological Approaches to the Treatment of Pain: New Concepts and Critical Issues, Progress in Pain Research and Management*, Vol. 1, ed. H. L. Fields and J. C. Liebeskind. Seattle: IASP Press, pp. 101–128.

Jänig, W. and McLachlan, E. M. (2002). Neurobiology of the autonomic nervous system. In *Autonomic Failure*, 4th edn., ed. C. J. Mathias and R. Bannister. New York, Oxford: Oxford University Press, pp. 3–15.

Jänig, W. and Morrison, J. F. B. (1986). Functional properties of spinal visceral afferents supplying abdominal and pelvic organs, with special emphasis on visceral nociception. *Prog. Brain Res.*, **67**, 87–114.

Jänig, W. and Räth, B. (1977). Electrodermal reflexes in the cat's paws elicited by natural stimulation of skin. *Pflügers Arch.*, **369**, 27–32.

Jänig, W. and Räth, B. (1980). Effects of anaesthetics on reflexes elicited in the sudomotor system by stimulation of Pacinian corpuscles and of cutaneous nociceptors. *J. Auton. Nerv. Syst.*, **2**, 1–14.

Jänig, W. and Spilok, N. (1978). Functional organization of the sympathetic innervation supplying the hairless skin of the hindpaws in chronic spinal cats. *Pflügers Arch.*, **377**, 25–31.

Jänig, W. and Stanton-Hicks, M. (eds.) (1996). *Reflex Sympathetic Dystrophy – a Reappraisal*. Seattle: IASP Press.

Jänig, W. and Szulczyk, P. (1980). Functional properties of lumbar preganglionic neurones. *Brain Res.*, **186**, 115–131.

Jänig, W. and Szulczyk, P. (1981). The organization of lumbar preganglionic neurons. *J. Auton. Nerv. Syst.*, **3**, 177–191.

Jänig, W., Krauspe, R. and Wiedersatz, G. (1982). Transmission of impulses from pre- to postganglionic vasoconstrictor and sudomotor neurons. *J. Auton. Nerv. Syst.*, **6**, 95–106.

Jänig, W., Krauspe, R. and Wiedersatz, G. (1983). Reflex activation of postganglionic vasoconstrictor neurones supplying skeletal muscle by stimulation of arterial chemoreceptors via non-nicotinic synaptic mechanisms in sympathetic ganglia. *Pflügers Arch.*, **396**, 95–100.

Jänig, W., Krauspe, R. and Wiedersatz, G. (1984). Activation of postganglionic neurones via non-nicotinic synaptic mechanisms by stimulation of thin preganglionic axons. *Pflügers Arch.*, **401**, 318–320.

Jänig, W., Schmidt, M., Schnitzler, A. and Wesselmann, U. (1991). Differentiation of sympathetic neurones projecting in the hypogastric nerves in terms of their discharge patterns in cats. *J. Physiol.*, **437**, 157–179.

Jänig, W., Levine, J. D. and Michaelis, M. (1996). Interactions of sympathetic and primary afferent neurons following nerve injury and tissue trauma. *Prog. Brain Res.*, **112**, 161–184.

Jänig, W., Khasar, S. G., Levine, J. D. and Miao, F. J. P. (2000). The role of vagal visceral afferents in the control of nociception. *Prog. Brain Res.*, **122**, 273–287.

Jänig, W., Chapman, C. R. and Green, P. G. (2006). Pain and body protection: sensory, autonomic, neuroendocrine and behavioral mechanisms in the control of inflammation and hyperalgesia. In *Proceeding of the 11th World Congress on Pain*, ed. H. Flor, E. Kalso and J. O. Dostrovsky. Seattle: IASP Press, in press.

Jankowska, E. (1992). Interneuronal relay in spinal pathways from proprioceptors. *Prog. Neurobiol.*, **38**, 335–378.

Jankowska, E. (2001). Spinal interneuronal systems: identification, multifunctional character and reconfigurations in mammals. *J. Physiol.*, **533**, 31–40.

Jankowska, E. and Lundberg, A. (1981). Interneurones in the spinal cord. *Trends Neurosci.*, **4**, 230–233.

Jansen, A. S. and Loewy, A. D. (1997). Neurons lying in the white matter of the upper cervical spinal cord project to the intermediolateral cell column. *Neuroscience*, **77**, 889–898.

Jansen, A. S., Ter Horst, G. J., Mettenleiter, T. C. and Loewy, A. D. (1992). CNS cell groups projecting to the submandibular parasympathetic preganglionic neurons in the rat: a retrograde transneuronal viral cell body labeling study. *Brain Res.*, **572**, 253–260.

Jansen, A. S., Farwell, D. G. and Loewy, A. D. (1993). Specificity of pseudorabies virus as a retrograde marker of sympathetic preganglionic neurons: implications for transneuronal labeling studies. *Brain Res.*, **617**, 103–112.

Jansen, A. S., Nguyen, X. V., Karpitskiy, V., Mettenleiter, T. C. and Loewy, A. D. (1995a). Central command neurons of the sympathetic nervous system: basis of the fight-or-flight response. *Science*, **270**, 644–646.

Jansen, A. S., Wessendorf, M. W. and Loewy, A. D. (1995b). Transneuronal labeling of CNS neuropeptide and monoamine neurons after pseudorabies virus injections into the stellate ganglion. *Brain Res.*, **683**, 1–24.

Jeske, I., Morrison, S. F., Cravo, S. L. and Reis, D. J. (1993). Identification of baroreceptor reflex interneurons in the caudal ventrolateral medulla. *Am. J. Physiol.*, **264**, R169–R178.

Jewett, D. L. (1964). Activity of single efferent fibres in the cervical vagus nerve of the dog, with special reference to possible cardio-inhibitory fibres. *J. Physiol.*, **175**, 321–357.

Jichia, D. and Frank, J. L. (2000). Spinal cord disease trauma and autonomic nervous system dysfunction. In *Handbook of Clinical Neurology*, Vol. 75, *The Autonomic Nervous System part II: Dysfunctions*, ed. O. Appenzeller, Amsterdam: Elsevier Science, pp. 567–587.

Jobling, P. and McLachlan, E. M. (1992). An electrophysiological study of responses evoked in isolated segments of rat tail artery during growth and maturation. *J. Physiol.*, **454**, 83–105.

Jobling, P., McLachlan, E. M., Jänig, W. and Anderson, C. R. (1992). Electrophysiological responses in the rat tail artery during reinnervation following lesions of the sympathetic supply. *J. Physiol.*, **454**, 107–128.

Jobling, P., Gibbins, I. L. and Morris, J. L. (2003). Functional organization of vasodilator neurons in pelvic ganglia of female guinea pigs: comparison with uterine motor neurons. *J. Comp. Neurol.*, **459**, 223–241.

Johnson, C. D. and Gilbey, M. P. (1996). On the dominant rhythm in the discharges of single postganglionic sympathetic neurones innervating the rat tail artery. *J. Physiol.*, **497**, 241–259.

Johnson, D. A. and Purves, D. (1981). Post-natal reduction of neural unit size in the rabbit ciliary ganglion. *J. Physiol.*, **318**, 143–159.

Johnson, D. A. and Purves, D. (1983). Tonic and reflex synaptic activity recorded in ciliary ganglion cells of anaesthetized rabbits. *J. Physiol.*, **339**, 599–613.

Jones, E. G. (2002). A pain in the thalamus. *J. Pain*, **3**, 102–104.

Jones, E. G. (2006). *The Thalamus,* 2nd edn., 2 volumes. Cambridge: Cambridge University Press.

Jones, J. F., Wang, Y. and Jordan, D. (1998). Activity of C fibre cardiac vagal efferents in anaesthetized cats and rats. *J. Physiol.*, **507**, 869–880.

Jordan, D. (1997a). Central nervous control of the airways. In *The Autonomic Nervous System*, Vol. 11, *Central Nervous Control of Autonomic Function*, ed. D. Jordan. Amsterdam: Harwood Academic Publishers, pp. 63–107.

Jordan, D. (ed.) (1997b). *The Autonomic Nervous System*, Vol. 11, *Central Nervous Control of Autonomic Function*, Amsterdam: Harwood Academic Press.

Joyner, M. J. and Dietz, N. M. (2003). Sympathetic vasodilation in human muscle. *Acta Physiol. Scand.*, **177**, 329–336.

Joyner, M. J. and Halliwill, J. R. (2000). Sympathetic vasodilatation in human limbs. *J. Physiol.*, **526**, 471–480.

Julé, Y. and Szurszewski, J. H. (1983). Electrophysiology of neurones of the inferior mesenteric ganglion of the cat. *J. Physiol.*, **344**, 277–292.

Juler, G. L. and Eltorai, I. M. (1985). The acute abdomen in spinal cord injury patients. *Paraplegia*, **23**, 118–123.

Kalia, M. and Richter, D. (1985a). Morphology of physiologically identified slowly adapting lung stretch receptor afferents stained with intra-axonal horseradish peroxidase in the nucleus of the tractus solitarius of the cat. I. A light microscopic analysis. *J. Comp. Neurol.*, **241**, 503–520.

Kalia, M. and Richter, D. (1985b). Morphology of physiologically identified slowly adapting lung stretch receptor afferents stained with intra-axonal horseradish peroxidase in the nucleus of the tractus solitarius of the cat. II. An ultrastructural analysis. *J. Comp. Neurol.*, **241**, 521–535.

Kalia, M. and Richter, D. (1988a). Rapidly adapting pulmonary receptor afferents: I. Arborization in the nucleus of the tractus solitarius. *J. Comp. Neurol.*, **274**, 560–573.

Kalia, M. and Richter, D. (1988b). Rapidly adapting pulmonary receptor afferents: II. Fine structure and synaptic organization of central terminal processes in the nucleus of the tractus solitarius. *J. Comp. Neurol.*, **274**, 574–594.

Kanosue, K., Hosono, T., Zhang, Y. H. and Chen, X. M. (1998). Neuronal networks controlling thermoregulatory effectors. *Prog. Brain Res.*, **115**, 49–62.

Karczmar, K., Koketsu, K. and Nishi, S. (eds.) (1986). *Autonomic and Enteric Ganglia*. New York: Plenum Press.

Karlsson, A. K. (1999). Autonomic dysreflexia. *Spinal Cord*, **37**, 383–391.

Katayama, Y. and Nishi, S. (1986). Peptidergic transmission. In *Autonomic and Enteric Ganglia*, ed. K. Karczmar, K. Koketsu and S. Nishi. New York, London: Plenum Press, pp. 181–200.

Katona, P. G., Poitras, J. W., Barnett, G. O. and Terry, B. S. (1970). Cardiac vagal efferent activity and heart period in the carotid sinus reflex. *Am. J. Physiol.*, **218**, 1030–1037.

Katter, J. T., Dado, R. J., Kostarczyk, E. and Giesler, G. J., Jr. (1996). Spinothalamic and spinohypothalamic tract neurons in the sacral spinal cord of rats. II. Responses to cutaneous and visceral stimuli. *J. Neurophysiol.*, **75**, 2606–2628.

Kazuyuki, K., Hosono, T., Zhang, Y. H. and Chen, X. M. (1998). Neuronal networks controlling thermoregulatory effectors. *Prog. Brain Res.*, **115**, 49–62.

Keast, J. R. (1995a). Pelvic ganglia. In *The Autonomic Nervous System*, Vol. 6, *Autonomic Ganglia*, ed. E. M. McLachlan. London: Harwood Academic Publishers GmbH, pp. 445–479.

Keast, J. R. (1995b). Visualization and immunohistochemical characterization of sympathetic and parasympathetic neurons in the male rat major pelvic ganglion. *Neuroscience*, **66**, 655–662.

Keast, J. R. (1999). Unusual autonomic ganglia: connections, chemistry, and plasticity of pelvic ganglia. *Int. Rev. Cytol.*, **193**, 1–69.

Keast, J. R., McLachlan, E. M. and Meckler, R. L. (1993). Relation between electrophysiological class and neuropeptide content of guinea pig sympathetic prevertebral neurons. *J. Neurophysiol.*, **69**, 384–394.

Keast, J. R., Luckensmeyer, G. B. and Schemann, M. (1995). All pelvic neurons in male rats contain immunoreactivity for the synthetic enzymes of either noradrenaline or acetylcholine. *Neurosci. Lett.*, **196**, 209–212.

Keay, K. A. and Bandler, R. (2001). Parallel circuits mediating distinct emotional coping reactions to different types of stress. *Neurosci. Biobehav. Rev.*, **25**, 669–678.

Keay, K. A. and Bandler, R. (2004). Periaqueductal gray. In *The Rat Nervous System*, 3rd edn., ed. G. Paxinos. San Diego: Academic Press, pp. 243–257.

Keay, K. A., Clement, C. I., Owler, B., Depaulis, A. and Bandler, R. (1994). Convergence of deep somatic and visceral nociceptive information onto a discrete ventrolateral midbrain periaqueductal gray region. *Neuroscience*, **61**, 727–732.

Keay, K. A., Feil, K., Gordon, B. D., Herbert, H. and Bandler, R. (1997). Spinal afferents to functionally distinct periaqueductal gray columns in the rat: an anterograde and retrograde tracing study. *J. Comp. Neurol.*, **385**, 207–229.

Keay, K. A., Clement, C. I., Matar, W. M. *et al.* (2002). Noxious activation of spinal or vagal afferents evokes distinct patterns of fos-like immunoreactivity in the ventrolateral periaqueductal gray of unanaesthetised rats. *Brain Res.*, **948**, 122–130.

Keef, K. D. and Kreulen, D. L. (1986). Venous mechanoreceptor input to neurones in the inferior mesenteric ganglion of the guinea-pig. *J. Physiol.*, **377**, 49–59.

Kellogg, D. L., Jr., Johnson, J. M. and Kosiba, W. A. (1989). Selective abolition of adrenergic vasoconstrictor responses in skin by local iontophoresis of bretylium. *Am. J. Physiol.*, **257**, H1599–H1606.

Kellogg, D. L., Jr., Pergola, P. E., Piest, K. L. *et al.* (1995). Cutaneous active vasodilation in humans is mediated by cholinergic nerve cotransmission. *Circ. Res.*, **77**, 1222–1228.

Kesler, B. S., Mazzone, S. B. and Canning, B. J. (2002). Nitric oxide-dependent modulation of smooth-muscle tone by airway parasympathetic nerves. *Am. J. Respir. Crit. Care*, **165**, 481–488.

Khasar, S. G., Green, P. G. and Levine, J. D. (1993). Comparison of intradermal and subcutaneous hyperalgesic effects of inflammatory mediators in the rat. *Neurosci. Lett.*, **153**, 215–218.

Khasar, S. G., Miao, F. J. P. and Levine, J. D. (1995). Inflammation modulates the contribution of receptor-subtypes to bradykinin-induced hyperalgesia in the rat. *Neuroscience*, **69**, 685–690.

Khasar, S. G., Miao, F. J. P., Jänig, W. and Levine, J. D. (1998a). Modulation of bradykinin-induced mechanical hyperalgesia in the rat by activity in abdominal vagal afferents. *Eur. J. Neurosci.*, **10**, 435–444.

Khasar, S. G., Miao, F. J. P., Jänig, W. and Levine, J. D. (1998b). Vagotomy-induced enhancement of mechanical hyperalgesia in the rat is sympathoadrenal-mediated. *J. Neurosci.*, **18**, 3043–3049.

Khasar, S. G., Green, P. G., Miao, F. J. P. and Levine, J. D. (2003). Vagal modulation of nociception is mediated by adrenomedullary epinephrine in the rat. *Eur. J. Neurosci.*, **17**, 909–915.

Kim, M., Chiego, D. J., Jr. and Bradley, R. M. (2004). Morphology of parasympathetic neurons innervating rat lingual salivary glands. *Auton. Neurosci.*, **111**, 27–36.

Kirchheim, H. R., Just, A. and Ehmke, H. (1998). Physiology and pathophysiology of baroreceptor function and neuro-hormonal abnormalities in heart failure. *Basic Res. Cardiol.*, **93**, **Suppl.1**, 1–22.

Kirkup, A. J., Brunsden, A. M. and Grundy, D. (2001). Receptors and transmission in the brain-gut axis: potential for novel therapies. I. Receptors on visceral afferents. *Am. J. Physiol. Gastrointest. Liver Physiol.*, **280**, G787–G794.

Klemm, M. F. (1995). Neuromuscular junctions made by nerve fibres supplying the longitudinal muscle of the guinea-pig ileum. *J. Auton. Nerv. Syst.*, **55**, 155–164.

Klemm, M., Hirst, G. D. and Campbell, G. (1992). Structure of autonomic neuromuscular junctions in the sinus venosus of the toad. *J. Auton. Nerv. Syst.*, **39**, 139–150.

Klemm, M. F., Van Helden, D. F. and Luff, S. E. (1993). Ultrastructural analysis of sympathetic neuromuscular junctions on mesenteric veins of the guinea pig. *J. Comp. Neurol.*, **334**, 159–167.

Koepchen, H. P. (1962). *Die Blutdruckrhythmik [The Rhythm of Blood Pressure]*. Darmstadt: Dr. Dietrich Steinkopff.

Koepchen, H. P. (1983). Respiratory and cardiovascular "centres": functional entirety or separate structures. In *Central Environment and the Control Systems of Breathing and Circulation*, ed. M. E. Schläfke, H. P. Koepchen and W. R. See. Berlin, Heidelberg, New York: Springer, pp. 221–237.

Koepchen, H. P. and Thurau, K. (1959). Über die Entstehungsbedingungen der atemsynchronen Schwankungen des Vagustonus (respiratorische Arrhythmie) [On the origin of the respiratory synchronous changes of the vagus tone (respiratory arrhythmia)]. *Pflügers Arch.*, **259**, 10–30.

Koepchen, H. P., Seller, H., Polster, J. and Langhorst, P. (1968). Über die Fein-Vasomotorik der Muskelstrombahn und ihre Beziehung zur Ateminnervation [Spontaneous vasomotor changes in the muscle and their relation to the respiratory rhythm]. *Pflügers Arch.*, **302**, 285–299.

Koepchen, H. P., Klüssendorf, D. and Sommer, D. (1981). Neurophysiological background of central neural cardiovascular-respiratory coordination: basic remarks and experimental approach. *J. Auton. Nerv. Syst.*, **3**, 335–368.

Koepchen, H. P., Abel, H.-H. and Klüssendorf, D. (1987). Brain stem generation of specific and non-specific rhythms. In *Organization of the Automatic Nervous*

System: Central and Peripheral Mechanisms, ed. J. Ciriello, F. R. Calaresu, L. P. Renaud and C. Polosa. New York: Alan R. Liss., pp. 179–188.

Koh, S. D., Ward, S. M., Ordog, T., Sanders, K. M. and Horowitz, B. (2003). Conductances responsible for slow wave generation and propagation in interstitial cells of Cajal. *Curr. Opin. Pharmacol.*, **3**, 579–582.

Koltzenburg, M., Häbler, H. J. and Jänig, W. (1995). Functional reinnervation of the vasculature of the adult cat paw pad by axons originally innervating vessels in hairy skin. *Neuroscience*, **67**, 245–252.

Kopin, I. J. (1989). Plasma levels of catecholamines and dopamine-beta-hydroxylase. In *Handbook of Experimental Pharmacology*, Vol. 90/II, *Catecholamines II*, ed. U. Trendelenburg and N. Weiner. Berlin: Springer-Verlag, pp. 211–275.

Kopp, U. C. and DiBona, G. F. (2000). The neural control of renal function. In *The Kidney: Physiology and Pathophysiology*, 3rd edn., ed. G. Seldin and G. Giebisch. New York: Raven Press, pp. 981–1006.

Korner, P. I. (1979). Central nervous control of autonomic cardiovascular function. In *Handbook of Physiology, The Cardiovascular System*, Vol. I, *The Heart*. ed. R. M. Barne. Bethesda: American Physiological Society, pp. 691–739.

Koshiya, N. and Guyenet, P. G. (1996). Tonic sympathetic chemoreflex after blockade of respiratory rhythmogenesis in the rat. *J. Physiol.*, **491**, 859–869.

Koshiya, N., Huangfu, D. and Guyenet, P. G. (1993). Ventrolateral medulla and sympathetic chemoreflex in the rat. *Brain Res.*, **609**, 174–184.

Kostarczyk, E., Zhang, X. and Giesler, G. J., Jr. (1997). Spinohypothalamic tract neurons in the cervical enlargement of rats: locations of antidromically identified ascending axons and their collateral branches in the contralateral brain. *J. Neurophysiol.*, **77**, 435–451.

Koushanpour, E. (1991). Baroreceptor discharge behavior and resetting. In *Baroreceptor Reflexes*, ed. P. B. Persson and H. R. Kirchheim. Berlin, Heidelberg: Springer-Verlag, pp. 9–44.

Krassioukov, A. V., Johns, D. G. and Schramm, L. P. (2002). Sensitivity of sympathetically correlated spinal interneurons, renal sympathetic nerve activity, and arterial pressure to somatic and visceral stimuli after chronic spinal injury. *J. Neurotrauma*, **19**, 1521–1529.

Kreis, M. E., Jiang, W., Kirkup, A. J. and Grundy, D. (2002). Cosensitivity of vagal mucosal afferents to histamine and 5-HT in the rat jejunum. *Am. J. Physiol. Gastrointest. Liver Physiol.*, **283**, G612–G617.

Krenz, N. R. and Weaver, L. C. (2000). Nerve growth factor in glia and inflammatory cells of the injured rat spinal cord. *J. Neurochem.*, **74**, 730–739.

Krenz, N. R., Meakin, S. O., Krassioukov, A. V. and Weaver, L. C. (1999). Neutralizing intraspinal nerve growth factor blocks autonomic dysreflexia caused by spinal cord injury. *J. Neurosci.*, **19**, 7405–7414.

Kress, M., Koltzenburg, M., Reeh, P. W. and Handwerker, H. O. (1992). Responsiveness and functional attributes of electrically localized terminals of cutaneous C-fibers in vivo and in vitro. *J. Neurophysiol.*, **68**, 581–595.

Kreulen, D. L. and Peters, S. (1986). Non-cholinergic transmission in a sympathetic ganglion of the guinea-pig elicited by colon distension. *J. Physiol.*, **374**, 315–334.

Kreulen, D. L. and Szurszewski, J. H. (1979a). Nerve pathways in celiac plexus of the guinea pig. *Am. J. Physiol.*, **237**, E90–E97.

Kreulen, D. L. and Szurszewski, J. H. (1979b). Reflex pathways in the abdominal prevertebral ganglia: evidence for a colo-colonic inhibitory reflex. *J. Physiol.*, **295**, 21–32.

Krier, J. and Hartman, D. A. (1984). Electrical properties and synaptic connections to neurons in parasympathetic colonic ganglia of the cat. *Am. J. Physiol.*, **247**, G52–G61.

Krier, J. and Szurszewski, J. H. (1982). Effect of substance P on colonic mechanoreceptors, motility, and sympathetic neurons. *Am. J. Physiol.*, **243**, G259–G267.

Krier, J., Schmalz, P. F. and Szurszewski, J. H. (1982). Central innervation of neurones in the inferior mesenteric ganglion and of the large intestine of the cat. *J. Physiol.*, **332**, 125–138.

Kuhn, R. A. (1950). Functional capacity of the isolated human spinal cord. *Brain*, **73**, 1–51.

Kumazawa, T. (1986). Sensory innervation of reproductive organs. *Prog. Brain Res.*, **67**, 115–131.

Kumazawa, T. (1990). Functions of the nociceptive primary neurons. *Jpn. J. Physiol.*, **40**, 1–14.

Kumazawa, T., Mizumura, K. and Sato, J. (1987). Response properties of polymodal receptors studied using in vitro testis superior spermatic nerve preparations of dogs. *J. Neurophysiol.*, **57**, 702–711.

Kümmel, H. (1983). Activity in sympathetic neurons supplying skin and skeletal muscle in spinal cats. *J. Auton. Nerv. Syst.*, **7**, 319–327.

Kunitake, T. and Kannan, H. (2000). Discharge pattern of renal sympathetic nerve activity in the conscious rat: spectral analysis of integrated activity. *J. Neurophysiol.*, **84**, 2859–2867.

Kuntz, A. (1940). The structural organization of the inferior mesenteric ganglia. *J. Comp. Neurol.*, **72**, 371–382.

Kuntz, A. (1954). *The Autonomic Nervous System*. Philadelphia: Lea & Febinger.

Kuntz, A. and Saccomanno, G. (1944). Reflex inhibition of intestinal motility mediated through decentralized prevertebral ganglia. *J. Neurophysiol.*, **7**, 163–171.

Kunze, D. L. (1972). Reflex discharge patterns of cardiac vagal efferent fibres. *J. Physiol.*, **222**, 1–15.

Kunze, W. A. and Furness, J. B. (1999). The enteric nervous system and regulation of intestinal motility. *Annu. Rev. Physiol.*, **61**, 117–142.

Kuo, D. C., Hisamitsu, T. and de Groat, W. C. (1984). A sympathetic projection from sacral paravertebral ganglia to the pelvic nerve and to postganglionic nerves on the surface of the urinary bladder and large intestine of the cat. *J. Comp. Neurol.*, **226**, 76–86.

LaBar, K. S. and LeDoux, J. E. (2001). Coping with danger: the neural basis of defensive behavior and fearful feelings. In *Handbook of Physiology. Section 7: The Endocrine System*, Vol. IV, *Coping with the Environment: Neural and Neuroendocrine Mechanisms*, ed. B. S. McEwen. Oxford, New York: Oxford University Press, pp. 139–154.

Lal, S., Kirkup, A. J., Brunsden, A. M., Thompson, D. G. and Grundy, D. (2001). Vagal afferent responses to fatty acids of different chain length in the rat. *Am. J. Physiol. Gastrointest. Liver Physiol.*, **281**, G907–G915.

Lamb, K., Kang, Y. M., Gebhart, G. F. and Bielefeldt, K. (2003). Gastric inflammation triggers hypersensitivity to acid in awake rats. *Gastroenterology*, **125**, 1410–1418.

Lammers, W. J. (2000). Propagation of individual spikes as "patches" of activation in isolated feline duodenum. *Am. J. Physiol. Gastrointest. Liver Physiol.*, **278**, G297–G307.

LaMotte, R. H., Lundberg, L. E. R. and Torebjörk, H. E. (1992). Pain, hyperalgesia and activity in nociceptive C units in humans after intradermal injection of capsaicin. *J. Physiol.*, **448**, 749–764.

Lange, C. S. (1920) [1887]. Über Gemüthsbewegungen [translated into English]. In *The Emotions*, ed. W. James and C. G. Lange. Baltimore: Williams and Wilkins.

Langley, J. N. (1891). On the course and connections of the secretory fibres supplying the sweat glands of the feet of the cat. *J. Physiol.*, **12**, 347–374.

Langley, J. N. (1892). On the origin from the spinal cord of the cervical and upper thoracic sympathetic fibres, with some observations on white and grey rami communicantes. *Phil. Trans. R. Soc. Lond. B*, **183**, 85–124.

Langley, J. N. (1894a). The arrangement of the sympathetic nervous system, based chiefly on observation upon pilo-motor nerves. *J. Physiol.*, **15**, 176–244.

Langley, J. N. (1894b). Further observations on the secretory and vaso-motor fibres of the foot of the cat, with notes on other sympathetic nerve fibres. *J. Physiol.*, **17**, 296–314.

Langley, J. N. (1895). Note on the regeneration of prae-ganglionic fibres of the sympathetic. *J. Physiol.*, **18**, 80–84.

Langley, J. N. (1897). On the regeneration of preganglionic and of postganglionic visceral nerve fibres. *J. Physiol.*, **22**, 215–230.

Langley, J. N. (1900). The sympathetic and other related systems of nerves. In *Textbook of Physiology*, ed. E. A. Schäfer. Edinburgh, London: Young J. Pentland, pp. 616–696.

Langley, J. N. (1903a). Das sympathische und verwandte nervöse System der Wirbeltiere (autonomes nervöses System) [The sympathetic and related nervous system of vertebrates (autonomic nervous system)]. *Ergeb. Physiol.*, **27/II**, 818–827.

Langley, J. N. (1903b). The autonomic nervous system. *Brain*, **26**, 1–26.

Langley, J. N. (1921). *The Autonomic Nervous System. Part I*. Cambridge: W. Heffer.

Langley, J. N. and Anderson, H. K. (1895a). On the innervation of the pelvic and adjoining viscera. Part I. The lower portion of the intestine. *J. Physiol.*, **18**, 67–105.

Langley, J. N. and Anderson, H. K. (1895b). The innervation of the pelvic and adjoining viscera. Part II. The bladder. *J. Physiol.*, **19**, 71–84.

Langley, J. N. and Anderson, H. K. (1895c). The innervation of the pelvic and adjoining viscera. Part III. The external generative organs. *J. Physiol.*, **19**, 85–121.

Langley, J. N. and Anderson, H. K. (1895d). The innervation of the pelvic and adjoining viscera. Part IV. The internal generative organs. *J. Physiol.*, **19**, 122–130.

Langley, J. N. and Sherrington, C. S. (1891). On pilo-motor nerves. *J. Physiol.*, **12**, 278–291.

Larsen, P. J., Enquist, L. W. and Card, J. P. (1998). Characterization of the multisynaptic neuronal control of the rat pineal gland using viral transneuronal tracing. *Eur. J. Neurosci.*, **10**, 128–145.

Lavidis, N. A. and Bennett, M. R. (1992). Probabilistic secretion of quanta from visualized sympathetic nerve varicosities in mouse vas deferens. *J. Physiol.*, **454**, 9–26.

Lawson, S. N. (1996). Neurochemistry of cutaneous nociceptors. In *Neurobiology of Nociceptors*, ed. C. Belmonte and F. Cervero. Oxford, New York, Tokyo: Oxford University Press, pp. 72–91.

Lawson, S. N. (2005). The peripheral sensory nervous system: dorsal root ganglion neurons. In *Peripheral Neuropathy*, 4th edn., ed. P. Dyck and P. K. Thomas. Amsterdam: W. B. Saunders, Elsevier, pp. 163–202.

Le Douarin, N. M. and Kalcheim, C. (1999). *The Neural Crest*, 2nd edn. Cambridge: Cambridge University Press.

LeDoux, J. E. (1996). *The Emotional Brain*. New York: Simon & Shuster.

Lee, B. Y., Karmakar, M. G., Herz, B. L. and Sturgill, R. A. (1995). Autonomic dysreflexia revisited. *J. Spinal Cord. Med.*, **18**, 75–87.

Lee, L. Y. and Pisarri, T. E. (2001). Afferent properties and reflex functions of bronchopulmonary C-fibers. *Respir. Physiol.*, **125**, 47–65.

Leite-Panissi, C. R., Coimbra, N. C. and Menescal-de-Oliveira, L. (2003). The cholinergic stimulation of the central amygdala modifying the tonic immobility response and antinociception in guinea pigs depends on the ventrolateral periaqueductal gray. *Brain Res. Bull.*, **60**, 167–178.

Levenson, R. W. (1993). Autonomic nervous system differences among emotions. *Psychol. Sci.*, **3**, 23–27.

Levenson, R. W., Ekman, P. and Friesen, M. V. (1990). Voluntary facial action generates emotion-specific autonomic nervous system activity. *Psychophysiology*, **27**, 363–384.

Levenson, R. W., Carstensen, L. L., Friesen, W. V. and Ekman, P. (1991). Emotion, physiology, and expression in old age. *Psychol. Aging*, **6**, 28–35.

Levenson, R. W., Ekman, P., Heider, K. and Friesen, W. V. (1992). Emotion and autonomic nervous system activity in the Minangkabau of West Sumatra. *J. Personal. Soc. Psychol.*, **62**, 972–988.

Levine, J. D., Fields, H. L. and Basbaum, A. L. (1993). Peptides and the primary afferent nociceptor. *J. Neurosci.*, **13**, 2273–2286.

Lew, M. J., Rivers, R. J. and Duling, B. R. (1989). Arteriolar smooth muscle responses are modulated by an intramural diffusion barrier. *Am. J. Physiol.*, **257**, H10–H16.

Lewin, G. R. and McMahon, S. B. (1993). Muscle afferents innervating skin form somatotopically appropriate connections in the adult rat dorsal horn. *Eur. J. Neurosci.*, **5**, 1083–1092.

Lewin, G. R., Ritter, A. M. and Mendell, L. M. (1993). Nerve growth factor-induced hyperalgesia in the neonatal and adult rat. *J. Neurosci.*, **13**, 2136–2148.

Lewin, G. R., Rueff, A. and Mendell, L. M. (1994). Peripheral and central mechanisms of NGF-induced hyperalgesia. *Eur. J. Neurosci.*, **6**, 1903–1912.

Lewis, D. I. and Coote, J. H. (1995). Chemical mediators of spinal inhibition of rat sympathetic neurones on stimulation in the nucleus tractus solitarii. *J. Physiol.*, **486**, 483–494.

Lewis, D. I. and Coote, J. H. (1996). Baroreceptor-induced inhibition of sympathetic neurons by GABA acting at a spinal site. *Am. J. Physiol.*, **270**, H1885–H1892.

Lewis, M. W., Hermann, G. E., Rogers, R. C. and Travagli, R. A. (2002). In vitro and in vivo analysis of the effects of corticotropin releasing factor on rat dorsal vagal complex. *J. Physiol.*, **543**, 135–146.

Li, Y. W. and Dampney, R. A. (1994). Expression of Fos-like protein in brain following sustained hypertension and hypotension in conscious rabbits. *Neuroscience*, **61**, 613–634.

Li, Y. W., Gieroba, Z. J., McAllen, R. M. and Blessing, W. W. (1991). Neurons in rabbit caudal ventrolateral medulla inhibit bulbospinal barosensitive neurons in rostral medulla. *Am. J. Physiol.*, **261**, R44–R51.

Lichtman, J. W. (1977). The reorganization of synaptic connexions in the rat submandibular ganglion during post-natal development. *J. Physiol.*, **273**, 155–177.

Lichtman, J. W., Purves, D. and Yip, J. W. (1979). On the purpose of selective innervation of guinea-pig superior cervical ganglion cells. *J. Physiol.*, **292**, 69–84.

Lichtman, J. W., Purves, D. and Yip, J. W. (1980). Innervation of sympathetic neurons in the guinea-pig thoracic chain. *J. Physiol.*, **298**, 285–299.

Liddell, E. G. T. (1960). *The Discovery of Reflexes*. Oxford: Clarendon Press.

Lindgren, I. and Olivecrona, H. (1947). Surgical treatment of angina pectoris. *J. Neurosurg.*, **4**, 19–39.

Lindh, B., Lundberg, J. M. and Hökfelt, T. (1989). NPY-, galanin-, VIP/PHI-, CGRP- and substance P-immunoreactive neuronal subpopulations in the cat autonomic and sensory ganglia and their projections. *Cell Tissue Res.*, **256**, 259–273.

Lindh, B., Risling, M., Remahl, S., Terenius, L. and Hökfelt, T. (1993). Peptide-immunoreactive neurons and nerve in lumbosacral sympathetic ganglia: selective elimination of a pathway-specific expression of immuno-reactivities following sciatic nerve resection in kittens. *Neuroscience*, **55**, 545–562.

Lipski, J., Kanjhan, R., Kruszewska, B. and Rong, W. (1996). Properties of presympathetic neurones in the rostral ventrolateral medulla in the rat: an intracellular study "in vivo'. *J. Physiol.*, **490**, 729–744.

Lipski, J., Kawai, Y., Qi, J., Comer, A. and Win, J. (1998). Whole cell patch-clamp study of putative vasomotor neurons isolated from the rostral ventrolateral medulla. *Am. J. Physiol.*, **274**, R1099–R1110.

Lipski, J., Lin, J., Teo, M. Y. and van Wyk, M. (2002). The network vs. pacemaker theory of the activity of RVL presympathetic neurons – a comparison with another putative pacemaker system. *Auton. Neurosci.*, **98**, 85–89.

Llewellyn-Smith, I. J. and Weaver, L. C. (2001). Changes in synaptic inputs to sympathetic preganglionic neurons after spinal cord injury. *J. Comp. Neurol.*, **435**, 226–240.

Llewellyn-Smith, I. J., Phend, K. D., Minson, J. B., Pilowsky, P. M. and Chalmers, J. P. (1992). Glutamate-immunoreactive synapses on retrogradely-labelled sympathetic preganglionic neurons in rat thoracic spinal cord. *Brain Res.*, **581**, 67–80.

Llewellyn-Smith, I. J., Minson, J. B., Pilowsky, P. M., Arnolda, L. F. and Chalmers, J. P. (1995). The one hundred percent hypothesis: glutamate or GABA in synapses on sympathetic preganglionic neurons. *Clin. Exp. Hypertens.*, **17**, 323–333.

Llewellyn-Smith, I. J., Cassam, A. K., Krenz, N. R., Krassioukov, A. V. and Weaver, L. C. (1997). Glutamate- and GABA-immunoreactive synapses on sympathetic preganglionic neurons caudal to a spinal cord transection in rats. *Neuroscience*, **80**, 1225–1235.

Llewellyn-Smith, I. J., Arnolda, L. F., Pilowsky, P. M., Chalmers, J. P. and Minson, J. B. (1998). GABA- and glutamate-immunoreactive synapses on sympathetic preganglionic neurons projecting to the superior cervical ganglion. *J. Auton. Nerv. Syst.*, **71**, 96–110.

Lloyd, D. P. C. (1960). Spinal mechanisms involved in somatic activities. In *Neurophysiology*, Vol. II, ed. J. Field, H. W. Magoun and V. E. Hall. Washington: American Physiological Society, pp. 929-949.

Loewy, A. D. (1990a). Central autonomic pathways. In *Central Regulation of Autonomic Functions*, ed. A. D. Loewy and K. M. Spyer. New York, Oxford: Oxford University Press, pp. 88-103.

Loewy, A. D. (1990b). Autonomic control of the eye. In *Central Regulation of Autonomic Functions*, ed. A. D. Loewy and K. M. Spyer. New York, Oxford: Oxford University Press, pp. 268-285.

Loewy, A. D. (1998). Viruses as transneuronal tracers for defining neural circuits. *Neurosci. Biobehav. Rev.*, **22**, 679-684.

Loewy, A. D. and Burton, H. (1978). Nuclei of the solitary tract: efferent projections to the lower brain stem and spinal cord of the cat. *J. Comp. Neurol.*, **181**, 421-449.

Loewy, A. D. and Haxhiu, M. A. (1993). CNS cell groups projecting to pancreatic parasympathetic preganglionic neurons. *Brain Res.*, **620**, 323-330.

Loewy, A. D. and Spyer, K. M. (1990a). Vagal preganglionic neurons. In *Central Regulation of Autonomic Functions*, ed. A. D. Loewy and K. M. Spyer. New York, Oxford: Oxford University Press, pp. 68-87.

Loewy, A. D. and Spyer, K. M. (eds.) (1990b). *Central Regulation of Autonomic Functions*. New York, Oxford: Oxford University Press.

Loewy, A. D., Franklin, M. F. and Haxhiu, M. A. (1994). CNS monoamine cell groups projecting to pancreatic vagal motor neurons: a transneuronal labeling study using pseudorabies virus. *Brain Res.*, **638**, 248-260.

Löfving, B. (1961). Cardiovascular adjustments induced from the rostral cingulate gyrus with special reference to sympatho-inhibitory mechanisms. *Acta Physiol. Scand.*, **53** (**Suppl 184**), 1-82.

Logan, S. D., Pickering, A. E., Gibson, I. C., Nolan, M. F. and Spanswick, D. (1996). Electrotonic coupling between rat sympathetic preganglionic neurones in vitro. *J. Physiol.*, **495**, 491-502.

Lombardi, F., Della Bella, P., Casati, R. and Malliani, A. (1981). Effects of intracoronary administration of bradykinin on the impulse activity of afferent sympathetic unmyelinated fibers with left ventricular endings in the cat. *Circ. Res.*, **48**, 69-75.

Longhurst, J. C. (1995). Chemosensitive abdominal visceral afferents. In *Visceral Pain. Progress in Pain Research and Management*, Vol. 5, ed. G. F. Gebhart. Seattle: IASP Press, pp. 99-132.

Lovén, C. (1866). Über die Erweiterung von Arterien in Folge einer Nervenerregung [On the vasodilation of arteries as a consequence of nerve stimulation]. *Ber. Verh. königl. -sächs. Ges. Wiss.: Math. -phys. Classe*, **18**, 85-110.

Lovick, T. A. (1987). Differential control of cardiac and vasomotor activity by neurons in nucleus paragigantocellularis lateralis in the cat. *J. Physiol.*, **389**, 23-35.

Low, P. (ed.) (1993). *Clinical Autonomic Disorders*. Boston, Toronto, London: Brown and Company.

Luckensmeyer, G. B. and Keast, J. R. (1998a). Characterisation of the adventitial rectal ganglia in the male rat by their immunohistochemical features and projections. *J. Comp. Neurol.*, **396**, 429-441.

Luckensmeyer, G. B. and Keast, J. R. (1998b). Projections of pelvic autonomic neurons within the lower bowel of the male rat: an anterograde labelling study. *Neuroscience*, **84**, 263-280.

Luff, S. E. and McLachlan, E. M. (1989). Frequency of neuromuscular junctions on arteries of different dimensions in the rabbit, guinea pig and rat. *Blood Vessels*, **26**, 95–106.

Luff, S. E., McLachlan, E. M. and Hirst, G. D. (1987). An ultrastructural analysis of the sympathetic neuromuscular junctions on arterioles of the submucosa of the guinea pig ileum. *J. Comp. Neurol.*, **257**, 578–594.

Luff, S. E., Hengstberger, S. G., McLachlan, E. M. and Anderson, W. P. (1991). Two types of sympathetic axon innervating the juxtaglomerular arterioles of the rabbit and rat kidney differ structurally from those supplying other arteries. *J. Neurocytol.*, **20**, 781–795.

Luff, S. E., Hengstberger, S. G., McLachlan, E. M. and Anderson, W. P. (1992). Distribution of sympathetic neuroeffector junctions in the juxtaglomerular region of the rabbit kidney. *J Auton. Nerv. Syst.*, **40**, 239–253.

Luff, S. E., Young, S. B. and McLachlan, E. M. (1995). Proportions and structure of contacting and non-contacting varicosities in the perivascular plexus of the rat tail artery. *J. Comp. Neurol.*, **361**, 699–709.

Luff, S. E., Young, S. B. and McLachlan, E. M. (2000). Ultrastructure of substance P-immunoreactive terminals and their relation to vascular smooth muscle cells of rat small mesenteric arteries. *J. Comp. Neurol.*, **416**, 277–290.

Lundberg, A. (1971). Function of the ventral spinocerebellar tract. A new hypothesis. *Exp. Brain Res.*, **12**, 317–330.

Lundberg, A. (1975). Control of spinal mechansims from the brain. In *The Basic Neurosciences*, ed. R. C. Brady. New York: Raven, pp. 253–265.

Lundberg, A. (1979). Multisensory control of spinal reflex pathways. *Prog. Brain Res.*, **50**, 11–28.

Lundberg, J. M. (1981). Evidence for coexistence of vasoactive intestinal polypeptide (VIP) and acetylcholine in neurons of cat exocrine glands. Morphological, biochemical and functional studies. *Acta Physiol. Scand.*, **496**, 1–57.

Lundberg, J. M. (1996). Pharmacology of cotransmission in the autonomic nervous system: integrative aspects on amines, neuropeptides, adenosine triphosphate, amino acids and nitric oxide. *Pharmacol. Rev.*, **48**, 113–178.

Lundberg, J. M., Hemsen, A., Rudehill, A. *et al.* (1988). Neuropeptide Y- and alpha-adrenergic receptors in pig spleen: localization, binding characteristics, cyclic AMP effects and functional responses in control and denervated animals. *Neuroscience*, **24**, 659–672.

Lundgren, O. (1988). Nervous control of intestinal transport. In *Bailliere's Clinical Gastroenterology*, Vol. 2/1, *Gastrointestinal Neurophysiology*, ed. D. Grundy and N. W. Read. London: Balliere Tindall, pp. 85–106.

Lundgren, O. (1989). Enteric nervous control of mucosal functions of the small intestine in vivo. In *Nerves and the Gastrointestinal Tract*, ed. M. V. Singer and A. Goebell. Dordrecht, The Netherlands: Kluwer Academic Publishers, pp. 275–285.

Lundgren, O. (2000). Sympathetic input into the enteric nervous system. *Gut*, **47**, **Suppl. 4**, iv33–iv35.

Luo, M., Hess, M. C., Fink, G. D. *et al.* (2003). Differential alterations in sympathetic neurotransmission in mesenteric arteries and veins in DOCA-salt hypertensive rats. *Auton. Neurosci.*, **104**, 47–57.

Lykken, D. T. (1998). *A Tremor in the Blood*. New York: Plenum Press.

Lynn, B. (1996a). Neurogenic inflammation caused by cutaneous polymodal receptors. *Prog. Brain Res.*, **113**, 361–368.

Lynn, B. (1996b). Efferent function of nociceptors. In *Neurobiology of Nociceptors*, ed. C. Belmonte and F. Cervero. Oxford, New York, Tokyo: Oxford University Press, pp. 418–438.

Lynn, B., Schütterle, S. and Pierau, F. K. (1996). The vasodilator component of neurogenic inflammation is caused by a special subclass of heat-sensitive nociceptors in the skin of the pig. *J. Physiol.*, **494**, 587–593.

Lynn, P. A. and Blackshaw, L. A. (1999). In vitro recordings of afferent fibres with receptive fields in the serosa, muscle and mucosa of rat colon. *J. Physiol.*, **518**, 271–282.

Lynn, P. A., Olsson, C., Zagorodnyuk, V., Costa, M. and Brookes, S. J. (2003). Rectal intraganglionic laminar endings are transduction sites of extrinsic mechanoreceptors in the guinea pig rectum. *Gastroenterology*, **125**, 786–794.

Macefield, V. G. and Wallin, B. G. (1995). Modulation of muscle sympathetic activity during spontaneous and artificial ventilation and apnoea in humans. *J. Auton. Nerv. Syst.*, **53**, 137–147.

Macefield, V. G. and Wallin, B. G. (1996). The discharge behaviour of single sympathetic neurones supplying human sweat glands. *J. Physiol.*, **61**, 277–286.

Macefield, V. G. and Wallin, B. G. (1999a). Firing properties of single vasoconstrictor neurones in human subjects with high level of muscle sympathetic activity. *J. Physiol.*, **516**, 293–301.

Macefield, V. G. and Wallin, B. G. (1999b). Respiratory and cardiac modulation of single sympathetic vasoconstrictor and sudomotor neurones to human skin. *J. Physiol.*, **516**, 303–314.

Macefield, V. G., Wallin, B. G. and Vallbo, A. B. (1994). The discharge behaviour of single vasoconstrictor motoneurones in human muscle nerves. *J. Physiol.*, **481**, 799–809.

Macefield, V. G., Elam, M. and Wallin, B. G. (2002). Firing properties of single postganglionic sympathetic neurons recorded in awake human subjects. *Auton. Neurosci.*, **95**, 146–159.

Macefield, V. G., Sverrisdottir, Y. B. and Wallin, B. G. (2003). Resting discharge of human muscle spindles is not modulated by increases in sympathetic drive. *J. Physiol.*, **551**, 1005–1011.

Madden, K. S. and Felten, D. L. (1995). Experimental basis for neural-immune interactions. *Physiol. Rev.*, **75**, 77–106.

Madden, C. J. and Sved, A. F. (2003). Cardiovascular regulation after destruction of the C1 cell group of the rostral ventrolateral medulla in rats. *Am. J. Physiol. Heart Circ. Physiol.*, **285**, H2734–H2748.

Madden, K. S., Sanders, V. M. and Felten, D. L. (1995). Catecholamine influences and sympathetic modulation of immune responsiveness. *Rev. Pharmacol. Toxicol.*, **35**, 417–448.

Maggi, C. A. (ed.) (1993). *The Autonomic Nervous System*, Vol. 3, *Nervous Control of the Urogenital System*. Chur, Switzerland: Harwood Academic Publishers.

Maggi, C. A. and Meli, A. (1988). The sensory-efferent function of capsaicin-sensitive sensory neurons. *Gen. Pharmacol.*, **19**, 1–43.

Maggi, C. A., Giachetti, A., Dey, R. D. and Said, S. I. (1995). Neuropeptides as regulators for airway function: vasoactive intestinal peptide and the tachykinins. *Physiol. Rev.*, **75**, 277–322.

Maier, S. F. and Watkins, L. R. (1998). Cytokines for psychologists: implications of bidirectional immune-to-brain communication for understanding behavior, mood, and cognition. *Psychol. Rev.*, **105**, 83–107.

Malliani, A. (1982). Cardiovascular sympathetic afferent fibers. *Rev. Physiol. Biochem. Pharmacol.*, **94**, 11–74.

Malliani, A. and Lombardi, F. (1982). Consideration of the fundamental mechanisms eliciting cardiac pain. *Am. Heart J.*, **103**, 575–578.

Malpas, S. C. (1998). The rhythmicity of sympathetic nerve activity. *Prog. Neurobiol.*, **56**, 65–96.

Mancia, G., Baccelli, G. and Zanchetti, A. (1972). Hemodynamic responses to different emotional stimuli in the cat: patterns and mechanisms. *Am. J. Physiol.*, **223**, 925–933.

Mano, T. (1999). Muscular and cutaneous sympathetic activity. In *Handbook of Clinical Neurology*, Vol. 74, *The Autonomic Nervous System, part I: Normal Functions*, ed. O. Appenzeller. Amsterdam: Elsevier, pp. 649–665.

Mantyh, P. W., Rogers, S. D., Honore, P. et al. (1997). Inhibition of hyperalgesia by ablation of lamina I spinal neurons expressing the substance P receptor. *Science*, **278**, 275–279.

Markakis, E. A. and Swanson, L. W. (1997). Spatiotemporal patterns of secretomotor neuron generation in the parvicellular neuroendocrine system. *Brain Res. Brain Res. Rev.*, **24**, 255–291.

Marsh, D. R. and Weaver, L. C. (2004). Autonomic dysreflexia, induced by noxious or innocuous stimulation, does not depend on changes in dorsal horn substance p. *J. Neurotrauma*, **21**, 817–828.

Marsh, D. R., Wong, S. T., Meakin, S. O. et al. (2002). Neutralizing intraspinal nerve growth factor with a trkA-IgG fusion protein blocks the development of autonomic dysreflexia in a clip-compression model of spinal cord injury. *J. Neurotrauma*, **19**, 1531–1541.

Marshall, J. M. (1994). Peripheral chemoreceptors and cardiovascular regulation. *Physiol. Rev.*, **74**, 543–594.

Marson, L. (1995). Central nervous system neurons identified after injection of pseudorabies virus into the rat clitoris. *Neurosci. Lett.*, **190**, 41–44.

Marson, L. (1997). Identification of central nervous system neurons that innervate the bladder body, bladder base, or external urethral sphincter of female rats: a transneuronal tracing study using pseudorabies virus. *J. Comp. Neurol.*, **389**, 584–602.

Marson, L. and McKenna, K. E. (1996). CNS cell groups involved in the control of the ischiocavernosus and bulbospongiosus muscles: a transneuronal tracing study using pseudorabies virus. *J. Comp. Neurol.*, **374**, 161–179.

Marson, L., Platt, K. B. and McKenna, K. E. (1993). Central nervous system innervation of the penis as revealed by the transneuronal transport of pseudorabies virus. *Neuroscience*, **55**, 263–280.

Martinez-Pena y Valencuela, I., Rogers, R. C., Hermann, G. E. and Travagli, R. A. (2004). Norepinephrine effects on identified neurons of the rat dorsal motor nucleus of the vagus. *Am. J. Physiol. Gastrointest. Liver Physiol.*, **286**, G333–G339.

Martínez-Pinna, J., Davies, P. J. and McLachlan, E. M. (2000). Diversity of channels involved in Ca^{2+} activation of K^+ channels during the prolonged AHP in guinea-pig sympathetic neurons. *J. Neurophysiol.*, **84**, 1346–1354.

Mason, P. (2001). Contributions of the medullary raphe and ventromedial reticular region to pain modulation and other homeostatic functions. *Annu. Rev. Neurosci.*, **24**, 737–777.

Mathias, C. J. and Bannister, R. (eds.) (2002). *Autonomic Failure*, 4th edn. Oxford: Oxford University Press.

Mathias, C. J. and Frankel, H. L. (2002). Autonomic disturbances and spinal cord lesions. In *Autonomic Failure*, 4th edn., ed. C. J. Mathias and R. Bannister. Oxford: Oxford University Press, pp. 494–513.

Matsuo, R. and Kang, Y. (1998). Two types of parasympathetic preganglionic neurones in the superior salivatory nucleus characterized electrophysiologically in slice preparations of neonatal rats. *J. Physiol.*, **513**, 157–170.

Matsuo, R. and Yamamoto, T. (1989). Gustatory-salivary reflex: neural activity of sympathetic and parasympathetic fibers innervating the submandibular gland of the hamster. *J. Auton. Nerv. Syst.*, **26**, 187–197.

Matsuo, R., Morimoto, T. and Kang, Y. (1998). Neural activity of the superior salivatory nucleus in rats. *Eur. J. Morphol.*, **36 Suppl**, 203–207.

Matthews, L. H. (1969). *The Life of Mammals*. London: Weidenfeld and Nicolson.

Matthews, M. R. and Cuello, A. C. (1984). The origin and possible significance of substance P immunoreactive networks in the prevertebral ganglia and related structures in the guinea-pig. *Philos. Trans. R. Soc. London B Biol. Sci.*, **306**, 247–276.

Matthews, M. R., Connaughton, M. and Cuello, A. C. (1987). Ultrastructure and distribution of substance P-immunoreactive sensory collaterals in the guinea pig prevertebral sympathetic ganglia. *J. Comp. Neurol.*, **258**, 28–51.

Mawe, G. M. (1995). Prevertebral, pancreatic and gallbladder ganglia: non-enteric ganglia that are involved in gastrointestinal function. In *The Autonomic Nervous System*, Vol. 6, *Autonomic Ganglia*, ed. E. M. McLachlan. Luxembourg: Harwood Academic Publishers, pp. 397–444.

Mawe, G. M. (1998). Nerves and hormones interact to control gallbladder function. *News Physiol. Sci.*, **13**, 84–90.

Mayer, E. A. and Raybould, H. E. (eds.) (1993). *Basic and Clinical Aspects of Chronic Abdominal Pain*. Amsterdam: Elsevier Science Publishers B.V.

Mayer, E. A., Munakata, J., Mertz, H., Lembo, T. and Bernstein, C. N. (1995). Visceral hyperalgesia and irritable bowel syndrome. In *Visceral Pain*, ed. G. F. Gebhart. Seattle: IASP Press, pp. 429–468.

McAllen, R. M. (1987). Central respiratory modulation of subretrofacial bulbospinal neurones in the cat. *J. Physiol.*, **388**, 533–545.

McAllen, R. M. (1992). Actions of carotid chemoreceptors on subretrofacial bulbospinal neurons in the cat. *J. Auton. Nerv. Syst.*, **40**, 181–188.

McAllen, R. M. and May, C. N. (1994a). Differential drives from rostral ventrolateral medullary neurons to three identified sympathetic outflows. *Am. J. Physiol.*, **267**, R935–R944.

McAllen, R. M. and May, C. N. (1994b). Effects of preoptic warming on subretrofacial and cutaneous vasoconstrictor neurons in anaesthetized cats. *J. Physiol.*, **481**, 719–730.

McAllen, R. M. and Spyer, K. M. (1978a). Two types of vagal preganglionic motoneurones projecting to the heart and lungs. *J. Physiol.*, **282**, 353–364.

McAllen, R. M. and Spyer, K. M. (1978b). The baroreceptor input to cardiac vagal motoneurones. *J. Physiol.*, **282**, 365–374.

McAllen, R. M., Häbler, H. J., Michaelis, M., Peters, O. and Jänig, W. (1994). Monosynaptic excitation of preganglionic vasomotor neurons by subretrofacial neurons of the rostral ventrolateral medulla. *Brain Res.*, **634**, 227–234.

McAllen, R. M., May, C. N. and Shafton, A. D. (1995). Functional anatomy of sympathetic premotor cell groups in the medulla. *Clin. Exp. Hypertens.*, **17**, 209–221.

McAllen, R. M., May, C. N. and Campos, R. R. (1997). The supply of vasomotor drive to individual classes of sympathetic neuron. *Clin. Exp. Hypertens.*, **19**, 607–618.

McCabe, P. M., Duan, Y. F., Winters, R. W. et al. (1994). Comparison of peripheral blood flow patterns associated with the defense reaction and the vigilance reaction in rabbits. *Physiol. Behav.*, **56**, 1101–1106.

McCall, R. B., Gebber, G. L. and Barman, S. M. (1977). Spinal interneurons in the baroreceptor reflex arc. *Am. J. Physiol.*, **232**, H657–H665.

McCrea, D. A. (1994). Can sense be made of spinal interneuron circuits? In *Movement Control*, ed. P. Cordo and S. Harnad. Cambridge: Cambridge University Press, pp. 31–41.

McCrea, D. A. (2001). Spinal circuitry of sensorimotor control of locomotion. *J. Physiol.*, **533**, 41–50.

McDonald, D. M. (1990). The ultrastructure and permeability of tracheobronchial blood vessels in health and disease. *Eur. Respir. J. Suppl.*, **12**, 572s–585s.

McDonald, D. M. (1997). Neurogenic inflammation in the airways. In *The Autonomic Nervous System*, Vol. 7, *Autonomic Control of the Respiratory System*, ed. P. J. Barnes. Amsterdam: Harwood Academic Publishers GmbH, pp. 249–289.

McDonald, D. M., Mitchell, R. A., Gabella, G. and Haskell, A. (1988). Neurogenic inflammation in the rat trachea. II. Identity and distribution of nerves mediating the increase in vascular permeability. *J. Neurocytol.*, **17**, 605–628.

McEwen, B. S. (1998). Protective and damaging effects of stress mediators. *New Engl. J. Med.*, **338**, 171–179.

McEwen, B. S. (2000). Protective and damaging effects of stress mediators: central role of the brain. *Prog. Brain Res.*, **122**, 25–34.

McEwen, B. S. (ed.) (2001a). *Handbook of Physiology. Section 7: The Endocrine System, Vol. IV, Coping with the Environment: Neural and Neuroendocrine Mechanisms*. Oxford, New York: Oxford University Press.

McEwen, B. S. (2001b). Neurobiology of interpreting and responding to stressful events: paradigmatic role of the hippocampus. In *Handbook of Physiology. Section 7: The Endocrine System, Vol. IV, Coping with the Environment: Neural and Neuroendocrine Mechanisms*, ed. B. S. McEwen, pp. 155–178. Oxford University Press, Oxford New York.

McEwen, B. S. and Wingfield, J. C. (2003). The concept of allostasis in biology and biomedicine. *Horm. Behav.*, **43**, 2–15.

McGaugh, J. L. (2000). Memory – a century of consolidation. *Science*, **287**, 248–251.

McGaugh, J. L. and Roozendaal, B. (2002). Role of adrenal stress hormones in forming lasting memories in the brain. *Curr. Opin. Neurobiol.*, **12**, 205–210.

McKenna, K. E. (1999). Central nervous system pathways involved in the control of penile erection. *Annu. Rev. Sex Res.*, **10**, 157–183.

McKenna, K. E. (2000). The neural control of female sexual function. *NeuroRehabilitation*, **15**, 133–143.

McKenna, K. E. (2001). Neural circuitry involved in sexual function. *J. Spinal Cord Med.*, **24**, 148–154.

McKenna, K. E. (2002). The neurophysiology of female sexual function. *World J. Urol.*, **20**, 93–100.

McKenna, K. E. and Marson, L. (1997). Spinal and brain stem control of sexual function. In *The Autonomic Nervous System*, Vol. 11, *Central Nervous Control of*

Autonomic Function, ed. D. Jordan. Amsterdam: Harwood Academic Publishers, pp. 151–187.

McKenna, K. E., Chung, S. K. and McVary, K. T. (1991). A model for the study of sexual function in anesthetized male and female rats. *Am. J. Physiol.*, **261**, R1276–R1285.

McKinley, M. J., Clarke, I. J. and Oldfield, B. J. (2004). Circumventricular organs. In *The Human Nervous System*, ed. G. Paxinos and J. K. Mai. Amsterdam: Elsevier Academic Press, pp. 562–591.

McLachlan, E. M. (1975). An analysis of the release of acetylcholine from preganglionic nerve terminals. *J. Physiol.*, **245**, 447–466.

McLachlan, E. M. (1985). The components of the hypogastric nerve in male and female guinea pigs. *J. Auton. Nerv. Syst.*, **13**, 327–342.

McLachlan, E. M. (ed.) (1995). *The Autonomic Nervous System*, Vol. 6, *Autonomic Ganglia*. Luxembourg: Harwood Academic Publishers.

McLachlan, E. M. and Hirst, G. D. (1980). Some properties of preganglionic neurons in upper thoracic spinal cord of the cat. *J. Neurophysiol.*, **43**, 1251–1265.

McLachlan, E. M. and Jänig, W. (1983). The cell bodies of origin of sympathetic and sensory axons in some skin and muscle nerves of the cat hindlimb. *J. Comp. Neurol.*, **214**, 115–130.

McLachlan, E. M. and Meckler, R. L. (1989). Characteristics of synaptic input to three classes of sympathetic neurone in the coeliac ganglion of the guinea-pig. *J. Physiol.*, **415**, 109–129.

McLachlan, E. M., Davies, P. J., Häbler, H. J. and Jamieson, J. (1997). Ongoing and reflex synaptic events in rat superior cervical ganglion cells. *J. Physiol.*, **501**, 165–181.

McLachlan, E. M., Häbler, H. J., Jamieson, J. and Davies, P. J. (1998). Analysis of the periodicity of synaptic events in neurones in the superior cervical ganglion of anaesthetized rats. *J. Physiol.*, **511**, 461–478.

McMahon, S. B. (1996). NGF as a mediator of inflammatory pain. *Philos. Trans. R. Soc. London B Biol. Sci.*, **351**, 431–440.

McMahon, S. B. and Morrison, J. F. (1982a). Spinal neurones with long projections activated from the abdominal viscera of the cat. *J. Physiol.*, **322**, 1–20.

McMahon, S. B. and Morrison, J. F. (1982b). Two groups of spinal interneurones that respond to stimulation of the abdominal viscera of the cat. *J. Physiol.*, **322**, 21–34.

Meckler, R. L. and Weaver, L. C. (1988). Characteristics of ongoing and reflex discharge of single splenic and renal sympathetic postganglionic fibres in the cat. *J. Physiol.*, **396**, 139–153.

Mei, N. (1983). Sensory structures in the viscera. *Prog. Sensory Physiol.*, **4**, 1–42.

Mei, N. (1985). Intestinal chemosensitivity. *Physiol. Rev.*, **65**, 211–237.

Meller, S. T. and Gebhart, G. F. (1992). A critical review of the afferent pathways and the potential chemical mediators involved in cardiac pain. *Neuroscience*, **48**, 501–524.

Melnitchenko, L. V. and Skok, V. I. (1970). Natural electrical activity in mammalian parasympathetic ganglion neurones. *Brain Res.*, **23**, 277–279.

Menendez, L., Bester, H., Besson, J. M. and Bernard, J. F. (1996). Parabrachial area: electrophysiological evidence for an involvement in cold nociception. *J. Neurophysiol.*, **75**, 2099–2116.

Mense, S. and Craig, A. D., Jr. (1988). Spinal and supraspinal terminations of primary afferent fibers from the gastrocnemius-soleus muscle in the cat. *Neuroscience*, **26**, 1023–1035.

Messenger, J. P., Anderson, R. L. and Gibbins, I. L. (1999). Neurokinin-1 receptor localisation in guinea pig autonomic ganglia. *J. Comp. Neurol.*, **412**, 693–704.

Meyer, R. A., Davis, K. D., Cohen, R. H., Treede, R. D. and Campbell, J. N. (1991). Mechanically insensitive afferents (MIAs) in cutaneous nerves of monkey. *Brain Res.*, **561**, 252–261.

Meyer, R. A., Ringkamp, M., Campbell, J. N. and Raja, S. N. (2005). Peripheral mechanisms of cutaneous nociception. In *Wall and Mezack's Textbook of Pain*, 5th edn., ed. S. B. McMahon and M. Koltzenburg. Edinburgh: Elsevier Churchill Livingstone, pp. 3–34.

Meyers, G. E. (1986). *William James, His Life and Thought*. New Haven: Yale University Press.

Miao, F. J. P., Green, P. G., Coderre, T. J., Jänig, W. and Levine, J. D. (1996a). Sympathetic-dependence in bradykinin-induced synovial plasma extravasation is dose-related. *Neurosci. Lett.*, **205**, 165–168.

Miao, F. J. P., Jänig, W. and Levine, J. D. (1996b). Role of sympathetic postganglionic neurons in synovial plasma extravasation induced by bradykinin. *J. Neurophysiol.*, **75**, 715–724.

Miao, F. J. P., Jänig, W., Green, P. G. and Levine, J. D. (1997a). Inhibition of bradykinin-induced synovial plasma extravasation produced by noxious cutaneous and visceral stimuli and its modulation by activity in the vagal nerve. *J. Neurophysiol.*, **78**, 1285–1292.

Miao, F. J. P., Jänig, W. and Levine, J. D. (1997b). Vagal branches involved in inhibition of bradykinin-induced synovial plasma extravasation by intrathecal nicotine and noxious stimulation in the rat. *J. Physiol.*, **498**, 473–481.

Miao, F. J. P., Jänig, W. and Levine, J. D. (2000). Nociceptive neuroendocrine negative feedback control of neurogenic inflammation activated by capsaicin in the rat paw: role of the adrenal medulla. *J. Physiol.*, **527**, 601–610.

Miao, F. J. P., Jänig, W., Jasmin, L. and Levine, J. D. (2001). Spino-bulbospinal pathway mediating vagal modulation of nociceptive-neuroendocrine control of inflammation in the rat. *J. Physiol.*, **532**, 811–822.

Miao, F. J. P., Green, P. G. and Levine, J. D. (2003b). Mechano-sensitive duodenal afferents contribute to vagal modulation of inflammation in the rat. *J. Physiol.*, **554**, 227–235.

Miao, F. J. P., Jänig, W., Jasmin, L. and Levine, J. D. (2003a). Blockade of nociceptive inhibition of plasma extravasation by opioid stimulation of the periaqueductal gray and its interaction with vagus-induced inhibition in the rat. *Neuroscience*, **119**, 875–885.

Michaelis, M., Göder, R., Häbler, H. J. and Jänig, W. (1994). Properties of afferent nerve fibres supplying the saphenous vein in the cat. *J. Physiol.*, **474**, 233–243.

Michaelis, M., Häbler, H. J. and Jänig, W. (1996). Silent afferents: a separate class of primary afferents? *Clin. Exp. Pharmacol. Physiol.*, **23**, 99–105.

Michl, T., Jocic, M., Heinemann, A., Schuligoi, R. and Holzer, P. (2001). Vagal afferent signaling of a gastric mucosal acid insult to medullary, pontine, thalamic, hypothalamic and limbic, but not cortical, nuclei of the rat brain. *Pain*, **92**, 19–27.

Mifflin, S. W., Spyer, K. M. and Withington-Wray, D. J. (1988). Baroreceptor inputs to the nucleus tractus solitarius in the cat: postsynaptic actions and the influence of respiration. *J. Physiol.*, **399**, 349–367.

Milner, T. A., Morrison, S. F., Abate, C. and Reis, D. J. (1988). Phenylethanolamine N-methyltransferase-containing terminals synapse directly on sympathetic preganglionic neurons in the rat. *Brain Res.*, **448**, 205–222.

Minson, J. B., Llewellyn-Smith, I. J., Chalmers, J. P., Pilowsky, P. M. and Arnolda, L. F. (1997). c-fos identifies GABA-synthesizing barosensitive neurons in caudal ventrolateral medulla. *NeuroReport*, **8**, 3015–3021.

Mitchell, J. H. (1985). Cardiovascular control during exercise: central and reflex neural mechanisms. *Am. J. Cardiol.*, **55**, 34D–41D.

Mitchell, R. A., Herbert, D. A., Baker, D. G. and Basbaum, C. B. (1987). In vivo activity of tracheal parasympathetic ganglion cells innervating tracheal smooth muscle. *Brain Res.*, **437**, 157–160.

Mitchell, S. W. (1872). *Injuries of Nerves and their Consequences*. Philadelphia: JB Lippincott.

Miura, M., Takayama, K. and Okada, J. (1994). Distribution of glutamate- and GABA-immunoreactive neurons projecting to the cardioacceleratory center of the intermediolateral nucleus of the thoracic cord of SHR and WKY rats: a double-labeling study. *Brain Res.*, **638**, 139–150.

Miyawaki, T., Goodchild, A. K. and Pilowsky, P. M. (2003). Maintenance of sympathetic tone by a nickel chloride sensitive mechanism in the rostral ventrolateral medulla of the adult rat. *Neuroscience*, **116**, 455–464.

Molander, C. and Grant, G. (1995). Spinal cord cytoarchitecture. In *The Rat Nervous System*, ed. G. Paxinos. San Diego: Academic Press, pp. 39–45.

Monnier, A., Alheid, G. F. and McCrimmon, D. R. (2003). Defining ventral medullary respiratory compartments with a glutamate receptor agonist in the rat. *J. Physiol.*, **548**, 859–874.

Moore, R. Y. (1996). Neural control of the pineal gland. *Behav. Brain Res.*, **73**, 125–130.

Moore, R. Y. (2003). Circadian timing. In *Fundamental Neuroscience*, 2nd edn., ed. L. R. Squire, F. E. Bloom, S. K. McConnell *et al.* San Diego: Academic Press, pp. 1067–1084.

Moran, T. H. and Schwartz, G. J. (1994). Neurobiology of cholecystokinin. *Crit. Rev. Neurobiol.*, **9**, 1–28.

Morgan, C. W., de Groat, W. C., Felkins, L. A. and Zhang, S. J. (1993). Intracellular injection of neurobiotin or horseradish peroxidase reveals separate types of preganglionic neurons in the sacral parasympathetic nucleus of the cat. *J. Comp. Neurol.*, **331**, 161–182.

Moriarty, M., Gibbins, I. L., Potter, E. K. and McCloskey, D. I. (1992). Comparison of the inhibitory roles of neuropeptide Y and galanin on cardiac vagal action in the dog. *Neurosci. Lett.*, **139**, 275–279.

Morris, J. L. (1995). Distribution and peptide content of sympathetic axons innervating different regions of the cutaneous venous bed in the pinna of the guinea pig ear. *J. Vasc. Res.*, **32**, 378–386.

Morris, J. L. (1999). Cotransmission from sympathetic vasoconstrictor neurons to small cutaneous arteries in vivo. *Am. J. Physiol.*, **277**, H58–H64.

Morris, J. and Dolan, R. (2004). Functional neuroanatomy of human emotion. In *Human Brain Function*, 2nd edn., ed. R. S. J. Frackowiak, K. J. Friston, C. D. Frith *et al.* Amsterdam: Elsevier Academic Press, pp. 365–396.

Morris, J. L. and Gibbins, I. L. (1992). Co-transmission and neuromodulation. In *The Autonomic Nervous System*, Vol. 1, *Autonomic Neuroeffector Mechanisms*, ed. G. Burnstock and C. H. V. Hoyle. Chur, Switzerland: Harwood, pp. 33–119.

Morris, J. L. and Gibbins, I. L. (eds.) (1997). *The Autonomic Nervous System*, Vol. 12, *Autonomic Innervation of the Skin*, Chur, Switzerland: Harwood Academic Publishers.

Morris, J. L., Gibbins, I. L. and Clevers, J. (1981). Resistance of adrenergic neurotransmission in the toad heart to adrenoceptor blockade. *Naunyn-Schmiedeberg's Arch. Pharmacol.*, **317**, 331–338.

Morris, M. J., Russell, A. E., Kapoor, V. et al. (1986). Increases in plasma neuropeptide Y concentrations during sympathetic activation in man. *J. Auton. Nerv. Syst.*, **17**, 143–149.

Morris, J. L., Gibbins, I. L. and Jobling, P. (2005). Post-stimulus potentiation of transmission in pelvic ganglia enhances sympathetic dilatation of guinea-pig uterine artery in vitro. *J. Physiol.*, **566**, 189–203.

Morrison, J. F. B. (1997). Central nervous control of the bladder. In *The Autonomic Nervous System*, Vol. 11, *Central Nervous Control of Autonomic Function*, ed. D. Jordan. Amsterdam: Harwood Academic Publishers, pp. 129–149.

Morrison, S. F. (1999). RVLM and raphe differentially regulate sympathetic outflows to splanchnic and brown adipose tissue. *Am. J. Physiol.*, **276**, R962–R973.

Morrison, S. F. (2001a). Differential control of sympathetic outflow. *Am. J. Physiol. Regul. Integr. Comp. Physiol.*, **281**, R683–R698.

Morrison, S. F. (2001b). Differential regulation of brown adipose and splanchnic sympathetic outflows in rat: roles of raphe and rostral ventrolateral medulla neurons. *Clin. Exp. Pharmacol. Physiol.*, **28**, 138–143.

Morrison, S. F. and Cao, W. H. (2000). Different adrenal sympathetic preganglionic neurons regulate epinephrine and norepinephrine secretion. *Am. J. Physiol. Regul. Integr. Comp. Physiol.*, **279**, R1763–R1775.

Morrison, S. F. and Gebber, G. L. (1985). Axonal branching patterns and funicular trajectories of raphespinal sympathoinhibitory neurons. *J. Neurophysiol.*, **53**, 759–772.

Morrison, S. F., Callaway, J., Milner, T. A. and Reis, D. J. (1991). Rostral ventrolateral medulla: a source of the glutamatergic innervation of the sympathetic intermediolateral nucleus. *Brain Res.*, **562**, 126–135.

Morrison, S. F., Sved, A. F. and Passerin, A. M. (1999). GABA-mediated inhibition of raphe pallidus neurons regulates sympathetic outflow to brown adipose tissue. *Am. J. Physiol.*, **276**, R290–R297.

Mosqueda-Garcia, R., Furlan, R., Tank, J. and Fernandez-Violante, R. (2000). The elusive pathophysiology of neurally mediated syncope. *Circulation*, **102**, 2898–2906.

Müller, L. R. (1902). Klinische und experimentelle Studien ueber die Innervation der Blase, des Mastdarmes und des Genitalapparates [Clinical and experimental studies on the innervation of the urinary bladder, hindgut and reproductive organs]. *Dtsch. Z. Nervenheilkd.*, **21**, 86–155.

Müller, L. R. (1906). Ueber die Exstirpation der unteren Haelfte des Rueckenmarkes und deren Folgeerscheinungen [On the consequences of the removal of the lower half of the spinal cord]. *Dtsch. Z. Nervenheilkd.*, **30**, 411–423.

Mulryan, K., Gitterman, D. P., Lewis, C. J. et al. (2000). Reduced vas deferens contraction and male infertility in mice lacking P2X1 receptors. *Nature*, **403**, 86–89.

Mulvany, M. J., Nilsson, H. and Flatman, J. A. (1982). Role of membrane potential in the response of rat small mesenteric arteries to exogenous noradrenaline stimulation. *J. Physiol.*, **332**, 363–373.

Murphy, S. M., Matthew, S. E., Rodgers, H. F., Lituri, D. T. and Gibbins, I. L. (1998). Synaptic organisation of neuropeptide-containing preganglionic boutons in lumbar sympathetic ganglia of guinea pigs. *J. Comp. Neurol.*, **398**, 551–567.

Murray, J. G. and Thompson, J. W. (1957). The occurrence and function of collateral sprouting in the sympathetic nervous system of the cat. *J. Physiol.*, **135**, 133–162.

Nadelhaft, I. and Vera, P. L. (1995). Central nervous system neurons infected by pseudorabies virus injected into the rat urinary bladder following unilateral transection of the pelvic nerve. *J. Comp. Neurol.*, **359**, 443–456.

Nadelhaft, I., Vera, P. L., Card, J. P. and Miselis, R. R. (1992). Central nervous system neurons labelled following the injection of pseudorabies virus into the rat urinary bladder. *Neurosci. Lett.*, **143**, 271–274.

Nagashima, K., Nakai, S., Tanaka, M. and Kanosue, K. (2000). Neuronal circuitries involved in thermoregulation. *Auton. Neurosci.*, **85**, 18–25.

Natarajan, M. and Morrison, S. F. (2000). Sympathoexcitatory CVLM neurons mediate responses to caudal pressor area stimulation. *Am. J. Physiol. Regul. Integr. Comp. Physiol.*, **279**, R364–R374.

Neff, R. A., Mihalevich, M. and Mendelowitz, D. (1998). Stimulation of NTS activates NMDA and non-NMDA receptors in rat cardiac vagal neurons in the nucleus ambiguus. *Brain Res.*, **792**, 277–282.

Ness, T. J. and Gebhart, G. F. (1990). Visceral pain: a review of experimental studies. *Pain*, **41**, 167–234.

Neuhuber, W. L. (1989). Vagal afferent fibers almost exclusively innervate islets in the rat pancreas as demonstrated by anterograde tracing. *J. Auton. Nerv. Syst.*, **29**, 13–18.

Niijima, A. (1996). The afferent dischargee from sensors for interleukin 1 beta in the hepatoportal system in the anesthetized rat. *J. Auton. Nerv. Syst.*, **61**, 287–291.

Nilsson, H., Jensen, P. E. and Mulvany, M. J. (1994). Minor role for direct adrenoceptor-mediated calcium entry in rat mesenteric small arteries. *J. Vasc. Res.*, **31**, 314–321.

Nilsson, S. (1983). *Autonomic Nerve Function in the Vertebrates*. Berlin: Springer-Verlag.

Nilsson, S. and Holmgren, S. (eds.) (1994). *The Autonomic Nervous System*, Vol. 4, *Comparative Physiology and Evolution of the Autonomic Nervous System*, Chur, Switzerland: Harwood Academic Publishers.

Nisida, I. and Okada, H. (1960). The activity of the pupilloconstrictor centers. *Jap. J. Physiol.*, **10**, 64–72.

Nja, A. and Purves, D. (1977a). Specific innervation of guinea-pig superior cervical ganglion cells by preganglionic fibres arising from different levels of the spinal cord. *J. Physiol.*, **264**, 565–583.

Nja, A. and Purves, D. (1977b). Re-innervation of guinea-pig superior cervical ganglion cells by preganglionic fibres arising from different levels of the spinal cord. *J. Physiol.*, **272**, 633–651.

Nolan, M. F., Logan, S. D. and Spanswick, D. (1999). Electrophysiological properties of electrical synapses between rat sympathetic preganglionic neurones in vitro. *J. Physiol.*, **519**, 753–764.

Noll, G., Elam, M., Kunimoto, M., Karlsson, T. and Wallin, B. G. (1994). Skin sympathetic nerve activity and effector function during sleep in humans. *Acta Physiol. Scand.*, **151**, 319–329.

Nordin, M. (1990). Sympathetic discharges in the human supraorbital nerve and their relation to sudo- and vasomotor responses. *J. Physiol.*, **423**, 241-255.

Nordin, M. and Fagius, J. (1995). Effect of noxious stimulation on sympathetic vasoconstrictor outflow to human muscles. *J. Physiol.*, **489**, 885-894.

Norman, R. A., Jr., Coleman, T. G. and Dent, A. C. (1981). Continuous monitoring of arterial pressure indicates sinoaortic denervated rats are not hypertensive. *Hypertension*, **3**, 119-125.

North, R. A. (1986). Receptors on individual neurones. *Neuroscience*, **17**, 899-907.

Oberle, J., Elam, M., Karlsson, T. and Wallin, B. G. (1988). Temperature-dependent interaction between vasoconstrictor and vasodilator mechanisms in human skin. *Acta Physiol. Scand.*, **132**, 459-469.

Obrist, P. A. (1981). *Cardiovascular Psychophysiology: A Perspective*. New York: Plenum Press.

Ochoa, J. and Torebjörk, E. (1983). Sensations evoked by intraneural microstimulation of single mechanoreceptor units innervating the human hand. *J. Physiol.*, **342**, 633-654.

Ochoa, J. and Torebjörk, H. E. (1989). Sensations evoked by intraneural microstimulation of C nociceptor fibres in human skin nerves. *J. Physiol.*, **415**, 583-599.

Oh, E. J., Mazzone, S. B., Canning, B. J. and Weinreich, D. (2006). Reflex regulation of airway sympathetic nerves in guinea-pigs. *J. Physiol.*, in press.

Oldfield, B. J. and McLachlan, E. M. (1981). An analysis of the sympathetic preganglionic neurons projecting from the upper thoracic spinal roots of the cat. *J. Comp. Neurol.*, **196**, 329-345.

Öngür, D. and Price, J. L. (2000). The organization of networks within the orbital and medial prefrontal cortex of rats, monkeys and humans. *Cereb. Cortex*, **10**, 206-219.

Öngür, D., An, X. and Price, J. L. (1998). Prefrontal cortical projections to the hypothalamus in macaque monkeys. *J. Comp. Neurol.*, **401**, 480-505.

Ootsuka, Y. and Terui, N. (1997). Functionally different neurons are organized topographically in the rostral ventrolateral medulla of rabbits. *J. Auton. Nerv. Syst.*, **67**, 67-78.

Orr, R. and Marson, L. (1998). Identification of CNS neurons innervating the rat prostate: a transneuronal tracing study using pseudorabies virus. *J. Auton. Nerv. Syst.*, **72**, 4-15.

Owens, N. C., Ootsuka, Y., Kanosue, K. and McAllen, R. M. (2002). Thermoregulatory control of sympathetic fibers supplying the rat's tail. *J. Physiol.*, **543**, 849-858.

Owsjannikow, P. (1874). Über einen Unterschied in den reflectorischen Leistungen des verlängerten und des Rückenmarkes der Kaninchen [On a difference between reflexes mediated by the medulla oblongata and spinal cord in rabbits]. *Sitzungsber. Akad. Wiss. Wien, Math.-Naturw., Abt. 2*, **26**, 457-464.

Ozaki, N. and Gebhart, G. F. (2001). Characterization of mechanosensitive splanchnic nerve afferent fibers innervating the rat stomach. *Am. J. Physiol. Gastrointest. Liver Physiol.*, **281**, G1449-G1459.

Page, A. J. and Blackshaw, L. A. (1998). An in vitro study of the properties of vagal afferent fibres innervating the ferret oesophagus and stomach. *J. Physiol.*, **512**, 907-916.

Page, A. J., Martin, C. M. and Blackshaw, L. A. (2002). Vagal mechanoreceptors and chemoreceptors in mouse stomach and esophagus. *J. Neurophysiol.*, **87**, 2095-2103.

Paintal, A. S. (1973). Vagal sensory receptors and their reflex effects. *Physiol. Rev.*, **53**, 159–227.

Paintal, A. S. (1986). The visceral sensations – some basic mechanisms. *Prog. Brain Res.*, **67**, 3–19.

Pan, H. L. and Longhurst, J. C. (1996). Ischaemia-sensitive sympathetic afferents innervating the gastrointestinal tract function as nociceptors in cats. *J. Physiol.*, **492**, 841–850.

Pan, H. L., Longhurst, J. C., Eisenach, J. C. and Chen, S. R. (1999). Role of protons in activation of cardiac sympathetic C-fibre afferents during ischaemia in cats. *J. Physiol.*, **518**, 857–866.

Panksepp, J. (1998). *Affective Neuroscience*. New York, Oxford: Oxford University Press.

Papka, R. E., Williams, S., Miller, K. E., Copelin, T. and Puri, P. (1998). CNS location of uterine-related neurons revealed by trans-synaptic tracing with pseudorabies virus and their relation to estrogen receptor-immunoreactive neurons. *Neuroscience*, **84**, 935–952.

Parr, E. J. and Sharkey, K. A. (1996). Immunohistochemically-defined subtypes of neurons in the inferior mesenteric ganglion of the guinea-pig. *J. Auton. Nerv. Syst.*, **59**, 140–150.

Paton, J. F. (1996a). The ventral medullary respiratory network of the mature mouse studied in a working heart–brainstem preparation. *J. Physiol.*, **493**, 819–831.

Paton, J. F. (1996b). A working heart–brainstem preparation of the mouse. *J. Neurosci. Methods*, **65**, 63–68.

Paton, J. F. (1999). The Sharpey-Schafer prize lecture: nucleus tractus solitarii: integrating structures. *Exp. Physiol.*, **84**, 815–833.

Paton, J. F. and Dutschmann, M. (2002). Central control of upper airway resistance regulating respiratory airflow in mammals. *J. Anat.*, **201**, 319–323.

Paton, J. F. and Kasparov, S. (2000). Sensory channel specific modulation in the nucleus of the solitary tract. *J. Auton. Nerv. Syst.*, **80**, 117–129.

Paton, J. F., Li, Y. W., Deuchars, J. and Kasparov, S. (2000). Properties of solitary tract neurons receiving inputs from the sub-diaphragmatic vagus nerve. *Neuroscience*, **95**, 141–153.

Paton, J. F., Deuchars, J., Li, Y. W. and Kasparov, S. (2001). Properties of solitary tract neurones responding to peripheral arterial chemoreceptors. *Neuroscience*, **105**, 231–248.

Paton, J. F., Deuchar, J., Wang, S. and Kasparov, S. (2005). Nitroxergic modulation in the NTS: implications for cardiovascular function. In *Advances in Vagal Afferent Neurobiology*, ed. B. Undem and D. Weinreich. Boca Raton: CRC Press, pp. 209–246.

Paxinos, G. and Watson, C. (eds.) (1998). *The Rat Brain*. San Diego: Academic Press.

Peng, H., Matchkov, V., Ivarsen, A., Aalkjaer, C. and Nilsson, H. (2001). Hypothesis for the initiation of vasomotion. *Circ. Res.*, **88**, 810–815.

Pernow, J. (1988). Co-release and functional interactions of neuropeptide Y and noradrenaline in peripheral sympathetic vascular control. *Acta Physiol. Scand. Suppl.*, **568**, 1–56.

Perry, M. J. and Lawson, S. N. (1998). Differences in expression of oligosaccharides, neuropeptides, carbonic anhydrase and neurofilament in rat primary afferent neurons retrogradely labelled via skin, muscle or visceral nerves. *Neuroscience*, **85**, 293–310.

Persson, P., Ehmke, H., Kirchheim, H. and Seller, H. (1988). Effect of sino-aortic denervation in comparison to cardiopulmonary deafferentiation on long-term blood pressure in conscious dogs. *Pflügers Arch.*, **411**, 160–166.

Petras, J. M. and Cummings, J. F. (1972). Autonomic neurons in the spinal cord of the Rhesus monkey: a correlation of the findings of cytoarchitectonics and sympathectomy with fiber degeneration following dorsal rhizotomy. *J. Comp. Neurol.*, **146**, 189–218.

Petrovich, G. D., Canteras, N. S. and Swanson, L. W. (2001). Combinatorial amygdalar inputs to hippocampal domains and hypothalamic behavior systems. *Brain Res. Brain Res. Rev.*, **38**, 247–289.

Petty, B. G., Cornblath, D. R., Adornato, B. T. et al. (1994). The effect of systemically administered recombinant human nerve growth factor in healthy human subjects. *Ann. Neurol.*, **36**, 244–246.

Phan, K. L., Wager, T., Taylor, S. F. and Liberzon, I. (2002). Functional neuroanatomy of emotion: a meta-analysis of emotion activation studies in PET and fMRI. *Neuroimage*, **16**, 331–348.

Phillips, J. K., Goodchild, A. K., Dubey, R. et al. (2001). Differential expression of catecholamine biosynthetic enzymes in the rat ventrolateral medulla. *J. Comp. Neurol.*, **432**, 20–34.

Phillips, R. J. and Powley, T. L. (2000). Tension and stretch receptors in gastrointestinal smooth muscle: re-evaluating vagal mechanoreceptor electrophysiology. *Brain Res. Brain Res. Rev.*, **34**, 1–26.

Phillips, R. J., Baronowsky, E. A. and Powley, T. L. (1997). Afferent innervation of gastrointestinal tract smooth muscle by the hepatic branch of the vagus. *J. Comp. Neurol.*, **384**, 248–270.

Pick, J. (1970). *The Autonomic Nervous System*. Philadelphia: Lippincott.

Pickering, A. E., Boscan, P. and Paton, J. F. (2003). Nociception attenuates parasympathetic but not sympathetic baroreflex via NK1 receptors in the rat nucleus tractus solitarii. *J. Physiol.*, **551**, 589–599.

Pierce, P. A., Xie, G. X., Peroutka, S. J., Green, P. G. and Levine, J. D. (1995). 5-Hydroxytryptamine-induced synovial plasma extravasation is mediated via 5-hydroxytryptamine2 A receptors on sympathetic efferent terminals. *J. Pharmacol. Exp. Ther.*, **275**, 502–508.

Pilowsky, P. M. and Goodchild, A. K. (2002). Baroreceptor reflex pathways and neurotransmitters: 10 years on. *J. Hypertens.*, **20**, 1675–1688.

Pilowsky, P. M. and Makeham, J. (2001). Juxtacellular labeling of identified neurons: kiss the cells and make them dye. *J. Comp. Neurol.*, **433**, 1–3.

Pilowsky, P. M., Jiang, C. and Lipski, J. (1990). An intracellular study of respiratory neurons in the rostral ventrolateral medulla of the rat and their relationship to catecholamine-containing neurons. *J. Comp. Neurol.*, **301**, 604–617.

Pilowsky, P., Llewellyn-Smith, I. J., Arnolda, L., Minson, J. and Chalmers, J. (1994). Intracellular recording from sympathetic preganglionic neurons in cat lumbar spinal cord. *Brain Res.*, **656**, 319–328.

Pinault, D. (1996). A novel single-cell staining procedure performed in vivo under electrophysiological control: morpho-functional features of juxtacellularly labeled thalamic cells and other central neurons with biocytin or Neurobiotin. *J. Neurosci. Methods*, **65**, 113–136.

Polgar, E., Gray, S., Riddell, J. S. and Todd, A. J. (2004). Lack of evidence for significant neuronal loss in laminae I-III of the spinal dorsal horn of the rat in the chronic constriction injury model. *Pain*, **111**, 144–150.

Polson, J. W., Halliday, G. M., McAllen, R. M., Coleman, M. J. and Dampney, R. A. (1992). Rostrocaudal differences in morphology and neurotransmitter content of cells in the subretrofacial vasomotor nucleus. *J. Auton. Nerv. Syst.*, **38**, 117–138.

Polson, J. W., Potts, P. D., Li, Y. W. and Dampney, R. A. (1995). Fos expression in neurons projecting to the pressor region in the rostral ventrolateral medulla after sustained hypertension in conscious rabbits. *Neuroscience*, **67**, 107–123.

Popescu, L. M., Ciontea, S. M., Cretoiu, D. *et al.* (2005a). Novel type of interstitial cell (Cajal-like) in human fallopian tube. *J. Cell Mol. Med.*, **9**, 479–523.

Popescu, L. M., Hinescu, M. E., Ionescu, N. *et al.* (2005b). Interstitial cells of Cajal in pancreas. *J. Cell Mol. Med.*, **9**, 169–190.

Possas, O. S., Campos, R. R., Jr., Cravo, S. L., Lopes, O. U. and Guertzenstein, P. G. (1994). A fall in arterial blood pressure produced by inhibition of the caudalmost ventrolateral medulla: the caudal pressor area. *J. Auton. Nerv. Syst.*, **49**, 235–245.

Potter, E. K. (1987). Guanethidine blocks neuropeptide-Y-like inhibitory action of sympathetic nerves on cardiac vagus. *J. Auton. Nerv. Syst.*, **21**, 87–90.

Potter, E. K. (1991). Neuropeptide Y as an autonomic neurotransmitter. In *Novel Peripheral Neurotransmitter*, ed. C. Bell. New York: Pergamon, pp. 81–112.

Powley, T. L., Berthoud, H. R., Fox, A. P. and Laughton, W. (1992). The dorsal vagal complex forms a sensory-motor lattice: the circuitry of gastrointestinal reflexes. In *Neuroanatomy and Physiology of Abdominal Vagal Afferents*, ed. S. Ritter, R. C. Ritter and C. D. Barnes. Boca Raton: CRC Press, pp. 55–79.

Prechtl, J. C. and Powley, T. L. (1985). Organization and distribution of the rat subdiaphragmatic vagus and associated paraganglia. *J. Comp. Neurol.*, **235**, 182–195.

Prechtl, J. C. and Powley, T. L. (1990a). The fiber composition of the abdominal vagus of the rat. *Anat. Embryol. (Berl.)*, **181**, 101–115.

Prechtl, J. C. and Powley, T. L. (1990b). B-afferents: a fundamental division of the nervous system mediating homeostasis [the article includes Open Peer Commentary]. *Behav. Brain Sci.*, **13**, 289–331.

Purves, D. (1988). *Body and Brain: A Tropic Theory of Neural Connection*. Cambridge: Harvard University Press.

Purves, D. and Hume, R. I. (1981). The relation of postsynaptic geometry to the number of presynaptic axons that innervate autonomic ganglion cells. *J. Neurosci.*, **1**, 441–452.

Purves, D. and Lichtman, J. W. (1978). Formation and maintenance of synaptic connections in autonomic ganglia. *Physiol. Rev.*, **58**, 821–862.

Purves, D. and Lichtman, J. W. (1985). Geometrical differences among homologous neurons in mammals. *Science*, **228**, 298–302.

Purves, D. and Wigston, D. J. (1983). Neural units in the superior cervical ganglion of the guinea-pig. *J. Physiol.*, **334**, 169–178.

Purves, D., Rubin, E., Snider, W. D. and Lichtman, J. (1986). Relation of animal size to convergence, divergence, and neuronal number in peripheral sympathetic pathways. *J. Neurosci.*, **6**, 158–163.

Purves, D., Snider, W. D. and Voyvodic, J. T. (1988). Trophic regulation of nerve cell morphology and innervation in the autonomic nervous system. *Nature*, **336**, 123–128.

Pyner, S. and Coote, J. H. (1994). Evidence that sympathetic preganglionic neurons are arranged in target-specific columns in the thoracic spinal cord of the rat. *J. Comp. Neurol.*, **342**, 15–22.

Pyner, S. and Coote, J. H. (1998). Rostroventrolateral medulla neurons preferentially project to target-specified sympathetic preganglionic neurons. *Neuroscience*, **83**, 617–631.

Qin, C., Chandler, M. J., Miller, K. E. and Foreman, R. D. (2001). Responses and afferent pathways of superficial and deeper c(1)–c(2) spinal cells to intrapericardial algogenic chemicals in rats. *J. Neurophysiol.*, **85**, 1522–1532.

Quigg, M., Elfvin, L. G. and Aldskogius, H. (1990). Anterograde transsynaptic transport of WGA-HRP from spinal afferents to postganglionic sympathetic cells of the stellate ganglion of the guinea pig. *Brain Res.*, **518**, 173–178.

Randall, W. C. (ed.) (1984). *Nervous Control of Cardiovascular Function*. New York, Oxford: Oxford University Press.

Randall, W. C., Wurster, R. D. and Lewin, R. J. (1966). Responses of patients with high spinal transection to high ambient temperatures. *J. Appl. Physiol.*, **21**, 985–993.

Randich, A. and Gebhart, G. F. (1992). Vagal afferent modulation of nociception. *Brain Res. Rev.*, **17**, 77–99.

Ranson, S. W. and Clark, S. L. (1959). *The Anatomy of the Nervous System*. 10th edn. Philadelphia, London: W. B. Saunders Company.

Ranson, S. W. and Magoun, H. W. (1939). The hypothalamus. *Ergebn. Physiol.*, **41**, 56–163.

Rathner, J. A. and McAllen, R. M. (1999). Differential control of sympathetic drive to the rat tail artery and kidney by medullary premotor cell groups. *Brain Res.*, **834**, 196–199.

Rathner, J. A., Owens, N. C. and McAllen, R. M. (2001). Cold-activated raphespinal neurons in rats. *J. Physiol.*, **535**, 841–854.

Reed, D. E. and Vanner, S. J. (2003). Long vasodilator reflexes projecting through the myenteric plexus in guinea-pig ileum. *J. Physiol.*, **553**, 911–924.

Reiner, A., Karten, H. J., Gamlin, P. D. R. and Erichsen, J. T. (1983). Parasympathetic ocular control. Functional subdivisions and circuitry of the avian nucleus Edinger–Westphal. *Trends Neurosci.*, **6**, 140–145.

Reis, D. J., Golanov, E. V., Ruggiero, D. A. and Sun, M. K. (1994). Sympathoexcitatory neurons of the rostral ventrolateral medulla are oxygen sensors and essential elements in the tonic and reflex control of the systemic and cerebral circulations. *J. Hypertens. Suppl.*, **12**, S159–S180.

Reis, D. J., Golanov, E. V., Galea, E. and Feinstein, D. L. (1997). Central neurogenic neuroprotection: central neural systems that protect the brain from hypoxia and ischemia. *Ann. New York Acad. Sci.*, **835**, 168–186.

Rekling, J. C. and Feldman, J. L. (1998). PreBotzinger complex and pacemaker neurons: hypothesized site and kernel for respiratory rhythm generation. *Annu. Rev. Physiol.*, **60**, 385–405.

Rexed, B. (1952). The cytoarchitectonic organization of the spinal cord in the cat. *J. Comp. Neurol.*, **96**, 414–495.

Rexed, B. (1954). A cytoarchitectonic atlas of the spinal cord in the cat. *J. Comp. Neurol.*, **100**, 297–379.

Richards, W., Hillsley, K., Eastwood, C. and Grundy, D. (1996). Sensitivity of vagal mucosal afferents to cholecystokinin and its role in afferent signal transduction in the rat. *J. Physiol.*, **497**, 473–481.

Richter, D. W. (1982). Generation and maintenance of the respiratory rhythm. *J. Exp. Biol.*, **100**, 93–107.

Richter, D. W. (1996). Neural regulation of respiration: rhythmogenesis and afferent control. In *Comprehensive Human Physiology*, ed. R. Greger and U. Windhorst. Berlin, Heidelberg, New York: Springer, pp. 2079–2095.

Richter, D. W. and Ballantyne, D. (1983). A three phase theory about the basic respiratory pattern generator. In *Central Neurone Environment*, ed. M. E. Schläfke, H. P. Koepchen and W. R. See. Berlin, Heidelberg, New York: Springer, pp. 164–174.

Richter, D. W. and Spyer, K. M. (1990). Cardiorespiratory control. In *Central Regulation of Autonomic Functions*, eds. A. D. Loewy and K. M. Spyer. New York, Oxford: Oxford University Press, pp. 189–207.

Richter, D. W. and Spyer, K. M. (2001). Studying rhythmogenesis of breathing: comparison of in vivo and in vitro models. *Trends Neurosci.*, **24**, 464–472.

Richter, D. W., Spyer, K. M., Gilbey, M. P. et al. (1991). On the existence of a common cardiorespiratory network. In *Cardiorespiratory and Motor Coordination*, ed. H. P. Koepchen and T. Huopaniemi. Berlin, Heidelberg, New York: Springer, pp. 118–130.

Rinaman, L., Card, J. P., Schwaber, J. S. and Miselis, R. R. (1989). Ultrastructural demonstration of a gastric monosynaptic vagal circuit in the nucleus of the solitary tract in rat. *J. Neurosci.*, **9**, 1985–1996.

Risold, P. Y., Thompson, R. H. and Swanson, L. W. (1997). The structural organization of connections between hypothalamus and cerebral cortex. *Brain Res. Brain Res. Rev.*, **24**, 197–254.

Ritter, S., Ritter, R. C. and Barnes, C. D. (eds.) (1992). *Neuroanatomy and Physiology of Abdominal Vagal Afferents*. Boca Raton: CRC Press.

Robertson, D. and Biaggioni, I. (eds.) (1995). *The Autonomic Nervous System*, Vol. 5, *Disorders of the Autonomic Nervous System*. Luxembourg: Harwood Academic Publishers.

Roddie, I. C. (1977). Human responses to emotional stress. *Irish J. Med. Sci.*, **146**, 395–417.

Roddie, I. C., Shepherd, J. T. and Whelan, R. F. (1957). The contribution of constrictor and dilator nerves to the skin vasodilation during body heating. *J. Physiol.*, **136**, 489–497.

Rogers, R. C. and Hermann, G. E. (1992). Central regulation of brainstem gastric vago-vagal control circuits. In *Neuroanatomy and Physiology of Abdominal Vagal Afferents*, ed. S. Ritter, R. C. Ritter and C. D. Barnes. Boca Raton: CRC Press, pp. 99–134.

Rogers, R. C., Kita, H., Butcher, L. L. and Novin, D. (1980). Afferent projections to the dorsal motor nucleus of the vagus. *Brain Res. Bull.*, **5**, 365–373.

Rogers, R. C., McTigue, D. M. and Hermann, G. E. (1995). Vagovagal reflex control of digestion: afferent modulation by neural and "endoneurocrine" factors. *Am. J. Physiol.*, **268**, G1–G10.

Rogers, R. C., Hermann, G. E. and Travagli, R. A. (1999). Brainstem pathways responsible for oesophageal control of gastric motility and tone in the rat. *J. Physiol.*, **514**, 369–383.

Rogers, R. C., Travagli, R. A. and Hermann, G. E. (2003). Noradrenergic neurons in the rat solitary nucleus participate in the esophageal-gastric relaxation reflex. *Am. J. Physiol. Regul. Integr. Comp. Physiol.*, **285**, R479–R489.

Rogers, R. F., Paton, J. F. and Schwaber, J. S. (1993). NTS neuronal responses to arterial pressure and pressure changes in the rat. *Am. J. Physiol.*, **265**, R1355–R1368.

Roman, C. and Gonella, J. (1994). Extrinsic control of digestive tract motility. In *Physiology of the Gastrointestinal Tract*, 3rd edn., ed. L. R. Johnson. New York: Raven Press, pp. 507-553.

Romano, T. A., Felten, S. Y., Felten, D. L. and Olschowka, J. A. (1991). Neuropeptide-Y innervation of the rat spleen: another potential immunomodulatory neuropeptide. *Brain Behav. Immun.*, **5**, 116-131.

Root, W. S. and Bard, P. (1947). The mediation of feline erection through sympathetic pathways with some remarks on sexual behavior after deafferentation of the genitalia. *Am. J. Physiol.*, **151**, 80-90.

Rosell, S. (1980). Neuronal control of microvessels. *Annu. Rev. Physiol.*, **42**, 359-371.

Rosell, S. and Belfrage, E. (1979). Blood circulation in adipose tissue. *Physiol. Rev.*, **59**, 1078-1104.

Rosicka, M., Krsek, M., Jarkovska, Z., Marek, J. and Schreiber, V. (2002). Ghrelin – a new endogenous growth hormone secretagogue. *Physiol. Res.*, **51**, 435-441.

Ross, C. A., Ruggiero, D. A., Park, D. H. et al. (1984b). Tonic vasomotor control by the rostral ventrolateral medulla: effect of electrical or chemical stimulation of the area containing C1 adrenaline neurons on arterial pressure, heart rate, and plasma catecholamines and vasopressin. *J. Neurosci.*, **4**, 474-494.

Ross, C. A., Ruggiero, D. A., Joh, T. H., Park, D. H. and Reis, D. J. (1984a). Rostral ventrolateral medulla: selective projections to the thoracic autonomic cell column from the region containing C1 adrenaline neurons. *J. Comp. Neurol.*, **228**, 168-185.

Rossiter, C. D., Norman, W. P., Jain, M. et al. (1990). Control of lower esophageal sphincter pressure by two sites in dorsal motor nucleus of the vagus. *Am. J. Physiol.*, **259**, G899-G906.

Rowell, L. B. (1993). *Human Cardiovascular Control*. New York, Oxford: Oxford University Press.

Ruggiero, D. A., Cravo, S. L., Arango, V. and Reis, D. J. (1989). Central control of the circulation by the rostral ventrolateral reticular nucleus: anatomical substrates. *Prog. Brain Res.*, **81**, 49-79.

Rushmer, R. F. and Smith, O. A., Jr. (1959). Cardiac control. *Physiol. Rev.*, **39**, 41-68.

Rybak, I. A., Paton, J. F. R., Rogers, R. F. and St.-John, W. M. (2002). Generation of the respiratory rhythm: state-dependency and switching. *Neurocomputing*, **44-46**, 605-614.

Rybak, I. A., Shevtsova, N. A., St.-John, W. M., Paton, J. F. and Pierrefiche, O. (2003). Endogenous rhythm generation in the pre-Bötzinger complex and ionic currents: modelling and in vitro studies. *Eur. J. Neurosci.*, **18**, 239-257.

Rybak, I. A., Shevtsova, N. A., Paton, J. F. et al. (2004). Modeling the pontomedullary respiratory network. *Respir. Physiol. Neurobiol.*, **143**, 307-319.

Sanders, K. M. and Smith, T. K. (2003). Neural regulation of colonic motor function. In *Textbook of Colonic Disease*, ed. T. Koch. Totowa, New Jersey, USA: Humana Press, Inc., pp. 35-52.

Sanders, K. M., Ordog, T., Koh, S. D., Torihashi, S. and Ward, S. M. (1999). Development and plasticity of interstitial cells of Cajal. *Neurogastroenterol. Motil.*, **11**, 311-338.

Sanders, K. M., Ordog, T., Koh, S. D. and Ward, S. M. (2000). A novel pacemaker mechanism drives gastrointestinal rhythmicity. *News Physiol. Sci.*, **15**, 291-298.

Santicioli, P. and Maggi, C. A. (1998). Myogenic and neurogenic factors in the control of pyeloureteral motility and ureteral peristalsis. *Pharmacol. Rev.*, **50**, 683–722.

Saper, C. B. (1995). Central autonomic system. In *The Rat Nervous System*, ed. G. Paxinos. San Diego: Academic Press, pp. 107–135.

Saper, C. B. (2002). The central autonomic nervous system: conscious visceral perception and autonomic pattern generation. *Annu. Rev. Neurosci.*, **25**, 433–469.

Saper, C. B. (2004). Anatomy of hypothalamus. In *The Human Nervous System*, ed. G. Paxinos and J. K. Mai. Amsterdam: Elsevier Academic Press, pp. 513–550.

Saphier, D. (1993). Psychoimmunology: the missing link. In *Hormonally Induced Changes in Mind and Brain*, ed. J. Schulkin. Boston, New York: Academic Press, pp. 191–224.

Sato, A. (1972). Somato-sympathetic reflex discharges evoked through supramedullary pathways. *Pflügers Arch.*, **332**, 117–126.

Sato, A. and Schmidt, R. F. (1971). Spinal and supraspinal components of the reflex discharges into lumbar and thoracic white rami. *J. Physiol.*, **212**, 839–850.

Sato, A. and Schmidt, R. F. (1973). Somatosympathetic reflexes: afferent fibers, central pathways, discharge characteristics. *Physiol. Rev.*, **53**, 916–947.

Sato, A., Sato, Y. and Schmidt, R. F. (1997). The impact of somatosensory input on autonomic functions. *Rev. Physiol. Biochem. Pharmacol.*, **130**, 1–328.

Schaible, H. G. and Schmidt, R. F. (1988). Time course of mechanosensitivity changes in articular afferents during a developing experimental arthritis. *J. Neurophysiol.*, **60**, 2180–2195.

Schelegle, E. S. and Green, J. F. (2001). An overview of the anatomy and physiology of slowly adapting pulmonary stretch receptors. *Respir. Physiol.*, **125**, 17–31.

Schläfke, M. E. and Loeschke, H. H. (1967). Lokalisation eines an der Regulation von Atmung und Kreislauf beteiligten Gebietes an der ventralen Oberfläche der Medulla oblongata durch Kälteblockade [Localization of a center on the ventral surface of the medulla oblongata involved in regulation of respiratory and cardiovascular system by cold blockade]. *Pflügers Arch.*, **297**, 201–220.

Schmelz, M., Michael, K., Weidner, C. et al. (2000). Which nerve fibers mediate the axon reflex flare in human skin? *NeuroReport*, **11**, 645–648.

Schmidt, R., Schmelz, M., Forster, C. et al. (1995). Novel classes of responsive and unresponsive C nociceptors in human skin. *J. Neurosci.*, **15**, 333–341.

Schmidt, R., Schmelz, M., Torebjörk, H. E. and Handwerker, H. O. (2000). Mechano-insensitive nociceptors encode pain evoked by tonic pressure to human skin. *Neuroscience*, **98**, 793–800.

Schmidt, R., Schmelz, M., Weidner, C., Handwerker, H. O. and Torebjörk, H. E. (2002). Innervation territories of mechano-insensitive C nociceptors in human skin. *J. Neurophysiol.*, **88**, 1859–1866.

Schmidt-Vanderheyden, W. and Koepchen, H. P. (1967). Zum Mechanismus der Adrenalindilatation der Skeletmuskelgefäße [On the mechanism of adrenaline dilation of skeletal muscle blood vessels]. *Pflügers Arch.*, **298**, 1–11.

Schomburg, E. D. (1990). Spinal sensorimotor systems and their supraspinal control. *Neurosci. Res.*, **7**, 265–340.

Schramm, L. P., Strack, A. M., Platt, K. B. and Loewy, A. D. (1993). Peripheral and central pathways regulating the kidney: a study using pseudorabies virus. *Brain Res.*, **616**, 251–262.

Schreihofer, A. M. and Guyenet, P. G. (1997). Identification of C1 presympathetic neurons in rat rostral ventrolateral medulla by juxtacellular labeling in vivo. *J. Comp. Neurol.*, **387**, 524–536.

Schreihofer, A. M. and Guyenet, P. G. (2000). Sympathetic reflexes after depletion of bulbospinal catecholaminergic neurons with anti-DβH-saporin. *Am. J. Physiol. Regul. Integr. Comp. Physiol.*, **279**, R729–R742.

Schreihofer, A. M. and Guyenet, P. G. (2002). The baroreflex and beyond: control of sympathetic vasomotor tone by GABAergic neurons in the ventrolateral medulla. *Clin. Exp. Pharmacol. Physiol.*, **29**, 514–521.

Schreihofer, A. M. and Guyenet, P. G. (2003). Baro-activated neurons with pulse-modulated activity in the rat caudal ventrolateral medulla express GAD67 mRNA. *J. Neurophysiol.*, **89**, 1265–1277.

Schreihofer, A. M., Stornetta, R. L. and Guyenet, P. G. (1999). Evidence for glycinergic respiratory neurons: Bötzinger neurons express mRNA for glycinergic transporter 2. *J. Comp. Neurol.*, **407**, 583–597.

Schreihofer, A. M., Stornetta, R. L. and Guyenet, P. G. (2000). Regulation of sympathetic tone and arterial pressure by rostral ventrolateral medulla after depletion of C1 cells in rat. *J. Physiol.*, **529**, 221–236.

Schulkin, J. (2003a). Allostasis: a neural behavioral perspective. *Horm. Behav.*, **43**, 21–27.

Schulkin, J. (2003b). *Rethinking Homeostasis. Allostatic Regulation in Physiology and Pathophysiology*. Cambridge Massachusetts: The MIT Press.

Schwartz, M. W. and Morton, G. J. (2002). Obesity: keeping hunger at bay. *Nature*, **418**, 595–597.

Schwarzacher, S. W., Wilhelm, Z., Anders, K. and Richter, D. W. (1991). The medullary respiratory network in the rat. *J. Physiol.*, **435**, 631–644.

Seagard, J. L., Gallenberg, L. A., Hopp, F. A. and Dean, C. (1992). Acute resetting in two functionally different types of carotid baroreceptors. *Circ. Res.*, **70**, 559–565.

Seals, D. R., Suwarno, N. O. and Dempsey, J. A. (1990). Influence of lung volume on sympathetic nerve discharge in normal humans. *Circ. Res.*, **67**, 130–141.

Seals, D. R., Suwarno, N. O., Joyner, M. J. *et al.* (1993). Respiratory modulation of muscle sympathetic nerve activity in intact and lung denervated humans. *Circ. Res.*, **72**, 440–454.

Seller, H., Langhorst, P., Richter, D. and Koepchen, H. P. (1968). Über die Abhängigkeit der pressoreceptorischen Hemmung des Sympathicus von der Atemphase und ihre Auswirkung in der Vasomotorik [Respiratory variations of baroreceptor reflex transmission and their effects on sympathetic activity and vasomotor tone]. *Pflügers Arch.*, **302**, 300–314.

Selyanko, A. A. (1992). Membrane properties and firing characteristics of rat cardiac neurones in vitro. *J. Auton. Nerv. Syst.*, **39**, 181–189.

Semans, J. H. and Langworthy, O. R. (1938). Observations on the neurophysiology of sexual function in the male cat. *J. Urol.*, **40**, 836–846.

Sengupta, J. N. and Gebhart, G. F. (1994a). Characterization of mechanosensitive pelvic nerve afferent fibers innervating the colon of the rat. *J. Neurophysiol.*, **71**, 2046–2060.

Sengupta, J. N. and Gebhart, G. F. (1994b). Mechanosensitive properties of pelvic nerve afferent fibers innervating the urinary bladder of the rat. *J. Neurophysiol.*, **72**, 2420–2430.

Sengupta, J. N. and Gebhart, G. F. (1995). Mechanosensitive afferent fibers in the gastrointestinal and lower urinary tracts. In *Visceral Pain*, ed. G. F. Gebhart. Seattle: IASP Press, pp. 75–98.

Sengupta, J. N., Saha, J. K. and Goyal, R. K. (1990). Stimulus–response function studies of esophageal mechanosensitive nociceptors in sympathetic afferents of opossum. *J. Neurophysiol.*, **64**, 796–812.

Sengupta, J. N., Su, X. and Gebhart, G. F. (1996). Kappa, but not mu or delta, opioids attenuate responses to distention of afferent fibers innervating the rat colon. *Gastroenterology*, **111**, 968–980.

Shade, R. E., Bishop, V. S., Haywood, J. R. and Hamm, C. K. (1990). Cardiovascular and neuroendocrine responses to baroreceptor denervation in baboons. *Am. J. Physiol.*, **258**, R930–R938.

Shafton, A. D., Oldfield, B. J. and McAllen, R. M. (1992). CRF-like immunoreactivity selectively labels preganglionic sudomotor neurons in cat. *Brain Res.*, **599**, 253–260.

Shah, S. D., Tse, T. F., Clutter, W. E. and Cryer, P. E. (1984). The human sympathochromaffin system. *Am. J. Physiol.*, **247**, E380–E384.

Shanahan, F. (1994). The intestinal immune system. In *Physiology of the Gastrointestinal Tract*, 3rd edn., ed. L. R. Johnson. New York: Raven Press, pp. 643–684.

Sharkey, K. A. and Mawe, G. M. (2002). Neuroimmune and epithelial interactions in intestinal inflammation. *Curr. Opin. Pharmacol.*, **2**, 669–677.

Sheehan, D. (1936). Discovery of the autonomic nervous system. *Arch. Neurol. Psychiat.*, **35**, 1081–1115.

Sheehan, D. (1941). The autonomic nervous system prior to Gaskell. *New Engl. J. Med.*, **224**, 457–460.

Shefchyk, S. J. (2001). Sacral spinal interneurons and the control of urinary bladder and urethral striated sphincter muscle function. *J. Physiol.*, **533**, 57–63.

Shefchyk, S. J. (2002). Spinal cord neural organization controlling the urinary bladder and striated sphincter. *Prog. Brain Res.*, **137**, 71–82.

Shepherd, J. Z. and Vatner, S. F. (eds.) (1996). *Nervous Control of the Heart.* Vol. 9, *The Autonomic Nervous System*, Amsterdam: Harwood Academic Publishers.

Sherbourne, C. D., Gonzales, R., Goldyne, M. E. and Levine, J. D. (1992). Norepinephrine-induced increase in sympathetic neuron-derived prostaglandins is independent of neuronal release mechanisms. *Neurosci. Lett.*, **139**, 188–190.

Sherrington, C. S. (1900). Cutaneous sensation. In *Textbook of Physiology*, Vol. 2, ed. E. A. Schäfer. Edinburgh, London: Young J. Pentland, pp. 920–1001.

Sherrington, C. (1947) [1906]. *The Integrative Action of the Nervous System*, 2nd edn. New Haven: Yale University Press.

Sillito, A. M. and Zbrozyna, A. W. (1970a). The localization of pupilloconstrictor function within the mid-brain of the cat. *J. Physiol.*, **211**, 461–477.

Sillito, A. M. and Zbrozyna, A. W. (1970b). The activity characteristics of the preganglionic pupilloconstrictor neurones. *J. Physiol.*, **211**, 767–779.

Silva-Carvalho, L., Paton, J. F., Goldsmith, G. E. and Spyer, K. M. (1991). The effects of electrical stimulation of lobule IXb of the posterior cerebellar vermis on neurones within the rostral ventrolateral medulla in the anaesthetised cat. *J. Auton. Nerv. Syst.*, **36**, 97–106.

Silverberg, A. B., Shah, S. D., Haymond, M. W. and Cryer, P. E. (1978). Norepinephrine: hormone and neurotransmitter. *Am. J. Physiol.*, **234**, E252–E256.

Simon, E. (1974). Temperature regulation: the spinal cord as a site of extrahypothalamic thermoregulatory functions. *Rev. Physiol. Biochem. Pharmacol.*, **71**, 1–76.

Simon, E., Pierau, F. K. and Taylor, D. C. (1986). Central and peripheral thermal control of effectors in homeothermic temperature regulation. *Physiol. Rev.*, **66**, 235–300.

Simone, D. A., Sorkin, L. S., Oh, U. et al. (1991). Neurogenic hyperalgesia: central neural correlates in responses of spinothalamic tract neurons. *J. Neurophysiol.*, **66**, 228–246.

Sjövall, H., Jodal, M. and Lundgren, O. (1987). Sympathetic control of intestinal fluid and electrolyte transport. *News Physiol. Sci.*, **2**, 214–217.

Skok, V. I. (1986). Spontaneous and reflex activities: general characteristics. In *Autonomic and Enteric Ganglia*, ed. K. Karczmar, K. Koketsu and S. Nishi. New York, London: Plenum Press, pp. 425–438.

Skok, V. I. (2002). Nicotinic acetylcholine receptors in autonomic ganglia. *Auton. Neurosci.*, **97**, 1–11.

Skok, V. I. and Ivanov, A. Y. (1983). What is the ongoing activity of sympathetic neurons? *J. Auton. Nerv. Syst.*, **7**, 263–270.

Skok, V. I. and Ivanov, A. Y. (1987). Organization of presynaptic input to neurones of a sympathetic ganglion. In *Organization of the Autonomic Nervous System: Central and Peripheral Mechanisms*, ed. J. Ciriello, F. R. Calaresu, L. P. Renaud and C. Polosa. New York: Alan Liss, pp. 37–46.

Skok, V. I. and Ivanov, A. Y. (1989). естественная активность вегетативных ганглиев [Natural activity in autonomic ganglia]. Kiev: Publisher Naukova Duma.

Sly, D. J., Colvill, L., McKinley, M. J. and Oldfield, B. J. (1999). Identification of neural projections from the forebrain to the kidney, using the virus pseudorabies. *J. Auton. Nerv. Syst.*, **77**, 73–82.

Smeyne, R. J., Klein, R., Schnapp, A. et al. (1994). Severe sensory and sympathetic neuropathies in mice carrying a disrupted Tr/NGF receptor gene. *Nature*, **368**, 246–249.

Smith, J. C., Ellenberger, H. H., Ballanyi, K., Richter, D. W. and Feldman, J. L. (1991). Pre-Bötzinger complex: a brain stem region that may generate respiratory rhythm in mammals. *Science*, **254**, 726–729.

Smith, J. C., Butera, R. J., Koshiya, N. et al. (2000). Respiratory rhythm generation in neonatal and adult mammals: the hybrid pacemaker-network model. *Respir. Physiol.*, **122**, 131–147.

Smith, J. E., Jansen, A. S., Gilbey, M. P. and Loewy, A. D. (1998). CNS cell groups projecting to sympathetic outflow of tail artery: neural circuits involved in heat loss in the rat. *Brain Res.*, **786**, 153–164.

Smith, O. A. and DeVito, J. L. (1984). Central neural integration for the control of autonomic responses associated with emotion. *Annu. Rev. Neurosci.*, **7**, 43–65.

Smith, O. A., Hohimer, A. R., Astley, C. A. and Taylor, D. J. (1979). Renal and hindlimb vascular control during acute emotion in the baboon. *Am. J. Physiol.*, **236**, R198–R205.

Smith, O. A., Astley, C. A., DeVito, J. L., Stein, J. M. and Walsh, K. E. (1980). Functional analysis of hypothalamic control of the cardiovascular responses accompanying emotional behavior. *Fed. Proc.*, **39**, 2487–2494.

Smith, O. A., Astley, C. A., Spelman, F. A. *et al.* (2000). Cardiovascular responses in anticipation of changes in posture and locomotion. *Brain Res. Bull.*, **53**, 69–76.

Smith, R. E. and Horwitz, B. A. (1969). Brown fat and thermogenesis. *Physiol. Rev.*, **49**, 330–425.

Smith, T. K., Oliver, G. R., Hennig, G. W. *et al.* (2003). A smooth muscle tone-dependent stretch-activated migrating motor pattern in isolated guinea-pig distal colon. *J. Physiol.*, **551**, 955–969.

Söderholm, J. D. and Perdue, M. H. (2001). Stress and gastrointestinal tract. II. Stress and intestinal barrier function. *Am. J. Physiol. Gastrointest. Liver Physiol.*, **280**, G7–G13.

Sonnenschein, R. R. and Weissman, M. L. (1978). Sympathetic vasomotor outflows to hindlimb muscles of the cat. *Am. J. Physiol.*, **235**, H482–H487.

Spencer, N. J. and Smith, T. K. (2001). Simultaneous intracellular recordings from longitudinal and circular muscle during the peristaltic reflex in guinea-pig distal colon. *J. Physiol.*, **533**, 787–799.

Spencer, N. J. and Smith, T. K. (2004). Mechanosensory S-neurons rather than AH-neurons appear to generate a rhythmic motor pattern in guinea-pig distal colon. *J. Physiol.*, **558**, 577–596.

Spencer, N., Walsh, M. and Smith, T. K. (1999). Does the guinea-pig ileum obey the 'law of the intestine'? *J. Physiol.*, **517**, 889–898.

Spencer, N. J., Hennig, G. W. and Smith, T. K. (2002). A rhythmic motor pattern activated by circumferential stretch in guinea-pig distal colon. *J. Physiol.*, **545**, 629–648.

Spencer, N. J., Hennig, G. W. and Smith, T. K. (2003a). Stretch-activated neuronal pathways to longitudinal and circular muscle in guinea pig distal colon. *Am. J. Physiol. Gastrointest. Liver Physiol.*, **284**, G231–G241.

Spencer, N. J., Sanders, K. M. and Smith, T. K. (2003b). Migrating motor complexes do not require electrical slow waves in the mouse small intestine. *J. Physiol.*, **553**, 881–893.

Spencer, S. E., Sawyer, W. B., Wada, H., Platt, K. B. and Loewy, A. D. (1990). CNS projections to the pterygopalatine parasympathetic preganglionic neurons in the rat: a retrograde transneuronal viral cell body labeling study. *Brain Res.*, **534**, 149–169.

Spike, R. C., Puskar, Z., Andrew, D. and Todd, A. J. (2003). A quantitative and morphological study of projection neurons in lamina I of the rat lumbar spinal cord. *Eur. J. Neurosci.*, **18**, 2433–2448.

Spillane, D. M. (1981). *The Doctrine of the Nerves*. Oxford: Oxford University Press.

Spyer, K. M. (1981). Neural organisation and control of the baroreceptor reflex. *Rev. Physiol. Biochem. Pharmacol.*, **88**, 24–124.

Spyer, K. M. (1994). Central nervous mechanisms contributing to cardiovascular control. *J. Physiol.*, **474**, 1–19.

Squire, L. R., Bloom, F. E., McConnell, S. R. *et al.* (eds.) (2003). *Fundamental Neuroscience*, 2nd edn. San Diego: Academic Press.

St.-John, W. M. (1998). Neurogenesis of patterns of automatic ventilatory activity. *Prog. Neurobiol.*, **56**, 97–117.

St.-John, W. M. and Paton, J. F. (2003). Defining eupnea. *Respir. Physiol. Neurobiol.*, **139**, 97–103.

St.-John, W. M. and Paton, J. F. (2004). Role of pontile mechanisms in the neurogenesis of eupnea. *Respir. Physiol. Neurobiol.*, **143**, 321–332.

Stanton-Hicks, M., Jänig, W., Hassenbusch, S. *et al.* (1995). Reflex sympathetic dystrophy: changing concepts and taxonomy. *Pain*, **63**, 127–133.

Staras, K., Chang, H. S. and Gilbey, M. P. (2001). Resetting of sympathetic rhythm by somatic afferents causes post-reflex coordination of sympathetic activity in rat. *J. Physiol.*, **533**, 537–545.

Stein, R. D. and Weaver, L. C. (1988). Multi- and single-fibre mesenteric and renal sympathetic responses to chemical stimulation of intestinal receptors in cats. *J. Physiol.*, **396**, 155–172.

Sterling, P. and Eyer, J. (1988). Allostasis: a new paradigm to explain arousal pathology. In *Handbook of Life Stress, Cognition and Health*, ed. S. Fisher and J. Reason. New York: Wiley, pp. 629–649.

Stjernberg, L. and Wallin, B. G. (1983). Sympathetic neural outflow in spinal man. A preliminary report. *J. Auton. Nerv. Syst.*, **7**, 313–318.

Stjernberg, L., Blumberg, H. and Wallin, B. G. (1986). Sympathetic activity in man after spinal cord injury. Outflow to muscle below the lesion. *Brain*, **109**, 695–715.

Stornetta, R. L., Schreihofer, A. M., Pelaez, N. M., Sevigny, C. P. and Guyenet, P. G. (2001). Preproenkephalin mRNA is expressed by C1 and non-C1 barosensitive bulbospinal neurons in the rostral ventrolateral medulla of the rat. *J. Comp. Neurol.*, **435**, 111–126.

Stornetta, R. L., Sevigny, C. P., Schreihofer, A. M., Rosin, D. L. and Guyenet, P. G. (2002). Vesicular glutamate transporter DNPI/VGLUT2 is expressed by both C1 adrenergic and nonaminergic presympathetic vasomotor neurons of the rat medulla. *J. Comp. Neurol.*, **444**, 207–220.

Stornetta, R. L., McQuiston, T. J. and Guyenet, P. G. (2004). GABAergic and glycinergic presympathetic neurons of rat medulla oblongata identified by retrograde transport of pseudorabies virus and in situ hybridization. *J. Comp. Neurol.*, **479**, 257–270.

Strack, A. M. and Loewy, A. D. (1990). Pseudorabies virus: a highly specific transneuronal cell body marker in the sympathetic nervous system. *J. Neurosci.*, **10**, 2139–2147.

Strack, A. M., Sawyer, W. B., Marubio, L. M. and Loewy, A. D. (1988). Spinal origin of sympathetic preganglionic neurons in the rat. *Brain Res.*, **455**, 187–191.

Strack, A. M., Sawyer, W. B., Hughes, J. H., Platt, K. B. and Loewy, A. D. (1989a). A general pattern of CNS innervation of the sympathetic outflow demonstrated by transneuronal pseudorabies viral infections. *Brain Res.*, **491**, 156–162.

Strack, A. M., Sawyer, W. B., Platt, K. B. and Loewy, A. D. (1989b). CNS cell groups regulating the sympathetic outflow to adrenal gland as revealed by transneuronal cell body labeling with pseudorabies virus. *Brain Res.*, **491**, 274–296.

Strauther, G. R., Longo, W. E., Virgo, K. S. and Johnson, F. E. (1999). Appendicitis in patients with previous spinal cord injury. *Am. J. Surg.*, **178**, 403–405.

Stricker, E. M. and Verbalis, J. G. (2003). Water intake and body fluids. In *Fundamental Neuroscience*, 2nd edn., ed. L. R. Squire, F. E. Bloom, S. K. McConnell *et al.* San Diego: Academic Press, pp. 1011–1029.

Strickland, J. H. and Calhoun, M. L. (1963). The integumentary system of the cat. *Am. J. Vet. Res.*, **24**, 1019–1029.

Su, X. and Gebhart, G. F. (1998). Mechanosensitive pelvic nerve afferent fibers innervating the colon of the rat are polymodal in character. *J. Neurophysiol.*, **80**, 2632–2644.

Su, X., Sengupta, J. N. and Gebhart, G. F. (1997a). Effects of kappa opioid receptor-selective agonists on responses of pelvic nerve afferents to noxious colorectal distension. *J. Neurophysiol.*, **78**, 1003–1012.

Su, X., Sengupta, J. N. and Gebhart, G. F. (1997b). Effects of opioids on mechanosensitive pelvic nerve afferent fibers innervating the urinary bladder of the bat. *J. Neurophysiol.*, **77**, 1566–1580.

Sugenoya, J., Iwase, S., Mano, T. et al. (1998). Vasodilator component in sympathetic nerve activity destined for the skin of the dorsal foot of mildly heated humans. *J. Physiol.*, **507**, 603–610.

Sugiura, Y., Terui, N. and Hosoa, Y. (1989). Differences in the distribution of central terminals between visceral and somatic unmyelinated primary afferent fibers. *J. Neurophysiol.*, **62**, 834–847.

Sun, M. K. (1995). Central neural organization and control of sympathetic nervous system in mammals. *Prog. Neurobiol.*, **47**, 157–233.

Sun, M. K. and Guyenet, P. G. (1985). GABA-mediated baroreceptor inhibition of reticulospinal neurons. *Am. J. Physiol.*, **249**, R672–R680.

Sun, M. K. and Reis, D. J. (1996). Medullary vasomotor activity and hypoxic sympathoexcitation in pentobarbital-anesthetized rats. *Am. J. Physiol.*, **270**, R348–R355.

Sun, M. K. and Spyer, K. M. (1991). Nociceptive inputs into rostral ventrolateral medulla-spinal vasomotor neurones in rats. *J. Physiol.*, **436**, 685–700.

Sun, M. K., Hackett, J. T. and Guyenet, P. G. (1988a). Sympathoexcitatory neurons of rostral ventrolateral medulla exhibit pacemaker properties in the presence of a glutamate-receptor antagonist. *Brain Res.*, **438**, 23–40.

Sun, M. K., Young, B. S., Hackett, J. T. and Guyenet, P. G. (1988b). Reticulospinal pacemaker neurons of the rat rostral ventrolateral medulla with putative sympathoexcitatory function: an intracellular study in vitro. *Brain Res.*, **442**, 229–239.

Sun, W. and Panneton, W. M. (2002). The caudal pressor area of the rat: its precise location and projections to the ventrolateral medulla. *Am. J. Physiol. Regul. Integr. Comp. Physiol.*, **283**, R768–R778.

Sun, W. and Panneton, W. M. (2004). Defining projections from the caudal pressor area of the caudal ventrolateral medulla. *J. Comp. Neurol.*, **482**, 273–293.

Sundlöf, G. and Wallin, B. G. (1977). The variability of muscle nerve sympathetic activity in resting recumbent man. *J. Physiol.*, **272**, 383–397.

Sundlöf, G. and Wallin, B. G. (1978). Effect of lower body negative pressure on human muscle nerve sympathetic activity. *J. Physiol.*, **278**, 525–532.

Sved, A. F. (2004). Tonic glutamatergic drive of RVLM vasomotor neurons? *Am. J. Physiol. Regul. Integr. Comp. Physiol.*, **287**, R1301–R1303.

Sved, A. F., Mancini, D. L., Graham, J. C., Schreihofer, A. M. and Hoffman, G. E. (1994). PNMT-containing neurons of the C1 cell group express c-fos in response to changes in baroreceptor input. *Am. J. Physiol.*, **266**, R361–R367.

Swanson, L. W. (1987). The hypothalamus. In *Handbook of Chemical Anatomy*, Vol. 5, *Integrated Systems of the CNS, part I: Hypothalamus, Hippocampus, Amygdala, Retina.*, ed. A. Björklund, T. Hökfelt and L. W. Swanson, Amsterdam, New York, Oxford: Elsevier, pp. 1–124.

Swanson, L. W. (1995). Mapping the human brain: past, present, and future. *Trends Neurosci.*, **18**, 471–474.

Swanson, L. W. (2000). Cerebral hemisphere regulation of motivated behavior. *Brain Res.*, **886**, 113–164.

Swanson, L. W. (2003). The architecture of the nervous system. In *Fundamental Neuroscience*, 2nd edn., ed. L. R. Squire, F. E. Bloom, S. K. McConnell *et al*. San Diego: Academic Press, pp. 15–45.

Swift, J. Q., Garry, M. G., Roszkowski, M. T. and Hargreaves, K. M. (1993). Effect of flurbiprofen on tissue levels on immunoreactive bradykinin and acute postoperative pain. *J. Oral. Maxillofac. Surg.*, **51**, 112–117.

Szurszewski, J. H. (1981). Physiology of mammalian prevertebral ganglia. *Annu. Rev. Physiol.*, **43**, 53–68.

Szurszewski, J. H. and King, B. F. (1989). Physiology of prevertebral ganglia in mammals with special reference to the inferior mesenteric ganglion. In *The Gastrointestinal System*, ed. J. D. Wood. Bethesda: American Physiological Society, pp. 519–592.

Taché, Y., Martinez, V., Million, M. and Wang, L. (2001). Stress and the gastrointestinal tract III. Stress-related alterations of gut motor function: role of brain corticotropin-releasing factor receptors. *Am. J. Physiol. Gastrointest. Liver Physiol.*, **280**, G173–G177.

Taiwo, Y. O. and Levine, J. D. (1988). Characterization of the arachidonic acid metabolites mediating bradykinin and noradrenaline hyperalgesia. *Brain Res.*, **458**, 402–406.

Tanaka, M., Nagashima, K., McAllen, R. M. and Kanosue, K. (2002). Role of the medullary raphe in thermoregulatory vasomotor control in rats. *J. Physiol.*, **540**, 657–664.

Tang, X., Neckel, N. D. and Schramm, L. P. (2003). Locations and morphologies of sympathetically correlated neurons in the T(10) spinal segment of the rat. *Brain Res.*, **976**, 185–193.

Tang, X., Neckel, N. D. and Schramm, L. P. (2004). Spinal interneurons infected by renal injection of pseudorabies virus in the rat. *Brain Res.*, **1004**, 1–7.

Tatarchenko, L. A., Ivanov, A. Y. and Skok, V. I. (1990). Organization of the tonically active pathways through the superior cervical ganglion of the rabbit. *J. Auton. Nerv. Syst.*, **30 Suppl.**, S163–S168.

Taylor, R. B. and Weaver, L. C. (1993). Dorsal root afferent influences on tonic firing of renal and mesenteric sympathetic nerves in rats. *Am. J. Physiol.*, **264**, R1193–R1199.

Ter Horst, G. J., Hautvast, R. W., De Jongste, M. J. and Korf, J. (1996). Neuroanatomy of cardiac activity-regulating circuitry: a transneuronal retrograde viral labelling study in the rat. *Eur. J. Neurosci.*, **8**, 2029–2041.

Thompson, R. H. and Swanson, L. W. (2003). Structural characterization of a hypothalamic visceromotor pattern generator network. *Brain Res. Brain Res. Rev.*, **41**, 153–202.

Thorén, P. (1979). Role of cardiac vagal C-fibers in cardiovascular control. *Rev. Physiol. Biochem. Pharmacol.*, **86**, 1–94.

Thrasher, T. N. (2002). Unloading of arterial baroreceptors causes neurogenic hypertension. *Am. J. Physiol. Regul. Integr. Comp. Physiol.*, **282**, R1044–R1053.

Thrasher, T. N. (2004). Baroreceptors and the long-term control of blood pressure. *Exp. Physiol.*, **89**, 331–335.

Thrasher, T. N. (2005). Baroreceptors, baroreceptor unloading, and the long-term control of blood pressure. *Am. J. Physiol. Integr. Comp. Physiol.*, **288**, R819–R827.

Thuneberg, L. (1982). Interstitial cells of Cajal: intestinal pacemaker cells? *Adv. Anat. Embryol. Cell Biol.*, **71**, 1–130.

Thuneberg, L. (1999). One hundred years of interstitial cells of Cajal. *Microsc. Res. Tech.*, **47**, 223–238.

Tokimasa, T. and Akasu, T. (1995). Biochemical gating for voltage-gated channels: mechanisms for slow synaptic potentials in autonomic ganglia. In *The Autonomic Nervous System*, Vol. 6, *Autonomic Ganglia*, ed. E. M. McLachlan. Chur Switzerland: Harwood Academic Publishers, pp. 259–295.

Tomita, T. (1975). Electrophysiology of mammalian smooth muscle. *Prog. Biophys. Mol. Biol.*, **30**, 185–203.

Tomori, Z. and Widdicombe, J. G. (1969). Muscular, bronchomotor and cardiovascular reflexes elicited by mechanical stimulation of the respiratory tract. *J. Physiol.*, **200**, 25–49.

Torebjörk, H. E., Lundberg, L. E. R. and LaMotte, R. H. (1992). Central changes in processing of mechanoreceptive input in capsaicin-induced secondary hyperalgesia in humans. *J. Physiol.*, **448**, 765–780.

Traube, L. (1865). Über periodische Thätigkeits-Aeusserungen des vasomotorischen und Hemmungs-Nervencentrums [On the periodic changes of the vasomotor and inhibitory neural centers]. *Centralbl. Medic. Wissensch.*, **56**, 881–885.

Travagli, R. A. and Rogers, R. C. (2001). Receptors and transmission in the brain-gut axis: potential for novel therapies. V. Fast and slow extrinsic modulation of dorsal vagal complex circuits. *Am. J. Physiol.*, **281**, G595–G601.

Travagli, R. A., Hermann, G. E., Browning, K. N. and Rogers, C. (2006). Brain stem circuits regulating gastric function. *Annu. Rev. Physiol.*, **68**, 279–305.

Treede, R. D., Kenshalo, D. R., Gracely, R. H. and Jones, A. K. (1999). The cortical representation of pain. *Pain*, **79**, 105–111.

Treede, R. D., Apkarian, A. V., Bromm, B., Greenspan, J. D. and Lenz, F. A. (2000). Cortical representation of pain: functional characterization of nociceptive areas near the lateral sulcus. *Pain*, **87**, 113–119.

Tsunoo, A., Konishi, S. and Otsuka, M. (1982). Substance P as an excitatory transmitter of primary afferent neurons in guinea-pig sympathetic ganglia. *Neuroscience*, **7**, 2025–2037.

Tucker, D. C., Saper, C. B., Ruggiero, D. A. and Reis, D. J. (1987). Organization of central adrenergic pathways: I. Relationships of ventrolateral medullary projections to the hypothalamus and spinal cord. *J. Comp. Neurol.*, **259**, 591–603.

Ulman, L. G., Potter, E. K. and McCloskey, D. I. (1992). Effects of sympathetic activity and galanin on cardiac vagal action in anaesthetized cats. *J. Physiol.*, **448**, 225–235.

Undem, B. and Weinreich, D. (eds.) (2005). *Advance in Vagal Afferent Neurobiology*. Boca Raton: CRC Press.

Unzicker, K. (ed.) (1996). *The Autonomic Nervous System* Vol. 10, *Autonomic–Endocrine Interactions*. Amsterdam: Harwood Academic Publishers.

Uvnäs, B. (1954). Sympathetic vasodilator outflow. *Pharmacol. Rev.*, **34**, 608–618.

Uvnäs, B. (1960). Sympathetic vasodilator system and blood flow. *Physiol. Rev.*, **40**, **Suppl**. 4, 68–75.

Vallbo, A. B., Hagbarth, K. E., Torebjörk, H. E. and Wallin, B. G. (1979). Somatosensory, proprioceptive and sympathetic activity in human peripheral nerves. *Physiol. Rev.*, **59**, 919–957.

Vallbo, A. B., Olsson, K. A., Westberg, K. G. and Clark, F. J. (1984). Microstimulation of single tactile afferents from the human hand. Sensory attributes related to unit type and properties of receptive fields. *Brain*, **107**, 727–749.

Vallbo, A. B., Hagbarth, K. E. and Wallin, B. G. (2004). Microneurography: how the technique developed and its role in the investigation of the sympathetic nervous system. *J. Appl. Physiol.*, **96**, 1262-1269.

van Bockstaele, E. J., Aston-Jones, G., Pieribone, V. A., Ennis, M. and Shipley, M. T. (1991). Subregions of the periaqueductal gray topographically innervate the rostral ventral medulla in the rat. *J. Comp. Neurol.*, **309**, 305-327.

van Dijk, J. G. (2003). Fainting in animals. *Clin. Auton. Res.*, **13**, 247-255.

van Helden, D. F. (1988a). An alpha-adrenoceptor-mediated chloride conductance in mesenteric veins of the guinea-pig. *J. Physiol.*, **401**, 489-501.

van Helden, D. F. (1988b). Electrophysiology of neuromuscular transmission in guinea-pig mesenteric veins. *J. Physiol.*, **401**, 469-488.

van Helden, D. F. (1991). Spontaneous and noradrenaline-induced transient depolarizations in the smooth muscle of guinea-pig mesenteric vein. *J. Physiol.*, **437**, 511-541.

Vecchiet, L., Albe-Fessard, D., Lindblom, U. and Giamberardino, M. A. (eds.) (1993). *New Trends in Referred Pain and Hyperalgesia*. Amsterdam: Elsevier Science.

Verberne, A. J., Stornetta, R. L. and Guyenet, P. G. (1999). Properties of C1 and other ventrolateral medullary neurons with hypothalamic projections in the rat. *J. Physiol.*, **517**, 477-494.

Victor, R. G., Leimbach, W. N. J., Seals, D. R., Wallin, B. G. and Mark, A. L. (1987). Effects of the cold pressor test on muscle sympathetic nerve activity in humans. *Hypertension*, **9**, 429-436.

Victor, R. G., Secher, N. H., Lyson, T. and Mitchell, J. H. (1995). Central command increases muscle sympathetic nerve activity during intense intermittent isometric exercise in humans. *Circ. Res.*, **76**, 127-131.

Villanueva, L. and Nathan, P. W. (2000). Multiple pain pathways. In *Proceedings of the 9th World Congress on Pain*, ed. M. Devor, M. C. Rowbotham and Z. Weisenfeld-Hallin. Seattle: IASP Press, pp. 371-386.

Vissing, S. F., Scherrer, U. and Victor, R. G. (1994). Increase of sympathetic discharge to skeletal muscle but not to skin during mild lower body negative pressure in humans. *J. Physiol.*, **481**, 233-241.

Vizzard, M. A., Erickson, V. L., Card, J. P., Roppolo, J. R. and de Groat, W. C. (1995). Transneuronal labeling of neurons in the adult rat brainstem and spinal cord after injection of pseudorabies virus into the urethra. *J. Comp. Neurol.*, **355**, 629-640.

Vizzard, M. A., Brisson, M. and de Groat, W. C. (2000). Transneuronal labeling of neurons in the adult rat central nervous system following inoculation of pseudorabies virus into the colon. *Cell Tissue Res.*, **299**, 9-26.

Vogt, B. A. (2005). Pain and emotion interactions in subregions of the cingulate gyrus. *Nat. Rev. Neurosci.*, **6**, 533-544.

Vogt, B. A., Rosene, D. L. and Pandya, D. N. (1979). Thalamic and cortical afferents differentiate anterior from posterior cingulate cortex in the monkey. *Science*, **204**, 205-207.

Vollmer, R. R. (1996). Selective neural regulation of epinephrine and norepinephrine cells in the adrenal medulla – cardiovascular implications. *Clin. Exp. Hypertens.*, **18**, 731-751.

von Euler, U. S. (1956). *Noradrenaline: Chemistry, Physiology, Pharmacology and Clinical Aspects*. Springfield Ill: Thomas.

von Euler, C. (1986). Brain stem mechanisms for generation and control of breathing pattern. In *Handbook of Physiology, The Respiratory System*, Vol. II, ed.

A. P. Fishman, N. S. Cherniack and J. G. Widdicombe. Bethesda: American Physiological Society, pp. 1–67.

von Holst, E. and St. Paul, U. (1960). Vom Wirkungsgefüge der Triebe [The Wirkungsgefüge of Drives]. *Die Naturwissenschaften*, **47**, 409–422.

von Holst, E. and St. Paul, U. (1962). Electrically controlled behavior. *Sci. Am.*, **206**, 50–60.

Voyvodic, J. T. (1989). Peripheral target regulation of dendritic geometry in the rat superior cervical ganglion. *J. Neurosci.*, **9**, 1997–2010.

Wallin, B. G. (1990). Neural control of human skin blood flow. *J. Auton. Nerv. Syst.*, **30 Suppl.**, S185–S190.

Wallin, B. G. (2002). Intraneural recordings of normal and abnormal sympathetic activity in humans. In *Autonomic Failure*, 4th edn., ed. C. J. Mathias and R. Bannister. Oxford: Oxford University Press, pp. 224–231.

Wallin, B. G. and Elam, M. (1994). Insights from intraneural recordings of sympathetic nerve traffic in humans. *News Physiol. Sci.*, **9**, 203–207.

Wallin, B. G. and Fagius, J. (1986). The sympathetic nervous system in man – aspects derived from microelectrode recordings. *Trends Neurosci.*, **9**, 63–67.

Wallin, B. G. and Fagius, J. (1988). Peripheral sympathetic neural activity in conscious humans. *Annu. Rev. Physiol.*, **50**, 565–576.

Wallin, B. G. and Stjernberg, L. (1984). Sympathetic activity in man after spinal cord injury. Outflow to skin below the lesion. *Brain*, **107**, 183–198.

Wallin, B. G., Batelsson, K., Kienbaum, P., Karlsson, T. and Gazelius, B. (1998). Two neural mechanisms for respiration-induced cutaneous vasodilatation in humans. *J. Physiol.*, **513**, 559–569.

Wang, F. B., Holst, M. C. and Powley, T. L. (1995). The ratio of pre- to postganglionic neurons and related issues in the autonomic nervous system. *Brain Res. Brain Res. Rev.*, **21**, 93–115.

Wang, H., Germanson, T. P. and Guyenet, P. G. (2002). Depressor and tachypneic responses to chemical stimulation of the ventral respiratory group are reduced by ablation of neurokinin-1 receptor-expressing neurons. *J. Neurosci.*, **22**, 3755–3764.

Wang, H., Weston, M. C., McQuiston, T. J., Stornetta, R. L. and Guyenet, P. G. (2003). Neurokinin-1 receptor-expressing cells regulate depressor region of rat ventrolateral medulla. *Am. J. Physiol. Heart Circ. Physiol.*, **285**, H2757–H2769.

Wank, M. and Neuhuber, W. L. (2001). Local differences in vagal afferent innervation of the rat esophagus are reflected by neurochemical differences at the level of the sensory ganglia and by different brainstem projections. *J. Comp. Neurol.*, **435**, 41–59.

Ward, S. M. and Sanders, K. M. (2001). Physiology and pathophysiology of the interstitial cell of Cajal: from bench to bedside. I. Functional development and plasticity of interstitial cells of Cajal networks. *Am. J. Physiol. Gastrointest. Liver Physiol.*, **281**, G602–G611.

Ward, S. M., Morris, G., Reese, L., Wang, X. Y. and Sanders, K. M. (1998). Interstitial cells of Cajal mediate enteric inhibitory neurotransmission in the lower esophageal and pyloric sphincters. *Gastroenterology*, **115**, 314–329.

Ward, S. M., Beckett, E. A., Wang, X. et al. (2000a). Interstitial cells of Cajal mediate cholinergic neurotransmission from enteric motor neurons. *J. Neurosci.*, **20**, 1393–1403.

Ward, S. M., Ordog, T., Koh, S. D. et al. (2000b). Pacemaking in interstitial cells of Cajal depends upon calcium handling by endoplasmic reticulum and mitochondria. *J. Physiol.*, **525**, 355–361.

Wasner, G., Heckmann, K., Maier, C. and Baron, R. (1999). Vascular abnormalities in acute reflex sympathetic dystrophy (CRPS I): complete inhibition of sympathetic nerve activity with recovery. *Arch. Neurol.*, **56**, 613–620.

Wasner, G., Schattschneider, J., Heckmann, K., Maier, C. and Baron, R. (2001). Vascular abnormalities in reflex sympathetic dystrophy (CRPS I): mechanisms and diagnostic value. *Brain*, **124**, 587–599.

Watkins, L. R. and Maier, S. F. (eds.) (1999). *Cytokines and Pain*. Basel, Boston, Berlin: Birkhäuser Verlag.

Watkins, L. R. and Maier, S. F. (2000). The pain of being sick: implications of immune-to-brain communication for understanding pain. *Annu. Rev. Psychol.*, **51**, 29–57.

Watts, A. G. and Swanson, L. W. (2002). Anatomy of motivational systems. In *"Stevens" Handbook of Experimental Psychology*, 3rd edn. Vol. 3, ed. G. R. Gallistel. New York: John Wiley, pp. 563–631.

Weaver, L. C. and Polosa, C. (1997). Spinal cord circuits providing control of sympathetic preganglionic neurons. In *The Autonomic Nervous System*, Vol. 11, *Central Nervous Control of Autonomic Function*, ed. D. Jordan. Amsterdam: Harwood Academic Publishers, pp. 29–61.

Weaver, L. C., Verghese, P., Bruce, J. C. *et al.* (2001). Autonomic dysreflexia and primary afferent sprouting after clip-compression injury of the rat spinal cord. *J. Neurotrauma*, **18**, 1107–1119.

Weaver, L. C., Marsh, D. R., Gris, D., Meakin, S. O. and Dekaban, G. A. (2002). Central mechanisms for autonomic dysreflexia after spinal cord injury. *Prog. Brain Res.*, **137**, 83–95.

Wesselmann, U. and McLachlan, E. M. (1984). The effect of previous transection on quantitative estimates of the preganglionic neurons projecting in the cervical sympathetic trunk of the guinea-pig and the cat made by retrograde labelling of damaged axons by horseradish peroxidase. *Neuroscience*, **13**, 1299–1309.

Westerhaus, M. J. and Loewy, A. D. (1999). Sympathetic-related neurons in the preoptic region of the rat identified by viral transneuronal labeling. *J. Comp. Neurol.*, **414**, 361–378.

Westerhaus, M. J. and Loewy, A. D. (2001). Central representation of the sympathetic nervous system in the cerebral cortex. *Brain Res.*, **903**, 117–127.

Weston, M., Wang, H., Stornetta, R. L., Sevigny, C. P. and Guyenet, P. G. (2003). Fos expression by glutamatergic neurons of the solitary tract nucleus after phenylephrine-induced hypertension in rats. *J. Comp. Neurol.*, **460**, 525–541.

White, J. C. and Bland, E. F. (1948). The surgical relief of severe angina pectoris. *Medicine*, **27**, 1–42.

Whyment, A. D., Wilson, J. M., Renaud, L. P. and Spanswick, D. (2004). Activation and integration of bilateral GABA-mediated synaptic inputs in neonatal rat sympathetic preganglionic neurons in vitro. *J. Physiol.*, **555**, 189–203.

Widdicombe, J. G. (1966). Action potentials in parasympathetic and sympathetic efferent fibers to the trachea and lungs of dogs and cats. *J. Physiol.*, **186**, 56–88.

Widdicombe, J. G. (1986). Sensory innervation of the lungs and airways. *Prog. Brain Res.*, **67**, 49–64.

Widdicombe, J. (2001). Airway receptors. *Respir. Physiol.*, **125**, 3–15.

Widdicombe, J. (2003). Functional morphology and physiology of pulmonary rapidly adapting receptors (RARs). *Anat. Rec.*, **270A**, 2–10.

Williams, C. L., Men, D., Clayton, E. C. and Gold, P. E. (1998). Norepinephrine release in the amygdala after systemic injection of epinephrine or escapable footshock: contribution of the nucleus of the solitary tract. *Behav. Neurosci.*, **112**, 1414–1422.

Williams, I. R. and Kupper, T. S. (1996). Immunity at the surface: homeostatic mechanisms of the skin immune system. *Life Sci.*, **58**, 1485–1507.

Williams, R. M., Berthoud, H. R. and Stead, R. H. (1997). Vagal afferent nerve fibers contact mast cells in rat small intestinal mucosa. *NeuroImmunoModulation*, **4**, 266–270.

Willis, W. D. J. (1985). *The Pain System*. Basel: Karger.

Willis, W. D., Jr. (1999). Dorsal root potentials and dorsal root reflexes: a double-edged sword. *Exp. Brain Res.*, **124**, 395–421.

Willis, W. D. J. and Coggeshall, R. E. (2004a). *Sensory Mechanisms of the Spinal Cord. Primary Afferent Neurons and the Spinal Dorsal Horn*, Vol. 1, 3rd edn. New York: Kluwer Academic/Plenum Publishers.

Willis, W. D. J. and Coggeshall, R. E. (2004b). *Sensory Mechanisms of the Spinal Cord. Ascending Sensory Tracts and Their Descending Control*, Vol. 2, 3rd edn. New York: Kluwer Academic/Plenum Publishers.

Willis, W. D., Al Chaer, E. D., Quast, M. J. and Westlund, K. N. (1999). A visceral pain pathway in the dorsal column of the spinal cord. *Proc. Natl. Acad. Sci. U. S. A.*, **96**, 7675–7679.

Willis, W. D., Jr., Zhang, X., Honda, C. N. and Giesler, G. J., Jr. (2001). Projections from the marginal zone and deep dorsal horn to the ventrobasal nuclei of the primate thalamus. *Pain*, **92**, 267–276.

Willis, W. D., Jr., Zhang, X., Honda, C. N. and Giesler, G. J., Jr. (2002). A critical review of the role of the proposed VMpo nucleus in pain. *J. Pain*, **3**, 79–94.

Wilson, L. B., Andrew, D. and Craig, A. D. (2002). Activation of spinobulbar lamina I neurons by static muscle contraction. *J. Neurophysiol.*, **87**, 1641–1645.

Winkler, H. (1988). Occurrence and mechanisms of exocytosis in adrenal medulla and sympathetic nerve. In *Handbook of Experimental Pharmacology*, Vol. 90/I *Catecholamines I*, ed. U. Trendelenburg and N. Weiner. Berlin, Heidelberg, New York: Springer-Verlag, pp. 43–118.

Wood, J. D. (1994). Physiology of the enteric nervous system. In *Physiology of the Gastrointestinal Tract*, 3rd edn., ed. L. R. Johnson. New York: Raven Press, pp. 423–482.

Wood, J. D. (2002). Enteric neuro-immunology. In *The Autonomic Nervous System*, Vol. 14, *Innervation of the Gastrointestinal Tract*, ed. S. Brookes and M. Costa. London, New York: Taylor and Francis, pp. 363–392.

Woods, S. C. and Stricker, E. M. (2003). Food intake and metabolism. In *Fundamental Neuroscience*, 2nd edn., eds. L. R. Squire, F. E. Bloom, S. K. McConnell *et al.* San Diego: Academic Press, pp. 991–1009.

Woolf, C. J. (1996). Phenotypic modification of primary sensory neurons: the role of nerve growth factor in the production of persistent pain. *Phil. Trans. R. Soc. London B*, **351**, 441–448.

Woolf, C. J., Safieh-Garabedian, B., Ma, Q.-P., Crilly, P. and Winter, J. (1994). Nerve growth factor contributes to the generation of inflammatory sensory hypersensitivity. *Neuroscience*, **62**, 327–331.

Woolf, C. J., Ma, Q. P., Allchorne, A. and Poole, S. (1996). Peripheral cell types contributing to the hyperalgesic action of nerve growth factor in inflammation. *J. Neurosci.*, **16**, 2716–2723.

Xi, X., Randall, W. C. and Wurster, R. D. (1991). Intracellular recording of spontaneous activity of canine intracardiac ganglion cells. *Neurosci Lett.*, **128**, 129–132.

Yang, H., Kawakubo, K., Wong, H. et al. (2000a). Peripheral PYY inhibits intracisternal TRH-induced gastric acid secretion by acting in the brain. *Am. J. Physiol. Gastrointest. Liver Physiol.*, **279**, G575–G581.

Yang, H., Yuan, P. Q., Wang, L. and Taché, Y. (2000b). Activation of the parapyramidal region in the ventral medulla stimulates gastric acid secretion through vagal pathways in rats. *Neuroscience*, **95**, 773–779.

Yang, Z. and Coote, J. H. (2003). Role of GABA and NO in the paraventricular nucleus-mediated reflex inhibition of renal sympathetic nerve activity following stimulation of right atrial receptors in the rat. *Exp. Physiol.*, **88**. 335–342.

Yang, Z., Smith, L. and Coote, J. H. (2004). Paraventricular nucleus activation of renal sympathetic neurones is synaptically depressed by nitric oxide and glycine acting at a spinal level. *Neuroscience*, **124**, 421–428.

Yeoh, M., McLachlan, E. M. and Brock, J. A. (2004a). Tail arteries from chronically spinal rats have potentiated responses to nerve stimulation in vitro. *J. Physiol.*, **556**, 545–555.

Yeoh, M., McLachlan, E. M. and Brock, J. A. (2004b). Chronic decentralization potentiates neurovascular transmission in the isolated rat tail artery, mimicking the effects of spinal transection. *J. Physiol.*, **561**, 583–596.

Yoshimura, M., Polosa, C. and Nishi, S. (1987a). Slow EPSP and the depolarizing action of noradrenaline on sympathetic preganglionic neurons. *Brain Res.*, **414**, 138–142.

Yoshimura, M., Polosa, C. and Nishi, S. (1987b). Slow IPSP and the noradrenaline-induced inhibition of the cat sympathetic preganglionic neuron in vitro. *Brain Res.*, **419**, 383–386.

Young, J. B. and Landsberg, L. (2001). Synthesis, storage, and secretion of adrenal medullary hormones: physiology and pathophysiology. In *Handbook of Physiology. Section 7. The Endocrine System*. Vol. IV, *Coping with the Environment: Neural and Neuroendocrine Mechanisms*, ed. B. S. McEwan. Oxford, New York: Oxford University Press, pp. 3–19.

Yu, L. C. and Perdue, M. H. (2001). Role of mast cells in intestinal mucosal function: studies in models of hypersensitivity and stress. *Immunol. Rev.*, **179**, 61–73.

Zagon, A. and Smith, A. D. (1993). Monosynaptic projections from the rostral ventrolateral medulla oblongata to identified sympathetic preganglionic neurons. *Neuroscience*, **54**, 729–743.

Zagorodnyuk, V. P. and Brookes, S. J. (2000). Transduction sites of vagal mechanoreceptors in the guinea pig esophagus. *J. Neurosci.*, **20**, 6249–6255.

Zagorodnyuk, V. P., Chen, B. N. and Brookes, S. J. (2001). Intraganglionic laminar endings are mechano-transduction sites of vagal tension receptors in the guinea-pig stomach. *J. Physiol.*, **534**, 255–268.

Zaretsky, D. V., Zaretskaia, M. V. and DiMicco, J. A. (2003a). Stimulation and blockade of GABA(A) receptors in the raphe pallidus: effects on body temperature, heart rate, and blood pressure in conscious rats. *Am. J. Physiol. Regul. Integr. Comp. Physiol.*, **285**, R110–R116.

Zaretsky, D. V., Zaretskaia, M. V., Samuels, B. C., Cluxton, L. K. and DiMicco, J. A. (2003b). Microinjection of muscimol into raphe pallidus suppresses tachycardia associated with air stress in conscious rats. *J. Physiol.*, **546**, 243–250.

Zhang, X. and Giesler, G. J., Jr. (2005). Response characterstics of spinothalamic tract neurons that project to the posterior thalamus in rats. *J. Neurophysiol.*, **93**, 2552–2564.

Zhang, X., Fogel, R. and Renehan, W. E. (1992). Physiology and morphology of neurons in the dorsal motor nucleus of the vagus and the nucleus of the solitary tract that are sensitive to distension of the small intestine. *J. Comp. Neurol.*, **323**, 432–448.

Zhang, X., Fogel, R. and Renehan, W. E. (1995a). Relationships between the morphology and function of gastric- and intestine-sensitive neurons in the nucleus of the solitary tract. *J. Comp. Neurol.*, **363**, 37–52.

Zhang, X., Kostarczyk, E. and Giesler, G. J., Jr. (1995b). Spinohypothalamic tract neurons in the cervical enlargement of rats: descending axons in the ipsilateral brain. *J. Neurosci.*, **15**, 8393–8407.

Zhang, X., Renehan, W. E. and Fogel, R. (1998). Neurons in the vagal complex of the rat respond to mechanical and chemical stimulation of the GI tract. *Am. J. Physiol.*, **274**, G331–G341.

Zhang, X., Honda, C. N. and Giesler, G. J., Jr. (2000a). Position of spinothalamic tract axons in upper cervical spinal cord of monkeys. *J. Neurophysiol.*, **84**, 1180–1185.

Zhang, X., Wenk, H. N., Honda, C. N. and Giesler, G. J., Jr. (2000b). Locations of spinothalamic tract axons in cervical and thoracic spinal cord white matter in monkeys. *J. Neurophysiol.*, **83**, 2869–2880.

Zhao, F. Y., Saito, K., Yoshioka, K. *et al.* (1996). Tachykininergic synaptic transmission in the coeliac ganglion of the guinea-pig. *Br. J. Pharmacol.*, **118**, 2059–2066.

Zhu, J. X., Zhu, X. Y., Owyang, C. and Li, Y. (2001). Intestinal serotonin acts as a paracrine substance to mediate vagal signal transmission evoked by luminal factors in the rat. *J. Physiol.*, **530**, 431–442.

Index

Acetylcholine
  autonomic ganglia 212
  muscarinic transmission, in vivo 240-247
  parasympathetic postganglionic neurons 15
  transmitter postganglionic 252-254, 254 (Tab. 7.1)
  vasodilator neurons 131, 133, 172 (Fig. 5.2), 173-174
Adenosine
  synaptic transmission to sympathetic preganglionic neurons 334
Adenosine triphosphate (ATP), transmitter
  enteric nervous system 176 (Tab. 5.1), 177
  peristalsis 186
  effector cells, autonomic 254 (Tab. 7.1), 262-263, 268-269
Adrenaline, see adrenal medulla, see catecholamines, circulating
  vasodilation in skeletal muscle 167
  sympathetic premotor neurons (C1 neurons) 318, 321 (Fig. 8.16), 333-334, 388 (Fig. 10.3), 389-391, 453-454
Adrenal medulla 19, 143-148
  Cannon's view of functioning 464-465
  inflammation and 146-147
  memory consolidation and 147
  nociceptor sensitization and 146-147
  neural regulation of 143-144
Airways
  parasympathetic innervation 159-161
  sympathetic innervation 141-142
Allostasis 2, 472
  allostatic load 6
  concept, definition 469-470
Amygdala
  autonomic responses during anger and fear, role of 495-498
  lamina I (spinal) neurons, parabrachial nuclei and 72 (Fig. 2.12), 74

Analysis, experimental, of the autonomic nervous system
  anesthesia, effects of 104
  axon tracer methods 93, 294, 295 (Fig. 8.1)
  closed-loop condition, studies in vivo 96, 105
  identification of neurons, neurophysiology 99-101
  immediate early gene c-fos (activity marker) 94, 188-190, 452
  intracellular labeling 94, 452
  microneurography, human 102-103, 103 (Fig. 3.6)
  neurochemistry 30 (Tab. 1.3), 94, 176 (Tab. 5.1)
  (see also neurochemical coding) 31 (Tab. 1.4)
  neurophysiology, animals in vivo 96-101
  reflexes, functional markers 95-96, 290
  suicide killing of neurons, specific, by a toxin 296, 452
  transneuronsal labeling, neurotropic virus 294-296
  working-heart-brain-stem (WHBS) preparation 93-94
Area postrema (AP)
  gastrointestinal tract and 440-441, 446-449
  integration of endocrine signals by 446-449, 447 (Fig. 10.27)
  neurohemal (circumventricular) organ 312
Arteries/arterioles, neural signal transmission to
  adenosine triphosphate, transmitter 268-269
  cotransmission 269
  hormones, effect on 274
  mesenteric artery 269
  myogenic activity 275
  neuroeffector transmission, mechanism 267-270
  nitric oxide (NO) 275-276
  peptidergic afferents, effect on (see axon reflex, neurogenic inflammation) 275
  tail artery, rat 269-270

Autonomic ganglia
  convergence, divergence in 214-216
  definition 14, 15
  parasympathetic 21-22 (Tab. 1.1), 23 (Fig. 1.4), 23-28, 238-240
  sympathetic 16-18, 21-22 (Tab. 1.1), 230-238
Autonomic motor cortex (see cingulate cortex, anterior)
Autonomic neural unit 223-225
Autonomic regulation, integrative
  behavior, brain and 2-5, 507-508
  cortical control of 5, 508-510, 4 (Fig. 2)
  endocrine regulation and 2
  failure of 5-6
  long-term 1
  future research questions 510-514
  range of 7
  short-term 1
  somatomotor system and 1
Autonomic and respiratory
  generator, integration 436-440
  (see also cardiovascular and respiratory regulation, coupling)
Axon reflex 83, 275

B-afferent neurons (small dark neurons)
  autonomic nervous system and 82-83
Bard, Philip (1898-1977)
  diencephalic cats, sham-rage behavior 462, 517
  mesencephalic, pontine cats 518
Baroreceptors, arterial
  blood pressure control 399-400
  functional types 399
  nucleus tractus solitarii, projection to 312 (Fig. 8.11), 313, 400
Baroreceptor pathways to
  cardiovascular neurons 401 (Fig. 10.7), 403-407
  caudal ventrolateral medulla, GABAergic interneurons 403-406, 408
  inhibition in spinal cord (alternative pathways) 406-407

modulation of 407–408, 409
(Fig. 10.11)
rostral ventrolateral medulla,
GABAergic inhibition 403–406
Baroreceptor reflexes 398–409
cutaneous vasoconstrictor
neurons 120
cardiomotor neurons,
parasympathetic 401 (Fig. 10.7),
402
detection of 398
functions of 399–400
muscle vasoconstrictor
neurons 118
reflex pathways 400,
401 (Fig. 10.7)
resetting of and nucleus tractus
solitarii 407–408
visceral vasoconstrictor neurons
(*see* muscle vasoconstrictor
neurons)
Bayliss, W. M. (1880–1924)
axon reflex 83
"The law of the intestine"
(peristalsis) 183–184
Behavioral state and autonomic
systems 4 (Fig. 2), 5
Behavioral patterns, autonomic
responses in 473 (Tab. 11.1)
conditioned emotional responses
(CER) 486–490
concept, general aspects, cortical
command 470–472, 471
(Fig. 11.1)
defense reactions 480–486
diving response 475–476
freezing reaction 491
skeletal muscle effort 477–480
syncope, neurally mediated
490–491
tonic immobility 490
vigilance reaction 491
Bernard, Claude (1813–1878)
cardiovascular center, medulla
oblongata 379
internal milieu 2, 460, 469
BDNF (brain-derived neurotrophic
factor)
sympathetic preganglionic
neurons 216
Blood vessels, integration of signals
273–276, 274 (Fig. 7.9)
"Brain of the gut", *see* enteric nervous
system 179

Brown adipose tissue
lipomotor neurons (sympathetic)
142, 417
Bursting activity in sympathetic
neurons 396–398
network oscillator hypothesis
396, 397
respiratory rhythmicity 396–397

Cajal, Ramón y (1852–1934)
interstitial cells of Cajal 188–190
Calcitonin gene-related peptide
(CGRP)
blood vessels, axon reflex 83, 253
enteric nervous system 176
(Tab. 5.1)
sympathetic neurons 29–31, 30
(Tab. 1.3), 31 (Tab. 1.4)
Cannon, Walter Bradford (1871–1945)
460–465
generalizing concept of the
autonomic nervous system,
consequences of 467–469
homeostasis 2
sympathico-adrenal system,
concept of 145, 460–465
The Wisdom of the Body 8,
463–465
Cardiac-somatic hypothesis 474
Cardiovascular center, medulla
oblongata
discovery of 379–381
Cardiovascular and respiratory
regulation, coupling 426–428
common cardiorespiratory
network, concept 437–439,
438 (Fig. 10.23)
common central oscillator model
436–437
irradiation model 436
respiratory sinus arrhythmia 428
Traube-Hering waves 426–428
Catecholamines, circulating
(adrenaline, noradrenaline)
concentration 144
functions 144–145, 148 (Fig. 4.18)
Caudal pressure area (CPA)
location 385 (Fig. 10.2), 383–386
Caudal ventrolateral medulla (CVLM)
baroreceptor pathways 403–406
location 385 (Fig. 10.2), 383
Chemoreceptors, arterial
nucleus tractus solitarii, projection
to 312 (Fig. 8.11), 410

Chemoreceptor reflexes, arterial
cutaneous vasoconstrictor
neurons, in 456–457
reflex pathways 412 (Fig. 10.13)
sympathetic vasoconstrictor
neurons, in 410–414
Cholecystokinin (CCK)
enteric hormone cells 172
enteric nervous system and defense
198 (Fig. 5.13), 200
Choline acetyltranferase (ChAT) 15
Cingulate cortex 74, 78
anterior cingulate cortex (ACC) 75,
76 (Fig. 2.13), 78
midcingulate cortex (MCC) 75, 76
(Fig. 2.13), 78
Common cardiorespiratory network,
concept 437–439, 438 (Fig. 10.23)
Conditioned emotional responses
(CER) 486–490
cardiovascular responses during
473 (Tab. 11.1), 487 (Fig. 11.8),
487–488
exercise, cardiovascular responses
during, comparison with CER
489–490
generation of, technique 517–518
hypothalamus, integration of
cardiovascular responses in CER
488–489
Conditioned fear response, rat
amygdala, role in 495–498, 497
(Fig. 11.12)
Confrontational defense, *see* defense
reactions
Convergence in autonomic ganglia
214–216
segmental origin of sympathetic
preganglionic neurons 216–217
Cortical control of autonomic
nervous system
conditioned emotional response
486–490
concept 4, 5, 471 (Fig. 11.1),
508–510 (Fig. 2)
defense reaction 484–486
diving response/reflex 475–476
muscle exercise 477–480
sensations and 471 (Fig. 11.1),
471–472
Corticotropin-releasing hormone
(CRH)
area postrema and gastrointestinal
functions 448

Corticotropin-releasing hormone (CRH) (cont.)
  sympathetic premotor neurons, in 321 (Fig. 8.16)
Craniosacral system 14, 15
  see parasympathetic system
Cutaneous vasoconstrictor (CVC) neurons
  chemoreceptor reflexes in 413–414
  emotional stimuli, reaction to 122, 165
  functional characteristics, animal 108 (Tab. 4.1), 113–117
  functional characteristics, human 119–123, 120 (Tab. 4.2)
  Lovén reflex 165
  non-nicotinic transmission (ganglia) 243–244, 246 (Fig. 6.15)
  nociceptive reflex, animal 115
  nociceptive reflex, human 121
  proportion of preganglionic neurons 116, 152 (Tab. 4.5)
  reflex pattern 114 (Fig. 4.4), 115–116
  respiratory rhythmicity in activity 116, 429–431, 432 (Fig. 10.21), 435–436
  spontaneous activity, animal 115, 231 (Tab. 6.2)
  spontaneous activity, human 117
  functional types of 116
Cutaneous vasoconstrictor pathway, central
  caudal raphe nuclei 415–417, 419 (Fig. 10.16)
  rostral ventrolateral medulla 416–417, 419 (Fig. 10.16)
Cytokines
  area postrema, tumor necrosis factor α & gastrointestinal functions 448–449
  defense of the gastrointestinal tract 197, 198 (Fig. 5.13)
  vagal afferents and 50

Deep dorsal horn projection neurons, spinal cord 70–71
  ascending pathways 70 (Fig. 2.11)
  functional classification 68 (Fig. 2.10), 70–71
  supraspinal projection sites 71, 74
Defense, of the gastrointestinal tract 198 (Fig. 5.13)
  cytokines, role of 197
  enteric nervous system and 196–200
  gut-associated lymphoid tissue (GALT) 196–197
  mast cells and 198 (Fig. 5.13), 199
  visceral afferents (spinal, vagal) 197, 199
Defense reactions, organized in the periaqueductal grey (PAG) 480–486
  afferent inputs from body tissues and 483–484
  autonomic responses in 473 (Tab. 11.1), 481
  cardiovascular centers in the medulla oblongata and 482–483, 484 (Fig. 11.6)
  hypothalamus, prefrontal cortex and 483–484
  muscle vasodilation and 132–135
  sensory adjustments in 481
  types of 480–481, 482 (Fig. 11.5)
Defensive (agonistic) behavior
  hypothalamus and 500, 506 (Tab. 11.2)
Divergence in autonomic ganglia 214–216
Diving response, cortically induced cardiovascular responses 473 (Tab. 11.1), 475–476
Domain theory (sympathetic ganglia) 215, 218
Dorsal motor nucleus of the vagus (DMNX)
  enteric nervous system 180, 200–201, 201 (Fig. 5.14)
  gastrointestinal tract, representation of 308–310, 311 (Fig. 8.10)
  heart 308, 311 (Fig. 8.10)
  intestino-intestinal reflex circuits 441–442
  numbers of neurons, rat 329
  preganglionic neurons to the gastrointestinal tract, functions 201, 201 (Fig. 5.14), 441
  viscerotopic organization 308–310, 311 (Fig. 8.10)
Dorsal vagal complex (DVC) 313
  forebrain, control of DVC by, concept 444, 446 (Fig. 10.26)
  integration of endocrine signals 446–449, 447 (Fig. 10.27)
  integration in the DVC, concept 443–444, 445 (Fig. 10.25)
  intestino-intestinal reflex circuits 441–442
Dynorphin (DYN)
  sympathetic neurons 29–31, 30 (Tab. 1.3), 31 (Tab. 1.4)
Dysreflexia, autonomic, after spinal cord transection 340–342
  mechanisms of, in spinal cord 340–341
  mechanisms of, in periphery 342

Edinger-Westphal nucleus 162, 310
Effector cells of the autonomic nervous system (see also neuroeffector transmission)
  cellular response to nerve stimulation 255–256
  endocrine cells (enteric nervous system) 174
  enteric nervous system 171
  functional syncytium 256
  intracellular pathways 255–256, 266–268, 272–273
  junctional/extrajunctional receptors see neuroeffector transmission
  neuroeffector junctions 257–259
  parasympathetic system 24, 25–27 (Tab. 1.2)
  sympathetic system 19, 25–27 (Tab. 1.1)
Emission, ejaculation 360–361
Emotions, basic
  autonomic responses, patterns of 492–494
  central representation of emotions and autonomic responses 494–498
  cingulated cortex and 495
  types of 492, 493 (Fig. 11.10)
Enkephalin (ENK)
  sympathetic neurons, in 31 (Tab. 1.4)
  sympathetic premotor neurons, in 321 (Fig. 8.16)
Enteric nervous system 169–205
  anatomy 169–171, 170 (Fig. 5.1)
  concept of functioning 178–181, 179 (Fig. 5.3)
  defense, gastrointestinal tract 196–200, 198 (Fig. 5.13)
  definition 15
  effector cells 171

interneurons 174-177,
  176 (Tab. 5.1)
intrinsic primary afferent neurons
  (IPANs) 171-172,
  176 (Tab. 5.1)
motility patterns, gastrointestinal
  tract 181-188
motor neurons 173-174,
  176 (Tab. 5.1)
neurochemical coding 176
  (Tab. 5.1), 177
parasympathetic control 200-203,
  201 (Fig. 5.14)
secretion, transmural transport
  194-196
sympathetic control 200-203, 203
  (Fig. 5.15)
transmitters 176 (Tab. 5.1)
types of neurons 171-177,
  172 (Fig. 5.2)
Erection
  mechanisms 374
  spinal reflex pathway 358-360
  transmitters 374
Exteroception 36-37, 75, 76 (Fig. 2.13)
Extraspinal (peripheral) reflex
  pathways
  gastrointestinal tract (prevertebral
    ganglia) 234-236
  heart (parasympathetic ganglia)
    231-234, 239 (Fig. 6.12)
  heart (stellate ganglion) 235
  visceral spinal afferents 237-238

Final autonomic (motor) pathway
  concept 88-89
  final common motor pathway and
    89 (Fig. 3.1)
  spinal autonomic circuits, and
    163 (Fig. 4.23), 367-369
Final common motor pathway 88
Flight, see defense reactions
Freezing reaction
  autonomic responses in 473
    (Tab. 11.1), 491
Functions of the autonomic nervous
  system
  hierarchical organization (levels of
    integration) 90-92, 91 (Tab. 3.1)
  organization of central circuits,
    concept 92-93

GABA (γ-aminobutyric acid),
  inhibitory transmitter

sympathetic premotor neurons
  333-334
interneurons, autonomic, in spinal
  cord 333-334, 391, 406-407
Rostral ventrolateral medulla
  baroreceptors reflexes 403-406
Galanin (GAL)
  sympathetic neurons 29-31,
    30 (Tab. 1.3)
Ganglia, autonomic
  definition 213
  postganglionic neurons,
    morphology 213-214
  convergence, divergence of
    preganglionic 214-216
Ganglia, parasympathetic 17,
  21-22 (Tab. 1.1), 238-240
  cranial ganglia 24-28 (Tab. 1.4),
    344 (Tab. 9.1)
  heart 239
  pelvic ganglia 239-240
Ganglia, paravertebral sympathetic
  230-234
  anatomy 16-18, 21-22 (Tab. 1.1)
  non-nicotinic transmission
    240-247
  relay function 225, 232 (Fig. 6.10)
Ganglia, prevertebral sympathetic
  234-238
  anatomy 16, 18, 21-22 (Tab. 1.1)
  integration, summation 232
    (Fig. 6.10), 236-237
  peripheral (extraspinal) reflexes
    234-236
  substance P in afferent collaterals
    to 43-44, 237-238
Gastrointestinal tract (see also enteric
    nervous system)
  afferent feedback to the brain and
    179-180
  defense of 196-200, 198 (Fig. 5.13)
  enteric nervous system and,
    concept 178-181, 179 (Fig. 5.3)
  neurogenic inflammation,
    mucosa 55
  parasympathetic neurons to,
    functional types 201, 201
    (Fig. 5.14), 441
  sympathetic neurons to, functional
    types 202-203, 203 (Fig. 5.15)
  vagal afferents and 46
Glucagon-like peptide (GLP-1)
  area postrema and gastrointestinal
    functions 448

Glutamate, transmitter
  baroreceptors reflex pathways
    401 (Fig. 10.7)
  chemoreceptor reflex pathways
    412 (Fig. 10.13)
  sympathetic premotor neurons
    333-334, 389-391
  vagal afferents, to nucleus tractus
    solitarii 442
Glycine, inhibitory transmitter
  interneurons, autonomic, spinal
    cord 333-334
  baroreceptors reflex pathway,
    spinal cord 381
Grey ramus 17-18
Gut-associated lymphoid tissue
  (GALT)
  enteric nervous system 171,
    196-197
  vagal afferents and 49

Hagbarth, Karl-Erik (1926-2005)
  microneurography 102, 426
Heart
  extraspinal reflex pathway to
    231-234, 235, 239, 239 (Fig. 6.12)
  neuroeffector transmission,
    parasympathetic 264-267
  neuroeffector transmission,
    sympathetic 267
  parasympathetic cardiomotor
    neurons 158
  sympathetic cardiomotor
    neurons 141
Hess, Walter Rudolf (1881-1973)
  ergotropic, trophotropic functions
    466-467
  organization of the autonomic
    nervous system, dichotomous
    465-467
  generalizing concept of the
    autonomic nervous system,
    consequences 467-469
Heymans, Corneille (1892-1968)
  baroreceptor reflexes 398
Hierarchical organization
  afferent feedback and 35
  autonomic motor system 4
  autonomic functions 90-92,
    91 (Tab. 3.1), 92-93, 291
Hindgut, parasympathetic (sacral)
  control 156-157
Homeostasis 2
  Cannon's concept 460-465

Homeostasis (cont.)
  concept, definition 469–470
5-Hydroxytryptamine (5-HT), transmitter
  enteric nervous system 176 (Tab. 5.1), 177
  enterochromaffin cells 171, 195, 195 (Fig. 5.12)
  sympathetic premotor neurons 321 (Fig. 8.16), 333–334
Hyperhydrosis, patients and vibration sudomotor reflex 166
Hypothalamus, integrative functions 498–507, 506 (Tab. 11.2)
  anatomy, functional of 499 (Fig. 11.13), 499–503
  behavior control column in 500–501, 502 (Fig. 11.14)
  circadian timing network in 502 (Fig. 11.14), 503
  functional model of, concept 503–507, 504 (Fig. 11.15)
  hypothalamic visceral pattern generator (HVPG) 501–503, 502 (Fig. 11.14)
  neuroendocrine motor zone in 500, 502 (Fig. 11.14)
Hypothalamo-sympathetic system, immune system and 150

Immune tissue
  sympathetic control of 148–151
Ingestive behavior
  hypothalamus and 501, 506 (Tab. 11.2)
Inspiratory neurons (sympathetic) 143, 152 (Tab. 4.5)
  respiratory rhythmicity in activity of 431, 432 (Fig. 10.21)
Insular cortex
  interoception 76 (Fig. 2.13)
  thalamus, projection to 76 (Fig. 2.13), 77, 79 (Fig. 2.14)
Interneurons, enteric
  peristalsis, role of 187 (Fig. 5.8)
  transmitter of 176 (Tab. 5.1), 177
  functional types of 176 (Tab. 5.1)
Interneurons, spinal autonomic 305–306
  criteria for existence 334–336
  micturition, continence 350–352, 354, 355

  propriospinal, thoracolumbar 306, 322–323 (Tab. 8.2), 335
  sacral spinal cord 308, 326 (Tab. 8.3), 335, 336
  segmental, thoracolumbar 305–306, 322–323 (Tab. 8.2), 335–336
  transmitters 333–334
Interneurons, spinal somatomotor
  control of movement, role of 363–366
Interoception 75
  cortical representation, insula 76 (Fig. 2.13)
  exteroception and, concept 36–37
  thalamic representation 76 (Fig. 2.13)
Interstitial cells of Cajal (ICC) 171
  criteria for ICC 205–206
  generation of pacemaker potentials by, mechanisms 206–207
  networks of ICC 189, 189 (Fig. 5.9)
  neuroeffector transmission and 173, 191–192
  pacemaker (gastrointestinal tract) and 190
  presence in organs other than the gastrointestinal tract 205–206
  slow waves of the gastrointestinal tract, role in 188–190
Intestinofugal neurons (enteric nervous system) 202–203, 203 (Fig. 5.15), 214, 235, 239
Intestino-intestinal reflexes
  (see also extraspinal [peripheral] reflex pathways)
  spinal 234–236, 238
  vagal (dorsal vagal complex, DVC) 441–442, 446 (Fig. 10.26)
Intrinsic primary afferent neurons (IPANs), see enteric nervous system

James, William (1842–1910)
  James–Lange theory of emotions 461–462, 491–492

Kidney
  sympathetic innervation 142

Lamina I projection neurons, spinal cord 67–70
  ascending pathways 70 (Fig. 2.11)
  functional classification 67, 68 (Fig. 2.10)

  supraspinal projection sites 69
  thalamic nuclei and 75–77, 76 (Fig. 2.13)
Langley, John Newport (1852–1925)
  enteric nervous system, definition 13, 168
  sympathetic, parasympathetic, definition 13–14, 15
  visceral afferent neurons, definition 40–41, 42–45
Limbic motor cortex, see anterior cingulate cortex
Limbic sensory cortex, see insula
Lovén, Otto Christian (1836–1904) 115
Lovén reflex 165
Lower brain stem
  baroreceptor reflexes 398–409
  cardiovascular-respiratory coupling 420–440
  chemoreceptor reflexes 410–414
  functional anatomy of 382–388, 385 (Fig. 10.2), 388 (Fig. 10.3)
  gastrointestinal reflexes 441–444
  general functions 377–378
Ludwig, Carl Friedrich (1916–1895) 115
  baroreceptors and cardiovascular control 398
  cardiovascular center 379–380
Lundberg, Anders (born 1923)
  motor control, role of spinal circuits 363

Micturition center, pontine 353, 353 (Fig. 9.9)
Micturition, continence, reflexes 352–354
  micturition reflex 352
  sacral visceral C- and Aδ-afferents, role in 373–374
  spinal reflexes 351 (Fig. 9.8), 354, 355–357
  supraspinal control 353, 353 (Fig. 9.9), 355–357, 373
Migrating myoelectric complex (MMC) 181, 193–194
Motility-regulating (MR) neurons to pelvic organs
  functional characteristics 137–138, 140, 152 (Tab. 4.5)
  reflex pattern 138 (Fig. 4.14)
  respiratory rhythmicity in activity 431, 432 (Fig. 10.21)

spontaneous activity 137,
231 (Tab. 6.2)
Motility patterns, gastrointestinal
tract
fed pattern of motility (see
peristalsis) 181
interdigestive motility pattern 181
(see migrating myoelectric
complex [MMC])
interstitial cells of Cajal and 192
rhythmoneuromuscular
apparatus 194
slow waves 181-183, 182 (Fig. 5.4),
183 (Fig. 5.5)
Muscle effort, cortically induced
cardiovascular responses 473
(Tab. 11.1), 477-480
Muscarinic (cholinergic) transmission
in autonomic ganglia, see non-
nicotinic transmission
Muscle vasoconstrictor (MVC)
neurons
functional characteristics, animal
107-112, 108 (Tab. 4.1)
functional characteristics, human
117-119, 120 (Tab. 4.2)
non-nicotinic transmission
(ganglia) 243-244, 246 (Fig. 6.15)
proportion of preganglionic
neurons 112, 152 (Tab. 4.5)
reflex pattern 109 (Fig. 4.1),
110-111
respiratory rhythmicity 429,
432 (Fig. 10.21), 433-435
spontaneous activity, animal 109,
231 (Tab. 6.2)
spontaneous activity, human 117
Myenteric plexus (Auerbach's plexus)
(enteric nervous system)
170 (Fig. 5.1)

NANC, non-adrenergic non-
cholinergic transmission
252-253
Nerve growth factor (NGF)
sympathetic postganglionic
neurons 216
sympathetic postganglionic fibers,
hyperalgesia and 274 (Fig. 7.9),
282-283
Neurochemical coding
concept of 32-33
enteric nervous system
176 (Tab. 5.1), 177

parasympathetic neurons 31-32
sympathetic neurons 29-31,
30 (Tab. 1.3), 31 (Tab. 1.4)
Neuroeffector junctions
functional characteristics
258-259
morphology 257-259,
258 (Tab. 7.2)
Neuroeffector transmission to
autonomic target cells
arteries, arterioles 267-270
cellular events 255-256
concept, principles 255-263
effector cells (see specific effector
cells)
electrophysiology 260-263
excitatory junction potential
261-262
heart 264-267
ileum 271-273
interstitial cells of Cajal and 273
intracellular pathways 255-256,
266-268, 272-273
morphology 257-259, 258 (Tab. 7.2)
quantal release of transmitter
260-262
receptors, postjunctional,
extrajunctional 255, 266, 267
ileum, smooth muscle 271-273
veins 270-271
Neurogenic inflammation (see axon
reflex) 83
sympathetic postganglionic fibers,
role of 277-281
Neurokinins
axon reflex, role of 83
Neuropeptides in autonomic neurons
see neurochemical coding
Neuropeptide Y (NPY)
arteries, modulation by 270
sympathetic neurons 29-31,
30 (Tab. 1.3), 31 (Tab. 1.4), 253
parasympathetic neurons 31-32
Neurturin
parasympathetic postganglionic
neurons 216
Nitric oxide (NO)
blood vessels 275-276
enteric motor neurons 173,
176 (Tab. 5.1), 177
erection 374
peristalsis 186
parasympathetic neurons 31-32
urinary tract 352, 374

Nociceptor sensitization
nerve growth factor 282-283
prostaglandin 281-282
sympathetic postganglionic fibers
281-283
Non-nicotinic (muscarinic, non-
cholinergic) transmission in
autonomic ganglia 240-247
LHRH-like peptide (bullfrog) 240
pelvic ganglia (to vasodilator
neurons), response to
preganglionic electrical
stimulation 247-248
reflexes in vasoconstrictor neurons
244-246
unmyelinated preganglionic
axons 241
Noradrenaline
adrenal medulla and 144, 158
adrenoceptors 25-27 (Tab. 1.2)
sympathetic premotor neurons 321
(Fig. 8.16), 333-334
transmitter, sympathetic 252-254
Nucleus ambiguus 308, 309 (Tab. 8.1)
Nucleus tractus solitarii (NTS)
311-317
baroreceptors, projection to
312 (Fig. 8.11), 400
chemoreceptors, projection to
312 (Fig. 8.11), 410
connections with brain centers 315,
316 (Fig. 8.13)
division of NTS nuclei 312 (Fig. 8.11)
functions of, concept 315-316,
317 (Fig. 8.14)
gastrointestinal tract afferents,
projections to 313,
314 (Fig. 8.12)
parabrachial nuclei and
72 (Fig. 2.12), 79 (Fig. 2.14)
pulse rhythmicity of activity of NTS
neurons 455
synaptic transmission to the
DMNX 442
thalamus, somatosensory and 78,
79 (Fig. 2.14)
visceroptopic organization of 312
(Fig. 8.11), 312-314

PACAP
enteric motor neurons 173, 176
(Tab. 5.1)
Parabrachial nuclei
hypothalamus and 72 (Fig. 2.12)

Parabrachial nuclei (cont.)
  lamina I neurons, projection to 72 (Fig. 2.12), 73
  nucleus tractus solitarii neurons, projection to 72 (Fig. 2.12), 73
Parasympathetic ganglia (see ganglia, parasympathetic)
Parasympathetic preganglionic neurons, cranial
  airways 308
  blood vessels (cranial) 310
  eye 310
  gastrointestinal tract 310
  glands (head) 310
  heart 308
  location 24–28, 308–310 (Fig. 1.4), 309 (Tab. 8.1), 326 (Tab. 8.3)
Parasympathetic preganglionic neurons, sacral morphology 306–307
Parasympathetic premotor neurons
  abdominal organs and 324 (Fig. 8.17), 326 (Tab. 8.3)
  location 319, 320–325, 324 (Fig. 8.17), 326 (Tab. 8.3)
  pelvic organs and 324 (Fig. 8.17), 326 (Tab. 8.3)
Parasympathetic systems innervating
  airways 159–161
  eye 161–162
  functional types 153–162, 155 (Tab. 4.6)
  gastrointestinal tract 161
  heart 158
  pelvic organs 154–158
  salivary glands 161
Parasympathetic system, peripheral
  anatomy, macroscopical 20–24
  definition 14, 15
  target organs, reactions to stimulation of 24–28, 25–27 (Tab. 1.2)
Paravertebral ganglia (sympathetic) 230–234 see ganglia, paravertebral
Pavlov, Ivan (1849-1936)
  conditioned autonomic reflex 472–474
Pelvic ganglia 19, 239–240, 250
Pelvic organs
  parasympathetic functional pathways 154–158
  sympathetic functional pathways 136–141

Peptide YY
  area postrema and gastrointestinal functions 448
Peptidergic transmission in autonomic ganglia, see non-nicotinic transmission
Periaqueductal grey (PAG)
  afferent input from body to 483–484
  defense reactions, organized in 480–481, 482 (Fig. 11.5), 517
  hypothalamic and cortical control of 484–486
  lamina I neurons (spinal), projection to 72 (Fig. 2.12)
  nucleus tractus solitarii (NTS), projection to 72 (Fig. 2.12)
  projection to cardiovascular centers in medulla oblongata 482–483, 484 (Fig. 11.6)
Peristalsis, gastrointestinal tract 183–188 (see also slow waves)
  colon, mechanism 186–188, 187 (Fig. 5.8)
  intrinsic primary afferent neurons (IPANs) 184–186
  motility patterns 181–183
  small intestine, mechanism 186, 187 (Fig. 5.8)
Pilomotor neurons, cat 128–129
Pineal gland, sympathetic innervation 143
Playing-dead response, see tonic immobility
Ponto-medullary respiratory network 386, 387
Postganglionic neurons
  morphology 213–214
  recording in vivo intracellular 220–221, 221 (Fig. 6.5)
  strong, weak synaptic input 217–223
  sympathetic 16–19, 20, 21–22 (Tab. 1.1)
  parasympathetic 21–22 (Tab. 1.1), 23–24
Postganglionic neurons, sympathetic 16–19
  electrophysiological properties 225–230, 228 (Tab. 6.1)
  LAH (long-afterhypolarizing) neurons 229, 230
  phasic neurons 226–227, 229
  neurogenic inflammation and 277–281

nociceptor sensitization and 281–283
  tonic neurons 227–229
Preganglionic neurons
  (see also parasympathetic or sympathetic preganglionic neurons)
  parasympathetic, anatomy 20–23
  parasympathetic, location 21–22 (Tab. 1.1)
  sympathetic, anatomy 19–20
  sympathetic, location 20, 21–22 (Tab. 1.1)
Premotor neurons, autonomic
  definition 317–318
  parasympathetic 319, 320–325, 324 (Fig. 8.17), 326 (Tab. 8.3)
  patterns of, labeled from distinct targets 322–323 (Tab. 8.2), 325–327, 326 (Tab. 8.3)
  sympathetic 318–320, 320 (Fig. 8.15), 322–323 (Tab. 8.2)
Prevertebral ganglia (sympathetic) see ganglia prevertebral
Purinoreceptor 263, 270

Quiescence, see defense reaction

Raphe nuclei
  cutaneous vasoconstrictor system and 415–417
  lipomotor neurons (brown adipose tissue) and 417–420
Reproductive behavior
  hypothalamus and 501, 506 (Tab. 11.2)
Reproductive organs, reflexes connected with
  central mechanisms 358
  emission, ejaculation 359–360
  erection 358–360
  female 357, 361
  parasympathetic (sacral) control 157–158
  sacral (parasympathetic) vs. thoracolumbar 357, 359–360
Research questions, future, integrative actions of the autonomic nervous system 510–514
Respiration
  eupnea (normal respiration) 420
  pacemaker in preBötzinger complex and 386, 396

respiratory phases 420,
425 (Fig. 10.18)
Respiratory network, ponto-
medullary 386, 420–422
cardiovascular network,
integration with 423–424,
424 (Fig. 10.17)
classification of respiratory
neurons 422, 457
respiratory rhythm, theories of
generation 421–422
Respiratory neurons, ventrolateral
medulla
cardiovascular neurons and 387
types of 386–387, 422, 457
location 385 (Fig. 10.2), 386–387
Respiratory rhythmicity in activity of
autonomic neurons
autonomic effector organs,
respiratory oscillations 423
cutaneous vasoconstrictor (CVC)
neurons 116
functionally identified autonomic
neurons, animals 428–433,
432 (Fig. 10.21)
functionally identified autonomic
neurons, humans 426, 433–436,
458
muscle vasoconstrictor (MVC)
neurons 118
species differences 433
sudomotor (SM) neurons 127
sympathetic nerves 423–426
Respiratory system, airways
vagal afferents and 46
neurogenic inflammation, mucosa
55
Rostral ventrolateral medulla (RVLM)
baroreceptor pathways 403–406
cardiovascular premotor nucleus
389–393, 391 (Fig. 10.5)
criteria as vasomotor nucleus 389
location 385 (Fig. 10.2), 385
numbers of bulbospinal neurons
454
sympathetic premotor neurons 318
Rhythmoneuromuscular apparatus,
enteric nervous system
gastrointestinal motility patterns
193 (Fig. 5.11), 194

Salivary glands
parasympathetic innervation 161
sympathetic innervation 142

Salivary nuclei, inferior, superior 309
(Tab. 8.1), 310
Secretion, transmural transport
(gastrointestinal tract)
reflex pathways, enteric nervous
system 194–196, 195 (Fig. 5.12)
Secretomotor neurons innervating
airways 159–161
enteric nervous system
172 (Fig. 5.2), 173–174
salivary glands 161
Sensorimotor programs
autonomic nervous system 3–5
behavior and 4
cortical control of 508–510
enteric nervous system 179–180
hierarchical organization of
4 (Fig. 2)
spinal autonomic motor programs
367
Sensory receptors, skin and
sympathetic innervation
131–132
Sensory receptors, skeletal muscle
and sympathetic innervation 135
Sham-rage behavior, see defense
behavior
Sherrington, Sir Charles Scott
(1857–1952)
common sensations, of the body 36
control of movement, role of spinal
motor circuits 362–363
final common motor path 88
interoception, exteroception 36–37
Silent (mechanoinsensitive) visceral
afferents 59–60
Skin potential
and sudomotor neurons 123–125
components of 165–166
Slow waves, gastrointestinal tract
(see also peristalsis)
interstitial cells of Cajal (ICC) and
188–190
mechanisms 181–182, 182 (Fig. 5.4),
183 (Fig. 5.5)
Somatostatin, transmitter
enteric interneurons 177
sympathetic neurons 29–31, 30
(Tab. 1.3)
Sphincter-detrusor dyssynergia 354
Spinal autonomic reflex pathway
concept 332–333, 333 (Fig. 9.1)
coordination of spinal autonomic
motor circuits 369–370

coordination of spinal autonomic
and spinal somatomotor circuits
370–371
integration of spinal and
supraspinal autonomic reflex
circuits, concept 367–369
(Fig. 9.15)
Spinal cord
deep dorsal horn projection
neurons 68 (Fig. 2.10)
dorsal horn neurons,
classification 84
integrative autonomic organ,
concept 366–370
integration of spinal and
supraspinal circuits in
autonomic control, concept
366–367, 369 (Fig. 9.15)
interneurons 66–67 (see
interneurons, spinal autonomic)
lamina I projection neurons 67–70,
68 (Fig. 2.10), 69
postsynaptic dorsal column tract
71–73
regulation of somatomotor system,
role of, concept 362–366
viscero-somatic convergent
neurons 67, 69
Spinal shock of sympathetic systems,
after spinal cord transection
339–340, 372
motility-regulating (MR) neurons
139–140
recovery of spontaneous activity in
autonomic neurons 339–340
recovery of spinal autonomic
reflexes 339–340, 343
vasoconstrictor systems 139–140
Spinal reflexes, parasympathetic
(sacral) 349–361
colon 354
colon–bladder interaction 355–357
reproductive organs 357–361
urinary bladder 350–354
Spinal reflexes, sympathetic
(thoracolumbar)
cardio-cardial reflexes 348
chronically after spinal cord
transection 336–349
cutaneous vasoconstrictor
neurons 343 (Fig. 9.4),
344 (Tab. 9.1), 345
intestino-intestinal reflexes
348–349

Spinal reflexes, sympathetic (thoracolumbar) (cont.)
  motility-regulating neurons 337–338, 345
  recovery after spinal cord transection, cat 339–340
  recovery after spinal cord transection, human 343
  reno-renal reflexes 347
  reflexes to physiological stimulation 342–349, 343 (Fig. 9.4), 344 (Tab. 9.1)
  reflexes to visceral stimuli 339–340, 347 (Fig. 9.6), 348 (Fig. 9.7)
  segmental, suprasegmental, to electrical stimulation of afferents 337–339
  sympathetic pathways, in 139–140
  visceral vasoconstrictor neurons 337
  viscero-cutaneous reflexes 349
Spleen and immune system, sympathetic control 149–150
Spontaneous activity in autonomic neurons
  caudal pressure area (CPA) and 394
  postganglionic neurons, sympathetic 231 (Tab. 6.2)
  postganglionic neurons, sympathetic after decentralization 250
  mechanism and origin 396–398
  preganglionic neurons, sympathetic 231 (Tab. 6.2)
  rostral ventrolateral medulla, role in 393–394
  rhythmic (bursting) changes of activity in "sympathetic nerves" 379, 396–398
  sympathetic premotor (C1) neurons 381
  sympathetic vasomotor "tone" 381, 396
Starling, Ernest Henry (1866–1927) see Bayliss, William M.
Stellate ganglion 17
Strong synapses in autonomic ganglia 217–223
  calcium channels 218, 220
  concept of functioning 222–223, 232 (Fig. 6.10)
  relay function 225
Submucosal plexus (Meissner's plexus) (enteric nervous system) 170 (Fig. 5.1)

Subretrofacial nucleus, cat, s. rostral ventrolateral medulla 391, 406–407
Substance P
  axon reflex, role in 83, 253
  sympathetic neurons 29–31, 30 (Tab. 1.3), 31 (Tab. 1.4)
  sympathetic premotor neurons, in 321 (Fig. 8.16)
  transmitter in prevertebral ganglia, see ganglia, prevertebral
Sudomotor (SM) neurons
  cutaneous vasoconstrictor neurons and 127
  electrodermal activity, human, role of 128
  functional characteristics, animal 123–127
  functional characteristics, human 127–128
  reflex pattern in 126 (Fig. 4.11)
  respiratory rhythmicity in activity of 127, 431, 432 (Fig. 10.21), 435–436
  skin potential, cat, and 123–125
  spontaneous activity in 231 (Tab. 6.2)
  vibration reflex, cat, in 126, 127
Superior cervical ganglion 17
Supersensitivity, autonomic effector cells to denervation, decentralization 167
Sympathetic outflow to pelvic organs 136–141
Sympathetic preganglionic neurons
  intermediate zone, spinal cord, subnuclei 297
  location of 297–305
  lumbar system, functional, topography 305 (Fig. 8.8), 305–306
  morphology of 297–305, 299 (Fig. 8.3)
  proportions of functional types of 151–153
  segmental organization of 299–300, 322–323 (Tab. 8.2)
  silent preganglionic neurons 152 (Tab. 4.5), 153
  spontaneous activity of 231 (Tab. 6.2)
  synchronization of activity of 230–232

  viscerotopic organization of 300–304, 303 (Fig. 8.6), 304 (Fig. 8.7)
Sympathetic premotor neurons
  caudal raphe nuclei 414–420
  cutaneous vasoconstrictor premotor neurons 415–417
  GABAergic transmission to 417
  lipomotor premotor neurons (brown adipose tissue) 417–420
  location of 318–320, 320 (Fig. 8.15), 322–323 (Tab. 8.2)
  rostral ventrolateral medulla, location in and functional topography 383, 391–393, 392 (Fig. 10.6)
  spontaneous activity in, origin 393–394
  synaptic input to 391, 391 (Fig. 10.5)
  transmitter, putative 318, 321 (Fig. 8.16), 333–334, 389–391
Sympathetic system, peripheral
  anatomy, macroscopical 16–20
  definition 14, 15
  target organs, reactions to stimulation of 24–28, 25–27 (Tab. 1.2)
Sympathetic trunk (chain) 16
Sympathico-adrenal system 463, 467
  Cannon's concept of 145, 460–465
  emergency function of the 464
Syncope, neurally mediated (see also tonic immobility)
  cardiovascular responses in 490–491
Syncytium of effector cells
  passive electrical behavior 256, 286

Tachykinins, see also substance P
  axon reflex 83
  enteric nervous system 176 (Tab. 5.1)
Thalamocortical system
  body sensations, representations in 74–78, 76 (Fig. 2.13)
Thalamus, somatosensory
  nuclei 76 (Fig. 2.13)
  lamina I neurons, projection to 74
  somatosensory cortex (SI, SII) 74–75
  ventral posterior nuclei (VPL, VPM) 74, 76 (Fig. 2.13)
  ventromedial nucleus, basal part (VMb) 74–78, 76 (Fig. 2.13), 78, 79 (Fig. 2.14)

ventromedial nucleus posterior (VMpo) 74–78, 76 (Fig. 2.13), 78, 79 (Fig. 2.14)
Thermoregulatory behavior hypothalamus and 501, 506 (Tab. 11.2)
Thoracolumbar system 14, 15 *see* sympathetic system
Thyrotropin-releasing hormone (TRH)
  area postrema and gastrointestinal functions 447–448
  sympathetic premotor neurons, in 321 (Fig. 8.16)
Tonic immobility
  autonomic responses in 473 (Tab. 11.1), 490
Transmitters
  interneurons, spinal 334–336
  postganglionic neurons 252–254, 254 (Tab. 7.1)
  sympathetic premotor neurons 318, 319, 321 (Fig. 8.16), 333–334
Traube-Hering waves, mechanism 426–428
Trigeminal afferent neurons 39

Urethro-genital reflex 357 (*see also* emission, ejaculation)
Urinary tract, lower
  parasympathetic (sacral) control of 157

Vagal afferents, *see* visceral afferent neurons, vagal
Vagus nerve
  abdominal, branches 330
  abdominal, numbers of fibers, rat 329
Vasoactive intestinal peptide (VIP)
  enteric motor neurons 173, 176 (Tab. 5.1)
  erection 374
  peristalsis 186
  parasympathetic neurons 31–32
  secretomotor neurons 253, 254 (Tab. 7.1)
  sympathetic neurons 29–31, 30 (Tab. 1.3), 31 (Tab. 1.4)
  sympathetic premotor neurons 321 (Fig. 8.16)
Vasoconstrictor neurons (*see under* cutaneous, muscle, visceral vasoconstrictor neurons)
  skeletal muscle 107–112, 117–119
  skin 113–117, 119–123
  viscera 112–113
Vasodilator neurons
  blood vessels, cranial 309 (Tab. 8.1), 310
  blood vessels, skeletal muscle, *see* Vasodilator neurons, skeletal muscle
  blood vessels, skin, *see* Vasodilator neurons, skin
  enteric nervous system 172 (Fig. 5.2), 173–174, 195 (Fig. 5.12)
  eye 161–162
  paracervical ganglia 247
  reproductive organs, parasympathetic 157–158
  reproductive organs, sympathetic 136
  salivary glands 161
Vasodilator neurons, skeletal muscle
  cholinergic 133
  functional characteristics, cat 132–135
  hypothalamic defense reaction and 132–135
  in humans 135
Vasodilator neurons, skin
  sympathetic, animals 129–130
  sympathetic, humans 130–131
  parasympathetic 131
Vasodilation, skin
  sweating and 130–131
  neural mechanisms of 130–131
  peptidergic afferents and 130
  role of vasoactive intestinal polypeptide 166
Vegetative nervous system 13
Veins, pulmonary artery
  neuroeffector transmission 270–271
Ventrolateral medulla oblongata (VLM) 381–388
  anatomy, functional 382–388, 385 (Fig. 10.2)
  adrenergic (C1) neurons 387–388, 388 (Fig. 10.3)
  cardiovascular and respiratory neurons 387
  noradrenergic (A1) neurons 387–388

Vigilance reaction
  autonomic responses in 473 (Tab. 11.1), 491
Visceral afferent neurons, general (*see also* visceral afferent neurons spinal, vagal)
  definition 40–42
  general functional characteristics 37–42, 38 (Fig. 2.1)
  interface between viscera and brain 40–41, 42–45
  receptive functions of 45–53
  sensation, organ regulation, and model of encoding 63–65, 64 (Fig. 2.9)
  specificity, function of visceral afferents, concept 62–63
Visceral afferent neurons, spinal 43 (Fig. 2.3)
  afferent functions of 43
  axon reflex and 83
  efferent functions of 44
  general characteristics 37–39, 40 (Fig. 2.2)
  peripheral (extraspinal) reflexes 43–44, 237–238
  projection to spinal cord 52–53
  prevertebral (sympathetic) ganglia and 237–238
  sacral visceral afferent neurons 53
  sensations and 48 (Tab. 2.1)
  silent (mechanoinsensitive) visceral afferents 59–60
  substance P in 43–44, 83, 237–238
  thoracolumbar visceral afferent neurons 53
  trophic functions of 44
  visceral pain and 56–60
Visceral afferent neurons, vagal
  body protection and 49–52, 54–55
  cardiovascular system 46
  cortical representation 78, 79 (Fig. 2.14)
  gastrointestinal tract 49–52, 440
  general characteristics of 39, 40 (Fig. 2.2)
  illness responses (sickness behavior) and 51 (Fig. 2.5)
  nociception and pain 54–56
  nucleus tractus solitarii, projection to 79 (Fig. 2.14)
  receptive functions of 46–52
  respiratory tract and 46

Visceral pain
  cardiac pain and vagal afferents 55–56
  spinal visceral afferents and 56–60
  theory, specificity vs. intensity 62–63
  vagal abdominal afferents and 54–55
  vagal thoracic afferents and 55–56
  visceral sensations, organ regulation and, concept 63–65, 64 (Fig. 2.9)
Visceral sensations
  spinal visceral afferents and 48 (Tab. 2.1)
  vagal visceral afferents and 48 (Tab. 2.1)
  organ regulation and 60–61
  specificity of visceral afferents and 62–63
Visceral sensory cortex, *see* insular cortex
Visceral vasoconstrictor (VVC) neurons
  functional characteristics, animal 112–113
  proportion of preganglionic neurons 113, 152 (Tab. 4.5)
  spontaneous activity, animal 231 (Tab. 6.2)
Volume conduction
  prevertebral ganglia 237

Weak synapses in autonomic ganglia 217–223
  calcium channels 218, 220
  concept of functioning 222–223, 232 (Fig. 6.10)
White ramus 17